Spermatology

Society for Reproduction and Fertility Volume 65

Proceedings of the 10th International Symposium
on Spermatology
held at El Escorial, Madrid, Spain
17-22 September 2006

Edited by
E.R.S. Roldan and M. Gomendio

NOTTINGHAM
University Press

First published by Nottingham University Press

This reissued original edition published 2023 by 5m Books Ltd www.5mbooks.com

British library Cataloguing in Publication Data
Spermatology
E.R.S. Roldan and M. Gomendio (Eds)

ISBN: 9781789183085

Disclaimer

Every reasonable effort has been made to ensure that the material in this book is true, correct, complete and appropriate at the time of writing. Nevertheless, the publishers and authors do not accept responsibility for any omission or error, or for any injury, damage, loss or financial consequences arising from the use of the book.

Typeset by Nottingham University Press, Nottingham
Printed and bound by PODWW, UK

Preface

The chapters in this book are based on presentations at the *10th International Symposium on Spermatology* («Spermadrid»), which took place at El Escorial (Madrid, Spain) from 17 to 22 September 2006. All invited speakers, as well as keynote lecturers, were asked to write an overview of the latest advances in their field. In addition, the authors of ten selected short communications were invited to present their results in extended form.

The Spermatology Symposia have a long history. The first meeting took place in 1969 at Rome and Siena and, since then, meetings have been held regularly every four years. Earlier symposia have taken place in Stockholm, Sweden (1973), Boston-Woods Hole, USA (1978), Seillac, France (1982), Fujiyoshida, Japan (1986), Siena, Italy (1990), Cairns, Australia (1994), Montreal, Canada (1998), and Cape Town, South Africa (2002). Early meetings focused mainly on comparative studies of sperm morphology and physiology. However, this initial emphasis rapidly expanded to encompass advances in the study of spermatozoa from different perspectives.

Spermadrid was an ambitious attempt to integrate different approaches to the study of spermatozoa: evolutionary, developmental and mechanistic. This was a huge task given the diverse and significant advances that have taken place in recent years. In addition, we tried to cover both basic research and the far-reaching applications that can be derived from the study of sperm cells. Including such a broad range of topics brought together researchers from different fields who seldom communicate with each other. This was perhaps the most exciting achievement of the meeting: to unite researchers with different conceptual frameworks and using different methodological tools under the umbrella of Spermatology. We very much hope that the book conveys the sense of excitement that was felt at Spermadrid.

Spermatozoa are highly specialised cells. Because reproduction ultimately depends on the ability of spermatozoa to fertilise ova, these cells have been under intense selective pressures leading to the evolution of an outstanding diversity of shapes, functions and behaviours. The widespread occurrence of sperm competition has also favoured specific adaptations designed to outcompete rival sperm.

To achieve fertilisation, spermatozoa face many hurdles. They have to actively reach the ovum by means of their motility and the regulation of the energy required to fuel it. Sperm cells need to interact with the ovum by exchanging appropriate recognition and activating signals and then have to penetrate the ovum vestments and deliver its DNA successfully. This multiplicity of tasks makes the spermatozoon unique among cells.

Understanding mechanisms underlying sperm formation and function represents a formidable challenge which has required the development of sophisticated methods that are gradually revealing a fascinating picture. The knowledge gained by the study of spermatozoa has allowed the manipulation of reproduction to an extent which was unforseen some years ago. Assisted reproductive techniques are presently used to overcome fertility problems in human patients, to avoid inbreeding and enhance reproduction among endangered species, and to maximise efficiency in domestic animals. At present, it is technically possible to manipulate reproductive processes to such an extent that new ethical issues are arising which need to be dealt with. The field is advancing so fast that new challenges will undoubtedly emerge.

We are grateful to the International Steering Committee for their guidance and, in particular, to Jim Cummins for his continued support and advice. The International Scientific Advisory Committee greatly assisted in the selection of topics and of speakers. The Local Organising

Committee had to deal with all aspects of the organisation and, most importantly, with fund raising. Julian Garde was pivotal in obtaining funds, as well as wine to cheer up the speakers. We would have been completely overwhelmed with organisational aspects without the professional help of Paloma Herrero and Belen Gomez from Unicongress. We are also very grateful to our students, E. Fernández, C. Crespo, A. del Olmo, J. Martin-Coello, N. Gañán, L. Arregui, R. González, M.J. Ruiz, J. Benavent-Corai, for their help before and during the meeting.

Spermadrid was well attended, with 341 participants from 35 countries in 5 continents. The meeting would not have been possible without the financial help received from many public and private institutions. We thank them for the support which was essential to allow the participation of invited speakers and many young investigators. A special mention should go to the Fundación BBVA, The Lalor Foundation, the International Society of Andrology and the Fundación Campollano who funded travel grants for 33 delegates (from 16 different countries) that were in the early stages of their scientific careers.

We wish to thank all the authors for the time and effort they have put into writing their reviews, and for the fascinating work that they do. Hopefully, the book will communicate the same feeling of adventure and enthusiasm that was present in so many of the talks. We are grateful to the Society for Reproduction and Fertility for the opportunity to publish this book. In addition, we sincerely appreciate the excellent work done by Sarah Keeling from Nottingham University Press.

Finally, we wish to express our sincerest thanks to Javier Hornedo and the staff at M.D. Anderson in Madrid (where most of the editing was done) who looked after Montse so well during many difficult months.

Eduardo R. S. Roldan
Montserrat Gomendio
Reproductive Ecology and Biology Group
Museo Nacional de Ciencias Naturales (CSIC)
28006-Madrid, Spain
roldane@mncn.csic.es
montseg@mncn.csic.es

Sponsors

- Ministerio de Educación y Ciencia, Spain
- Consejo Superior de Investigaciones Científicas, Spain
- Instituto Nacional de Investigación y Tecnología Agraria y Alimentaria, Spain
- Consejería de Educación, Comunidad de Madrid
- Junta de Comunidades de Castilla-La Mancha
- Consejería de Agricultura, Junta de Comunidades de Castilla-La Mancha
- Universidad de Castilla-La Mancha
- Museo Nacional de Ciencias Naturales, Madrid
- Consejo Regulador Denominación de Origen «La Mancha»
- Canadian Institutes of Health Research
- CONRAD
- Society for Reproduction and Fertility
- European Academy of Andrology
- Fundación BBVA
- Fundación Campollano
- The Lalor Foundation
- International Society of Andrology
- XY Inc
- SCSA Diagnostics
- Committee of the 9th International Symposium on Spermatology

Exhibitors

- Telstar (Planer, CHART, Custom BioGenic Systems)
- Microptic SL
- Poiser Projectes i Serveis R+D
- Minitub Ibérica
- Hamilton Thorne Biosciences
- Izasa-Nikon
- Society for Reproduction and Fertility
- European Academy of Andrology

10th International Symposium on
Spermatology Madrid 2006
17-22 September 2006

Contents

ACCESSORY GLANDS AND ROLE OF SECRETORY FLUIDS IN SPERM FUNCTION

SIGNALLING AND SPERM FUNCTION

SPERM MOTILITY

SPERM-EGG INTERACTION

EPIGENETICS OF FERTILISATION

SPERMATOLOGY AND WILDLIFE CONSERVATION

MANIPULATION OF SPERM CELLS

SPERMATOZOA AND HUMAN FERTILITY

Spermatology. SRF Vol. 65. ERS Roldan and M Gomendio (eds) Nottingham University Press, Nottingham

Origin and evolution of spermatogenesis genes on the human sex chromosomes

Margaret L. Delbridge and Jennifer A. Marshall Graves

Comparative Genomics Group, Research School of Biological Sciences, The Australian National University, Canberra ACT 2601, Australia

Both the X and Y chromosomes have a remarkable enrichment of genes involved in gonadogenesis and gametogenesis. The small Y chromosome contains the sex determining gene *SRY*, as well as many genes that are critical for spermatogenesis and are expressed exclusively in the testis. The X chromosome, too, is enriched for genes involved in sex and reproduction. This biased gene content can be best understood in terms of the origin and evolution of our sex chromosomes. The Y chromosome can be seen as the relic of the ancient autosome, on which only a few genes survive by virtue of their critical male-specific role. The X is more complicated - it has evolved male-advantage genes because of its representation as a single copy in males, where it is exposed to selection for male-advantage genes.

Introduction

Many genes are involved in the complex process of gamete formation, and we have known for decades that these genes are found all around the genome. However, new information from sequencing the genomes of humans and other mammals makes it clear that genes involved in gonadogenesis and gametogenesis are remarkably enriched on the sex chromosomes. The small Y chromosome is almost completely specialised for sex and spermatogenesis. The gene content of the X chromosome, which had been thought to be unremarkable, also appears to be greatly biased toward genes that affect gonads and gametes.

To understand how and why sex chromosomes became enriched for sex and reproduction genes, and discover where these overabundant genes came from, we need to understand the origin and evolution of sex chromosomes. Here we describe the make-up of the human X and Y chromosomes, discuss their origin from an ordinary pair of autosomes, and present evidence that special evolutionary forces have been at work to accumulate new sex and reproduction genes, and to evolve new functions for existing genes on the sex chromosomes.

Mammal sex chromosomes

All the peculiarities of mammalian sex chromosomes, including their remarkable enrichment for genes involved in gonads and gametes, result from their different representation in males and females.

Corresponding author E-mail: Margaret.Delbridge@anu.edu.au

Mammals have an XX female:XY male chromosomal sex determination system, in which the Y chromosome controls male development. The Y chromosome is uniquely male specific, and contains the *SRY* gene that triggers male determination. The X chromosome is also unique in the mammalian genome because it is present in two doses in females, but only a single dose in males.

The X chromosome is a medium-sized chromosome (about 155 megabase pairs (Mb)), containing about 1300 genes with a variety of housekeeping and specialised functions. In eutherian ("placental") mammals, the X chromosome is almost entirely conserved between species, and contains the same suite of genes. The X dosage difference between males and females has major genetic consequences. Firstly, an X-linked mode of inheritance is immediately recognisable (e.g. for haemophilia and colour blindness) because male offspring receive their single X chromosome only from their mother. Secondly, the X dosage difference between males and females is compensated for by X chromosome inactivation, whereby one or the other randomly chosen X is stably inactivated in the somatic cells of females.

The Y chromosome is much smaller - only about 57 Mb. Most of the Y chromosome does not pair with the X, so is male-specific (male-specific region of the Y chromosome, MSY). However, there is a 2.7 Mb region shared by the tips of the X and Y short arms (called the pseudoautosomal region, PAR1) and a second, very small (0.4 Mb) pseudoautosomal region at the other end of the X and Y chromosomes (PAR2) (Fig. 1). The Y chromosome contains only 45 genes that code for unique proteins. Many of these genes are packed into the pseudoautosomal regions, leaving only 27 genes in the MSY. The number of transcribed sequences on the Y chromosome is higher (172), but this includes amplified copies of some genes, as well as several pseudogenes that are transcribed but cannot be translated to produce a protein product. The Y chromosome has a high content of repetitive non-coding sequence, which is concentrated in a heterochromatic (that is, differently staining) region on the distal half of the long arm that is apparently bereft of genes. Although the gene content of Y chromosomes of different mammals overlaps to a considerable extent, the copy number and the activity state is very variable between species.

Spermatogenesis genes on the human Y chromosome

Perhaps the most remarkable thing about the human Y chromosome is that it seems to be specialised for male sex and reproduction. It contains the *SRY* gene that triggers the differentiation of testis, and leads to the hormone cascade that differentiates the male soma and germline. The human Y chromosome also contains several regions, deletions of which cause azoospermia. The azoospermia factor (AZF) locus was first identified as microscopically visible deletions or rearrangements in the Y chromosome of sterile patients (Tiepolo and Zuffardi, 1976). Mapping microdeletions by PCR amplification of a standard panel of STS markers revealed three distinct AZF regions within Yq11, AZFa, AZFb, and AZFc, but the borders of these regions were not always clear because the molecular reference markers against which they were mapped turned out to be variable. Atypically for human gene maps, there does not seem to be a consistent order for these anonymous markers. Deletion of individual markers was found also in fertile men.

The AZFa locus is located near the centromere on the long arm of the Y (in band Yq11), and encompasses three single copy genes, *USP9Y*, *DBY*, and *UTY*. Most infertile patients with AZFa deletions have lost the entire interval, so it is not clear whether the AZFa phenotype is caused by loss of one or more of these genes. These three genes all have homologues on the X chromosome and are all expressed ubiquitously; however, *DBY* also produces an alternative

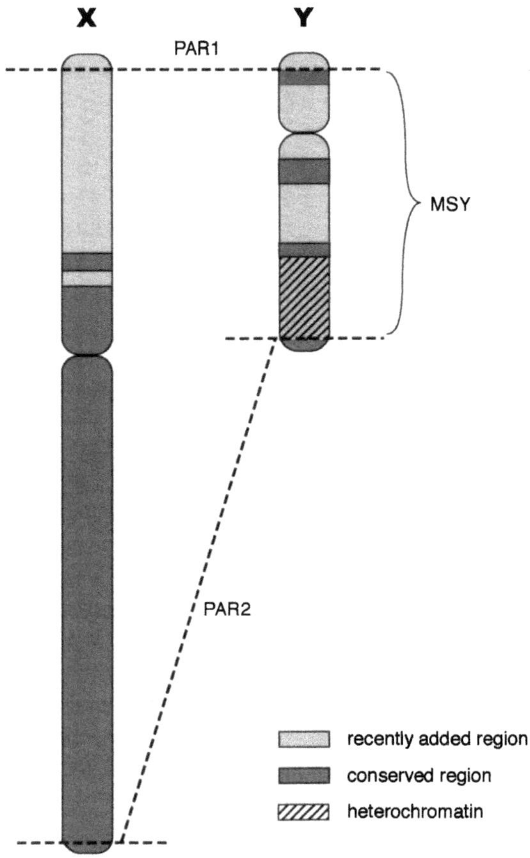

Figure 1. Regions of the human X and Y chromosomes and their evolutionary origins. The pseudoautosomal regions (PAR1 and PAR2) at both ends of the long and short arms of the human X and Y chromosomes pair and undergo recombination. The remainder of the Y is male-specific (MSY). The long arm of the X chromosome, and part of the short arm, is conserved on the X in all mammals and relics of this region occur on the Y. Most of the short arm of the X chromosome is autosomal in distantly related mammals, so was added to the eutherian X chromosome after the divergence of the marsupials and monotremes from placental mammals; relics of this region make up most of the human Y chromosome.

transcript that is expressed only in the testis. *DBY* and *USP9Y* are the major AZFa candidates as deletion of *DBY* only is associated with a complete absence of germ cells in the testis (Sertoli cell only syndrome), and a small (4 base pairs(bp)) deletion within *USP9Y* results in severe oligospermia (Sun, Skaletsky, Birren, Devon, Tang, Silber, Oates and Page, 1999; Foresta, Ferlin and Moro, 2000).

A little further down band Yq11 lie the AZFb and AZFc azoospermia regions, whose mapping was even more fraught. Sequencing of the entire human Y chromosome has cleared up some of the inconsistencies. This region contains eight large palindromic sequences (that is they are the same read forward and backward), their arms ranging from 9 kilobase pairs (kb) to 1.45 Mb in size. The palindromes are imperfect, the arms are separated by a spacer of unique sequence, ranging in size from 2 – 170 kb. The two arms of the palindromes are more than 99.95% identical (Skaletsky *et al.*, 2003), accounting for difficulties in localising molecular markers and defining the extent of the AZF intervals. We now know that the original AZFb and AZFc intervals overlap by 1.5 Mb (Repping, Skaletsky, Lange, Silver, Van Der Veen, Oates, Page and Rozen, 2002, Vogt, 2004), and five of the eight palindromes in distal Yq11 lie within this 'AZFb + AZFc' region. Several active testis specific gene family members lie within the 'AZFb + AZFc' region, including *XKRY*, *CDY*, *HSFY*, *RBMY*, *PRY*, *BPY2* and *DAZ*, and all are candidate AZF genes (Skaletsky *et al.*, 2003). Active members of the gene families are found within different palindromes. Presumably, therefore, each is responsible for a critical step in spermatogenesis. It is difficult to pinpoint the copy whose loss causes infertility because no point mutations within the AZF regions associated with a particular testicular pathology have yet been identified, and deletions characterised to date usually involve a number of genes.

Thus the Y chromosome is unique in its structure, as well as its coherence of male-specific function.

Origin of the X and Y chromosomes

The unique male-specific function of the human Y is best explained by the selective forces that lead to degradation of one member of an originally autosomal chromosome pair. There is a lot of evidence that the mammal X and Y both evolved from an ordinary chromosome pair that had nothing to do with sex. Understanding this proto-X and -Y is a basis for understanding how the gene content of our sex chromosomes became biased.

The theory that sex chromosomes derived ultimately from an ordinary autosome pair was first proposed for insects nearly a century ago (Muller, 1914), and was applied to vertebrates when it was observed that different families of snakes had sex chromosomes that showed intermediate degrees of differentiation (Ohno, 1967). Although the human X and Y are very different in size and gene content, homology within the pseudoautosomal regions, and also between many of the genes that have copies shared by the X and Y, supports the hypothesis that the mammalian X and Y chromosomes also differentiated from a homologous autosomal pair.

Differentiation of the X and Y chromosomes in mammals must have begun early in mammal evolution because the mammal sex chromosome pair shows no homology with the sex chromosomes of other vertebrates. Birds and many reptiles have a ZZ male:ZW female sex determining system, in which Z dosage and the presence of the W chromosome interact to control sex determination. Although the evolution of the large gene-rich Z and the small heterochromatic W parallels that of the mammal X and Y, genes on the mammalian X chromosome are found on chicken chromosomes 1 and 4, and most genes on the chicken Z chromosome are found on human chromosome 9 (Nanda *et al.*, 1999; Shetty, Griffin and Graves, 1999; Kohn, Kehrer-Sawatzki, Vogel and Graves, 2004). The XX:XY and ZZ:ZW systems are therefore thought to have evolved independently from different autosomal chromosome pairs in a reptilian ancestor (Graves and Shetty, 2001).

Evidence from comparative gene mapping shows that the human X and Y originated as at least two, and probably three, completely different ancient autosomal regions that were put together at different times during our evolution. Sex and reproduction genes within the ancient and added regions have a different origin and a different evolutionary trajectory.

Mapping human X-borne genes in distantly related mammals showed that part of the X chromosome is conserved in all three major groups of mammals: placental mammals, marsupials (which diverged from placentals approximately 180 million years ago), and monotremes, which diverged from therians (placentals and marsupials) approximately 210 million years ago (Graves, 1995). Cross-species chromosome painting (hybridization of DNA from isolated chromosomes) confirms that the long arm and pericentric region of the human X is equivalent to the marsupial X (Glas, De Leo, Delbridge, Reid, Ferguson-Smith, O'Brien, Westerman and Graves, 1999) and much of the largest monotreme X (Waters, Delbridge, Deakin, El-Mogharbel, Kirby, Carvalho-Silva and Graves, 2005). This ancient conserved region of the mammalian X chromosome must be at least 210 million years old. However, genes on the short arm of the human X, including the pseudoautosomal region on the short arm (PAR1), were found to be autosomal in marsupials and monotremes, implying addition of a large region to the sex chromosomes between 180 and 100 million years ago (Fig. 1). These two blocks are also distinct in chicken, where they lie on chicken chromosomes 4 and 1 respectively, and a third block of intermediate age is identified (Kohn *et al.*, 2004).

Comparative mapping of human Y-borne genes in distantly related mammals shows that the Y chromosome, too, contains the same ancient and added regions (Waters, Duffy, Frost, Delbridge and Graves, 2001). Most of the MSY derives from the added region, and this region has been subjected to the same evolutionary forces that have shaped the content of the original MSY region (Fig. 1).

Thus the human sex chromosomes originated largely from two autosomal regions that were put together at different stages during mammal evolution. These regions therefore have had different time periods to accumulate sex and reproduction genes and to evolve functions in spermatogenesis.

Degradation of the Y chromosome

If the male-specific Y chromosome was once an ordinary autosome that had nothing to do with sex, what led to the enrichment of the sex and reproduction genes on it? The evidence is that the original proto-Y chromosome has been progressively whittled away, and nothing survived except genes that are selected for a critical male-specific function.

Theory, backed up with many observations of intermediate sex chromosome differentiation plus some experimental observation from *Drosophila*, predicts that once a chromosome acquires a male-determining gene, it is doomed. This is because of two different evolutionary forces – heightened mutation on the Y chromosome, and its genetic isolation because it can no longer recombine with the X chromosome.

The high number of mutations suffered by the Y results from its limitation to males. It therefore spends all its reproductive time in a testis, whereas autosomes are in an ovary half the time (and the X more than half the time). Making sperm requires many more cell divisions than making eggs, so the testis provides more opportunities for mutation during DNA replication (Miyata, Hayashida, Kuma, Mitsuyasu and Yasunaga, 1987); indeed we now know that most *de novo* mutations in human patients originated in the father's reproductive tract (Makova and Li, 2002). In addition, the highly oxidative biochemistry of sperm presents a very mutagenic environment, and there are no DNA repair enzymes to mop up the damage (Aitken and Graves, 2002).

The acquisition of a male-determining locus immediately confines the proto-Y to the males, and this sets up strong selection for genes on the Y that have an advantage in males. The Y chromosome is an especially good repository for genes that are advantageous in males but detrimental to females. These male-advantage genes accumulated near the sex-determining locus, and it was advantageous to suppress recombination to keep a male-specific gene package together. Mutations and deletions rapidly accumulated on the Y as a result of the higher mutation rate, and were not eliminated efficiently by selection because of the lack of recombination (Charlesworth, 1991). Beneficial new mutations that arose on a poor Y were dragged to extinction, whereas detrimental mutations and deletions could hitchhike on a Y that contained other beneficial genes. The lower number of Y chromosomes in the population exacerbated problems of accidental loss of relatively intact Y chromosomes by genetic drift, and in the absence of recombination, they could not be regenerated. Further degradation of the proto-Y resulted in the expansion of the non-recombining region independently in different lineages, explaining why the mammalian Y chromosome is unusually variable between species in its gene content, number and activity.

The ancient and added regions of the X and Y have had different times to differentiate, and this means that sequence will have diverged much more in the more ancient strata. Sequence comparisons between X and Y paralogues confirm that genes fall into different clusters that represent ancient, intermediate and recently added regions, and even subdivide the added region to three or more layers that diverged at different times, perhaps as the result of inversions that eliminated pairing between regions of the X and Y chromosomes (Lahn and Page, 1999).

Thus the mammalian Y chromosome is a degraded version of the X. It has lost about 1250 of the 1300 genes it started with, and at this rate it is projected to completely disappear in 15 million years or so (Graves, 2006). The extraordinary evolutionary history of the Y can explain its unique specialisation for male reproduction.

Origin of male-specific genes on the mammalian Y chromosome

Is the concentration of spermatogenesis genes on the Y explained by selective retention of male-specific genes? Or were genes with other functions moulded to perform new functions? Or were male-specific genes recruited to the Y from other chromosomes? The answer is probably all three.

The male-specific region of the mammalian Y chromosome contains many genes with a male-specific role, or at least with testis-specific expression. Some, like the testis determining gene *SRY*, are present in a single copy. However, nine of the 27 MSY genes coding for unique proteins (*TSPY, VCY, XKRY, CDY, HSFY, RBMY, PRY, BPY2* and *DAZ*) are present on the Y in multiple copies, between them accounting for approximately 60 of the 78 active gene copies on the MSY. Four MSY genes (*DBY, USP9Y, RBMY* and *DAZ*) have been demonstrated to have functions in spermatogenesis, but others that are expressed specifically in the testis (e.g. *TSPY, TGIF2LY, VCY*) may also prove to function in spermatogenesis.

The MSY contains 16 single copy genes (e.g. *ZFY, USP9Y, DBY, UTY, TMSB4Y*), which appear to have retained the ubiquitous expression pattern of a housekeeping gene, or a specialised function in a somatic tissue (e.g. the tooth enamel gene, amelogenin, *AMELY* and the brain gene, protocadherin 11 Y, *PCDH11Y*). These Y genes that are ubiquitously expressed retain both sequence and functional homology with X borne partners. The X homologues of these genes are not subject to X inactivation in females because there is equal dosage of these genes in males and females. These genes are all relics of X-Y homology on the ancient proto-X and proto-Y chromosomes. In addition, there are a large number of inactive pseudogene sequences on the human Y chromosome, which seem to have been generated by degeneration of ancient X-Y gene pairs, and amplified copies of active genes.

Genes on the Y chromosome with testis-specific expression were initially grouped into a Class II of male-specific genes, distinct from the relics of X-Y homology (Class I). They were originally thought to have no X partners, and therefore were proposed to have been acquired by the Y chromosome from autosomes because they had functions that conferred a selective advantage in males (Lahn and Page, 1997), as has been documented in *Drosophila* (Carvalho, 2002). Indeed, this does seem to describe two genes that function in spermatogenesis, *DAZ* and *CDY*. *DAZ* is a multicopy testis-specific gene, with no X homologue (Saxena, Brown, Hawkins, Alagappan, Skaletsky, Reeve, Reijo, Rozen, Dinulos, Disteche and Page, 1996), but is closely related to the autosomal *DAZL* gene with an ancient function in reproduction (Burgoyne, 1996, Seboun, Barbaux, Bourgeron, Nishi, Agulnik, Egashira, Nikkawa, Bishop, Fellous, McElreavey, Kasahara and Algonik,

1997). *DAZ* was evidently copied and transposed to the Y chromosome early in the primate lineage, whereupon it was amplified (Saxena *et al.*, 1996). Another arrival to the human Y chromosome is the intronless *CDY* gene, which is a testis-specific retrotransposed copy of a ubiquitously transcribed autosomal gene, *CDYL*. The presence of *CDY* on the Y in non-primate mammals implies that *CDY* arrived on the Y earlier than *DAZ* (Dorus, Gilbert, Forster, Barndt and Lahn, 2003).

However, these two genes are exceptional. We now know that several testis-specific Y genes with roles in male sex determination and fertility have widely expressed X homologues. The spermatogenesis gene *RBMY* was found to have a closely related homologue on the X (*RBMX*) and the testis-specific *TSPY* has a partner *TSPX* (Delbridge, Lingenfelter, Disteche and Graves, 1999; Delbridge, Longepied, Depetris, Mattei, Disteche, Marshall Graves and Mitchell, 2004). Even the sex-determining gene *SRY* has a homologue on the X, *SOX3*, from which it obviously evolved (Foster and Graves, 1994). The subdivision of Y genes into Class I and Class II therefore has no foundation; most sex and spermatogenesis genes on the Y are relics of the ancient autosome or the recent addition to the proto-sex chromosomes.

Is this simply a result of selection for genes that already had a function in male sex and spermatogenesis? To survive the inexorable Y degradation a gene must confer a selective advantage on its bearer. Since the Y chromosome is confined to males, it cannot contain genes essential to survival, so the only critical functions that can be selected are male reproduction. For instance, retention of an ancient function in sex determination may explain the origin of one gene that lies on the Y in marsupials but not placentals. *ATRY* is a testis-specific homologue of the widely expressed *ATRX* gene on the X chromosome. Mutations in human *ATRX* cause sex reversal as well as mental retardation and other abnormalities. The Y homologue appears to have been modified from the original, with amino acid changes in several protein domains that would alter the binding of this protein to factors involved in chromatin condensation (Park, Pask, Huynh, Renfree Harley and Graves, 2004).

How did the testis-specific functions of Y genes arise, given that their X homologues from which they evolved seem to be widely transcribed and code for proteins involved in general functions in both sexes. Curiously, several of these genes appear to be involved in brain differentiation and function. For instance, *RBMX* codes for a protein involved in RNA processing. It is highly expressed in brain, and knock-down experiments in zebrafish suggest a brain phenotype (Tsend-Ayush, O'Sullivan, Grutzner, Onnebo, Lewis, Delbridge, Marshall Graves and Ward, 2005). Similarly, *SOX3* belongs to a gene family that codes for a DNA-binding and bending protein that modifies chromatin architecture. It is highly expressed in the developing central nervous system and plays a role in X-linked mental retardation and pituitary function (Laumonnier, Ronce, Hamel, Thomas, Lespinasse, Raynaud, Paringaux, Van Bokhoven, Kalscheuer, Fryns, Chelly, Moraine and Briault, 2002). The general functions in both sexes of the X progenitor genes imply that the testis-specific Y homologues specialised, at least in their expression profile.

How did ubiquitously expressed brain-differentiating genes on the original autosome transmogrify into testis specific genes? Sequence comparisons show that the coding regions of the testis-specific *RBMY* and *TSPY* genes underwent changes in exon number compared to their more conserved widely expressed X counterparts *RBMX* and *TSPX* (Delbridge *et al.*, 1999; Delbridge *et al.*, 2004). In several placental mammals, the entire *RBMY* and *TSPY* genes have also been subject to amplification, which is probably mediated by duplication between the palindromic sequences within which they lie.

We know most about the sex determining gene *SRY* and its X partner *SOX3*. Alignment with *SOX3* sequence shows that *SRY* retains only the DNA binding HMG box domain of its ancestor, and has traded the transactivating domains of *SOX3* for flanking regions, which are quite different in human, mouse and kangaroo (Foster and Graves, 1994). Thus *SRY* is essentially a truncated *SOX3* gene. Given that truncation of the related *SOX9* gene turns a transcriptional activator into a repressor, this led to the proposal that the role of *SRY* in the testis was to interfere with the DNA binding of another HMG box containing gene, rather than to activate testis-differentiating genes (Graves, 1998).

Thus most genes on the mammalian Y chromosome evolved from homologous gene pairs on the proto-X and proto-Y chromosomes, with the exception of two more recent recruits. Many of these genes still complement the functions of their X partners, so are in an early stage of Y differentiation. Several have recently undergone changes that render them inactive, suggesting that this class of gene is readily lost from the Y. In contrast, genes that are retained long-term on the Y are those that have an important male-specific role. This role could be an original function of the gene when it was autosomal, or could have been acquired during evolution. Genes on the MSY therefore represent a spectrum of evolution of male-specific genes from widely expressed genes on an ancestral autosome (Fig. 2).

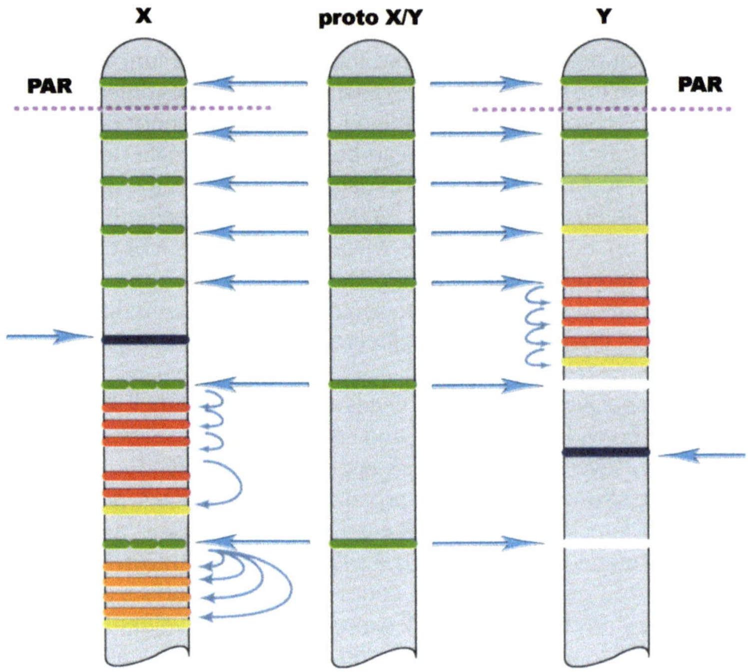

Figure 2. Fate of genes on the human sex chromosomes. Initially all genes were active (green lines), but as the X and Y chromosomes differentiated from the homologous proto-X/Y, genes on the Y near the testis determining factor *SRY* became isolated from recombination with their partners on the X chromosome. They accumulated deleterious mutations, became partially active (pale green) or completely inactivated (yellow) and were eventually lost from the Y chromosome (white). Their newly unpartnered active X homologues were then recruited into the X inactivation system (dashed green), to equalise the dosage of these X genes in males and females. Genes that remain on the Y are either in an early stage of degradation, or have acquired a male-specific role. Genes with functionally equivalent X and Y copies escape inactivation on the X (solid green on X and Y). These genes are clustered in or near the pseudoautosomal region (PAR), and probably represent an early stage of degradation. Genes with a male-specific function (red) no longer complement their partners on the X, which are inactivated. A few male-specific genes (blue) have been transposed to both the X and Y chromosomes. On both the X and Y chromosome, genes with a testis-specific expression pattern are located within large inverted repeat regions and have undergone amplification, by duplication (red) and retrotransposition (orange). Some of these copies have become inactive due to mutation.

Sex and reproduction genes on the mammalian X chromosome

It was assumed that throughout this process of Y degradation, the X remained steadfastly invariant. Indeed, the X is by far the most conserved region of the eutherian genome, with a size and gene content that is almost identical between eutherian species (Wakefield and Graves, 2003), perhaps because it is protected from rearrangement by its involvement in the chromosome-wide X chromosome inactivation (Ohno, 1967). Gene order on the X has also been conserved between eutherian groups as divergent as primates, equids and even insectivores (Raudsepp, Lee, Kata, Brinkmeyer, Mickelson, Skow, Womack and Chowdhary, 2004). Sequence divergence is typically less for X-linked than autosomal genes (Ellegren and Fridolfsson, 1997; McVean and Hurst, 1997), perhaps because it is more frequently in the ovary than the testis, so it is subject to a lower mutation pressure. Thus the mammalian X chromosome was considered to be constant in gene content, arrangement and sequence.

However, genome sequencing projects have now confirmed growing suspicion that the mammalian X chromosome too has an atypical gene content. The proposal that the X is enriched in genes that are involved in sex and reproduction was first made on the basis of comparisons of the functions of X-borne versus autosomal genes (Saifi and Chandra, 1999), and supported by database analyses that showed that the mammalian X chromosome is over-endowed with genes that are involved in sex and reproduction in both sexes, as well as in brain development and function (Wilda, Bachner, Zechner, Kehrer-Sawatzki, Vogel and Hameister, 2000; Wang, McCarrey, Yang and Page, 2001; Zechner, Wilda, Kehrer-Sawatzki, Vogel, Fundele and Hameister, 2001). Consistent with this proposal was the observation that nearly half the cDNAs isolated from mouse spermatocytes mapped to the X (Wang *et al.*, 2001). The mammalian X chromosome seems to be enriched also for genes expressed in the ovary and placenta (Vallender and Lahn, 2004; Khil, Smirnova, Romanienko and Camerini-Otero, 2004).

The over-representation of genes involved in spermatogenesis was surprising in view of evidence that such genes were entirely absent for the X in fruit flies and nematodes (Reinke, Smith, Nance, Wang, Van Doren, Begley, Jones, Davis, Scherer, Ward and Kim, 2000; Thoemke, Yi, Ross, Kim, Reinke and Zarkower, 2005), and there was initial confusion in the literature about whether genes involved in spermatogenesis were over or underrepresented on the human and mouse X chromosomes. Detailed examination of expression patterns showed that the subset of genes expressed in the early stages of spermatogenesis are over-represented on the human and mouse X chromosomes, whereas genes expressed in the later stages of spermatogenesis are poorly represented (Lercher Urrutia and Hurst, 2003; Khil *et al.*, 2004). This is not surprising, since transcription is repressed when the sex chromosomes condense and inactivate at the onset of male meiosis.

Early spermatogonial genes include the testis-specific sex and reproduction related (SRR) genes, and the cancer-testis (CT) antigen genes. These two groups are not mutually exclusive and several genes have been classified as both. The CT antigen genes are normally expressed in human spermatogonia (Fig. 3), but are also expressed in cancer cells, and are the targets of research into tumour-specific antigens for the development of cancer vaccines. Some X-borne CT antigen genes (e.g. *FTHL17*, *NXF2*, *TAF2Q* and *FATE1*) are single copy genes, but several are present on the human X chromosome as vastly expanded gene families. For example, the *MAGE* (melanoma antigen) gene family consists of 52 members, the *GAGE/PAGE/XAGE* (G antigen/prostate-associated) gene family is made up of 19 members, the *SSX* (sarcoma, synovial, X breakpoint) gene family has 9 members and the *CT47* gene family has 13 members. The 5 member *SPANX* (sperm protein associated with the nucleus, X chromosome) gene family is a unusual CT antigen gene on the X chromosome that is not expressed in early spermatogenesis, but later, in the spermatids following meiosis (Fig. 3) (Khil *et al.*, 2004; Ross *et al.*, 2005; Simpson, Caballero, Jungbluth, Chen and Old, 2005).

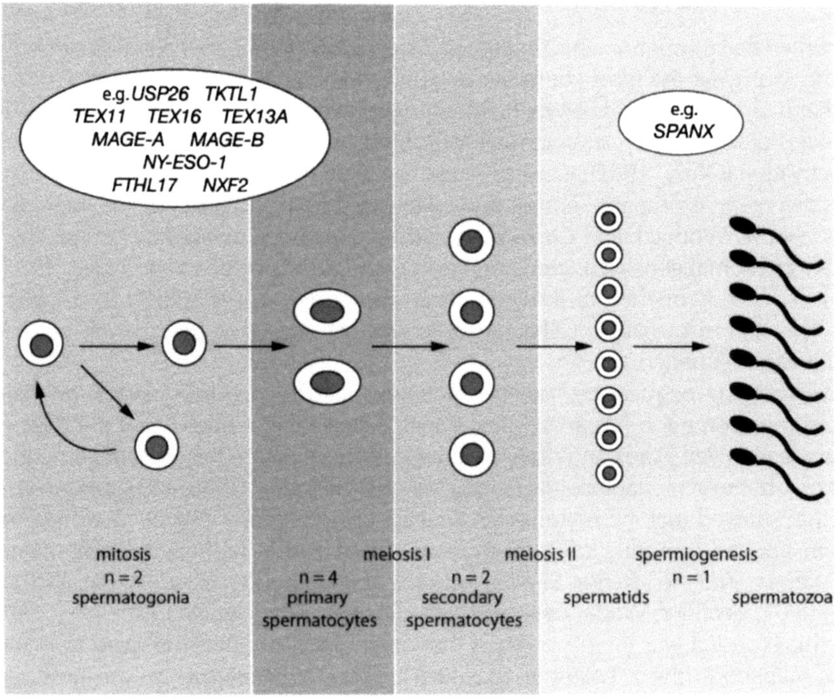

Figure 3. Roles of X-borne testis specific genes in spermatogenesis. Most CT antigen and SRR genes are expressed in the early stages of spermatogenesis prior to meiosis. The exception is the *SPANX* CT antigen gene family that is expressed in spermiogenesis.

What is the evolutionary explanation for this over-representation of sex and reproduction genes on the X chromosome? As early as 1931 it was proposed that the presence of two X chromosomes in females and only one in males favours the accumulation of male-advantage genes on the X chromosome (Fisher, 1931). A new allele that confers an advantage on males - such as more, or faster sperm - will be immediately exposed to selection in males, since they have only a single X. However, if it is recessive, it will not be expressed in females, and as long as it is rare there will be no homozygotes to worry about. Thus, male-advantage genes will be rapidly selected if they are on the X as well as the Y, even if they have no effect, or even a detrimental effect, on female carriers.

The over-representation of female-advantage genes, such as those expressed in ovary and placenta, can also be explained by the differences in X copy number between males and females, since the probability of a new mutation is higher in XX females than in XX males.

Mechanism for accumulation of SRR genes on sex chromosomes

Duplication of intron-containing genes, duplication of large sections between palindromic repeat sequences, and retrotransposition has all contributed to the functional bias of the human X and Y chromosomes. The mechanism for the accumulation of SRR genes (including many testis-specific genes that are required for spermatogenesis) is similar on the X and Y chromosomes.

When did accumulation of SRR genes on the X chromosome occur? Several of the SRR genes on the human and mouse X are single copy, intron-containing genes and were probably located on the ancient proto-X and -Y chromosomes. For instance, the testis-specific *TEX11*, *NXF2*, *TKTL1* genes on the X chromosome are present on the orthologous chromosomes in chicken, and so must be very ancient (Kohn *et al.*, 2004). However, although these are ancient genes, it is not certain that their function in sex determination or spermatogenesis is ancient, since selection for male-advantage traits may have selected for testis-specific variants once the autosome bearing them became proto-sex chromosomes. Just such neofunctionalisation seems to have occurred repeatedly to the Y copies of X-Y shared genes, as discussed above.

Many X-borne single copy testis-specific genes contain no introns, so probably arose by retrotransposition from an autosomal source gene. For instance the intronless *ATXN3L* and *YY2* genes have autosomal, but not X-borne paralogues, and were therefore probably derived from the insertion of a DNA copy of an (intronless) mRNA transcript (Loriot, Boon and De Smet, 2003; Emerson, Kaessmann, Betran and Long, 2004). More than 75% of the retrogenes (genes generated by retrotransposition) on the human X chromosome are also found on the mouse X chromosome, implying that the retrotransposition events that generated these gene copies occurred prior to the divergence of primates and rodents approximately 80 million years ago (Emerson *et al.*, 2004).

The X chromosome appears to be a frequent target for retrotranspositions, recruiting a disproportionately high number of functional retroposed genes from autosomal origins (there are 13 X located retroposons, compared with 3.4 expected by chance alone). The autosomal genes which give rise to the retroposons are widely expressed, but about half the retrotransposed copies on the X chromosome are testis-specific (Emerson *et al.*, 2004; Simpson *et al.*, 2005).

Some gene families have been greatly expanded by gene amplification and retrotransposition. For instance, the CT antigen gene families on the human X chromosome, which represent a remarkable 10% of the genes on the human X (Simpson *et al.*, 2005) consist of expanded gene families. Comparison of CT antigen gene sequences suggests that the CT gene families duplicated and evolved rapidly so that identification of orthologues is difficult. Intron-containing genes were evidently amplified, and then other members of the family were generated by retrotransposition. Many members of these testis-specific gene families on the human X appear to have no orthologues in other species, implying that they were generated recently. For most families, only one or two members of a gene family can be found in other mammals, identifying the founder members of the families, from which all other family members arose.

One of the best studied CT antigen gene families on the human X chromosome is the *MAGE* family. Retrotransposition has featured strongly in the expansion of the *MAGE* gene family (Chomez *et al.*, 2001). Comparisons of gene structure have revealed that the *MAGE* gene family has expanded on the human X chromosome largely by retrotransposition from the *MAGE-D* subfamily located at Xp11. The *MAGE-D* subfamily consists of four ubiquitously expressed genes that arose by duplication prior to the divergence of primates and rodents. These four genes then generated the numerous members of the testis-specific *MAGE-A*, *MAGE-B* and *MAGE-C* family in the human lineage by retrotransposition (Chomez, De Backer, Bertrand, De Plaen, Boo and Lucas, 2001; Bertrand, Huijbers, Chomez and De Backer, 2004).

Retrotransposition goes both ways, for many genes on the X chromosome have generated copies on autosomes. This has given rise to autosomal testis-specific paralogues that substitute for source housekeeping genes that are inactivated in the late stages of spermatogenesis (e.g. *PGK1*, *PDHA1* and *RBMX* (Dahl, Brown, Hutchinson, Maragos and Brown, 1990; McCarrey *et al.*, 1996; Lingenfelter, Delbridge, Thomas, Hoekstra, Mitchell, Graves and Disteche, 2001). This was probably the result of selection to retain male germline expression of genes (Emerson *et al.*, 2004).

Retrogenes (functional processed genes) on the X are heavily biased toward testis-specific function, as predicted by the hypothesis that there is strong selection for male-advantage genes on the X. Fewer than expected X-borne retrogenes are expressed in female tissues, consistent with the hypothesis that for male-advantage genes that are detrimental in females, there is strong selection for limitation of their function to male tissue. These sexually antagonistic (male-advantage/female disadvantage) retrogenes have therefore accumulated on the X, as well as the Y chromosome (Emerson *et al.*, 2004).

The *SSX* (synovial sarcoma breakpoint, X chromosome) gene family is a CT antigen gene family that expanded on the X chromosome via gene duplication, rather than retrotransposition. It comprises many copies on both the human and mouse X. The *SSX* genes are likely to encode transcriptional repressors, and expression of all the family members is normally restricted to the testis, suggesting a normal role in mammalian spermatogenesis. The 9 members of the human *SSX* gene family lie in two clusters approximately 3 Mb apart on the short arm of the X chromosome. Expansion of the family occurred via duplication of an original intron-containing gene, and the two clusters arose by duplication of a region of approximately 100 kb (Gure, Wei, Old and Chen, 2002).

The human *SPANX* (human sperm protein associated with the nucleus on the X chromosome) gene family is the only CT antigen gene family on the X chromosome that is expressed in late spermatogenesis. Expression of the mouse gene is implied by the existence of a number of ESTs. They are also amplified on the X in great apes, and a single copy of *SPANX-N* has been identified from the sequence databases in both the mouse and rat. This implies that the *SPANX-N* genes are the ancestral members of the *SPANX* gene family, from which the other members of the family evolved early in the hominid lineage. The human *SPANX* gene family contains nine genes clustered at Xq27, and one at Xp11. The family arose largely by tandem duplication. Four *SPANX-N* genes are at located Xq27 and a single *SPANX-N* gene lies at Xp11. The *SPANX-N* genes are the founder members of the family, the five other members of the family also cluster within 800 kb in Xq27. All *SPANX* genes reside within a series of 20 kb tandem duplications in this region (Kouprina, Pavlicek, Noskov, Solomon, Otstot, Isaacs, Carpten, Trent, Schleutker, Barrett, Jurka and Larionov, 2004; Kouprina, Mullokandov, Rogozin, Collins, Solomon, Otstot, Risinger, Koonin, Barrett and Larionov, 2005).

The repetitive structure of the X might be responsible for the amplification and relocation of some of the multiple copy CT antigen gene families (e.g. *MAGE, GAGE/PAGE/XAGE, SSX,* and *SPANX*), which appear to be located in several distinct clusters on the human X chromosome (Fig. 4). A genome-wide analysis of inverted repeat structure in the human genome has revealed that the human X chromosome contains several large inverted repeats (or imperfect palindromes) near the centromere on the short arm, and towards the telomere of the long arm of the X chromosome. The inverted repeats in these regions contain genes with a testis-specific expression pattern, including members of the CT antigen gene families (Warburton, Giordano, Cheung, Gelfand and Benson, 2004). This pattern is analogous to the palindromic repeats found on the human Y chromosome in Yq11, which also contain amplified testis specific genes.

The inverted repeats on the X chromosome predominantly contain genes expressed in the testis, suggesting a role for the structure in gene expression during spermatogenesis. The mammalian X and Y chromosomes undergo male germline sex chromosome inactivation. This prevents expression of X and Y linked genes during and after the first meiotic division of the primary spermatocytes. This inactivation of many essential X-linked genes appears to be compensated for by the testis-specific expression of autosomal retrotransposed copies (e.g. *PGK, PDHA1* and *RBMX*). There is a requirement for at least some X- and Y-linked testis genes to be expressed in the late stages of spermatogenesis (e.g. *SPANX*). It has been suggested that formation of cruciforms or other structures by the inverted repeat sequences might permit escape from meiotic inactivation, and allow expression of crucial genes in late spermatogenesis (Skaletsky *et al.*, 2003; Warburton *et al.*, 2004).

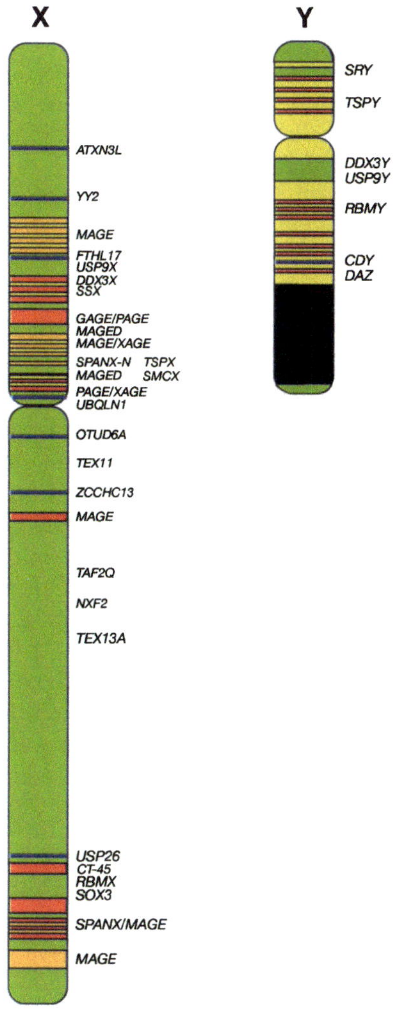

Figure 4. Clusters of testis-specific genes on the human X and Y chromosomes, colour coded as in Figure 2. Most genes on the X are active (green) and subject to X inactivation; only a few regions of the Y contain active ubiquitously expressed genes (green). Most of the Y chromosome is littered with inactive remnants of X-Y shared genes and inactive pseudogenes (yellow). Many genes, including *SRY,* have differentiated from their partners to have a male specific function in sex determination or spermatogenesis. Members of the testis-specific CT antigen gene families are clustered on the short arm of the X near the centromere, and the long arm near to the telomere. Family members have been generated by duplications (red) or retrotransposition (orange). Some family members have become inactive (yellow). The known founder members of some of the CT antigen gene families are shown in bold type. Genes that are the result of retrotranspositions onto the sex chromosomes are shown in blue.

In contrast, it has also been suggested that the formation of inverted repeat secondary structures might suppress expression of the testis genes within them. The accumulation of recessive alleles beneficial to males (XY) but detrimental to females (XX) on the X chromosome would not be selected against because the detrimental effect in females would be masked in heterozygous females. As these genes accumulate on the X, it would be expected that modifiers would arise to limit the expression of these genes to males. Formation of inverted repeat structures on the X chromosome might be one mechanism by which expression of testis genes might be limited to the male germ line, to prevent deleterious expression of these X-linked genes in females (Wu and Xu, 2003).

The functional bias of genes on both the X and Y chromosomes therefore has been generated by similar mechanisms including duplication of intron-containing genes, duplication of large sections between palindromic repeat sequences, and retrotransposition.

Conclusions

Suspicions of a bias in gene content of the human X have been confirmed by sequencing. Thus the X, as well as the Y chromosome contains many genes that are expressed only in the testis. Concentration of male-advantage genes on the Y resulted from loss the genes that conferred no selective advantage in males. Accumulation of such genes on the X chromosome probably results from the immediate exposure of new X-borne alleles to selection in the hemizygous (XY) male. Clearly on both sex chromosomes there has been recruitment, retrotransposition, duplication and neofunctionalisation. This has resulted in the striking bias of X-borne genes toward functions in gonadogenesis and spermatogenesis (Fig. 2). Analysis of the intergenic sequences of both the X and Y chromosomes has revealed that they are more alike than first thought, both containing large palindromic sequences which contain many of the testis specific genes on these chromosomes. These palindromic sequences on the X and Y probably contributed to amplification of the testis specific genes within them, and they may play an important regulatory role in controlling testis specific gene expression during meiosis.

References

Aitken, R. J. and Graves, J. A. M. (2002). The future of sex. *Nature* **415:** 963

Bertrand, M., Huijbers, I., Chomez, P. and De Backer, O. (2004). Comparative expression analysis of the *MAGED* genes during embryogenesis and brain development. *Developmental Dynamics* **230:** 325-334

Burgoyne, P. S. (1996). Fruit(less) flies provide a clue. *Nature* **381:** 740-741

Carvalho, A. B. (2002). Origin and evolution of the Drosophila Y chromosome. *Current Opinion in Genetics and Development* **12:** 664-668

Charlesworth, B. (1991). The evolution of sex chromosomes. *Science* **251:** 1030-1033

Chomez, P., De Backer, O., Bertrand, M., De Plaen, E., Boon, T. and Lucas, S. (2001). An overview of the *MAGE* gene family with the identification of all human members of the family. *Cancer Research* **61:** 5544-5551

Dahl, H. H., Brown, R. M., Hutchison, W. M., Maragos, C. and Brown, G. K. (1990). A testis-specific form of the human pyruvate dehydrogenase E1 alpha subunit is coded for by an intronless gene on chromosome 4. *Genomics* **8:** 225-232

Delbridge, M. L., Lingenfelter, P. A., Disteche, C. M. and Graves, J. A. (1999). The candidate spermatogenesis gene *RBMY* has a homologue on the human X chromosome. *Nature Genetics* **22:** 223-224

Delbridge, M. L., Longepied, G., Depetris, D., Mattei, M. G., Disteche, C. M., Marshall Graves, J. A. and Mitchell, M. J. (2004). *TSPY*, the candidate gonadoblastoma gene on the human Y chromosome, has a widely expressed homologue on the X - implications for Y chromosome evolution. *Chromosome Research* **12:** 345-356

Dorus, S., Gilbert, S. L., Forster, M. L., Barndt, R. J. and Lahn, B. T. (2003). The *CDY*-related gene family: coordinated evolution in copy number, expression profile and protein sequence. *Human Molecular Genetics* **12:** 1643-1650

Ellegren, H. and Fridolfsson, A. K. (1997). Male-driven evolution of DNA sequences in birds. *Nature Genetics* **17:** 182-18

Emerson, J. J., Kaessmann, H., Betran, E. and Long, M. (2004). Extensive gene traffic on the mammalian X chromosome. *Science* **303:** 537-540

Fisher, R. A. (1931). The evolution of dominance. *Biological Reviews* **6:** 345-368

Foresta, C., Ferlin, A. and Moro, E. (2000). Deletion and expression analysis of AZFa genes on the human Y chromosome revealed a major role for *DBY* in male infertility. *Human Molecular Genetics* **9:** 1161-1169

Foster, J. W. and Graves, J. A. M. (1994). An SRY-related sequence on the marsupial X chromosome: implications for the evolution of the mammalian testis-determining gene. *Proceedings of the National Academy of Sciences USA* **91:** 1927-1931

Glas, R., De Leo, A. A., Delbridge, M. L., Reid, K., Ferguson-Smith, M. A., O'brien, P. C., Westerman, M. and Graves, J. A. (1999). Chromosome painting in marsupials: genome conservation in the kangaroo family. *Chromosome Research* **7:** 167-176

Graves, J. A. (2006). Sex chromosome specialization and degeneration in mammals. *Cell* **124:** 901-914

Graves, J. A. M. (1995). The origin and function of the mammalian Y chromosome and Y-borne genes - an evolving understanding. *Bioessays* **17:** 311-320

Graves, J. A. M. (1998). Interactions between *SRY* and *SOX* genes in mammalian sex determination. *Bioessays* **20:** 264-269

Graves, J. A. M. and Shetty, S. (2001). Sex from W to Z - evolution of vertebrate sex chromosomes and sex determining genes. *Journal of Experimental Zoology* **281:** 472-481

Gure, A. O., Wei, I. J., Old, L. J. and Chen, Y. T. (2002). The *SSX* gene family: characterization of 9 complete genes. *International Journal of Cancer* **101:** 448-453

Khil, P. P., Smirnova, N. A., Romanienko, P. J. and Camerini-Otero, R. D. (2004). The mouse X chromosome is enriched for sex-biased genes not subject to selection by meiotic sex chromosome inactivation. *Nature Genetics* **36:** 642-642

Kohn, M., Kehrer-Sawatzki, H., Vogel, W., Graves, J. A. and Hameister, H. (2004). Wide genome comparisons reveal the origins of the human X chromosome. *Trends in Genetics* **20:** 598-603

Kouprina, N., Mullokandov, M., Rogozin, I. B., Collins, N. K., Solomon, G., Otstot, J., Risinger, J. I., Koonin, E. V., Barrett, J. C. and Larionov, V. (2004). The *SPANX* gene family of cancer/testis-specific antigens: rapid evolution and amplification in African great apes and hominids. *Proceedings of the National Academy of Sciences USA* **101:** 3077-3082

Kouprina, N., Pavlicek, A., Noskov, V. N., Solomon, G., Otstot, J., Isaacs, W., Carpten, J. D., Trent, J. M., Schleutker, J., Barrett, J. C., Jurka, J. and Larionov, V. (2005). Dynamic structure of the SPANX gene cluster mapped to the prostate cancer susceptibility locus HPCX at Xq27. *Genome Research* **15:** 1477-1486

Lahn, B. T. and Page, D. C. (1997). Functional coherence of the human Y chromosome. *Science* **278:** 675-680

Lahn, B. T. and Page, D. C. (1999). Four evolutionary strata on the human X chromosome. *Science* **286:** 964-967

Laumonnier, F., Ronce, N., Hamel, B. C., Thomas, P., Lespinasse, J., Raynaud, M., Paringaux, C., Van Bokhoven, H., Kalscheuer, V., Fryns, J. P., Chelly, J., Moraine, C. and Briault, S. (2002). Transcription factor SOX3 is involved in X-linked mental retardation with growth hormone deficiency. *American Journal of Human Genetics* **71:** 1450-1455

Lercher, M. J., Urrutia, A. O. and Hurst, L. D. (2003). Evidence that the human X chromosome is enriched for male-specific but not female-specific genes. *Molecular Biology and Evolution* **20:** 1113-1116

Lingenfelter, P. A., Delbridge, M. L., Thomas, S., Hoekstra, H. E., Mitchell, M. J., Graves, J. A. M. and Disteche, C. M. (2001). Expression and conservation of processed copies of the *RBMX* gene. *Mammalian Genome* **12:** 538-545

Loriot, A., Boon, T. and De Smet, C. (2003). Five new human cancer-germline genes identified among 12 genes expressed in spermatogonia. *International Journal of Cancer* **105:** 371-376

Makova, K. D. and Li, W. H. (2002). Strong male-driven evolution of DNA sequences in humans and apes. *Nature* **416:** 624-626

McCarrey, J. R., Kumari, M., Aivaliotis, M. J., Wang, Z., Zhang, P., Marshall, F. and Vandeberg, J. L. (1996). Analysis of the cDNA and encoded protein of the human testis-specific *PGK-2* gene. *Developmental Genetics* **19:** 321-332

Mcvean, G. T. and Hurst, L. D. (1997). Evidence for a selectively favourable reduction in the mutation rate of the X chromosome. *Nature* **386:** 388-392

Miyata, T., Hayashida, H., Kuma, K., Mitsuyasu, K. and Yasunaga, T. (1987). Male-driven molecular evolution: a model and nucleotide sequence analysis. *Cold Spring Harbor Symposia of Quantative Biology* **52:** 863-867

Muller, H. J. (1914). A gene for the fourth chromosome of *Drosophila*. *Journal of Experimental Zoology* **17:** 325-336

Nanda, I., Shan, Z., Schartl, M., Burt, D. W., Koehler, M., Nothwang, H., Grutzner, F., Paton, I. R., Windsor, D., Dunn, I., Engel, W., Staeheli, P., Mizuno, S., Haaf, T. and Schmid, M. (1999). 300 million years of conserved synteny between chicken Z and human chromosome 9. *Nature Genetics* **21:** 258-259

Ohno, S. (1967). *Sex Chromosomes and Sex Linked Genes*. Springer-Verlag, Berlin

Park, D. J., Pask, A. J., Huynh, K., Renfree, M. B., Harley, V. R. and Graves, J. A. (2004). Comparative analysis of *ATRX*, a chromatin remodeling protein. *Gene* **339:** 39-48

Raudsepp, T., Lee, E. J., Kata, S. R., Brinkmeyer, C., Mickelson, J. R., Skow, L. C., Womack, J. E. and Chowdhary, B. P. (2004). Exceptional conservation of horse-human gene order on X chromosome revealed by high-resolution radiation hybrid mapping. *Proceedings of the National Academy of Sciences USA* **101:** 2386-2391

Reinke, V., Smith, H. E., Nance, J., Wang, J., Van Doren, C., Begley, R., Jones, S. J., Davis, E. B., Scherer, S., Ward, S. and Kim, S. K. (2000). A global profile of germline gene expression in C. elegans. *Molecular Cell* **6:** 605-616

Repping, S., Skaletsky, H., Lange, J., Silber, S., Van Der Veen, F., Oates, R. D., Page, D. C. and Rozen, S. (2002). Recombination between palindromes P5 and P1 on the human Y chromosome causes massive deletions and spermatogenic failure. *American Journal of Human Genetics* **71:** 906-922

Ross, M. T., Grafham, D. V., Coffey, A. J., Scherer, S., Mclay, K., Muzny, D., Platzer, M., Howell, G. R., Burrows, C., Bird, C. P., Frankish, A., Lovell, F. L., Howe, K. L., Ashurst, J. L., Fulton, R. S., Sudbrak, R., Wen, G., Jones, M. C., Hurles, M. E., Andrews, T. D., Scott, C. E., Searle, S., Ramser, J., Whittaker, A., Deadman, R., Carter, N. P., Hunt, S. E., Chen, R., Cree, A., Gunaratne, P., Havlak, P., Hodgson, A., Metzker, M. L., Richards, S., Scott, G., Steffen, D., Sodergren, E., Wheeler, D. A., Worley, K. C., Ainscough, R., Ambrose, K. D., Ansari-Lari, M. A., Aradhya, S., Ashwell, R. I., Babbage, A. K., Bagguley, C. L., Ballabio, A., Banerjee, R., Barker, G. E., Barlow, K. F., Barrett, I. P., Bates, K. N., Beare, D. M., Beasley, H., Beasley, O., Beck, A., Bethel, G., Blechschmidt, K., Brady, N., Bray-Allen, S., Bridgeman, A. M., Brown, A. J., Brown, M. J., Bonnin, D., Bruford, E. A., Buhay, C., Burch, P., Burford, D., Burgess, J., Burrill, W., Burton, J., Bye, J. M., Carder, C., Carrel, L., Chako, J., Chapman, J. C., Chavez, D., Chen, E., Chen, G., Chen, Y., Chen, Z., Chinault, C., Ciccodicola, A., Clark, S. Y., Clarke, G., Clee, C. M., Clegg, S., Clerc-Blankenburg, K., Clifford, K., Cobley, V., Cole, C. G., Conquer, J. S., Corby, N., Connor, R. E., David, R., Davies, J., Davis, C., Davis, J., Delgado, O., Deshazo, D., et al. (2005). The DNA sequence of the human X chromosome. *Nature* **434:** 325-337

Saifi, G. M. and Chandra, H. S. (1999). An apparent excess of sex- and reproduction-related genes on the human X chromosome. *Proceedings of the Royal Society of London, series B* **266:** 203-209

Saxena, R., Brown, L. G., Hawkins, T., Alagappan, R. K., Skaletsky, H., Reeve, M. P., Reijo, R., Rozen, S., Dinulos, M. B., Disteche, C. M. and Page, D. C. (1996). The *DAZ* gene cluster on the human Y chromosome arose from an autosomal gene that was transposed, repeatedly amplified and pruned. *Nature Genetics* **14:** 292-299

Seboun, E., Barbaux, S., Bourgeron, T., Nishi, S., Agulnik, A., Egashira, M., Nikkawa, N., Bishop, C., Fellous, M., McElreavey, K., Kasahara, M. and Algonik, A. (1997). Gene sequence, localization, and evolutionary conservation of *DAZLA*, a candidate male sterility gene. *Genomics* **41:** 227-235

Shetty, S., Griffin, D. K. and Graves, J. A. M. (1999). Comparative painting reveals strong chromosome homology over 80 million years of bird evolution. *Chromosome Research* **7:** 289-295

Simpson, A. J., Caballero, O. L., Jungbluth, A., Chen, Y. T. and Old, L. J. (2005). Cancer/testis antigens, gametogenesis and cancer. *Nature Reviews. Cancer* **5:** 615-625

Skaletsky, H., Kuroda-Kawaguchi, T., Minx, P. J., Cordum, H. S., Hillier, L., Brown, L. G., Repping, S., Pyntikova, T., Ali, J., Bieri, T., Chinwalla, A., Delehaunty, A., Delehaunty, K., Du, H., Fewell, G., Fulton, L., Fulton, R., Graves, T., Hou, S. F., Latrielle, P., Leonard, S., Mardis, E., Maupin, R., Mcpherson, J., Miner, T., Nash, W., Nguyen, C., Ozersky, P., Pepin, K., Rock, S., Rohlfing, T., Scott, K., Schultz, B., Strong, C., Tin-Wollam, A., Yang, S. P., Waterston, R. H., Wilson, R. K., Rozen, S. and Page, D. C. (2003). The male-specific region of the human Y chromosome is a mosaic of discrete sequence classes. *Nature* **423:** 825-837

Sun, C., Skaletsky, H., Birren, B., Devon, K., Tang, Z., Silber, S., Oates, R. and Page, D. C. (1999). An azoospermic man with a de novo point mutation in the Y-chromosomal gene *USP9Y*. *Nature Genetics* **23:** 429-432

Thoemke, K., Yi, W., Ross, J. M., Kim, S., Reinke, V. and Zarkower, D. (2005). Genome-wide analysis of sex-enriched gene expression during C. elegans larval development. *Developmental Biology* **284:** 500-508

Tiepolo, L. and Zuffardi, O. (1976). Localization of factors controlling spermatogenesis in the nonfluorescent portion of the human Y chromosome long arm. *Human Genetics* **34:** 119-124

Tsend-Ayush, E., O'Sullivan, L. A., Grutzner, F. S., Onnebo, S. M., Lewis, R. S., Delbridge, M. L., Marshall Graves, J. A. and Ward, A. C. (2005). *RBMX* gene is essential for brain development in zebrafish. *Developmental Dynamics* **234:** 682-688

Vallender, E. J. and Lahn, B. T. (2004). How mammalian sex chromosomes acquired their peculiar gene content. *Bioessays* **26:** 159-169

Vogt, P. H. (2004). Genomic heterogeneity and instability of the AZF locus on the human Y chromosome. *Molecular and Cellular Endocrinology* **224:** 1-9

Wakefield, M. J. and Graves, J. A. (2003). The kangaroo genome. Leaps and bounds in comparative genomics. *EMBO Reports* **4:** 143-147

Wang, P. J., McCarrey, J. R., Yang, F. and Page, D. C. (2001). An abundance of X-linked genes expressed in spermatogonia. *Nature Genetics* **27:** 422-426

Warburton, P. E., Giordano, J., Cheung, F., Gelfand, Y. and Benson, G. (2004). Inverted repeat structure of the human genome: the X-chromosome contains a preponderance of large, highly homologous inverted repeats that contain testes genes. *Genome Research* **14:** 1861-1869

Waters, P. D., Delbridge, M. L., Deakin, J. E., El-Mogharbel, N., Kirby, P. J., Carvalho-Silva, D. R. and Graves, J. A. (2005). Autosomal location of genes from the conserved mammalian X in the platypus (Ornithorhynchus anatinus): implications for mammalian sex chromosome evolution. *Chromosome Research* **13:** 401-410

Waters, P. D., Duffy, B., Frost, C. J., Delbridge, M. L. and Graves, J. A. M. (2001). The human Y chromosome derives largely from a single autosomal region added to the sex chromosomes 80-130 million years ago. *Cytogenetics and Cell Genetics* **92:** 74-79

Wilda, M., Bachner, D., Zechner, U., Kehrer-Sawatzki, H., Vogel, W. and Hameister, H. (2000). Do the constraints of human speciation cause expression of the same set of genes in brain, testis, and placenta? *Cytogenetics and Cell Genetics* **91:** 300-302

Wu, C. I. and Xu, E. Y. (2003). Sexual antagonism and X inactivation—the SAXI hypothesis. *Trends in Genetics* **19:** 243-247

Zechner, U., Wilda, M., Kehrer-Sawatzki, H., Vogel, W., Fundele, R. and Hameister, H. (2001). A high density of X-linked genes for general cognitive ability: a run-away process shaping human evolution? *Trends in Genetics* **17:** 697-701

Spermatology. SRF Vol. 65. ERS Roldan and M Gomendio (eds) Nottingham University Press, Nottingham

From stem cells to spermatozoa and back

Harry Moore and Behrouz Aflatoonian

Centre for Stem Cell Biology, University of Sheffield, Sheffield S10 2UH, UK

As a strategy for survival, the metazoa segregated immortal germ cells from the mortal soma by an early differentiation process and thereby reproduced asymmetrically in a specialised tissue. The legacy of this process is seen in the generation of mammalian gametes which are derived from a founder population of primordial germ cells (PGCs) that are determined early in embryogenesis and set aside for unique development. Understanding the mechanisms of PGC determination and differentiation is important for elucidating causes of infertility and how endocrine disrupting chemicals may increase susceptibility to congenital reproductive abnormalities and conditions such as testicular cancer in adulthood (testicular dysgenesis syndrome). Primordial germ cells are closely related to embryonic stem (ESCs) cells and embryonic germ (EG) cells and comparisons between these various cell types are providing new information about pluripotency and epigenetic processes. Murine embryonic stem cells can differentiate to PGCs, gametes and even blastocysts. Recently live mouse pups were born from sperm generated from ESCs. Although investigations are still preliminary, human embryonic stem cells (hESCs) apparently display a similar developmental capacity to generate PGCs and subsequently gametes. The findings indicate hESCs have the potential to differentiate to PGCs (as determined with surface markers and gene expression profiles) although these phenotypes represented a small proportion of the total cell population (~0.1%). Post-meiotic spermatids have been identified using specific markers, while primordial follicular structures have been observed in culture although not characterised definitely. Germ cells will enter meiosis autonomously and develop as oocytes unless meiosis is blocked and cells are induced into a spermatogenic pathway. During hESC differentiation both pathways seem to occur, regardless of the sex karyotype. Exactly how gamete-like cells are generated during stem cell culture remains unclear especially as conditions are currently ill defined. The findings are discussed in relation to the mechanisms of human PGC and gamete development and the biotechnology of hESCs and hEG cells.

Introduction

At some early stage of life on earth, primitive unicellular organisms deprived of nutrients evolved a strategy of survival by differentiating into 'fruiting bodies' and germ cell-like spores.

Corresponding author E-mail: h.d.moore@shef.ac.uk

The metazoa segregated immortal germ cells from the mortal soma by an early differentiation process and in so doing reproduced asymmetrically in a specialised tissue. Germ cells evolved as haploid gametes faithfully carrying the genome from one generation to another and thus becoming vehicles of the immortal DNA of an individual and providing a totipotent blueprint to create soma. However, the soma lost immortality, resulting in ageing and death as the inevitable consequence. The evolutionary legacy of this archaic 'germ–soma conflict' (Heininger, 2002) is manifest in mammals today in the foetal segregation of primordial germ cells (Donovan and de Miguel, 2003; McLaren, 2003) - a process that recapitulates the fate of the fruiting body. Thus many genes involved in germ cell development are highly conserved (Donovan and de Miguel, 2003) although subtle but important differences occur between relatively closely related species.

In mammals, primordial germ cell development has been characterised the most in the mouse. By contrast, much less is known of the earliest stages of germ cell development in the human due to the practical and ethical difficulties associated with obtaining human tissue samples. Hence, an *in vitro* model system that can recapitulate the development of human germ cells and gametes would be an particularly valuable research tool which in due course could lead to novel reproductive toxicological assays and ultimately new clinical applications to overcome infertility. In our laboratory we have investigated the use of human embryonic stem cells (hESCs) to generate germ cells and gametes as an *in vitro* system to study human germ cell development (Aflatoonian and Moore, 2005). ESCs are defined as pluripotent stem-cell lines derived from early embryos before formation of the tissue germ layers (Smith, 2006). They are derived usually from the pre-implantation blastocyst and exhibit indefinite proliferative capacity under appropriate conditions *in vitro* (Kaufman and Evans, 1981; Martin, 1981; Thomson, Itskovitz-Eldor, Shapiro, Waknitz, Swiergiel, Marshall and Jones, 1998; Draper, Moore, Ruban, Gokhale and Andrews, 2004). Several milestone investigations have demonstrated that murine embryonic stem cells (mESCs) can differentiate to primordial germ cell (PGCs) and subsequently early gametes and blastocysts (Hubner Fuhrmann, Christenson, Kehler, Reinhold, De La Fuente, Wood, Strauss, Boiani and Scholer, 2003; Geijsen, Horoschak, Kim, Gribnau, Eggan and Daley, 2004). Recently, immature sperm cells derived from mESCs in culture have generated live offspring (Nayernia, Nolte, Michelmann, Lee, Rathsack, Drusenheimer, Dev, Wulf, Ehrmann, Elliott, Okpanyi, Zechner, Haaf, Meinhardt and Engel, 2006a).

Preliminary but persuasive data indicate that human embryonic stem cells (hESCs) most likely display a similar developmental capacity (Clark, Bodnar, Fox, Rodriguez, Abeyta, Firpo and Pera, 2004; Aflatoonian, Fazeli, Ruban, Andrews and Moore, 2005). Thus, this hESC technology offers great potential for new types of reproductive investigations including a readily accessible system to investigate the very earliest stages of human gametogenesis including epigenetic modifications of the germline (Allegrucci, Thurston, Lucas and Young, 2005). Indeed, a range of pluripotent stem cell lines have now been derived from different stages of germ cell development in the mouse and human (Fig. 1). Here, we discuss the derivation of germ cells and their formation from embryonic stem cells *in vitro*. These investigations are providing fresh insight into some of the mechanisms of human germ cell development.

Primordial germ cells

Mammalian gametes are derived from a small founder pluripotent population which when segregated early in embryogenesis in the developing yolk sac become primordial germ cells (PGCs). A detailed knowledge of the mechanisms controlling human PGC and gamete development is crucial for understanding the aetiology of various aspects of infertility, including the

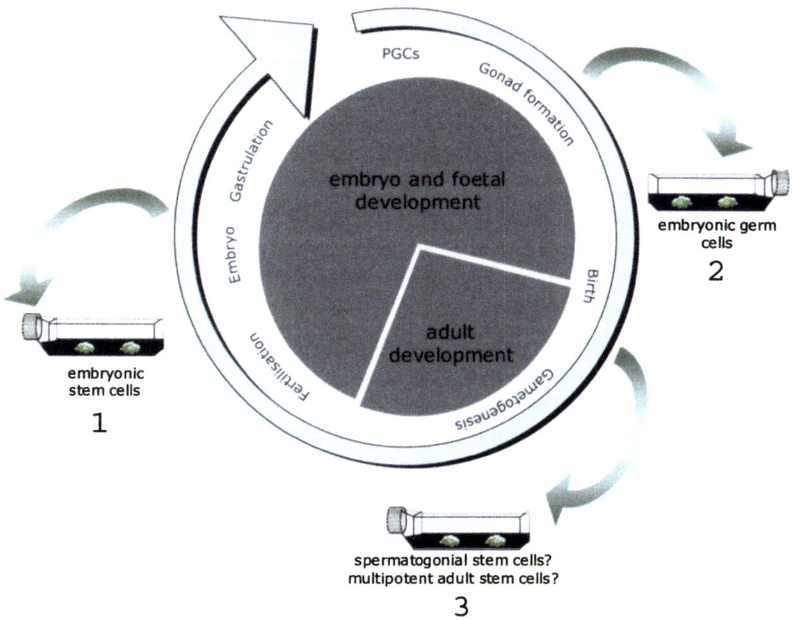

Figure 1. The germ cell cycle and stages where pluripotent cell lines have been reported to be derived. 1. Embryonic stem cells are derived from the inner cell mass of pre-implantation embryos in a number of mammalian species including the mouse and human. 2. Embryonic germ cells are derived from the genital ridge of the developing gonad. 3. Mouse spermatogonia stem cells from foetal, neonatal and adult testis may have multipotent potential.

influence environmental chemicals may have on reproductive capacity. For instance, it has been argued that endocrine disrupting chemicals in the environment or the workplace may interfere with early male germ cell development thereby increasing susceptibility to testicular cancer and infertility in adulthood; the so-called 'testicular dysgenesis syndrome' (Skakkebaek, Rajpert-De Meyts and Main, 2001). The specification of PGCs, their proliferation, maintenance and then differentiation to primary oocytes and prospermatogonia (precursor spermatogonia stem cells) have a profound bearing on the number and function of germ cells that are available subsequently for gametogenesis. The evidence is growing for effects of environmental chemicals on the various earliest stages of these processes (Baillie, Pacey and Moore, 2003). For example, organochlorines were implicated recently in affecting the ratio of X and Y sperm in human semen indicating a potential effect on early meiosis (Tiido, Rignell-Hydbom, Jonsson, Giwercman, Rylander, Hagmar and Giwercman, 2005). Moreover, transient exposure of pregnant rats in early gestation to the fungicides and endocrine disrupting chemicals vinclozolin (antiandrogenic compound) or methoxychlor (an oestrogenic compound) increased the incidence of male infertility in offspring (Anway, Cupp, Uzumcu and Skinner, 2005). Disturbingly, this infertility was transferred through the male germ line to subsequent generations possibly indicating modification to epigenetic pathways.

Over many years, the activity of tissue non-specific alkaline phosphatase (TNAP) has been utilised to mark PGCs and observe their transition from the base of the allantois, through the hindgut to the dorsal body wall where they enter into the genital ridges of the gonadal anlagen (McLaren, 2003). The high expression of this enzyme in PGCs is a feature also shared with

ESCs, embryonic germ (EG) cells and embryonal carcinoma (EC) cells, all of which have pluripotent capabilities. PGC migration (whether active or passive) occurs at about 7-10 days post conception (d.p.c.) in the mouse and between weeks 5-8 of human gestation (Freeman, 2003; Molyneaux and Wylie, 2005). Difficulties of undertaking detailed investigations of early human foetal development has meant that relatively little is known of the specification of human PGCs, although it is likely that common signalling pathways occur in mammals and possibly all vertebrates (see Donovan and de Miguel, 2003). In the mouse, PGCs originate in the proximal epiblast, a region of the embryo that also contributes to the first blood lineages of the embryonic yolk sac (Ginsburg, Snow and McLaren, 1990). As early as 6.25 d.p.c, germline competence can be identified in a founder population of as few as six epiblast cells that express the protein Blimp1 (B-lymphocyte-induced maturation protein 1; Ohinata, Payer, O'Carroll, Ancelin, Ono, Sano, Barton, Obukhanych, Nussenzweig, Tarakhosvsky, Saitou and Surani, 2005; Vincent, Dunn, Sciammas, Shapiro-Shalef, Davis, Calame, Bikoff and Robertson, 2005; McLaren and Lawson, 2005). Blimp1 was first identified as a transcriptional repressor that enables the further differentiation of immunoglobulin-secreting plasma cell by inhibiting the expression of genes involved in alternative B-cell development. Mutant null-allele mice lacking *Blimp1* generate very few PGCs, and those that do develop lack normal migratory behaviour (Vincent *et al.*, 2005). In the female, the gonocyte surrounded by a cortical interstitial layer initiates meiosis and becomes a primary oocyte and follicle, thereby ending precursor proliferative potential. In the male, the gonocyte surrounded by the foetal sex cord of the gonadal ridge (pre-seminiferous tubules) arrests in G0/G1 of mitosis as a prospermatogonium but retains a proliferative precursor potential. Following birth, prospematogonia migrate to the basement membrane of the seminiferous tubule and differentiate into spermatogonial stem cells (SSCs). Like adult stem cells, SSCs can both self-renew and provide daughter cells, which differentiate into one or more terminal cell types (Brinster, 2002). In this respect there is link relating the self-renewal/totipotent capacity of embryonic stem cell, primordial germ cells, embryonic germ cells and spermatogonial stem cells (Fig. 1).

Germ cell competence is induced in the murine epiblast in response to signals secreted by extra embryonic ectoderm including the synergistic action of the growth factors; particularly bone morphogenic proteins (BMP) 4 and 8b; both members of the transforming growth factor-beta (TGF-ß) superfamily of secreted proteins (Shimasaki, Moore, Otsuka and Erickson, 2004). BMP4 in its mature form is a dimer that binds and then signals through heteromeric receptor complexes and downstream SMAD proteins. On their own BMP4 or BMP8b are unable to induce PGCs from cultured epiblast, but can do so in combination. This indicates that signalling for various BMPs may occur through separate receptor complexes. Interestingly, BMP2, a close relative of BMP4, is expressed in visceral endoderm at the time of PGC specification, and inactivation of BMP2 results in a reduction in PGC number, revealing a function of visceral endoderm in PGC generation in the mouse at least (Lawson, Dunn, Roelen, Zeinstra, Davis, Wright, Korving and Hogan, 1999; Ying, Qi and Zhao, 2001; Shimasaki *et al.*, 2004; de Sousa Lopes, Roelen, Monteiro, Emmens, Lin, Li, Lawson and Mummery, 2004).

The genes *fragilis* and *stella* also have key roles in germ cell competency and development. Fragilis, a trans-membrane protein and part of a larger interferon- inducible family of genes, is evolutionarily conserved and has human homologues. Interferon-inducible proteins have an anti-proliferative function and fragilis may serve to increase the length of the cell cycle in PGCs. During induction of germ cell fate, there is only transient expression of *fragilis* but this gene is also expressed in ESCs and embryonic germ (EG) cells signifying a potential role in pluripotency status (Saitou, Barton and Surani, 2002). Similarly, the gene *stella* may have a function during the development of pluripotency. It is expressed in the oocyte, through preimplantation embryo development and in germ cell tumours (Payer, Saitou, Barton, Thresher, Dixon, Zahn, Colledge, Carlton, Nakano and Surani, 2003). *Stella*-positive germ cells exhibit repression of homeobox genes, which may go some way to explaining how germ cells escape from a somatic cell fate and the retention of pluripotency (Saitou *et al.*, 2002). Recently, transgenic mice have been generated that express a green fluorescent protein (GFP)-*stella* reporter transgene which appear to accurately follow PGC development (Payer, de Sousa Lopes, Barton, Lee, Saitou and Surani, 2006).

A range of other factors has been implicated in PGC derivation and maintenance. These include: (1) Mouse vasa homologue (mvh) protein is exclusively expressed in primordial germ cells just after their colonization of embryonic gonads and in germ cells undergoing gametogenic processes until the post-meiotic stage in both males and females (Toyooka, Tsunekawa, Takahashi, Matsui, Satoh and Noce, 2000). (2) The tyrosine-kinase receptor c-kit and its ligand, stem cell factor (SCF), are also essential for the maintenance of PGCs in both sexes. In the adult testis, the c-kit receptor is re-expressed in differentiating spermatogonia, but not in spermatogonial stem cells, whereas SCF is expressed by Sertoli cells under FSH stimulation (Rossi, Sette, Dolci and Geremia, 2000). (3) Men with deletions encompassing the Y-chromosome *DAZ* (Deleted in AZoospermia) genes have few or no germ cells, indicating they are defective in the formation or maintenance of germ cells. A *DAZ* homolog, *DAZL* (*DAZ-Like*), found in diverse organisms, including humans is required for germ cell development in males and females. Significantly, *PUM2*, a human homologue of pumilio, a protein required to maintain germ line stem cells in *Drosophila* and *Caenorhabditis elegans*, forms a stable complex with *DAZ* through the same functional domain required for RNA binding, protein-protein interactions (Moore, Jaruzelska, Fox, Urano, Firpo, Turek, Dorfman and Pera, 2003) suggesting mechanisms of germ cell development are highly conserved. (4) The POU domain transcription factor Oct4 has been shown to have a role in PGC survival. This transcription factor is crucial for maintaining pluripotency in the inner cell mass of the blastocyst and in ESCs. By use of conditional gene targeting with the Cre/LoxP system, Kehler and co-workers showed that loss of Oct4 leads to apoptosis (Kehler, Tolkunova, Koschorz, Pesce, Gentile, Bioani, Lomeli, Nagy, McLaughlin, Scholer and Tomilin, 2004). In human foetal tissue, Oct4 is strongly expressed in migrating PGCs as well as in human germ cell tumours and EG cells (Looijenga, Stoop, de Leeuw, de Gouveia Brazao, Gillis, van Roozendaal, van Zoelen, Weber, Wolffenbuttel, van Dekken, Honecker, Bokemeyer, Perlman, Schneider, Kononen, Sauter and Oosterhuis, 2003; Rajpert-De Meyts, Hanstein, Jorgensen, Graem, Vogt and Skakkebaek, 2004). Oct4 expression is down regulated rapidly in the human female gonad and silenced as oocytes enter the first meiotic prophase. In contrast, the same process occurs much more gradually in the male with Oct4 expression often persisting in some gonocytes and infantile spermatogonia. This observation has prompted the suggestion that differential germ cell expression of Oct4 between the sexes might contribute to the fact that germ-cell derived cancer is much higher in men than women (Rajpert-De Meyts *et al.*, 2004).

Hence, some of the genes that play a crucial role in germ cell differentiation are *Blimp1, stella, fragilis, c-Kit, vasa,* and *DAZL*. Their expression is stage-specific, thereby allowing identification of germ cells at different developmental phases. In addition to the expression of these genes, other markers associated with germ cell development are non-specific alkaline phosphatase activity, the stage-specific embryonic antigen 1 (SSEA1), the transcription factor Oct4 and beta1- and alpha6-integrins (Lacham-Kaplan, 2004; Fig. 2).

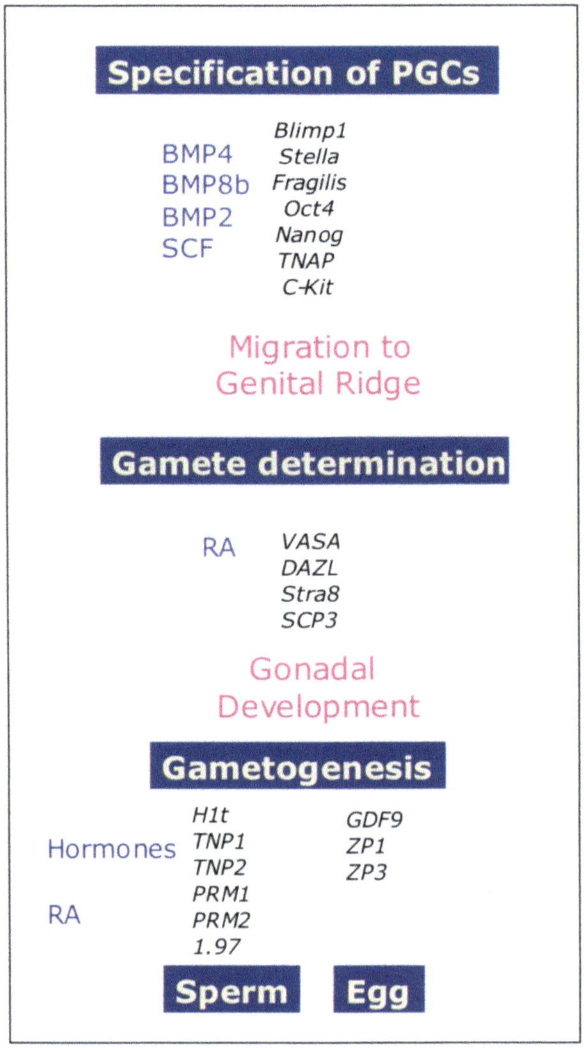

Figure 2. Developmental stages of mammalian germ cell development and some of the factors and genes (italics) involved. The genes have been used as markers of PGC and gamete development from ES cells in culture.

From embryonic stem cells to germ cells

ESCs have been defined as pluripotent (Smith, 2006) as they differentiate along somatic cell lineages but were not believed to form trophoblast. However, for hESCs at least it may be more appropriate to use the term totipotent as they spontaneously generate trophoblast and trophoblast stem cells, while mESCs form trophoblast if Oct4 expression is conditionally down regulated (Xu, Chen, Li, Li, Addicks, Glennon, Zwaka and Thomson, 2002; Hay, Sutherland, Clark and Burdon, 2004; Harun, Ruban, Matin, Draper, Jenkins, Liew, Andrews, Li, Laird and Moore, 2006). Moreover, ESCs can apparently form germ cells which have a totipotent potential. In the last three years it has become clear that murine ESCs can develop into PGCs *in vitro* (Hubner *et al.*, 2003; Geijsen *et al.*, 2004; Payer *et al.*, 2006) and with extended culture occasionally form early spermatids or oogonia. Ovarian-like structures and oocytes (i.e. having morphological similarities to follicles and gametes from gonads) may form that subsequently develop to blastocyst-like structures, presumably due to parthenogenetic activation (Hubner *et al.*, 2003; Geijsen *et al.*, 2004; Lacham-Kaplan, Chy and Trounson, 2006). Since these germ cell derivatives have not originated from true gonadal structures it is imperative that biological function is properly characterised and studies have increased in their sophistication. First mESC-derived spermatids were tested *in vitro* by using ESCs initially carrying an Oct 4-promoter driven green fluorescent protein (GFP) reporter gene which enabled the tracking of these cells (Hubner *et al.*, 2003; Geijsen *et al.*, 2004). When haploid spermatids generated from these GFP-ESCs were isolated by flow-cytometry and injected intracytoplasmically into oocytes the diploid chromosome complement was restored and green fluorescent blastocysts generated. At about the same time, Noce and coworkers (Toyooka, Tsunekawa, Akasu and Noce, 2003) showed that *Mvh-LacZ* murine germ cells (transgene with germ cell-specific mouse vasa homolog and *LacZ* reporter) generated *in vitro* from mESCs could also develop to some extent *in vivo*. These cells were aggregated with day 12.5-13.5 d.p.c gonadal tissue which formed seminiferous tubules within 6-8 weeks when co-transplanted under the testis capsule. Such tubules (separate from host tubules) participated in spermatogenesis to produce elongated ß-gal-stained spermatids. This demonstrated that such *Mvh-LacZ* ESCs formed PGCs with developmental potential to integrate into a somatic epithelium and undergo meiosis to form early sperm cells. Following these reports there still remained considerable uncertainty as to the biological fidelity of these various culture systems but Nayernia and co-workers (Nayernia *et al.*, 2006a) recently produced viable transgenic offspring from sperm produced from mESCs using a two-stage culture system. First mESCs were transfected with the reporter gene *Stra8-EGFP*, a retinoic acid (RA) responsive gene expressed in pre-meiotic mouse germ cells and fused with enhanced green fluorescent protein gene construct. After an initial ESC culture, green cells expressing the gene and therefore marking pre-meiotic germ cells were isolated by fluorescent activated flow cytometry (FACS) and cultured with RA as a proliferative cell line. These cells were subsequently transfected with a second fusion reporter gene *Prm1-dsred* (protamine 1 gene fused to red fluorescent protein gene construct) to identity post-meiotic sperm cells. After further RA induction for just a few days, red post-meiotic sperm cells were isolated and microinjected into mouse eggs to generate viable offspring. However, offspring generated in this way exhibited abnormalities and died prematurely (5 days to 5 months), probably due to inappropriate epigenetic control mechanisms during *in vitro* development of the germ cells.

Although investigations with hESCs are more preliminary they also show that spontaneous or induced differentiation in culture can generate cells displaying mRNA expression profiles and cell surface markers consistent with PGCs (Clark *et al.*, 2004, Clark and Reijo Pera, 2006) and of subsequent meiosis. In our laboratory, PGCs (and on rare occasions early spermatid cells but so far not oocytes) have been identified after the appropriate culture conditions (Aflatoonian *et al.*, 2005).

Gamete determination and differentiation

Markers of female germ cells are expressed in both XX and XY cells of murine and human ESC culture (Toyooka *et al.*, 2003; Geijsen *et al.*, 2004). The prevailing hypothesis to date has been that pathways intrinsic to the cell rather than inductive signals regulate PGC development. Thus, whatever the complement of sex chromosomes, germ cells are cell-autonomous and intrinsically programmed to undergo meiosis, enter prophase and develop as oocytes, unless otherwise prevented from doing so by a meiosis inhibiting factor(s) (McLaren, 2003). Under these circumstances it is perhaps unsurprising for both male and female ESC lines to display female germ cell markers since culture conditions may be sub-optimal and lack meiosis inhibition. However, recent studies (Bowles, Knight, Smith, Wilhelm, Richman, Mamiya, Yashiro, Chawengsaksophak, Wilson, Rossant, Hamada and Koopman, 2006; Koubova, Menke, Zhou, Capel, Griswold and Page, 2006) indicate that instead of an intrinsic programme to enter meiosis, germ cells respond to the external signal of RA and its metabolism. Thus, in the embryonic mouse ovary, RA induces germ cells to express the pre-meiotic marker *Stra-8* (stimulated by retinoic acid 8) and initiate meiosis. On the other hand, in the embryonic mouse testis RA is metabolised and inactivated by the P450 enzyme CYP26 (B1) thereby preventing early germ cell entry into meiosis with down-regulation of genes such as SCP3 (synaptonemal complex protein; associated with meiotic events) and the induction into the alternative pathway of mitotic arrest as G0/G1 prospermatogonia. In this case the induction of presumptive PGCs into meiosis in culture medium containing RA might be expected although local concentrations within cell aggregates could differ significantly and affect the timing. Concentrations of RA in plasma or serum (or serum replacement) that supplement ESC culture medium are normally in the 2 – 8 nM range (Eckhoff and Nau, 1990). This level is much lower than the 1 –5 μM concentration often used with induction protocols but the sensitivity of cells to RA can vary considerably depending on composition of medium.

This process of sex determination seems even more dysregulated with hESCs. Markers of both male and female germ cell development have been detected regardless of the sex of the cell line (Clark *et al.*, 2004; Aflatoonian *et al.*, 2005). The inference from these findings is that RA or possibly other factors affecting meiosis and gamete determination are present in the culture conditions possibly generated by male and female hESCs. For male germ cell development *in vivo,* seminiferous epithelium with Sertoli cells is necessary and the development of these cells is dependent normally on expression of the transcription factor SRY, the gene of which is located on the Y chromosome. Therefore, the generation of presumptive male germ cells with XX hESCs will require either an artifactual differentiation of Sertoli cells, for example, by female supporting cells differentiating to a Sertoli-like phenotype (Adams and McLaren, 2002) and/or production of meiosis inhibitor factors (or metabolised in the case of RA) from a different origin. In vitamin A deficient adult mice, where only spermatogonia and Sertoli cells populate the seminiferous epithelium, it is of interest that the addition of retinoids such as RA and retinol may induce the resumption of spermatogenesis (Baleato, Aitken and Roman, 2005).

The germ cell niche and ESC culture

A fascinating aspect of the generation of ESC-derived germ cells *in vitro* is how normal checkpoints of the developmental process which would naturally span a timeline from early foetal development to puberty and lead to arrest of cells at different stages is apparently overcome and compressed into a relatively much short culture period. In the absence of environmental cues within the developing gonad germ cells may develop according to an intrinsic clock. Payer and co-workers (2006) used *stella*-GFP murine ESCs to show that ESC colonies display a subpopulation of *stella*-positive cells. Significantly, there was overlap of expression of Oct4 and *stella* in some cells which might indicate a population of germ cell-like ESCs (Payer *et al.*, 2006). Whether *stella*-positive ESCs are destined to form PGCs in the mouse is not clear but these results suggest that there is perhaps an inherent propensity of ESC culture systems to form PGCs. Thereafter, these cells may create a stem cell niche continuum for induction of germ cell development although how ESC cultures may mimic the somatic environment that encapsulates either developing oogonia (the follicle) or the sperm stem cell (seminiferous epithelium) is unknown. The appropriate growth factor and hormonal microenvironment required to support and sustain such complex niches probably depends on the type of culture system adopted for the differentiation process.

Usually ESCs are cultured in combinations of two basic methods. Monolayer adherent cultures of ESCs can be allowed to differentiate directly to form an appropriate niche or ESCs can be induced to aggregate to form embryoid bodies that form a more three-dimensional micro-environment. In conjunction with these two protocols, ESCs can be initially co-cultured with feeder layers (i.e. mouse embryonic fibroblasts, neonatal gonadal cell culture) or conditioned-medium from feeder cells and various growth factor and serum supplementation. Generally, these culture conditions attempt to create an environment conducive for germ cell proliferation and differentiation but they are not specific. Aside from the absence of a defined culture medium, the various procedures for passaging ESCs (i.e. to single cells or as clumps), the influence of feeder cells and extracellular matrix (e.g. Matrigel) and the cell line itself can all contribute to proportion of germ cells specified, if at all. If a phenotype is not selected (i.e. using fluorescent reporter systems) then only a small proportion of cells (usually less than 5%) will normally display primordial germ cell or gonocyte/spermatogonia markers but the fact that some cells progress further to a post-meiotic stage seems extraordinary given the complex hormonal requirements of gamete development. It is of note that oestradiol production (50-100 pg/ml) was detected after 12 days in mESC cultures that generated follicle-like structures (Hubner *et al.*, 2003) while appreciable levels (60 pg/ml) of dihydrotestosterone can be detected in hESCs cultures that generate PGCs and spermatids (Aflatoonian and Moore, unpublished). Presently, the source of steroidogenesis is unclear but it may originate from additives to the culture medium (i.e. serum or serum replacement supplement) or gonadal cell types (granulosa, Leydig cells) that co-differentiate with germ cells. The latter appeared to be case for oocytes generated from mESCs since increased oestradiol levels correlated with increased proportion of cells resembling a follicular morphology (Hubner *et al.*, 2003). Alternatively, steroids may be produced from non-reproductive cell phenotypes (e.g. liver or adrenal cells) that could conceivably differentiate from ESCs in culture at the same time as germ cells.

The proliferative capacity of germ cells in culture will also depend on their fate. Since *in vivo*, at least, germ cells are programmed by default to enter meiosis, the efficiency at which ESC cultures will generate germ cells will depend on the initial rate of differentiation to PGCs and subsequently whether factors are present to divert development down spermatogenesis when mitotic proliferation of prospermatogonia can potentially occur. The self-renewal of spermatogonial stem cells (SSCs) in rodents is dependent on glial cell line-derived neurotrophin factor (GDNF), the GDNF-family receptor a-1 and basic fibroblast growth factor (bFGF) (Hamra, Chapman, Nguyen, Williams-Stephens, Hammer and Garbers, 2005; Ryu, Kubota, Avarbock and Brinster, 2005).

In the mouse, pluripotency of ESCs is mirrored also by EG cells (Matsui, Zsebo and Hogan, 1992; Resnick, Bixler, Cheng and Donovan, 1992) and seem to include a population of SSCs in the neonatal mouse testis (Kinatsu-Shinohara, Inoue, Lee, Yoshimoto, Ogonuki, Miki, Baba, Kato, Kazuki, Toyokuni, Toyoshima, Niwa, Oshimura, Heike, Nakahata, Ishino, Ogura and Shinohara, 2004); and perhaps more remarkably to cells in adult mouse testis (Guan, Nayernia, Maier, Wagner, Dressel, Lee, Nolte, Wolf, Li, Engel and Hasenfuss, 2006). These adult SSCs can be transformed in culture to ES-like cell lines displaying pluripotency both *in vitro* and significantly *in vivo*. When injected into mouse pre-implantation blastocysts, adult germline stem cells (maGSCs) contributed to the development of all the major organs and displayed germline transmission (Guan *et al.*, 2006). The obvious conservation in many of the mechanisms for germ cell development in mammals suggests that it might be possible to develop similar cell lines from human adult spermatogonial populations although it may not be straightforward to translate from the murine to the human system. While human EG cell lines have been reported (Shamblott, Axelman, Wang, Bugg, Littlefield, Donovan, Blumenthal, Huggins and Gearhart, 1998), their proliferative and pluripotent capacity currently as true EG cells has been questioned (Turnpenny, Brickwood, Spalluto, Piper, Cameron, Wilson and Hanley, 2003; Aflatoonian and Moore, 2005).

While the evidence is becoming more compelling that in culture ESCs can generate PGCs and germ cells, there remains the question as to whether germ cells can be derived (or transdifferentiated) from adult stem cells residing outside the gonad. There are reports that germ cells can be derived from bone marrow (BM) and peripheral blood cells (Johnson, Canning, Kaneko, Pru and Tilly, 2004, 2005; Nayernia, Lee, Drusenheimer, Nolte, Wulf, Dressel, Gromoll and Engel, 2006b) and BM-derived germ cells can repopulate the ovarian follicular reserve (Johnson *et al.*, 2004; Johnson, Bagley, Skaznik-Wikiel, Lee, Adams, Niikura, Tschudy, Tilly, Cortes, Forkert, Spitzer, Iacomini, Scadden and Tilly, 2005). However, the latter investigations have not been substantiated (Eggan, Jurga, Gosden, Min and Wagers, 2006). Germ cell derivation from pig foetal skin cells *in vitro* has also been reported (Dyce, Wen and Li, 2006) with formation of morphological structures resembling follicles, that secrete oestradiol and progesterone; and oocyte-like cells displaying markers of germ cells (oct4, vasa, GDF9b) or exhibiting the meiosis marker SCP3 and displaying apparent parthenogenetic embryo structures. Provided these report of transdifferentiation are fully corroborated, they indicate that PGC and germ cell formation *in vitro* and possibly *in vivo* is much more plastic than previously believed (Fig. 3). However, it is very difficult to reconcile some of these observations with known reproductive physiology in mammals and an alternative explanation is that detection of many germ cell markers is due to aberrant expression or detection in culture and does not correspond to true germ cell phenotype. A wide battery of specific protein and functional phenotypic assays will be necessary to resolve this issue.

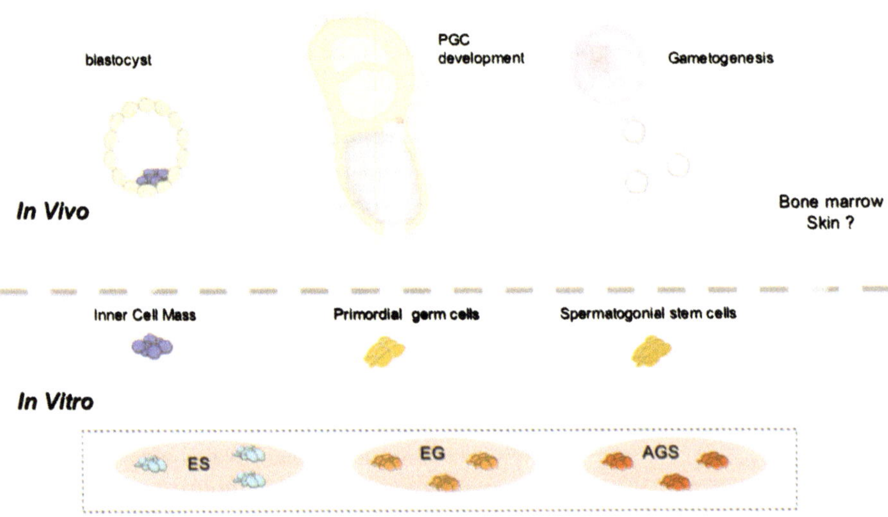

Figure 3. Diagram showing the relationship between *in vivo* development of germ cells and the *in vitro* pluripotent stem cell lines that have been reported to be derived from mouse and human cells (ES and EG) and their developmental potential. The relationship between ES, EG and AGS still remains unclear although they express similar markers of pluripotency.

Epigenetic mechanisms

Notwithstanding that embryonic stem cells can undergo differentiation to PGCs and possibility early gametes there is the added complexity of a range of epigenetic processing required for germ cell development (Allegrucci *et al.*, 2005). Female PGCs reactivate their silenced X chromosome thereby allowing each oocyte to have an active X chromosome. However this characteristic is not exclusively confined to female PGCs as XXY male germ cells also may have this capability (Mroz, Carrel and Hunt, 1999). Moreover, as PGCs enter the gonads they become hypomethylated resulting in a loss of imprinting and biallelic gene expression. Recent evidence suggests that cell demethylation at this stage may control the expression of genes required for gametogenesis (Maatouk, Kellam, Mann, lei, Li, Bartolomei and Resnick, 2006) but exactly how such processes might operate in embryonic stem cell differentiation to germ cells in culture remains unclear.

Conclusions

Currently ESCs differentiation remains difficult to control and accurately monitor *in vitro*. However with the advent of highly specific transgene reporter systems to monitor germ cell differentiation, evidence has accumulated that PGCs can be derived readily from mESCs. Furthermore, there is growing data that very similar processes are likely to occur with hESCs. Occasionally postmeiotic gametes, albeit immature, can be generated although appropriate developmental and epigenetic competence remains uncertain. Considering the complex hormone-dependent somatic cell niche that normally surrounds germ cells *in situ*, this spontaneous development seems remarkable and may indicate a high degree of intrinsic regulatory mechanisms in germ cells. However, the promiscuous nature of spontaneous ESC differentiation in culture makes it especially difficult to determine at present whether PGC and germ cell development follow similar programmes to those occurring *in vivo* or whether the conditions merely provide suitable conducive conditions that permit an intrinsic developmental process but not necessarily the same as the gonadal environment. For example, the findings that the development of both haemopoietic plasma cells and PGCs depend on expression of the same protein (Blimp1) indicate conservation in these differentiation systems and therefore there is potential for interference between pathways during *in vitro* development. With careful selection of cells with the appropriate markers, it should be feasible to devise stage-specific protocols (with specific growth factors) for differentiation of ESCs to PGCs and subsequently PGCs to gonocytes and male or female germ cell fate. However, there remains major uncertainty about the genetic/epigenetic processing of germ cells *in vitro* and very careful consideration would need to be given to safety aspects of such cells if they were ever to be used for clinical applications.

References

Adams, I.R. and McLaren, A. (2002). Sexually dimorphic development of mouse primordial germ cells: switching from oogenesis to spermatogenesis. *Development* **129:** 1155-1164

Aflatoonian, B. and Moore, H. (2005). Human primordial germ cells and embryonic germ cells, and their use in cell therapy. *Current Opinions in Biotechnology* **16:** 530-535

Aflatoonian, B., Fazeli, A., Ruban, L., Andrews, P. and Moore, H. (2005). Human embryonic stem cells differentiate to primordial germ cells as determined by gene expression profiles and antibody markers. Proc. 21st Annual Meeting of the European Society for Human Reproduction and Embryology, Copenhagen. *Human Reproduction* **20** (suppl): 1 i6

Allegrucci, C., Thurston, A., Lucas, E. and Young, L. (2005). Epigenetics and the germline. *Reproduction* **129:** 137-149

Anway, M.D., Cupp, A.S., Uzumcu, M. and Skinner, M.K. (2005). Epigenetic transgenerational actions of endocrine disruptors and male fertility. *Science* **308:** 1466-1469

Baillie, H., Pacey, A. and Moore, H. (2002) Environmental chemicals and the threat to male fertility in mammals: Evidence and perspective. In: *Conservation Biology 8. Reproductive Science and Integrated Conservation*. Edited by Holt, W.V., Pickard, A.R., Rodger, J.C. and Wildt, D.E., Cambridge University Press, Cambridge, pp. 57-66

Baleato, R.M., Aitken, R.J. and Roman, S.D. (2005). Vitamin A regulation of BMP4 expression in the male germ line. *Developmental Biology* **286:** 78-90

Bowles, J., Knight, D., Smith, C., Wilhelm, D., Richman, J., Mamiya, S., Yashiro, K., Chawengsaksophak, K., Wilson, M.J., Rossant, J., Hamada, H. and Koopman, P. (2006). Retinoid signaling determines germ cell fate in mice. *Science* 312: 596-600

Brinster, R.L. (2002). Germline stem cell transplantation and transgenesis. *Science* **296:** 2174-2176

Clark, A.T., Bodnar, M.S., Fox, M., Rodriguez, R.T., Abeyta, M.J., Firpo, M.T. and Pera, R.A. (2004). Spontaneous differentiation of germ cells from human embryonic stem cells in vitro. *Human Molecular Genetics* **13:** 727-739

Clark, A.T. and Reijo Pera, R.A. (2006). Modeling human germ cell development with embryonic stem cells. *Regenerative Medicine* **1:** 85-93

de Sousa Lopes, S.M., Roelen, B.A., Monteiro, R.M., Emmens, R., Lin, H.Y., Li, E., Lawson, K.A. and Mummery, C.L. (2004). BMP signaling mediated by ALK2 in the visceral endoderm is necessary for the generation of primordial germ cells in the mouse embryo. *Genes and Development* **18:** 1838-1849

Donovan, P.J. and de Miguel, M.P. (2003). Turning germ cells into stem cells. *Current Opinions in Genetics and Development* **13:** 463-471

Draper, J.S., Moore, H.D., Ruban, L.N., Gokhale, P.J. and Andrews, P.W. (2004). Culture and characterization of human embryonic stem cells. *Stem Cells and Development* **13**: 325-336

Dyce, P.W., Wen, L. and Li, J. (2006). In vitro germline potential of stem cells derived from foetal porcine skin. *Nature Cell Biology* **8:** 384-390

Eckhoff, C. and Nau, H. (1990). Identification and quantitation of all-*trans*- and 13-*cis*-retinoic acid and 13-*cis*-4-oxoretinoic acid in human plasma. *Journal of Lipid Research* **31:** 1445-1454

Eggan, K., Jurga, S., Gosden, R., Min, I.M. and Wagers A.J. (2006). Ovulated oocytes in adult mice derive from non-circulating germ cells. *Nature* **441:** 1109-1114

Evans, M.J. and Kaufman, M.H. (1981). Establishment in culture of pluripotential cells from mouse embryos. *Nature* **292:** 154-156

Freeman, B. (2003). The active migration of germ cells in the embryos of mice and men is a myth. *Reproduction* **125:** 635-643.

Geijsen, N., Horoschak, M., Kim, K., Gribnau, J., Eggan, K. and Daley, G.Q. (2004). Derivation of embryonic germ cells and male gametes from embryonic stem cells. *Nature* **427:** 106-107

Ginsburg, M., Snow, M.H. and McLaren, A. (1990). Primordial germ cells in the mouse embryo during gastrulation. *Development* **110:** 521-528

Guan, K., Nayernia, K., Maier, M.S., Wagner, S., Dressel, R., Lee, J.H., Nolte, J., Wolf, F., Li, M., Engel, W. and Hasenfuss, G. (2006). Pluripotency of spermatogonial stem cells from adult mouse testis. *Nature* **440:** 1199-1203

Hamra, F.K., Chapman, K.M., Nguyen, D.M., Williams-Stephens, A.A., Hammer, R.E. and Garbers, D.L. (2005). Self renewal, expansion, and transfection of rat spermatogonial stem cells in culture. *Proceedings of the National Academy of Sciences USA* **102:** 17430-17435

Harun, R., Ruban, L., Matin. M., Draper, J., Jenkins, N.M., Liew, G.C., Andrews, P.W., Li, T.C., Laird, S.M. and Moore, H.D. (2006). Cytotrophoblast stem cell lines derived from human embryonic stem cells and their capacity to mimic invasive implantation events. *Human Reproduction* **21**: 1349-1358

Hay, D.C., Sutherland, L., Clark, J. and Burdon, T. (2004). Oct-4 knockdown induces similar patterns of endoderm and trophoblast differentiation markers in human and mouse embryonic stem cells. *Stem Cells* **2:** 225-235

Heininger, K. (2002). Aging is a deprivation syndrome driven by a germ-soma conflict. *Ageing Research Review* **1**: 481-536

Hubner, K., Fuhrmann, G., Christenson, L.K., Kehler, J., Reinbold, R., De La Fuente, R., Wood, J., Strauss, J.F. 3rd, Boiani, M. and Scholer, H.R. (2003). Derivation of oocytes from mouse embryonic stem cells. *Science* **300:** 1251-1256

Johnson, J., Canning, J., Kaneko, T., Pru, J.K. and Tilly, J.L. (2004). Germline stem cells and follicular renewal in the postnatal mammalian ovary. *Nature* **428:** 145-150

Johnson, J., Bagley, J., Skaznik-Wikiel, M., Lee, H.J., Adams, G.B., Niikura, Y., Tschudy, K.S., Tilly, J.C., Cortes, M.L., Forkert, R., Spitzer, T., Iacomini, J., Scadden, D.T. and Tilly, J.L. (2005). Oocyte generation in adult mammalian ovaries by putative germ cells in bone marrow and peripheral blood. *Cell* **122:** 303-315

Kanatsu-Shinohara, M., Inoue, K., Lee, J., Yoshimoto, M., Ogonuki, N., Miki, H., Baba, S., Kato, T., Kazuki, Y., Toyokuni, S., Toyoshima, M., Niwa, O., Oshimura, M., Heike, T., Nakahata, T., Ishino, F., Ogura, A. and Shinohara, T. (2004). Generation of pluripotent stem cells from neonatal mouse testis. *Cell* **119:** 1001-1012

Kehler, J., Tolkunova, E., Koschorz, B., Pesce, M., Gentile, L., Boiani, M., Lomeli, H., Nagy, A., McLaughlin, K.J., Scholer, H.R. and Tomilin, A. (2004). Oct4 is required for primordial germ cell survival. *European Molecular Biology Organisation Report* **5:** 1078-1083

Lacham-Kaplan, O. (2004). *In vivo* and *in vitro* differentiation of male germ cells in the mouse. *Reproduction* **128:** 147-152

Lacham-Kaplan, O., Chy, H. and Trounson, A. (2006).

Testicular cell conditioned medium supports differentiation of embryonic stem cells into ovarian structures containing oocytes. *Stem Cells* **24:** 266-273

Lawson, K.A., Dunn, N.R., Roelen, B.A., Zeinstra, L.M., Davis, A.M., Wright, C.V., Korving, J.P. and Hogan, B.L. (1999). Bmp4 is required for the generation of primordial germ cells in the mouse embryo. *Genes and Development* **13:** 424-436

Looijenga, L.H., Stoop, H., de Leeuw, H.P., de Gouveia Brazao, C.A., Gillis, A.J., van Roozendaal, K.E., van Zoelen, E.J., Weber, R.F., Wolffenbuttel, K.P., van Dekken, H., Honecker, F., Bokemeyer, C., Perlman, E.J., Schneider, D.T., Kononen, J., Sauter, G. and Oosterhuis, J.W. (2003). POU5F1 (OCT3/4) identifies cells with pluripotent potential in human germ cell tumors. *Cancer Research* **63:** 2244-2250

Koubova, J., Menke, D.B., Zhou, Q., Capel, B., Griswold, M.D. and Page, D.C. (2006). Retinoic acid regulates sex-specific timing of meiotic initiation in mice. *Proceedings of the National Academy of Science USA* **103**: 2474-2479

Maatouk, D.M., Kellam, L.D., Mann, M.R., Lei, H., Li, E., Bartolomei, M.S. and Resnick, J.L. (2006). DNA methylation is a primary mechanism for silencing postmigratory primordial germ cell genes in both germ cell and somatic cell lineages. *Development* **133**: 3411-3418

Martin, G.R. (1981). Isolation of a pluripotent cell line from early mouse embryos cultured in medium conditioned by teratocarcinoma stem cells. *Proceedings of the National Academy of Science USA* **78:** 7634-7638

Matsui, Y., Zsebo, K. and Hogan, B.L. (1992). Derivation of pluripotential embryonic stem cells from murine primordial germ cells in culture. *Cell* **70:** 841-847

McLaren, A and Lawson, K.A. (2005). How is the mouse germ-cell lineage established? *Differentiation* **73:** 435-437

McLaren, A. (2003). Primordial germ cells in the mouse. *Developmental Biology* **262:** 1-15

Molyneaux, K. and Wylie, C. (2005). Primordial germ cell migration. *International Journal of Developmental Biology* **48**: 537-543

Moore, F.L., Jaruzelska. J., Fox, M.S., Urano, J., Firpo, M.T., Turek, P.J., Dorfman, D.M. and Pera, R.A. (2003). Human Pumilio-2 is expressed in embryonic stem cells and germ cells and interacts with DAZ (Deleted in AZoospermia) and DAZ-like proteins. *Proceedings of the National Academy of Science USA* **100:** 538-543

Mroz, K., Carrel, L. and Hunt, P.A. (1999). Germ cell development in the XXY mouse: Evidence that X chromosome reactivation is independent of sexual differentiation. *Developmental Biology* **207**: 229-238

Nayernia, K., Nolte, J., Michelmann, H.W., Lee, J.H., Rathsack, K., Drusenheimer, N., Dev, A., Wulf, G., Ehrmann, I.E., Elliott, D.J., Okpanyi, V., Zechner, U., Haaf, T., Meinhardt, A. and Engel, W. (2006a). In vitro-differentiated embryonic stem cells give rise to male gametes that can generate offspring mice. *Developmental Cell* **11:** 125-132

Nayernia, K., Lee, J.H., Drusenheimer, N., Nolte, J., Wulf, G., Dressel, R., Gromoll, J. and Engel, W. (2006b). Derivation of male germ cells from bone marrow stem cells. *Laboratory Investigation* **86:** 654-663

Ohinata, Y., Payer, B., O'Carroll D., Ancelin, K., Ono, Y., Sano, M., Barton, S.C., Obukhanych, T., Nussenzweig, M., Tarakhovsky, A., Saitou, M. and Surani, M.A. (2005). Blimp1 is a critical determinant of the germ cell lineage in mice. *Nature* **436**: 207-213

Payer, B., Saitou, M., Barton, S.C., Thresher, R., Dixon, J.P., Zahn, D., Colledge, W.H., Carlton, M.B., Nakano, T. and Surani, M.A. (2003). *Stella* is a maternal effect gene required for normal early development in mice. *Current Biology* **13:** 2110-2117

Payer, B., Chuva de Sousa Lopes, S.M., Barton, S.C., Lee, C., Saitou, M. and Surani M.A. (2006). Generation of stella-GFP transgenic mice: A novel tool to study germ cell development. *Genesis* **44:** 75-83

Rajpert-De Meyts, E., Bartkova, J., Samson, M., Hoei-Hansen, C.E., Frydelund-Larsen, L., Bartek, J. and Skakkebaek, N.E. (2003). The emerging phenotype of the testicular carcinoma in situ germ cell. *Acta Pathologica Microbiologica et Immunologica Scandinavica* **111:** 267-278

Rajpert-De Meyts, E., Hanstein, R., Jorgensen, N., Graem, N., Vogt, P.H. and Skakkebaek, N.E. (2004). Developmental expression of POU5F1 (OCT-3/4) in normal and dysgenetic human gonads. *Human Reproduction* **19:** 1338-1344

Resnick, J.L., Bixler, L.S., Cheng, L. and Donovan, P.J. (1992). Long-term proliferation of mouse primordial germ cells in culture. *Nature* **359:** 550-551

Rossi, P., Sette, C., Dolci, S. and Geremia, R. (2000). Role of c-kit in mammalian spermatogenesis. *Journal of Endocrinological Investigation* **23:** 609-615

Ryu, B.Y., Kubota, H., Avarbock, M.R. and Brinster, R.L. (2005). Conservation of spermatogonial stem cell self-renewal signaling between mouse and rat. *Proceedings of the National Academy of Science USA* **102:** 14302-14307

Saitou, M., Barton, S.C. and Surani, M.A. (2002). A molecular programme for the specification of germ cell fate in mice. *Nature* **418:** 293-300

Shamblott, M.J., Axelman, J., Wang, S., Bugg, E.M., Littlefield, J.W., Donovan, P.J., Blumenthal, P.D., Huggins, G.R. and Gearhart, J.D. (1998). Derivation of pluripotent stem cells from cultured human primordial germ cells. *Proceedings of the National Academy of Science USA* **95:** 13726-13731

Shimasaki S., Moore, R.K., Otsuka, F. and Erickson, G. F. (2004). The bone morphogenetic protein system in mammalian reproduction. *Endocrine Reviews* **25:** 72-101

Skakkebaek, N.E., Rajpert-De Meyts, E. and Main, K.M. (2001). Testicular dysgenesis syndrome: an increasingly common developmental disorder with environmental aspects. *Human Reproduction* **16:** 972-978

Smith, A. (2006). A glossary for stem-cell biology. *Nature* **441:** P1060

Thomson, J.A., Itskovitz-Eldor, J., Shapiro, S.S., Waknitz, M.A., Swiergiel, J.J., Marshall, V.S. and Jones, J.M. (1998). Embryonic stem cell lines derived from human blastocysts. *Science* **282:** 1145-1147

Tiido, T., Rignell-Hydbom, A., Jonsson, B,, Giwercman, Y.L., Rylander, L., Hagmar, L. and Giwercman, A. (2005). Exposure to persistent organochlorine pollutants associates with human sperm Y:X chromosome ratio. *Human Reproduction* **20:** 1903-1909

Toyooka, Y., Tsunekawa, N., Takahashi, Y., Matsui, Y., Satoh, M. and Noce, T. (2000). Expression and intracellular localization of mouse Vasa-homologue protein during germ cell development. *Mechanisms of Development* **93:** 139-149

Toyooka,Y., Tsunekawa, N., Akasu, R. and Noce, T. (2003). Embryonic stem cells can form germ cells in vitro. *Proceedings of the National Academy of Science USA* **100:** 11457-11462

Turnpenny, L., Brickwood, S., Spalluto, C.M., Piper, K., Cameron, I.T., Wilson, D.I. and Hanley, N.A. (2003). Derivation of human embryonic germ cells: an alternative source of pluripotent stem cells. *Stem Cells* **21:** 598-609

Vincent, S.D., Dunn, N.R., Sciammas, R., Shapiro-Shalef, M., Davis, M.M., Calame, K., Bikoff, E.K. and Robertson, E.J. (2005). The zinc finger transcriptional repressor Blimp1/Prdm1 is dispensable for early axis formation but is required for specification of primordial germ cells in the mouse. *Development,* **132:** 1315-1325

Xu, R.H., Chen, X., Li, D.S., Li, R., Addicks, G.C., Glennon, C., Zwaka, T.P. and Thomson, J.A. (2002). BMP4 initiates human embryonic stem cell differentiation to trophoblast. *Nature Biotechnology* **20:** 1261-1264

Ying, Y., Qi, X. and Zhao, G.Q. (2001). Induction of primordial germ cells from murine epiblasts by synergistic action of BMP4 and BMP8b signaling pathways. *Proceedings of the National Academy of Science USA* **98:** 7858-7862

Spermatology. SRF Vol. 65. ERS Roldan and M Gomendio (eds) Nottingham University Press, Nottingham

Molecular biology of sperm head shaping

Abraham L. Kierszenbaum, Eugene Rivkin and Laura L. Tres

Department of Cell Biology and Anatomical Sciences, The Sophie Davis School of Biomedical Education/The City University of New York Medical School, New York, NY 10031, USA

The shaping of the mammalian sperm involves the elongation and condensation of the spermatid nucleus, the development of the acrosome, and the transient appearance of the microtubular manchette. F-actin-containing ectoplasmic hoops of Sertoli cells embrace the upper third of the spermatid head during elongation. During acrosomal biogenesis, proacrosomal vesicles derived from the Golgi apparatus, dock and fuse along the acroplaxome, an F-actin/keratin 5-containing cytoskeletal plate. The acroplaxome consists of a bent plate and a marginal ring encircling the spermatid nucleus. It anchors the developing acrosome to the spermatid nucleus. The manchette, consisting of a perinuclear rings with inserted microtubules, lies subjacent to the marginal ring of the acroplaxome. During spermatid elongation, the two overlapping rings reduce their diameter to fit, in a sleeve-like fashion, the decreasing diameter of the spermatid nucleus. The acroplaxome may provide a planar scaffold to modulate exogenous constriction forces generated by Sertoli cell F-actin hoops during spermatid head elongation. The dynamics of the F-actin cytoskeleton, one of the components of the acroplaxome and Sertoli cell hoops, can be regulated by tyrosine kinases, which target cortactin, an F-actin-associated protein. Tyrosine phosphorylation of cortactin correlates with a reduction in the crosslinking properties of F-actin. Phosphorylated cortactin and tyrosine kinase Fer are present in the acroplaxome, thus supporting a role of this F-actin remodelling pathway during spermatid head shaping. Keratin 5, an additional component of the acroplaxome, may also undergo dynamic reorganization during spermatid head elongation. We postulate that the F-actin/keratin 5 cytoskeleton in the acroplaxome may undergo a dynamic reorganization to modulate exogenous shear forces exerted by Sertoli cell F-actin hoops during spermatid head shaping. The acroplaxome-manchette perinuclear rings may reduce their diameter to balance exogenous constriction forces generated by the embracing Sertoli cell F-actin hoops and guide nuclear elongation.

Introduction

Abnormalities in sperm head shaping are a severe cause of male infertility (Escalier, 1990) but the mechanism of sperm head shaping remains unknown. Approaches to elucidate the molecu-

Corresponding author E-mail: kier@med.cuny.edu

lar mechanism of spermatid head shaping include mouse mutants with defective spermiogenesis, mutants lacking the expression of molecules judged important for spermatogenesis, and the fractionation of structures involved in spermatid head morphogenesis (e.g., the manchette; Mochida, Tres and Kirszenbaum, 1998). One unexpected difficulty with the use of some of the mutants has been a modest change in fertility despite the null expression of genes considered significant in reproductive biology (Craig, Zirngibl, Williams, Cole and Greer, 2001; Greer, 2002). It is possible that molecular redundancy may account for the unexpected outcomes of transgenic mice.

Three examples of mutants leading the way to an understanding of sperm head shaping are the acrosome-deficient Hrb mouse (Kang-Decker, Mantchev, Juneja, McNiven and van Deursen, 2001; Kierszenbaum, Tres, Rivkin, Kang-Decker and van Deursen, 2004), the GOPC-deficient mouse (Yao, Ito, Natsume, Sugitani, Yamanaka, Kuretake, Yanagida, Sato, Toshimori and Noda, 2002), and the abnormal head-shaped azh mouse mutant (Meistrich, 1993; Mochida, Tres and Kierszenbaum, 1999; Mendoza-Lujambio, Burfeind, Dixkens, Meinhardt, Hoyer-Fender, Engel and Neesen, 2002). Hrb and GOPC spermatids display pseudoacrosomes (flat sacs aligned along the acroplaxome) and spherical shaped heads. The major defect is a lack of fusion of proacrosomal vesicles. In wild-type rodents, proacrosomal vesicles, derived from the Golgi apparatus, are transported during acrosome biogenesis along actin filaments (Kierszenbaum, Rivkin and Tres, 2003b) and microtubule tracks (Yang and Sperry, 2003) by myosin Va and kinesin motor proteins, respectively. Proacrosomal vesicles are recruited to the acroplaxome, an F-actin-containing cytoskeletal plate attached to the spermatid nuclear envelope (Kierszenbaum, Rivkin and Tres, 2003a). During enlargement and descent of the acrosomal sac, the acroplaxome develops a marginal ring that secures the caudal recess of the acrosomal sac to the spermatid nuclear envelope. The marginal ring of the acroplaxome can be regarded as a desmosomal-like anchoring complex encircling the spermatid nucleus. The ring is a tripartite structure consisting of an acrosome-associated plaque with attached bundles of F-actin and the intermediate filament protein-containing keratin 5, and a nuclear envelope-associated plaque (Fig. 1). Additional proteins may be present to secure the adhesion properties of the marginal ring plaques to the corresponding membranes. The components of the adhesion plaques are presently unknown.

Three events occur during acrosome biogenesis. First, the axoneme of the future sperm tail initiates its assembly from the centrosomal region located at the pole opposite to the acrosome. The centrosome consists of a pair of centrioles surrounded by a pericentriolar matrix containing a number of proteins known to participate in nucleocytoplasmic transport and microtubule assembly (Ran GTPase; Kierszenbaum, Gil, Rivkin and Tres, 2002) and in the degradation of ubiquitinylated proteins (the 26S proteasome; Mochida, Tres and Kierszenbaum, 2000).

Second, the Sertoli cell ectoplasmic region adjacent to a developing spermatid displays bundles of actin filaments, forming a stack of hoops embracing the spermatid region where the acrosome-acroplaxome complex is developing. The close Sertoli cell-spermatid apposition is stabilized by the afadin-nectin-2/3 complex, integrin subunit ß1, and presumably by other cell-cell adhesion molecules. The afadin-nectin complex interacts with F-actin cytoskeleton. Nectin-2- and nectin-3-deficient mice display a severe disorganization of Sertoli cell-spermatid contact sites resulting in abnormal spermiogenesis and sterility (Bouchard, Dong, McDermott, Lam, Brown, Shelanski, Bellve and Racaniello, 2000; Ozaki-Kuroda, Nakanishi, Ohta, Tanaka, Kurihara, Mueller, Irie, Ikeda, Sakai, WImmer, Nishimune and Takai, 2002).

Third, rat spermatids (step 8) begin to assemble the manchette. The manchette consists of a perinuclear ring separated from the perinuclear marginal ring of the acroplaxome by a narrow gap. The manchette ring nucleates microtubules extending, like a dense perinuclear screen,

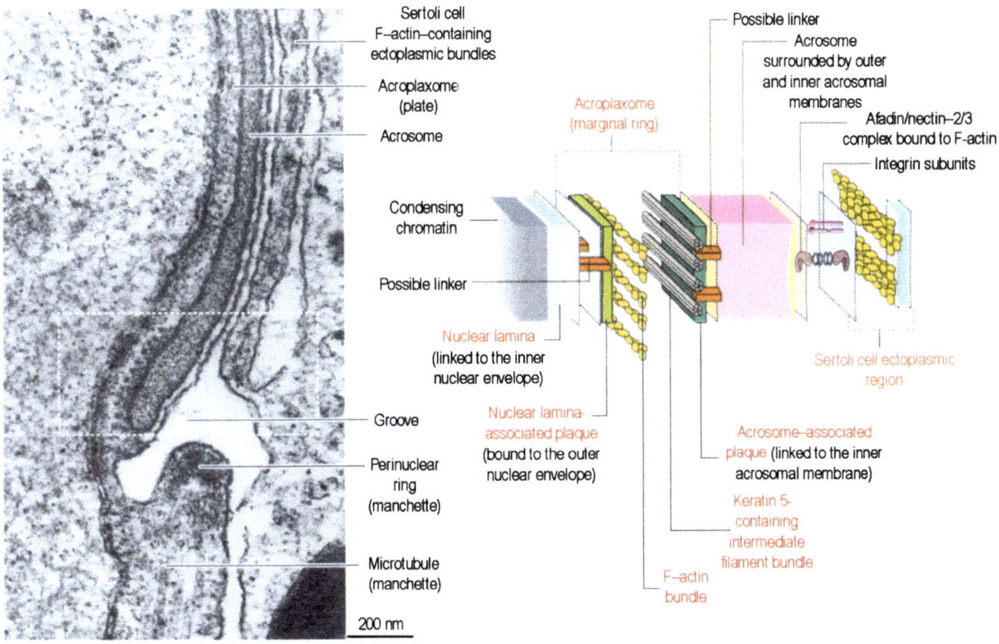

Figure 1. The dashed box in the electron micrograph (left) points to rat spermatid components depicted in the accompanying diagram, with particular emphasis on the structure of the marginal ring of the acroplaxome and adjacent structures. Note that (1) the nuclear lamina associated to the spermatid nuclear envelope is prominent in the acroplaxome region and decreases in density beyond the circular groove. (2) The bundle of intermediate filaments, shown in cross section, is firmly attached to the acrosome-associated plaque. (3) The perinuclear ring of the manchette is separated from the marginal ring of the acroplaxome by the circular groove. (4) Sertoli cell F-actin bundles along the ectoplasmic region are adjacent to a longitudinal channel. The ectoplasmic region terminates abruptly at the circular groove region. Diagram of the Sertoli-spermatid relationship (right). The diagram (drawn not to scale) denotes the components and localization of the acroplaxome, Sertoli cell F-actin-containing ectoplasmic bundles, and adhesion molecules (afadin-nectin-2/3 complex and integrin subunits) linking the spermatid to a contiguous Sertoli cell.

toward the centrosomal region and beyond, and contains keratin 9 (Mochida, Rivkin, Gil and Kierszenbaum, 2000) and other proteins. It is associated with microtubules consisting of several post-translationally modified tubulin molecules (reviewed in Kierszenbaum, 2002a) and F-actin (Kierszenbaum *et al.*, 2003b). A number of microtubule- and F-actin associated motor proteins, involved in intramanchette transport (Kierszenbaum, 2002b), are present in the microtubular component of the manchette. Observations in the acrosome-deficient Hrb mutant indicate that the position of the manchette is dependent on acrosomal biogenesis (Kierszenbaum *et al.*, 2004).

Two structural features are noteworthy: a circular groove develops at the boundary of the acroplaxome-manchette rings, and the stacks of Sertoli cell F-actin hoops stop at the groove (Fig. 1). From a hypothetical mechanical point of view, the groove appears to provide a flexible zone between exogenous constriction forces, exerted by Sertoli cell F-actin-containing hoops embracing the apical one-third of the spermatid nucleus, and the overlapping acroplaxome-manchette encircling rings. As spermiogenesis progresses, both rings reduce their diameter, adjusting in a sleeve-like manner, to the perimeter of the elongating spermatid nucleus. The elongating spermatid head is kept relatively stationary with respect to the adjacent Sertoli cell by the actin-linked afadin-nectin complex and the tubulobulbar complex extending from the concave face of the elongating spermatid into cytoplasmic microcrypts in the adjacent Sertoli cell (Guttman, Takai and Vogl, 2004; reviewed in Kierszenbaum and Tres, 2004).

F-actin, the primary template for the assembly of the acroplaxome

How can the acroplaxome contribute to spermatid nuclear shaping? During the early stages of acroplaxome assembly, F-actin precedes the appearance of keratin 5 (previously designated Sak57; Kierszenbaum *et al.*, 2003a). This suggests that F-actin may provide a template for the assembly of keratin 5-containing filaments, including the development of the marginal ring. The acroplaxome undergoes a dynamic change during spermiogenesis: it increases its diameter in parallel with the corresponding enlargement and spreading of the acrosome over the elongating spermatid nucleus; then it reduces the diameter of its marginal ring when nuclear elongation is underway. These changes may occur at the expense of F- actin and keratin 5 polymerization and depolymerization.

The actin regulatory testis-expressed protein profilin III (Braun, Aszodi, Hellerbrand, Berna, Fassler and Brandau, 2002) and profilin IV (Obermann, Raabe, Balvers, Brunswig, Schulze and Kirchhoff, 2005), a protein family that regulates the assembly of actin filaments by catalyzing the exchange of G-actin bound to ADP for ATP and transferred to the barbed (polymerization end) of actin, is present in the acrosome-acroplaxome region (Obermann *et al.*, 2005). Phosphorylation of cofilin, an actin depolymerization factor that stimulates the dissociation of ADP-bound actin, by LIM kinase, prevents cofilin binding and depolymerization of actin. Targeted disruption of the testicular isoform of LIM kinase 2 (LIMK2) present in spermatids disrupts the progression of spermatogenesis (Takahashi, Koshimizu, Myzakai and Nakamura, 2002). A variant of Wiskott-Aldrich syndrome protein-interacting actin polymerization src homology 3 protein (designated WISH) was found in testis (Fukuoda, Suetsugu, Miki, Fukami, Endo and Takenawa, 2001). WISH enhances Wiskott-Aldrich syndrome protein-induced actin-related protein 2/3 (Arp 2/3) complex activation, resulting in actin nucleation and branching.

Fer tyrosine kinase and cortactin, two components of the acroplaxome

Texas Red-X phalloidin, a fluorescent probe for filamentous actin, is visualised in the acroplaxome plate and marginal ring of round and elongated spermatids (Fig. 2, A-C). We have detected by immunocytochemistry the presence in the acroplaxome of the tyrosine-phosphorylated form of the F-actin-binding protein cortactin (anti-p (Y421) cortactin serum, Stressgen, B.C., Canada) (Fig. 2, D-F). Cortactin is phosphorylated by non-receptor tyrosine kinases. Stable F-actin prevents cortactin tyrosine phosphorylation; depolymerizing actin enables phosphorylation of unbound cortactin by Fer tyrosine kinase (Fan, Di Ciano-Oliveira, Weed, Craig and Greer, 2004). Cortactin emerges as a probable feedback regulator of F-actin remodeling in the acroplaxome and the Sertoli cell hoops.

Cortactin (~ 80 kDa) is an F-actin associated protein and a substrate for several non-receptor protein tyrosine kinases such as Src, Fyn and Fer (Greer, 2002; Lua and Low, 2005). Cortactin consists of an N-terminal with an NTA region (N-terminal acidic region) serving as the binding site for the actin-related protein (Arp)2/3 complex, and six and one half tandem repeats (cortactin repeats) of a 37-amino-acid motif, involved in F-actin binding. NTA and cortactin repeat do-

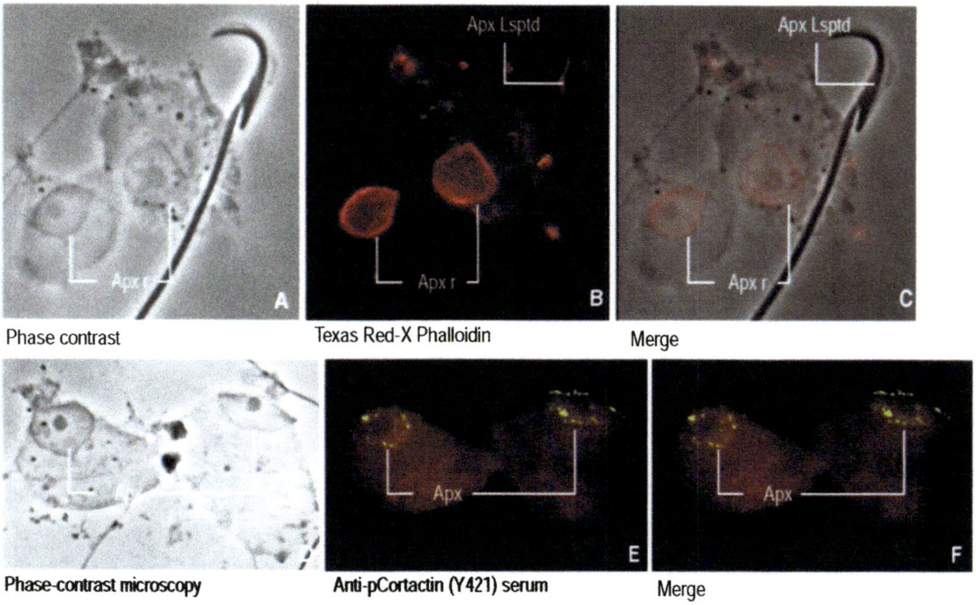

Figure 2. (**A-C**) Localization of F-actin in the acroplaxome and its marginal ring (Apx r) of rat round spermatids (step 8-9 of spermiogenesis) and a late spermatid (Apx LSptd) detected with Texas Red-X Phalloidin. (**D-F**) Localization of phosphorylated cortactin (pCortactin) in the acroplaxome (Apx) of rat round spermatids (step 8 of spermiogenesis) using a polyclonal antibody generated against a cortactin synthetic peptide phosphorylated at tyrosine 421 (Y421; Working dilution: 1:50 first antibody; Catalog number: KAP—TK001; Stressgen, Victoria, B.C., Canada). Tyrosine 421 (Y421) is present in the tyrosine phosphorylation Src homolog 3 domain. The antibody does not detect non-phosphorylated cortactin. Panels A-C reprinted from Kierszenbaum, Rivkin and Tres, 2003b, with permission of the publisher (Karger AG).

mains promote F-actin nucleation, *de novo* actin polymerization and Arp2/3-mediated actin branching and stability. The C-terminal of cortactin contains an a-helical domain followed by a PRR (Pro-rich region), which is targeted by various protein kinases, and a Src homology 3 (SH3) domain, which can be the binding site of different proteins.

We have found that the non-receptor (cytoplasmic) Fer tyrosine kinase is present in the acroplaxome. Two forms of Fer tyrosine kinase exist in testis: a somatic form, FerS, and a truncated testis-specific form, FerT (Keshet, Itin, Fischman and Nir, 1990; Fischman, Edman, Shackleford, Turner, Rutter and Nir, 1990). FerS tyrosine kinase (Fig. 3A) consists of an N-terminus FCH (Fsp/Fes/Fer/CIP4 homology) domain, implicated in the regulation of cytoskeletal rearrangements, followed by coiled-coil domains, that mediate the formation of monotrimers, a SH2 domain, involved in protein-protein interactions, and a kinase domain with a role in tyrosine phosphorylation (Greer, 2002). FerS is ubiquitously expressed in somatic tissues (designated FerS). The truncated testis-specific form, FerT (Fischman *et al.*, 1990; Keshet *et al.*, 1990), lacks the N-terminus FCH domain and coiled-coil domains, but retains SH2 and kinase domains. An antibody produced against the N-terminal amino acid region of FerT, not present in FerS, demonstrates antigenic sites in the acroplaxome and the manchette (Fig. 3, B-F). Therefore, there is a correlation in the localization sites of FerT and phosphorylated cortactin during the transition of round spermatids to elongated spermatids.

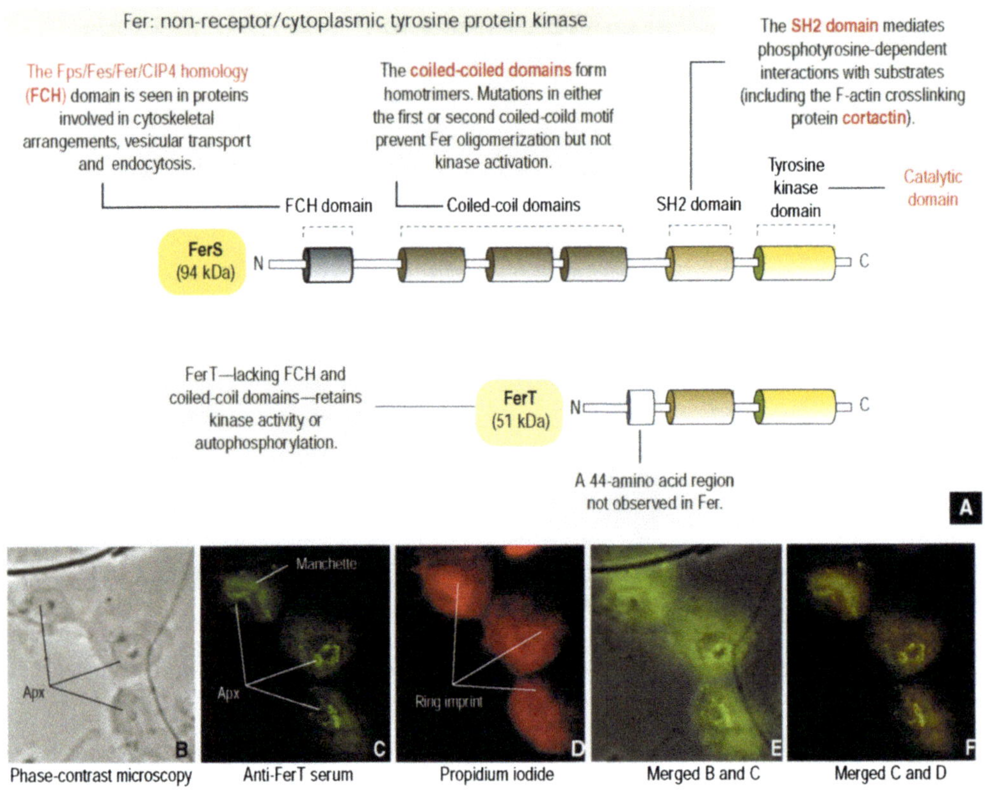

Figure 3. (**A**) Diagram of the non-receptor/cytoplasmic tyrosine protein kinase Fer somatic (FerS) and truncated testis-specific Fer (FerT). (**B-F**) Localization of FerT in the acroplaxome and the manchette of rat spermatids (step 9). A punctuated FerT immunoreactive pattern is associated with the acroplaxome and with a portion of the manchette adjacent to its perinuclear ring. The red propidium iodide staining represents nuclear and cytoplasmic nucleic acids. Note a sharp red band corresponding to the imprint of the marginal ring of the acroplaxome. Polyclonal antisera to Fer (data not shown) and FerT were generated against synthetic peptides. Anti-FerT was produced against a 12-amino acid peptide corresponding to a region of the 44-amino acid region not present in Fer. Figure reprinted with modifications from Kierszenbaum, 2006, with permission of the publisher (John Wiley and Sons).

Fer tyrosine kinase coexists with Fyn tyrosine kinase in spermatids. In the male, Fyn and Fer colocalise with F-actin in Sertoli cell ectoplasmic specializations (Fyn: Maekawa, Toyama, Yasuda, Yagi and Yasa, 2002; Fer: Chen, Lee, Mruk, Lee and Cheng, 2003). These observations are particularly relevant to our hypothesis that Sertoli cell F-actin hoops, as components of Sertoli cell ectoplasmic specializations, contribute exogenous constriction forces during the elongation of the spermatid head and possibly during spermiation. Fyn and Fer might control in part the function of the F-actin-containing hoops but this possibility needs to be determined. Fer and Fyn immunoreactivity in the acroplaxome can be seen in the illustrations of Maekawa *et al.* (2002; Fig. 8,B) and Chen *et al.* (2003, Fig. 3,D).

A puzzling observation is that Fer- and Fyn-deficient male mice are fertile (Craig *et al.*, 2001; Maekawa *et al.*, 2002). Yet, Fyn- and Fer- deficient mutants display degenerating spermatogenic cells (Maekawa et *al.*, 2002). Fertility in male Fyn- and Fer-deficient mutants has obviously raised the question of biological redundancy (substitution of Fer deficiency by Fyn tyrosine kinase). It is not known whether Fyn deficiency is compensated for by Fer kinase or vice versa.

Role of keratin 5 in the acroplaxome

Coincident with the simultaneous presence of phosphorylated cortactin and Fer tyrosine kinase in the acroplaxome is the visualization of a keratin 5 discontinuous immunoreactive pattern in the acroplaxome of elongating spermatids (Fig. 4). In contrast, the acroplaxome in round spermatids displays a continuous linear immunofluorescent pattern. F-actin is not a strong cytoskeletal scaffold when compared to the viscoelastic properties of keratins. The coexistence of F-actin and keratin 5 in the acroplaxome is consistent with the possibility that the keratin cytoskeleton transmits mechanical signals from the cell surface to all portions of the cell. The F-actin-keratin 5-containing acroplaxome may gently modulate exogenous forces generated by the contractile Sertoli cell F-actin-containing hoops encircling the upper region of the elongating spermatid head harbouring the condensing nucleus. If keratin 5 is absent, as in the acrosome-less *Hrb* mutant mouse, the spermatid nucleus does not elongate and unbalanced forces generate nucleopodes, protrusions of nuclear material restricted to the acroplaxome site (Kierszenbaum *et al.*, 2004).

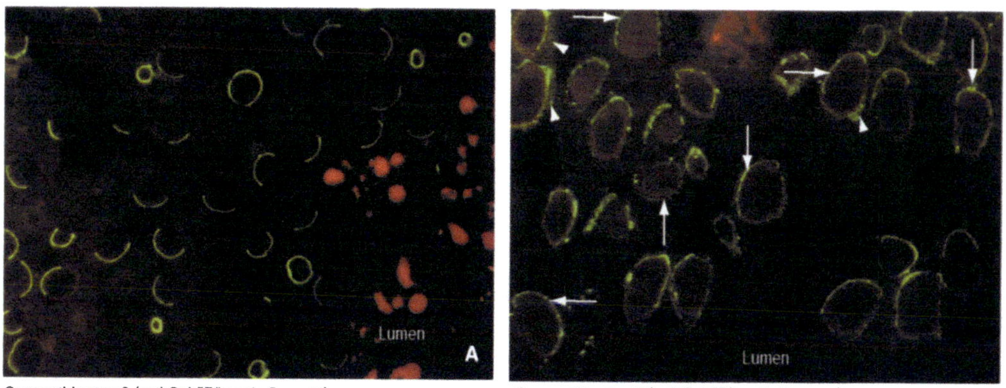

Spermatids step 6 (anti-Sak57/keratin 5 serum) Spermatids step 9 (anti-Sak57/keratin 5 serum)

Figure 4. (**A**) Localization of Sak57/keratin 5 in the acroplaxome of rat spermatids (step 6) shown in a section of the seminiferous tubule. Note the continuous green fluorescent linear pattern and the red propidium iodide staining of residual bodies near the lumen. (**B**) Localization of the same antigen in step 9 elongating spermatids seen in an adjacent seminiferous tubule (same section). The arrows indicate a discontinuous green fluorescent pattern regarded as a breakdown of the keratin 5 cytoskeleton in the acroplaxome. Immunorectivity in the manchette, representing intramanchette transport of keratin 5 to the developing spermatid tail, can also be seen (arrowheads). Tissue embedded in a water soluble resin (Lowicryl) and processed as described by Kierszenbaum, Rivkin and Tres (2003b).

Similar to all keratins, testis keratin 5 consists of a central rod domain flanked by a head domain at the N-terminus and a tail at the C-terminus (see Kierszenbaum *et al.*, 2003a, for detailed molecular data). Three aspects related to the function of keratin 5 in the acroplaxome are relevant: (1) keratin 5 contains abundant Ser residues on the head (25 residues) and tail (26 residues) domains. Ser residues are the target of serine/threonine protein kinases. (2) Phosphorylation of keratin 5 may target Ser residues in the head and tail domains and affect their assembly state (Helfand, Chang and Goldman, 2003). Shear stress exerted by exogenous constriction forces (Sertoli cell hoops) may induce a shift in the equilibrium between filamentous keratin (unphosphorylated) toward

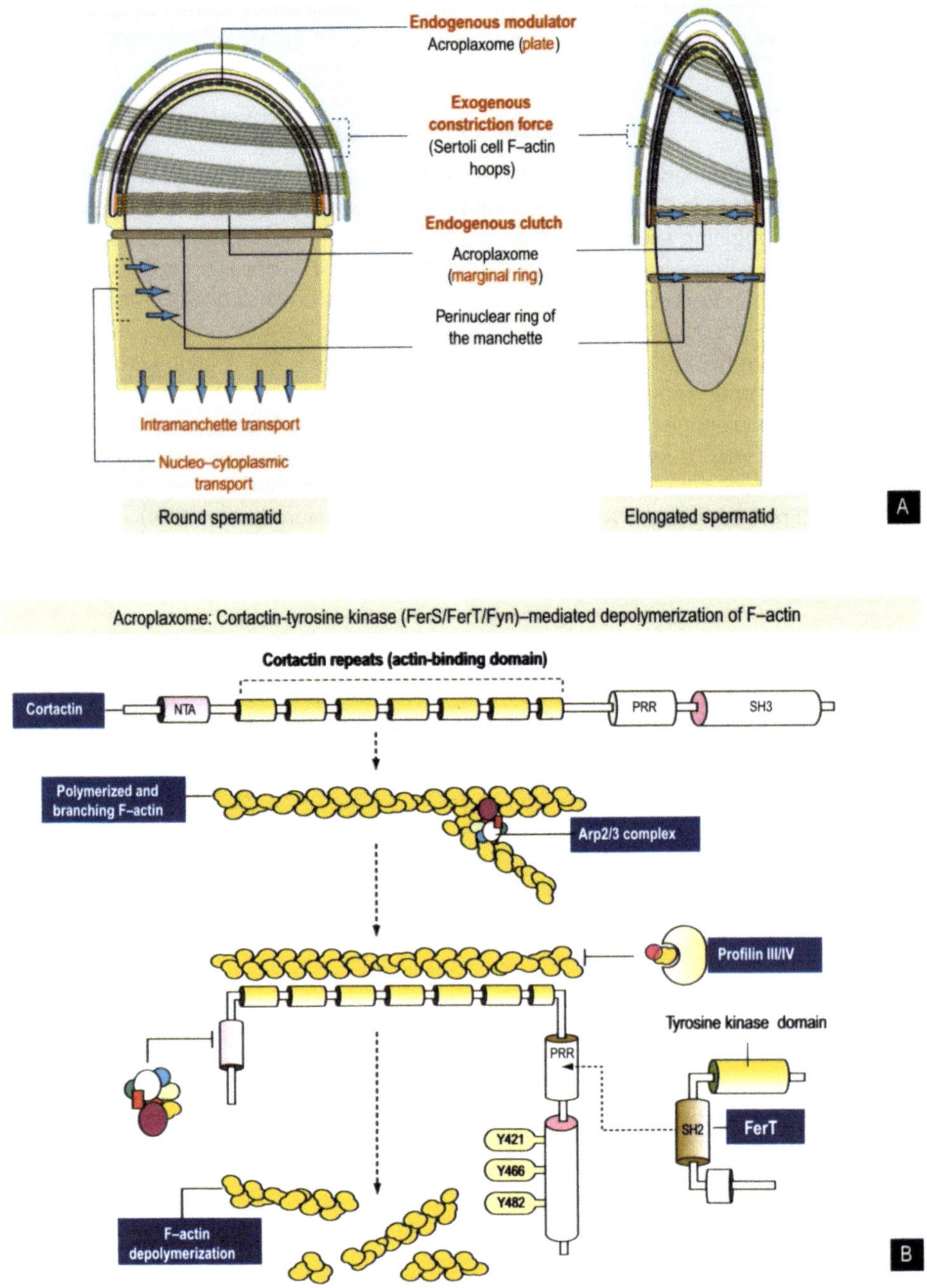

Figure 5. (**A**) Summary diagram (drawn not to scale) of a round spermatid and an elongated spermatid illustrating the exogenous constriction-endogenous modulator-endogenous clutch hypothesis. Sertoli cell F-actin hoops, embracing the apical region of the spermatid, represent the hypothetical exogenous constriction force. Note the change in the angular orientation and width of the Sertoli cell hoops in the elongated spermatid with respect to the round spermatid. A similar angular change is observed in the acroplaxome and manchette rings;

Figure 5 (contd) of exogenous constriction forces. The endogenous clutch involves the participation of the two rings (the marginal ring of the acroplaxome and the perinuclear ring of the manchette) acting as constricting sleeve-like structures adjusting to the decreasing diameter of the elongating nucleus. (**B**) Diagram illustrating the hypothesis that binding of cortactin to branching and polymerizing F-actin determines FerT-mediated tyrosine phosphorylation of cortactin. The SH2 domain of FerT (or FerS) binds to the PPR domain of cortactin and tyrosine phosphorylation of the SH3 domain of cortactin occurs as a result of the catalytic activity of the tyrosine kinase domain of FerT (or FerS). Phosphorylated cortactin prevents branching and further polymerization of G-actin monomers carried out by profilin III and/or IV, followed by disassembly of F-actin. The diagram of the structure of cortactin (drawn not to scale) indicates the presence of the N-terminal acidic region (NTA), the binding site of the actin-related protein (Arp) 2/3 complex, corresponding to the actin branching site along a pre-existing actin filament, and six-and-one-half tandem repeats of a 37-amino acid motif, involved in F-actin binding. The C-terminal of cortactin contains a proline-rich region (PPR) targeted by various protein kinases, and a Src homology 3 (SH3) domain with tyrosine phosphorylation sites (Y421, Y466, and Y482), which also has a domain for protein-protein interaction.

the soluble form (phosphorylated) under conditions in which the maintenance of the F-actin network can also be affected by the tyrosine kinase induced phosphorylation of cortactin. (3) Protein kinase Ca, a serine protein kinase, is present around the elongating spermatid (Guttman *et al.*, 2004). Future studies should determine additional details of the mechanism responsible for the apparent breakdown of the keratin scaffold in the acroplaxomal plate.

The summary diagram in Figure 5A emphasises the highlights of the hypothesis derived from observations in mutant mice with abnormal sperm head shaping in relation to the three major participants in spermatid head morphogenesis: Sertoli cell F-actin hoops (representing the exogenous constriction force), the acroplaxome F-actin-keratin 5-containing plate (acting as an endogenous modulator of exogenous constriction forces), and the marginal ring of the acroplaxome and perinuclear ring of the manchette (corresponding to the endogenous clutch operating at the descending end of the elongating spermatid nucleus). It should be stressed that additional roles of the manchette are nucleocytoplasmic transport (Kierszenbaum *et al.*, 2002) and intramanchette transport of cargos required for spermatid tail development (Kierszenbaum, 2002b).

How can cortactin phosphorylation modulate the integrity of F-actin in the acroplaxome and presumably the Sertoli cell hoops? Figure 5B illustrates the hypothesis that cortactin binding of F-actin blocks F-actin branching and polymerization by preventing the activities of the Arp 2/3 complex and testis-expressed profilin III (Braun *et al.*, 2002) and/or profilin IV (Obermann et *al.*, 2005). Depolymerization may occur when the SH2 domain of FerT (or FerS) binds to the PRR domain of cortactin and induces the phosphorylation of tyrosine residues in cortactin SH3 domain. The tyrosine kinase domain is available in FerS and truncated FerT to carry out cortactin phosphorylation. An alternative mechanism (not shown) was proposed by Fan *et al.* (2004): a stable F-actin network prevents phosphorylation of F-actin bound cortactin. When F-actin depolymerizes, cortactin detaches and becomes phosphorylated by non-receptor tyrosine kinases (Fer/Fyn tyrosine kinases). Essentially, the integrity of F-actin in the acroplaxome and Sertoli cell hoops may be protected by bound unphosphorylated cortactin. Whichever mechanism may occur, both alternatives attribute an important role to cortactin phosphorylation in the regulation of the F-actin cytoskeleton by enabling transient changes in F-actin architecture during spermatid head shaping.

References

Bouchard, M.J., Dong, Y., McDermott, B.M. Jr., Lam, D.H., Brown, K.R., Shelanski, M., Bellve, A.R. and Racaniello, V.R. (2000). Defects in nuclear and cytoskeletal morphology and mitochondrial localization in spermatozoa of mice lacking nectin-2, a component of cell-cell adherent junctions. *Molecular Cell Biology* **20:** 2865-2873

Braun, A., Aszodi, A., Hellerbrand, H., Berna, A., Fassler, R. and Brandau, O. (2002). Genomic organization of profilin-III and evidence for a transcript expressed exclusively in testis. *Gene* **283:** 219-225

Chen, Y-M., Lee, N.P.Y., Mruk, D.D., Lee, W.M. and Cheng, C.Y. (2003). Fer kinase/Fer T and adherens junctions dynamics in the testis: an in vitro and in vivo study. *Biology of Reproduction* **69:** 656-672

Craig, A.W.B., Zirngibl, R., Williams, K., Cole, L.A. and Greer, P.A. (2001). Mice devoid of Fer protein tyrosine kinase activity are viable and fertile but display reduced cortactin phosphorylation. *Molecular Cell Biology* **21:** 603-613.

Escalier, D. (1990). Failure of differentiation of the nuclear-perinuclear skeletal complex in the round-headed human spermatozoa. *International Journal of Developmental Biology* **34:** 287-297

Fan, L., Di Ciano-Oliveira, C., Weed, S.A., Craig, A.W.B. and Greer, P.A. (2004). Actin depolymerization-induced tyrosine phosphorylation of cortactin: the role of Fer kinase. *Biochemistry Journal* **380:** 581-591

Fischman, K., Edman, J.C., Shackleford, M., Turner, J.A., Rutter, W.J. and Nir, U. (1990). A murine fer testis-specificc transcript (ferT) encodes a truncated Fer protein. *Molecular Cell Biology* **10:** 146-153

Fukuoka, M., Suetsugu, S., Miki, H., Fukami, K., Endo, T. and Takenawa T. (2001). A novel neural Wiskott-Aldrich syndrome protein (N-WASP) binding protein, WISH, induces Arp2/3 complex activation independent of Cdc42. *Journal of Cell Biology* **152:** 471-482

Greer, P. (2002). Closing in on the biological functions of FBS/FES and FER. *Nature Reviews Molecular Cell Biology* **3:** 278-289

Guttman, J.A., Takai, Y. and Vogl, A.W. (2004). Evidence that tubulobulbar complexes in the seminiferous epithelium are involved with internalization of adhesion junctions. *Biology of Reproduction* **71:** 548-559

Helfand, B.T., Chang, L. and Goldman, R.D. (2003). The dynamic and motile properties of intermediate filaments. *Annual Review of Cell and Developmental Biology* **19:** 445-467

Kang-Decker, N., Mantchev, G.T., Juneja, S.C., McNiven, M.A. and van Deursen, J.M. (2001). Lack of acrosome formation in Hrb-deficient mice. *Science* **294:** 1531-1533

Keshet, E., Itin, A., Fischman,. K. and Nir, U. (1990). The testis-specific transcript (ferT) of the tyrosine kinase FER is expressed during spermatogenesis in a stage-specific manner. *Molecular Cell Biology* **10:** 5021-5025

Kierszenbaum, A.L. (2002a). Sperm axoneme: a tale of tubulin posttranslation diversity. *Molecular Reproduction and Development* **62:** 1-3

Kierszenbaum, A.L. (2002b). Intramanchette transport (IMT): Managing the making of the spermatid head, centrosome, and tail. *Molecular Reproduction and* Development **63:** 1-4

Kierszenbaum, A.L. (2006). Tyrosine protein kinases and spermatogenesis: truncation matters. *Molecular Reproduction and Development* **73:** 399-403

Kierszenbaum, A.L., Gil, M., Rivkin, E. and Tres, L.L. (2002). Ran, a GTP-binding protein involved in nucleocytoplasmic transport and microtubule nucleation, relocates from the manchette to the centrosome region during rat spermiogenesis. *Molecular Reproduction and Development* **63:** 131-140

Kierszenbaum, A.L., Rivkin, E. and Tres, L.L. (2003a). Acroplaxome, an F-actin-keratin-containing plate, anchors the acrosome to the nucleus during shaping of the spermatid head. *Molecular Biology of the Cell* **14:** 4628-4540

Kierszenbaum, A.L., Rivkin, E. and Tres, L.L. (2003b). The actin-based motor myosin Va is a component of the acroplaxome, an acrosome-nuclear envelope junctional plate, and of manchette-associated vesicles. *Cytogenetics and Genome Research* **103:** 337-344

Kierszenbaum, A.L. and Tres L.L. (2004). The acrosome-acroplaxome-manchette complex and the shaping of the spermatid head. *Archives of Histology and Cytology* **67:** 271-284

Kierszenbaum, A.L., Tres, L.L.,, Rivkin E., Kang-Decker, N. and van Deursen, J.M. (2004). The acroplaxome is the docking site of Golgi-derived myosin Va/Rab27a/b-containing proacrosomal vesicles in wild-type and Hrb mutant mouse spermatids. *Biology of Reproduction* **70:**1400-1410

Lua, B.L. and Low, B.C. (2005). Cortactin phosphorylation as a switch for actin cytoskeletal network and cell dynamics control. *FEBS Letters* **579:** 577-585

Maekawa, M., Toyama, Y., Yasuda, M., Yagi, T. and Yuasa, S. (2002). Fyn tyrosine kinase in Sertoli cells is involved in mouse spermatogenesis. *Biology of Reproduction* **66:** 211-221

Meistrich, M. L, (1993). Nuclear morphogenesis during spermiogenesis. In: *Molecular Biology of the Male Reproductive System*. Edited by de Kretser, D. Academic Press, New York, pp. 67-97

Mendoza-Lujambio, I., Burfeind, P., Dixkens, C., Meinhardt, A., Hoyer-Fender, S., Engel, W. and Neesen,. J (2002). The Hook1 gene is non-functional in the abnormal spermatozoon head shape (azh) mutant mouse. *Human Molecular Genetics* **11:** 1647-1658

Mochida, K., Tres, L.L. and Kierszenbaum, A.L. (1998). Isolation of the rat spermatid manchette and its perinuclear ring. *Developmental Biology* **200**: 46-56

Mochida, K., Tres, L.L. and Kierszenbaum, A.L. (1999). Structural and biochemical features of fractionated spermatid manchettes and sperm axonemes of the azh/azh mutant mouse. *Molecular Reproduction and Development* **52**: 434-444

Mochida, K., Rivkin, E., Gil, M. and Kierszenbaum, A.L. (2000). Keratin 9 is a component of the perinuclear ring of the manchette of rat spermatids. *Developmental Biology* **227:** 510-519

Mochida, K., Tres, L.L. and Kierszenbaum, A.L. (2000). Structural features of the 26S proteasome complex isolated from rat testis and sperm tail. *Molecular Reproduction and Development* **57:**176-184

Obermann, H., Raabe, I., Balvers, M., Brunswig, B., Schulze, W. and Kirchhoff, C. (2005). Novel testis-expressed profilin IV associated with acrosome biogenesis and spermatid elongation. *Molecular Human Reproduction* **11:** 53-64

Ozaki-Kuroda, K., Nakanishi, H., Ohta, H., Tanaka, H., Kurihara, H., Mueller, S., Irie, K., Ikeda, W., Sakai, T., Wimmer, E., Nishimune, Y. and Takai, Y. (2002). Nectin couples cell-cell adhesion and actin scaffold at heterotypic testicular junctions. *Current Biology* **12:** 1145-1150

Takahashi, H., Koshimizu, U., Myzakai, J. and Nakamura, T. (2002). Impaired spermatogenic ability of testicular germ cells in mice deficient in the LIM-kinase 2 gene. *Developmental Biology* **241:** 259-272

Yang, W.X. and Sperry, A.O. (2003). C-terminal kinesin motor KIFC1 participates in acrosome biogenesis and vesicle transport. *Biology of Reproduction* **69:** 1719-1729

Yao, R., Ito, C., Natsume, Y., Sugitani, Y., Yamanaka, H., Kuretake, S., Yanagida, K., Sato, A., Toshimori, K. and Noda, T. (2002). Lack of acrosome formation in mice lacking Golgi protein, GOPC. *Proceedings of the National Academy of Sciences USA* **99:** 11211-11216

Spermatology. SRF Vol. 65. ERS Roldan and M Gomendio (eds) Nottingham University Press, Nottingham

The scaffold role of the fibrous sheath

E. M. Eddy

Gamete Biology Section, Laboratory of Reproductive and Developmental Toxicology, National Institute of Environmental Health Sciences, National Institutes of Health, 111 T.W. Alexander Drive, Research Triangle Park, NC 27709, USA

The fibrous sheath (FS) is a novel structural feature in the principal piece region of the sperm flagellum in marsupial and eutherian mammals. It begins anteriorly at the annulus, is composed of two longitudinal columns connected by circumferential ribs, and surrounds the outer dense fibers in the principal piece. The formation of the FS was described eloquently nearly twenty-five years ago, but the identification of its components has occurred only during the last few years. Most of the genes encoding the known FS components are expressed during the postmeiotic period of spermatogenesis, are expressed only in spermatogenic cells, and are either novel genes or paralogues of genes expressed in somatic cells. The proteins of the FS can be classified in nine categories based on their functional or structural roles. Most of these are resistant to solubilization and remain in the FS during rigorous isolation procedures. Mouse sperm with a knock out of the gene encoding GAPDHS are immotile, which demonstrates that most of the ATP needed for sperm motility is generated by glycolysis. Most of the glycolytic enzymes have been localised to the principal piece and some are tightly associated with the FS. However, the nature of the protein-protein interactions involved in assembly and maintenance of FS are known for only a few of the component proteins.

Introduction

Many of the transcripts produced during male germ cell development are found only in spermatogenic cells. Some are products of genes that are paralogues of genes expressed in somatic cells and some are products of novel genes. Others are products of variant transcripts produced by alternative exon usage or splicing events that occur during transcription from genes also expressed in somatic cells. Genes expressed in somatic cells usually are down-regulated as their germ cell-specific counterparts are up-regulated. Most of the proteins that form the novel structural and functional features of spermatozoa are products of genes expressed from the haploid genome during the post-meiotic phase of spermatogenesis (Eddy, 2002, 2006). Although progress has been made in identifying proteins involved in determining structural and functional features of spermatozoa, there are considerable gaps in our knowledge. These include a need to identify genes essential for the development and function of spermatozoa, to gain an understanding of how their expression is regulated, to determine the roles of their encoded proteins, and to identify the mechanisms that regulate the function of these proteins. The fibrous sheath (FS) provides an excellent example of a sperm-specific structure for which a growing number of constituent and associated proteins are known, but with much remaining to

Corresponding author E-mail: eddy@niehs.nih.gov

be learned about the composition, assembly, and role of this unique structure in sperm function. Most studies on the FS have been carried out on human or rodent sperm. A large majority of these have been in the mouse because of the genetic and genomic information available and the ability to selectively disrupt individual genes by homologous recombination in this species. The gene symbols and other genomic information reported here are based on information currently available in the Mouse Genome Informatics (MGI) database (http://www.informatics.jax.org/).

Fibrous sheath components

The principal piece region of the flagellum is defined by the presence of the FS, which consists of two longitudinal columns connected by closely arrayed circumferential ribs (Fig. 1). The FS surrounds the axoneme and outer dense fibers, the other major cytoskeletal structures in the mammalian sperm flagellum. FS development begins in round spermatids soon after completion of the second meiotic division and is not completed until the final step of remodeling spermatids into spermatozoa. An unusual aspect of FS formation is that it proceeds from distal to proximal, beginning after the axoneme is formed and when the outer dense fibers are partially formed. The longitudinal columns first appear and extend proximally and the ribs begin to appear at the distal end of the FS in the early part of spermatid elongation (Fig. 2). The definitive study of FS formation was in the rat (Irons and Clermont, 1982). The classical view was that the FS provides the sperm tail with mechanical support that modulates flagellar bending and defines the shape of the flagellar beat (Fawcett, 1975). This is likely to be one of its roles, but more recent studies have established that an essential role of the FS is to serve as a scaffold for proteins with indispensable roles in sperm motility.

The proteins currently known to be associated with the FS can be grouped in nine categories based on their roles: PKA signalling, Rho pathway, chemoprotective, structural, calcium binding, citric acid cycle, myosin-like, role unknown, and glycolysis (Table 1). Based on what is known currently, several of these components can be placed in more than one category. They have been identified using a variety of biochemical and molecular biology approaches, with a majority of these proteins being identified in the last six years. It is likely that there are numerous other components to be identified, considering that the apparently simple cilium from the green alga *Chlamydomonas reinhardtii* contains over 650 proteins (Pazour, Agrin, Leszky and Whitman, 2005).

PKA signalling pathway

The activity of the PKA signalling pathway (Table 1) is triggered by cyclic adenosine monophosphate (cAMP) activation of protein kinase A (PKA). PKA is a tetramer containing two cAMP-activated regulatory (R) subunits and two catalytic subunits which phosphorylate serine and threonine residues of proteins. The N-terminal domains of the two R subunits dimerise and bind to an amphipathic helix region characteristic of PKA anchoring proteins (AKAPs) (Pawson and Scott, 1997). Three AKAPs are found only in association with the FS in spermatids and sperm. One of these, AKAP4, is the most abundant protein in the FS. Transcription of the mouse *Akap4* gene begins during the early part of round spermatid development, but protein synthesis is delayed until spermatids become elongated (Carrera, Gerton, and Moss, 1994; Fulcher, Mori, Welch, O'Brien, Klapper and Eddy, 1995). The FS anlagen forms in round spermatids, but AKAP4 is not deposited on this framework until late in spermiogenesis (Fig. 2).

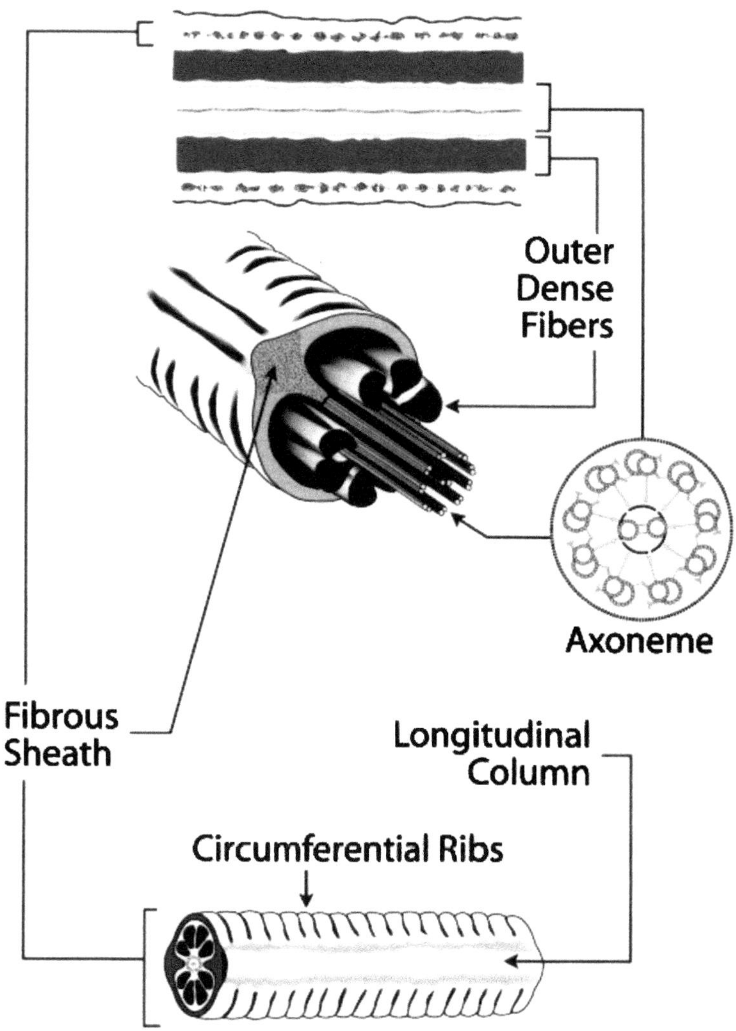

Figure 1. Structural features of the fibrous sheath. A longitudinal section of the fibrous sheath is shown at the top of the figure and surface views are seen in the center and lower parts of the figure. The fibrous sheath surrounds the outer dense fibers throughout the length of the principal piece of the flagellum. The two longitudinal columns of the fibrous sheath are connected by closely arrayed circumferential ribs. The columns attach to microtubule doublets 3 and 8 of the axoneme and the plane of the flagellar beat appears to be perpendicular to the columns (From Eddy, 2006).

Two variant transcripts that arise by alternative exon utilization encode precursor forms of AKAP4 that differ by 9 amino acids at the N-terminus. Processing removes the prodomains from both forms to produce a single ~80kDa mature protein. The AKAP4 protein is present in both the longitudinal columns and the ribs of the FS of mouse sperm (Johnson, Foster, Haig-Ladewig, VanScoy, Rubin, Moss and Gerton, 1997). An unusual feature of AKAP4 is the presence of two PKA-anchoring domains, one with dual binding ability for RIα or RIIα subunits of PKA and the other binding specifically the RIα subunit (Miki and Eddy, 1998).

Table 1. Fibrous sheath associated proteins

Protein	Gene	References
PKA Signalling		
PKA-binding protein 4 (AKAP4)	*Akap4*	Carrera et *al*., 1994; Fulcher et *al*., 1995a.
PKA-binding protein 3 (AKAP3)	*Akap3*	Vijayaraghavan et *al*., 1999; Mandel et *al*., 1999
PKA-binding protein 14 (TAKAP-80)	*Akap14*	Mei et *al*., 1997
Sperm autoantigen protein 17 (SP17)	*Spa17*	Lea et *al*., 2004
Fibrous sheath interacting protein 2 (FSIP2)	*Fsip2*	Brown et *al*., 2003
Rhophilin	*Rophn*	Nakamura et *al*., 1999
ASP (ropporin 1-like)	*Ropn1l*	Carr et *al*., 2001
CABYR	*Cabyr*	Naaby-Hansen et *al*., 2002
Rho Pathway		
Rhophilin	*Rophn*	Nakamura et *al*., 1999
Ropporin (ropporin 1)	*Ropn1*	Fujita et *al*., 2000
ASP (ropporin 1-like)	*Ropn1l*	Carr et *al*., 2001
Chemoprotective		
Glutathine S-transferase mu 5 (GSTM5)	*Gstm5*	Fulcher et *al*., 1995b
Thioredoxin domain containing protein 2, spermatozoa (TXNDC2; previously SPTRX-1)	*Txndc2*	Yu et *al*., 2002
Thioredoxin domain containing protein 3, spermatozoa (TXNDC3; previously SPTRX-2)	*Txndc3*	Miranda-Vizuete et *al*., 2002
Structural		
PKA-binding protein 4 (AKAP4)	*Akap4*	Carrera et *al*., 1994; Fulcher et *al*., 1995a
PKA-binding protein 3 (AKAP3)	*Akap3*	Vijayaraghavan et *al*., 1999; Mandel et *al*., 1999
PKA-binding protein 14 (TAKAP-80)	*Akap14*	Mei et *al*., 1997
Sperm autoantigen protein 17	*Spa17*	Lea et *al*., 2004
Testis specific 10 (TSGA10)	*Tsga10*	Modaressi et *al*., 2004
DDC8 (FS39)	*Ddc8*	Catalano et *al*., 2001
Rhophilin	*Rophn*	Nakamura et *al*., 1999
Calcium Binding		
(CABYR)	*Cabyr*	Naaby-Hansen et *al*., 2002
Citric Acid Cycle		
Pyruvate dehydrogenase E1 component B (PDHB, 36K-A, 36K-B)	*Pdhb*	Fujinoki et *al*., 2003, 2004
Myosin-like Features		
TSGA10	*Tsga10*	Modaressi et *al*., 2004
Role Unknown		
Fibrous sheath-interacting protein 1	*Fsip1*	Brown et *al*., 2003
Glycolysis		
Hexokinase type 1, spermatogenic cell-specific isoform (HK1S)	*Hk1*	Mori et *al*., 1993, 1998
Glyceraldehyde 3-phosphate dehydrogenase, sperm-specific (GAPDHS)	*Gapdhs*	Bunch et *al*., 1998
Lactate dehydrogenase C (LDHC)	*Ldhc*	Beyler et *al*., 1985
Aldolase A, two testis-specific expressed retrogenes	*Aldoa*	Vemagunti et *al*., 2006

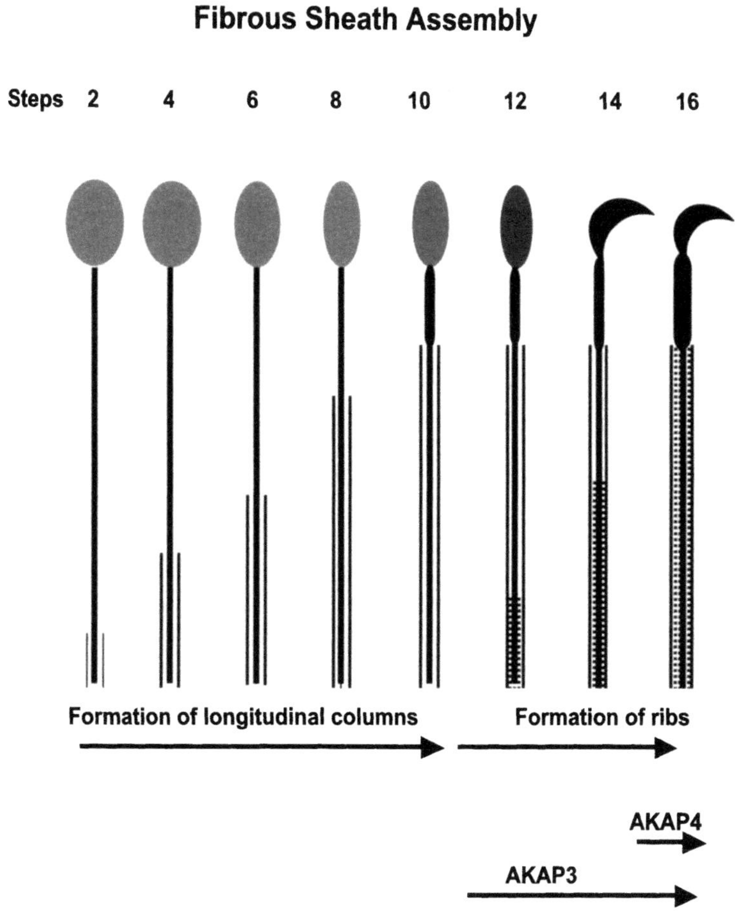

Figure 2. Fibrous sheath assembly in mouse sperm. The fibrous sheath assembles from distal to proximal along the flagellum. The longitudinal columns begin to form in the early part of the postmeiotic phase of spermatogenesis and the ribs form later. Two of the major structural proteins, AKAP3 and AKAP4, become part of the fibrous sheath only during the latter part of its development.

The AKAP3 protein also is a major component of the FS. It has an RIIα-binding domain in the N-terminal region and is 35% similar at the amino acid level to AKAP4 (Vijayaraghavan, Liberty, Mohan, Winfrey, Olson and Carr, 1999). Transcription of the *Akap3* gene and synthesis of AKAP3 protein begins in round spermatids, soon after the second meiotic division. The protein remains in the cytoplasm until spermatid elongation is underway and then is integrated into the FS as its formation progresses (Fig. 2). This suggests that AKAP3 contributes to the formation of the framework of the FS well before AKAP4 is integrated. A yeast two-hybrid screen with AKAP4 found that binding occurs between the C-terminal end of AKAP4 and the central region of AKAP3 (Brown, Miki, Harper and Eddy, 2003).

A third AKAP was identified in the FS of rat sperm that originally was referred to as AKAP-80 (Mei, Singh, Erlichman and Orr, 1997) and now is called AKAP14. Transcripts of *Akap14* were not detectable in testis before day 30 and were seen by Northern blotting in testis and not in seven other tissues. The MGI database indicates that mouse and human have *Akap14* orthologues, but they remain to be characterised.

Possible regulators of RII subunit binding to AKAPs that have been identified include SP17, CABYR, FSIP2, ropporin, and ASP. The SP17 protein bears an N-terminal domain 45% identical to the PKA binding region of human RII. It binds to the RII anchoring site on AKAP3 *in vitro* (Lea, Widgren and O'Rand, 2004), but does not bind to AKAP4 under the same conditions. The SP17 protein is present mainly in the principal piece, with some also being found in the middle piece and in the equatorial region of the mouse sperm head. While AKAP4 and AKAP3 are unique to spermatids, SP17 is present in a variety of normal and neoplastic cell types (Lea *et al.*, 2004). The testis-specific CABYR protein is present in multiple isoforms, some of which contain an N-terminal sequence with similarity to the RII dimerization and AKAP-binding domain (Naaby-Hansen, Mandal, Wolkowicz, Sen, Westbrook, Shetty,Coonrod, Klotz, Him, Bush, Flickinger and Herr, 2002). Antisera to the full-length isoform and to C-terminal isoforms detect the protein in the ribs and longitudinal columns of the FS (Naaby-Hansen *et al.*, 2002; Kim, Jha, Mandal, Vanage, Farris, Snow, Klotz, Naaby-Hansen, Flickinger and Herr, 2005). The binding site on AKAP4 for the testis-specific FSIP2 protein overlies the RIα/RIIα dual binding domain and this protein might also serve to regulate interaction between PKA and AKAP4 at this site (Brown *et al.*, 2003).

Ropporin and ASP (ropporin-1-like) also are able to bind to the PKA-anchoring site on AKAP3, but in addition are putative components of the Rho pathway (Table 1). Both proteins have an N-terminal region with strong similarity to the PKA binding region of RII. Transcription of ropporin begins near the end of the meiotic phase (Nakamura, Fujita, Murata, Watanabe, Mori, Fujita, Watanabe, Ishizaki, Yoshida and Narumiya, 1999). The ropporin protein is bound tightly to the outer surface of the FS of sperm from the cauda epididymis (Fujita, Nakamura, Kato, Watanabe, Ishizaki, Kimura, Mizoguchi and Narumiya, 2000), suggesting that it is incorporated into the FS late in its formation. ASP and ropporin are 39% identical at the amino acid level, but highly similar in their 80 amino acid N-terminal regions (Carr, Fujita, Stentz, Liberty, Olson and Narumiya, 2001).

Rho signalling pathway

Rhophilin, a small GTPase Rho-binding protein, is highly expressed in the testis and present at lower levels in several other tissues. The RNA was not detected in the mouse testis on postnatal day 15, but was present on day 18 and later, suggesting that it is transcribed in late pachytene spermatocytes and in round spermatids. It was localised to the principal piece of mouse sperm. Rhophilin binds Rho, but the lack of a catalytic domain suggests that it is an adaptor protein linking Rho to other proteins (Nakamura *et al.*, 1999).

Ropporin was isolated in a yeast two-hybrid screen using the PZD domain of rhophilin as bait. The *Ropn1* gene is transcribed in round spermatids and the protein is localised to the principal piece and end piece regions of the flagellum. By electron microscopy ropporin is seen mostly on the inner surface of the FS, while rhophilin appears to be more abundant on the outer surface of the FS (Fujita *et al.*, 2000). The N-terminal 40-amino acid regions of ropporin form dimers, similar to the association between RII proteins and an AKAP. While rhophilin was insoluble in 6 M urea (Nakamura *et al.*, 1999), ropporin was solubilised in 4 M urea (Fujita *et al.*, 2000). The *Ropn1l* transcripts that encode ASP (ropporin 1-like) were detected only in the testis and encode a novel protein. However, ASP and ropporin are similar only in their N-terminal sequences and ability to bind to the PKA anchoring domain of AKAP3. This suggests that ASP is not associated with the Rho pathway.

The association of PKA and Rho signalling components with the FS suggests a possible interaction between these two pathways. Interaction between ropporin, ASP and RII at the PKA binding domain on AKAP3 might modulate the association between RIIα and AKAP3 and FSIP2 might modulate the binding of RIα or RIIα to AKAP4. It also was proposed that a role of sperm AKAPs might be to serve as scaffolds for the Rho-GTPase pathway, and that Rho might regulate protein phosphatases through activation of Rho kinase (Carr *et al.*, 2001).

Chemoprotective

Sperm are highly susceptible to damage by oxidative stress due to peroxidation of unsaturated fatty acids in the plasma membrane. Three proteins associated with the FS might protect proteins in the principal piece from damage resulting from oxidative stress. One of these is GSTM5, a member of the mu-class glutathione S-transferase family. It is a major protein in the FS of mouse sperm (Fulcher, Welch, Klapper, O'Brien and Eddy, 1995). The *Gstm5* gene is expressed at high levels in pachytene spermatocytes and spermatids, lower levels in brain, and trace levels in several other tissues. The GST proteins are scavengers of cytotoxic substances, including electrophilic substances and peroxides. The superoxide anion responsible for the sperm oxidative stress apparently is not from sperm mitochondria, but the source has not been established (De Iuliis, Wingate, Koppers, McLaughlin and Aitken, 2006).

Thioredoxins have similar capabilities and two members of the thioredoxin protein family, TXNDC2 and TXNDC3, are products of spermatogenic cell-specific genes and are associated with the FS. The thioredoxins can reduce disulfide bonds if thioredoxin reductase is present and are reported to regulate DNA synthesis, apoptosis, and binding of transcription factors to DNA. The thioredoxin motif of TXNDC2 lies in the C-terminal region while the N-terminal region of the protein contains a novel 15-amino-acid sequence repeated 23 times (Yu, Oko, and Miranda-Vizuete, 2002). This protein is seen in the longitudinal columns of the FS during formation of the flagellum, but it is not detected by Western analysis or immunostaining in epididymal sperm. This suggests that TXNDC2 is a transient contributor to the formation of the longitudinal columns of the FS (Yu *et al.*, 2002). The N-terminal region of TXNDC3 also contains a thioredoxin domain and the C-terminal region contains three tandem-repeat NDP-kinase domains. Synthesis of the TXNDC3 protein begins in spermatocytes and peaks in spermatids. The TXNDC3 protein is present mainly in the cytoplasm of round spermatids and becomes confined to the principal piece region of the flagellum in elongating spermatids. The protein is present in the longitudinal columns and ribs of the FS and remains as an integral component of the FS in mature sperm (Miranda-Vizuete, Tsang, Yu, Jiménez, Pelto-Huikko, Flickinger, Sutovsky and Oko, 2003).

Structural

There are not distinct criteria for classifying proteins as structural component of the FS. For the current discussion, proteins that can only be extracted from the FS with 4 M urea or higher in the presence of low concentrations of a disulfide-reducing reagent are considered structural.

Although AKAP4 is the most abundant protein in the FS, it becomes a part of the FS relatively late in spermiogenesis (Fig. 2). Nevertheless, sperm from *Akap4* gene knockout mice have reduced amounts of AKAP3 and GAPDHS and the principal piece is thin compared to that of sperm from wild type mice (Miki, Willis, Brown, Goulding, Fulcher and Eddy, 2002). Both of these proteins require rigorous extraction procedures for their solubilization from wild type sperm. This suggests that AKAP4 plays an important role in cementing these and probably other proteins (Lea *et al.*, 2004) together to complete formation of the FS.

The AKAP3 protein was detected by immunostaining in the cytoplasm of early round spermatids, but was not found in the flagellum until spermatid elongation was under way (Fig. 2) (Brown *et al.*, 2003). The protein was localised to the ribs in human sperm by electron microscopic immunolabelling (Johnson *et al.*, 1997), but the precise location of AKAP3 in other species has not been determined. Because AKAP3 is a relatively abundant protein in the isolated FS, it presumably helps form the framework upon which AKAP4 is later deposited. The *Akap14* mRNA and AKAP14 protein were present at 30 days after birth in the rat. It remained associated with FS after extraction in 4.5 M urea and 20 mM dithiothreitol (Mei *et al.*, 1997) and is likely to have a structural role similar to that of the other two AKAP proteins.

Proteins present in other categories that can be considered structural components of the FS as well include SP17, GSTM5, DDC8, rhophilin, and PDHA (Table 1). The majority of SP17 remains in the insoluble fraction of rat sperm and in fibrous sheaths isolated from mouse sperm, but is reduced in amount in sperm from *Akap4* knockout mice. Furthermore, SP17 showed a strong level of association with AKAP3 and only a weak association with AKAP4 (Lea *et al.*, 2004). GSTM5 was identified by microsequencing tryptic digests of rigorously purified preparations of FSs (Fulcher *et al.*, 1995). The GSTM5 protein was present in the mouse FS after an isolation procedure that involved passage through solutions containing a nonionic detergent, potassium thiocyanate, DTT, and 4 M urea. Transcripts for *Ddc8* are first seen in round spermatids and the protein is present in the FS in elongating spermatids. The DDC8 protein is not homologous to any other proteins, is highly insoluble, and has an intermediate filament-like amino acid composition and predicted secondary structure (Catalano, Hillhouse, and Vlad, 2001). Rhophilin was found to be insoluble in 6 M urea and minimally extracted from mouse sperm using Laemmli buffer, unless pretreated with 10% trichloroacetic acid (Nakamura *et al.*, 1999). PDHA was present in FS isolated from hamster sperm after extraction in 6 M urea (Fujinoki, Kawamura, Toda, Ohtake, Ishimoda-Takagi, Shimizu, Yamaoka and Okuno, 2004).

Calcium-binding

The only FS protein identified so far that binds calcium is CABYR (calcium-binding tyrosine phosphorylation-regulated protein). Gel overlay methods demonstrated that CABYR binds calcium and it is present in the testis in multiple isoforms that result from alternative transcript splicing. Some isoforms lack the RII-like domain or a calcium-binding EF-hand domain in the N-terminal portion (Naaby-Hansen *et al.*, 2002). CABYR becomes phosphorylated during sperm capacitation and treatment of CABYR with alkaline phosphatase abolished calcium binding, strongly suggesting the calcium binding is phosphorylation dependent. Immunostaining of human testis demonstrated that the CABYR protein is present in round and elongating spermatids. The protein was found by electron microscopic immunocytochemistry staining to be present in the longitudinal columns and ribs of the FS of human sperm (Naaby-Hansen *et al.*, 2002).

Citric acid cycle

Two 36kDa proteins are present in hamster sperm that are phosphorylated during capacitation. They were subjected to proteomic analysis and both were found to be the pyruvate dehydrogenase E1 component ß (PDHB) subunit (Fujinoki, Kawamura, Toda, Ohtake, Ishimoda-Takagi, Shimizu, Yamaoka, and Okuno, 2003; Fujinoki *et al.*, 2004). However, one of the proteins (36K-B) was present in the middle piece (Fujinoki *et al.*, 2003) and the other (36K-A) was present in the principal piece and in isolated FSs (Fujinoki *et al.*, 2004). Pyruvate dehydrogenase is a complex of seven subunits and links glycolysis to the citric acid cycle during energy production under aerobic conditions. It is not obvious what role PDHB might have in the principal piece where most of the ATP required for sperm motility is produced anaerobically (Miki *et al.*, 2004).

Myosin-like

The *Tsga10* gene is first expressed in pachytene spermatocytes and is present at high levels in the testis, but it also was detected by PCR in several foetal tissues and a variety of tumour cells. The TSGA10 protein is synthesised in late spermatids and is present in the FS. A 67 kDa form of the protein was seen in total testis and sperm tails, while a 27 kDa form was present in an isolated FS preparation. An antibody to the N-terminal portion of the protein reacted with the 27kDa form and the protein apparently is processed around the time it becomes part of the FS (Modarressi, Behnam, Cheng, Taylor, Wolfe and van der Hoorn, 2004). The 67 kDa TSGA10 protein contains a domain that is similar to the myosin tail and highly similar to the ERM (Ezrin/radixin/moesin) family domain. Transient transfection studies in heterologous cells indicated that TSGA10 can form short, thick filaments, lending support to the suggestion that it might function in the formation of filaments in elongating spermatids (Modarressi *et al.*, 2004).

Unknown role

The FSIP1 protein was isolated in yeast two-hybrid screens using AKAP4 as bait. It has a predicted size of 49.9 kDa and was found in yeast two-hybrid assays to bind to regions of AKAP4 located between amino acids 351 and 359 and between amino acids 692 and 721. However, the protein does not contain recognizable motifs that suggest its function and the role of FSIP1 is yet to be determined (Brown *et al.*, 2003).

Glycolytic enzymes

A gene knock out of the spermatogenic cell-specific paralogue of glyceraldehyde-3-phosphate dehydrogenase demonstrated that glycolysis is essential for the production of ATP in sperm. Mice that were homozygous for a mutation in the *Gapdhs* gene were infertile and their sperm were incapable of progressive motility (Miki *et al.*, 2004). GAPDHS is located in the FS suggesting that the other glycolytic enzymes must be in close proximity for glycolysis to occur. Indeed, several other glycolytic enzymes have been identified in the principal piece and the majority of these are either paralogues of genes expressed in somatic cells or spermatogenic cell-specific variant transcripts of ubiquitously expressed genes.

Hexokinase. The first enzyme in the glycolytic pathway (Fig. 3) utilises an ATP to phosphorylate glucose, producing glucose-6-phosphate (G6P). Three of the four genes in the hexokinase family (*Hk1, Hk2,* and *Hk3*) encode ~100 kDa isozymes that contain two homologous sequences in tandem and are inhibited by G6P (HK1, HK2, HK3). The other gene (*Gck*) encodes glucokinase (GCK), a ~50kDa isozyme containing one HK domain and this isozyme is not inhibited by G6P. Earlier studies using starch gel electrophoresis found a sperm-type hexokinase activity in homogenates of testes from 20 day-old, but not 15 day-old rats (Sosa, Altamirano, Hernandez and Rosado, 1972). It was shown more recently that three spermatogenic cell-specific splice variants of type 1 hexokinase (*Hk1*) are present. The three variants (*Hk1_v1, Hk1_v2* and *Hk1_v3*) have different 5′ untranslated regions and encode a protein

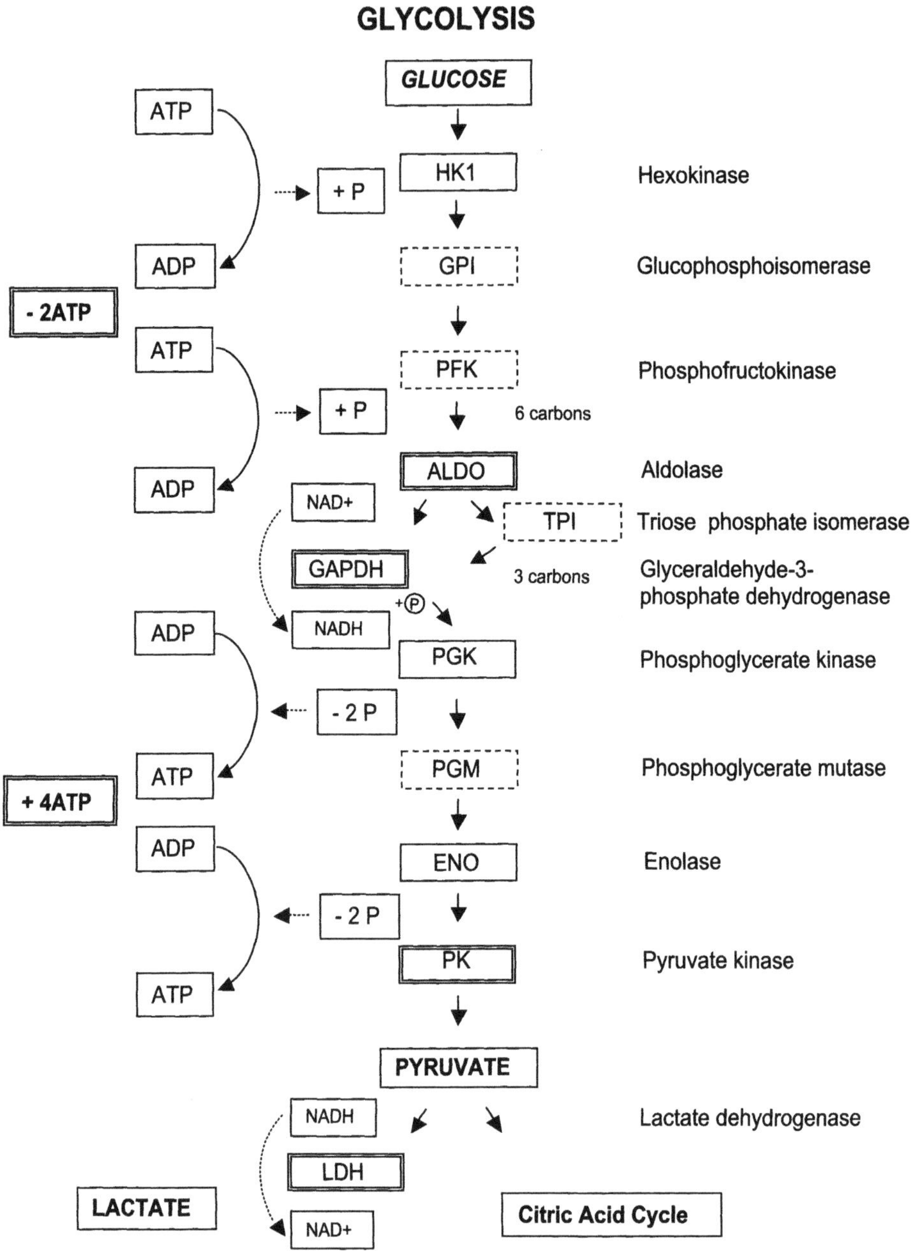

Figure 3. The glycolytic pathway. Most of the ATP required for sperm motility is generated by glycolysis. This process utilises two ATP molecules and produces four ATP molecules for each glucose molecule metabolised to pyruvate. Four of the glycolytic enzymes are present in the FS (double lines) and three others are present in the principal piece region of the flagellum (solid lines). It remains to be shown that the other four (dashed line) are present in this region of the flagellum.

(HK1S) in which the HK1 N-terminal porin-binding domain (PBD) is replaced with a 24 amino acid N-terminal spermatogenic cell-specific region (SSR) by alternative exon utilization during transcription (Mori, Welch, Fulcher, O'Brien and Eddy, 1993). The PBD binds HK1 to porins (also known as voltage-dependent anion channels) present in the outer mitochondrial membrane. The HK1S protein is found in the principal piece region of the flagellum (Mori, Nakamura, Welch, Gotoh, Goulding, Fajioka and Eddy, 1998), but 0.5% Triton X-100 released hexokinase from mouse sperm (Travis, Foster, Rosenbaum, Visconti, Gerton and Moss, 1998), indicating that this enzyme is not tightly bound to the FS.

Glucose phosphate isomerase. The second enzyme in the glycolytic pathway (GPI; also called phosphoglucose isomerase, PGI) converts the six-carbon ring into the linear fructose 6-phosphate structure. An electrophoretically distinct form of GPI was identified in mouse sperm (Beuhr and McLaren, 1981) suggesting that a different or modified GPI might be present. In addition, a monoclonal antibody to human sperm that caused agglutination was used to isolate cDNA clones with high identity to human GPI (Yakirevich and Naot, 2000). However, a novel *Gpi* gene or transcript has not been identified in spermatogenic cells. Further studies are needed to determine if some or all of the GPI associated with human sperm is the product of a *Gpi* paralogue or of the ubiquitously expressed *Gpi* gene.

Phosphofructokinase. The next enzyme in the glycolytic pathway (PFK; also called phosphohexokinase, PHK) uses another ATP to phosphorylate the sugar moiety. Enzymatically active PFK was released from boar sperm by hypotonic treatment with a potassium phosphate buffer (Jones and Piccolo, 1999), demonstrating that this enzyme is present, but not bound to the FS in boar sperm. There are four know genes in the PFK family in mouse, referred to as *Pfkl* (liver), *Pfkm* (muscle), *Pfkp* (platelet), and *Pfkx* (about which little is known). It has not been determined which gene encodes the PFK found in sperm.

Aldolase. This enzyme splits the six-carbon sugar into two three-carbon moieties, each bearing a high energy phosphate. There are three known aldolase genes in the mouse, referred to as *Aldoa, Aldob,* and *Aldoc,* and two additional pseudogenes (*Aldo-ps1* and *Aldo-ps2*) in the MGI database. An unusual aldolase was reported to be present in bovine sperm, the majority of which was relatively insoluble. The activity of this aldolase for the precursor substrate was much lower than for the liver or muscle forms and it was concluded that a different aldolase isozyme was present in sperm (Gillis and Tamblyn, 1984). Proteomic analyses demonstrated recently that aldolase A (ALDOA) is present in the mouse sperm flagellum (Cao, Gerton and Moss, 2006) and in the FS isolated from mouse sperm (Krisfalusi, Miki, Magyar and O'Brien, 2006). Further studies determined that two *Aldo*-like sequences previously identified as pseudogenes are instead functional retrogenes. These two retrogenes are expressed only in the testis and encode aldolase isoforms present in the FS (Vemuganti, Bell, Scarlett, Parker, Pardo-Manuel de Villena and O'Brien, 2006).

Triose phosphate isomerase. The role of this enzyme (TPI) is to interconvert the two three-carbon moieties to form glyceraldehyde-3-phosphate, the substrate for GAPDHS. Two TPI transcripts were identified in rat testis, one of which was detected only in spermatids. The difference between the two transcripts was a 100-bp deletion in one (Russell and Kim, 1996). A recent proteomics analysis detected TPI in the flagellum of mouse sperm (Cao *et al.*, 2006). Although only one functional TPI gene (*Tpi1*) in known in the mouse, the MGI database indicates that there are ten TPI-related sequences (*Tpi-rs2-11*) in the mouse genome that remain to be characterised.

Glyceraldehyde-3-phosphate dehydrogenase, spermatogenic. The first enzyme in the three-carbon portion of the glycolysis pathway is glyceraldehyde-3-phosphate dehydrogenase (GAPDH). It is NAD+ cofactor-dependent and adds a phosphate to glyceraldehyde 3-phosphate (G3P) to form 1,3-phosphoroglycerate. A testis-specific paralogue of GAPDH is transcribed in round spermatids and the protein (GAPDHS) is first seen in elongating spermatids (Bunch, Welch, Magyar, Eddy and O'Brien, 1998). The protein is highly resistant to extraction from sperm and is tightly bound to the FS. As noted before, the sperm from mice lacking functional GAPDHS are unable to generate progressive motility, resulting in infertility (Miki *et al.*, 2004). The ATP levels in these sperm were greatly reduced while oxygen consumption was comparable to that of sperm from wild-type mice, indicating that mitochondrial function was not altered. These results indicate that the energy required for sperm motility is derived primarily from ATP produced by glycolysis and not by oxidative phosphorylation in mitochondria (Miki *et al.*, 2004).

Phosphoglycerate kinase-2. This enzyme catalyses the first ATP-generating step during glycolysis in spermatozoa. It converts one ADP to an ATP per three-carbon moiety, resulting in production of two ATP molecules per glucose equivalent. The *Pgk1* gene is expressed constitutively in all cell types except spermatocytes and spermatids. While *Pgk1* is considered to be a single copy gene, seven *Pgk1* pseudogenes are listed in the MGI database. The *Pgk1* gene down-regulates at the beginning of spermatocyte development and there is a concomitant up-regulation of *Pgk2*. This a functional, intron-less gene that evolved by retroposition from the intron-containing *Pgk1* gene (McCarrey and Thomas, 1987; Boer, Adra, Lau and McBurney, 1987). Although *Pgk2* transcripts are present in primary spermatocytes, PGK2 protein is found only in spermatids, indicating both transcriptional and translational control of expression occurs. PGK2 protein accumulates in condensing spermatids, and is the only form of phosphoglycerate kinase found in spermatozoa (McCarrey, Kumari, Aivaliotis, Wang, Zhang, Marshall and Vandeberg, 1996). A knockout of the *Pgk2* gene resulted in markedly reduced sperm motility, low sperm ATP levels, and greatly impaired fertility (Danshina, Geyer, Dai, Goulding, Willis, McCarrey, Eddy and O'Brien, 2006). These results confirm that the ATP essential for sperm motility is produced by glycolysis.

Phosphoglycerate mutase. Four members of the phosphoglycerate mutase (PGAM) family are listed in the MGI database, one of which is a pseudogene. The *Pgam2* gene that encodes the phosphoglycerate mutase M (muscle) isozyme (PGAM2), is expressed in pachytene spermatocytes and spermatids in mouse (Fundele, Winking, Illmensee and Jagerbauer, 1987) and rat (Briceño, Ruiz, Reina, Vilaró and Pons, 1995). PGAM activity is present in rat sperm (Brooks, 1976), and PGAM2 was found in mouse sperm (Fundele *et al.*, 1987).

Enolase. There are three recognised members of the enolase gene family (*Eno1, Eno2,* and *Eno3*). An early study reported the presence of a fourth enolase (ENOS) in human, ram, and mouse sperm based on electrophoretic mobility, thermostability, and ability to undergo structural alteration at high temperatures. The ENOS protein was present only in testicular cell fractions containing elongating spermatids and sperm (Edwards and Grootegoed, 1983). It was reported more recently that the α-isoform of enolase (ENO1) is present in elongating spermatids and in the sperm flagellum in the rat (Gitlits, Toh, Loveland and Sentry, 2000). In addition, electrophoretic studies demonstrated that human sperm contain ENO1 and a sperm-specific isoform suggested to be the human ENOS (Force, Viallard, Grizard and Boucher, 2002; Force, Viallard, Saez, Grizard and Boucher, 2004). Disruption of mouse *Eno1* by a retroviral gene trap resulted in early embryonic lethality (Couldrey, Carlton, Ferrier, Colledge and Evans, 1998), but the beta-galactosidase reporter in the gene trap vector was expressed in testis and not in other adult tissues examined. Because *Eno1* is reported to be expressed in many tissues, there is a reasonable doubt that the gene disrupted was the ubiquitously expressed *Eno1* gene. It instead may have been the gene encoding ENOS.

Pyruvate kinase. Genes for two pyruvate kinase (PK) family member are curated in the MGI database, *Pklr* (pyruvate kinase, liver and red blood cell; pyruvate kinase 1) and *Pkm2* (pyruvate kinase, muscle; pyruvate kinase 2). However, two splice variants of *Pkm2*, referred to as M1 and M2, are expressed in the mouse testis, with the M1 variant being more abundant in the latter part of spermatogenesis (de Luis and del Mazo, 1998). Pyruvate kinase was identified by Western blotting in FS isolated from mouse sperm, indicating that it is tightly bound to the FS (Krisfalusi *et al.*, 2006). Additional studies are needed to determine if the PK associated with the FS is the product of one or both of these variants.

Lactate dehydrogenase. The three genes in the lactate dehydrogenase family (*Ldh1, Ldh2, Ldh3*) encode the LDHA, LDHB, and LDHC isozymes. Under anaerobic conditions, LDH converts pyruvate to lactate in a NADH cofactor-dependent reaction. All three isozymes are expressed in the testis, but LDHC is found in spermatogenic cells (Hintz and Goldberg, 1977; Li, O'Brien, Hou, Versola, Rockett and Eddy, 1989), and oocytes (Coonrod, Vitale, Duan, Bristol-Gould, Herr and Goldberg, 2006). An antibody to LDHC bound primarily to the principal piece region of mouse and human sperm (Beyler, Wheat and Goldberg, 1985) and proteomic studies confirmed that LDHC is present in the mouse sperm flagellum (Cao *et al.*, 2006). However, another proteomic study found that LDHA, but not LDHC was present in the FS isolated from mouse sperm (Krisfalusi *et al.*, 2006). The presence of LDHA in the FS was confirmed by Western blotting in this study.

The phenotype of the *Gapdhs* gene knock out mouse provided convincing evidence that glycolysis is required for sperm motility (Miki *et al.*, 2004). This was *prima facie* evidence that all of the glycolytic enzymes must be present in sperm. The studies referred to above also provide evidence that all of the glycolytic enzymes are present in sperm. Furthermore, seven of the twelve enzymes are known to be present in the principal piece region of the flagellum. Four are these are bound tightly to the isolated FS (ALDOA, GAPDHS, PK, and LDHA) and immunostaining studies have shown that the other three are present in the principal piece (HK1, ENO1, and LDHC). While there is evidence that the other enzymes (GPI, PFK, TPI, PGK2, and PGAM) are present in sperm, it remains to be demonstrated that they are resident in the principal piece.

A notable feature of the sperm glycolytic enzymes is that several are products of spermatogenic cell-specific paralogues of genes expressed ubiquitously (ALDOA, GAPDHS, PGK2, and LDHC) or are products of germ cell-specific variant transcripts (HK1S). Two of the other glycolytic enzymes (GPI, PK) show features suggesting that they are products of a spermatogenic cell-specific genes or transcripts. This is a relatively common theme for genes expressed in the testis, particularly during the post-meiotic phase (Eddy, 2002, 2006). In some cases, it is clear that these genes encode proteins required for the unique structural and functional properties of sperm, but it is less obvious why germ cells require their own set of genes for a process so highly conserved as glycolysis.

Towards a fibrous sheath interactome

The studies described above have demonstrated that an array of proteins are present in the principal piece and associated with the FS. However, it is largely unknown how these proteins bind to each other or are anchored to the FS. Some of these proteins can be released in part from mouse sperm by sonication (HK1S, PGK2, ENO1, LDHC, RIα, RIIα). Additional amounts of some of these (HK1S, PGK2, LDHC) were extracted with solutions containing relatively low concentrations of nonionic detergents (e.g., 1% Triton X-100, 0.2 mM DTT) (Krisfalusi *et al.*,

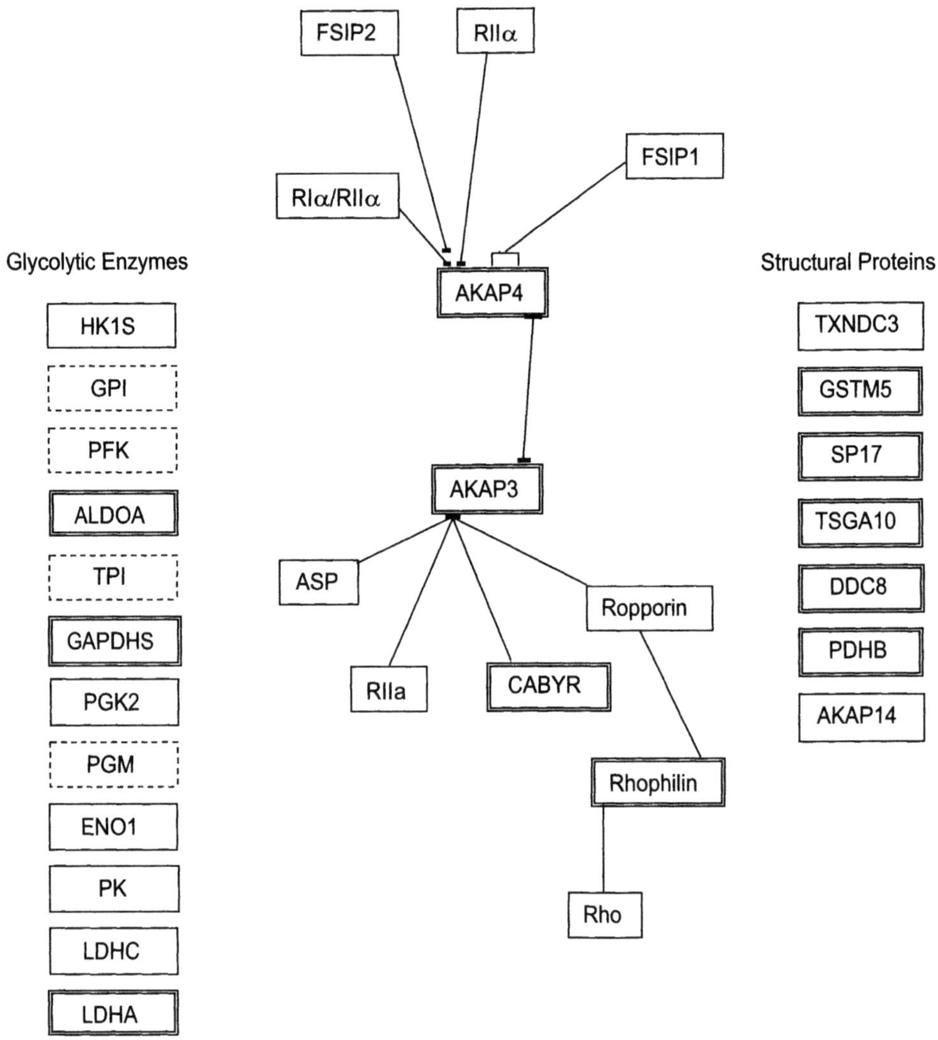

Figure 4. The fibrous sheath interactome. AKAP4 interacts with PKA through its RIα and RIIα subunits, with FSIP1 and FSIP2, and with AKAP3. The sites of these interactions have been mapped to specific regions of the AKAP4 sequence. AKAP3 interacts with RIIa, and ASP, ropporin, and CAYBR can interact with AKAP3 at the same site. Ropporin interacts with rhophilin and rhophilin interacts with Rho. Some glycolytic enzymes are known to be tightly associated with the FS (double lines), while others are released by sonication or treatment with a relatively low concentration of a nonionic detergent (single lines). Additional studies are needed to determine how the glycolytic enzymes are secured in the principal piece or FS. In addition, five of the structural proteins remain associated with the isolated FS (double lines) and two others appear to be tightly bound to the fibrous sheath (single lines).

2006). These proteins probably are associated with the plasma membrane or bind with low affinity to cytoplasmic or FS components. Subsequent extraction in a solution containing 0.6M potassium thiocyanate and 0.2mM DTT released the remaining fraction of some of these proteins (LDHC, PGK2, ENO1). Other FS proteins were found to be mostly insoluble (AKAP4) or partly soluble in 6M urea and 20 mM DTT (ALDOA, GAPDHS, LDHA) (Krisfalusi *et al.*, 2006) or 6M urea and 2mM DTT (PDHB) (Fujinoki *et al.*, 2004). Other studies have indicated that additional FS proteins are insoluble in 4M urea (AKAP3, AKAP14, GSTM5, SP17, DDC8, rhophilin, CABYR) (see Table 1 for references). The low solubility of these structural proteins is due at least in part to the formation of intermolecular disulfide bonds as sperm move down the epididymis (Calvin and Bedford, 1971; Shalgi, Seligman and Kosower, 1989).

Based on what currently is known, AKAP4 and AKAP3 appear to have important roles in linking together some of the FS proteins (Fig. 4). AKAP4 anchors PKA at two sites, binds to AKAP3, and directly interacts with FSIP1 and FSIP2. AKAP3 has one PKA anchoring site and also can bind ASP, ropporin or CABYR at the same site. Through ropporin, AKAP3 can link to rophilin and in turn to Rho (Fig. 4). However, it is not known how the three insoluble glycolytic enzymes (ALDOA, GAPDHS, LDHA) and seven other structural proteins are anchored to the FS. Nor is it understood how relatively soluble glycolytic enzymes (HK1S, LDHC, ENO1, PGK2) remain localised in the principal piece. While it is known that one protein (TXNDC2) has a transient association with the FS during its formation (Yu *et al.*, 2002), it remains to be learned if two other FS associated proteins are present in mature sperm (FSIP1, FSIP2) (Brown *et al.*, 2003).

There has been considerable progress in the last few years in identifying proteins associated with the FS. However, it remains to be determined: 1) which proteins form the nascent FS, 2) when during the formation of the FS most of the proteins are incorporated, 3) how these proteins are delivered to the FS as it forms from distal to proximal along the flagellum, 4) how these proteins assemble and bind to each other to complete FS formation, and 5) how the complex process of FS assembly is regulated. In addition, it is unknown when the different FS proteins arose during vertebrate evolution, what similarities and differences exist between FS proteins of the spectrum of mammalian species, and what the FS contributes to shaping of the flagellar wave form. Progress towards a better understanding of these topics will benefit from the comparative genomic data and bioinformatics tools that are becoming available. It is likely that a substantially better understanding of the composition, protein interactions and functions of the FS and other novel sperm structure will be available by the next Spermatology meeting.

Acknowledgement

This research was supported by the Intramural Research Program, NIH, National Institute of Environmental Health Sciences.

References

Beyler, S.A., Wheat, T.E. and Goldberg, E. (1985). Binding of antibodies against antigenic domains of murine lactate dehydrogenase-C_4 to human and mouse spermatozoa. *Biology of Reproduction* **32:** 1201-1210

Boer, P.H., Adra, C.N., Lau., Y-F. and McBurney, M.W. (1987). The testis-specific phosphoglycerate kinase gene Pgk-2 is a recruited retroposon. *Molecular Cell Biology* **7:** 3107-3112

Briceño, C., Ruiz, P., Reina, M., Vilaró, S. and Pons, G. (1995). The muscle-specific phosphoglycerate mutase gene is specifically expressed in testis during spermatogenesis. *European Journal of Biochemistry* **227:** 629-635

Brooks, D.E. (1976). Activity and androgenic control of glycolytic enzymes in the epididymis and epididymal spermatozoa in the rat. *Biochemistry Journal* **156:** 527-537

Brown, P.R., Miki, K., Harper, D.B. and Eddy, E.M. (2003). AKAP4 binding proteins in the fibrous sheath of the sperm flagellum. *Biology of Reproduction* **68:** 2241-2248

Buehr, M. and McLaren, A. (1981). An electrophoretically detectable modification of glucosephosphate isomerase in mouse spermatozoa. *Journal of Reproduction and Fertility* **63:** 169-173

Bunch, D.O., Welch, J.E., Magyar, P.L. Eddy, E.M. and O'Brien, D.A. (1998). Glyceraldehyde 3-phosphate dehydrogenase-S is bound to the fibrous sheath of mammalian spermatozoa. *Biology of Reproduction* **58:** 834-841

Calvin, H. I. and Bedford, J. M. (1971). Formation of disulfide bonds in the nucleus and accessory structures of mammalian spermatozoa during maturation in the epididymis. *Journal of Reproduction and Fertility Supplement* **13:** 65-75

Cao, W., Gerton, G.L. and Moss, S.B. (2006). Proteomic profiling of accessory structures from the mouse sperm flagellum. *Molecular and Cellular Proteomics* **5:** 801-810

Carr, D.W., Fujita, A., Stentz, C.L., Liberty, G.A., Olson, G.E. and Narumiya, S. (2001). Identification of sperm-specific proteins that interact with A-kinase anchoring proteins in a manner similar to the type II regulatory subunit of PKA. *Journal of Biological Chemistry* **276:** 17332-17338

Carrera, A., Gerton, G.L. and Moss, S.B. (1994). The major fibrous sheath polypeptide of mouse sperm: structural and functional similarities to the A-kinase anchoring proteins. *Developmental Biology* **165:** 272-284

Catalano, R.D., Hillhouse, E.W. and Vlad, M. (2001). Developmental expression and characterization of FS39, a testis complementary DNA encoding an intermediate filament-related protein of the sperm fibrous sheath. *Biology of Reproduction* **65:** 277-287

Coonrod, S., Vitale, A., Duan, C., Bristol-Gould, S., Herr, J. and Goldberg, E. (2006). Testis-specific lactate dehydrogenase (LDH-C_4; Ldh3) in murine oocytes and preimplantation embryos. *Journal of Andrology* **27:** 502-509

Couldrey, C., Carlton, M.B.L., Ferrier, J., Colledge, W.H. and Evans, M.J. (1998). Disruption of murine α-enolase by a retroviral gene trap results in early embryonic lethality. *Developmental Dynamics* **212:** 284-292

Danshina, P., Geyer, C.B., Dai, Q., Goulding, E.H., Willis, W.D., McCarrey, J.R., Eddy, E.M. and O'Brien, D.A. (2006). Male fertility and sperm function are severely impaired in mice lacking phosphoglycerate kinase-2. *Journal of Andrology Supplement* **27:**35

De Iuliis, G.N., Wingate, J.K., Koppers, A.J., McLaughlin, E.A. and Aitken, R.J. (2006). Definitive evidence for the nonmitochondrial production of superoxide anion by human spermatozoa. *Journal of Clinical Endocrinology and Metabolism* **91:** 1968-1975

de Luis, O. and del Mazo, J. (1998). Gene expression of mouse M1 and M2 pyruvate kinase isoenzymes correlates with differential poly(A) tract extension of their mRNAs during the development of spermatogenesis. *Biochemica et Biophysica Acta – Gene Structure and Expression* **1396:** 294-305

Eddy, E.M. (2002). Male germ cell gene expression. In: *Recent Progress in Hormone Research*. Volume 57. Edited by Means, A.R., Academic Press, New York, pp. 103-128

Eddy, E.M. (2006). The spermatozoon. In: *Knobil and Neill's Physiology of Reproduction*, 3rd Edition. Edited by Neill, J.D., Raven Press, New York, pp. 3-54

Edward, Y.H. and Grootegoed, A. (1983). A sperm-specific enolase. *Journal of Reproduction and Fertility* **68:** 305-310

Fawcett, D.W. (1975). The mammalian spermatozoon. *Developmental Biology* **44:** 394-436

Force, A., Viallard, J-L., Grizard, G. and Boucher, D. (2002). Enolase isoforms activities in spermatozoa from men with normospermia and abnormospermia. *Journal of Andrology* **23:** 202-210

Force, A., Viallard, J-L., Saez, F., Grizard, G. and Boucher, D. (2004). Electrophoretic characterization of the human sperm-specific enolase at different stages of maturation. *Journal of Andrology* **25:** 824-829

Fujinoki, M., Kawamura, T., Toda, T., Ohtake, H., Ishimoda-Takagi, T., Shimizu, N., Yamaoka, S. and Okuno, M. (2003). Identification of 36-kDa flagellar phosphoproteins associated with hamster sperm motility. *Journal of Biochemistry* **133:** 361-369

Fujinoki, M., Kawamura, T., Toda, T., Ohtake, H., Ishimoda-Takagi, T., Shimizu, N., Yamaoka, S. and Okuno, M. (2004). Identification of 36 kDa phosphoprotein in fibrous sheath of hamster spermatozoa. *Comparative Biochemistry and. Physiology, Part B: Biochemistry and Molecular Biology* **137:** 509-520

Fujita, A., Nakamura, K., Kato, T., Watanabe, N., Ishizaki, T., Kimura, K., Mizoguchi, A. and Narumiya, S. (2000). Ropporin, a sperm-specific binding protein of rhophilin, that is localized in the fibrous sheath of sperm flagella. *Journal of Cell Science* **113:** 103-112

Fulcher, K.D., Mori, C., Welch, J.E., O'Brien, D.A., Klapper, D.G. and Eddy, E.M. (1995). Characterization of Fsc1 cDNA for a mouse sperm fibrous sheath component. *Biology of Reproduction* **52:** 41-49

Fulcher, K.D., Welch, J.E., Klapper, D.G., O'Brien, D.A. and Eddy, E.M. (1995). Identification of a unique m-class glutathione S-transferase in mouse spermatogenic cells. *Molecular Reproduction and Development* **42:** 415-424

Fundele, R., Winking, H., Illmensee, K. and Jagerbauer, E-M. (1987). Developmental activation of phosphoglycerate mutase-2 in the testis of the mouse. *Developmental Biology* **124:** 562-566

Gillis, B.A. and Tamblyn, T.A. (1984). Association of bovine sperm aldolase with sperm subcellular components. *Biology of Reproduction* **31:** 25-35

Gitlits, V.K., Toh, B.H., Loveland, K.L. and Sentry, J.W. (2000). The glycolytic enzyme enolase is present in sperm tail and displays nucleotide-dependent association with microtubules. *European Journal of Cell Biology* **79:** 104-111

Hintz, M. and Goldberg, E. (1977). Immunohistochemical localization of LDH-X during spermatogenesis in mouse testes. *Developmental Biology* **57:** 375-384

Irons, M.J. and Clermont, Y. (1982). Kinetics of fibrous sheath formation in the rat spermatid. *American Journal of Anatomy* **165:** 121-130

Johnson, L.R., Foster, J.A., Haig-Ladewig, L., VanScoy, H., Rubin, C., Moss, S.B. and Gerton, G.B. (1997). Assembly of AKAP82, a protein kinase A anchor protein, into the fibrous sheath of mouse sperm. *Developmental Biology* **192:** 340-350

Jones, A.R. and Piccolo F. (1999). Glycolytic enzyme activity in hypotonically treated boar spermatozoa. *Reproduction Fertility and Development* **11:** 409-413

Kim, Y-H., Jha, K.N., Mandal, A., Vanage, G., Farris, E., Snow, P.L., Klotz, K., Naaby-Hansen, S., Flickinger, C.J. and Herr, J.C. (2005). Translation and assembly of CABYR coding region B in fibrous sheath and restriction of calcium binding to coding region A. *Developmental Biology* **286:** 46-56

Krisfalusi, M., Miki, K., Magyar, P.L. and O'Brien, D.A. (2006). Multiple glycolytic enzymes are tightly bound to the fibrous sheath of mouse spermatozoa. *Biology of Reproduction* **75:** 270-278

Lea, I.A., Widgren, E.E. and O'Rand, M.G. (2004). Association of sperm protein 17 with A-kinase anchoring protein 3 in flagella. *Reproductive Biology and Endocrinology* **2:** 1-7

Li, S.-L., O'Brien, D.A., Hou, E.W., Versola, J., Rockett, D.L. and Eddy, E.M. (1989). Differential activity and synthesis of lactate dehydrogenase isozymes A (muscle), B (heart) and C (testis) in mouse spermatogenic cells. *Biology of Reproduction* **40:** 173-180

McCarrey, J.R., Kumari. M., Aivaliotis, M.J., Wang, Z., Zhang, P., Marshall, F. and Vandeberg, J.L. (1996). Analysis of the cDNA and encoded protein of the human testis-specific PGK-2 gene. *Developmental Genetics* **19:** 321-332

McCarrey, J.R. and Thomas, K. (1987). Human testis-specific PGK gene lacks introns and possesses characteristics of a processed gene. *Nature* **326:** 501-505

Mei, X., Singh, I.S., Erlichman, J. and Orr, G.A. (1997). Cloning and characterization of a testis-specific, developmentally regulated A-kinase-anchoring protein (TAKAP-80) present on the fibrous sheath of rat sperm. *European Journal of Biochemistry* **246:** 425-432

Miki, K. and Eddy, E.M. (1998). Identification of tethering domains for protein kinase A type Iα regulatory subunits on sperm fibrous sheath protein FSC1. *Journal of Biological Chemistry* **273:** 34384-34390

Miki, K., Qu, W., Goulding, E.H, Willis, W.D., Bunch, D.O., Strader, L.F., Perreault, S.D., Eddy, E.M. and O'Brien D.A. (2004). Glyceraldehyde 3-phosphate dehydrogenase-S, a sperm specific glycolytic enzyme, is required for sperm motility and male fertility. *Proceedings of the National Academy of Sciences USA* **101:** 16501-16506

Miki, K., Willis, W.E., Brown, P.R., Goulding, E.H., Fulcher, K.D. and Eddy, E.M. (2002). Targeted disruption of the *Akap4* gene causes defects in sperm flagellum and motility. *Developmental Biology* **248:** 331-342

Miranda-Vizuete, A, Tsang, K., Yu, Y., Jiménez, A., Pelto-Huikko, M., Flickinger, C.J., Sutovsky, P. and Oko, R. (2003). Cloning and developmental analysis of murid spermatid-specific thioredoxin-2 (SPTRX-2), a novel sperm fibrous sheath protein and autoantigen. *Journal of Biological Chemistry* **278:** 44874-44885

Modarressi, M. H., Behnam, B., Cheng, M., Taylor, K. E., Wolfe, J. and van der Hoorn, F. A. (2004). *Tsga10* encodes a 65-kilodalton protein that is processed to the 27-kilodalton fibrous sheath protein. *Biology of Reproduction* **70:** 608-615

Mori, C., Nakamura, N., Welch, J.E., Gotoh, H., Goulding, E.H., Fajioka, M. and Eddy, E.M. (1998). Mouse spermatogenic-cell specific type 1 hexokinase (*mHk1-s*) transcripts are expressed by alternative splicing from the *mHk1* gene and the HK1-S protein is localized mainly in the sperm tail. *Molecular Reproduction and Development* **49:**374-385

Mori, C., Welch, J.E., Fulcher, K.D., O'Brien, D.A. and Eddy, E.M. (1993). Unique hexokinase messenger ribonucleic acids lacking the porin-binding domain are developmentally expressed in mouse spermatogenic cells. *Biology of Reproduction* **49:** 191-203

Naaby-Hansen, S., Mandal, A., Wolkowicz, M.J., Sen, B., Westbrook, V.A., Shetty, J., Coonrod, S.A., Klotz, K.L., Him, Y-H., Bush, L.A, Flickinger, C.J. and Herr, J.C. (2002). CABYR, a novel calcium-binding tyrosine phosphorylation-regulated fibrous sheath protein involved in capacitation. *Developmental Biology* **242:** 236-254

Nakamura, K., Fujita, A., Murata, T., Watanabe, G., Mori, C., Fujita, J., Watanabe, N., Ishizaki, T., Yoshida, O. and Narumiya, S. (1999). Rhophilin, a small GTPase Rho-binding protein, is abundantly expressed in the mouse testis and localized in the principal piece of the sperm tail. *FEBS Letters* **445:** 9-13

Pawson, T. and Scott, J.D. (1997). Signaling through scaffold, anchoring, and adaptor proteins. *Science* **278:** 2075-2080

Pazour, J.G., Agrin, N., Leszyk, J. and Whitman, G.B. (2005). Proteomic analysis of a eukaryotic cilium. *Journal of Cell Biology* **170:** 103-113

Russell, D.L. and Kim, K.H. (1996). Expression of triosephosphate isomerase transcripts in rat testis: evidence for retinol regulation and a novel germ cell transcript. *Biology of Reproduction* **55:** 11-18

Shalgi, R., Seligman, J. and Kosower, N.S. (1989). Dynamics of the thiol status of rat spermatozoa during maturation: analysis with the fluorescent labeling agent monobromobimane. *Biology of Reproduction* **40:** 1037-1045

Sosa, A., Altamirano, E., Hernandez, P. and Rosado, A. (1972). Developmental pattern of rat testis hexokinase. *Life Science* **11:** 499-510

Travis, A.J., Foster, J.A., Rosenbaum, N.A., Visconti, P.E., Gerton, G.L. and Moss, S.B. (1998). Targeting of a germ cell-specific type 1 hexokinase lacking a porin-binding domain to the mitochondria as well as to the head and fibrous sheath of murine spermatozoa. *Molecular Biology of the Cell* **9:** 263-276

Vemuganti, S.A., Bell, T.A., Scarlett, C.O., Parker, C.E., Pardo-Manuel de Villena, F. and O'Brien, D.A. (2006). Genomic and proteomic analysis of aldolase A retrogenes expressed during spermatogenesis. *Journal of Andrology Supplement* **27:** 39

Vijayaraghavan, S., Liberty, G., Mohan, J., Winfrey, V., Olson, G. and Carr, D. (1999). Isolation and molecular characterization of AKAP110, a novel, sperm-specific protein kinase A-anchoring protein. *Molecular Endocrinology* **13:** 705-717

Yakirevich, E. and Naot, Y. (2000). Cloning of a glucose phosphate isomerase/neuroleukin-like sperm antigen involved in sperm agglutination. *Biology of Reproduction* **62:** 1016-1023

Yu, Y., Oko, R. and Miranda-Vizuete, A. (2002). Developmental expression of spermatid-specific thioredoxin-1 protein: transient association to the longitudinal columns of the fibrous sheath during sperm tail formation. *Biology of Reproduction* **67:** 1546-1554

Spermatology. SRF Vol. 65. ERS Roldan and M Gomendio (eds) Nottingham University Press, Nottingham

Evolution of vertebrate chromosomal sperm proteins: implications for fertility and sperm competition

Juan Ausió[1], José María Eirín-López[1,2] and Lindsay J. Frehlick[1]

[1]*Department of Biochemistry and Microbiology, University of Victoria, Victoria, British Columbia, Canada V8W 3P6;* [2]*Departamento de Biología Celular y Molecular, Universidade da Coruña, Campus de A Zapateira s/n, 15071-A Coruña, Spain*

The three major types of sperm nuclear basic proteins (SNBPs), histone (H type), protamine-like (PL type) and protamine (P type), are well represented in vertebrates. The three groups are evolutionarily related through a vertical evolutionary process (H → PL → P) that involves a transition from lysine to arginine-rich proteins and results in a sporadic but non-random distribution that can be phylogenetically traced. The arginine-rich P type has been selected in the course of evolution of the vertebrates, probably due to constraints imposed by internal fertilisation. Protamines are subject to a positive Darwinian selection process that results in the characteristic fast evolutionary rate shown by these proteins. This makes their use very suitable for the reconstruction of phylogenies of the different vertebrate groups. In mammals, two different types of protamines (P1 and P2) are present which, in the course of the evolution of this vertebrate group, have undergone a further transition to cysteine-rich proteins which further enhanced their DNA packing efficiency. From a functional perspective, protamines provide the most efficient packaging of sperm chromatin and can probably influence the shape of the sperm nucleus and chromatin stability, both of which have direct implications for fertility. In mammals, alterations of the ratio between P1 and P2 protamines as well as the ratio between histones and protamines are important determinants of sperm fertility. All of this suggests a potential involvement of protamines in sperm competition which is discussed in this paper.

"[Ramon y Cajal] would envisage the sperm-cells as activated by a sort of passionate urge in their rivalry for penetration into the ovum cell."
(Sherrington, 1949)

Corresponding author E-mail: jausio@uvic.ca

Evolution of vertebrate sperm nuclear basic proteins

Chromatin, the complex of histones, non-histone proteins and DNA, is a dynamic structure. The core subunit of chromatin, known as the nucleosome core particle, is composed of 146 base pairs of DNA wrapped around an octamer of core histones (H3, H4, H2A and H2B). A linker histone (H1) protects an additional 20 base pairs to form the chromatosome (van Holde, 1988). Altering or remodeling of this structure is important for regulation of many nuclear metabolic events such as, regulation of gene expression, DNA replication and DNA repair. One of the most dramatic examples of chromatin remodeling occurs within male germ cells, where changes in both the protein composition and compaction level of the chromatin take place (Caron *et al.*, 2005). During the postmeiotic maturation of sperm (spermiogenesis), chromatin becomes supercondensed and transcriptionally inert. High compaction of the sperm haploid genome is needed to allow for a more hydrodynamic sperm head and may also protect the DNA from physical and chemical damage (Braun, 2001).

The chromosomal proteins involved in the organization of the mature sperm chromatin are generically known with the name of sperm nuclear basic proteins (SNBPs). They can be grouped into three categories: protamine (P type), protamine-like (PL type) and histones (H type) (Ausió, 1995, 1999). All three types of SNBPs are structurally analogous and condense DNA into chromatin fibers of 300-500 Å (Casas *et al.*, 1993), regardless of the structure of the individual nucleoprotein complexes (Eirín-López *et al.*, 2006a).

Organisms containing SNBPs of the P type are well represented amongst vertebrates and have been found in all classes with the exception of Agnatha (see Fig. 1). In these organisms the histones from the progenitor germ cells at the onset of spermatogenesis are replaced by protamine in the late stages of spermiogenesis (Oliva and Dixon, 1991). In mammals, somatic histones undergo post-translational modifications as well as replacement with specialised histone variants during meiotic prophase [reviewed in Lewis *et al.* (2003a); Govin *et al.* (2004); Kimmins and Sassone-Corsi (2005)]. Immediately after meiosis, histones are replaced by transition proteins (TNPs) (Meistrich *et al.*, 1978, 2003), which are unique to mammals. During spermiogenesis, the transition proteins are replaced by protamines, which vary greatly in amino acid sequence, but are all small and have a very high arginine content, which is often much greater than 30%. The high charge density of these arginine-rich proteins allows them to bind DNA with high affinity and to more efficiently shield the charges on the DNA phosphate backbone, compared to somatic histones, resulting in maximal compaction of the genome. Many mammalian protamines have cysteine residues which are involved in disulfide crosslinking that further contributes to the compaction of chromatin fibers (Lewis *et al.*, 2003b).

Some amphibians and fish have structurally intermediate protamine-like proteins in their sperm (PL type) (Fig. 1). These are highly heterogeneous but generally have a high combined arginine + lysine content (35-50%) and are closely related to histone H1 (Ausió, 1995, 1999). Alternatively, some amphibians and fish retain histones in their mature sperm (H type). Often sperm-specific histone variants of either linker histones or core histones or both are present in these instances. For example, frogs of the genus *Rana* have sperm containing sperm-specific histone H1s, which have higher lysine contents than the *Rana* somatic H1s, as well as a full somatic type histone complement (Itoh, Ausió and Katagiri, 1997).

From the examination of SNBPs from many invertebrates and vertebrates it was proposed that SNBPs had undergone a vertical evolution H → PL → P (Saperas *et al.*, 1994; Ausió, 1995, 1999). Although a link between PL proteins and linker histone H1 had been proposed (Ausió, 1995, 1999), it remained unclear how a high lysine content, which is characteristic of the members of the histone H1 family, could have given rise to arginine rich PL and P proteins.

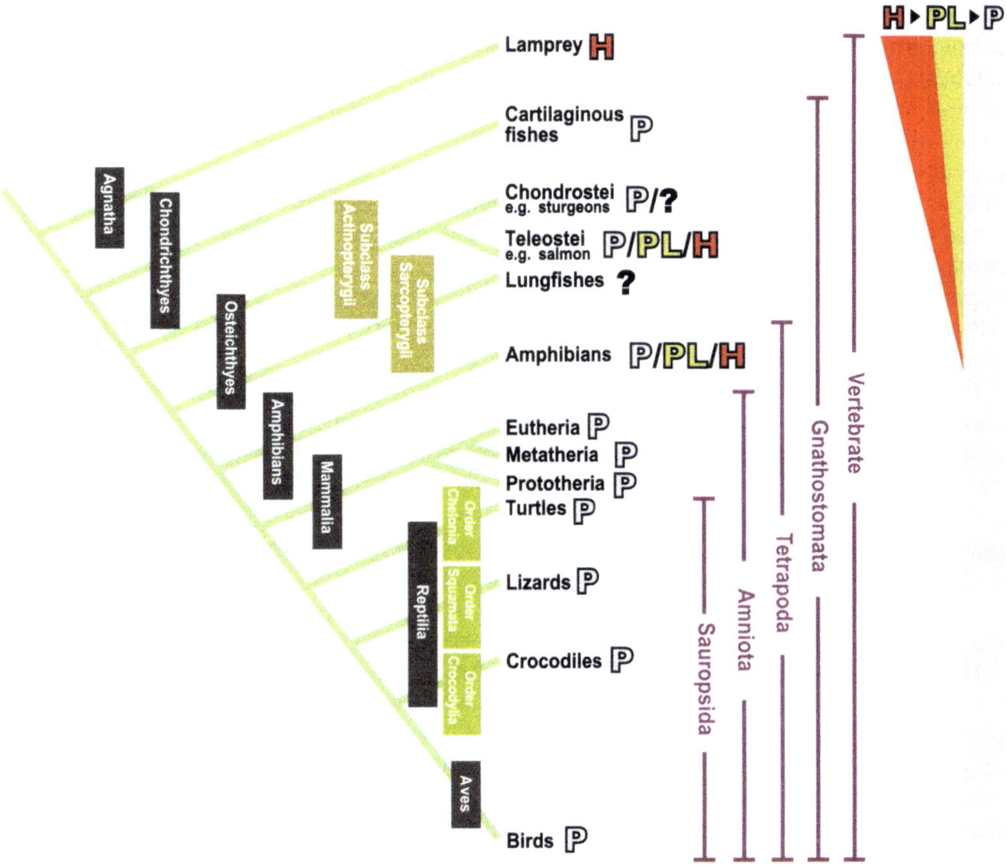

Figure 1. Cladogram of the subphylum Vertebrata showing the currently accepted relationships of monophyletic groups making up the subphylum. Each branch represents a monophyletic group. The arrow shown on the right hand side depicts the direction and approximate distribution of the three main types of SNBPs (in different colours) during the course of evolution of this group. H: histone type; PL: protamine-like type; P: protamine type. It has been proposed that protamines (P type) may have evolved from a histone H1-related protein as indicated at the base of the large arrow (Ausió, 1999; Lewis *et al.*, 2004; Eirin-Lopez *et al.*, 2006).

Two recent studies have shed light on the possible mechanisms underlying this change, lending direct support for a link between H1s, PLs, and protamines (Lewis *et al.*, 2004; Eirín-López *et al.*, 2006b). In the first study, examination of the SNBPs of two closely related species of tunicates, *Styela montereyensis* and *Ciona intestinalis*, revealed the presence of proteins of the PL type in their sperm (Lewis *et al.*, 2004). However, while in *S. montereyensis* the PLs were arginine rich, in *C. intestinalis* the PLs were lysine-rich. Detailed analyses of the coding nucleotide sequences suggested that a single frameshift mutation may have allowed lysine-rich clusters in the C-terminal tail of *C. intestinalis* to convert to arginine-rich clusters in *S. montereyensis*, establishing the evolutionary link between PL and protamine type SNBPs (Lewis *et al.*, 2004). A second report, focused on the evolutionary relationships shown by members of the histone H1 family and the SNBPs of the PL type, revealed the presence of a common origin for both groups of proteins (Eirín-López *et al.*, 2006b). In this regard, it now seems clear that H1 histones and PLs are descendants of an ancient group of 'orphon' H1 replication-dependent histones, which were excluded to solitary genomic regions of the genome as early in metazoan evolution as before the differentiation of bilaterians. This 'orphon' lineage was ultimately responsible for the origin of the replication-independent somatic H1 lineage (as histone H5 and H1°), as well as SNBP lineage. Both of these lineages were subject to a purifying process, that shifted towards adaptive selection in the case of the SNBP lineage at the time of the lysine to arginine transition (Eirín-López *et al.*, 2006b). Due to the more efficient DNA condensation properties of arginine (Puigdoménech *et al.*, 1976; Hélène and Lancelot, 1982; Ausió *et al.*, 1984), proteins with a high global content of this amino acid would have been positively selected for during the course of evolution.

The evolution of SNBPs through a H → PL → P process involving a primordial replication independent histone H1 (Ausió, 1995, 1999; Eirín-López *et al.*, 2006a; Eirín-López *et al.*, 2006b) lends support to the old concept of a potential relationship between ontogeny and phylogeny (Bloch, 1976; Ausió, 1995, 1999). Indeed, during mammalian spermiogenesis a gradual transition in the composition of chromosomal proteins is observed that involves highly specialized H1 histones [H1t (Seyedin and Kistler, 1980), HILS1 (Iguchi *et al.*, 2003; Yan *et al.*, 2003), Hanp1/H1T2 (Martianov *et al.*, 2005; Tanaka *et al.*, 2005)] that precede and/or coexist with TNPs (Meistrich, 1989; Meistrich *et al.*, 2003) which are ultimately replaced by protamines (Lewis *et al.*, 2003b) in the mature spermatozoa.

The different SNBP types display a heterogeneous distribution across the different vertebrate groups (Fig. 1). Nevertheless, as it was shown in fish, the sporadic appearance of the different types is not random and follows the vertical evolutionary phylogeny of the groups where these proteins are present (Saperas et *al.*, 1994). As can be seen from Figure 1, the occurrence of protamines is seen in vertebrates as ancient as sharks and other cartilaginous fishes. Agnatha, which are a superclass of primitive jawless fish (lamprey and hagfish) have sperm that only contain histones and thus are of the H type (Saperas *et al.*, 1994, 1997). In contrast, all the species examined within the class Chondrichthyes (cartilaginous fish), which includes the sharks, skates and rays, are of the P type. Within the subclass Actinopterygii, or ray-finned fish, representatives from all SNBP types are present (Chiva, 1995). The examined sturgeons and paddle fish which are within the subclass Chondrostei contain protamines, whereas the teleost fish or bony fish (the largest group of living fish) are more diverse containing organisms of the H, PL and P types, even though they are all within the subclass Sarcopterygii. A similar sporadic SNBP distribution is seen within the class Amphibia (Kasinsky, 1989). It is apparent from the analysis of the SNBP composition of both fishes and amphibians that a lysine to arginine transition (or divergence from H type to PL type to P type) likely occurred multiple times in the course of their evolution. It has been suggested that the driving force behind this evolution in fish and amphibians may be differing constraints placed on the sperm by internal versus external fertilisation (Kasinsky, 1989, 1995). There is a correlation that suggests that the harsh and viscous environment that sperm are subjected to within female reproductive tracts during internal fertilisation may select for protamines (Mann *et al.*, 1982; Kasinsky *et al.*, 1985; Kasinsky, 1989, 1995).

Following the appearance of Amniota, all organisms contain protamines in their sperm (P type) (Fig. 1). This may suggest an evolutionary trend towards the use of protamines to package sperm DNA in 'higher' vertebrate organisms. The replacement of histones with protamines is typical of taxa located at the uppermost tips of evolutionary branches (Ausió, 1999). In mammals, two types of protamines have been identified: protamine P1 and the protamine P2 family. For the P2 family, proteolytic cleavage of the N-terminus of the P2 precursor protein yields the form of the protein that is present in the mature sperm (Sautiere et *al.*, 1988; Hecht, 1989; Lewis *et al.*, 2003b). Although P1 has been found in all species studied, P2 is exclusively expressed in only a few eutherian organisms, including human and mouse (Oliva, 2006; and references therein). An additional compositional transition in the course of vertebrate SNBP evolution took place in mammals where some of their protamines became rich in cysteine (Oliva and Dixon, 1991; Lewis *et al.*, 2003b). This residue is absent from metatherian protamines and is uncommon in other chromosomal proteins (van Holde, 1988). Cysteine first appears in the protamines of placental (eutherian) mammals and it is well established in both the P1 and P2 protamine lineages. Cysteine also appeared later in the marsupial P1 protamines of the genus *Planigales* through a process of convergent evolution (Retief *et al.*, 1995a). The acquisition of cysteine, which can form inter- and intramolecular disulfide bonds (Vilfan et *al.*, 2004), adds stability and increases the compaction of the nucleoprotamine complexes. The lysine to arginine conversion in the transition from H to PL SNBPs and the acquisition of cysteine by the P type during the course of evolution of vertebrate SNBPs is highly reminiscent of the similar compositional transitions that have been observed in the equivalent invertebrate SNBP types (Lewis *et al.*, 2003b).

The importance of being a protamine and the evolution of vertebrate protamines

Despite the continued effort to characterise protamines from different representative groups of vertebrates, there is still an important lack of information on these proteins. Although protamines appear to be compositionally simple proteins (small and arginine rich), the amino acid distribution in the primary structure of protamines does not seem to be random. In fact, as it will be described in the following sections, the amino acid sequences of these proteins appear to be critical for their function and can be used to trace their evolution.

Protamines exhibit a large structural heterogeneity, as can be seen from the large extent of variation observed in their electrophoretic mobility and primary structure (Fig. 2). Despite this diversity, their amino acid sequence stores very valuable phylogenetic information (Oliva, 1995; Lewis *et al.*, 2003b). Figure 2A provides an electrophoretic analysis of protamines of representative organisms from different vertebrate groups (see Fig. 1) and Figure 2B outlines their corresponding amino acid sequences. We have shown in the past that it is possible to use this kind of sequence information to produce phylogenetic trees that closely resemble those obtained using other molecular markers and/or those based on paleontological/anatomical traits (Lewis *et al.*, 2003b). For instance, we were able to generate a phylogeny for the closely related species of sticklebacks that was indistinguishable from that generated from morphological and behavioural analysis of this group of Gasterosteoidei fish (Giménez-Bonafé *et al.*, 2000). Using a broader range of teleost fish species, for which there is a relatively abundant amount of sequence information for protamines, it was possible to reconstruct a phylogenetic tree which was almost identical to that currently available from the use of conventional markers (Frehlick et *al.*, 2006b). The protamine sequence information has also been useful in confirming the close relationship between birds and reptiles (Sauropsida) (Hunt et *al.*, 1996; Ausió et *al.*, 1999) and has additionally provided support to the monophyletic origin of Paleognathae birds (ostriches, emus, tinamou) (Ausió *et al.*, 1999). Similar phylogenetic results have been obtained with different groups of mammals (also see below) (Retief *et al.*, 1993; Queralt *et al.*, 1995; Retief et *al.*, 1995c).

The fidelity with which protamines work as phylogenetic markers is primarily due to the rapid evolution of these proteins (Oliva and Dixon, 1991; Oliva, 1995; Lewis *et al.*, 2003b), a trait they share with other reproductive proteins (Swanson and Vacquier, 2002). This fast evolu-

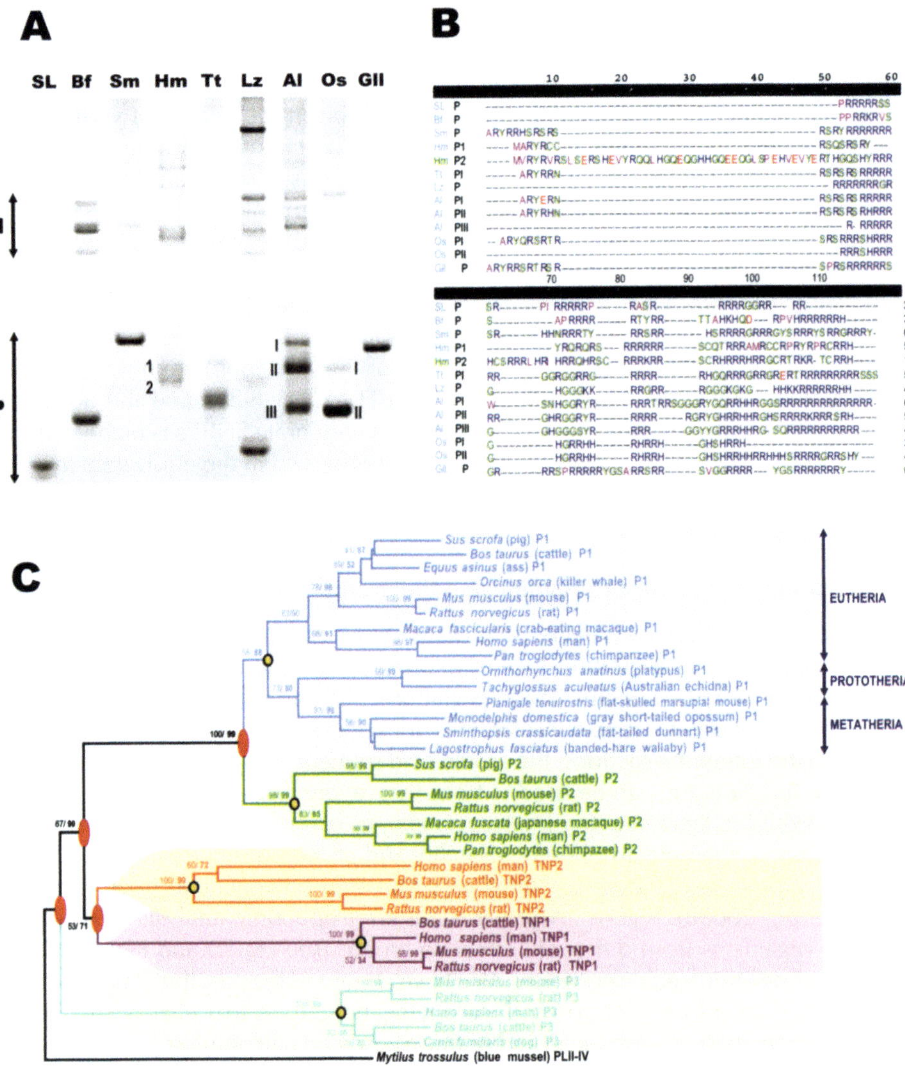

Figure 2. (**A**) Acetic acid-urea polyacrylamide gel electrophoresis (AU-PAGE) of protamines from several representative species of the main monophyletic groups of the subphylum Vertebrata shown in Figure 1. Al: *Alligator mississipiensis* (alligator); Bm: *Bufo marinus* (cane toad); Gll: *Gallus gallus* (rooster); Hm: *Homo sapiens* (human); Lz : *Anolis carolinensis* (green anole, lizard); Os: *Struthio camelus* (ostrich); SL *Oncorhyinchus keta* (chum salmon, salmine); Sm: *Sminthopsis crassicaudata* (fat-tailed dunnart); Tt: *Chrysemys picta* (painted turtle). H: histones; P: protamines. Amino acids are colour-coded according to their side-chain properties: Blue, basic; green, polar uncharged; pink, non-polar hydrophobic; red, acid. (**B**) Amino acid sequences of the protamines from the species shown in (A). The sequences have been aligned using the BIOEDIT (Hall, 1999) and CLUSTAL_X (Thompson *et al.*, 1997) programs. (**C**) Phylogenetic relationships between different mammalian chromosomal sperm proteins (P: protamines; TNP: transition proteins; PL: protamine-like proteins). The amino acid sequences were aligned as in B and the phylogenetic tree derived from this alignment was constructed as described elsewhere (Eirin-Lopez *et al.*, 2006). Numbers for the interior branches represent the BS (bootstrap probability) values followed by the CP (confidence probability, internal-branch test) values based on 1000 replications and are only shown when their value is higher than 50 % (Eirin-Lopez *et al.*, 2006).

tion has been ascribed to positive Darwinian selection (Rooney and Zhang, 1999; Clark and Civetta, 2000; Wyckoff *et al.*, 2000; Swanson and Vacquier, 2002) and involves an unusual form of purifying selection that is responsible for maintaining the overall high arginine composition of these proteins regardless of the specific position of the residues in the molecule (Rooney *et al.*, 2000).

We have taken advantage of all this and revisited here (Fig. 2C) the phylogenetic analysis of mammals using the SNBP sequence information available to date for this vertebrate group. The phylogenetic relationships between mammalian protamines and transition proteins bear an enormous resemblance to those obtained for these proteins when analysed within a broader phylogenetic tree that included SNBPs from all types belonging to both deuterostome and protostome organisms (Eirín-López *et al.*, 2006c). Accordingly, mammalian protamine P3 encoded by the *PRM3/GENE4* gene of the human protamine locus (Nelson and Krawetz, 1994; Kramer and Krawetz, 1998) appears to be more closely related to invertebrate PL proteins than to any other protamines (including mammalian P1 and P2 protamines encoded by *PRM1/PRM2* genes) (Eirín-López *et al.*, 2006b), a fact that could be taken as indicative of the more primitive origin of this mammalian protamine type. As pointed out by Oliva (2006) the name protamine for the protein encoded by *PRM3* may be misleading as it contains a large amount of glutamic acid which is uncharacteristic of a DNA binding protein. Interestingly however, the phylogenetic analysis suggests that a P3 ancestor may have been the precursor of mammalian TNP and P1 and P2 protamines (see Fig. 2C), in contrast to earlier propositions which suggested that *PRM3* was derived from *PRM1* (Kramer and Krawetz, 1998). It is then not surprising that, *PRM3* (Schluter and Engel, 1995; Schluter *et al.*, 1996; Kramer and Krawetz, 1998) and the *PRM1*, *PRM2* and *TNP2* genes are found in the same vicinity within the same protamine locus (Nelson and Krawetz, 1993).

As can be seen in Figure 2C, the analysis not only provides useful information about the potential evolutionary links between the SNBPs of mammals but their protamines (Oliva, 1995), in particular protamine P1, accurately define the phylogeny of the group. The Metatheria (Marsupialia) (Retief *et al.*, 1995a,b), Prototheria (Monotremata) (Retief *et al.*, 1993) and Eutheria (Oliva, 1995; Queralt *et al.*, 1995) appear well defined, thus providing yet another example of the usefulness of protamines as highly reliable phylogenetic markers.

In addition to all this, the tree shown in Figure 2C also suggests that transition proteins arose earlier in evolution than mammalian P1/P2 protamines and that P1 and P2 are closely related, with the slightly more recent appearance of the P2 type likely arising by gene duplication of a P1 precursor (Krawetz and Dixon, 1988). The evolutionary path followed by these mammalian SNBPs is reminiscent of the sequential order in which these proteins appear during spermatogenesis. Together with the close relation of TNPs to invertebrate PLs (Eirín-López *et al.*, 2006c) all this lends additional support to the notion of a relationship between the ontogeny and phylogeny of the three main SNBP types (Bloch, 1976; Ausió, 1995, 1999).

Do SNBPs affect sperm shape?

Amphibians provide an excellent system for a first approximation to the answer of this question as this vertebrate group contain species that are representative of each of the three main SNBP types (Frehlick *et al.*, 2006a) (Fig. 3).

A few preliminary considerations about chromatin organization need to be introduced here before further discussion on the topic. The association of histones, PL proteins and protamines with DNA results in nucleoprotein complexes with a fiber-like organization of 300-500 Å (van Holde, 1988; Casas *et al.*, 1993; Saperas *et al.*, 2006) that is ultimately determined by an overall energy minimization of the complex and not by the specific nature of the protein–DNA interactions described next (Subirana, 1992). In the H type, such as in *Rana*, the chromatin is organised in a 300 Å fiber consisting of approximately six nucleosomes per turn (see Fig. 3D, H). In contrast, the invertebrate (Ausió and Subirana, 1982) and vertebrate (Saperas *et al.*, 2006) PL type chromatin consists of irregular parallel DNA bundles. In mammalian protamines, chromatin loops are organised in toroidal structures containing similar parallel nucleoprotamine bundles (Hud *et al.*, 1995; Ward and Zalensky, 1996; Balhorn, 1999; Brewer *et al.*, 1999) (see Fig. 3D PL/P).

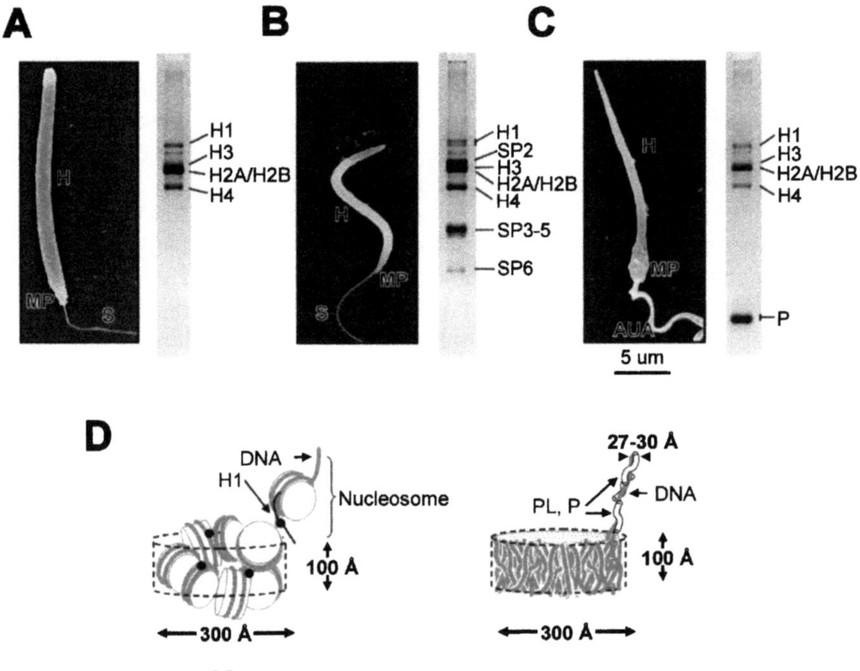

Figure 3. Scanning electron microscopy micrographs of the sperm heads (lanes 1) and AU-PAGE analysis of the SNBPs (lanes 2) of three different species of amphibians: (**A**) *R. catesbeiana* (bullfrog); (**B**) *X. laevis* (African clawed frog); (**C**) *B. marinus* (Cane toad). AUA: axoneme-undulating membrane-axial rod; H: head, MP: midpiece and S: single flagellum only with axoneme. (**D**) Schematic representation of a 100 Å cross-section of a 300 Å fiber of a histone-containing (H) or a P/PL-containing sperm chromatin fiber (see text for more details).

With the structural information currently available for these different types of chromatin organizations, it is possible to theoretically calculate the extent of DNA compaction achieved by each of them (see Fig. 3D). If we consider a 110 Å thick section and the DNA retains its B conformation (Ausió and Subirana, 1982; van Holde, 1988) with an average raise of 3.4 Å per base pair, it is possible to fit 100 nucleoprotein (P/PL) bundles (of approximately 30 Å in diameter) into a PL/P chromatin fiber with a 300 Å diameter. If the DNA was fully stretched this would amount to approximately (100 / 3.4) x 100 (complexes) = 2900 base pairs of DNA compacted within this cross section. In the H type, chromatin is organized in discrete subunits (nucleosomes) each consisting of approximately 210 base pairs of DNA. In the presence of histone H1, the nucleosome arrays can fold into a higher order structure consisting of approximately 6 nucleosomes per turn and a 300 Å diameter. Since the nucleosome is approximately 100 Å tall, this implies that about 1300 base pairs of DNA are compacted within a 100 Å by 300 Å section. Thus, the PL/P– DNA complexes can compact DNA about twice (2.2) as densely as the complexes of the H type.

We have used several scanning electron microscopy (SEM) pictures of *Rana catesbeiana* (H type), *Xenopus laevis* (PL type) and *Bufo marinus* (P type) sperm whose haploid C-value ranges are: 6.63-9.00, 3.00-3.85 and 3.98-5.65 picograms respectively (Gregory, 2006) to calculate the approximate sperm head volume/C-value ratios (v/C) and in doing so we observed that both *X. laevis* and *B. marinus* apparently exhibit the same v/C ratio, whereas the v/C ratio is approximately 1.7 times larger in *R. catesbeiana* (Fig. 3A-C). This higher value for the H type chromatin is in good agreement with the value theoretically calculated from the corresponding chromatin conformations, although it is a bit lower (1.7 vs. 2.2). The lower than expected compaction value could be attributed to the fact that the 300 Å nucleosomally organised chromatin fibers (Fig. 3 D, H) can interdigitate to some extent (Daban, 2003; Robinson *et al.*, 2006) increasing the theoretically calculated value.

While all this does not answer completely the question raised at the beginning of this section, it does demonstrate that, in general, histones are less efficient in packaging sperm chromatin than P and PL proteins. In addition to better compacting the sperm chromatin P and PL proteins erase the epigenetic contribution of histones (Caron *et al.*, 2005; Rousseaux *et al.*, 2005). These are most likely two of the important reasons why these protein types, especially protamines, have been selected throughout evolution (see Fig.1). In terms of sperm head shape it is difficult from just the three amphibian examples shown in Figure 3 to make any predictions other than the observation that while PL/P types seem to lead to more conical streamlined head shapes (Fig. 3 B,C), histones appear to result in rounder shapes (Fig. 3A). The morphology of the sperm head for those organisms containing exclusively protamines can be extremely heterogeneous across different taxa (Baccetti and Afzelius, 1976; Oliva and Dixon, 1991) and no SNBP-related rule appears to exist other than the previous generalization for H versus PL/P type.

It is very likely that SNBPs are not the sole determinants of the sperm nuclear shape. Mouse sperm co-expressing different amounts of a heterologous rooster protamine does not exhibit significant changes in its gross morphology (Maleszewski *et al.*, 1998). It has been shown that SNBP composition and regulation of the contraction of the perinuclear microtubule matrix may have co-evolved to produce the highly unique structural characteristics of the sperm nucleus and head, such as occurs with the cysteine-rich protamines of the nucleus of the sperm of the marine octopod *Eledone cirrhosa* (Giménez-Bonafé *et al.*, 2002). A synergistic effect between multiple protein components would explain the enormous sperm head shape diversity, when compared to the three main groups of SNBPs. However, at the level of closely related species or taxa is possible that protamines and their relative composition (when several molecular species are present, as is the case in mammals) may result in alterations of the overall sperm architecture, as has been observed in protamine deficient human spermatozoa (Blanchard *et al.*, 1990) and in protamine 2 deficient mice (Cho *et al.*, 2003). As the shape of the head may play an important role in sperm competition (Roldan, Gomendio and Vitullo, 1992) it would be worth analyzing the protamine composition of closely related species in relation to the sperm head morphology to determine the existence of any potential correlation.

Mammalian SNBPs and infertility

The multiplicity of SNBPs involved in mammalian spermiogenesis [H1t (Seyedin and Kistler, 1980), HILS1 (Iguchi *et al.*, 2003; Yan *et al.*, 2003), Hanp1/H1T2 (Martianov *et al.*, 2005;

Tanaka et al., 2005), TNPs (Meistrich, 1989; Meistrich et al., 2003) and P1/P2 protamines (Lewis et al., 2003b)] suggests that alterations in any of these proteins could affect male fertility. In this regard, homozygous *Hanp1/H1T2* disrupted male mice are infertile (Tanaka et al., 2005), *Tnp1/Tnp2* deficient mice exhibit an altered chromatin structure and reduced fertility (Yu et al., 2000; Zhao et al., 2001; Zhao et al., 2004) and haplo-insufficiency of P1 or P2 causes infertility in the same organism (Cho et al., 2001, 2003). In the last example, the effect is system-dependent as bull and boar lack protamine P2 as a result of mutations on this gene (Maier et al., 1990) and, in contrast to mice, P2 expression is suppressed in rats (Bunick et al., 1990).

The role of vertebrate SNBPs and protamines in male infertility has been the subject of two recent extensive reviews (Nishimune and Tanaka, 2006; Oliva, 2006). In humans, which retain some (ca. 15%) sperm-specific residual histones in the mature sperm (Gatewood et al., 1987; Wykes and Krawetz, 2003; Churikov et al., 2004), three main types of alterations that have been described so far which can affect fertility are: altered histone/protamine ratio, altered P1/P2 ratio and mutated P1or P2 (Fig. 4 A, B).

Alterations in the relative amount of histones with respect to the protamine content can alter the sperm morphology (Blanchard et al., 1990) and have been described in infertile men (Hofmann and Hilscher, 1991; Foresta et al., 1992; de Yebra et al., 1993; Carrell and Liu, 2001; Zhang et al., 2006). Although this type of protein imbalance is usually related to protamine deficiency (Zhang et al., 2006), in some instances it may alternatively be the result of an incomplete replacement of histones during spermiogenesis (Oliva, 2006), as is observed in males with antecedents of cryptorchidism (Fernandez Valades et al., 2001).

An increase in the P1/P2 ratios as a source of infertility has long been recognized (Chevaillier et al., 1987; Balhorn et al., 1988; Bach et al., 1990; Blanchard et al., 1990; Belokopytova et al., 1993; de Yebra et al., 1993; Bench et al., 1998; de Yebra et al., 1998; Mengual et al., 2003; Aoki et al., 2005). Interestingly, an increased level of P2 precursors has also been described, suggesting a potential role of protamine processing in infertility (Bench et al., 1998; de Yebra et al., 1998; Torregrosa et al., 2006).

The potential role of SNBP mutations in infertility has been recently reviewed (Nishimune and Tanaka, 2006). Although this does not appear to be a common source of the problem, a few mutations have been described in both the *PRM1* and *PRM2* genes of infertile males. A single nucleotide polymorphism (SNP) (*c248t*) in the *PRM2* gene that results in early mRNA termination has been detected in an infertile patient that presumably results in haplo-insufficiency of P2 and is responsible for infertility (Tanaka et al., 2003). More recently, an amino acid transition from R to S at position 34 of human P1 (see Fig. 4B) in a group of infertile patients has been described (Iguchi et al., 2006). This mutation was shown to be the result of a heterozygous single nucleotide polymorphism and there are reasons to believe that it may be involved in infertility. The replacement of arginine by a serine in the context of the region of P1 where the transition takes place is interesting as it generates an RSR motif amenable to phosphorylation by SR-protein-specific kinase 1 (Papoutsopoulou et al., 1999). Phosphorylation of the repetitive N-terminal "RS" domains and of protamine serine residues in general has been shown to play a critical role in the proper interaction and deposition of these proteins onto DNA (Bode et al., 1977; also see Oliva and Dixon, 1991, and Raukas and Mikelsaar, 1999, for reviews) and hence are important for the proper organization of sperm chromatin. Thus, the appearance of an new internal serine in P1 could potentially have a very disruptive effect. In this regard, and as shown in Figure 4B, phosphorylation of S(34) introduces two negative charges that have the potential to neutralise the adjacent positive charge of the surrounding R(33)/R(35) leading to a significant modification of the protamine-DNA interaction and/or to changes in the local secondary structure of the protein, both of which can result in an anomalous chromatin organization. Alterations of sperm chromatin structure have the potential to lead to infertility and altered binding of protamines to DNA may affect the proper formation of a male pronucleus after fertilisation.

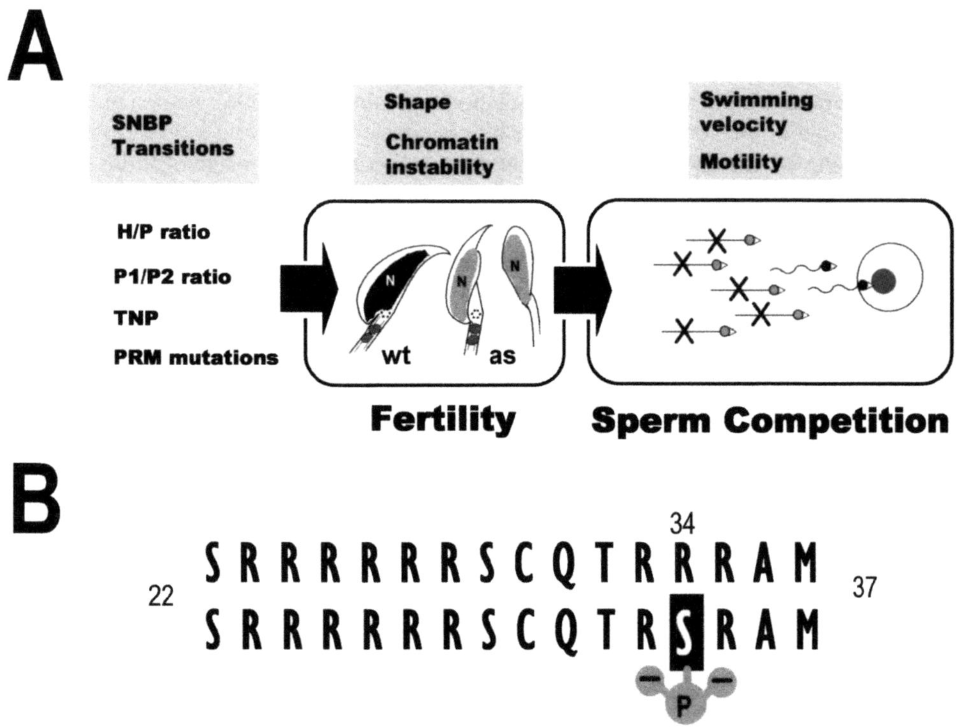

Figure 4. (**A**) Transitions in SNBPs can affect sperm morphology (shape) and chromatin instability, affecting fertility. Schematically shown is a representation of mouse sperm heads of the wild type (wt) or exhibiting an altered shape (as) resulting from a protamine 2 deficiency (Cho *et al.*, 2003) [redrawn from pictures shown in Fig. 2 from Cho *et al.* (2003)]. In conjunction with other cellular changes it is possible that the SNBP-mediated conformational changes may have a significant effect on sperm mobility (in this representation X indicates reduced mobility) which may play an important role in sperm competition (see text for a detailed discussion). H: histones; N: nucleus, P: protamines; TNP: transition proteins. (**B**) A recently discovered single nucleotide polymorphism leads to an amino acid transition (R to S) at position 34 of human protamine 1 and may participate in infertility (Iguchi *et al.*, 2006).

Can protamines play a role in sperm competition?

Support for the notion that protamines may have a role in sperm competition comes from evolutionary studies carried out with these proteins. It has been proposed that the positive Darwinian selection of protamine P1 in primates may have been driven by sperm competition (Rooney and Zhang, 1999; Wyckoff *et al.*, 2000; Swanson and Vacquier, 2002). However, the structural determinants involved in the process are not known. If protamines can have some direct or indirect effect on sperm shape and chromatin stability (Fig. 4A), altering the overall mobility and having implications for infertility (Cho et *al.*, 2003), the exciting possibility exists that protamines may also play a role in sperm competition (Fig. 4A). It has also been shown that expression of a heterologous avian protamine from rooster in transgenic mice (Rhim *et al.*, 1995) resulted in reduced mobility and lower fertility (Maleszewski et *al.*, 1998). Therefore, it appears that protamines may have a significant effect in sperm competition, however further research in this area is still needed.

Recent work with Iberian red deer (*Cervus elaphus hispanicus*) has shown that the hydrodynamic shape of the sperm head in conjunction with the force generated by the flagellum play a critical role in determining sperm swimming velocity (Malo *et al.*, 2006). Also, a unique morphological conformation has been described for the falciform head of the wood mouse (*Apodemus sylvaticus*) sperm which results in sperm cooperation where distinctive aggregations of genetically related sperm results in the increase of the sperm's progressive motility (Moore et *al.*, 2002). As sperm swimming velocity and motility are very important determinants of male fertilisation success (Birkhead *et al.*, 1999; Froman et *al.*, 1999; Levitan, 2000; Gage et *al.*, 2004) the shape of the sperm head may be a component in sperm competition (Roldan et *al.*, 1992). In this hypothetical setting the influence of the protamine P1/P2 composition in the shape of the mice sperm heads is quite revealing and lends further support to the notion that protamines may play a role in sperm competition (Fig. 4A).

It has been pointed out that sperm competition is a potent evolutionary force among mammals (Roldan et *al.*, 1992) which influences sperm morphology (Gomendio and Roldan, 1991). It could be that the selection for protamines in the course of SNBP evolution was not exclusively determined by the transition from external to internal fertilisation but may have also been influenced by sperm competition.

Concluding remarks

The rapid evolution of sperm proteins, in particular protamines, is allowing us to observe the evolution of proteins at its best. The transition from the H type to the P type in the course of evolution may have ultimately been driven by the enhanced ability of these proteins to compact the genome, while efficiently erasing the epigenetic histone component inherited from the stem cells at the onset of spermatogenesis.

In general, the sperm head shape appears to be highly variable and independent of the SNBP composition. However, protamines and PL proteins compact DNA much more efficiently and the possibility exists that protamine variation may affect the shape of the sperm head in relatively closely related species and hence have a role in fertility and even perhaps in sperm competition. Increased chromatin compaction by protamines decreases the volume and streamlines the shape of the sperm head, providing better protection against externally damaging agents and enhancing sperm mobility. Ultimately, differences in sperm swimming velocity in internally fertilising organisms are an important component of sperm competition. Also, there is now a significant amount of accumulated data that indicates that protamines play an important direct role in fertility. This involvement suggests again that the SNBP transitions responsible may also affect sperm competition.

In closing, we would like to go back to Sherrington's quote at the beginning of this chapter on Cajal's anthropomorphic views of cell systems (Sherrington, 1949). Evolution of SNBPs together with other male reproductive proteins appears to have an important role in optimizing the cell parameters that allow sperm to accomplish its "passionate urge" (fertility) and successful "rivalry" (competition). This later issue, sperm competition, is an important selective force whose role in the evolution of SNBPs has yet to be explored.

Acknowledgements

This work was supported by Natural Science and Engineering Research Council of Canada Grant (NSERC) OGP 0046399-02 to J.A. J.M.E-L is a recipient of a Postdoctoral Marie Curie International Fellowship within the 6th European Community Framework Programme. L.J.F. is a recipient of an NSERC postgraduate doctoral scholarship.

References

Aoki, V.W., Liu, L. and Carrell, D.T. (2005). Identification and evaluation of a novel sperm protamine abnormality in a population of infertile males. *Human Reproduction* **20:** 1298-1306

Ausió J. (1995). Histone H1 and the evolution of the nuclear sperm specific proteins. In: *Advances in Spermatozoal Phylogeny and Taxonomy*, Volume 166. Edited by Jamieson, B.G.M., Ausió, J. and Justine, J.L., Memoires de Museum National d'Histoire Naturelle, Paris, pp 501-514

Ausió, J. (1999). Histone H1 and evolution of sperm nuclear basic proteins. *Journal of Biological Chemistry* **274:** 31115-31118

Ausió, J., Greulich, K.O., Haas, E. and Wachtel, E. (1984). Characterization of the fluorescence of the protamine thynnine and studies of binding to double-stranded DNA. *Biopolymers* **23:** 2559-2571

Ausió, J., Soley, J.T., Burger, W., Lewis, J.D., Barreda, D. and Cheng, K.M. (1999). The histidine-rich protamine from ostrich and tinamou sperm. A link between reptile and bird protamines. *Biochemistry* **38:** 180-184

Ausió, J. and Subirana, J.A. (1982). Nuclear proteins and the organization of chromatin in spermatozoa of *Mytilus edulis*. *Experimental Cell Research* **141:** 39-45

Baccetti, B. and Afzelius, B.A. (1976). *The Biology of the Sperm Cell.* Monographs in developmental biology, Volume 10, Karger, Basel, Switzerland

Bach, O., Glander, H.J., Scholz, G. and Schwarz, J. (1990). Electrophoretic patterns of spermatozoal nucleoproteins (NP) in fertile men and infertility patients and comparison with NP of somatic cells. *Andrologia* **22:** 217-224

Balhorn, R., Cosman, M., Thornton, K., Krishnan, V.V., Corzett, M., Bench, G., Kramer, C., Lee IV, J., Hud, N.V., Allen, M., Prieto, M., Meyer-Ilse, W., Brown, T.J., Kirz, J., Zhang, X., Bradbury, E.M., Maki, G., Braun, R.E. and Breed, W. (1999). Protamine mediated condensation of DNA in mammalian sperm. *In: The Male Gamete: From Basic Knowledge to Clinical Applications.* Edited by Gagnon, C., Cache River Press, Vienna , IL , pp. 55-70

Balhorn, R., Reed, S. and Tanphaichitr, N. (1988). Aberrant protamine 1/protamine 2 ratios in sperm of infertile human males. *Experientia* **44:** 52-55

Belokopytova, I.A., Kostyleva, E.I., Tomilin, A.N. and Vorob'ev, V.I. (1993). Human male infertility may be due to a decrease of the protamine P2 content in sperm chromatin. *Molecular Reproduction and Development* **34:** 53-57

Bench, G., Corzett, M.H., De Yebra, L., Oliva, R. and Balhorn, R. (1998). Protein and DNA contents in sperm from an infertile human male possessing protamine defects that vary over time. *Molecular Reproduction and Development* **50:** 345-353

Birkhead, T.R., Martinez, J.G., Burke, T. and Froman, D.P. (1999). Sperm mobility determines the outcome of sperm competition in the domestic fowl. *Proceedings of the Royal Society of London, series B* **266:** 1759-1764

Blanchard, Y., Lescoat, D. and Le Lannou, D. (1990). Anomalous distribution of nuclear basic proteins in round-headed human spermatozoa. *Andrologia* **22:** 549-555

Bloch, D.P. (1976). Histones of sperm. In: *Handbook of Genetics.* Edited by King, R.C., Plenum Press. New York, NY, pp 139-167

Bode, J., Willmitzer, L. and Opatz, K. (1977). On the competition between protamines and histones: studies directed towards the understanding of spermiogenesis. *European Journal of Biochemistry* **72:** 393-403

Braun, R.E. (2001). Packaging paternal chromosomes with protamine. *Nature Genetics* **28:** 10-12

Brewer, L.R., Corzett, M. and Balhorn, R. (1999). Protamine-induced condensation and decondensation of the same DNA molecule. *Science* **286:** 120-123

Bunick, D., Balhorn, R., Stanker, L.H., and Hecht, N.B. (1990). Expression of the rat protamine 2 gene is suppressed at the level of transcription and translation. *Experimental Cell Research* **188:** 147-152

Caron, C., Govin, J., Rousseaux, S., and Khochbin, S. (2005). How to pack the genome for a safe trip. *Progress in Molecular and Subcellular Biology* **38:** 65-89

Carrell, D.T. and Liu, L. (2001). Altered protamine 2 expression is uncommon in donors of known fertility, but common among men with poor fertilizing capacity, and may reflect other abnormalities of spermiogenesis. *Journal of Andrology* **22:** 604-610

Casas, M.T., Ausió, J. and Subirana, J.A. (1993). Chromatin fibers with different protamine and histone compositions. *Experimental Cell Research* **204:** 192-197

Chevaillier, P., Mauro, N., Feneux, D., Jouannet, P. and David, G. (1987). Anomalous protein complement of sperm nuclei in some infertile men. *The Lancet* **2:** 806-807

Chiva, M., Saperas, N., Caceres, C. and Ausió, J. (1995). Nuclear basic proteins from the sperm of tunicates, cephalochordates, agnathans and fish. In: *Advances in Spermatozoal Phylogeny and Taxonomy*, Volume 166. Edited by Jamieson, B.G.M., Ausió, J. and Justine J.L., Memoires de Museum National d'Histoire Naturelle, Paris, pp. 501-514

Cho, C., Jung-Ha, H., Willis, W.D., Goulding, E.H., Stein, P., Xu, Z., Schultz, R.M., Hecht, N.B. and Eddy, E. M. (2003). Protamine 2 deficiency leads to sperm DNA damage and embryo death in mice. *Biology of Reprodution* **69:** 211-217

Cho, C., Willis, W.D., Goulding, E.H., Jung_Ha, H., Choi, Y. C., Hecht, N.B. and Eddy, E.M. (2001). Haploinsufficiency of protamine-1 or -2 causes infertility in mice. *Nature Genetics* **28:** 82-86

Churikov, D., Zalenskaya, I.A. and Zalensky, A.O. (2004). Male germline-specific histones in mouse and man. *Cytogenetics and Genome Research* **105:** 203-214

Clark, A.G. and Civetta, A. (2000). Evolutionary biology. Protamine wars. *Nature* **403:** 261-263

Daban, J.R. (2003). High concentration of DNA in condensed chromatin. *Biochemistry and Cell Biology* **81:** 91-99

de Yebra, L., Ballesca, J.L., Vanrell, J.A., Bassas, L. and Oliva, R. (1993). Complete selective absence of protamine P2 in humans. *Journal of Biological Chemistry* **268:** 10553-10557

de Yebra, L., Ballesca, J.L., Vanrell, J.A., Corzett, M., Balhorn, R. and Oliva, R. (1998). Detection of P2 precursors in the sperm cells of infertile patients who have reduced protamine P2 levels. *Fertility and Sterility* **69:** 755-759

Dixson, A.F. and Anderson, M.J. (2004). Sexual behavior, reproductive physiology and sperm competition in male mammals. *Physiology and Behavior* **83:** 361-371

Eirín-López, J.M., Frehlick, L.J. and Ausió, J. (2006a). Protamines, in the footsteps of linker histone evolution. *Journal of Biological Chemistry* **281:** 1-4

Eirín-López, J.M., Lewis, J.D., Howe, L.A. and Ausió, J. (2006b). Common phylogenetic origin of protamine-like (PL) proteins and histone H1: Evidence from bivalve PL genes. *Molecular Biology and Evolution* **23:** 1304-1317

Fernandez Valades, R., Hortas Nieto, M.L., Castilla, J.A., Lopez de la Torre Casares, M., Valladares Mendias, J.C., Alaminos Mingorance, M., Ruiz Montes, A., Castejon Casado, F.J. and Sanchez Lopez-Tello, C. (2001). [Alterations of nuclear maturity in spermatozoids of males with antecedents of cryptorchidism]. *Cirugía pediátrica: organo oficial de la Sociedad Española de Cirugía Pediátrica* **14:** 95-97

Foresta, C., Zorzi, M., Rossato, M. and Varotto, A. (1992). Sperm nuclear instability and staining with aniline blue: abnormal persistence of histones in spermatozoa in infertile men. *International Journal of Andrology* **15:** 330-337

Frehlick, L.J., Eirín-López, J.M., Field, E.D., Hunt, D.F. and Ausió, J. (2006a). The characterization of amphibian nucleoplasmins yields new insight into their role in sperm chromatin remodeling. *BMC Genomics* **7:** 99

Frehlick, L.J., Eirín-López, J.M., Prado, A., Su, H.W., Kasinsky, H.E. and Ausió, J. (2006b). Sperm nuclear basic proteins of two closely related species of scorpaeniform fish (*Sebastes maliger, Sebastolobus* sp.) with different sexual reproduction and the evolution of fish protamines. *Journal of Experimental Zoology A Comparative Experimental Biology* **305:** 277-287

Froman, D.P., Feltmann, A.J., Rhoads, M.L. and Kirby, J.D. (1999). Sperm mobility: A primary determinant of fertility in the domestic fowl (*Gallus domesticus*). *Biology of Reprodution* **61:** 400-405

Gage, M.J., Macfarlane, C.P., Yeates, S., Ward, R.G., Searle, J.B. and Parker, G.A. (2004). Spermatozoal traits and sperm competition in Atlantic salmon: relative sperm velocity is the primary determinant of fertilization success. *Current Biology* **14:** 44-47

Gatewood, J.M., Cook, G.R., Balhorn, R., Bradbury, E.M. and Schmid, C.W. (1987). Sequence-specific packaging of DNA in human sperm chromatin. *Science* **236:** 962-964

Giménez-Bonafé, P., Ribes, E., Zamora, M.J., Kasinsky, H.E and Chiva, M. (2002). Evolution of octopod sperm I: Comparison of nuclear morphogenesis in Eledone and Octopus. *Molecular Reproduction and Development* **62:** 357-362

Giménez-Bonafé, P., Laszczak, M., Kasinsky, H.E., Lemke, M.J., Lewis, J.D., Iskandar, M., He, T., Ikonomou, M.G., White, F.M., Hunt, D.F., Chiva, M. and Ausió, J. (2000). Characterization and evolutionary relevance of the sperm nuclear basic proteins from stickleback fish. *Molecular Reproduction and Development* **57:** 185-193

Gomendio, M. and Roldan, E.R.S. (1991). Sperm competition influences sperm size in mammals. *Proceedings of the Royal Society of London, series B* **243**: 181-185

Govin, J., Caron, C., Lestrat, C., Rousseaux, S. and Khochbin, S. (2004). The role of histones in chromatin remodelling during mammalian spermiogenesis. *European Journal of Biochemistry* **271:** 3459-3469

Gregory, T.R. (2006). Animal Genome Size Database. http://www.genomesize.com

Hecht, N.B. (1989). Mammalian protamines and their expression. In: *Histones and Other Basic Nuclear Proteins*. Edited by Hnilica, L.S., Stein, G.S. and Stein J.L., CRC Press, Boca Raton, FL, pp. 347-373

Hélène, C. and Lancelot, G. (1982). Interactions between functional groups in protein-nucleic acid associations. *Progress in Biophysics & Molecular Biology* **39:** 1-68

Hofmann, N. and Hilscher, B. (1991). Use of aniline blue to assess chromatin condensation in morphologically normal spermatozoa in normal and infertile men. *Human Reproduction* **6:** 979-982

Hud, N.V., Downing, K.H. and Balhorn, R. (1995). A constant radius of curvature model for the organization of DNA in toroidal condensates. *Proceedings of* the National Academy of Sciences USA **92:** 3581-3585

Hunt, J.G., Kasinsky, H.E., Elsey, R.M., Wright, C.L., Rice, P., Bell, J.E., Sharp, D.J., Kiss, A.J., Hunt, D.F., Arnott, D.P., Russ, M.M., Shabanowitz, J. and Ausió, J. (1996). Protamines of reptiles. *Journal of Biological Chemistry* **271:** 23547-23557

Iguchi, N., Tanaka, H., Yomogida, K. and Nishimune, Y. (2003). Isolation and characterization of a novel cDNA encoding a DNA-binding protein (Hils1) specifically expressed in testicular haploid germ cells. *International Journal of Andrology* **26:** 354-365

Iguchi, N., Yang, S., Lamb, D. J. and Hecht, N. B. (2006). An SNP in protamine 1: a possible genetic cause of male infertility? *Journal of Medical Genetics* **43:** 382-384

Itoh, T., Ausió, J. and Katagiri, C. (1997). Histone H1 variants as sperm-specific nuclear proteins of *Rana catesbeiana*, and their role in maintaining a unique condensed state of sperm chromatin. *Molecular Reproduction and Development* **47:** 181-190

Kasinsky, H.E. (1989). Specificity and distribution of sperm basic proteins. In: *Histones and Other Basic Nuclear Proteins*. Edited by Hnilica, L.S., Stein G.S. and Stein J.L., CRC Press Inc., Boca Raton, FL, p 73-163

Kasinsky, H.E. (1995). Evolution and origins of sperm basic proteins. In: *Advances in Spermatozoal Phylogeny and Taxonomy*, Volume 166. Edited by Jamieson, B.G.M., Ausió, J. and Justine, J.L., Memoires de Museum National d'Histoire Naturelle, Paris, pp. 447-462

Kasinsky, H.E., Mann, M., Lemke, M. and Huang, S.Y. (1985). Diversity of sperm basic chromosomal proteins in vertebrates: a phylogenetic point of view. In: *Chromosomal Proteins and Gene Expression*. Edited by Reeck, G.R., Goodwin, G.H. and Puigdomenech, P., Plenum Press, New York, NY, pp. 333-359

Kimmins, S. and Sassone-Corsi, P. (2005). Chromatin remodelling and epigenetic features of germ cells. *Nature* **434:** 583-589

Kramer, J.A. and Krawetz, S.A. (1998). Genesis of a novel human sequence from the protamine PRM1 gene. *Comparative Biochemistry and Physiology. C, Pharmacology, Toxicology & Endocrinology* **120:** 467-473

Krawetz, S.A. and Dixon, G.H. (1988). Sequence similarities of the protamine genes: implications for regulation and evolution. *Journal of Molecular Evolution* **27:** 291-297

Levitan, D.R. (2000). Sperm velocity and longevity trade off each other and influence fertilization in the sea urchin Lytechinus variegatus. *Proceedings of the Royal Society of London, series B* **267:** 531-534

Lewis, J.D., Abbott, D.W. and Ausió, J. (2003a). A haploid affair: core histone transitions during spermatogenesis. *Biochemistry and Cell Biology* **81:** 131-40

Lewis, J.D., Saperas, N., Song, Y., Zamora, M.J., Chiva, M. and Ausió, J. (2004). Histone H1 and the origin of protamines. *Proceedings of the National Academy of Sciences USA* **101:** 4148-4152

Lewis, J.D., Song, Y., de Jong, M.E., Bagha, S.M. and Ausió, J. (2003b). A walk though vertebrate and invertebrate protamines. *Chromosoma* **111:** 473-482

Maier, W.M., Nussbaum, G., Domenjoud, L., Klemm, U. and Engel, W. (1990). The lack of protamine 2 (P2) in boar and bull spermatozoa is due to mutations within the P2 gene. *Nucleic Acids Research (Online)* **18:** 1249-1254

Maleszewski, M., Kuretake, S., Evenson, D., Yanagimachi, H., Bjordahl, J. and Yanagimachi, R. (1998). Behavior of transgenic mouse spermatozoa with galline protamine. *Biology of Reproduction* **58:** 8-14

Malo, A.F., Gomendio, M., Garde, J., Lang-Lenton, B., Soler, A.J. and Roldan, E.R.S. (2006). Sperm design and sperm function. *Biology Letters* **2:** 246-249

Mann, M., Risley, M.S., Eckhardt, R.A. and Kasinsky, H.E. (1982). Characterization of spermatid/sperm basic chromosomal proteins in the genus *Xenopus* (Anura, Pipidae). *Journal of Experimental Zoology* **222:** 173-186

Martianov, I., Brancorsini, S., Catena, R., Gansmuller, A., Kotaja, N., Parvinen, M., Sassone-Corsi, P. and Davidson, I. (2005). Polar nuclear localization of H1T2, a histone H1 variant, required for spermatid elongation and DNA condensation during spermiogenesis. *Proceedings of the National Academy of Sciences USA* **102:** 2808-2813

Meistrich, M.L. (1989). Histone and basic nuclear prtein transitions in mammalian spermatogenesis. In: *Histones and Other Basic Nuclear Proteins*. Edited by Hnilica, L.S., Stein, G.S. and Stein, J.L., CRC Press Inc., Boca Raton, FL, pp. 165-182

Meistrich, M.L., Brock, W.A., Grimes, S.R., Platz, R.D. and Hnilica, L.S. (1978). Nuclear protein transitions during spermatogenesis. *Federation Proceedings* **37:** 2522-2525

Meistrich, M.L., Mohapatra, B., Shirley, C.R. and Zhao, M. (2003). Roles of transition nuclear proteins in spermiogenesis. *Chromosoma* **111:** 483-488

Mengual, L., Ballesca, J.L., Ascaso, C. and Oliva, R. (2003). Marked differences in protamine content and P1/P2 ratios in sperm cells from percoll fractions between patients and controls. *Journal of Andrology* **24:** 438-447

Moore, H., Dvorakova, K., Jenkins, N. and Breed, W. (2002). Exceptional sperm cooperation in the wood mouse. *Nature* **418:** 174-177

Nelson, J.E. and Krawetz, S.A. (1993). Linkage of human spermatid-specific basic nuclear protein genes. Definition and evolution of the P1—>P2—>TP2 locus. *Journal of Biological Chemistry* **268:** 2932-2936

Nelson, J.E. and Krawetz, S.A. (1994). Characterization of a human locus in transition. *Journal of Biological Chemistry* **269:** 31067-31073

Nishimune, Y. and Tanaka, H. (2006). Infertility caused by polymorphisms or mutations in spermatogenesis-specific genes. *Journal of Andrology* **27:** 326-334

Oliva, R. (1995). Sequence, evolution and transcriptional regulation of avian-mammalian P1 type protamines. In: *Advances in Spermatozoal Phylogeny and Taxonomy*, Volume 166, Edited by Jamieson, B.G.M., Ausió, J. and Justine, J.L., Memoires du Museum national d'Histoire naturelle, Paris, pp. 537-548

Oliva, R. (2006). Protamines and male infertility. *Human Reproduction Update* **12:** 417-435

Oliva, R. and Dixon, G.H. (1991). Vertebrate protamine genes and the histone-to-protamine replacement reaction. *Progress in Nucleic Acid Research and Molecular Biology* **40:** 25-94

Papoutsopoulou, S., Nikolakaki, E., Chalepakis, G., Kruft, V., Chevaillier, P. and Giannakouros, T. (1999). SR protein-specific kinase 1 is highly expressed in testis and phosphorylates protamine 1. *Nucleic Acids Research* **27:** 2972-2980

Puigdoménech, P., Martinez, P., Palau, J., Bradbury, E. M. and Crane-Robinson, C. (1976). Studies on the role and mode of operation of the very-lysine-rich histones in eukaryote chromatin. Nuclear-magnetic-resonance studies on nucleoprotein and histone phi 1-DNA complexes from marine invertebrate sperm. *European Journal of Biochemistry* **65:** 357-363

Queralt, R., Adroer, R., Oliva, R., Winkfein, R.J., Retief, J.D. and Dixon, G.H. (1995). Evolution of protamine P1 genes in mammals. *Journal of Molecular Evolution* **40:** 601-607

Raukas, E. and Mikelsaar, R.H. (1999). Are there molecules of nucleoprotamine? *Bioessays: News and Reviews in Molecular, Cellular and Developmental Biology* **21:** 440-448

Retief, J.D., Krajewski, C., Westerman, M. and Dixon, G.H. (1995a). The evolution of protamine P1 genes in dasyurid marsupials. *Journal of Molecular Evolution* **41:** 549-555

Retief, J.D., Krajewski, C., Westerman, M., Winkfein, R.J. and Dixon, G.H. (1995b). Molecular phylogeny and evolution of marsupial protamine P1 genes. *Proceedings of the Royal Society of London, series B* **259:** 7-14

Retief, J.D., Rees, J.S., Westerman, M. and Dixon, G. H. (1995c). Convergent evolution of cysteine residues in sperm protamines of one genus of marsupials, the Planigales. *Molecular Biology and Evolution* **12:** 708-712

Retief, J.D., Winkfein, R.J. and Dixon, G.H. (1993). Evolution of the monotremes. The sequences of the protamine P1 genes of platypus and echidna. *European Journal of Biochemistry* **218:** 457-61

Retief, J.D., Winkfein, R.J., Dixon, G.H., Adroer, R., Queralt, R., Ballabriga, J. and Oliva, R. (1993). Evolution of protamine P1 genes in primates. *Journal of Molecular Evolution* **37:** 426-434

Rhim, J.A., Connor, W., Dixon, G.H., Harendza, C.J., Evenson, D.P., Palmiter, R.D. and Brinster, R.L. (1995). Expression of an avian protamine in transgenic mice disrupts chromatin structure in spermatozoa. *Biology of Reproduction* **52:** 20-32

Robinson, P.J., Fairall, L., Huynh, V.A. and Rhodes, D. (2006). EM measurements define the dimensions of the "30-nm" chromatin fiber: Evidence for a compact, interdigitated structure. *Proceedings of the National Academy of Sciences USA* **103:** 6506-6511

Roldan, E.R.S., Gomendio, M. and Vitullo, A.D. (1992). The evolution of eutherian spermatozoa and underlying selective forces: female selection and sperm competition. *Biological Reviews* **67:** 551-593

Rooney, A.P. and Zhang, J. (1999). Rapid evolution of a primate sperm protein: relaxation of functional constraint or positive Darwinian selection? *Molecular Biology and Evolution* **16:** 706-710

Rooney, A.P., Zhang, J. and Nei, M. (2000). An unusual form of purifying selection in a sperm protein. *Molecular Biology and Evolution* **17:** 278-283

Rousseaux, S., Caron, C., Govin, J., Lestrat, C., Faure, A. K. and Khochbin, S. (2005). Establishment of male-specific epigenetic information. *Gene* **345:** 139-153

Saperas, N., Ausió, J., Lloris, D. and Chiva, M. (1994). On the evolution of protamines in bony fish: alternatives to the "retroviral horizontal transmission" hypothesis. *Journal of Molecular Evolution* **39:** 282-295

Saperas, N., Chiva, M., Casas, M. T., Campos, L., Eirín-López, J. M., Prieto, C., Subirana, J. A. and Ausió, J. (2006). A unique vertebrate histone H1-related protamine-like (PL) protein results in an unusual sperm chromatin organization. *FEBS Letters* **273:** 9548-9561

Saperas, N., Chiva, M., Pfeiffer, D. C., Kasinsky, H. E. and Ausió, J. (1997). Sperm nuclear basic proteins (SNBPs) of agnathans and chondrichthyans: variability and evolution of sperm proteins in fish. *Journal of Molecular Evolution* **44:** 422-431

Sautiere, P., Martinage, A., Belaiche, D., Arkhis, A. and Chevaillier, P. (1988). Comparison of the amino acid sequences of human protamines HP2 and HP3 and of intermediate basic nuclear proteins HPS1 and HPS2. Structural evidence that HPS1 and HPS2 are pro-protamines. *Journal of Biological Chemistry* **263:** 11059-11062

Schluter, G., Celik, A., Obata, R., Schlicker, M., Hofferbert, S., Schlung, A., Adham, I. M. and Engel, W. (1996). Sequence analysis of the conserved protamine gene cluster shows that it contains a fourth expressed gene. *Molecular Reproduction and Development* **43:** 1-6

Schluter, G. and Engel, W. (1995). The rat Prm3 gene is an intronless member of the protamine gene cluster and is expressed in haploid male germ cells. *Cytogenetics and Cell Genetics* **71:** 352-355

Seyedin, S.M. and Kistler, W.S. (1980). Isolation and characterization of rat testis H1t. An H1 histone variant associated with spermatogenesis. *Journal of Biological Chemistry* **255:** 5949-5954

Sherrington, C. (1949). A memoir of Dr. Cajal. In: *Explorer of the Human Brain. The Life of Santiago Ramón y Cajal (1852-1934)*, Edited by Canon, D.F., H. Schuman, New York, NY. pp. 14

Subirana, J.A. (1992). Order and disorder in 30 nm chromatin fibers. *FEBS Letters* **302:** 105-107

Swanson, W.J. and Vacquier, V.D. (2002). The rapid evolution of reproductive proteins. *Nature Reviews.*

Genetics **3:** 137-144

Tanaka, H., Iguchi, N., Isotani, A., Kitamura, K., Toyama, Y., Matsuoka, Y., Onishi, M., Masai, K., Maekawa, M., Toshimori, K., Okabe, M. and Nishimune, Y. (2005). HANP1/H1T2, a novel histone H1-like protein involved in nuclear formation and sperm fertility. *Molecular and Cellular Biology* **25:** 7107-7119

Tanaka, H., Miyagawa, Y., Tsujimura, A., Matsumiya, K., Okuyama, A. and Nishimune, Y. (2003). Single nucleotide polymorphisms in the protamine-1 and -2 genes of fertile and infertile human male populations. *Molecular Human Reproduction* **9:** 69-73

Torregrosa, N., Dominguez-Fandos, D., Camejo, M.I., Shirley, C.R., Meistrich, M.L., Ballesca, J.L. and Oliva, R. (2006). Protamine 2 precursors, protamine 1/protamine 2 ratio, DNA integrity and other sperm parameters in infertile patients. *Human Reproduction* **21:** 2084-2089

van Holde, K.E. (1988). *Chromatin*. Springer-Verlag, NY.

Vilfan, I.D., Conwell, C.C. and Hud, N.V. (2004). Formation of native-like mammalian sperm cell chromatin with folded bull protamine. *Journal of Biological Chemistry* **279:** 20088-20095

Ward, W.S. and Zalensky, A.O. (1996). The unique, complex organization of the transcriptionally silent sperm chromatin. *Critical Reviews in Eukaryotic Gene Expression* **6:** 139-147

Wyckoff, G.J., Wang, W. and Wu, C.I. (2000). Rapid evolution of male reproductive genes in the descent of man. *Nature* **403:** 304-309

Wykes, S.M. and Krawetz, S.A. (2003). The structural organization of sperm chromatin. *Journal of Biological Chemistry* **278:** 29471-29477

Yan, W., Ma, L., Burns, K.H. and Matzuk, M.M. (2003). HILS1 is a spermatid-specific linker histone H1-like protein implicated in chromatin remodeling during mammalian spermiogenesis. *Proceedings of the National Academy of Sciences USA* **100:** 10546-105451

Yu, Y.E., Zhang, Y., Unni, E., Shirley, C.R., Deng, J.M., Russell, L.D., Weil, M.M., Behringer, R.R. and Meistrich, M. L. (2000). Abnormal spermatogenesis and reduced fertility in transition nuclear protein 1-deficient mice. *Proceedings of the National Academy of Sciences USA* **97:** 4683-4688

Zhang, X., San Gabriel, M and Zini, A. (2006). Sperm nuclear histone to protamine ratio in fertile and infertile men: evidence of heterogeneous subpopulations of spermatozoa in the ejaculate. *Journal of Andrology* **27:** 414-420

Zhao, M., Shirley, C.R., Hayashi, S., Marcon, L., Mohapatra, B., Suganuma, R., Behringer, R.R., Boissonneault, G., Yanagimachi, R. and Meistrich, M. L. (2004). Transition nuclear proteins are required for normal chromatin condensation and functional sperm development. *Genesis* **38:** 200-213

Zhao, M., Shirley, C.R., Yu, Y.E., Mohapatra, B., Zhang, Y., Unni, E., Deng, J.M., Arango, N.A., Terry, N.H., Weil, M.M., Russell, L.D., Behringer, R.R. and Meistrich, M.L. (2001). Targeted disruption of the transition protein 2 gene affects sperm chromatin structure and reduces fertility in mice. *Molecular and Cellular Biology* **21:** 7243-7255

Spermatology. SRF Vol. 65. ERS Roldan and M Gomendio (eds) Nottingham University Press, Nottingham

Value of DNA integrity assays for fertility evaluation

R. John Aitken and Geoffry N. De Iuliis

ARC Centre of Excellence in Biotechnology and Development and Discipline of Biological Sciences, University of Newcastle, NSW 2308, Australia

DNA damage in the male germ line is associated with failed fertilisation, impaired preimplantation development and poor pregnancy outcomes, whether the insemination is natural or artificial. It is also apparent that DNA damage in spermatozoa is associated with adverse impacts on the health and wellbeing of children. This may be particularly important in the case of conceptions involving intracytoplasmic sperm injection. With this technique DNA damaged spermatozoa that would, under physiological circumstances, be excluded from the conception process are able to initiate pregnancies. Although the oocyte actively repairs the DNA damage brought into the zygote by the fertilising spermatozoon, errors in this repair process would generate mutations that might in turn be linked to the increased incidence of dominant genetic disease and childhood cancer seen in the offspring of fathers possessing DNA damaged spermatozoa. Significantly, the incidence of such mutations is so low that it might be several generations before the full consequences of using DNA damaged spermatozoa in assisted conception cycles are realised. The factors modulating DNA damage in the male germ line are complex and largely unresolved. They involve advanced paternal age, exposure to xenobiotics and male genital tract infection. The types of damage being measured by the assays used in most laboratories (SCSA, Comet and TUNEL) are also uncertain. Molecular characterization of this DNA damage might provide insights into the underlying aetiologies, facilitate the development of optimised diagnostic methods for its detection and suggest logical avenues to pursue for its correction and ultimate prevention.

Introduction

One of the most significant advances in assisted conception over the past twenty years has been the introduction of intracytoplasmic sperm injection (ICSI) as a therapeutic technique for the treatment of infertile males (Palermo, Joris, Devroey and Van Steirteghem, 1992). With this technique conception can be achieved with a single gamete recovered from the ejaculate, the epididymis or even the seminiferous tubules. In general, fertilisation and pregnancy rates for ICSI and IVF are similar for male patients with borderline semen quality (van der Westerlaken, Naaktgeboren, Verburg, Dieben and Helmerhorst, 2006), although in severe male infertility

---

Corresponding author E-mail: jaitken@mail.newcastle.edu.au

cases, where IVF is not an option, decreases in fertilisation and implantation rates have been reported (Shveiky, Simon, Gino, Safran, Lewin, Reubinoff, Laufer and Revel, 2006). One of the most remarkable aspects about ICSI, is how forgiving this process is of the quality of the spermatozoon used for fertilisation. It is important that such spermatozoa are viable (Sallam, Farrag, Agameya, El-Garej and Ezzeldin, 2005) but not essential that they are competent to exhibit normal motility. Poorly motile ejaculated spermatozoa and even completely immotile sperm recovered from the testes are still able to fertilise the oocyte when ICSI is used as the insemination technique (Katsoff, Check, Summers-Chase and Check, 2005; Kovacic, Vlaisavljevic and Reljic, 2006). Similarly, although normal sperm morphology and a high level of sperm maturity are certainly beneficial for ICSI (Stalf, Mehnert, Hajimohammad, Manolopoulos, Shen, Schuppe, Diemer, Schill, Weidner and Tinneberg, 2005; Vural, Sofuoglu, Caliskan, Delikara, Aksoy, Uslu and Karan, 2005; Hazout, Dumont-Hassan, Junca, Cohen Bacrie and Tearik, 2006), it is also evident that imperfectly formed spermatozoa and even round spermatids can be successfully used to fertilise oocytes using this approach (Dittrich, Maltaris, Dragonas, Huber, Beckmann and Kiesewetter, 2005; Hikichi, Kishigami, Thuan, Ohta, Mizutani, Wakayama and Wakayama, 2005). An analysis of the impact of pathological leukocytospermia (median 2.68 million/ml) on ICSI also revealed little impact on fertilisation rates (82% vs 87% in the controls), estimates of embryo quality (88% vs 92% for good quality embryos) cleavage (50% vs 50% normal cleavage) or pregnancy rates (Yilmaz, Koyuturk, Kilic, Alpak and Aytoz, 2005), despite the oxidative stress associated with this condition. These results are in keeping with early studies indicating that high rates of oxidative DNA damage artificially induced in spermatozoa using a variety of mechanisms, including direct exposure to hydrogen peroxide, were no contra-indication to successful fertilisation using ICSI (Twigg, Irvine and Aitken, 1998). In a similar vein, clinical cases have been cited in which successful pregnancies have been generated using spermatozoa from men exhibiting extremely high rates of DNA damage (Gandini, Lombardo, Paoli, Caruso, Eleuteri, Leter, Ciriminna, Culasso, Dondero, Lenzi and Spano, 2004).

Such results lend support to the concept that ICSI is not only an extremely efficient means of generating human pregnancies, it is also extremely safe. This conclusion has been echoed in analyses of birth defects following assisted conception. These studies have tended to indicate that although assisted reproduction techniques are associated with a significantly elevated incidence of birth defects, this increase is largely a consequence of the multiple births associated with this technology and no higher in ICSI babies than those conceived by conventional IVF (Hansen, Kurinczuk, Bower and Webb, 2002). Notwithstanding these data, it may yet be too early to determine the relative safety of ICSI and IVF because not enough time has passed to determine the full extent of the phenotypic and behavioural abnormalities that might be present in these children and not enough babies have been born. This point is exemplified by considering the relationships between paternal factors such as age and smoking, DNA damage in the spermatozoa and morbidity in the offspring.

Consequences of DNA damage in the germ line

As men age, their fertility appears to undergo a progressive but variable decline (Kidd, Eskenazi and Wyrobek, 2001) and the integrity of DNA in their spermatozoa deteriorates. In one recent study for example, the percentage of spermatozoa exhibiting DNA damage according to the Comet assay was found to be significantly higher in men aged 36-57 years than in those aged 20-35 years, (Singh, Muller and Berger, 2003; Fig. 1A). Similarly, using a Sperm Chromatin Structure Assay (SCSA) to monitor DNA integrity a significant relationship was detected between age and DNA damage in spermatozoa, such that around 40% of the variance in logged

SCSA measurements could be accounted for on the basis of age (Wyrobek, Eskenazi, Young, Arnheim, Tiemann-Boege, Jabs, Glser, Pearson and Evenson, 2006). The increased DNA damage seen in the spermatozoa of ageing males is correlated with a range of diseases in their offspring that show a dependency on paternal age at the time of conception. These conditions range from miscarriage (de La Rochebrochard and Thonneau, 2002) to dominant genetic mutations, such as achondroplasia (Crow, 2000; Wyrobek *et al.*, 2006) and neurological disorders such as schizophrenia (Sipos, Rasmussen, Harrison, Tynelius, Lewis, Leon and Gunnell, 2004).

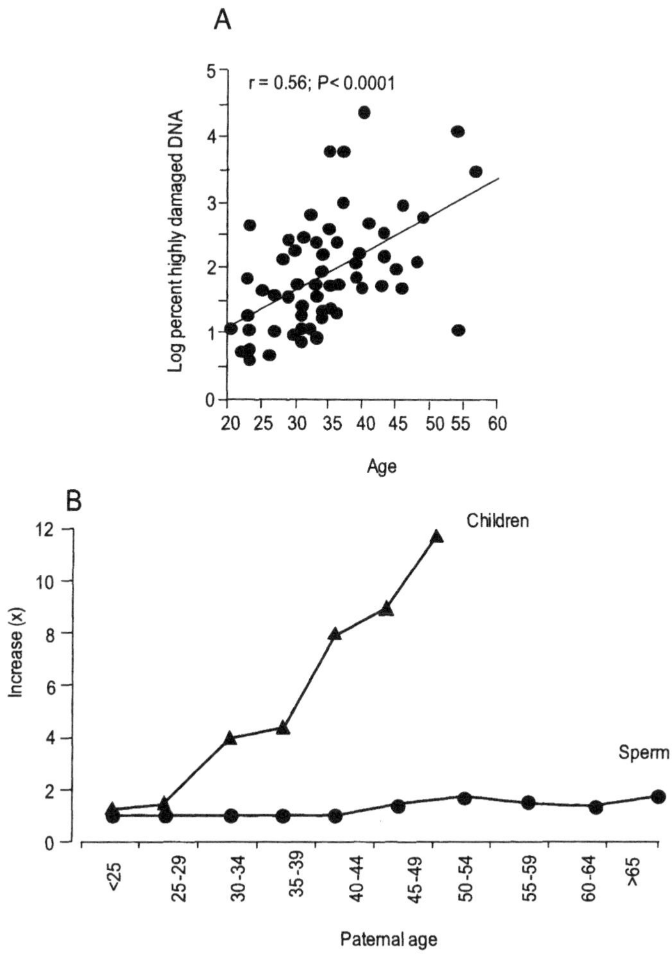

Figure 1. Impact of age on DNA damage and mutation frequency in the male germ line. (**A**) Relationship between age and DNA damage in spermatozoa as measured with the Comet assay (after Singh *et al.*, 2003). (**B**) Relationship between the frequency of GB1134A mutations in spermatozoa (●), the mutation associated with achondroplasia, and the observed vs expected ratio for sporadic achondroplasia in the offspring of ageing males (▲) (after Hurst and Ellegren, 2002).

The link between DNA damage in the spermatozoa on the one hand and an increased incidence of dominant genetic diseases on the other is not direct. Although it is traditionally believed that the dominant genetic mutations associated with paternal age arise as a consequence of replication errors in the male germ line (Crow, 2000), this view does not completely accord with the facts. Firstly, studies with the Big Blue transgenic mouse, which contains genetic inserts capable of reporting on mutation frequencies, suggest that the spontaneous mutation frequency in the male germ line is one of the lowest in the body and does not increase significantly with age (Hill, Halangoda, Heinmoeller, Gonzalez, Chitaphan, Longmate, Scaringe, Wang and Sommer, 2005). Secondly, although paternal ageing is associated with an increased incidence of the fibroblast growth factor 3 (FGFR3) mutations responsible for achondroplasia in spermatozoa, the magnitude of this increase does not account for absolute incidence of this disease *in vivo* (Fig. 1B; Tiemann-Boege, Navidi, Grewal, Cohn, Eskenazi, Wyrobek and Arnheim, 2002). So while ageing does increase FGFR3 mutation frequencies in the male germ line (Wyrobek *et al.*, 2006), we have to look elsewhere to explain the exponential increase in achondroplasia observed in the offspring of ageing males (Crow, 2000). Several explanations for this situation have been advanced. It is possible that spermatozoa carrying this mutation would have some selective advantage in terms of fertilisation. However no plausible hypothesis has been advanced as to why this would be the case, particularly when we remember that spermatozoa develop as a syncytium sharing a common plasma membrane until late in their development. An alternative possibility that we have advanced (Aitken and Krausz, 2001; Aitken, Koopman and Lewis, 2004; Baker and Aitken, 2005) is that the correlation with paternal age reflects an age-dependent increase in oxidative DNA damage in male germ line and it is the aberrant repair of this DNA damage in the fertilised egg that results in the achondroplasia mutation. In keeping with this explanation, the most common mutation observed in achondroplasia patients is a CG-AT transition in the FGFR3 gene. Since this is the most common base substitution observed following oxidative damage to DNA (Kreutzer and Essigmann, 1998) we hypothesize that ageing is associated with increased oxidative stress in the male gamete and that the FGF3 mutation arises as a consequence of the defective repair of this damage by the oocyte in the period that elapses between fertilisation and the initiation of S-phase prior to the first mitotic division (Fig. 2). In this scheme, oxidative deamination of cytosine 1138 in the sperm is proposed to lead to the creation of a uracil derivative (Kreutzer and Essigmann, 1998) that is then aberrantly paired with adenine and subsequently replaced with thymine in an aberrant CG to UG to UA to TA repair cycle in order to create the pathological CG-AT transition (G1138A) (Fig. 2).

Another important facet of these correlations between paternal DNA damage and adverse outcomes in the offspring is their rare incidence. Conditions such as achondroplasia, Apert syndrome and schizophrenia are observed with an incidence of less than 10 per 100,000 of the population. Similarly, men who smoke heavily produce spermatozoa that contain high levels of DNA fragmentation associated with oxidative lesions to DNA bases (Fraga, Motchnik, Wyrobek, Rempel and Ames, 1996; Sepaniak, Forges, Gerard, Foliguet, Bene and Monnier-Barbarino, 2006). The consequence of this damage is not seen so much in the fertility of these men but in the health and wellbeing of their children. Heavy paternal smoking is associated with an increased risk of childhood disease according to a Chinese study that examined the incidence of cancer in the offspring of parents characterized by a non-smoking mother and a smoking father (Ji, Shu, Linet, Zheng, Wacholder, Gao, Ying and Jin, 1997). The results of this study revealed that paternal preconception smoking was related to a significantly elevated risk of childhood cancers, particularly acute leukaemia and lymphoma but, again, the incidence of these types of cancer is typically 2-4 per 100,000 children less than14 years old (McNeil, Cote, Clegg and Mauer, 2002). From these examples, it must be clear that until hundreds of thousands of children have been born as a consequence of ICSI, we shall not see an overt, detectable phenotype in the F1 generation, if genetic lesions have been created as a result of this therapeutic procedure.

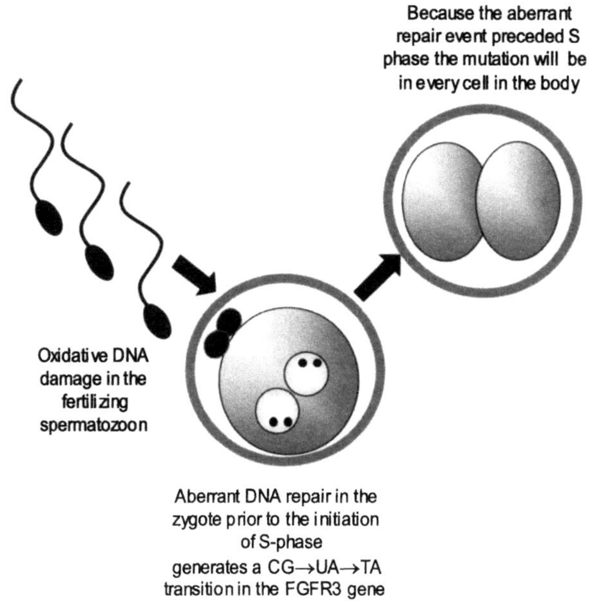

Figure 2. Hypothesis for the origins of mutations in the embryo subsequent to DNA damage in the fertilising sperm, as exemplified by achondroplasia. In the fertilising spermatozoon a key cytosine in the FGFR 3 gene is oxidatively damaged to create a uracil derivative. Defective repair of this damage in the fertilised egg results in the uracil derivative aberrantly paired with adenine followed by replacement of the former with thymine to create the final G1138A transition. This series of events must precede S-phase otherwise a genetic mosaic will be formed.

Thus, because a significantly increased incidence of genetic (or epigenetic) defects has not yet been reported following this type of therapy, we cannot conclude, as yet, that ICSI does not result in the trans-generational propagation of genetic damage. The fact is that with this technique the natural safeguards designed to ensure that defective human spermatozoa do not participate in the process of conception have been removed. Under these circumstances, it is inevitable that conceptions are occurring using DNA damaged spermatozoa. Under normal circumstances such DNA damage will be repaired by the oocyte. Indeed DNA synthesis in both the male and female pronuclei is held in abeyance until the oocyte has completed an immediate post fertilisation period of DNA repair (Shimura, Inoue, Taga, Shiraishi, Uematsu, Takei, Yuan, Shinohara and Niwa, 2002). As indicated above, if the oocyte makes a mistake during this repair process a mutation may be induced that, because it precedes S-phase, will be in every cell in the body, including the germ line. In this way factors that impact upon DNA integrity in human spermatozoa have the potential to create mutations in the embryo that will influence the health and wellbeing of children for generations to come. We are probably all carrying around in our genes the legacy of our great grand fathers pipe smoking habit; it did not end with his death from lung cancer.

The ageing and smoking models presented above demonstrate that DNA damage in the male germ line is associated with a statistical increase in the incidence of childhood diseases including dominant genetic mutations and cancer. This mutagenesis model is supported by a wealth of toxicology literature demonstrating that the induction of DNA damage in the male germ line with xenobiotics can have a disruptive effect on embryonic development. This, after all, is the basis of the dominant lethal assay; a routine toxicological test that identifies compounds capable of generating mutations in the male germ line that disrupt the viability and developmental normality of the embryo (Adler, 2000). Given the potential importance of DNA damage in the male germ line, it is of interest to consider the type of DNA damage in spermatozoa, its probable cause and the tests that are available to detect its presence.

Detection of DNA damage in the male germ line

To a large extent the molecular nature of DNA damage incurred in the male germ line is unknown and further, many of the commonly used assays for detecting DNA damage in spermatozoa are unable to directly provide this information. It can be predicted that this damage involves both single and double DNA strand breaks, generated via either oxidative or hydrolytic pathways, as well as abasic sites and DNA base modifications arising as a result of adduct formation or oxidation. Conceivable cellular pathways by which DNA damage might be generated include: abortive apoptosis, aberrant recombination, faulty remodelling, defective chromatin packing, formation of adducts or oxidative stress. Clearly the type of DNA damage occurring in human spermatozoa will depend upon the cause. Damage arising from toxicants or xenobiotics may include base- or phosphate-alkylation, base-adduct formation, intercalation, base oxidation and oxidative stand breakage. DNA fragmentation arising from the endonuclease activation associated with abortive apoptosis should result in free 3′- or 5′- OH groups and the coincident expression of apoptotic markers such as Fas or annexin V binding. DNA damage arising from oxidative stress will be associated with oxidative base formation and fragmentation. The information currently available on DNA in human spermatozoa suggests that oxidative stress is a more likely cause of DNA fragmentation than abortive apoptosis (Kodama, Yamaguchi, Fukuda, Kasi and Tanaka, 1997; Aitken, Gordon, Harkiss, Twigg, Milne, Jennings and Irvine, 1998; Aitken and Krausz, 2001; Muratori, Maggi, Spinelli, Filimberti, Forti and Baldi, 2003; Aitken, 2004; Moustafa, Sharma, Thornton, Mascha, Abdel-Hafez, Thomas and Agarwal, 2004; McVicar, McClure, Williamson, Dalzell and Lewis, 2004; Smith, Kaune, Parodi, Madariaga, Rios, Morales and Castro, 2006). However, detailed analyses of DNA adduct formation in patients' spermatozoa has not yet been performed.

Assays of DNA damage

The three main DNA damage assays used to assess spermatozoa are the Comet assay (Irvine, Twigg, Gordon, Fulton, Milne and Aitken, 2000), TUNEL assay (terminal transferase dUTP nick end labelling) (Lopes, Jurisicova, Sun and Casper, 1998) and SCSA (Virro, Larson-Cook and Evenson, 2004). All these tests detect the integrity of the chromatin structure by visualizing DNA strand breaks. However it is important to note that these assays do not directly measure DNA modifications that do not lead to stand breaks.

Comet assay

The Comet assay is a single cell electrophoretic method that depends upon the migration of DNA fragments in an electric field. Under the conditions of the assay, uncompromised DNA is maintained as a single coherent entity, while smaller DNA fragments arising from single and double strand breaks stream away from the main mass of DNA to create a structure that resembles a "comet" with distinct head and tail features that can be quantified (Fig. 3). Following staining with a DNA sensitive fluorochrome, individual Comets can be examined and the relative intensity of the signal in the tail versus the head of these structures taken as a measure of DNA damage. The assay exists in two forms comprising the alkaline or neutral pH versions. The alkaline method uses high pH (pH 12) to denature DNA fragments, thereby facilitating their release from the nucleus when the electric field is applied. At such a high pH, alkali labile sites within the chromatin can be converted to strand breaks and will contribute to the DNA damage observed. Some researchers consider this approach too artificial and hence use the neutral comet assay. However, alkali labile sites typically arise because DNA bases (or indeed the backbone) have been modified. Thus, if the objective of the assay is to measure total DNA damage and not just strand breaks, then including alkali labile sites in the assay (alkaline comet assay) is beneficial. When applied to human spermatozoa both the alkaline and the neutral versions of the Comet assay are sensitive, repeatable and detect high levels of DNA damage in patients' spermatozoa that are generally reflective of semen quality (Irvine *et al.*, 2000; Trisini, Singh, Duty and Hauser, 2004). To the authors' knowledge no direct comparison of the neutral and alkaline versions of the assay have been conducted in a diagnostic setting.

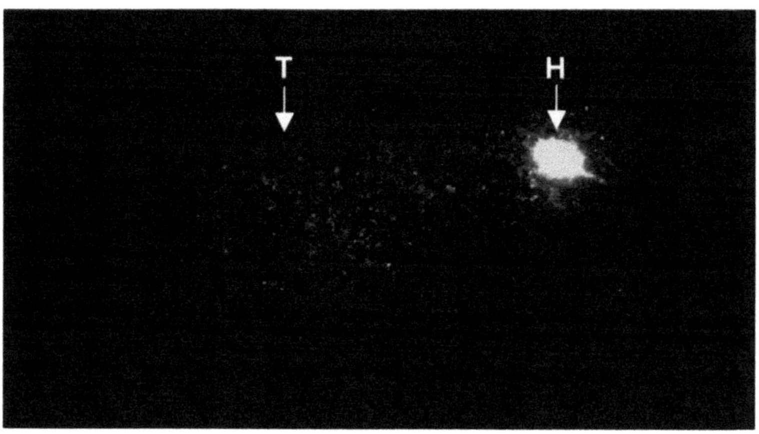

Figure 3. Typical Comet image generated following the electrophoresis of human spermatozoa showing head (H) and tail (T) components of the 'Comet'. DNA damage is proportional to the relative size and length of the tail.

SCSA

The SCSA method, involves denaturing sperm chromatin by exposure to low pH (pH 1.2) and then staining the treated cells with acridine orange. This stain fluoresces green when bound to double stranded (native) DNA and orange when bound to single stranded (denatured) DNA. Therefore, appropriately packaged and undamaged chromatin will only stain green, while damage will be associated with an increase in orange staining. Single stranded DNA fragments generated by the acid treatment are then measured by flow cytometry and a DNA fragmentation index (DFI) calculated. This assay has been shown to be highly correlated to male fertility (Evenson and Wixon, 2006). The SCSA method, unlike the Comet assay, only measures one type of DNA damage: strand breaks that yield fragments of single stranded DNA. The low pH may hydrolyse purine bases, leading to assay-derived DNA damage, but low pH will not hydrolyse the DNA phosphodiester backbone. Notwithstanding this difference, SCSA test results appear to be highly correlated with those obtained with the Comet assay (Aravindan, Bjordahl, Jost, and Evenson, 1997).

TUNEL assay

The TUNEL assay works on a different principle to the SCSA and Comet assays. Initially used for somatic cells, it has been adapted for use with germ cells. By utilising terminal deoxynucleotidyl transferase (TdT) to ligate fluorescently-stained dUTP residues onto the 3'-protruding -OH end of a DNA initiator, both double and single DNA strand breaks can be conveniently measured by flow cytometry. Since TdT will preferentially ligate dUTP onto the 3'-OH phosphate ends of exposed DNA termini, the TUNEL assay is specific for phosphodiester strand breaks or enzymatically induced breaks. For this reason, the TUNEL assay is typically used for apoptotic studies were endonucleases digest the chromatin leaving largely 3'-OH phosphate ends. The TUNEL assay has been successfully applied to the analysis of DNA damage in human spermatozoa and correlations noted between the outcome of this assay and fertilisation *in vitro* (Sun, Jurisicova and Casper, 1997). The DNA damage revealed by similar nick translation assays has also been shown to correlate with the ability of spermatozoa to bind chromomycin 3 (CMA3), a dye that competes with protamines for binding sites on the DNA and readily identifies poorly protaminated chromatin (Manicardi, Bianchi, Pantano, Azzoni, Bizzaro, Bianchi, and Sakkas, 1995). More recent data also suggest a positive link between DNA damage detected by TUNEL and the protamine composition of human spermatozoa (Torregrosa, Dominguez-Fandos, Camejo, Shirley, Meistrich, Ballesca and Oliva, 2006). However, it is not clear how cause and effect can be differentiated in this situation. Normally, the sperm nucleus is extremely compact and it may be difficult for the enzymes used in TUNEL (terminal transferase) or nick translation (DNA polymerase 1) assays to physically gain access to the sites of DNA damage, under these circumstances. Whether poor chromatin compaction is responsible for rendering sperm more vulnerable to DNA damage, as recently suggested in a comparative analysis of sperm protamine structure and susceptibility to DNA damage (Bennetts and Aitken, 2005), or simply allows these enzymes access to the sites of pre-existing nicks, is still an open question.

Another curiosity with the TUNEL assay is that it is frequently used in situations where oxidative stress is thought to be involved in the aetiology of DNA damage. For example, the DNA damage associated with heavy smoking is known to be oxidatively induced (Fraga *et al.*, 1996) and yet a recent report claims to have detected this damage with the TUNEL assay (Sepaniak *et al.*, 2006). Similarly, stimulation of oxidative damage in human spermatozoa using a redox cycling xenobiotic, menadione, resulted in the generation of DNA damage that could be detected with the TUNEL assay (De Iuliis, Wingate, Koppers, McLaughlin and Aitken, 2006). Furthermore in varicocoele patients, there is a great deal of data to indicate that the spermatozoa are under severe oxidative stress and exhibit correspondingly high levels of DNA damage, which can be readily detected with both the SCSA and TUNEL assays (Smith *et al.*, 2006). More generally, the spontaneous DNA damage observed in male infertility patients, which appears to be largely, if not exclusively, due to oxidative stress (Aitken, 2004; Aitken *et al.*, 2004, Aitken, Roman, Baker and De Iuliis, 2006; Baker and Aitken, 2005), can be detected with equal facility by SCSA or TUNEL (Chohan, Griffin, Lafromboise, De Jonge and Carrell, 2006). The ability of the TUNEL assay to detect oxidatively induced DNA damage in human spermatozoa is not readily explained unless the spermatozoa possess a redox regulated endonuclease. Human spermatozoa certainly possess endonuclease activity (Sotolongo, Huang, Isenberger and Ward, 2005). Moreover, in other cell types the induction of oxidative stress through the suppression of catalase and glutathione peroxidase activities can induce endonuclease mediated DNA fragmentation that is detectable with the

TUNEL assay (Ishihara and Shimamoto, 2006). In this case, the redox-regulated effector is endonuclease G, which is known to be present in the male germ line (Zhang, Dong, Li, Fan, Pathre, Dong, Lou, Wells, Olivares-Villagomez, Van Kaer, Wang and Xu, 2003).

Concluding comments

There can be no doubt that DNA damage in the male germ line is associated with adverse reproductive outcomes in terms of fertilisation rate, embryo quality, pregnancy and childhood heath (Lewis and Aitken, 2005). It is known that during the physiological remodeling of sperm chromatin during spermiogenesis, DNA nicks induced by topoisomerase may be required to relieve torsional stress (Laberge and Boissonneault, 2005). These double strand breaks are functional and clearly do not constitute DNA damage. Pathological DNA strand breaks can occur by hydrolysis of a phosphodiester bond (typically catalysed by nucleases) leaving 3′ or 5′ phosphate ends. Redox processes, typically directed toward the more chemically vulnerable bases and sugar moieties, can also break DNA strands. Of these processes, H-abstraction is a common mechanism resulting in DNA strand breaks. This involves intimate interaction with radical species (for example, the hydroxyl radical, HO•). The process of H-abstraction leaves the effected group (for example the ribose moiety) electron deficient, which subsequently leads to the rearrangement of electrons, ring opening and ultimately strand breakage. Oxidative damage to purine bases can also lead to apurinic sites that leave the affected region quite vulnerable to chemical strand breakage. For DNA damage assays that measure physical DNA breaks (Comet and SCSA), the two types of strand breaks outlined above (hydrolytic and oxidative) will appear as one in the same, both adding to the total measurable DNA damage, even though the aetiology of the damage is completely different. The interpretation of TUNEL data is also complex because this assay appears to be detecting DNA damage even when the source of damage should not theoretically involve the creation of free 3′ OH groups, as in the case of oxidative stress. We suggest that this enigma might be resolved if spermatozoa possessed a stress activated endonuclease. Progress in this area will only be made once we have resolved the precise nature of the DNA damage extant in the spermatozoa of male subjects. This information will not only provide insights into the origins of the DNA damage but also point us towards rational, sensitive methods for detecting this damage and managing its severity.

Acknowledgements

We are grateful to the ARC Centre of Excellence in Biotechnology and Development for financial support. Figure 1A reprinted from Fertility and Sterility Volume 80, Singh, N.P., Muller, C.H. and Berger, R.E., Effects of age on DNA double-strand breaks and apoptosis in human sperm, pages 1420-1430, Copyright (2003), with permission from the American Society of Reproductive Medicine.

References

Adler, I.D. (2000). Spermatogenesis and mutagenicity of environmental hazards: extrapolation of genetic risk from mouse to man. *Andrologia* **32:** 233-237

Aitken, R.J. (2004). Founders' Lecture. Human spermatozoa: fruits of creation, seeds of doubt. *Reproduction, Fertility and Development* **16:** 655-664

Aitken, R.J. and Krausz, C.G. (2001). Oxidative stress, DNA damage and the Y chromosome. *Reproduction* **122:** 497-506

Aitken, R.J., Gordon, E., Harkiss, D., Twigg, J.P., Milne, P., Jennings, Z. and Irvine, D.S. (1998). Relative impact of oxidative stress on the functional compo tence and genomic integrity of human spermatozoa. *Biology of Reproduction* **59:** 1037-1046

Aitken, R.J., Koopman, P. and Lewis S.E. (2004). Seeds of concern. *Nature* **432:** 48-52

Aitken, R.J., Roman, S.D., Baker, M.A. and De Iuliis, G. (2006). Redox regulation of DNA damage in the male germ line. In: *Male Mediated Developmental Toxicity*. Edited by Anderson, D. and Brinkworth, M. Royal Society of Chemistry (in press)

Aravindan, G.R., Bjordahl, J., Jost, L..K and Evenson, D.P. (1997). Susceptibility of human sperm to in situ DNA denaturation is strongly correlated with DNA strand breaks identified by single-cell electrophoresis. *Experimental Cell Research* **236:** 231-237

Baker, M.A. and Aitken, R.J. (2005). Reactive oxygen species in spermatozoa: methods for monitoring and significance for the origins of genetic disease and infertility. *Reproductive Biology and Endocrinology* **3:** 67-76

Bennetts, L.E. and Aitken, R.J. (2005). A comparative study of oxidative DNA damage in mammalian spermatozoa. *Molecular Reproduction and Development* **71:** 77-87

Chohan, K.R., Griffin, J.T., Lafromboise, M., De Jonge, C.J. and Carrell, D.T. (2006). Comparison of chromatin assays for DNA fragmentation evaluation in human sperm. *Journal of Andrology* **27:** 53-59

Crow, J.F. (2000). The origins, patterns and implications of human spontaneous mutation. *Nature Reviews Genetics* **1:** 40-47

De Iuliis, G.N., Wingate, J.K., Koppers, A.J., McLaughlin, E.A. and Aitken, R.J. (2006). Definitive evidence for the non-mitochondrial production of superoxide anion by human spermatozoa. *Journal of Clinical Endocrinology and Metabolism* **91:** 1968-1975

de la Rochebrochard, E. and Thonneau, P. (2002). Paternal age and maternal age are risk factors for miscarriage; results of a multicentre European study. *Human Reproduction* **17:** 1649-1656

Dittrich, R., Maltaris, T., Dragonas, C., Huber, C., Beckmann, M.W. and Kiesewetter, F. (2005). Individual assessment of sperm morphology of single spermatozoa used for intracytoplasmic sperm injection. *Andrologia* **37:** 53-56

Evenson, D.P. and Wixon, R. (2006). Clinical aspects of sperm DNA fragmentation detection and male infertility. *Theriogenology* **65:** 979-999

Fraga, C. G., Motchnik, P. A., Wyrobek, A. J., Rempel, D. M., and Ames, B. N. (1996). Smoking and low antioxidant levels increase oxidative damage to DNA. *Mutation Research* **351:** 199-203

Gandini, L., Lombardo, F., Paoli, D., Caruso, F., Eleuteri, P., Leter, G., Ciriminna, R., Culasso, F., Dondero, F., Lenzi, A. and Spano, M. (2004). Full-term pregnancies achieved with ICSI despite high levels of sperm chromatin damage. *Human Reproduction* **19:** 1409-1417

Hansen, M., Kurinczuk, J.J., Bower, C. and Webb, S. (2002). The risk of major birth defects after intracytoplasmic sperm injection and *in vitro* fertilization. *New England Journal of Medicine* **346:** 725-730

Hazout, A., Dumont-Hassan, M., Junca, A.M., Cohen Bacrie, P. and Tesarik, J. (2006). High-magnification ICSI overcomes paternal effect resistant to conventional ICSI. *Reproductive Biomedicine Online* **12:** 19-25

Hikichi, T., Kishigami, S., Thuan, N.V., Ohta, H., Mizutani, E., Wakayama, S. and Wakayama, T. (2005). Round spermatids stained with mitotracker can be used to produce offspring more simply. *Zygote* **13:** 55-61

Hill, K.A., Halangoda, A., Heinmoeller, P.W., Gonzalez, K., Chitaphan, C., Longmate, .J, Scaring,e W.A., Wang, J.C. and Sommer, S.S. (2005). Tissue-specific time courses of spontaneous mutation frequency and deviations in mutation pattern are observed in middle to late adulthood in Big Blue mice. *Environmental and Molecular Mutagenesis* **45:** 442-454

Hurst, L.D. and Ellegren, H. (2002). Human genetics: mystery of the mutagenic male. *Nature* **420:** 365-366

Irvine, D.S., Twigg, J.P., Gordon, E.L., Fulton, N., Milne, P.A. and Aitken, R.J. (2000). DNA Integrity in human spermatozoa: relationships with semen quality. *Journal of Andrology* **21:** 33-44

Ishihara, Y. and Shimamoto, N. (2006). Involvement of endonuclease G in nucleosomal DNA fragmentation under sustained endogenous oxidative stress. *Journal of Biological Chemistry* **281:** 6726-6733

Ji, B. T., Shu, X. O., Linet, M. S,, Zheng, W., Wacholder, S., Gao, Y. T., Ying, D. M. and Jin, F. (1997). Paternal cigarette smoking and the risk of childhood cancer among offspring of nonsmoking mothers. *Journal of the National Cancer Institute* **89:** 238-244

Katsoff, B., Check, M.L., Summers-Chase, D. and Check, J.H. (2005). Absence of sperm with rapid motility is not detrimental to IVF outcome measures when ICSI is used. *Archives of Andrology* **51:** 413-417

Kidd, S.A., Eskenazi, B. and Wyrobek, A.J. (2001). Effects of male age on semen quality and fertility: a review of the literature. *Fertility and Sterility* **75:** 237-248

Kodama, H., Yamaguchi, R., Fukuda, J., Kasi, H. and Tanaka, T. (1997). Increased deoxyribonucleic acid damage in the spermatozoa of infertile male patients. *Fertility and Sterility* **68:** 519-524

Kovacic, B., Vlaisavljevic, V., Reljic, M. (2006). Clinical use of pentoxifylline for activation of immotile testicular sperm before ICSI in patients with azoospermia. *Journal of Andrology* **27:** 45-52

Kreutzer, D.A. and Essigmann, J.M. (1998). Oxidized, deaminated cytosines are a source of C—T transitions *in vivo*. *Proceedings of the National Academy of Sciences USA* **95:** 3578-3582

Laberge, R.M. and Boissonneault, G. (2005). On the nature and origin of DNA strand breaks in elongating spermatids. *Biology of Reproduction* **73:** 289-96

Lewis, S.E. and Aitken, R.J. (2005). DNA damage to spermatozoa has impacts on fertilization and pregnancy. *Cell and Tissue Research* **322:** 33-41

Lopes, S., Jurisicova, A., Sun, J.G. and Casper, R.F. (1998). Reactive oxygen species: potential cause for

DNA fragmentation in human spermatozoa. *Human Reproduction* **13:** 896-900

Manicardi, G.C., Bianchi, P.G., Pantano, S., Azzoni, P., Bizzaro, D., Bianchi, U. and Sakkas, D. (1995). Presence of endogenous nicks in DNA of ejaculated human spermatozoa and its relationship to chromomycin A3 accessibility. *Biology of Reproduction* **52:** 864-867

McNeil, D.E., Cote, T.R., Clegg, L. and Mauer, A. (2002). SEER update of incidence and trends in pediatric malignancies: acute lymphoblastic leukemia. *Medical and Pediatric Oncology* **39:** 554-557

McVicar, C.M., McClure, N., Williamson, K., Dalzell, L.H. and Lewis, S.E. (2004). Incidence of Fas positivity and deoxyribonucleic acid double-stranded breaks in human ejaculated sperm. *Fertility and Sterility* **81** (Supplement 1): 767-774

Moustafa, M.H., Sharma, R.K., Thornton, J., Mascha, E., Abdel-Hafez, M.A., Thomas, A.J. Jr, Agarwal, A. (2004). Relationship between ROS production, apoptosis and DNA denaturation in spermatozoa from patients examined for infertility. *Human Reproduction* **19:** 129-138

Muratori, M., Maggi, M., Spinelli, S., Filimberti, E., Forti, G. and Baldi, E. (2003). Spontaneous DNA fragmentation in swim-up selected human spermatozoa during long term incubation. *Journal of Andrology* **24:** 253-262

Palermo, G., Joris, H., Devroey, P. and Van Steirteghem, A.C. (1992). Pregnancies after intracytoplasmic injection of single spermatozoon into an oocyte. *Lancet* **340:** 17-18

Sallam, H.N., Farrag, A., Agameya, A.F., El-Garem, Y. and Ezzeldin, F. (2005). The use of the modified hypo-osmotic swelling test for the selection of immotile testicular spermatozoa in patients treated with ICSI: a randomized controlled study. *Human Reproduction* **20:** 3435-3440

Sepaniak, S., Forges, T., Gerard, H., Foliguet, B., Bene, M.C. and Monnier-Barbarino, P. (2006). The influence of cigarette smoking on human sperm quality and DNA fragmentation. *Toxicology* **223:** 54-60

Shimura, T., Inoue, M., Taga, M., Shiraishi, K., Uematsu, N., Takei, N., Yuan, Z.M., Shinohara, T and Niwa, O. (2002). p53-dependent S-phase damage checkpoint and pronuclear cross talk in mouse zygotes with X-irradiated sperm. *Molecular and Cellular Biology* **22:** 2220-2228

Shveiky, D., Simon, A., Gino, H., Safran, A., Lewin, A., Reubinoff, B., Laufer, N. and Revel, A. (2006). Sibling oocyte submission to IVF and ICSI in unexplained infertility patients: a potential assay for gamete quality. *Reproductive Biomedicine Online* **12:** 371-374

Singh, N.P., Muller, C.H. and Berger, R.E. (2003). Effects of age on DNA double-strand breaks and apoptosis in human sperm. *Fertility and Sterility* **80:** 1420-1430

Sipos, A., Rasmussen, F., Harrison, G., Tynelius, P., Lewis, G., Leon, D.A. and Gunnell D. (2004). Paternal age and schizophrenia: a population based cohort study. *British Medical Journal* **329:** 1070

Smith, R., Kaune, H., Parodi, D., Madariaga, M., Rios, R., Morales, I. and Castro, A. (2006). Increased sperm DNA damage in patients with varicocele: relationship with seminal oxidative stress. *Human Reproduction* **21:** 986-993

Sotolongo, B., Huang, T.T., Isenberger, E. and Ward, W.S. (2005). An endogenous nuclease in hamster, mouse, and human spermatozoa cleaves DNA into loop-sized fragments. *Journal of Andrology* **26:** 272-280

Stalf, T., Mehnert, C., Hajimohammad, A., Manolopoulos, K., Shen, Y., Schuppe, H.C., Diemer, T., Schill, W.B., Weidner, W. and Tinneberg, H.R. (2005). Influence of motility and vitality in intracytoplasmic sperm injection with ejaculated and testicular sperm. *Andrologia* **37:** 125-130

Sun, J.G., Jurisicova, A. and Casper, R.F. (1997). Detection of deoxyribonucleic acid fragmentation in human sperm: correlation with fertilization *in vitro*. *Biology of Reproduction* **56:** 602-607

Tiemann-Boege, I., Navidi, W., Grewal, R., Cohn, D., Eskenazi, B., Wyrobek, A.J. and Arnheim, N. (2002). The observed human sperm mutation frequency cannot explain the achondroplasia paternal age effect. *Proceedings of the National Academy of Sciences USA* **99:** 14952-14957

Torregrosa, N., Dominguez-Fandos, D., Camejo, M.I., Shirley, C.R., Meistrich, M.L., Ballesca, J.L. and Oliva, R. (2006). Protamine 2 precursors, protamine 1/protamine 2 ratio, DNA integrity and other sperm parameters in infertile patients. *Human Reproduction* **21:** 2084-2089

Trisini, A.T., Singh, N.P., Duty, S.M. and Hauser, R. (2004). Relationship between human semen parameters and deoxyribonucleic acid damage assessed by the neutral comet assay. *Fertility and Sterility* **82:** 1623-1632

Twigg, J.P., Irvine, D.S. and Aitken, R.J. (1998). Oxidative damage to DNA in human spermatozoa does not preclude pronucleus formation at intracytoplasmic sperm injection. *Human Reproduction* **13:** 1864-1871

van der Westerlaken, L., Naaktgeboren, N., Verburg, H., Dieben, S. and Helmerhorst, F.M. (2006). Conventional *in vitro* fertilization versus intracytoplasmic sperm injection in patients with borderline semen: a randomized study using sibling oocytes. *Fertility and Sterility* **85:** 395-400

Virro, M.R., Larson-Cook, K.L. and Evenson, D.P. (2004). Sperm chromatin structure assay parameters are related to fertilization, blastocyst development, and ongoing pregnancy in in vitro fertilization and intracytoplasmic sperm injection cycles. *Fertility and Sterility* **81:** 1289-1295

Vural, B., Sofuoglu, K., Caliskan, E., Delikara, N., Aksoy, E., Uslu, H. and Karan, A. (2005). Predictors of intracytoplasmic sperm injection (ICSI) outcome in couples with and without male factor infertility. *Clinical and Experimental Obstetrics and Gynecology* **32:** 158-162

Wyrobek, A.J., Eskenazi, B., Young, S., Arnheim, N.,

Tiemann-Boege, I., Jabs, E.W., Glaser, R.L., Pearson, F.S. and Evenson, D. (2006). Advancing age has differential effects on DNA damage, chromatin integrity, gene mutations, and aneuploidies in sperm. *Proceedings of the National Academy of Sciences USA* **103**: 9601-9606

Yilmaz, S., Koyuturk, M., Kilic, G., Alpak, O. and Aytoz, A. (2005). Effects of leucocytospermia on semen parameters and outcomes of intracytoplasmic sperm injection. *International Journal of Andrology* **28:** 337-342

Zhang, J., Dong, M., Li, L., Fan, Y., Pathre, P., Dong, J., Lou, D., Wells, J.M., Olivares-Villagomez, D., Van Kaer, L., Wang, X. and Xu, M. (2003). Endonuclease G is required for early embryogenesis and normal apoptosis in mice. *Proceedings of the National Academy of Sciences USA* **100:** 15782-15787

Spermatology. SRF Vol. 65. ERS Roldan and M Gomendio (eds) Nottingham University Press, Nottingham

Analysis of sperm DNA fragmentation using flow cytometry and other techniques

Donald P. Evenson, Kay Kasperson and Regina L. Wixon

Department of Biology and Microbiology, ASC 136, South Dakota State University, Brookings, SD 57007, USA

The goal of a rapid, precise and objective sperm measure of sperm DNA fragmentation, *in situ*, was realised a quarter century ago by the pioneer development of the sperm chromatin structure assay (SCSA). Since then, measurements on many thousands of sperm samples from a variety of animals and humans exposed to reproductive toxicants and having different diseases or infertility pathologies have solidly established the utility of sperm DNA fragmentation in experimental and clinical settings. New methods have been developed that measure sperm DNA fragmentation by different means such as the enzymatic labelling of broken DNA strands (Tunel assay by light microscopy or flow cytometry) and the light microscope observations of DNA fragments in the Comet assay and the negative loops of sperm chromatin in the Halo sperm assay.These assays have proven useful in the areas of livestock and captive wildlife reproduction efficiency, toxicology and more recently in the human infertility clinic. A significant problem is the lack of quality control and standards of threshold values between laboratories. At this time, only the SCSA has been deemed sufficiently rigorous in methodology and established clinical thresholds for utility in the human infertility clinic.

Introduction

Sperm quality determination, as related to both fertility and toxicology, has for decades relied on the classical semen analyses of sperm count, motility and morphology. These criteria have served useful; however, it is well recognised that the overlap in semen values between fertile and infertile, as well as toxic-exposed and non-exposed subjects, has prompted the development of new sperm integrity assays for both of these fields. Especially in this era of genomic characterization and intracytoplasmic sperm injection (ICSI), it is obvious that the integrity of the male genome is of primary interest.

In the 1970's a research team at the Lawrence Livermore National Laboratory attempted to use the newly invented flow cytometer to measure genetic alterations of sperm DNA content in mice (Pinkel, 1984). However, the flow cytometer being used had the standard orthogonal axes of sample flow direction at 90 degrees to the angle of laser beam, and 90 degrees to the angle of sample fluorescent beam collection. While this orientation is not a problem for measuring DNA content in typical round somatic cells, sperm cells with asymmetrical shape and a high refractive index, due to the highly condensed chromatin, presented a daunting problem

Corresponding author E-mail: scsa@brookings.net

(Pinkel, 1984). Since the asymmetrically shaped sperm can flow in the sample stream at any angle, it became obvious that asymmetrical shaped mammalian sperm had fluorescent light amplification at certain angles that caused an artifact of differential fluorescent amplitudes even though all sperm had identical DNA content. These investigations did show, among a population of sperm, obtained from irradiated mice, a dose response relationship between dosage and the CV of the DNA fluorescence signals (Pinkel, 1984).

In that same time frame, a new approach was taken by D. Evenson and colleagues at the Memorial Sloan Kettering Cancer Center in New York. In a quest to learn whether cancer treatment caused sperm DNA damage, a method was developed to determine if sperm DNA strand breaks could be detected in a flow cytometer. Sperm were treated with heat to denature DNA at the sites of DNA strand breaks followed by staining with acridine orange (AO) that produced green fluorescence when intercalated into double stranded (ds) DNA and red fluorescence when associated with single stranded (ss) DNA. In a pioneering publication (Evenson, Darzynkiewicz and Melamed, 1980) the term "altered chromatin structure" was used since it was not known whether only DNA strand breaks were involved or whether the differential fluorescence patterns were also due to nuclear protein alterations. The original protocol was altered later to an acid DNA denaturation procedure and named the sperm chromatin structure assay (SCSA®) (Evenson, Higgins, Grueneberg and Ballachey, 1985).

The Sperm Chromatin Structure Assay (SCSA)

The SCSA has been validated for over a quarter century with many thousands of sperm samples from many species. It is not possible to extrapolate SCSA data to validate the other DNA fragmentation assays. The SCSA relies on small AO molecules being in strict equilibrium staining with DNA (~2 AO molecules/DNA base pair). In contrast, the terminal uridine end labeling (Tunel) assay requires large enzymes, and in some cases antibodies, to detect the presence of 3′OH broken ends of DNA. In addition, there is some concern regarding the complete washing out of the non-specific labelled fluorochrome from the complex maze of sperm chromatin. Sperm, especially human sperm, have a great amount of heterogeneity in chromatin structure; thus, the labelling pattern may not be the same in different nuclei having an equivalent number of DNA strand breaks. Even so, correlations between data from these two assays using the same samples strongly suggest a similar pattern of detection of sperm DNA strand breaks (Gorczyca, Melamed and Darzynkiewicz, 1993; Sailer, Jost and Evenson, 1995). However, data on the threshold for placing a man at risk for infertility have been dissimilar for the Tunel and Comet assay (Evenson and Wixon 2006).

The original heating technique for the SCSA was later switched to using a pH 1.2 buffer for 30 seconds to accomplish DNA denaturation at the sites of DNA strand breaks with the same results as with the heating method (Evenson *et al.*, 1985). However, the acid technique is much easier than the more cumbersome heating method. SCSA analysis of sonicated and sucrose gradient purified sperm nuclei ruled out the idea that any RNA in cytoplasmic remnants of sperm added to a significant increase in red fluorescence above that due to ss DNA (Figs. 1 and 2) (Evenson *et al.*, 1985).

In our pioneering studies (Evenson *et al.*, 1980), bull fertility ranking was predicted from the SCSA data. Likewise, men presenting at an infertility clinic had a mean two fold higher DNA fragmentation index (DFI) than samples from proven donors. Finally, mice fed a zinc-deficient diet showed a significant increase in sperm DNA fragmentation as was confirmed later for rats (Evenson, Emerick, Jost, Kayong-Male and Stewart, 1993). Zinc is required for normal repackaging of sperm DNA with disulfide bonds between cysteine residues of sperm nuclear protamines. Thus, the stage was set for determining whether the SCSA was a sensitive marker for dose response toxicology studies and for men at risk for infertility.

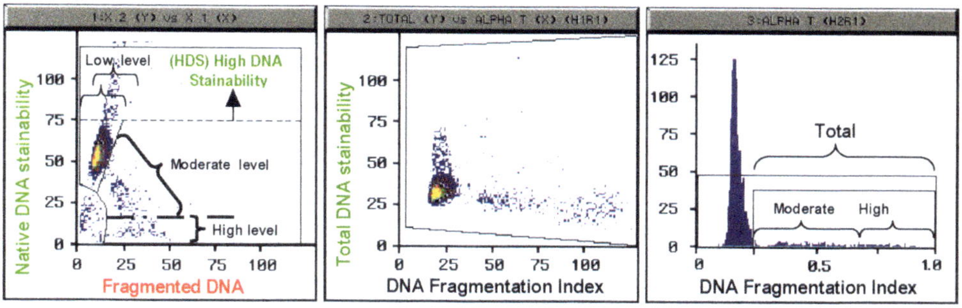

Patient	Date	Measurement	X DFI	SD DFI	DFI (%)	HDS (%)
7272-87	####	1	213.5	111.6	6.8	5.0
		2	221.2	118.1	8.3	5.4
		mean	217.4	114.8	7.5	5.2
		sd	5.4	4.6	1.1	0.2

Figure 1. SCSA cytograms from a man with excellent sperm DNA integrity. Each dot represents one of 5000 sperm cells.

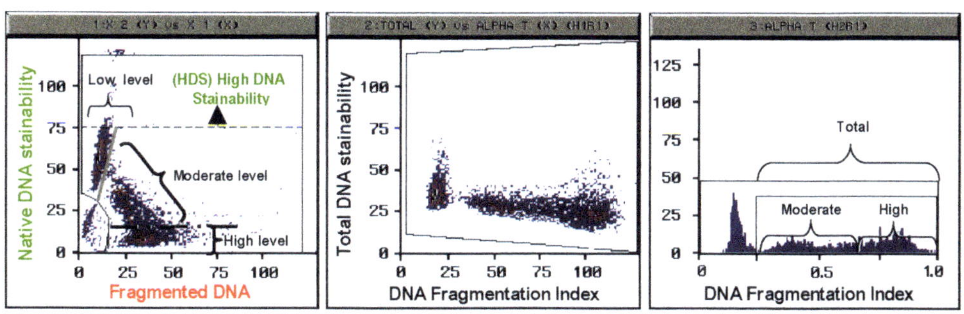

Patient	Date	Measurement	X DFI	SD DFI	DFI (%)	HDS (%)
7272-113	####	1	563.7	307.0	64.9	6.4
		2	561.4	304.8	64.9	7.2
		mean	562.6	305.9	64.9	6.8
		sd	1.2	1.1	0.0	0.4

Figure 2. SCSA cytograms from a man with fair to poor sperm DNA integrity. Each dot represents one of 5000 sperm cells. Note the 64.9% of sperm characterised by high red fluorescence due to fragmented DNA.

Animal toxicology

The first approach to validate the biochemistry and flow cytometry protocol for SCSA toxicology measurements was to repeat the experiments of Wyrobek (1979) that showed a high correlation between genotoxic chemical dose and abnormal sperm head morphology. Interestingly, for mice, many agents that caused sperm head morphology changes also showed a highly significant increase in DNA damage as measured by the SCSA. Chemical agents were selected that were known, or suspected, to target germ cells at various stages of spermatogenesis or spermiogenesis. Thus, in general, agents that affected spermatocytes (Sailer, Jost, Erickson, Tajiran and Evenson, 1995), mitotic/meiotic spindles and epididymal sperm (Evenson, Jost and Baer, 1993) all produced changes in mouse sperm morphology and DNA integrity as measured by the SCSA. The great value of the studies with mouse toxicology was the fact that the SCSA data were shown to be very highly repeatable not only between individual mice and repeated experiments, but also between fresh and frozen sperm. Data on epididymal sperm obtained from mice exposed to three different toxicants are described below.

Triethylmelamine (TEM). SCSA data obtained on fresh sperm were highly correlated with measurements of aliquots of the same samples collected over 44 weeks, frozen, and measured on the same day (SD of DFI, 0.88 and DFI 0.93, $p<0.001$). SCSA data remained abnormal at 44 weeks after triethylenemelamine (TEM) dosage suggesting that the abnormalities, present long after the initial toxic response, may be a result of mutation (Evenson, Baer and Jost, 1989). SCSA data and sperm head morphology abnormalities were highly correlated (0.87-0.93, $p<0.001$) with dose of TEM and with each other.

Methyl methane sulfonate (MMS). Epididymal DNA damage by exposure to MMS was detected three days post injection (dpi) (and likely occurring earlier), however, but dominant lethal mutations were not detected until 8 dpi mice were bred (Evenson *et al.*, 1993). Therefore, the SCSA was apparently detecting molecular lesions days before MMS caused sufficient chromosome damage to become lethal to the developing embryo. In a similar vein, bull testes exposed to elevated temperatures caused DNA damage detectable by the SCSA at three days post heat while alterations in sperm morphology/morphometry were not detected until 12 days post heat exposure (Fig. 3) (Karabinus, Vogler, Saacke and Evenson, 1997).

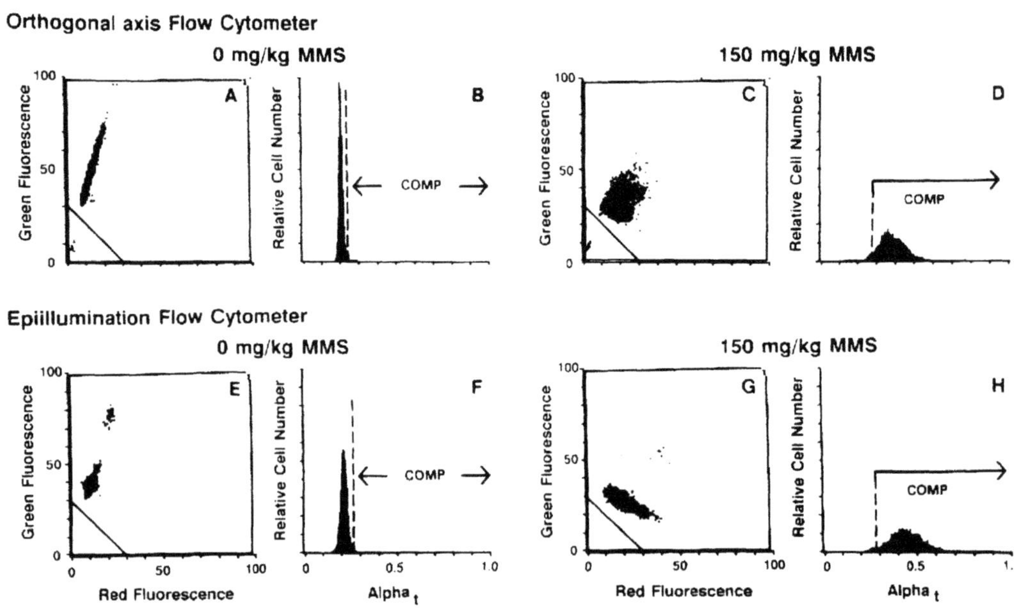

Figure 3. Green vs red fluorescence cytograms of SCSA data obtained by measuring caudal mouse sperm with a Cytofluorograf (**A,C**) or ICP22A (**E,G**) flow cytometer. Sperm were obtained from mice treated with 0 or 150 mg/kg and killed at 13 days. Panels **B,D,F** and **H** are alpha-t frequency histograms; cells to the right of the vertical lines are cells with DFI.

X-radiation. The studies on effects of X-ray to mouse testes showed that the SCSA was the most sensitive assay for low dose radiation damage. If more than three mice had been used per time point, the dose of 12 rads would likely have shown to be significant instead of a trend, while the dose of 25 rads produced a statistically significant increase in DNA fragmentation. It is of interest that the X-ray damage was detected in epididymal sperm 40 days after treatment; thus, the damage done to spermatocytes persisted into epididymal sperm (Sailer *et al.*, 1995). One can question why the biological monitors/checkpoints of sperm quality did not signal for these sperm to be destroyed by apoptosis. Or, were these sperm actually apoptotic, i.e., genetically dead, in the epididymis?

Male infertility

Bull. In collaboration with Eastern Artificial Insemination, Inc. (EAI, Ithaca, NY), semen was obtained from 49 bulls for a total of 191 collections. The SCSA data consisted of both the % of sperm with denatured DNA (%DFI, % sperm with fragmented DNA) and the SD of DFI (red/red+green fluorescence). Of interest, the SD of DFI had a higher correlation (-0.65, $p<0.01$) than the % DFI (-0.53, $p<0.05$) (Ballachey, Hohenboken and Evenson, 1987). Thus, this increased sensitivity was similar to data from the mouse toxicology experiments. This parameter is of significant power and is a parameter that can not be calculated by those who use Tunel, Comet or SCSA without SCSAsoft® software.

While bovine fertility studies are much less complicated than human, even so, the calculations of a fertility index for bulls are extensive. To minimise the fertility variations of the cows, a more significant experiment would be to take out the cow variations by using heterospermic insemination, i.e., mixing equal numbers of motile sperm from bulls of different phenotypes and simply determining the percent of calves born with the different phenotypes (Ballachey, Evenson and Saacke, 1988). In this case, the correlations between % DFI and SD DFI were -0.94 and -0.74, respectively (Fig. 4).

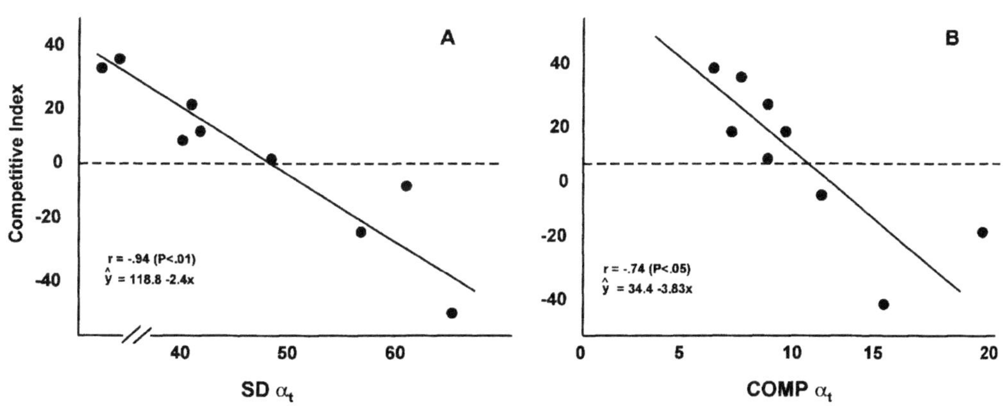

Figure 4. Relationship of Competitive Index with (**A**) SD of DFI (SDα_t) and (**B**) DFI (cells outside the main population, COMPα_t). Modified from Ballachey *et al.* (1988).

Pig. The same type of heterospermic index was also derived from pigs. An advantage of utilizing pigs is that with a multiparous animal, not only can one calculate the fertility index but also relate the SCSA data with number of piglets born (Evenson, Thompson and Jost, 1994). Semen from six boars was mixed in equal sperm numbers in six 3 way combinations and inseminated into at least three Duroc gilts per combination. Four of the six combinations yielded two litters, while the remaining two combinations yielded three litters. The SCSA correctly predicted both the high and low fertility boars based on a ratio of offspring as deviated from the theoretical percentage.

Human. With the validation of the SCSA for both toxicology and animal fertility, it was obvious that the next frontier should be human infertility, a much more complicated investigation for obvious reasons. The pioneering study to investigate this relationship was a highly controlled male factor infertility study at Georgetown University led by Michael Zinaman and Eric Clegg. This study enrolled 165 couples attempting natural pregnancy with the female partner having no known infertility factors (this is a very important point relative to later publications not taking this factor into account). These couples were carefully screened and instructed on how to maximise pregnancy by timed intercourse. Monitoring was extensive for the first three months with monthly urine analyses to determine if a biochemical pregnancy had occurred. Semen samples were collected monthly for the first three months and measured for sperm count, morphology, motility and concentration. An aliquot of raw semen was flash frozen and sent to our laboratory for the SCSA test and the blinded data were then sent to an independent statistician.

From these data, it was learned that couples where the man with a %DFI of less than 15% had the highest percent pregnancy in the first three months. Those taking a longer time to pregnancy (4-12 months) were in the group of 15-30% DFI and the poorest pregnancy rate was with men who had a >30% DFI. No man with a >30% DFI had a pregnancy in the first three months. However, a very important point, and in sharp contrast to many published misconceptions (Evenson, Larson and Jost, 2002), there was not an absolute cut off that said that no pregnancies were possible with a DFI > 30%. Some men with a DFI > 30% did initiate successful pregnancies, but it simply took a longer time, i.e., in months 4-9 rather than months 1-3, while others did not initiate a pregnancy. It would not be expected to have a sharp cut off at exactly 30.0% DFI for lack of pregnancies; biology does not work that way. The 30% threshold is a statistical threshold only.

These data were corroborated by a "first pregnancy planner" study in Denmark. Likewise, the pregnancy rate fell to a very low level when the %DFI was above 30-40% (Spano, Bonde, Hjollund, Kolstatd, Cordelli and Leter, 2000). Using the SCSA, *in vivo* patients (n=215) (with no previous knowledge of their fertility status) showed a declining pregnancy outcome when their DFI > 20% and negligible pregnancies when the DFI approached 40%. Patients were about ten times more likely to achieve a pregnancy when DFI was <40%.

Given the advances of routine *in vitro* fertilisation (IVF), and especially the advent of ICSI, an all-important question was the predictability of the SCSA for patients undergoing assisted reproductive techniques (ART). The first published study unfortunately had only 24 participants. Fourteen patients had DFI values above 30% and none of those couples had a successful pregnancy after one cycle of ICSI (Larson, DeJonge, Barnes, Jost and Evenson, 2000). This manuscript added to the incorrect interpretation that no pregnancies were possible when the DFI was >30%.

A recently published meta-analysis for SCSA data and pregnancy outcomes showed that couples have a 6.5-10X, 7.0 - 8.7X, ~2 X, and ~1.5 X greater probability of a successful pregnancy by *in vivo*, intra-uterine insemination (IUI), routine IVF and ICSI fertilisations, respectively, if the semen samples contains < 27-30% DFI (Evenson and Wixon, 2006).

Tunel assay

In the Tunel assay (Gorczyca *et al.*, 1993; Sailer *et al.*, 1995), sperm are incubated with proteases to provide access to DNA for labelled markers enzymatically added to 3′OH ends of broken DNA. These labelled sperm can be measured by light microscopy or by flow cytometry. The obvious advantage of flow cytometry is that thousands of sperm can be measured with great precision in a short time thus providing excellent statistical power to the analysis. The Tunel assay is used by a variety of investigators. The primary current problem is that, as was stated by an U.S. Environmental Protection Agency (EPA) workshop review team a few years ago (Perreault, Aitken, Baker, Evenson, Huszar, Irvine, Morris, Morris Robbins, Sakkas, Spano and Wyrobek, 2003) there are about as many subtle variations of the technique as there are individuals using it. Consequently, statistical thresholds for pregnancy outcomes differ between laboratories. Duran, Morshedi, Taylor and Oehninger (2002) found no pregnancies occurring in patients having greater than 10% Tunel positive sperm while the statistical threshold found by Henkel, Kierspel, Hajimohammad, Stalf, Hoogendijk, Mehnert, Menkveld, Schill and Kruger (2003) with a receiver operating characteristic (ROC) analysis was 36.5%. In the latter case, the pregnancy rate was reduced by 40% compared to patients with less than 36.5% Tunel positive sperm (<36.5% DFI). A concern for the Tunel assay is that the sperm nucleus is very compact and even after digestion there exists a tangle of DNA/chromatin that may inhibit both the entrance of large molecules, i.e., TdT enzyme and antibodies to BrdU tagged nucleotides, and furthermore, the washing out of the non-specific tags prior to fluorometric analysis. The Tunel assay supposedly tags ds DNA breaks primarily which are lethal for any required gene. However, comparison of data from Tunel and SCSA measures have shown a 0.9 correlation between the two assays (Gorczyca *et al.*, 1993). Since the SCSA is thought to measure both ss- and ds-DNA breaks, the picture is not clear. A disadvantage of Tunel is that it can not detect the immature sperm population, known as the High DNA Stainable sperm (HDS) in the SCSA. This factor together with the %DFI has shown to be of importance for assessing male infertility (Virro, Larson-Cook and Evenson, 2004) and reproductive toxic effects (Jeffay, Strader, Buus, Evenson, Olshan, Herring, Bradley, Smith and Perreault, 2006).

Comet assay

The single cell gel electrophoretic assay (Comet assay) initially developed for somatic cells (Tice, Agurell, Anderson, Burlinson, Hartmann, Kobayashi, Miyamae, Rojas, Ryu and Sasaki, 2000) was later adapted for use with sperm (Singh, Danner, Tice, McCoy, Collins and Schneider, 1989). This method has been done under both high alkaline conditions (pH 12.5) that denatures all DNA and also under neutral pH (pH10, i.e., neutral relative to not causing DNA denaturation). The alkaline condition causes an artifact in measurements for sperm since its DNA possess alkaline sensitive sites. Under neutral conditions DNA with a double strand break in close proximity to another ds DNA break allows a segment of DNA to migrate out in the gel electrophoresis system. The pieces that migrate out form a tail, making the image look like a celestial comet. While this assay is useful for providing a "yes" or "no" answer to the existence of ds DNA strand breaks, it is difficult to quantitate the extent of the DNA damage. Researchers have attempted to quantitate the extent of damage by image analysis measures of the length and width of the tail (relative to sizes of DNA segments). Sperm DNA is labelled with a fluorochrome such as YOYO or propidium iodide (PI). Since YOYO is 100 times brighter than PI the data from each type of staining can not be compared with the other. Another problem is that whereas researchers report a % of DNA in the tail relative to the nuclear core, these measures are likely not correct. Work done in our laboratory, not yet reported, show that the core contains compacted DNA/protein that does not give stochiometric fluorescence intensity relative to the free DNA in the tail. In addition, Comet lacks the statistical power of measuring thousands of sperm as done using a flow cytometer.

Some laboratories are using the Comet for clinical assessment; this may be appropriate for a single laboratory to conduct the assay exactly the same way. However, given the variations that can occur with factors such as exact amperage in the electrophoresis set up, it is considered difficult to compare computer image values from one lab to another. One of the first reports on the comparison between SCSA and Comet came from our laboratory, showing a very high correlation between the two assays based on % of positivity in both assays (Aravindan, Bjordahl, Jost and Evenson, 1997); unfortunately, we have not been able to repeat those results and are unclear of the relationship between the two assays.

Acridine Orange (AO) assay

The photomicrograph of red and green sperm on the cover of *Science* (Evenson *et al.*, 1980) prompted Tejada, Cameron, Norman, Marik and Friedman (1984) to pursue measuring by light microscopy human clinical DNA fragmentation of acid treated/AO stained sperm. The initial report was very optimistic that this technique could be used for prognosis of male infertility. However, other investigators who tried this technique would say that on a single slide one could find areas where all the sperm were green and nearby areas where all the sperm were red. This is caused by the fact that the surfaces of the glass slide and cover slips are not smooth at the molecular level. The metachromatic properties of AO depend on exact equilibrium conditions. At those sites where the cover slip and slide are smooth, the AO is squeezed out and all the sperm may be green; at those sites were there are micro dips in the glass surfaces, an excess of AO is present and forces DNA denaturation resulting in all red sperm. From some extensive unpublished studies in our lab, we do not find this AO method suitable for any clinical studies.

Halo sperm assay

The Halo sperm method for assessing sperm DNA damage was ingeniously developed by Fernandez, Muriel, Rivero, Goyanes, Vazquez and Alvarez (2003). In this methodology, sperm on a microscope slide are treated with 0.08 N HCl and lysing solutions containing 0.8 M dithiothreitol (DTT), 1% sodium dodecyl sulfate (SDS), and 2 M NaCl. Then, the slides were sequentially stained with DAPI (4',6-diamidino-2-phenylindole) and Diff-Quik reagent, and the percentages of sperm with nondispersed and dispersed chromatin loops were monitored by fluorescence and bright field microscopy, respectively. Sperm with intact DNA form chromatin loops. DNA strand breaks prevent this looping and the degree of DNA damage is proportional to the decreasing size of the loops ranging down to the nuclear core with no loops. Practically, several hundred sperm are scored by light microscopy as plus or minus having a decreased halo size and then five levels of sizes. A recent publication by Fernandez, Muriel, Goyanes, Segrelles, Gosalvez, Enciso, LaFromboise and De Jonge (2005) compared the equivalency of the "gold standard" SCSA with the newly developed Halosperm kit commercialised as the poor man's method to do the SCSA equivalent for clinical samples. This study reported a correlation of 0.85 between the %DFI and % halo positive sperm and the conclusion was made that the Halo test was "equivalent to the SCSA". The data showed a good correlation at the low DFI values and the very high DFI values. However, in the clinically critical range of 20-40% DFI, the points for the Halo sperm assay have little confidence; for example, a sample with a SCSA DFI of 30-35% could have a Halo result ranging between 20% and 50% (i.e. -10 to +15%). Moreover, given that the inter-observer error was up to 12%, the real biological variability could be either negligible or almost double what is reported in Fernandez *et al.* (2005). Current attempts are to quantitate the percent and the size of the halos with computer image analysis. It has been shown for those species that have a very low DFI and a concomitant low statistical threshold value for reduced pregnancy success (~8% for the pig) that the Halo sperm kit does not posses sufficient sensitivity for being clinically useful.

Sperm DNA fragmentation in a variety of animal species

The SCSA has been used to measure DNA/chromatin alterations in numerous species including bull, stallion, ram, boar, gorilla, monkey, rabbit, mouse, rat, hamster, whale, porpoise, fish, fowl, deer and others. Technically, sperm from all of these species present an interpretable cytogram and DFI values. The very large tail on rat sperm often clogs the flow cytometer sample channel and we sonicate the whole sperm to break up the tails prior to SCSA analysis. Each species presents a minor to major difference in the scattergram due to the artifact derived from using a flow cytometer with orthoganol axes as cited above. This problem can be overcome by using an epifluorescence flow cytometer, such as a Partec (Partec, Germany). Equivalent species comparisons have not been done with the other DNA fragmentation test. The many thousands of sperm measurements from multiple species lend a very strong credence to the biochemical and instrumental validity of the SCSA.

Bulls

Thousands of bull semen samples have been measured by the SCSA for the purpose of deriving correlations between DFI values and reproductive outcomes. Since bull studs store huge numbers of extended semen samples around the world, these samples are easy to obtain. Furthermore, detailed records typically exist for breeding outcomes on all batches of processed and shipped semen batches. Furthermore, since samples stored in liquid nitrogen (LN_2) for at least a decade, and perhaps many more, will have the same DNA fragmentation values over time, retrospective studies can be done. Our laboratory has used hundreds of aliquots of single bull semen samples over a number of years. The cytogram pattern remained an exact duplicate so that one can "recognise the bull" when the cytogram comes up on the oscilloscope screen. Likewise, we obtained repeat ejaculates from a single stallion over a ~9 year period. This highly fertile stallion always produced a near identical, very recognizable, cytogram pattern with nearly identical DFI values.

Implantation of a growth hormone, such as zeranol (a resorcylic acid lactone) has been shown to cause a significant increase in abnormal spermatogenesis and sperm DNA fragmentation (Ballachey, Miller, Jost and Evenson, 1986).

Boars

The SCSA was used to evaluate boar fertility as measured by farrowing rate (FR) and average pigs born (APB) born per litter. The %DFI and SD of DFI showed the following significant negative correlations *vs* FR and APB: DFI *vs* FR, $r = -0.55$, $p<0.01$; SD *vs* FR, $r = -0.67$, $p<0.002$; DFI *vs* APB, $r = -0.54$, $p<0.01$; SD *vs* APB, $r = -0.54$, $p<0.02$. Odds ratio for DFI and SD DFI were 1.5 times ($p=0.0003$, CI 1.21, 1.94), and 2.5 times ($p=0.001$, CI 1.87, 3.32), respectively (Didion, in preparation).

The SCSA correctly predicted the three high and three low fertility boars in a heterspermic trial (Evenson *et al.*, 1994). The effect on semen quality and sperm DNA fragmentation was investigated in four breeds of boars including Duroc, Hampshire, Landrace and Danish Large White (Fig. 5).

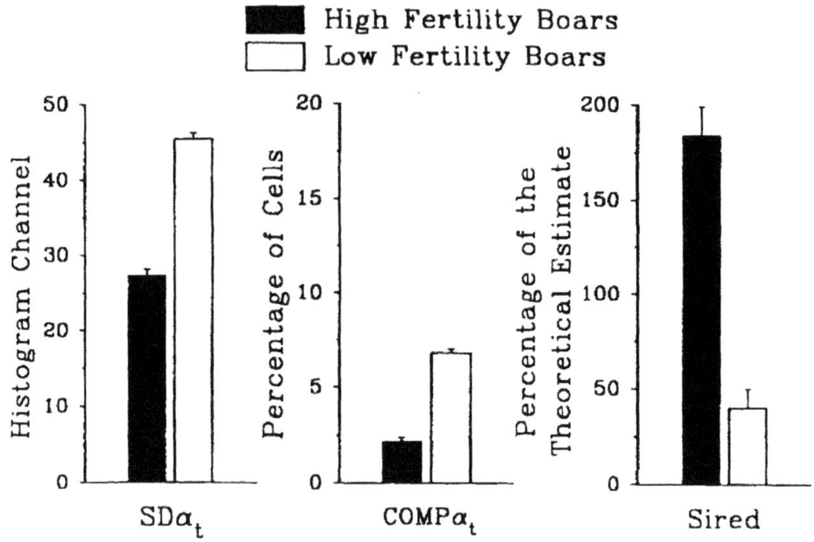

Figure 5. Relationship between sperm chromatin structure assay data and heterospermic fertiltiy ptoential expressed as a percentage of the theoretical estimate (each boar sired 1/3 of the piglets). SDα_t = standard deviation and Compα_t = percentage of cells outside the main population of the α_t distribution [Reprinted from Evenson, D.P., Thompson, L. and Jost, L. (1994) Flow cytometric evaluation of boar semen by the Sperm Chromatin Structure Assay as related to cryopreservation and fertility. *Theriogenology* **41**: 637-651, with permission from Elsevier Inc.].

Zoo and endangered species

As human activities encroach into wildlife habitat, there is an ever increasing number of animals that are placed on the endangered species list and those that are now gone from the wild and living only in zoos. Thus, breeding in captivity becomes ever more important and for those situations where it is not practical to move animals from one place of captivity to another, the use of frozen and shipped semen to another site to be used in an artificial insemination site is efficient. Obviously, for zoo breeding programmes, the best quality sperm is desired and one independent assay for such quality assurance is to conduct sperm DNA fragmentation measures.

Penfold, Jost, Evenson and Wildt (2003) examined the influence of teratospermia on sperm nuclear integrity in felids. Teratospermic ejaculates expressed more elevated levels of sperm DNA fragmentation than the normospermic counterparts both in unprocessed and swim-up processed samples ($p<0.05$). Fertilisation success *in vitro* was higher from normospermic compared with teratospermic inseminates ($p<0.05$). In contrast, similar proportions of oocytes were fertilised after ICSI using spermatozoa from normo- and teratospermic cats ($p<0.05$) (Fig. 6).

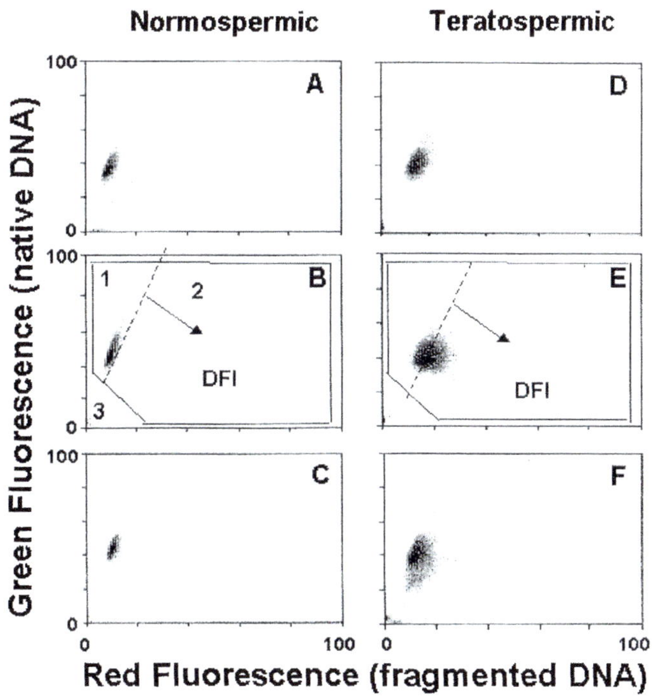

Figure 6. Native DNA versus fragmented DNA cytograms of selected normo- and teratospermic felid individuals. Each cytogram contains 5000 sperm, each represented by a dot. The location of each sperm represents the amount of native and fragmented DNA each sperm contains. Sperm falling further to the right are more heterogeneous and contain larger amounts of fragmented DNA. From Penfold *et al.* (2003), reprinted with permission.

Whales and porpoises

The SCSA pattern resembles that of other mammals. Figure 7 shows a good specimen of porpoise semen (Fig. 7A) as well as one of relatively poorer quality (Fig. 7B) and Figure 8 shows the pattern of a killer whale.

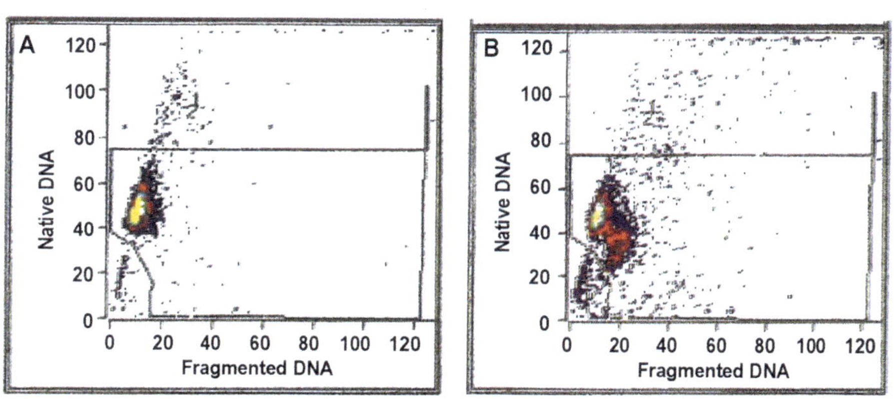

Figure 7. SCSA data from porpoises showing excellent (**A**) and poor (**B**) sperm DNA patterns.

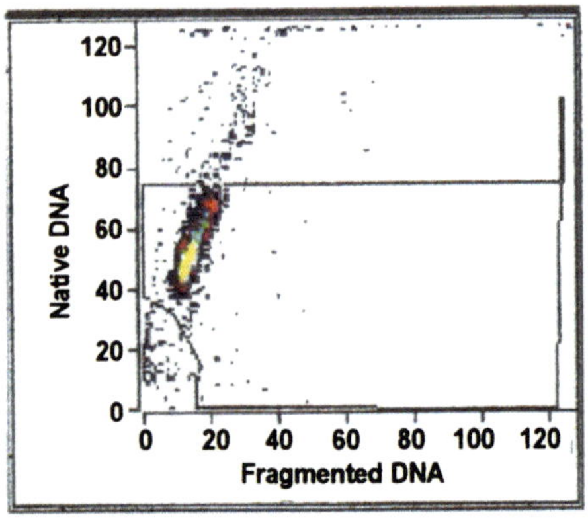

Figure 8. SCSA pattern from a killer whale.

Animal Toxicology

Many thousands of animals per year are used for examining potential toxicity to various body organs as a screening devise for human commercial products. Likewise, animals are used as sentinels in environmental studies to determine safe levels of potentially hazardous substances. Since the mammalian testis is characterised by a very high rate of cell proliferation and a unique and complex pattern of differentiation, examination of the ratio of testis cell types as well as sperm for DNA fragmentation/protein alterations, makes the flow cytometer an excellent tool for rapid determination of toxic response. For example, the testis differentiation pattern of pubertal mice exposed during gestation and post partum to genetically modified corn and soybean diets showed that no effect was detectable over several generations (Brake, Thaler and Evenson, 2004; Brake and Evenson, 2004).

Methyl methane sulfonate is known to alkylate the free -SH groups in epididymal sperm that leads to DNA fragmentation. Intraperitoneal injections of MMS into mice (Evenson *et al.*, 1993) lead to dramatic SCSA defined DNA fragmentation by 3 days post injection (dpi). However, females bred to these mice treated the same way did not show dominant lethal mutations until 5-8 dpi of the male rat. Thus, we likely observed the damage at the molecular level of DNA strand breaks accumulating until gross chromosome alterations occurred leading to embryo death.

The SCSA appears to be the most sensitive assay for the detection of DNA fragmentation in mouse epididymal sperm obtained from mice whose testes were exposed to X radiation from 5 to 400 rads 40 days earlier. The lowest level of detection was at 12.5 rads with a dose response increase (Sailer *et al.*, 1995) (Fig. 9).

Human toxicology

Insecticides

Polychlorinated biphenyls (PCBs) and the insecticide dichlorodiphenyltrichloroethane (DDT) and its major metabolite, dichlorodiphenyldichloroethylene (p,p′-DDE) are a concern in the environment due to their resistance to degradation and their ability to bioaccumulate with negative effects on male reproduction. Rignell-Hydborn and colleagues found a significantly lower %DFI in the lowest CB-153 quintile compared with the other quintiles (Rignell-Hydbom,

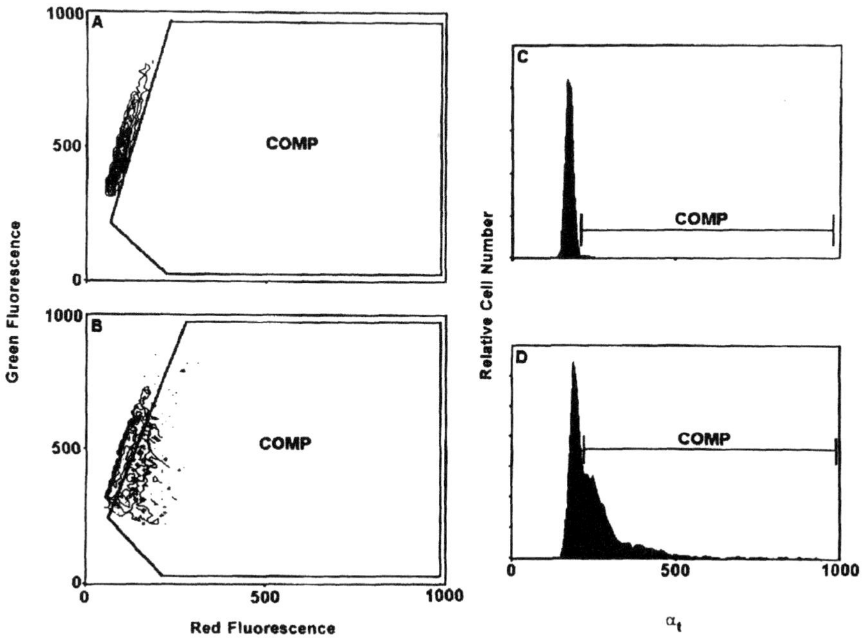

Figure 9. Sample shown in (**A**) is from a control mouse, while (**B**) is from a mouse which received 250 rads of radiation exposure. (**C,D**) are computer generated histograms corresponding to samples presented in (**A**) and (**B**). From Sailer *et al.* (1995), reprinted with permission.

Rylander, Giwercman, Jonsson, Lindh, Eleuteri, Rescia, Leter, Cordelli, Spano and Hagmar, 2005). In a related study, Spano, Toft, Hagmar, Eleuteri, Rescia, Rignell-Hydbom, Tyrkiel, Zvyezday and Bonde (2005) found a strong relationship between increasing DNA fragmentation and increasing serum levels of CB-153 among European men, reaching a 60% higher average level in the highest exposure group.

Men exposed to insecticides such as carbaryl and pesticides such as Fenvalerate, showed significantly increased levels of sperm DNA fragmentation (Xia, Cheng, Bian, Xu, Collins, Chang, Song, Liu, Wang and Wang, 2005). A dramatic effect of organophosphorous was reported by Sanchez-Pena, Reyes, Lopez-Carrillo, Recio, Moran-Martinez, Cebrian and Quintanilla-Vega (2004) where nearly three-fourths of these operators had DFI values >30%. Significantly higher levels of sperm DNA fragmentation were found in factory workers exposed to styrene which is used to make plastics, rubber, and resins (Spano, Kolstad, Larsen, Cordelli, Leter, Giwercman and Bonde, 1998).

Air pollution

Residents of Teplice, Czech Republic, a town with a history of heavy winter time air pollution generated by burning soft brown coal experienced a higher than normal rate of infertility and spontaneous miscarriages (Rubes, Selevan, Evenson, Zudova, Vozdova, Zudova, Robbins and Perreault, 2005). Czech army conscripts, 18 years of age, provided semen samples for a cross sectional study and then a two year longitudinal study that went through periods of clean and polluted air. Sperm DNA fragmentation as measured by the SCSA was the only measure to detect a correlation between air pollution and semen quality in 18-year old army conscripts. One-fourth of these young men had %DFI >30, placing them in a statistical group known to be at an increased risk for infertility. This is the first report on sperm DNA fragmentation caused by air pollution.

Medication

We have reported a case study where a man had a semen analysis as well as an SCSA evaluation for 8 consecutive months. The SCSA analysis for seven of the eight months showed sperm DNA fragmentation consistent with excellent fertility. After a back injury, the man was given cortisone injections and the following month the SCSA analysis showed very poor sperm DNA integrity (Evenson, Jost, Corzett and Balhorn, 2000).

Fever

In one case study (Evenson *et al.*, 2000) a man had a one-day influenza induced fever of 104°F and then recovered quickly. The first sample obtained at 18 days after the fever had an approximate 50% DFI and then that factor improved. However, at day 33 there was a high percent of HDS sperm. These sperm were isolated and the proteins analysed by gel electrophoresis. An unknown protein was seen on the gel that was isolated and partially sequenced and determined to be the precursor to protamine 2. Thus, the fever apparently disrupted spermiogensis in a way that prevented the cleavage of protamine 2 that replaced the histones but the chromatin could not come to full condensation thereby providing more DNA sites to be stained.

Age of man

Age of man is highly correlated with SCSA defined %DFI (Wyrobek, Eskenazi, Young, Arnheim, Tiemann-Boege, Jabs, Glaser, Pearson and Evenson, 2006). In this study, ~15 men per decade of life were enrolled and characterised by no known infertility issues, non-smokers, similar socio-economic standing and had worked or were working at the Lawrence Livermore National Laboratory. SCSA data with regard to both %DFI and SD DFI were significantly correlated with age. Men in their early 20's had an average of ~5% DFI and this value rose 5 % per year. The DFI values were more heterogeneous with age than a woman's loss of fertility. However, regression analysis showed that a man at age 50 has a one-third chance of being above the 30% DFI threshold that places him into the statistical chance of risk, especially by natural pregnancy and IUI fertilisation. Thus, even though a man fathered children in his 20's, age alone, on the average, will increase the risk for infertility. This observation is not surprising. Reactive oxygen species (ROS) damage is an age related phenomenon and the testis is not an exception. It is of interest that a small study (Rubes *et al.*, in preparation) measuring SCSA defined %DFI showed that men who were positive for the gene glutathione S transferase had on the average a lower %DFI than men who were negative.

SCSA and human fertility

Human fertility in vivo

To date, there are two published studies regarding sperm DNA fragmentation and pregnancy outcome using *in vivo* fertilisation. The Georgetown Male Factor Infertility Study established the SCSA statistical ranges of excellent (<15% DFI), good (>15 to <30% DFI), and fair to poor (>30% DFI) sperm DNA integrity (Evenson *et al.*, 1999). This study included 144 couples trying to conceive naturally. Odds ratios from this study indicated that couples were 6.5 times more likely to become pregnant when their DFI was <30%. Spano *et al.* (2000), using the SCSA analysis, found that *in vivo* patients (n=215) (with no previous knowledge of their fertility status) showed a declining pregnancy outcome when their DFI was >20% and negligible pregnancies when the DFI approached 40%. Patients were 10 times more likely to achieve a pregnancy when DFI was >40%.

IUI fertilisation

In a recent study (Bungum, Humaidan, Spano, Jepson, Bungum and Giwercman, 2004), the relationship between SCSA data and outcomes of IUI, IVF and ICSI fertilisation were investigated. Two SCSA parameters, DFI and high HDS, gave a high predictive value regarding the outcome of IUI. HDS sperm are known (Tritle and Evenson, unpublished data) to have immature chromatin and other characteristics of immature spermatozoa. IUI patients were grouped by <27% DFI and HDS <10% (group 1) and >27% DFI or HDS >10% (group 2). Using odds ratios, group 1 with DFI <27% were 8.7 times more likely to deliver a baby than group 2 with DFI >27%. In a later study by (Bungum, 2005, personal communication), similar results were found. Patients undergoing IUI were 7.0 times more likely to become pregnant when their DFI was <27% (n=387, p=0.0001).

Routine IVF

When SCSA defined levels of sperm DNA fragmentation were <30%, routine IVF patients were twice as likely to carry a pregnancy >12 weeks (Virro *et al.*, 2004). In this study, high levels of DNA fragmentation were clearly shown to be predictive for reduced pregnancy in routine IVF. (Bungum *et al.*, 2004) found IVF patients (n=109) in the SCSA defined <27% DFI group were 1.5 times more likely to deliver as the >27% DFI group.

ICSI

In ICSI patients (n=66) the SCSA defined <27% DFI group were 2.5 times more likely to deliver a baby as the >27% DFI group (Bungum *et al.*, 2004). Comparison of ICSI and routine IVF in the >27% group indicated that the ICSI patients were 3.6 times more likely to deliver than the routine IVF group. These results were significant using the chi-square test. Benchaib, Braun, Lornage, Hadj, Salle, Lejeune and Guerin (2003) found sperm DNA fragmentation was significantly lower when a pregnancy was obtained using ICSI (p<0.05, n=54).

The relationship between sperm DNA damage and pregnancy rate was investigated in 52 couples undergoing routine IVF and ICSI procedures (Chohan, Fiffin, LaFramboise, DeJonge and Carrell, 2006). The authors concluded that while there was no significant relationship with IVF outcome, the negative trend for increasing DFI and pregnancy outcome may become significant with a larger study population. Odds ratio calculations showed that a patient was 2 times more likely to become pregnant when the DFI was <30% (not statistically significant). Check, Graziano, Cohen, Krotec and Check (2005) found pregnancy rates in the >30% DFI group to be about half of the <30% group though the differences were not significant. Zini, Blumenfeld, Libman and Willis (2005) also found a trend for spontaneous abortions in couples with DFI >30%.

Tunel and human fertility

SCSA data on human sperm have been shown to be closely related to Tunel data ($r=0.859$, $p<0.001$) (Gorzczyca *et al.*, 1993), strongly suggesting that the SCSA technique adds acridine orange fluorochrome markers at the sites of DNA strand breaks which are also the sites of the enzymatic addition of a fluorochrome with the Tunel assay. Studies on animal spermatozoa have shown similar correlations between SCSA and Tunel data for bulls ($r=0.78$, $p<0.001$), stallions ($r=0.65$, $p<0.001$) and rams ($r=0.84$, $p<0.001$) (Sailer *et al.*, 1995). At present, however, the SCSA is a more reliable and repeatable technique for precise repeatable determination of DNA integrity. This is in agreement with Erenpreiss, Jepson, Giwercman, Tsarev and Erenpreiss (2004), who showed significant correlations between SCSA and Tunel data ($r=063$, $p=0.005$). SCSA and Tunel data were positively related ($p<0.05$) when homozygous ß-thalassaemia patients with iron overload were evaluated for sperm DNA damage (Perera, Pizzey, Campbell, Katz, Porter, Petrou, Irvine and Chatterje, 2002). Although Tunel and SCSA tests likely measure the same sites of sperm DNA damage, the threshold for Tunel from various laboratories appear to differ significantly. Duran *et al.* (2002) found no pregnancies when the Tunel threshold was >12%, while Benchaib *et al.* (2003) found a significant decrease in pregnancy rate at a >20% threshold. While thresholds may differ between laboratories, there may be consensus within the same laboratory as shown by (Henkel, Kierspel, Hajimohammad, Stalf, Hoogendijk, Mehnert, Menkveld, Schill and Kruger, 2003, 2004), where a significant decrease in pregnancies was found when the Tunel threshold was >36.5%.

IUI fertilisation

Duran *et al.* (2002) found no pregnancies were initiated when sperm DNA fragmentation was >12% by Tunel analysis in 119 IUI patients.

Routine IVF

Henkel *et al.* (2004) demonstrated that DNA fragmentation, as determined by the Tunel assay, is significantly predictive for pregnancy rate in IVF. Patients with a lower percentage of DNA fragmentation (Tunel) had an almost 2x pregnancy rate (34.65 *vs* 19.05% $p=0.0344$), compared with patients with higher levels of DNA fragmentation. In this study, the ROC established Tunel threshold was set at >36.5%.

Comet and human fertility

Using Comet analysis, Morris, Ilott, Dixon and Brison (2002) found that elevated sperm DNA damage was predictive of embryo development failure in a group of IVF/ICSI men. Three of the nine pregnancies (33%) in the elevated sperm DNA damaged group resulted in an early miscarriage, while all six of the pregnancies in the low sperm DNA damaged group yielded live births.

Recurrent pregnancy loss

Recurrent pregnancy loss is a devastating issue for infertility patients. Current research indicates a trend towards increased spontaneous abortions when the DFI is >30%. In the Georgetown study, where human SCSA statistical thresholds were first derived, combined DFI and HDS scores predicted 39% of miscarriages (Evenson, Jost, Marshall, Zinaman, Clegg, Purvis, de Angelis and Claussen, 1999; Zinaman, Brown, Selevan and Clegg, 2000). Check *et al.* (2005) found slightly lower clinical pregnancies and higher miscarriage rates in the >30% DFI patient group resulting in an ongoing pregnancy rate that was almost half of the <15 and 15-30% groups though the differences were not significant. (Zini *et al.*, 2005), also found a trend for spontaneous abortions in couples with DFI >30%. A >30% DFI score was associated with a higher rate of spontaneous abortion at 12 weeks of gestation (p<0.01) in comparison to the <30% group (Virro *et al.*, 2004). Men of couples with repetitive (>2) spontaneous abortions had a 4-fold increase in Tunel positive spermatozoa relative to sperm donors (Carrell, Liu, Peterson, Jones, Hatasaka, Erickson and Campbell, 2003).

Aetiology of sperm DNA fragmentation

Sperm DNA damage is multifactorial and may be due to many environmental conditions such as chemotherapy, radiation, some prescription medications, air pollution, smoking, pesticides, chemicals, heat, assisted reproduction preparation protocols (Evenson *et al.*, 1980; Evenson *et al.*, 1985; Evenson *et al.*, 1989; Evenson, Jost, Baer, Turner and Schrader 1991; Evenson *et al.*, 1993; Evenson and Jost, 1993; Sailer *et al.*, 1995; Fossa, DeAngelis, Kraggerud, Evenson, Theodorsen and Clausen, 1997; Potts, Newbury, Smith, Notarianni and Jefferies, 1999; Rignell-Hydbom *et al.*, 2005; Rubes *et al.*, 2005; Adams, personal communication) and various pathological conditions including cryptorchidism, cancer, fever, age, infection, leukocytospermia and varicocele among others (Evenson, Jost, Corzett and Balhorn, 2000; Zini *et al.*, 2001; Alvarez, Sharma, Ollero, Saleh, Lopez, Thomas, Evenson and Agarwal, 2002; Aitken, Baker and Sawyer, 2003; Burrello, Calogero, Perdichizzi, Salmeri, D'Agata and Vicari, 2004; Stahl, Eberhard, Jepson, Spano, Cwikiel, Cavallin-Stahl and Giwercman, 2004). It would appear that elevated sperm fragmentation, damage due to reactive oxygen species and apoptotic events may all play a part in the pathogenesis of male infertility (Aitken *et al.*, 2003). Protamine-deficient spermatozoa may also contribute to elevated sperm DNA fragmentation (Nasr-Esfahani, Razavi, Anjomshoa, Rozbahani, Moulavi and Mardani, 2005). Several studies have shown a correlation between elevated protamine-1/protamine-2 (P1/P2) ratio and male infertility (de Yebra, Ballesca, Vanrell, Bassas and Oliva, 1993; Carrell and Liu, 2001). Aoki, Moskovtsev, Willis, Liu, Mullen and Carrell (2005) found the correlation between elevated levels of sperm DNA fragmentation and low P1/P2 levels significant, but biologically weak. Decreased P1/P2 ratios were found in a population of infertile males undergoing IVF and/or ICSI. The overall fertilisation rate (IVF and ICSI) was significantly decreased in patients with a low P1/P2 ratio compared with patients with normal and elevated P1/P2 ratios. Fertilisation has shown weakly significant to non-significant correlations with elevated sperm DNA fragmentation (Henkel *et al.*, 2004; Virro *et al.*, 2004). Reports have been conflicting regarding the correlations between SCSA DFI values and standard semen analysis measures. Results from several studies show r-values between DFI and classical semen measurements ranging from -0.42 to 0.12 for sperm density and -0.60 to 0.25 for morphology and for motility from -0.59 to 0.18. While many of the DFI values and semen analysis measures are statistically significant, the correlations may have little biological meaning. The great variability of the r values is probably due to variations in the measurement of the standard semen analysis. Furthermore, DFI showed no significant relationships to fertilisation (Henkel *et al.*, 2003, 2004; Virro *et al.*, 2004), implantation rates (Bungum *et al.*, 2004), and embryo quality (Benchaib *et al.*, 2003). Since the male genome is not activated until day three, embryo quality at the time of transfer may not show future negative male genome effects after fertilisation.

Conclusions

The development of the SCSA came about as a confluence of three factors in a single laboratory at the Memorial Sloan Kettering Cancer Center in New York: (1) invention of the flow cytometer and its interfacing to a computer (L. Kametsky and M. Melamed), (2) expertise of AO biochemistry (Z. Darzynkiewicz) and, (3) an NIH grant on human sperm chromatin structure (D. Evenson). When the % DFI threshold for risk of infertility is low (e.g., ~8% for pigs) or a very narrow range around a threshold exists, as for SCSA-defined IUI data, only flow cytometry measures of sperm stained with highly exacting conditions for chromatin integrity can be useful for clinical diagnosis and prognosis. The Tunel assay has promise, but it may not match the precision of using a small molecule like AO for marking DNA fragmentation rather than large enzyme molecules in heterogeneous sperm nuclear chromatin. The Halo sperm kit does not appear at this time to have the necessary precision for human clinical use.

At this time, the aetiology and mechanisms of sperm DNA fragmentation are not clear. Numerous studies have pursued the thought that apoptosis is the prime candidate. However, other studies put that idea into question due to the lack of known apoptotic markers not appearing concomitantly with fragmented DNA. Consensus at this time strongly favours oxidative stress as the primary factor causing sperm DNA fragmentation (Alvarez, 2003; Greco, Romano, Iacobelli, Ferrero, Baroni, Minasi, Ubaldi, Rienzi and Tesarik, 2005; Aitken, 2006). Sperm DNA fragmentation is a clinical test whose time has come. While andrologists are rapidly realising this, the infertile couple often is referred first to the gynecologist. This group of physicians must become more aware of the fact that nearly half of the couple infertility problem is due to the male and that 20 to 30% of the couples coming to an infertility clinic have a male factor of high sperm DNA fragmentation. The male partner should be examined by a urologist for factors such as a variocoele, excess heat to the testis, hormone balance, life style factors and antioxidant therapy. All of these factors need to be explored to potentially bring down a high level of DNA fragmentation prior to sending the couple for ICSI fertilisation.

The SCSA is a powerful tool which can be used for statistical predictions on negative pregnancy outcome with *in vivo*, IUI and routine IVF fertilisation. As stated in a current review by Erenpreiss, Spano, Erenpreisa, Bungum and Giwercman (2006) "At the moment SCSA is the only method which has demonstrated clear clinically useful cut-off levels between fertile and infertile men and its prognostic value for ART has also been shown. The undisputed advantages of this technique are its robustness and small intra- and inter-assay variation".

Acknowledgements

Research supported by Grant EPA R827019 and the South Dakota State University Experiment Station.

References

Aitken, R.J. (2006). Sperm function tests and fertility. *International Journal of Andrology* **29**: 69-75

Aitken, R.J., Baker, M.A., Sawyer, D., Aitken, R.J., Baker, M.A. and Sawyer, D. (2003). Oxidative stress in the male germ line and its role in the aetiology of male infertility and genetic disease. *Reproductive Biomedicine Online* **7**: 65-70

Alvarez, J.G., (2003). DNA fragmentation in human spermatozoa: significance in the diagnosis and treatment of infertility. *Minerva Ginecologica* **3**: 233-239

Alvarez, J.G., Sharma, R.K., Ollero, M., Saleh, R.A., Lopez, M.C., Thomas, A.J. Jr, Evenson, D.P. and Agarwal, A. (2002). Increased DNA damage in sperm from leukocytospermic semen samples as determined by the sperm chromatin structure assay. *Fertility and Sterility* **78**: 319-329

Aoki, V.W., Moskovtsev, S.I., Willis, J., Liu, L., Mullen,

J.B. and Carrell, D.T. (2005). DNA integrity is compromised in protamine-deficient human sperm. *Journal of Andrology* **26**: 741-748

Aravindan, G.R., Bjordahl, J., Jost, J.K. and Evenson, D.P. (1997). Susceptibility of human sperm to *in situ* DNA denaturation is strongly correlated with DNA strand breaks identified by single-cell electrophoresis. *Experimental Cell Research*. **236**: 231-237

Ballachey, B.E., Hohenboken, W.D. and Evenson, D.P. (1987). Heterogeneity of sperm nuclear chromatin structure and its relationship to bull fertility. *Biology of Reproduction* **36**: 915-925

Ballachey, B.E., Evenson, D.P. and Saacke, R.G. (1988). The sperm chromatin structure assay relationship with alternate tests of semen quality and heterospermic performance of bulls. *Journal of Andrology* **9**: 109-115

Ballachey, B.E., Miller, H.L., Jost, L.K., and Evenson, D.P. (1986). Flow cytometry evaluation of testicular and sperm cells obtained from bulls implanted with zeranol. *Journal of Animal Science* **63**: 995-1004

Benchaib, M., Braun, V., Lornage, J., Hadj, S., Salle, B., Lejeune, H. and Guerin, J.F. (2003). Sperm DNA fragmentation decreases the pregnancy rate in an assisted reproductive technique. *Human Reproduction* **18**: 1023-1028

Brake, D.G. and Evenson, D.P. (2004). A generational study of glyphosate-tolerant soybeans on mouse fetal, postnatal, pubertal and adult testicular development. *Food and Chemical Toxicology* **42**: 29-36

Brake, D.G., Thaler, R. and Evenson, D.P. (2004). Evaluation of Bt (*Bacillus thuringiensis*) corn on mouse testicular development by dual parameter flow cytometry. *Journal of Agricultural Food Chemistry* **52**: 2097-2102

Bungum, M., Humaidan, P., Spano, M., Jepson, K., Bungum, L. and Giwercman A. (2004). The predictive value of sperm chromatin structure assay (SCSA®) parameters for the outcome of intrauterine insemination, IVF and ICSI. *Human Reproduction* **19**: 1401-1408

Burrello, N., Calogero, A.E., Perdichizzi, A., Salmeri, M., D'Agata, R. and Vicari E. (2004). Inhibition of oocyte fertilization by assisted reproductive techniques and increased sperm DNA fragmentation in the presence of *Candida albicans*: a case report. *Reproductive Biomedicine Online* **8**: 569-573

Carrell, D.T. and Liu, L. (2001) Altered protamine 2 expression is uncommon in donors of known fertility, but common among men with poor fertilizing capacity, and may reflect other abnormalities of spermiogenesis. *Journal of Andrology* **22**: 604-610

Carrell, D.T., Liu, L., Peterson, C.M., Jones, K.P., Hatasaka, H.H., Erickson, L. and Campbell, B. (2003) Sperm DNA fragmentation is increased in couples with unexplained recurrent pregnancy loss. *Archives of Andrology* **49**: 49-55

Check J.H., Graziano V., Cohen R., Krotec J. and Check M.L. (2005) Effect of an abnormal sperm chromatin structural assay (SCSA®) on pregnancy outcome following (IVF) with ICSI in previous IVF failures. *Archives of Andrology* **51**: 121-124

Chohan, K.R., Griffin, J.T., Lafromboise, M., De Jonge, C.J. and Carrell, D.T. (2006). Comparison of chromatin assays for DNA fragmentation evaluation in human sperm. *Journal of Andrology* **27**: 53-59

de Yebra L., Ballesca J.L., Vanrell J.A., Bassas L. and Oliva R. (1993). Complete selective absence of protamine P2 in humans. *Journal of Biological Chemistry* **15**: 10553-10557

Duran E.H., Morshedi M., Taylor S. and Oehninger S. (2002). Sperm DNA quality predicts intrauterine insemination outcome: a prospective cohort study. *Human Reproduction* **17**: 3122-3128

Erenpreiss J., Jepson K., Giwercman A., Tsarev I., Erenpreisa J. and Spano M. (2004). Toluidine blue cytometry test for sperm DNA conformation: comparison with the flow cytometric sperm chromatin structure and TUNEL assays. *Human Reproduction* **19**: 2277-2282

Erenpreiss J., Spano M., Erenpreisa J., Bungum M. and Giwercman A. (2006). Sperm chromatin structure and male fertility: biological and clinical aspects. *Asian Journal of Andrology* **8**: 11-29

Evenson, D.P., Baer, R.K. and Jost L.K. (1989). Long-term effects of triethylenemelamine exposure on mouse testis cells and sperm chromatin structure assayed by flow cytometry. *Environmental and Molecular Mutagenesis* **14**: 79-89

Evenson, D.P., Baer, R.K., Jost, L.K. and Gesch, R.W. (1986). Toxicity of thiotepa on mouse spermatogenesis as determined by dual-parameter flow cytometry. *Toxicology and Applied Pharmacology* **82**: 151-163

Evenson, D.P., Darzynkiewicz, Z. and Melamed, M.R. (1980). Relation of mammalian sperm chromatin heterogeneity to fertility. *Science* **240**: 1131-1133

Evenson, D.P., Emerick, R.J., Jost, L.K. Kayongo-Male, H. and Stewart, S.R. (1993). Zinc-silicon interactions Influencing sperm chromatin integrity and testicular cell development in the rat as measured by flow cytometry. *Journal of Animal Science* **71**: 955-962

Evenson, D.P., Jost, L.K., Corzett, M. and Balhorn, R. (2000). Characteristics of human sperm chromatin structure following an episode of influenza and high fever: a case study. *Journal of Andrology* **21**: 739-746

Evenson, D.P., Higgins, P.J., Grueneberg, D. and Ballachey, B.E .(1985). Flow cytometric analysis of mouse spermatogenic function following exposure to ethylnitrosourea. *Cytometry* **6**: 238-253

Evenson, D.P. and Jost, L.K. (1993). Hydroxyurea exposure alters mouse testicular kinetics and sperm chromatin structure. *Cell Proliferation* **26**: 147-159

Evenson, D.P., Jost, L.K. and Baer, R.K. (1993). Effects of methyl methanesulfonate on mouse sperm chromatin structure and testicular cell kinetics. *Environmental and Molecular Mutagenesis* **21**: 144-153

Evenson, D.P., Jost, L.K., Baer, R.K., Turner, T.W. and Schrader, S.M. (1991) Individuality of DNA denaturation patterns in human sperm as measured by the sperm chromatin structure assay. *Reproductive Toxicology* **5**: 115-125

Evenson, D.P., Jost, L.K., Corzett, M. and Balhorn R. (2000). Characteristics of human sperm chromatin structure following an episode of influenza and high fever: a case study. *Journal of Andrology* **21**: 739-746

Evenson, D.P., Jost, L.K., Zinaman, M.J., Clegg, E., Purvis K., de Angelis P. and Claussen O.P. (1999). Utility of the sperm chromatin structure assay (SCSA®) as a diagnostic and prognostic tool in the human fertility clinic. *Human Reproduction* **14**: 1039-1049

Evenson, D.P., Larson, K.L. and Jost, L.K. (2002). Sperm chromatin structure assay: its clinical use for detecting sperm DNA fragmentation in male infertility and comparison with other techniques. *American Society of Andrology* **23**: 25-43

Evenson, D.P., Thompson, L. and Jost, L. (1994). Flow cytometric evaluation of boar semen by the sperm chromatin structure assay as related to cyropreservation and fertility. *Theriogenology* **41**: 637-651

Evenson, D.P. and Wixon, R.L. (2006) Meta-analysis of sperm DNA fragmentation using the sperm chromatin structure assay. *Reproductive BioMedicine Online*. **12**: 466-472

Fernandez, J.L., Muriel, L., Rivero, M.T., Goyanes, V., Vazquez, R and Alvarez, J.G. (2003). The sperm chromatin dispersion test: a simple method for the determination of sperm DNA fragmentation. *Journal of Andrology* **24**: 59-66

Fernandez, J.L., Muriel, L., Goyanes, V., Segrelles, E., Gosalvez, J., Enciso, M., LaFromboise, M. and De Jonge, C. (2005). Halosperm is an easy, available, and cost-effective alternative for determining sperm DNA fragmentation. *Fertility and Sterility* **84**: 860

Fossa, S.D., De Angelis, P., Kraggerud, S.M., Evenson, D., Theodorsen, L. and Clausen O.P. (1997). Prediction of posttreatment spermatogenesis in patients with testicular cancer by flow cytometric sperm chromatin structure assay. *Cytometry* Aug **15**: 192-196

Gorczyca, W., Melamed, M.R. and Darzynkiewicz, Z. (1993). Apoptosis of S-phase HL-60 cells induced by DNA topoisomerase inhibitors: detection of DNA strand breaks by flow cytometry using the in situ nick translation assay. *Toxicology Letters* **67**: 249-258

Greco, E., Romano, S., Iacobelli, M., Ferrero, S., Baroni, E., Minasi, M.G., Ubaldi, F., Rienzi, L. and Tesarik, J. (2005). ICSI in cases of sperm DNA damage: beneficial effect of oral antioxidant treatment. *Human Reproduction* **20**: 2590-2594

Henkel, R., Hajimohammad, M., Stalf, T., Hoogendijk, C., Mehnert, C., Menkveld, R., Gips, H., Schill, W.B. and Kruger, T.F. (2004). Influence of deoxyribonucleic acid damage on fertilization and pregnancy. *Fertility and Sterility* **81**: 965-972

Henkel, R., Kierspel, E., Hajimohammad, M., Stalf, T., Hoogendijk, C., Mehnert, C., Menkveld, R., Schill, W.B. and Kruger, T.F.(2003). DNA fragmentation of spermatozoa and assisted reproduction technology. *Reproductive Biomedicine Online* **7**: 477-484

Jeffay, S.C., Strader, L.F., Buus, R.M., Evenson, D.P.. Olshan, A.F., Herring, A.H., Bradley, L.E., Smith, J.C. and Perreault S.D. (2006). Relationships among semen endpoints used as indicators of sperm nuclear integrity. American Society of Andrology 2005, Chicago

Karabinus, D.S., Vogler, C.J., Saacke, R.G., Evenson, D.P. (1997). Chromatin structural changes in sperm after scrotal insulation of Holstein bulls. *Journal of Andrology* **18**: 549-555

Larson, K.L., DeJonge, C.J., Barnes, A.M., Jost, K.L. and Evenson, D.P. (2000). Sperm chromatin structure assay parameters as predictors of failed pregnancy following assisted reproductive techniques. *Human Reproduction* **15**: 1717-1722

Morris, I.D., Ilott, S., Dixon, L. and Brison, D.R. (2002). The spectrum of DNA damage in human sperm assessed by single cell gel electrophoresis (Comet assay) and its relationship to fertilization and embryo development. *Human Reproduction* **17**: 990-998

Nasr-Esfahani, M.H., Razavi, S., Mozdarani, H., Mardani, M. and Azvagi, H. (2004). Relationship between protamine deficiency with fertilization rate and incidence of sperm premature chromosomal condensation post-ICSI. *Andrologia*. **36**: 95-100

Penfold, L.M., Jost, L., Evenson, D.P. and Wildt, D.E. (2003). Normospermic versus teratospermic domestic cat sperm chromatin integrity Eevaluated by flow cytometry and intracytoplasmic sperm injection. *Biology of Reproduction* **69**: 1730-1735

Perera, D., Pizzey, A., Campbell, A., Katz, M., Porter, J., Petrou, M., Irvine, D.S. and Chatterjee, R.(2002). Sperm DNA damage in potentially fertile homozygous beta-thalassaemia patients with iron overload. *Human Reproduction* **17**: 1820-1825

Perreault, S.D., Aitken, R.J., Baker, H.W., Evenson, D.P., Huszar, G., Irvine, D.S., Morris, I.D., Morris, R.A., Robbins, W.A., Sakkas, D., Spano, M. and Wyrobek, A.J. (2003). Integrating new tests of sperm genetic integrity into semen analysis: breakout group discussion. *Advances in Experimental Medicine and Biology* **518**: 253-268

Pinkel, D. (1984). Cytometric analysis of mammalian sperm for induced morphologic and DNA content errors. In: *Biological Dosimetry: Cytometric Approaches to Mammalian Systems*. Edited by Eisert, W.G. and Mendelsohn, M.L., Springer-Verlag, Berlin, pp. 111-126

Potts, R.J., Newbury, C.J., Smith, G., Notarianni, L.J. and Jefferies, T.M. (1999). Sperm chromatin damage associated with male smoking. *Mutation Research* **423**: 103-111

Rignell-Hydbom, A., Rylander, L., Giwercman, A., Jonsson, B.A., Lindh, C., Eleuteri, P., Rescia, M., Leter, G., Cordelli, E., Spano, M. and Hagmar, L. (2005). Exposure to PCBs and p,p'-DDE and human sperm chromatin integrity. *Environmental Health Perspectives* **113**: 175-179

Rubes, J., Selevan, S.G., Evenson, D.P., Zudova, D., Vozdova, M., Zudova, Z., Robbins, W.A. and

Perreault, S.D. (2005). Episodic air pollution is associated with increased DNA fragmentation in human sperm without other changes in semen quality. *Human Reproduction* **20**: 2776-2783

Sailer, B.L., Jost, L.K., Erickson, K.R., Tajiran and Evenson, D.P. (1995). Effects of X-irradiation on mouse testicular cells and sperm chromatin structure. *Environmental and Molecular Mutagenesis* **25**: 23-30

Sailer, B.L., Jost, L.K. and Evenson, D.P. (1995). Mammalian sperm DNA susceptibility to *in situ* denaturation associated with the presence of DNA strand breaks as measured by the terminal deoxynucleotidyl transferase assay (TDTA). *Journal of Andrology* **16**: 80-87

Sanchez-Pena, L.C., Reyes, B.E., Lopez-Carrillo, L., Recio, R., Moran-Martinez, J., Cebrian, M.E. and Quintanilla-Vega B. (2004). Organophosphorous pesticide exposure alters sperm chromatin structure in Mexican agricultural workers. *Toxicology and Applied Pharmacology* **196**: 108–113

Singh, N.P., Danner, D.B., Tice, R.R., McCoy, M.T., Collins, G.D. and Schneider, E.L. (1989). Abundant alkali-sensitive sites in DNA of human and mouse sperm. *Experimental Cell Research* **184**: 461-470

Spano, M., Bonde, J., Hjollund, H.I., Kolstatd, H.A., Cordelli, E. and Leter, G. (2000). Sperm chromatin damage impairs human fertility. *Fertility and Sterility* **73**: 43-50

Spano, M., Kolstad, A.H., Larsen, S.B., Cordelli, E., Leter, G., Giwercman, A. and Bonde, J.P. (1998). The applicability of the flow cytometric sperm chromatin structure assay in epidemiological studies. Asclepios *Human Reproduction* **13**: 2495-2505

Spano, M., Toft, G., Hagmar, L., Eleuteri, P., Rescia, M., Rignell-Hydbom, A., Tyrkiel, E., Zvyezday, V. and Bonde, J.P.; INUENDO. (2005). Exposure to PCB and p, p'-DDE in European and Inuit populations: impact on human sperm chromatin integrity. *Human Reproduction* **20**: 3488-3499

Stahl, O., Eberhard, J.. Jepson, K., Spano, M., Cwikeiel, M., Cavallin-Stahl, E. and Giewercman, A. (2004). The impact of testicular carcinoma and its treatment on sperm DNA integrity. *Cancer* **100**: 1137-1144

Tejada, R.I., Cameron, J., Norman, A., Marik, J. and Friedman, S. (1984) A test for the practical evaluation of male fertility by acridine orange (AO) fluorescence. *Fertility and Sterility* **42**: 87-91

Tice, R.R., Agurell, E., Anderson, D., Burlinson, B., Hartmann, A., Kobayashi, H., Miyamae, Y., Rojas, E., Ryu, J.C. and Sasaki, Y.F. (2000). Single cell gel/comet assay: guidelines for in vitro and in vivo genetic toxicology testing. *Environmental and Molecular Mutagenesis* **35**: 206-221

Virro, M.R., Larson-Cook, K.L. and Evenson, D.P. (2004). Sperm chromatin structure assay (SCSA) parameters are related to fertilization, blastocyst development, and ongoing pregnancy in in vitro fertilization and intracytoplasmic sperm injection cycles. *Fertility and Sterility* **81**: 1289-1295

Wyrobek, AJ. (1979). Changes in mammalian sperm morphology after X-ray and chemical exposures. *Genetics* **92**: 105-119

Wyrobek, A.J., Eskenazi, B., Young, S., Arnheim, N., Tiemann-Boege, I., Jabs, E.W., Glaser, R.L., Pearson, F.S. and Evenson, D. (2006) Advancing age has differential effects on DNA damage, chromatin integrity, gene mutations, and aneuploidies in sperm. *Proceedings of the National Academy of Sciences USA* **103**: 9601-9606

Xia, Y., Cheng, S., Bian, Q., Xu, L., Collins, M.D., Chang, H.C., Song, L., Liu, J., Wang, S. and Wang, X. (2004). Genotoxic effects on spermatozoa of carbaryl-exposed workers. *Toxicological Sciences* **85**: 615-623

Zinaman, M.J., Brown, C.C., Selevan, S.G. and Clegg, E.D. (2000). Semen quality and human fertility: a prospective study with healthy couples. *Journal of Andrology* **21**: 145-153

Zini, A., Bielecki, R., Phang, D. and Zenzes, M.T. (2001). Correlations between two markers of sperm DNA integrity, DNA denaturation and DNA fragmentation, in fertile and infertile men. *Fertility and Sterility* **75**: 674-677

Zini, A., Blumenfeld, A., Libman, J. and Willis, J., (2005). Beneficial effect of microsurgical varicocelectomy on human sperm DNA integrity. *Human Reproduction* **20**: 1018-1021

Spermatology. SRF Vol. 65. ERS Roldan and M Gomendio (eds) Nottingham University Press, Nottingham

Sperm competition and ejaculate evolution

Nina Wedell

Centre for Ecology and Conservation, School of Biosciences, University of Exeter, Cornwall Campus, Penryn TR10 9EZ, UK

Under sperm competition, production of large numbers of sperm is often advantageous, either due to numerical superiority or by displacing a larger volume of rival sperm. However, sperm and ejaculate production are inevitably associated with costs, which at times can directly limit male mating rates. As a consequence of sperm limitation, males are expected to strategically allocate their ejaculates to maximise fertilisations. Here I review the evidence for sperm production costs and male ejaculate tailoring showing that males are sensitive to cues relating to varying levels of sperm competition and female fecundity and modulate their ejaculates accordingly. Variation in ejaculation production costs are also shown to influence male ejaculates in relation to outbreeding, cost of inbreeding, and reproductive parasites. There is evidence that females suffer reduced fertility as a consequence of male strategic ejaculation, potentially resulting in sexual conflict over number of sperm delivered. The cost of ejaculate production and the necessity of producing large quantities of sperm may have promoted the evolution of 'manipulative' ejaculates, (e.g. mating plugs, non-fertile sperm, cooperative sperm, and seminal peptides). Many of these ejaculate components are designed to reduce or remove the risk of sperm competition altogether and hence the need to transfer large numbers or elaborate sperm to the female. Some of these adaptations are costly to females since they interfere with their reproductive rate or physiology.

Introduction

Sperm competition is the phenomenon when ejaculates from more than one male overlap and compete for a given set of ova. It is virtually ubiquitous throughout the animal kingdom and is found in most taxa (Smith, 1984; Birkhead and Møller, 1992, 1998; Simmons, 2001). With the advent of new molecular techniques such as microsatellite analyses, the majority of studies of females in the wild show multiple paternities within broods, indicating the female has mated with more than one male during her receptive period. In fact, monogamy, when females only mate with one male, seems to be uncommon in the animal kingdom. It is even possible that in some of these putative cases of monogamy sperm competition may also have occurred, as determining paternity in offspring only measures the outcome of sperm competition and not necessarily sperm in competition.

The research area of sperm competition is only about 40 years old, and was more or less conceived and developed single handedly by Geoff Parker in the late 1960ies after studying

Corresponding author E-mail: N.Wedell@Exeter.ac.uk

the love life of yellow dung flies (Parker, 1970, 2001). He has developed a rich theoretical basis for the study of evolution of sperm size and number and how males should allocate their ejaculates strategically. Parker's work has provided a solid theoretical foundation for a research field that has exploded in recent time. There are now text books in the area (Smith, 1984; Birkhead and Møller, 1992, 1998; Simmons, 2001) and several leading evolutionary journals regularly feature articles on sperm competition. From a mere 13 papers being published on the topic until 1979, sperm competition studies increased to around 50 and 700 in the 1980s and 1990s respectively, with now >1000 papers being published on sperm competition since the 21st century (ISI Web of Science)!

One of the underlying assumptions of Parker's sperm competition models is the importance of producing large numbers of sperm (Parker, 1998). In fact, sperm competition has been suggested as being partly responsible for the evolution of anisogamy and hence the two sexes (Parker, Baker and Smith, 1972; Parker, 1978; Bulmer and Parker, 2002). When the survival of the zygote is strongly dependent on zygote size, disruptive selection will ensure where large gametes are favoured in terms of high viability, whereas small gametes are numerically superior. Competition between gametes over fusion partners will result in strong disruptive selection leading to either large viable gametes (eggs), or small numerous gametes (sperm). There are alternative hypothesis for the evolution of anisogamy such as small sperm avoiding transmitting selfish genetic elements (e.g. Cosmides and Tooby, 1981; Hurst, 1990; Randerson and Hurst, 1999), but this does not preclude the importance of sperm competition playing a role for shaping the initial size divergence between egg and sperm.

The theoretical reason for why producing large numbers of sperm is advantageous in sperm competition relies in its simplest form on numerical superiority. According to the 'Lottery Principle' where all sperm have an equal chance of fertilising the egg, males that ejaculate proportionally more sperm have a higher chance of being successful (Parker, 1998). The idea is analogous to a lottery where the more tickets (sperm) you buy, the greater your chances of winning the prize (fertilising the egg). The pressure on males to produce increasingly larger quantities of sperm has promoted the evolution of larger testes size, as bigger testes have larger sperm producing capacity (Smith, 1984; Birkhead and Møller, 1992, 1998; Simmons, 2001). An indication of the general importance of sperm competition for promoting increasing testes size comes from comparative studies in numerous animal groups ranging from insects to fishes, frogs and mammals, showing that across species males have relatively bigger testes size when they face an increased risk of sperm competition. In other words, in highly promiscuous species where females mate multiply, males tend to have larger testes for their body size.

Within species, males should also be sensitive to the varying risk of sperm competition. For example, if males are able to discriminate the mating status of females, they should provide virgin females with fewer sperm since the risk of sperm competition is reduced. The male is not certain his ejaculate will face sperm competition in the future. In contrast, when mating with a previously inseminated female the male 'knows' that his sperm will face competition, and hence should increase the number of sperm ejaculated. There is plenty of empirical evidence that males are sensitive to cues revealing female mating status and vary the number of sperm ejaculated accordingly (Wedell, Gage and Parker, 2002, and see below).

However, as is now clear, not all sperm have an equal chance of fertilising the egg. Males produce sperm that vary in swimming speed and longevity for example, that directly affect their chances of fertilisation (Smith, 1984; Birkhead and Møller, 1992, 1998; Simmons 2001). In particular, the role in which males inseminate the female has a profound impact on their likelihood of fertilisation. This may partly be due to variation in time since the first mating, number of sperm being stored by females, and the number of eggs or offspring produced, all of which will affect sperm competitive success. This led to the construction of increasingly more sophisticated models, where a loading factor was applied to account for males frequently mating in either favoured or unfavoured roles (Parker, 1998, 1990a). The outcome of these models show that if these roles are occupied randomly, males should expend equally on ejaculates, irrespective of whether competing in a fair or loaded raffle (Parker, 1990b). Things are a bit more complicated when males of a particular phenotype (e.g. big or small) typically occupy a particular role (e.g. first or second to mate with the female) in a loaded raffle. The outcome depends on whether the types of males or the role they mate in are considered (Ball and Parker, 2000). When comparing types of males (i.e. small or large), the type that is usually mating in the disfavoured role should expend more on their ejaculates. On the other hand, across roles, a larger ejaculate investment should be made when mating in the role that is usually occupied by that type of male. For example, if mating first is favoured and is usually occupied by males of a particular phenotype (say large males), and small males typically mate second then, overall, small males should invest more sperm than bigger males. Bigger males should increase their ejaculate investment when mating first, whereas small males should spend more when mating second.

Lately, it has become increasingly apparent that males frequently suffer from varying degrees of sperm limitation, for example as a consequence of a recent ejaculation. As a consequence females may also suffer from limited sperm reducing their fertility. In the circumstances of female sperm limitation, recent models predict that when mating in the favoured role males should ejaculate more sperm if this role is randomly occupied as this will result in higher fertilisation returns for the male (Mesterton-Gibbons, 1999). Things are a bit more unpredictable when roles are occupied non-randomly (Ball and Parker, 2000).

Males not only face differing degree of risk of sperm competition, which vary between 0 (no sperm competition) and 1 (competition is inevitable). The intensity of sperm competition also varies between matings, and relates to the number of competitors at any given mating. These models tend to be of particular relevance for species experiencing high levels of sperm competition (Parker, Ball, Stockley and Gage, 1996, 1997; Ball and Parker, 1997). These models show that, across species, ejaculate expenditure per mating opportunity should increase with the number of competing ejaculates for the eggs of a given female. However, the opposite is true for ejaculate investment within a species. Providing males can assess the number of competitors, it is predicted males should decrease the number of sperm ejaculated once the number of competitors rise above 2. Again a fair lottery is assumed where all sperm have equal probability of fertilisation, resulting in males ejaculating very little when there are no competitors present, most in the presence of one competitor, and reduce the number of sperm when there are two or more competitors. The reason being that the benefit of increasing the number of sperm ejaculated decreases with increasing number of competitors due to diminishing returns. In other words analogous to the lottery principle: the more lottery tickets being sold in total, the chances of winning by buying additional tickets are reduced.

The realization that males frequently strategically vary the number of sperm provided at mating has stimulated additional models exploring the possibility of coevolution between degree of sperm competition and males' sperm allocation strategies (Williams, Day and Cameron, 2005). The rationale being that since the ejaculate allocation strategy adopted by males will directly influence the degree of sperm competition, the two are coupled and should be considered simultaneously. Variation in environmental parameters such as resource availability is assumed to affect the cost of mating and thereby affecting both the degree of sperm competition and sperm allocation strategies. For example, variation in resource availability can result in evolution of greater ejaculate expenditure per mating because it influences both the number of matings and the paternity achieved per mating. The model also shows that the distinction between sperm competition risk and intensity is no longer necessary, since variation in the distribution of sperm competition risk across females at the population level will simultaneously alter the intensity of sperm competition. Similarly, another recent model shows that both sperm competition and male mating rate can directly affect relative testes size (an index of ejaculate expenditure) across species (Parker and Ball, 2005).

Sperm competition

The process of sperm competition can be divided into two broad classes: behaviours and morphological adaptations designed to reduce the risk of future sperm competition once a male has mated with the female, or increase the first male's fertilisation success when the female remates, termed *'defense'* mechanisms. The other form represent adaptations involved in increasing male fertilisation success when a male mates with a female that is already inseminated and hence sperm will compete, termed *'offence'* mechanisms (Simmons and Siva-Jothy, 1998).

Defense mechanisms

These are adaptations that tend to be associated with preventing the female from remating and hence reducing the risk of sperm competition for the first male to mate. There are three main ways in which males can prevent or reduce the likelihood of female remating. Males can transfer material together with the sperm in the ejaculate that form a mating plug making it more difficult for a rival male to inseminate the female shortly after her first mating. Secondly, males of many animals may also guard the female, most commonly after inseminations thereby preventing her from mating again before fertilising her eggs. In some species males also guard the female before mating (e.g. *Gammarus*). In these circumstances females typically only have a very short period of receptivity, or may only mate once in their lifetime. Here there is no sperm competition, instead males monopolise females for future mating. Pre-copulatory mate guarding is also expected in species where the first male sires the majority of offspring, as copulating with non-virgin females does not result in any fertilisation returns. Finally, males of the majority of polyandrous species frequently transfer compounds in the ejaculate that suppress female receptivity or render the female unattractive to rival males following mating, thereby reducing the risk of sperm competition (see below). Suppression of receptivity is often associated with sperm storage in females (Gillott, 2003). However, the extent to which males are able to influence storage of their sperm is far from clear, as females have several sophisticated ways to manage sperm supplies (Eberhard, 1996).

Offence mechanisms

Commonly when we think of male adaptations to sperm competition we refer to the situation when males mate with an already inseminated female and his sperm will be in direct competition with sperm from previous matings. Male animals have devised a suite of morphological and behavioural means to increase their paternity when in competition.

Mechanical sperm removal. The first documented case of mechanical sperm removal comes from studies of dragonflies, where males frequently possess a modified intromittent organ that scoop out sperm present in the females' bursa (where sperm are received at mating) before depositing his own sperm (Waage, 1979). This behaviour seems to be widespread in this insect group (e.g. Siva-Jothy and Tsubaki, 1989; Cordero and Miller, 1992). However, later studies show that males are unable to remove sperm already transported to the females' spermatheca (sperm stores) ensuring some control by females over sperm storage (Siva-Jothy and Hooper, 1995). Interestingly, there appears to be intra-specific variation in sperm manipulation. In *Calopteryx haemorrhoidalis* damselflies, males of a Spanish population stimulate females to empty sperm stored in their spermathecae, whereas males of an Italian population physically remove sperm. Associated with these differences, genitalia differed in the two populations, indicating geographic variation in post-mating sexual selection in this species (Cordero-Rivera, Andrés, Córdoba-Aguilar and Utzeri, 2004). Sperm removal has also been reported in bush-crickets where the male even consume rival male's sperm (von Helversen and von Helversen, 1991). In other species males flush out rival males sperm with their own ejaculate, as is the case in tree crickets (Ono, Siva-Jothy and Kato, 1989). In some beetles males have a spiny intromittent organ that can remove rival males' sperm. There is some suggestion that males can transfer sperm removed from rival males to females they later inseminate, and these sperm may even end up fertilising eggs despite the male never mating with the female (termed 'fertilisation by proxy' by the authors, Haubruge, Arnaud, Mignon and Gage, 1999). Similarly, spiny intromittent organs are widespread in many other animal groups such as rodents, snakes and spiders (Eberhard, 1985), which may play a role in sperm removal, although alternative suggestions that they may function to induce ovulation and stimulate the females have been proposed (Eberhard, 1996). For example, copulation-induced ovulation has been demonstrated in mammals (e.g. Lariviere and Ferguson, 2003), and marsupials (Smith, 1992), which may be beneficial to both sexes. Males increase the chances of siring offspring and females receive a reliable cue that successful insemination has occurred.

Sperm stratification. The last male to mate with the female can also increase paternity by depositing their sperm in the most favourable place inside the female thereby increasing the chances it will be used in fertilisation. Males can push previous ejaculates to the back of the females' sperm storage, placing his ejaculate closer to the site of fertilisation (Siva-Jothy, 1988). Sperm stratification is present in snow crabs, for example (Urbani, Saint-Marie, Sévigny, Zadworny and Kuhnlein, 1998). In the red flour beetle *Tribolium castaneum*, sperm stratification results in differential sperm loss by subsequent males' sperm, which increases paternity for the first male to mate (Lewis, Kobel, Fedina and Beeman, 2005). There is also some evidence that sperm stratification may play a role in determining paternity in the marsupial *Anthechinus agilis*, where it is observed that the lower storage crypts in the female oviduct receive sperm from the first mating, whereas sperm from subsequent matings occupy crypts closer to the ovary (Taggart and Temple-Smith, 1991), which may explain the apparent second male precedence in this species (Kraaijeveld-Smit, Ward, Temple-Smith and Paetkau, 2002). Packing and repositioning of rival sperm is another possibility that seems to operate in some insects (e.g. Siva-Jothy and Tsubaki, 1989).

Sperm numbers. As mentioned above, one of the most important determinants for the outcome of sperm competition is the relative number of sperm transferred by the two males and the number of sperm that compete with each other. Several factors can influence the number of sperm in the female reproductive tract. Not all sperm that a male inseminates will end up in the female tract and hence be available for fertilisation. Females of many species frequently eject sperm from males following mating. This can be a consequence of insemination by un-preferred males, as seem to be the case in fowl where females eject sperm after mating to subordinate males (Pizzari and Birkhead, 2000). Many mammals eject the majority of sperm following copulation, as do several bird and insect species (e.g. Eberhard, 1996; Cordoba-Aguilar,

2006). This may be one way in which females could influence the outcome of sperm competition. Another important impact on the number of sperm in competition is the timing between the two inseminations. If there is a long time interval between copulations, the majority of sperm from the first mating may have been lost, or have died due to old age, or have been used by the female in fertilisation. All these factors will directly influence the relative number of sperm from different males that will end up competing with each other. If males can influence any of these factors then they could enjoy greater paternity without necessarily having to expend a lot of resources on sperm. For example, guarding males frequently have smaller testes size and deliver substantially smaller ejaculates than males trying to sneak matings with the female (Gage, Stockley and Parker, 1995; Fu, Neff and Gross, 2001). This is because the risk of sperm competition for guarding males is substantially lower compared to sneaking males whose sperm is always in competition and hence need to transfer large ejaculates to obtain paternity.

It has been suggested that the frequently observed last male sperm precedence in many animals is due to a combination of factors where males are able to remove, displace or inactivate sperm from previous matings (Birkhead and Møller, 1998; Simmons, 2001). This indicates there may be stronger selection pressure on males to evolve effective sperm offence mechanisms compared to investing in sperm defense mechanisms when in sperm competition. On the other hand, a positive genetic correlation has recently been found in *D. melanogaster* between males sperm defense and offense abilities, suggesting the same or linked genes are involved (Friberg, Lew, Byrne and Rice, 2005, but see Civetta and Clark, 2000). However, as pointed out, there may be even stronger selection on males to reduce the risk of sperm competition altogether, which would preclude the need to invest in sperm offence mechanisms.

Sperm are costly

There are several lines of evidence indicating ejaculate production carries a cost to males. In several species, ejaculate production can directly constrain male mating rate promoting male mate choice (Dewsbury, 1982). For example, males of most animals frequently require a recuperation period after mating before they are able to produce a new ejaculate, which tend to have successively lower sperm numbers. Fertilisation rate in lemon tetras, for example, decreases with increasing spawning frequency and females prefer to pair with males that have not recently spawned (Nakatsuru and Kramer, 1982). There are also some direct indications that producing sperm is energetically expensive. In adders, for example, there is data to suggest sperm production is as energetically costly as the more active phase of searching, fighting over and copulating with females (Olsson, Madsen and Shine, 1997). In *C. elegans* nematodes sperm production may also reduce male life span (van Voorhies, 1992). In other animals, like male water-mites, sperm production capacity is directly related to food availability or the amount of resources allocated, with males given less food producing fewer spermatophores (sperm packages) than males given more food (Proctor, 1992). Similarly, in copepods spermatophore output is reduced during food shortage (Ianora, Miralto and Buttino, 1999), and food restriction in the Indian meal moth results in reduced sperm numbers (Gage and Cook, 1994). Additional circumstantial evidence that sperm production can be energetically costly comes from observations that males of several animals accumulate nutrients before the onset of spermatogenesis. Sea urchins, for example, even have specialised nutritive phagocytes situated both in the gut and the gonadal wall that mobilise nutrients used in spermatogenesis (Walker, Harrington, Lesser and Fagerberg, 2005).

Empirical data for ejaculate tailoring

The pressure on males to deliver large, and potentially costly, ejaculates has promoted strategic sperm allocation across matings. Contrary to what was once thought, males' reproductive success can be directly limited by the number of sperm they are able to produce. Since males may suffer sperm limitation it is important they are economical with the number of sperm they ejaculate at any given mating and should tailor their ejaculate to generate the greatest fertilisation returns for a minimal investment. As predicted by models, males are expected to be economical with their sperm and only ejaculate large numbers of sperm when this results in increased fertilisation returns. There is ample evidence that males are sensitive both to the risk and intensity of sperm competition and tailor their ejaculates accordingly. For example, males ranging from slugs to man increase the number of sperm delivered when there is an increased risk of sperm competition, either in the form of the presence of rival males, but also in relation to female mating status and expected fecundity (see Wedell *et al.*, 2002 for a review).

Males of several species have been shown to be sensitive to the risk of sperm competition and provide more sperm in the presence of rival males (see Wedell *et al.*, 2002). It is apparent that males are very sensitive to a variety of cues relating to risk of sperm competition. For example, male sticklebacks ejaculate more sperm when viewing videos of rival males courting females than of a male tending eggs (Zbinden, Largiader and Bakker, 2004). Similarly, human males viewing images depicting sperm competition ejaculate higher quality semen (Kilgallon and Simmons, 2005). As mentioned above, there is no point wasting many sperm when sperm are not in competition, and, as predicted, males often provide virgin females with fewer sperm (e.g. Cook and Wedell, 1999). Conversely, there is plenty of empirical evidence showing that males are sensitive to information relating to the risk of sperm competition and provide non-virgin females with more sperm resulting in higher sperm competitive success (e.g. Gage and Barnard, 1996; Nicholls, Burke and Birkhead, 2001; Wedell *et al.*, 2002). However, this is not always the case. Males may also discriminate against mated females, and preferentially copulate with virgins and experience reduced sperm competition (e.g. Saether, Fiske and Kalas, 2001; Arnaud and Haubruge, 1999). Interestingly, the pattern is influenced by sperm precedence. As predicted, in species where the first male to mate has higher paternity, as is the case in many spiders, males preferentially mate with and provide virgin females with more sperm (e.g. Bukowski and Christensen, 1997). Males of many species have also been shown to vary the number of sperm provided to females in relation to female quality. For example, female size often covaries with female fecundity, and males of many species provide larger, more fecund females with more sperm resulting in higher fertilisation returns. However, bigger females often mate more frequently and hence may also represent an increased risk of sperm competition (Wedell and Cook, 1999). Female age is another aspect of female quality directly affecting male fertilisation returns. Older females are often less fecund as they have fewer eggs left to lay, and are more likely to have already mated therefore representing a higher risk or even higher intensity of sperm competition (e.g. Simmons, Craig, Llorens, Schinzig and Hosken, 1993). As the models described above predict, the number of sperm ejaculated at any given mating by a particular male will be a complex decision based on the expected fertilisation return in relation to the amount invested.

Other aspects relating to maximising male fertilisation returns can also influence ejaculate tailoring. In sailfin mollies (*Poecilia latipinna*), males are sexually parasitised by the Amazon molly (*P. formosa*), a unisexual gynogenetic species where clonal females require sperm from the related bisexual sailfin molly males to initiate embryogenesis. Male sailfin mollies not only prefer to mate with conspecific females, but males from both allopatric and sympatric populations produce more sperm when in the presence of conspecific females than heterospecific females (Aspbury and Gabor, 2004). These results indicate that costs of sperm production resulting in strategic sperm allocation may promote species isolation in these fish. Similarly, Amazon mollies (*P. formosa*), which require sperm to trigger embryogenesis, also frequently mate with *P. mexicana* males. Interestingly, mating with parasitic females makes *P. mexicana* males more attractive to conspecific females, but they seem to reduce the cost of these matings by providing *P. formosa* females with less sperm (Schlupp and Plath, 2005). Preferential conspecific sperm allocation is likely to be advantageous, not only due to reducing the cost of producing hybrid offspring, but since conspecific sperm precedence seems to be common (e.g. Howard, 1999). Hence males will risk substantial sperm loss when providing hetero-specific females with large quantities of sperm.

At the other end of the spectrum, males may be sensitive not to the costs of out-breeding, but risk of mating with relatives potentially resulting in reduced offspring production due to the costs of inbreeding. There are several examples of individuals avoiding mating with relatives (Pusey and Wolf, 1996). So far, few studies have examined the possibility that males may also strategically allocate their sperm in relation to mate relatedness. There is some suggestion that male moths provide sisters with fewer sperm than unrelated females (Lewis and Wedell, unpublished). Theory predicts that males should be less sensitive to the risk of mating with kin compared to females in terms of potential loss due to costs of inbreeding. In addition, and somewhat counterintuitive, incestuous mating can even under certain circumstances be beneficial to males (Parker, 1979; Kokko and Ots, 2006). This is because mating with a relative can increase males' inclusive fitness by helping that individual to spread shared genes. This will depend on the cost to males of inbreeding depression relative to the cost of ejaculate production. So far the potential for variation in inbreeding costs to influence males' ejaculate expenditure has not been examined.

Males have been shown to be sensitive to the expected fecundity returns and preferentially provide more sperm when the expected fecundity returns are greater (Wedell *et al.*, 2002). One recent realization is that males are sensitive to the presence of reproductive parasites. For example, populations of the isopod *Armadillium vulgare*, are often infected by the endosymbiont *Wolbachia pipientis*, which feminises males turning them into functional females. This is advantageous to the parasite as it can only be transmitted in the cytoplasm of eggs and hence males represent an evolutionary dead end. Feminised males are less fecund, partly due to inefficient sperm use than true females, and male pill bugs provide low fecundity feminised males with fewer sperm (Moreau, Bertin, Caubert and Rigaud, 2001; Rigaud and Moreau, 2004). A similar situation seems to be the case in the shrimp *Gammarus duebeni* infected with a feminising microsporidium, which is also maternally inherited. Male shrimps suffer severe sperm depletion after only two matings and provide feminised males with fewer sperm. Again this is advantageous as feminised males produce fewer eggs (Dunn, Andrews, Ingrey, Riley and Wedell, 2006). A further indication that reproductive parasites can affect male ejaculation strategies comes from recent work showing that *Wolbachia*-infected male *D. simulans* males suffer reduced paternity when competing with uninfected males' sperm (Champion de Crespigny and Wedell, 2006). Strategic sperm allocation may thus reduce sperm competitive ability, and be an important, but underestimated, mechanism undermining the transmission advantage to reproductive parasites and hence affect their population dynamics.

Sperm evolution

Sperm size

There are other ways, apart from ejaculating many sperm, that can influence a males' fertilisation success. One way is to produce more elaborate sperm, either sperm that live longer or bigger and faster sperm that have a higher likelihood of succeeding in sperm competition. For example the biggest known sperm is found in the fly *D. bifurca*, where they can reach more than 58 mm in length, which is 20 times the size of the male! Needless to say, producing gargantuan sperm comes at a cost. In *Drosophila* species, delayed maturity is the cost males pay for producing large sperm (Pitnick, Markow and Spicer, 1995). Males cannot produce a vast number of these large sperm, instead *D. bifurca* males produce almost the same number of sperm as the female produces eggs (Bjork and Pitnick, 2006). Only a few of these big sperm are transferred per mating. Interestingly, in this species sperm competition is not intense, as females do not remate frequently unless they initially received few sperm (Luck and Joly, 2005). There is some evidence that big sperm have an advantage in sperm competition. In mites for example, males with relatively longer sperm gain more fertilisations (Radwan, 1996). Similarly, in polyandrous butterfly species where sperm competition is pronounced, males have relatively longer sperm than less polyandrous species (Gage, 1994). On the other hand, the opposite is found in the field cricket, *Gryllus bimaculatus*, where males with the shortest sperm have higher paternity (Morrow and Gage, 2001). In insects in general, there is a negative correlation between sperm size and sperm number among related taxa, suggesting that production of large sperm is expensive (Pitnick *et al.*, 1995). In birds, sperm length co-varies with degree of extra pair paternity, but this seems due to the relationship between female sperm tubule length (where females store sperm) and sperm length (Briskie and Montgomerie, 1993). In mammals, production of large numbers of sperm is metabolically constrained, as suggested by the fact that in species where sperm morphology and anatomy of the female tract increase the probability of fertilisation, the number of sperm produced is lower (Cohen, 1977; Harcourt, 1991). According to Harcourt (1991), sperm competition in mammals occurs through scramble competition (i.e. high sperm numbers), rather than contest competition (i.e. through sperm elaboration). This may explain why mammal sperm are small and plentiful, (c.f. external fertilisers), rather then few and elaborate like in fruit flies. In many species sperm size appears to be optimised, as manipulation of diet that affect sperm number seem to have no influence on sperm size (Gage and Cook, 1994), indicating sperm size may be under stabilising selection.

Sperm motility

Longer sperm in rodents, primates and birds have been suggested to have a motility advantage (Cohen, 1977; Roldan and Gomendio, 1991). Long sperm may swim faster than short sperm and so out-compete shorter sperm for access to ova. In the fowl, for example, motility directly influences sperm competitive success (Froman, Pizzari, Feltmann, Castillo-Juares and Birkhead, 2002). However, this cannot apply for the majority of insects, as sperm may even be longer than the distance they travel, and only partially fill the female sperm storage (Pitnick *et al.*, 1995). Large sperm also do not seem to represent a nutrient investment to offspring (e.g. Pitnick, Spice and Markow, 1997). It is possible they have evolved in response to sperm competition. Across fish species, relative testes size increases with increasing degree of sperm competition, but sperm length decreases (Stockley, Gage, Parker and Møller, 1997). This may be because sperm length increases its speed, but at the same time decreases its longevity, indicating a trade off between sperm size and numbers. In general, sperm motility appears to be dependent on the amount of mitochondrial DNA in the sperms' midpiece. In primates for example, the volume of the sperm midpiece is correlated with relative testes size and degree of promiscuity (Anderson and Dixon, 2002).

Sperm longevity

Another important factor affecting male fertilisation success is the longevity of sperm. This is particularly important for external fertilisers such as many aquatic animals that shed their sperm into the water. Sperm longevity may also be important in species where females store the sperm for extended periods of time without supporting sperm while in storage or where the sperm have a time consuming journey to the site of fertilisation. As yet, it is not known whether species with sperm with larger mid-piece, which may swim faster, also have shorter lived sperm. In fish, longer sperm (which may have bigger mid-pieces) do have shorter longevity (see above).

Sperm cooperation

Increased ejaculate efficiency may, instead of producing larger sperm, be achieved by grouping sperm together into multicellular units. In the silverfish *Termobius,* sperm are paired and only motile when coiled around each other (Bawa, 1964). Paired sperm are also found in possums, millipedes, beetles and mollusks (Sivinski, 1980). The *Colymbetes* beetles even have groups of three or more sperm (Mackie and Walker, 1974). Sperm grouped in bundles often have proteinaceous caps that may serve as a nutrient source or contain enzymes that break down and assimilate secretions from the genital tract (e.g. Cantacuzene, 1968), which may increase sperm longevity or motility. In whirligig beetles for example, ~100 sperm are attached by their heads to a rod like object called a spermatostyle, the movement of which is due to the swimming sperm tails which are moved by joint effort, although individual sperm are motile on their own (Mackie and Walker, 1974; Bretland and Simmons, 1970). It is possible that this increases the swimming speed. In fishflies, males also produce sperm in bundles of several hundred, which swim by synchronous flagellate movements, and larger sperm bundles swim at a higher speed than smaller bundles of sperm (Hayashi, 1998). A fascinating example of sperm cooperation was recently shown in the wood mouse, *Apodemus sylvaticus,* where hundreds or thousands of sperm link together as a train using hooked structures on their heads, which enables them to swim at almost twice the speed of a single sperm. These trains break up prior to fertilisation as only one sperm can fertilise the egg, requiring the majority of sperm to sacrifice themselves by undergoing the acrosome reaction (Moore, Dvorakova, Jenkins and Breed, 2002). Similarly, in some sessile immobile prosobranch marine mollusks, thousands of sperm are carried on giant plate-like cells on a thickened tail, which, according to the authors, is 'a functional equivalent to a penis' (Sivinski, 1980, citing Purchon, 1977).

Sperm polymorphism

Instead of clumping together more than one sperm into units that may be more efficient in transporting the sperm either to the female or inside the female, males of several species also produce different sperm types, often referred to as sperm heteromorphism. Sperm heteromorphism occurs in a variety of animals from molluscs, spiders, insects, millipedes, nematodes, centipedes, platyhelminths, echinoderms and fish, and appear to have evolved independently many times (Jamieson 1987, 2000; Swallow and Wilkinson, 2002; Till-Bottraud, Joly, Lachaise and Snook, 2004). Sperm types can be morphologically distinct and have different developmental pathways (e.g. butterflies and moths; Friedländer, Seth and Reynolds, 2005). Most commonly, only one of the sperm types in sperm heteromorphic species is able to fertilise the egg, although in some species all sperm morphs appear to be capable of fertilisation. In mammals it has been suggested that sperm morphs may be a consequence of errors during meiosis (Harcourt, 1991), but this seems unlikely to be the case in invertebrates, where different developmental pathways are involved. Non-fertile sperm are likely to have a function, as they have been show to respond to selection in predictable ways (e.g. Crudgington, Beckerman, Brustle, Green and Snook, 2005).

There are several hypotheses for the function of non-fertile sperm. They have been suggested to aid the successful transfer of the fertile sperm either during insemination or to the site of fertilisation or sperm storage in the female (e.g. Sahara and Takemura, 2003). Sperm can be of distinct types like in the externally fertilising siboglinid pogonophores where the sperm packet (spermatophore) contains two types of sperm. The normal fertilising sperm and 'atypical sperm', which are smaller, but contain a nucleus. Once the spermatophore is released into the water it opens immediately at one end. Some 10-30 non-fertile sperm then protrude from the opening, and by vigorous movements they propel the spermatophore through the water. The sperm survive for at least 10 min (Adiyodi and Adiyodi, 2000). This may increase the chances of bumping into eggs before the sperm dies and hence increase male fertilisation prospects. It has been suggested that non-fertile sperm may represent a nutrient source either for the female, the developing zygote or the fertile sperm while in storage. There is no real support for this idea at least in insects (Swallow and Wilkinson, 2002; Till-Bottraud *et al.*, 2004). This is mainly because sperm do not appear to be digested, but can remain in the female reproductive tract for extended periods of time. In addition, in species where males provide nutrients in the ejaculate this can often form a separate part of the ejaculate (Vahed, 1998). In mollusks there is some suggestion non-fertile sperm may serve a nutritive role benefiting the female rather then the fertile sperm (Buckland-Nicks, 1998), although there is, as yet, no experimental evidence.

Non-fertile sperm have also been implicated to play a role in sperm competition (Silbergleid, Shepherd and Dickinson, 1984). This can be achieved either by reducing female receptivity and thereby reducing the risk of sperm competition, or by interfering with rival males' sperm from previous matings by inactivating, destroying or removing sperm. There is some evidence in support of this hypothesis. For example, in butterflies males produce a large number of anucleate sperm that cannot fertilise the females' eggs. These are typically produced in great quantity representing >90% of total sperm numbers (Cook and Wedell, 1996). In the green-veined white butterfly *Pieris napi*, these sperm appear to function to fill the female's sperm storage organ, as females that refrain from remating store more non-fertile sperm, whereas no difference in number of fertile sperm stored was found (Cook and Wedell, 1999). There is genetic variation between females in the number of non-fertile sperm stored, which covaries with their remating rate (Wedell, 2001). This indicates that the non-fertile sperm in this butterfly species may function to fill the females' sperm storage organ thereby reducing her willingness to remate. Female remating rate appears to be influence by number of sperm stored in many insects (Gillott, 2003). It is possible males have exploited a female system that evolved to allow females to monitor the number of sperm stored to ensure maximum fertility levels. Males may be using their non-fertile sperm rather than the larger fertile sperm to switch off female receptivity. Crucial to the function of non-fertile sperm as cheap fillers is that they are cheaper to produce than fertile sperm. In the Lepidoptera this may be the case as they are often smaller in size than fertile sperm and their nuclear material is lost during development (Kawamura, Yamashiki, Saitoh and Sahara, 2000), which may be recoverable by the male. In addition, non-fertile sperm are accompanied by less extracellular material than fertile sperm during transfer to the female (Friedländer *et al.*, 2005). Finally, unlike fertile sperm, non-fertile sperm in the Lepidoptera are highly motile upon ejaculation and may therefore be more easily 'detected' by females while in storage. However, the fact that there is genetic variation in females' tendency to store non-fertile sperm indicate females have regained some control over sperm storage, at least in *P. napi*. Comparative studies of moths also suggest non-fertile sperm may function as cheap filler. There is a positive correlation between the number of non-fertile sperm and female sperm storage organ size, whereas no such relationship was found for fertile sperm (Morrow and Gage, 2000).

A filler function on non-fertile sperm has not been corroborated in flies. In *D. pseudoobscura* males produce about equal number of short and long sperm, but only long sperm are used in fertilisation (Snook and Karr, 1998). Short sperm appear to reach the site of sperm storage, but there is no association between sperm storage and female remating tendency, indicating they are unlikely to be involved in suppressing female receptivity in this species (Snook, 1998). Similarly in stalk-eyed flies, it is unlikely that non-fertile sperm function to fill the females sperm storage as they are produced in low numbers relative to fertile sperm (Lorch, Wilkinson, and Reillo, 1993; Swallow and Wilkinson, 2002).

Non-fertile sperm have also been suggested to actively incapacitate rival males sperm (Silberglied *et al.*, 1984; Baker and Bellis, 1988). In Tanniode mollusks, non-fertile sperm carry lysosomes, which could potentially be detrimental to rival sperm although this has yet to be shown (Buckland-Nicks, 1998). If non-fertile sperm primarily function to interfere with rival sperm, they should reach the female sperm storage killing rival sperm from previous matings, prior to the arrival of the inseminating males' fertile sperm. Non-fertile sperm reach the sperm storage organ before the fertile sperm in butterflies and moths (Tschudi-Rein and Benz, 1990), but whether they directly interfere with sperm stored from previous matings is not known. On the other hand, in *Drosophila*, both short and long sperm reach the sperm storage site at the same time (Snook and Markow, 2001). There is some indication that male butterflies increase the number of non-fertile sperm provided to non-virgin females, as predicted if they serve to interfere with previous males' sperm resulting in higher fertilisation success (Wedell and Cook, 1998). However, males also increase the number of fertile sperm to mated females (Wedell and Cook, 1999). In marine sculpine fish, non-fertile sperm may enhance the delivery of sperm to the site of fertilisation, but may also hamper subsequent rival males' sperm from reaching the egg. Both sperm types are shed in the water where the non-fertile sperm form clumps which 'swallow up' fertile sperm. This may result in rival males' sperm being prevented from getting access to the eggs (Hayakawa, Munehara and Komaru, 2002a; Hayakawa, Akiyama, Munehara and Komaru, 2002b).

Ejaculate evolution

An alternative way to reduce the need for producing costly sperm is to evolve other aspects of the ejaculate that either reduce the risk of sperm competition, or increase the chances of fertilisation in sperm competition. Reducing the risk of sperm competition can be achieved either by preventing the female from remating, or suppressing her receptivity and stimulating oviposition. Males have evolved various factors transferred in the ejaculate that function to achieve this aim.

Mating plugs

Mating plugs are common in several animals (e.g. Eberhard, 1985). Butterflies, for example, have mating plugs (so called sphraga) that can be remarkably elaborate and at times so large they directly interfere with female egg laying. There is some evidence they are effective in reducing female remating: the number of matings by females decreases with increasing elaboration or size of the mating plug across butterfly species (Simmons, 2001). These plugs may also function as a visual deterrent to rival males, as the larger the plug the less likely males are to attempt to mate with the female in some butterfly species (Orr and Rutowski, 1991). However, it is not completely effective as females are capable of remating despite the presence of a plug (Dickinson and Rutowski, 1989; Orr and Rutowski, 1991; Matsumoto and Suzuki, 1992).

Plugs are also common in many other species of insects (Parker, 1970), spiders (Jackson, 1980), nematodes (Barker, 1994), snakes (Shine, Olsson and Mason, 2000), and mammals including rodents and primates (Gomendio, Harcourt and Roldan, 1998; Kingan, Tatar and Rand, 2003). For example, in red-sided garter snakes, the mating plug is effective in preventing further matings (Shine *et al.*, 2000). Male *Nephila fenestra* spiders plug the female reproductive opening by sacrificing one of their two pedipalps (which are used to transfer sperm), thereby protecting their sperm and hence increase paternity (Fromhage and Schneider, 2006). Similarly, male bumblebees cement shut the female reproductive opening (Sauter, Brown, Baer and Schmidt-Hempel, 2001). This is potentially costly to females, as multiple mating is advantageous by increasing the genetic diversity of offspring, which decreases the susceptibility towards parasites and pathogens (Baer and Schmidt-Hempel, 1999). Mating plugs may have had their origin in making sure sperm is not lost from the female reproductive tract following insemination, and later become more elaborate when taking on the role of physically forming a barrier against further copulation attempts by rival males.

Non-sperm components

One way in which males can reduce female receptivity is by rendering them unattractive to rival males following mating. In some butterfly and moth species, males release pheromones that repel potential rivals (e.g. Hirai, Shorey and Gaston, 1978; Lecompte, Thibout, Pierre and Auger, 1998). Male butterflies and moths also transfer chemical compounds, or anti-aphrodisiacs, to females at mating that reduce their attractiveness. These are frequently synthesised by males and appear to be species-specific and differ even between closely related species (Gilbert, 1976; Andersson, Borg-Karlson and Wiklund, 2000, 2003). These anti-aphrodisiacs are emitted by the mated female and act as a strong deterrent to courting males, reducing potentially costly harassment. However, these compounds only have a transient effect.

Males of the majority of polyandrous species frequently transfer compounds in the ejaculate that suppress female receptivity and stimulate oviposition and egg maturation rate (Gillott, 2003). In insects, these compounds frequently involve juvenile hormones (JH), either by direct transfer (e.g. Park and Ramaswamy, 1998; Shirk, Bhaskaran and Roller, 1980, 1983), or by stimulating the females' own JH production (Park, Ramaswamy and Srinivasan, 1998). Male insects also transfer other compounds apart from JH that affect female receptivity and reproductive physiology. The best-studied ejaculate compounds are the accessory gland proteins (Acps) transferred in the ejaculate to the female at mating in *D. melanogaster*. These compounds reduce female receptivity and stimulate egg maturation and oviposition (Wolfner, 2002; Kubli, 2003; Chapman and Davies, 2004). The most intensively studied and characterised Acp is the sex peptide (Acp70A), which reduces female receptivity and increases egg-laying by stimulating the release of JH (Soller, Bownes and Kubli, 1997; Chapman and Davies, 2004). Interestingly, there appears to be functional similarity between moth and *D. melanogaster* sex peptides. For example, the sex peptide of *D. melanogaster* has been found to increase JH production when injected in female *Helicoverpa armigera* moths (Fan, Rafaeli, Moshitzky, Kubli, Choffat and Applebaum, 2000). The exciting possibility of shared functionality of seminal peptides in these two disparate insect groups requires further examination.

Seminal peptides are common also in other animals (e.g. Swanson and Vacquier, 2002; Wolfner, 2002). Males may have exploited the females' endocrine system by hijacking the very molecules (e.g. JH, ecdysteriods, neuropeptides) that are involved in regulating key reproductive processes to manipulate females' reproductive physiology (c.f. sensory exploitation; Ryan, Fox, Wilczynsk and Rand, 1990). This may result in the use of different, although ancient, physiological pathways that may be largely conserved across insect groups (e.g. Diptera and Lepidoptera). Male-derived female receptivity suppressing factors may be under positive selection promoting rapid divergence among species and populations. This possibility is suggested for seminal proteins that manipulate female reproductive physiology in insects and vertebrates (Wyckoff, Wang and Wu, 2000; Swanson and Vacquier, 2002; Dorus, Evans, Wyckoff, Choi and Lahn, 2004).

Cost to females

Whenever females mate multiply and hence sperm competition occurs, there is a conflict between the sexes over female remating, the number of eggs laid before remating, and sperm use amongst other things. Males and females frequently disagree over all these aspects, and males have evolved several means aimed at manipulating females into making a larger investment in the males' offspring, even if this is directly detrimental to female fitness (Chapman, Arnqvist, Bangham and Rowe, 2002; Arnqvist and Rowe, 2005; Tregenza, Wedell and Chapman, 2006; Wedell, Kvarnemo, Lessells and Tregenza, 2006). This is because males are selected to extract as much investment by the female as possible, as he will gain no or little fitness returns once the female remates. As mentioned, male insects frequently transfer compounds in the ejaculate that manipulate female reproductive physiology to increase immediate egg output and suppress receptivity or prevent female remating by mate guarding. This is beneficial for the male as it results in higher paternity. However, this often comes at a cost to the female (Rice, 1996; Stockley, 1997). For example, in poecilid fish with internal fertilisation, males' intromittent organ frequently damages the female genitalia, which may serve to reduce female remating due to cost of mating to females (Constanz, 1984). It is possible that a high cost of copulation in terms of enduring spiny intromittent organs, taking on board seminal products, etc, all which reduce female longevity (Rice, 1996; Crudgington and Siva-Jothy, 2000), may promote females to make a greater investment in current reproduction as she has reduced residual reproductive value (Lessells, 2005, 2006). This assumes that the main function of these male manipulative actions is to increase the cost of mating to females, thereby reducing the benefit to female of remating due to high mating costs (Lessells, 1999). Alternatively, the cost to females of male manipulation may be an incidental side effect (Chapman and Davies, 2004).

Male ejaculate tailoring can also represent a cost to females in terms of reduced fertility. For example, females are often not able to discriminate between males in terms of their mating status (e.g. Kaitala and Wiklund, 1995). This can result in extended copulations as males can require a recuperation time before he is able to produce a new ejaculate up to a week or more in some cases, during which time he remains in copula before transferring his sperm (e.g. brimstone butterfly; Labitte, 1919). In monogamous species, females may risk reduced fertility if she mates with a recently mated male (e.g. Jones, 2001). Females mating to particularly popular males may also suffer sperm limitation, as these males experience dramatically higher copulation rates than less attractive males and hence can run out of sperm. In these cases females must trade-off reduced fertility against the benefit of mating with popular males. For example, dominant male Soay sheep run out of sperm at the end of the rutting season leading to subordinate males having a chance of gaining some paternity in sperm competition (Preston, Stevenson, Pemberton and Wilson, 2001). Similarly, in blue-headed wrasse fish, males that enjoy the highest mating success confer the lowest fertility to females (Warner, Shapiro, Marconato and Petersen, 1995). In external fertilisers, group-spawning can at times result in reduced number of sperm released as a consequence of males reducing the number of sperm ejaculated in response to increased sperm intensity (Marconato, Tessari and Marin, 1995; Wedell *et al.*, 2002). In some species of fish, males are sensitive to cues relating to expected fertilisation returns and release varying number of sperm in direct response to variation in female fecundity (Shapiro, Marconato and Yoshikawa, 1994; Marconato *et al.*, 1995).

The production of non-fertile sperm may potentially also pose a cost to females by compromising their fertility. For example, transfer and storage of large number of non-fertile sperm may interfere with successful storage and utilization of fertile sperm. There is individual variation in the proportion of fertile sperm in sperm heteromorphic species (e.g. Cook and Wedell, 1996; Crudgington *et al.*, 2005). However, there is as yet no experimental examination of the possibility that high proportion of non-fertile sperm carries fertility costs to females. In a survey of the incidence of infertility across insects, ~22% of matings involved insemination failures (García-González, 2004). Some of these failures may be due to transfer of large quantities of non-fertile sperm.

However, not all paternity assurance mechanisms are costly to females. It has been suggested that the evolution of paternal care in fish, for example, may have evolved as a means to increase male fertilisation success (Ah-King, Kvarnemo and Tullberg, 2005). The fact that male care is often correlated with paternity assurance both across and within species (e.g. Møller and Birkhead, 1993; Freeman-Gallant, 1996; Smith, 1997; Wright, 1998), has led to the proposal that caring males may enjoy a higher paternity (Tallamy, 2000; Kvarnemo, 2006). The idea being that male care can increase paternity either by being preferred as mating partners because they make good fathers (e.g. Owens, Burke and Thompson, 1994), or because males simply enjoy higher paternity by caring since this activity may directly reduce sperm competition (Kvarnemo, 2006).

Summary and future prospects

It is clear that the cost to males of producing many sperm, which is required to succeed in sperm competition, has promoted the evolution of sperm shape and form. In many species males have evolved impressive ejaculate production capacities as one way of gaining higher paternity. In other species, males have evolved more elaborate sperm. In this review I have argued that the cost of ejaculate production and the necessity of producing vast quantities of sperm, have promoted the evolution of 'manipulative' ejaculates, such as mating plugs, non-fertile sperm, cooperative sperm, and seminal peptides. Many of these ejaculate components are designed to reduce or remove the risk of sperm competition altogether and hence the need to transfer large number or elaborate sperm to the female. Some of these adaptations are costly to females since they interfere with their reproductive rate or physiology.

Sexual conflict over reproduction is rife in animals. Theory predicts males will have more to gain from inseminating hetero-specific females because of a higher ratio of potential gains to cost per mating than females, and hence tend to break down barriers against gene-flow (Parker and Partridge, 1998). However, high cost of ejaculate production may act to reduce gene flow and hence reinforce such genetic barriers, since males should strategically allocate their precious ejaculate to maximise their reproductive output. There is evidence showing that males modulate their ejaculates to cues relating to expected fertilisation returns, including reducing sperm numbers in hetero-specific matings. There may therefore be a hitherto largely unrecognised conflict between the sexes over the number of sperm ejaculated. It is known that in species where males' investment in reproduction approaches that of females (e.g. Simmons, 1992), sex-role reversal can occur, with males discriminating and rejecting low fecundity females as mates (Gwynne and Simmons, 1990). Similarly, male mate choice in the form of ejaculate tailoring may directly compromise female reproduction and hence generate sexual conflict.

Although sperm generally greatly outnumber eggs at the gamete production level (Bateman, 1948), this is not always true at the point of fertilisation where competition between gametes occurs, resulting in both variation and reduction in sperm-egg ratios. For the female, the obvious consequence of sperm limitation is that producing an egg is no longer equivalent of producing a zygote, since sperm prudence by males can result in sperm limitation for eggs. Females' reproductive success may under these situations be limited by the number of matings performed rather than by amount of resources, thus being similar to that of males. We then have a situation of similar near-linear returns of gametic investment for both sexes and a convergence of male and female life histories. Moreover, in species with gargantuan sperm the cost of production is extreme. In *D. bifurca*, for example, males and females are nearly isogamous with respect to gamete size and production rate, and the high production cost of the giant sperm directly determine variation in sexual selection intensity between the sexes (Bjork and Pitnick, 2006). Cost of gamete production therefore has direct implications for life history theory, for example when determining the optimal time for emergence/eclosion and mating rate.

Acknowledgements

I thank E.R.S. Roldan and M. Gomendio for kindly inviting me to contribute to the 10th International Symposium of Spermatology, and D. Hosken and T. Tregenza for discussion. This work was supported by a NERC research grant, the European Social Fund, and a Royal Society University Research Fellowship.

References

Ah-King, M., Kvarnemo. C. and Tullberg, B.S. (2005). The importance of sperm competition and territoriality for the evoution of male care: a phylogenetic study of fish. *Journal of Evolutionary Biology* **18**: 371-382

Anderson, M.J. and Dixon, A.F. (2002). Sperm competition: motility and the midpiece in primates. *Nature* **416**: 367-381

Andersson, J., Borg-Karlson, A-K. and Wiklund, C. (2000). Sexual cooperation and conflict in butterflies: A male transferred anti-aphrodisiac reduces harassment of recently mated females. *Proceedings of the Royal Society Series B* **267**: 1271-1275

Andersson, J., Borg-Karlson, A-K. and Wiklund, C. (2003). Functional significance of citral as an aphrodisiac in the butterfly *Pieris napi*: Behavioural experiments and EAG tests. *Journal of Chemical Ecology* **29**: 1489-1499

Arnaud, L. and Haubruge, E. (1999). Mating behaviour and male mate choice in *Tribolium castaneum* (Coleoptera, Tenebrionidae). *Behaviour* **136**: 67-77

Arnqvist, G. and Rowe, L. (2005). *Sexual Conflict*. Princeton University Press, Princeton

Aspbury, A.S. and Gabor, C.R. (2004). Discriminating males alter sperm production between species. *Proceedings of the National Academy of Sciences USA* **101**: 15970-15973

Adiyodi, K.G. and Adiyodi, R.D. (2000). *Reproductive Biology of Invertebrates*, vol IX. Wiley

Baer, B. and Schmid-Hempel, P. (1999). Experimental variation in polyandry affects parasite loads and fitness in a bumble-bee. *Nature* **397**: 151-154

Baker, R.R. and Bellis, M.A. (1988). "Kamikaze' sperm in mammals? *Animal Behaviour* **36**: 936-939

Ball, M.A. and Parker, G.A. (1997). Sperm competition games: Inter- and intra-species results of a continuous fertilization model. *Journal of Theoretical Biology* **186**: 459-466

Ball, M.A. and Parker, G.A. (2000). Sperm competition games: A comparison of loaded raffle models and their biological implications. *Journal of Theoretical Biology* **206**: 487-506

Barker, D.M. (1994). Copulatory plugs and paternity assurance in the nematode *Caenorhabtitis elegans*. *Animal Behaviour* **48**: 147-156

Bateman, A.J. (1948). Intra-sexual selection in *Drosophila*. *Heredity* **2**: 349-368

Bawa, S.R. (1964). Electron microscope study of spermiogenesis in a firebrat insect *Thermobia domestica* Pack. *Journal of Cell Biology* **23**: 431-444

Birkhead, T.R. and Møller, A.P. (1992). *Sperm Competition in Birds*, Academic Press, London

Birkhead, T.R. and Møller, A.P. (1998). *Sperm Competition and Sexual Selection*. Academic Press, London

Bjork, A. and Pitnick, S. (2006). Intensity of sexual selection along the anisogamy-isogamy continuum. *Nature* **441**: 742-745

Bretland, O.P. and Simmons, E. (1970). Preliminary studies of the spermatozoa and the male reproductive system of some whirligig beetles (Coleoptera: Gyrinidae). *Entomology News* **81**: 101-110

Briskie, J.V. and Montgomerie, R. (1993). Patterns of sperm storage in relation to sperm competition in passerine birds. *Condor* **95**: 442-454

Buckland-Nicks, J. (1998). Prosopbranch parasperm: sterile germ cells that promote paternity? *Micron* **29**: 267-280

Bukowski, T.C. and Christenson, T.E. (1997). Determinants of sperm release and storage in a spiny orbweaving spider. *Animal Behaviour* **53**: 381-395

Bulmer, M.G. and Parker, G.A. (2002). The evolution of anisogamy: a game-theoretic approach. *Proceedings of the Royal Society Series B* **269**: 2381-2388

Cantacuzene, A.M. (1968). On types of sperm association in Orthoptera. *Zeitschrift für Zellforschnung und Microscopie* **90**: 113

Champion de Crespigny, F.E. and Wedell, N. (2006). *Wolbachia* infection reduces sperm competitive ability in an insect. *Proceedings of the Royal Society Series B* **273**: 1455-1458

Chapman, T. and Davies, S.J. (2004). Functions and analysis of the seminal fluid proteins of male *Drosophila melanogaster* fruit flies. *Peptides* **25**: 1477-1490

Chapman, T., Arnqvist, G., Bangham, J. and Rowe, L. (2003). Sexual conflict. *Trends in Ecology and Evolution* **18**: 41-47

Civetta, A. and Clark, A.G. (2000). Chromosomal effects on male and female components of sperm precedence in Drosophila. *Genetical Research* **75**: 143-151

Cohen, J. (1977). *Reproduction*. Butterworths, London

Constantz, G.D. (1984). Sperm competition in Poecilid fishes. In: *Sperm Competition and the Evolution of Animal Mating Systems*. Edited by Smith, R.L., Academic Press, London, pp 465-485

Cook, P.A. and Gage, M.J.G. (1995). Effects of risks of sperm competition on the numbers of eupyrene and apyrene sperm ejaculated by the male moth *Plodia interpunctella* (Lepidoptera: Pyralidae). *Behavioral Ecology and Sociobiology* **36**: 261-268

Cook, P.A. and Wedell, N. (1996). Ejaculate dynamics in butterflies: a strategy for maximizing fertilization success? *Proceedings of the Royal Society Series B* **263**: 1047-1051

Cook, P.A. and Wedell, N. (1999). Non-fertile sperm delay female remating. *Nature* **397**: 486

Cordero, A. and Miller, P.L. (1992). Sperm transfer, displacement and precedence in *Ischnura graellsii* (Odontata: Coenagrionidae). *Behavioral Ecology and Sociobiology* **30**: 261-267

Cordero-Rivera, A., Andrés, J.A., Córdoba-Aguilar, A. and Utzeri, C. (2004). Postmating sexual selection: allopatric evolution of sperm competition mechanisms and genital morphology in Calopterygig damselflies (Insecta: Odonata). *Evolution* **58**: 349-359

Cordoba Aguilar, A. (2006). Sperm ejection as a possible cryptic female choice mechanism in Odonata (Insecta). *Physiological Entomology* **31**: 146-153

Cosmides, L.M. and Tooby, J. (1981). Cytoplasmic inheritance and intragenomic conflict. *Journal of Theoretical Biology* **89**: 83-129

Crudgington, H.S. and Siva-Jothy, M.T. (2000). Genital damage, kicking and early death. *Nature* **407**: 855-856

Crudgington, H.S., Beckerman, A.P., Brustle, L. Green, K. and Snook, R.R. (2005). Experimental removal and elevation of sexual selection: Does sexual selection generate manipulative males and resistant females? *American Naturalist* **165**: S72-S87

Dewsbury, D.A. (1982). Ejaculate cost and male choice. *American Naturalist* **119**: 601-610

Dickinson, J.L. and Rutowski, R.L. (1989). The function of the mating plug in the chalcedon checkerspot butterfly. *Animal Behaviour* **38**: 154-162

Dorus, S., Evans, P.D., Wyckoff, G.J., Choi, S.S. and Lahn, B.T. (2004). Rate of molecular evolution of the seminal protein gene *SEMG2* correlates with levels of female promiscuity. *Nature Genetics* **36**: 1326-1329

Dunn, A.M., Andrews, T., Ingrey, H., Riley, J. and Wedell, N. (2006). Strategic sperm allocation under parasitic sex ratio distortion. *Biology Letters* **2**: 78-80

Eberhard, W.G. (1985). *Sexual Selection and Animal Genitalia*. Harvard University Press, Cambridge

Eberhard, W.G. (1996). *Female Control: Sexual Selection by Cryptic Female Choice*. Princeton, Princeton University Press, Princeton

Fan, Y., Rafaeli, A., Moshitzky, P., Kubli, E., Choffat, Y. and Applebaum, S.W. (2000). Common functional elements of *Drosophila melanogaster* sex peptide stimulates juvenile hormone biosynthesis and depresses sex pheromone production in *Helicoverpa armigera*. *Insect Biochemistry and Molecular Biology* **30**: 805-812

Freeman-Gallant, C.R. (1996). DNA fingerprinting reveals female preference for male parental care in Savannah Sparrows. *Proceedings of the Royal Society Series B* **263**: 157-160

Friberg, U., Lew, T.A., Byrne, P.G. and Rice, W.R. (2005). Assessing the potential for an ongoing arms race within and between the sexes: selection and heritable variation. *Evolution* **59**: 1540-1551

Friedländer, M., Seth, R.K. and Reynolds, S.E. (2005). Eupyrene and apyrene sperm: Dichotomous spermatogenesis in Lepidoptera. *Advances in Insect Physiology* **32**: 206-308

Froman, D.P., Pizzari, T., Feltmann, A.J., Castillo-Juares, H. and Birkhead, T.R. (2002). Sperm mobility: mechanisms of fertilizing efficiency, genetic variation and phenotypic relationship with male status in the domestic fowl, *Gallus gallus domesticus*. *Proceedings of the Royal Society Series B* **269**: 607-612

Fromhage, L. and Schneider, J.M. (2006). Emasculation to plug up females: the significance of pedipalp damage in *Nephila fenestrata*. *Behavioral Ecology* **17**: 353-357

Fu, P., Neff, B.D. and Gross, M.R. (2001). Tactic-specific success in sperm competition. *Proceedings of the Royal Society Series B* **268**: 1105-1112

Gage, A.R. and Barnard, C.J. (1996). Male crickets increase sperm number in relation to competition and female size. *Behavioral Ecology and Sociobiology* **38**: 227-237

Gage, M.J.G. (1994). Associations between body size, mating pattern, testes size, and sperm lengths across butterflies. *Proceedings of the Royal Society Series B* **258**: 247-254

Gage, M.J.G. and Cook, P.A. (1994). Sperm size or numbers? Effects of nutritional stress upon eupyrene and apyrene sperm production strategies in the moth *Plodia interpunctella* (Lepidoptera, Pyralidae). *Functional Ecology* **8**: 594-599

Gage, M.J.G. and Morrow, E. (2003). Experimental evidence for the evolution of many tiny sperm. *Current Biology* **13**: 754-757

Gage, M.J.G., Stockley, P. and Parker, G.A. (1995). Effects of alternative male mating strategies on characteristics of sperm production in the Atlantic salmon (*Salmo salar*): Theoretical and empirical investigations. *Philosophical Transactions of the Royal Society of London Series B* **350**: 391-399

García-González, F. (2004). Infertile matings and sperm competition: The effect of 'nonsperm representation' on intraspecific variation in sperm precedence patterns. *American Naturalist* **164**: 457-472

Gilbert, L.E. (1976). Post-mating female odour in Heliconius butterflies: a male contributed anti-aphrodisiac? *Science* **193**: 419-420

Gillott, C. (2003). Male accessory gland secretions: Modulators of female reproductive physiology and behavior. *Annual Review of Entomology* **48**: 163-184

Gomendio, M., Harcourt, A.H. and Roldan E.R.S. (1998). Sperm competition in mammals. In: *Sperm Competition and Sexual Selection*. Edited by T.R. Birkhead and A.P. Møller, Academic Press, London, pp. 667-752

Gomendio, M. and Roldan, E.R.S. (1991). Sperm competition influences sperm size in mammals. *Proceedings of the Royal Society series B* **243**: 181-185

Gwynne, D.T. and Simmons, L.W. (1990). Experimental reversal of courtship role in an insects. *Nature* **346**: 172-174

Harcourt, A.H. (1991). Sperm competition and evolution of nonfertilizing sperm in mammals. *Evolution* **45**: 314-328

Haubruge, E., Arnaud, L., Mignon, J. and Gage, M.J,G. (1999). Fertilization by proxy: rival sperm removal and translocation in a beetle. *Proceedings of the Royal Society Series B* **266**: 1183-1187

Hayakawa, Y., Munehara, H. and Komaru, A. (2002a). Obstructive role of the dimorphic sperm in a non-copulatory marine sculpin, *Hemilepidotus gilberti,* to prevent other males' eusperm from fertilization. *Environmental Biology of Fishes* **64**: 419-427

Hayakawa, Y., Akiyama, R., Munehara, H. and Komaru, A. (2002b). Dimorphic sperm influence semen distribution in a non-copulatiry sculpin *Hemilepidotus gilberti. Environmental Biology of Fishes* **65**: 311-317

Hayashi, F. (1998). Sperm co-operation in the fishfly, *Parachauliodes japonicus. Functional Ecology* **12**: 347-350

Hirai, K., Shorey, H.H. and Gaston, L.K. (1978). Competition among courting male moths: male-to-male inhibitory pheromone. *Science* **202**: 644-645

Howard, D.J. (1999). Conspecific sperm and pollen precedence and speciation. *Annual Review of Ecology and Sysematics* **30**: 109-132

Hurst, L.D. (1990). Parasite diversity and the evolution of diploidy, multicellularity and anisogamy. *Journal of Theoretical Biology* **144**: 429-443

Ianora, A., Miralto, A., and Buttino, I. (1999). First evidence of some dinoflagellates reducing male copepod fertilization capacity. *Limnology and Oceanography* **44**: 147-153

Jackson, R.R. (1980). The mating strategy of *Phidippus johnsoni* (Aranae, Salticidae), II. Sperm competition and the function of copulation. *Journal of Arachnology* **8**: 217-240

Jamieson, B.G.M. (1987). *The Ultrastructure and Phylogenies of Insect Spermatozoa*. Cambridge University Press, Cambridge

Jamieson, B.G.M. (2000). *Reproductive Biology of Invertebrates. Volume IX. Progress in Male Gamete Ultrastructure and Phylogeny. Part C.* John Wiley and Sons, Chichester

Jones, TM. (2001). A potential cost of monandry in the lekking sandfly *Lutzomyia longipalpis. Journal of Insect Behavior* **14**: 385-399

Kaitala, A. and Wiklund, C. (1995). Female mate choice and mating costs in the polyandrous butterfly *Pieris napi* (Lepidoptera, Pieridae). *Journal of Insect Behavior* **8**: 355-363

Kawamura, N., Yamashiki, N., Saitoh, H. and Sahara, K. (2000). Peristaltic squeezing of sperm bundles at the later stage of spermatogenesis in the silkworm, *Bombyx mori. Journal of Morphology* **246**: 53-58

Kilgallon, S.J. and Simmons. L.W. (2005). Image content influences men's semen quality. *Biology Letters* **1**: 253-255

Kingan, S.B., Tatar, M. and Rand, D.M. (2003). Reduced polymorphism in the chimpanzee semen coagulating protein, semenogelin I. *Journal of Molecular Evolution* **57**: 159-169

Kokko, H. and Ots, I. (2006). When not to avoid inbreeding. *Evolution* **60**: 467-475

Kraaijevel-Smith, F.J.L., Ward, S.J., Temple-Smith, P.D. and Paetkau, D. (2002). Factors influencing paternity success in *Anthechinus agilis*: last-male sperm precedence, timing of mating and genetic compatibility. *Journal of Evolutionary Biology* **15**: 100-107

Kubli, E. (2003). Sex-peptides: seminal peptides of the *Drosophila* male. *Cellular and Molecular Life Sciences* **60**: 1689-1704

Kvarnemo, C. (2006). Evolution and maintenance of male care: is increased paternity a neglected benefit of care? *Behavioral Ecology* **17**: 144-148

Labitte, M.A. (1919). Observations sur *Rhodocera rhamni. Bulletin of the Museum of Natural History of Paris* **25**: 624-625

Lariviere, S. and Ferguson, S.H. (2003). Evolution of

induced ovulation in North American carnivores. *Journal of Mammology* **84**: 937-947

Lecompte, C., Thibout, E., Pierre, D. and Auger, J. (1998). Transfer, perception, and activity of male pheromone of *Acrolepiopsis assectella* with special reference to conspecific male sexual inhibition. *Journal of Chemical Ecology* **24**: 655-671

Lessells, C.M. (1999). Sexual conflicts in animals. In: *Levels of Selection*. Edited by Keller, L., Princeton University Press, Princeton, pp. 75-99

Lessells, C.M. (2005). Why are males bad for females? Models for the evolution of damaging male mating behavior. *American Naturalist* **165**: S46-S63

Lessells, C.M. (2006). The evolution of sexual conflict. *Philosophical Transactions of the Royal Society of London Series B* **361**: 301-317

Lewis, S.M., Kobel, A., Fedina, T. and Beeman, R.W. (2005). Sperm stratification and paternity success in red flour beetles. *Physiological Entomology* **30**: 303-307

Lorch, P.D., Wilkinson, G.S. and Reillo, P.R. (1993). Copulation duration and sperm precedence in the stalk-eyed fly *Cyrtodiopsis whitei* (Diptera: Diopsidae). *Behavioral Ecology and Sociobiology* **32**: 775-781

Luck, N. and Joly, D. (2005). Sexual selection and mating advantage in the giant sperm species, *Drosophila bifurca*. *Journal of Insect Sciences* **5**: Art. No.10

Mackie, J.B. and Walker, M.H. (1974). A study of the conjugate sperm of the dytiscid water beetles *Dytiscus marginalis* and *Colymbetes fuscus*. *Cell and Tissue Research* **148**: 505-519

Marconato, A., Tessari, V. and Marin, G. (1995). The mating system of *Xyrichthys novacula*: sperm economy and fertilization success. *Journal of Fish Biology* **47**: 292-301

Matsumoto, K. and Suzuki, N. (1992). Effectiveness of the mating plug in *Atrophaneura alcinous* (Lepidoptera: Papilionidae). *Behavioral Ecology and Sociobiology* **30**: 157-163

Mesterton-Gibbons, M. (1999). On sperm competition games: incomplete fertilization risk and the equity paradox. *Proceedings of the Royal Society Series B* **266**: 269-274

Møller, A.P. and Birkhead, T.R. (1993). Certainty of paternity covaries with paternal care in birds. *Behavioral Ecology and Sociobiology* **33**: 631-368

Moore, H., Dvorakova, K., Jenkins, N. and Breed, W. (2002). Exceptional sperm cooperation in the wood mouse. *Nature* **418**: 174-177

Moreau, J., Bertin, A., Caubert, Y. and Rigaud, T. (2001). Sexual selection in an isopod with *Wolbachia*-induced sex reversal: males prefer real females. *Journal of Evolutionary Biology* **14**: 388-394

Morrow, E. and Gage, M.J.G. (2001). Sperm competition experiments between lines of crickets producing different sperm lengths. *Proceedings of the Royal Society Series B* **268**: 2281-2286

Nakatsuru, K. and Kramer, D.L. (1982). Is sperm cheap? Limited male fertility and female choice in the lemon tetra (Pisces, Characidae). *Science* **216**: 753-755

Nicholls, E.H., Burke, T. and Birkhead T.R. (2001) Ejaculate allocation by male sand martins, *Riparia riparia*. *Proceedings of the Royal Society Series B* **268**: 1265-1270

Olsson, M., Madsen, T. and Shine, R. (1997). Is sperm really so cheap? Costs of reproduction in male adders, *Vipera berus*. *Proceedings of the Royal Society Series B* **264**: 455-459

Ono,T., Siva-Jothy, M.T. and Kato, A. (1989). Removal and subsequent inestion of rivals' semen during copulation in a tree cricket. *Physiological Entomology* **14**: 195-202

Orr, A.G. and Rutowski, R.L. (1991). The function of the sphragis in *Cressida cressida* (Fab) (Lepidoptera, Papilionidae): A visual deterrent to copulation attempts. *Journal of Natural History* **25**: 703-710

Owens, I.P.F., Burke, T. and Thompson, D.B.A. (1994). Extraordinary sex roles in the Eurasian dotterel: female mating arenas, female-female competition, and female mate choice. *American Naturalist* **144**: 76-100

Park, Y.I. and Ramaswamy, S.B. (1998). Role of brain, ventral nerve cord, and corpora cardiaca-corpora allata complex in the reproductive behavior of female tobacco budworm (Lepidoptera: Noctuidae). *Annual Entomological Society of America* **91:** 329-334

Park, Y.I., Ramaswamy, S.B. and Srinivasan, A. (1998). Spermatophore formation and regulation of egg maturation and oviposition in female *Heliothis virescens* by the male. *Journal of Insect Physiology* **44**: 903-908

Parker, G.A. (1970). Sperm competition and its evolutionary consequences in insects. *Biological Reviews* **45**: 525-567

Parker, G.A. (1978). Selection on nonrandom fusion of gametes during evolution of anisogamy. *Journal of Theoretical Biology* **73**: 1-28

Parker, G.A. (1979). Sexual selection and sexual conflict. In: *Sexual Selection and Reproductive Competition in Insects*. Edited by Blum, M.S. and Blum, N.A., Academic Press, New York, pp. 123-166

Parker, G.A. (1990a). Sperm competition games – sneaks and extra-pair copulations. *Proceedings of the Royal Society Series B* **242**: 127-133

Parker, G.A. (1990b). Sperm competition games - raffles and roles. *Proceedings of the Royal Society Series B* **242**: 120-126

Parker, G.A. (1998). Sperm competition and the evolution of ejaculates. In: *Sperm Competition and Sexual Selection*. Edited by Birkhead, T.R. and Møller, A.P., Academic Press, London, pp. 3-54

Parker, G.A. (2001). Golden flies, sunlit meadows: a tribute to the yellow dungfly. In: *Model Systems in Behavioral Ecology: Integrating Conceptual, Theoretical, and Empirical Approaches*. Edited by Dugatkin, L.A., Princeton University Press, Princeton, pp. 3-26

Parker, G.A., Baker, R.R. and Smith, V.G.F. (1972). The origin of evolution of gamete dimorphism and the male-female phenomenon. *Journal of Theoretical*

Biology *36: 529-553*

Parker, G.A. and Ball, M.A. (2005). Sperm competition, mating rate and the evolution of testis and ejaculation sizes: a population model. *Biology Letters* **1**: 235-238

Parker, G.A. and Partridge, L. (1998). Sexual conflict and speciation. *Philosophical Transactions of the Royal Society of London Series B* **353**: 261-274

Parker, G.A., Ball, M.A., Stockley, P. and Gage, M.J.G. (1996). Sperm competition games: assessment of sperm competition intensity by group spawners. *Proceedings of the Royal Society Series B* **263**: 1291-1297

Parker, G.A., Ball, M.A., Stockley, P. and Gage, M.J.G. (1997). Sperm competition games: a prospective analysis of risk assessment. *Proceedings of the Royal Society Series B* **264**: 1793-1802

Pitnick, S., Markow, T.A. and Spicer, G.S. (1995). Delayed male maturity is a cost of producing large sperm in Drosophila. *Proceedings of the National Academy of Sciences USA* **92**: 10614-10618

Pitnick, S., Spicer, G.S. and Markow, T.A. (1997). Phylogenetic examination of female incorporation of ejaculates in Drosophila. *Evolution* **51**: 833-845

Pizzari, T. and Birkhead, T.R (2000). Female feral fowl eject sperm of subdominant males. *Nature* **405**: 787-789

Preston, B.T., Stevenson, I.R., Pemberton, J.M. and Wilson, K. (2001). Dominant rams lose out by sperm depletion. *Nature* **409**: 681-682

Proctor, H.C. (1992). Effect of food deprivation on mate searching and spermatophore production in male water mites (Acari, Unionicolidae). *Functional Ecology* **6**: 661-665

Purchon, R.D. (1977). *The Biology of the Mollusca, 2nd* ed. Pergamon Press, Oxford

Pusey, A. and Wolf, M. (1996). Inbreeding avoidance in animals. *Trends in Ecology and Evolution* **11**: 201-206

Radwan, J. (1996). Intraspecific variation in sperm competition success in the bulb mite: A role for sperm size. *Proceedings of the Royal Society Series B* **263**: 855-859

Randerson, J.P. and Hurst, L.D. (1999). Small sperm, uniparental inheritance and selfish cytoplasmic elements: a comparison of two models. *Journal of Evolutionary Biology* **12**: 1110-1124

Rice, W.R. (1996). Sexually antagonistic male adaptation triggered by experimental arrest of female evolution. *Nature* **381**: 232-243

Rigaud, T. and Moreau, J. (2004). A cost of *Wolbachia*-induced sex reversal and female-biased sex ratios: decrease in female fertility after sperm depletion. *Proceedings of the Royal Society Series B* **271**: 1941-1946

Ryan, M.J., Fox, J.H., Wilczynski, W. and Rand, A.S. (1990). Sexual selection for sensory exploitation in the frog *Physalaemus pustulosus*. *Nature* **343**: 66-67

Sæther, S.A., Fiske, P. and Kalas, JA. (2001). Male mate choice, sexual conflict and strategic allocation of copulations in a lekking bird. *Proceedings of the Royal Society Series B* **268**: 2097-2102

Sahara, K. and Takemura, N. (2003). Application of artificial insemination technique to eupyrene and/or apyrene sperm in *Bombyx mori*. *Journal of Experimental Zoology* **297A**: 196-200

Sauter, A., Brown, M.J.F., Baer, B. and Schmidt-Hempel, P. (2001). Males of social insects can prevent queens from multiple mating. *Proceedings of the Royal Society Series B* **268**: 1449-1454

Schlupp, I. and Plath, M. (2005). Male mate choice and sperm allocation in a sexual/asexual mating complex of *Poecilia* (Poeciliidae, Teleostei). *Biology Letters* **1**: 169-171

Shapiro, D.Y., Marconato, A. and Yoshikawa, T. (1994). Sperm economy in a coral reef fish, *Thalassoma bifasciatum*. *Ecology* **75**: 1334-1344

Shine, R., Olsson, M.M. and Mason, R.T. (2000). Chastity belts in gartersnakes: the functional significance of mating plugs. *Biological Journal of the Linnean Society* **70**: 377-390

Shirk, P.D., Bhaskaran, G. and Roller, H. (1980). The transfer of juvenile hormone from male to female during mating in the *Hyalophora cecropia* silkmoths. *Experientia* **36**: 682-683

Shirk, P.D., Bhaskaran, G. and Roller, H. (1983). Developmental physiology of corpora allata and accessory glands in the *Cecropia* silkmoths. *Journal of Experimental Zoology* **227**: 69-79

Silberglied, R.E., Shepherd, J.G. and Dickinson, J.L. (1984). Eunuchs – the role of apyrene sperm in Lepidoptera. *American Naturalist* **123**: 255-265

Simmons, L.W. (1992). Quantification of role reversal in relative parental investment in a bush cricket. *Nature* **358**: 61-63

Simmons, L.W. (2001). *Sperm Competition and its Evolutionary Consequences in the Insects*. Princeton University Press, Princeton

Simmons, L.W. and Siva-Jothy, M.T. (1998). Sperm competition in insects. In: *Sperm Competition and Sexual Selection*. Edited by Birkhead, T.R. and Møller, A.P., Academic Press, London, pp. 341-434

Simmons, L.W., Craig, M., Llorens, T., Schinzig, M. and Hosken, D. (1993). Bushcricket spermatophores vary in accord with sperm competition and parental investment theory. *Proceedings of the Royal Society Series B* **251**: 183-186

Siva-Jothy, M.T. (1988). Sperm repositioning in *Crocothermis erythraea*, a Libellulid dragonfly with a breif copulation. *Journal of Insect Behavior* **1**: 235-245

Siva-Jothy, M.T. and Hooper, R.E. (1995). The disposition and genetic diversity of stored sperm in females of the damselfly *Calopteryx splendens xanthostoma* (Charpentier). *Proceedings of the Royal Society Series B* **259**: 313-318

Siva-Jothy, M.T. and Tsubaki, Y. (1989). Variation in copulation duration in *Mnais pruinosa* Selys (Odonata: Calypterygidae). 1. Alternative male-securing tactics and sperm competition. *Behavioral Ecology and Sociobiology* **24**: 39-45

Sivinski, J. (1980). Sexual selection and insect sperm. *Florida Entomologist* **63**: 99-111

Smith, M.J. (1992). Evidence from the estrus cycle for male-induced ovulation in *Bettongia penicillata* (Marsupialia). *Journal of Reproduction and Fertility* **95**: 283-289

Smith, R.L. (1984). *Sperm Competition and Evolution of Animal Mating Systems*, Academic Press, London

Smith, R.L. (1997). Evolution of paternal care in giant water bugs (Heteroptera: Belostomatidae). In: *The Evolution of Social Behavior in the Insects and Arachnids.* Edited by Choe, J.C. and Crespi, B.J., Cambridge University Press, Cambridge, pp 116-149

Snook, R.R. (1998). The risk of sperm competition and the evolution of sperm heteromorphism. *Animal Behavior* **56**: 1497-1507

Snook, R.R. and Karr, T.L. (1998). Only long sperm are fertilization-competent in six sperm-heteromorphic Drosophila species. *Current Biology* **8**: 291-294

Snook, R.R. and Markow, T.A. (2001). Mating system evolution in sperm heteromorphic Drosophila. *Journal of Insect Physiology* **47**: 957-964

Soller, M., Bownes, M. and Kubli, E. (1997). Mating and sex peptide stimulate the accumulation of yolk in oocytes of *Drosophila melanogaster. European Journal of Biochemistry* **243**: 732-738

Stockley, P. (1997). Sexual conflict resulting from adaptations to sperm competition. *Trends in Ecology and Evolution* **12**: 154-159

Stockley, P., Gage, M.J.G., Parker, G.A. and Møller, A.P. (1997). Sperm competition in fishes: the evolution of testis size and ejaculate characteristics. *American Naturalist* **149**: 933-954

Swallow, J.G. and Wilkinson, G.S. (2002). The long and the short of sperm polymorphism in insects. *Biological Reviews* **77**: 153-182

Swanson, W.J. and Vacquier, V.D. (2002). Reproductive protein evolution. *Annual Reviews of Ecology and Systematics* **33**: 161-179

Tallamy, D.W. (2000). Sexual selection and the evolution of exclusive paternal care in arthropods. *Animal Behavior* **60**: 559-567

Taggart, D.A. and Temple-Smith, P.D. (1991). Transport and storage of spermatozoa in the female reproductive tract of the brown marsupial mouse, *Anthechinus stuartii* (Dasyuridae). *Journal of Reproduction and Fertility* **93**: 97-110

Till-Bottraud, I., Joly, D., Lachaise, D. and Snook, R.R. (2004). Pollen and sperm heteromorphism: convergence across kingdoms? *Journal of Evolutionary Biology* **18**: 1-18

Tregenza, T., Wedell, N. and Chapman, T. (2006). Sexual conflict: a new paradigm? *Philosophical Transactions of the Royal Society of London Series B* **361**: 229-234

Tschudi-Rein, K. and Benz, G. (1990). Mechanisms of sperm transfer in female *Pieris brassicae* (Lepidoptera, Pieridae). *Annals of the Entomological Society of America* **83**: 1158-1164

Urbani, N., Saint-Marie, B., Sévigny, J.-M., Zadworny, D. and Kuhnlein, U. (1998). Sperm competition and paternity assurance during the first breeding period of female snow crab (*Chionoecetes opilio*) (Brachyura: Majidae). *Canadian Journal of Fisheries and Aquatic Sciences* **55**: 1104-1113

Vahed, K. (1998). The function of nuptial feeding in insects: a review of empirical studies. *Biological Reviews* **73**: 43-78

van Voorhies, W.A. (1992). Production of sperm reduces nematode life-span. *Nature* **360**: 456-458

von Helversen, D. and von Helversen, O. (1991). Premating sperm removal in the bush-cricket *Metaplastes ornatus* Ramme 1931 (Orthoptera, Tettigoniidae, Phaneropteridae). *Behavioral Ecology and Sociobiology* **28**: 391-396

Waage, J.K. (1979). Dual function of the damselfly penis: sperm removal and transfer. *Science* **203**: 916-918.

Walker, C.W., Harrington, L.M., Lesser, M.P. and Fagerberg, W.R. (2005). Nutritive phagocyte incubation chambers provide a structural and nutritive microenvironment for germ cells of *Strongylocentrotus droebachiensis*, the green sea urchin. *Biological Bulletin* **209**: 31-48

Warner, R.R., Shapiro, D.Y., Marconato, A. and Petersen, C.W. (1995). Sexual conflict – males with the highest mating success convey the lowest fertilization benefits to females. *Proceedings of the Royal Society Series B* **262**: 135-139

Wedell, N. (2001). Female remating in butterflies: interaction between female genotype and non-fertile sperm. *Journal of Evolutionary Biology* **14**: 746-754

Wedell, N. and Cook, P.A. (1998). Determinants of paternity in a butterfly. *Proceedings of the Royal Society Series B* **265**: 625-630

Wedell, N. and Cook, P.A. (1999). Butterflies tailor their ejaculate in response to sperm competition risk and intensity. *Proceedings of the Royal Society Series B* **266**: 1033-1039

Wedell, N., Gage, M.J.G. and Parker, G.A. (2002). Sperm competition, male prudence and sperm limited females. *Trends in Ecology and Evolution* **17**: 313-320

Wedell, N., Kvarnemo, C., Lessells, C.K. and Tregenza, T. (2006). Sexual conflict and life histories. *Animal Behavior* **71**: 999-1011

Williams, P.D., Day, T. and Cameron, E. (2005). The evolution of sperm allocation strategies and the degree of sperm competition. *Evolution* **59**: 492-499

Wright, J. (1998). Paternity and paternal care. In: *Sperm Competition and Sexual Selection.* Edited by Birkhead, T.R. and Møller, A.P., Academic Press, London, pp. 117-145

Wolfner, M.F. (2002). The gifts that keep on giving: physiological functions and evolutionary dynamics of male seminal proteins in *Drosophila*. *Heredity* **88**: 85-93

Wyckoff, G.J., Wang, W. and Wu, C-I. (2000). Rapid evolution of male reproductive genes in the descent of man. *Nature* **403**: 304-309

Zbinden, M., Largiader, C.R. and Bakker, T.C.M. (2004). Body size of virtual rivals affects ejaculate size in sticklebacks. *Behavioral Ecology* **15**: 137-140

Spermatology. SRF Vol. 65. ERS Roldan and M Gomendio (eds) Nottingham University Press, Nottingham

Post-insemination sexual selection in birds

Tommaso Pizzari

Edward Grey Institute of Ornithology, Department of Zoology, University of Oxford, Oxford OX1 3PS, UK

In many avian species females obtain sperm from multiple males during a single reproductive event, setting the scene for post-insemination sexual selection through the competition of the ejaculates of different males over fertilisation (sperm competition) and female biased utilisation of sperm (cryptic female choice). The use of poultry techniques in combination with molecular tools is catalysing an interest in birds as vertebrate model systems to study the mechanisms of post-insemination sexual selection. This chapter: (i) outlines the main mechanisms of avian sperm competition, (ii) introduces methodological approaches to study post-insemination sexual selection in birds, (iii) reviews recent evidence of multiple mechanisms of strategic sperm allocation by males, and (iv) discusses mechanisms of cryptic female choice. Post-insemination variance in paternity in birds, appears to be determined by the interactions between complex male and female strategies of differential sperm utilisation. It is argued that a better understanding of the operation of post-insemination sexual selection in birds may be achieved through a two-pronged approach which, on the one hand, investigates behavioural and physiological mechanisms applying poultry techniques and molecular tools to domestic model species, and on the other, verifies these mechanisms and tests their adaptive significance in more natural populations.

Introduction

Sexual selection is one of the most powerful agents of evolutionary change arising from variance in the reproductive success among members of the same sex (Shuster and Wade, 2003).

Darwin (1871) proposed the idea of sexual selection to explain the evolution of male secondary sexual characters that were inconsistent with natural selection, as an evolutionary process that "depends on the advantage which certain individuals have over others of the same sex and species solely in respect of competition over mating and fertilization". Therefore, sexual selection is a consequence of sexual reproduction that promotes traits conferring an advantage in reproductive competition. Sex-specific investment in gametes typically results in higher potential reproductive rates - and thus variance in reproductive success and sexual selection intensity - in males (Bateman, 1948; Trivers, 1972). Variation in paternity is determined by competition between males for access to females (intra-sexual selection), and selection by females of males (inter-sexual selection; Andersson, 1994). Traditionally, sexual selection was thought to act exclusively on variation in copulation success. However, in most species females obtain sperm from multiple males within a reproductive event (Birkhead and Pizzari,

Corresponding author E-mail: tommaso.pizzari@zoology.oxford.ac.uk

2002). This promiscuity promotes both intra- and inter-sexual selection after insemination through the competition of the ejaculates of different males for the fertilisation of a set of eggs (sperm competition; Parker, 1970, 1998) and female preferential utilisation of the sperm of certain partners (cryptic female choice; Eberhard, 1996).

Sperm competition influences the variance in male reproductive success and prolongs intra-sexual selection beyond insemination (Birkhead and Pizzari, 2002; Shuster and Wade, 2003) by targeting the fertilising efficiency of an insemination (Parker, 1970; Simmons, 2001; Birkhead and Pizzari, 2002; Pizzari and Birkhead, 2002). Theory predicts that the all else being equal, the outcome of sperm competition is determined by the number of sperm that a male inseminates into a female relative to the number of sperm inseminated by his competitors (Parker, 1998). Consistent with this prediction, substantial evidence across a number of invertebrate and vertebrate taxa indicates that large ejaculates convey an advantage in sperm competition (e.g. Parker, 1984; Dziuk, 1996; Simmons and Siva-Jothy, 1998; Gage and Morrow, 2003; but see Gage, McFarlane, Yeates, Ward, Searle and Parker, 2006). In some species with internal fertilisation, where females store sperm, sperm are lost from the sperm storage sites over time, resulting in a progressive reduction in the number of sperm of an ejaculate available to fertilise successive eggs and hence the ejaculate's fertilising efficiency (e.g. Colegrave, Birkhead and Lessells, 1995). Traits that prolong the fertilising efficiency of the ejaculates stored by females are thus likely to be favoured by sperm competition. Seminal fluid, the non-cellular components of an ejaculate, may also have a profound impact on the outcome of sperm competition in some species. In principle, seminal fluid may influence sperm competition in three different, non-mutually exclusive ways. First, seminal fluid products may influence the reproductive decisions of an inseminated female in a way that is advantageous to the reproductive success of the inseminating male. For example, seminal peptides produced by the accessory glands (Acps) in *Drosophila melanogaster* manipulate female oviposition rate and propensity to re-mate (Chapman, 2001; Wolfner, 2002; Chapman, Bangham, Vinti, Seifried, Lung, Wolfner, Smith and Partridge, 2003), thus reducing the risk of sperm competition. Second, seminal fluid products may increase the fertilising efficiency of the sperm contained in the same ejaculate (self). For example, *D. melanogaster* sex peptide (Acp70A) increases the fertilising efficiency of self sperm (Liu and Kubli, 2003; Chapman *et al.*, 2003), while Acp36DE facilitates self sperm storage in the female sperm storage organs (Qazi and Wolfner, 2003). Similarly, seminal fluid may mediate the rate of sperm release and thus the fertilising performance of 'sperm trails' in the grass goby, *Zosterisessor ophiocephalus* (Mazzoldi, Scaggiante, Ambrosin and Rasotto, 2000). Third, seminal fluid products may reduce the fertilising efficiency of the sperm from competing ejaculates (non-self). Unidentified *D. melanogaster* Acps have been suggested to displace sperm previously stored by a female (Harshman and Prout, 1994; Clark, Aguadé, Prout, Harshman and Langley, 1995; Prout and Clark, 1996; Price, Kim, Posluszny and Coyne, 1999). Although seminal fluid and specialised accessory glands excreting seminal fluid products appear phylogenetically widespread, the mechanisms through which seminal fluid influences sperm quality and potentially neutralises competing ejaculates have been explored in few species and warrant further investigation (Chapman, 2001; Pizzari, 2004).

Cryptic female choice is expected to bias the outcome of sperm competition. However, the signature of female bias is difficult to predict. Directional cryptic female choice is a female-driven bias in paternity caused by a response to a male or ejaculate trait that is consistent across females. Non-directional cryptic female choice, on the other hand, is caused by female-driven biases in paternity driven by a response that is consistent within females but different across females. Depending on the experimental design, non-directional cryptic female choice can be detected by an individual female effect or by an individual male by individual female interaction effect (for discussion see: Birkhead, 2000; Pitnick and Brown, 2000; Birkhead and Pizzari, 2002).

Birds as a model system for mechanisms of post-insemination sexual selection

Modern understanding of the dynamics of sperm competition in birds is based mostly on studies of domestic galliforms (fowl, *Gallus gallus domesticus*, turkey, *Meleagris gallopavo*, Japanese quail, *Coturnix japonica)*, and passerine model systems such as the zebra finch (*Taenopygia guttata*). The way the ejaculates of different males interact over fertilisation within the female reproductive tract has a crucial impact on variation in paternity among males. The basic dynamics of sperm competition are fairly well established (Birkhead and Møller, 1992; Birkhead and Pizzari, 2002). Following insemination, sperm are taken up through peristaltic vaginal contractions and stored in the female's sperm storage tubules (SSTs) at the utero-vaginal junction of the female oviduct. Sperm are lost from the SSTs at a constant rate and passively migrate to the infundibulum, where fertilisation occurs, through peristaltic contractions of the female duct (Brillard, 1993; Etches, 1996).

Number of sperm inseminated

Artificial insemination experiments have revealed that the fertilising efficiency of an ejaculate in sperm competition depends on the number of sperm inseminated (Martin, Reimers, Lodge and Dziuk, 1974; Froman, Pizzari, Feltmann, Castillo-Juarez and Birkhead, 2002). The way different ejaculates are stored in the SSTs and the mechanisms through which sperm egress from the SSTs remain unresolved (Froman, 2003). Three main models have been proposed to explain the reproductive advantage gained by the last male inseminating a female (last male precedence): (a) passive sperm loss (Birkhead, Wishart and Biggins, 1995), (b) sperm stratification (Compton, Van Krey and Siegel, 1978; Van Krey, Balander and Compton, 1981), and (c) sperm displacement (Lessells and Birkhead, 1990). The passive sperm loss model proposes that ejaculates mix in the SSTs and the constant loss of sperm from the SSTs translates into an advantage of the insemination that is numerically most represented in the SSTs at the onset of ovulation. This results in a constant advantage gained by the last male inseminating a female (Birkhead and Parker, 1997). The stratification model proposes that different ejaculates are stratified in the SST with the most recent ejaculate layered on top of previous inseminations, and closer to the entrance of the SST, thus generating a 'last in, first out' pattern of sperm egression from the SST. The stratification model therefore predicts that the second insemination will have an initial fertilising advantage over the first. However the first ejaculate will acquire a fertilising advantage as the second ejaculate becomes progressively depleted. Finally, the sperm displacement model proposes that a new ejaculate will flush out sperm previously stored in the SSTs, resulting in loss of the previous insemination. While most experimental evidence is consistent with the passive sperm loss model (Birkhead *et al.*, 1995; Birkhead and Biggins, 1998), other work indicates potential for sperm stratification (Van Krey *et al.*, 1981; but see King, Brillard, Garrett, Bakst and Donoghue, 2002), and yet other studies have provided results which may be consistent with sperm displacement (Lessells and Birkhead, 1990; but see Birkhead *et al.*, 1995). It is important to note here that these models are not necessarily mutually exclusive. For example, King *et al.* (2002) observed that, following artificial insemination of the ejaculates of two males, on average only 4% of the SSTs of a female turkey contained sperm from both ejaculates. If this observation reflected a general pattern, it would suggest that the three models presented above have limited scope in explaining sperm competition dynamics in birds. It is indeed possible that some SSTs may remain empty despite the insemination of large ejaculates. This differential 'switching on and off' of SSTs may result in different inseminations being stored separately. This scenario has not been explicitly addressed theoretically. However, segregation of different ejaculates in different SSTs is likely to result in a scenario consistent with the passive sperm loss with direct interactions between ejaculates (predicted by the other models) only possible in SSTs where multiple ejaculates co-occur. Importantly, the testing of these models is based on the analysis of variation in paternity over successive eggs. However, so far studies of sperm competition dynamics in birds have not considered the temporal distribution of progeny during the laying period following insemination.

Sperm fertilising efficiency

Another important component of the success of an ejaculate in sperm competition is mediated by its efficiency to reach, be stored and remain competitive in the SST until and throughout the production of a clutch of eggs (Pizzari and Birkhead, 2002). In the domestic fowl sperm mobility (an assay which measures the net movement of a population of spermatozoa in an inert medium) is an important predictor of fertility (Froman, Feltmann, Rhoads and Kirby, 1999) particularly under sperm competition (Birkhead, Martinez, Burke and Froman, 1999). Recent experimental work indicates that while the number of sperm inseminated determines how many sperm reach the SST, less mobile ejaculates are lost from the SST at a faster rate (Froman *et al.*, 2002). The mechanism through which sperm are released from the SST is unresolved (Froman, 2003). A parsimonious -albeit untested- possibility is that sperm remain in the SST by swimming against a current generated by the epithelial cells that line the SST (Froman, 2003). As sperm swimming velocity declines below a threshold, sperm are flushed out of the SST. Because the mobility of an ejaculate is highly correlated with the velocity of individual sperm, this hypothesis provides a mechanistic explanation for how sperm mobility mediates sperm performance in the SST. This hypothesis predicts that ejaculates of low sperm mobility will be lost at a faster rate from the SSTs and thus, paradoxically, retain a fertilising advantage over the first few eggs ovulated following insemination. This prediction has been recently tested by inseminating a common set of domestic fowl hens with semen from distinct male pairs (Pizzari, Worley, Burke and Froman, unpublished). In each case, two insemination doses were used, a total of either 30 or 50×10^6 sperm per hen. In either case, the dose contained 10×10^6 sperm from the high mobility male. Thus, the ratios of sperm used for each male pair were 2:1 and 4:1 low to high. This design enabled to distinguish the independent effects of sperm number and sperm quality on time-dependent variation in paternity. Overall, high mobility ejaculates retained a fertilising advantage despite a strong numerical disadvantage, confirming the crucial role of sperm quality in avian sperm competition dynamics. Importantly, consistent with theory, low mobility ejaculates appeared to have a fertilising advantage over the first few eggs produced following insemination which was progressively lost over successive days.

These results indicate that males producing low mobility ejaculates are selected to invest large ejaculates and re-mate frequently with the same female to compensate for the higher rate of senescence of their sperm. This scenario may predict preferential investment in social dominance by males of low sperm quality to retain control over female re-mating rates. Increasing evidence suggests a physiological trade-off between social status and ejaculate fertilising efficiency in male fowl. First, controlling for copulation rates, males of a random-bred population of domestic fowl showed a weak but significant, negative relationship between status and the mobility of their extragonadal sperm reserves (Froman *et al.*, 2002). Second, in male feral fowl, a measure of sperm quality closely correlated with sperm mobility (average path swimming velocity, VAP) declined over successive natural ejaculates in socially dominant males but remained constant in subdominant males (Cornwallis, 2004). This pattern was reversed within individual males when their status was experimentally changed (Cornwallis, 2004). Third, a recent study revealed rapid and longer term fluctuations in sperm mobility within males, following a general pattern consistent with within-male trade-offs between sperm quality and male status (Pizzari, Cornwallis and Froman, in preparation). Clearly, more research is needed to understand the independent effect of sperm number and sperm quality in determining sperm competition dynamics and variation in paternity in birds.

Seminal fluid

Seminal fluid may also contribute to the competitive efficiency of an insemination. While many species produce dense ejaculates (e.g. passerines), others such as some galliforms produce specific seminal fluid products (Fujihara, 1992). Little is known about the effect of such seminal fluid products on sperm competition. Male *Coturnix* quails possess a well-developed proctodeal gland which produces a foamy substance that contributes to the inseminated ejaculate (Adkins-Reagan, 1999). Experimental evidence indicates that the fertilising efficiency of an insemination close to oviposition is improved by the presence of this foam, suggesting that this highly derived seminal product may help ejaculates by-pass a hard-shelled egg in the oviduct (Adkins-Reagan, 1999).

In the fowl, males lack specialised glands and seminal fluid comprises: (i) glutamate-rich seminal plasma, originating from the testis and the excurrent duct, and (ii) lymph-like transparent fluid, constituted by lymphatic exudates originating form the phallic folds of the male cloaca (Lake, 1971; Bahr and Johnson, 1991; Fujihara, 1992; Etches, 1996; Kirby and Froman, 2000). The effect of seminal plasma and transparent fluid on ejaculate mobility and the swimming velocity of individual sperm in the fowl is unresolved and ambiguous (Fujihara, 1992). For example, a seminal plasma protein (sperm motility inhibiting factor, SMIF) which has antibacterial properties, inhibits sperm motility, although this effect disappears in the presence of seminal plasma (Mohan, Saini and Joshi, 1995). Similarly, male fowl carrying the *sd* (sperm degeneration) allele are characterised by malformed proximal efferent ducts and reduced sperm longevity (Alhagbari, Engel and Froman, 1993). *sd*-mutants lack at least one non-albumin seminal plasma protein and their seminal fluid is characterised by an imbalance of electrolytes, amino acids, and protein (Alhagbari *et al.*, 1992; 1993). Fertility of *sd*-mutants is rescued when their spermatozoa are mixed with albumin-depleted seminal plasma protein from fertile roosters before artificial insemination (Alhagbari *et al.*, 1992). Transparent fluid has been involved in *in vitro* reduction in sperm motility over exposure time but this effect appears mediated by temperature and sperm density (Fujihara, 1992). Similarly, prolonged *in vitro* exposure to transparent fluid appears to result in spermatozoa bending at the mid-piece region (Fujihara, 1992). In contrast, other studies suggest that brief exposure to transparent fluid may confer a fertilising advantage when sperm were artificially deposited on the female cloaca but not when sperm were inseminated in the vagina (Lake, Ravie and Waddington, 1985). A transient stimulatory effect on sperm velocity may be determined by transparent fluid promoting alkalinity and high glucose-level in semen (Fujihara, 1992). The way interactions between sperm and seminal fluid are mediated by genetic similarity of the competing males is also unknown. One possibility is that genetic similarity may operate through immunoglobulins contained in both, seminal plasma and transparent fluid of the fowl, and seminal fluid - sperm antigenicity (Fujihara, 1992). Allocation of seminal fluid to individual ejaculates appears non-random in feral fowl (T. Pizzari, unpublished data). Strategic utilisation of seminal fluid would substantiate the idea that seminal fluid is important in sperm competition.

Methodological approaches to the study of sperm competition in birds

The study of sperm competition dynamics in internally fertilising species such as birds, presents considerable methodological challenges. Two breakthroughs have catalysed the study of sperm competition mechanisms in birds: (i) the collection of natural ejaculates, and (ii) artificial insemination of ejaculates into females.

The study of male sperm allocation requires the collection of natural ejaculates. This is particularly difficult to achieve non-invasively in internally fertilising species, such as birds. Recently, a variety of ingenious methods has been developed to collect natural ejaculates from different species, catalysing the study of strategic sperm allocation by male birds. First, a number of techniques based on the use of female 'dummies' fixed in a soliciting position have provided crucial information on ejaculate traits in wild species of birds. In particular, the 'freeze-dried' female technique developed by Pellatt and Birkhead (1994) uses a dead female fixed through 'freeze-drying' in a soliciting position and provided with a removable 'false cloaca' filled with PBS, to collect and preserve semen. This technique appears to work only in some of the species studied, mostly passerines. Interestingly, only a subsample of a male population typically responds to this technique by producing ejaculates consistently (T. R. Birkhead, personal communication). Second, Hunter, Harcourt, Wright and Davis (2000) swiped microscope slides on the cloaca of female Adélie penguins, *Pygoscelis adeliae*, to detect the presence of sperm following copulation. The use of this technique is limited to species or populations that are tame enough to enable researchers to be in close proximity of copulating pairs. In addition, this technique cannot provide a quantitative measure of sperm allocation but simply an indication as to whether sperm transfer occurred. This basic information however, can be very informative (see below). Third, attempts to collect ejaculates produced in natural copulations were first made in domestic fowl by attaching a 'semen collector' under the male cloaca which would interject the ejaculate before it is transferred to the female cloaca (e.g. Parker, McKenzie and Kempster, 1940). Fourth, recently, Pizzari and colleagues have developed a technique to obtain natural ejaculates by means of fitting a live female with a harness that covers the female cloaca and prevents semen transfer (Pizzari and Birkhead, 2000; Pizzari, Cornwallis, Løvlie, Jakobsson and Birkhead, 2003; Pizzari, Løvlie and Cornwallis, 2004; Løvlie, Cornwallis and Pizzari, 2005). By presenting a male fowl with a female held in a soliciting position, it is possible to reduce variation in ejaculate traits due to interactions between male and female (e.g. female resistance). The ejaculate is then collected with a pipette from the harness and the waterproof mat upon which the female is rested (Fig. 1). The advantage of this technique is that all males of a population can be trained to copulate with a harnessed female, however, its main limitation is that its use is restricted to tame birds.

Molecular tools to assign paternity have also played an important role in the demonstration and study of sperm competition. Indeed, molecular studies demonstrated multiple paternity in single clutches, leading to the realisation that sperm competition is typical across many species of birds, and to the quantification of variation in multiple paternity within and across species (Birkhead, 1998; Griffith, Thuman and Owens, 2002). Paternity studies remain crucial to explore variation in the level of promiscuity and measure the intensity of post-insemination sexual selection on different male traits. However, it is important to realise that because they measure the outcome of competitive fertilisation rather than the mechanisms of sperm competition and cryptic female choice, paternity studies alone have limited scope to unravel the male and female mechanisms whose interaction determine variation in paternity at a post-insemination stage. However, combined with an experimental approach that controls number and quality of sperm inseminated into a female, molecular tools are key to study the mechanisms of post-insemination sexual selection in birds.

Figure 1. Collection of natural ejaculates through the harness technique. (**a**) Live females (feral domestic fowl) are fitted with a harness of light elastic bands holding a thin film of soft plastic over the female cloaca, which prevents ejaculates from being inseminated. (**b**) The female is held by a researcher in a soliciting (crouching) position while the male mounts her and ejaculates on the harness. The ejaculate is then collected with a pipette from the harness or from the plastic mat upon which the female is rested. (**c**) This technique has been extensively used in recent years to study male sperm investment in natural ejaculates in populations of fowl. The use of this technique enables the experimental manipulation of several important variables, including a male's knowledge of the level of sperm competition associated with a female (Photographs, T. Pizzari, H. Løvlie).

Strategic sperm allocation by males

Sperm competition exerts a strong selective pressure on males to defend the paternity of a set of eggs from competitors, through increased investment in individual ejaculate size (i.e. number of sperm inseminated) and quality. Because the rate of spermatogenesis is limited, increased sperm investment in the current insemination may jeopardise investment in future inseminations (Wedell, Gage and Parker, 2002). Gametic production is costly and limits the reproductive success of an individual. This was first noted in females where the costs of reproduction are typically higher than in males (Bateman, 1948), while the fact that male reproductive success can also be constrained by sperm production was realised only recently (Dewsbury, 1982; Wedell *et al.*, 2002). Therefore, individual males are expected to respond to sperm competition by strategically allocating sperm according to the reproductive value of individual copulation opportunities (see Wedell chapter in this volume). The idea that males allocate sperm economically in order to inseminate multiple females has received surprisingly little attention, considering that male sperm reserves have been shown to be limited (Parker, McKenzie and Kempster, 1942; Dewsbury, 1982; Nakatsuru and Kramer, 1982; Levitan and Petersen, 1995).

Below, I review the evidence of strategic sperm allocation by male birds in relation to: (a) female sexual familiarity, (b) female reproductive quality, (c) female promiscuity (i.e. level of sperm competition associated with a copulation or a female), and (d) genetic compatibility between a male and female.

Sexual familiarity, the Coolidge effect and aspermic copulations

Consider a situation where a male has the opportunity of copulating again with a female he has already inseminated recently and/or wait for a new copulation partner. Provided that both copulation opportunities occur within a time scale which does not allow the replenishment of his sperm reserves, the male faces a trade-off between sperm investment in the sexually familiar and the sexually novel female. This scenario is plausible in several avian mating systems. For example, in socially monogamous species, a male may have the option of copulating with his social partner and/or with an extra-pair female (Birkhead and Møller, 1992; Birkhead, 1998). Similarly, in polygynandrous systems a male may have access to multiple females simultaneously (e.g. Sæther, Fiske and Kålås, 2001). All else being equal, there will be an optimal sperm investment by the male in each female which maximises the overall number of eggs fertilised. If he has already inseminated an optimal number of sperm in the familiar female, a male is expected to preferentially invest sperm in the novel female and enter competition for a new set of eggs. Preferential sperm investment in sexually novel females may be mediated by the 'Coolidge effect', renewed sexual interest in terms of likelihood of copulation, by males to a novel female (Wilson, Kuehn and Beach, 1963). Despite much anecdotal information concerning a minority of species, particularly mammals, the extent to which the Coolidge effect drives male copulatory behaviour is unresolved (Dewsbury, 1981). More importantly, in the absence of strategic sperm allocation, the Coolidge effect may result in males renewing copulation with novel females but not necessarily inseminating any sperm. Because the number of sperm inseminated by a male into different females has seldom been quantified, it is not known whether the Coolidge effect mediates male sperm allocation and the biological relevance of the Coolidge effect has remained uncertain until recently. Furthermore, although the Coolidge effect was originally inspired by the behaviour of poultry (Wilson *et al.*, 1963), until now there was no evidence of its occurrence in birds.

By studying semen transfer in Adélie penguins, Hunter *et al.* (2000) provided the first experimental evidence consistent with the idea of preferential sperm investment in novel females by male birds. Adélie penguins are socially monogamous but engage in extra-pair copulations. Hunter *et al.* (2000) showed that extra-pair copulations (EPCs) were more likely to result in traces of sperm on the female cloaca than within-pair copulations (WPCs), thus male Adélie penguins were more likely to transfer sperm during EPCs than during WPCs.

More recently, Pizzari *et al.* (2003) demonstrated that the Coolidge effect mediates preferential male sperm investment in sexually novel females in two populations of fowl: a feral population of domestic fowl, and a captive population of red junglefowl. A male was exposed to a female fitted with a harness and allowed to copulate *ad libitum*. When the male stopped copulating, the first female was replaced with a new female to test whether copulations had stopped because of sperm depletion or preferential investment in a new female. In both populations copulation rate declined with successive copulations with a female and was rescued by the presence of a new female. Importantly, in both populations, this Coolidge effect was mirrored by differential sperm allocation: the first ejaculates invested in a female contained typically more sperm than the last ejaculate produced with the same female, controlling for female order (i.e. regardless of whether a female was the first, second or third that a male encountered in an experimental trial). These results strongly suggest that when male fowl cease to inseminate a female they may not necessarily have depleted their sperm reserves but they may be saving sperm for new partners. The Coolidge effect may have far-reaching evolutionary implications. In a mating system where males typically encounter multiple females before being able to replenish their sperm reserves, as in the fowl (Pizzari and Birkhead, 2000), the economical allocation of sperm through the Coolidge effect may have important repercussions for a male's reproductive success. In addition, in socially monogamous species, extra-pair copulations may have a disproportionate fertilising success. Many factors can account for this advantage. For example, female reproductive asynchrony would enable a male to copulate with his social partner until her clutch is laid, following which he can allocate all of his sperm reserves to copulate exclusively with extra-pair females who are yet to lay (e.g. Dunn, Afton, Gloutney and Alisauskas, 1999). A non-mutually exclusive alternative is that even when females are synchronous, the Coolidge effect may enable a male to invest more sperm in less familiar females (e.g. extra-pair). The results by Hunter *et al.* (2000) are consistent with this idea. Finally, the Coolidge effect may contribute to the evolution of female polyandry if a female obtains more sperm from a novel male than from a male with whom she has copulated before. Because copulation is costly to females, one would expect females to minimise the number of copulations required to obtain a sperm supply sufficient to fertilise all the eggs. Because of the Coolidge effect, the number of sperm obtained by a female by copulating *n* times with one male is likely lower than the number of sperm she would obtain by copulating once with *n* males (Pizzari, 2002).

In several taxa, including a number of bird species, some copulations do not to result in the transfer of semen ('aspermic' copulations, e.g. Parker *et al.*, 1942; Westneat, Rambo, Fraterrigo, Birkhead and Fletcher, 1998; see also Garcia-Gonzalez, 2004). Both Hunter *et al.* (2000) and Pizzari *et al.* (2003) support the idea that aspermic copulations are not randomly distributed across partners and may in fact be a form of extreme male sperm allocation. The probability of male fowl failing to inseminate any semen into a female increased significantly with the number of times a male had previously inseminated the same female, regardless of whether she was his first, second or third copulation partner (Pizzari *et al.* 2003). Why do males copulate if they do not deliver any semen? The adaptive significance of aspermic copulations remains unclear. A recent study in feral fowl has shed some light onto the adaptive significance of this male behaviour. Løvlie *et al.* (2005) tested the idea that mounting *per se*, independently of insemination products, may play a role in sperm competition by inhibiting female propensity to copulate again. The authors measured the effect of mounting *per se* on female sexual behaviour, through experimental manipulations afforded by the harness technique. Four treatments were used: (a) hens exposed to natural copulations (mounting + insemination) by two males, (b) hens fitted with the harness and exposed to natural copulations by the same two males (i.e. mounting but no insemination), (c) hens exposed to the same two males which were prevented from mating, (d) hens fitted with the harness and exposed to males as in (c). Following these treatments the sexual response of hens (one from each treatment) to a third focal male was monitored over four successive days. Mounting alone caused a transient but drastic reduction in female propensity to re-mate, and – crucially – the number of sperm that a female obtained from a new male. These results indicate that the inhibition of female propensity to copulate again, observed across many taxa, may be at least partly explained by male mounting alone. In particular, the possibility that female birds respond to mounting alone by reducing their mating propensity is suggested by two studies. First, female Japanese quails, *Coturnix japonica*, producing fertilised eggs following natural mating associated with males less than females producing eggs that were not fertilised (Persaud and Galef, 2004). A potential explanation of this result may be that females that produced fertilised eggs, had accumulated more copulations than the control females. Second, female zebra finches paired to vasectomised males (that mounted but could not deliver semen to their partners), were not more likely to re-mate with a new male than females paired to intact males (Birkhead and Fletcher, 1995).

Together, these studies suggest that aspermic copulations may play an important role in the sperm competition strategies of male birds. In particular, males may perform aspermic copulations on females they recently inseminated in order to reduce the risk of sperm competition faced by their ejaculates with minimal additional investment in a female.

Female quality

Whenever females vary in reproductive investment (i.e. number and quality of eggs produced), males may gain by preferentially investing sperm in females of higher reproductive value. In principle, two selective mechanisms may operate on male preferential sperm investment in high quality females: (a) direct selection, determined by the higher number of offspring that a male can produce, and (b) indirect selection, determined by the superior reproductive performance of the offspring which may be conveyed by superior maternal investment in sons and daughters and superior reproductive quality inherited by daughters.

The evolution of male preferential investment in relation to female quality assumes reliable predictors of female reproductive investment the expression of which should form the basis of male sperm allocation decisions. In several bird species, females display sexual ornaments that may covary with different aspects of reproductive investment and there is increasing evidence that these ornaments may play an important role in male selection of partners (Hill, 1993; Jones and Hunter ,1993). It is therefore plausible that female ornaments covarying with female reproductive investment may also mediate cryptic male choice of partners in the form of differential sperm allocation across different females. Evidence of this comes from a recent study of feral fowl (Pizzari *et al.*, 2003). In this species, relative comb size is positively correlated with different measures of female reproductive investment: fecundity, egg size and relative yolk mass (Tufvesson *et al.*, 1999; Pizzari *et al.*, 2003). When simultaneously presented with females that differed in relative comb size male feral fowl preferred to copulate with the large-combed female first and with the less ornamented female second. By reversing the order with which a male copulated with the two females in a second trial (by preventing the male from copulating first with the female with whom he had copulated first in the previous trial), the authors were able to determine the relative sperm investment of a male in both types of females, controlling for copulation order. On average, large combed females obtained a higher proportion of a male's sperm expenditure over both trials. Interestingly, this male response was status-specific (Pizzari *et al.*, 2003). These results have been confirmed by a recent experimental follow up of the same population (Cornwallis and Birkhead, 2006). Sperm expenditure by dominant males was more biased in favour of large-combed females than that of subdominant males. One explanation for this response is that dominant male fowl have higher re-mating rates than subdominants and this higher risk of sperm depletion may exert a stronger selective pressure for prudent sperm allocation.

Levels of sperm competition

An extensive theoretical framework has been developed by Parker and colleagues to study optimal ejaculate expenditure by males under different levels of sperm competition (Parker, 1998; Wedell *et al.*, 2002). Models analysing optimal sperm allocation under varying levels of sperm competition generate multiple predictions. In general, within a species, optimal ejaculate investment by a male is predicted to increase with the probability that his sperm will compete with the sperm of another male (risk of sperm competition). Intraspecific models of sperm investment in relation to the number of other males that will inseminate the same female (or intensity of sperm competition) predict that a male should maximise sperm expenditure when his sperm compete with the sperm of only another male, and should progressively reduce sperm investment as the number of competitors inseminating the same female increases beyond one (Wedell *et al.*, 2002).

There is some evidence that the level of sperm competition associated with a female or copulation may influence male sperm investment in an ejaculate. The use of freeze-dried female dummies in colonies of sand martins, *Riparia riparia*, where males often seek and compete over access to extra-pair females, showed that ejaculates transferred by a male in the presence of a rival male contained significantly more sperm than ejaculates transferred in the absence of a rival male (Nicholls, Burke and Birkhead, 2001). These results are generally consistent with the predictions generated by sperm competition risk models. A recent study of feral fowl produced more complex results. The study compared the number of sperm invested by a male in a female when alone, in the presence of one competitor, and in the presence of three other competitors (Pizzari *et al.*, 2003). The sperm investment of individual males changed across treatments in a status-specific manner. Consistent with theory, all males minimised sperm investment in the absence of competition. However, when competitors were added, dominant males (i.e. the top-ranking male in the male pairs, and the top two ranking males in groups of four) increased sperm investment as competitors increased from one to three, whereas the sperm investment of subdominant males was maximised in the presence of just one competitor. The functional significance of this differential response remains unclear. It is possible that status-specific access to copulation opportunities in this species may underlie this difference. Because dominant males have more control over females, the addition of three more competitors may result in an increased risk that a female will copulate again with another male. Consistent with models of optimal sperm investment under sperm competition risk, dominant males increase their investment as the risk increases. Subdominant males on the other hand, have limited access to females and, in the presence of other males, their ejaculates have a high probability of facing sperm competition. Subdominant males may therefore reduce sperm investment in copulations associated with relatively intense sperm competition, as predicted by sperm competition intensity models. The decision-making process of such sperm allocation strategies remains unclear. Males appear to respond to the number of competitors, while the number of times they observe a competitor copulate with the same female does not appear to have an influence. This may make sense in a species where ejaculates are stored by a female for over two weeks, which may make information on the immediate promiscuity of a female little informative. Similarly, the response of dominant males needs further investigation. It is expected to decline when enough competitors are added so that sperm competition becomes intense for a dominant male.

Genetic compatibility

Reproductive success may also vary due to the genetic compatibility of two partners (Tregenza and Wedell, 2002). Given the choice, both males and females are therefore expected to prefer genetically compatible partners. In addition, a male is expected to copulate with a non-compatible female whenever this does not jeopardise his reproductive success with more compatible partners. A captive population of red junglefowl provides two examples consistent with differential sperm allocation in relation to two mechanisms of genetic compatibility: (a) Inbreeding, and (b) Major Histocompatibility Complex (MHC). (a) Inbreeding promotes homozygosity which can depress fitness due to the expression of deleterious recessive alleles and loss of heterozygote advantage (Charlesworth and Charlesworth, 1987; Lynch and Walsh, 1998). Red junglefowl live in small, sexually promiscuous populations where reproductive skew, limited dispersal by both sexes (Collias and Collias, 1996) and male sexual coercion (Pizzari, Froman and Birkhead, 2002) create potential for inbreeding: Around 4% of the copulations observed in free-ranging, unconstrained groups of red junglefowl occurred between brothers and sisters and mothers and sons (Collias and Collias, 1996). Furthermore, in domestic fowl inbreeding depresses different fitness components (e.g. Craig and Baruth, 1965; Cheng, Burns and Shoffner, 1985; Abplanalp, Sato, Napolitano and Reid, 1992). Males are expected to inseminate related females whenever this investment does not constraint investment in copulations with unrelated (i.e. compatible) females and the reproductive success gained exceeds the inclusive costs of reducing the reproductive success of a relative (Parker, 1979; Kokko and Otz, 2006). When presented with the opportunity to copulate with a single female, male red junglefowl readily copulated with full-sib sisters (Pizzari *et al.*, 2004). In fact there was a trend for males to invest more sperm in their sisters than in unrelated females (Pizzari *et al.*, 2004), which may be explained as a response to female selection against inseminations by relatives (see below). (b) The MHC is a highly polymorphic gene complex encoding receptors for antigen presentation to T lymphocytes and thus is a crucial component of the vertebrate acquired immune system, playing a fundamental role in resistance to parasites, infectious and autoimmune disease. Overdominance selection due to heterozygosity advantages on MHC performance may result in disassortative mating and fertilisation between genetically compatible partners (Tregenza and Wedell, 2000). MHC-disassortative mating is expected as this would result in MHC-diverse offspring, which respond to a greater variety of pathogens (Potts and Slev, 1995; Hedrick, 2002; Penn, Damjanovich and Potts, 2002). This idea is being tested in a captive population of red junglefowl which has been MHC-typed through Reference Strand Mediated Conformation Analysis (RSCA). Preliminary results suggest that, when presented with the choice of an MHC-similar (55-100% alleles shared) and an MHC-dissimilar female (0-14% alleles shared), males were more likely to initiate copulation with the MHC-dissimilar female. Importantly, controlling for copulation order (i.e. by forcing the male to copulate with least preferred female first in a second trial, see above), males appear to invest more semen in MHC-dissimilar females (Gillingham, Worley, Richardson and Pizzari, unpublished).

Strategic sperm utilisation by females (cryptic female choice)

Evidence that female birds bias the outcome of sperm competition through differential utilisation of different ejaculates is ambiguous. A careful experimental study of over 200 copulation trials in Japanese quails, showed that although many copulations were forced by the male, fewer eggs were fertilised when the female ran away from the male (i.e. resisted, Adkins-Reagan, 1995). While many factors can contribute to explain it, this pattern is consistent with the general idea that female birds may be able to reduce the fertilising efficiency of 'unwanted' inseminations.

Much research of strategic sperm utilisation by females has focused on mechanisms driven by genetic compatibility. A recent study of sedge warblers, *Acrocephalus schoenobaenus*, raised the intriguing possibility that female birds may be able to bias sperm utilisation in favour of specific haplotypes within the sperm of a single male. This mechanism would enable females to manipulate the genetic diversity of the young even under monogamy. The mallard, *Anas platyrhynchos*, has been the subject of several studies because intense sexual coercion by males would suggest that females may be selected to control paternity cryptically, after insemination, rather than through partner choice. Cunningham and Cheng (1999) tested the hypothesis that female mallards consistently bias paternity in favour of specific male genotypes. Following heterospermic artificial inseminations, the study found that paternity was biased in favour of one of the male genotypes, however these biases were not consistent across multiple insemination events within individual females, which is not consistent with the idea of strategic sperm utilisation by females. More recently, Denk, Holzmann, Peters, Vermeirssen and Kempenaers (2005) conducted a similar study in which female mallards were exposed to heterospermic artificial inseminations comprising the sperm of a genetically related and a genetically unrelated male. While the sperm traits (sperm swimming speed and motility) of a male were important factors in determining paternity shares, the authors failed to detect an effect of genetic relatedness, suggesting no female sperm selection based on genetic compatibility in this species. Similar studies have been conducted in a number of fowl populations with somewhat different results. Birkhead, Chaline, Biggins, Burke and Pizzari (2004) performed a large scale heterospermic artificial insemination experiment and found evidence for female-driven heterogeneity in the paternity of individual males. In six out of the nine male pairs tested across multiple females, individual females were found to bias paternity in favour of one or the other male, consistently, over multiple insemination events. Female cryptic response to inbreeding was experimentally tested in red junglefowl (Pizzari *et al.*, 2004, see above). Females were exposed to natural inseminations under standardised conditions by an unrelated male or by a full-sib brother, and the order was reversed after females had depleted the sperm reserves stored from the previous insemination. By analysing the number of sperm contained on the perivitelline layer of eggs laid by a female over successive days following an insemination, the authors found that females retained fewer sperm following an insemination by a brother than by an unrelated male. The study controlled for the possibility that this trend was due to males allocating more sperm to unrelated females, through similar trials in which ejaculates were collected through the harness technique (see above). Preliminary results of successive follow-ups in the same population appear to confirm such female bias against sperm by related males. However, this bias is only obvious with natural copulations and disappears when artificial inseminations are performed (Løvlie and Pizzari, unpublished). If confirmed, this result would suggest that phenotypic cues may be needed for female fowl to bias sperm utilisation in favour of genetically compatible males. This may explain why artificial insemination studies may fail to detect sperm selection by females. Future studies should aim to disentangle the role of social familiarity and genetic relatedness *per se*. In addition, a female response to genetic relatedness may be driven by MHC similarity. In the future it will be important to determine the independent influence of MHC and genetic relatedness.

Social cues mediate strategic sperm utilisation by females in response to male social status. In feral fowl, females show a preference to copulate with socially dominant males. However, sexual coercion enables subdominant males to impose inseminations on females. Females are able to eject an ejaculate (or part of it) immediately following inseminations and subdominant males suffer a higher probability of sperm ejection than dominant males (Pizzari and Birkhead, 2000). Experimental manipulation of male status elicited a change in the probability of sperm ejection, suggesting an active female mechanism based on male social status (Pizzari and Birkhead, 2000). A large amount of variance in sperm ejection remained unexplained by male status. Future work needs to focus on other mechanisms that may mediate female decision of sperm utilisation. These factors include: (a) male genetic relatedness, (b) whether a female will lay eggs in time for the current insemination to have the opportunity to achieve fertilisation, and (c) the recent copulation history of a female. For example, female kittiwakes, *Rissa tridactyla*, are more likely to eject sperm following copulations at the beginning of the breeding season than closer to egg-laying time (Wagner, Helfenstein and Danchin, 2005). This time-dependent female selection against insemination by the social partner is puzzling as extra-pair copulations are unlikely in this species. However, because fertilisations by old sperm may result in genetic disorders and impaired offspring viability (Tarín, Perez-Albala and Cano, 2000), it is possible that sperm ejection may enable female kittiwakes to reduce the risk of fertilisation by old sperm. Consistent with this idea, Wagner *et al.* (2005) provide evidence that when females eject early ejaculates at relatively low frequency, they produce chicks of inferior quality.

The mechanisms mediating differential sperm utilisation in female birds remain unresolved. Sperm ejection is likely determined by muscular contractions of the cloaca, triggered by environmental stimuli (e.g. social dominance of the male, genetic relatedness with the male, insemination time relative to ovulation). However, sperm selection may also occur at subsequent stages. A candidate arena for female selection of sperm is the SSTs. First, SSTs may be more than passive reservoirs for sperm. By using fluorescent stained sperm, Bakst and Vinyard (2002) showed that sperm storage in the SSTs of female turkeys is non-random. Following intra-magnal artificial inseminations of large ejaculates, the authors found that SSTs in the middle region of the utero-vaginal junction which are longer and more pleiomorphic than SSTs at either end of the junction, are more likely to store sperm. However, even in this middle region over 40% of the SSTs remained empty. Second, SSTs are innervated and this may provide female birds with neural mechanisms to control sperm storage in the SSTs (Freedman, Akuffo and Bakst, 2001). Finally, it is possible that female responses to genetic relatedness may be mediated by the MHC. Zheng, Nishibori, Isobe and Yoshimura (2001) demonstrated that MHC class II mRNA is expressed in cells of the oviductal mucosal stroma in the infundibulum (but not in the uterovaginal junction where SSTs are) of female domestic fowl following insemination. Interestingly, in this region, MHC class II expression was significantly lower in control hens that were sham inseminated with saline. This may indicate that sperm or seminal fluid may trigger anti sperm immune response.

Conclusions

Complex mechanisms of male and female differential sperm utilisation interact to influence the fertilising success of individual inseminations. The application of poultry techniques in combination with molecular tools to avian model study systems enables a non-invasive, quantitative and experimental approach to study these mechanisms in vertebrates. While poultry and molecular techniques are not new, their application to the evolutionary biology of male and female reproductive strategies is more recent. This novel approach has catalysed the use of birds as model systems to study post-insemination mechanisms of sexual selection. The emerging pattern of these studies reveals sperm allocation by males and females as a phenotypically plastic trait, under the influence of a complex network of environmental, social and genetic factors. However, we have only begun to scratch the surface. A concerted effort to understand the mechanisms of post-insemination sexual selection in birds should branch out along two trajectories: (a) In-depth studies to pinpoint the physiological and molecular mechanisms of male and female differential sperm utilisation and investment in model systems; and (b) studies aiming to quantify the fitness consequences of male and female reproductive strategies in natural or semi-natural populations.

References

Abplanalp, H., Sato, K., Napolitano, D. and Reid, J. (1992). Reproductive performance of inbred congenic leghorns carrying different haplotypes for the major histocompatibility complex. *Poultry Science* **71**: 9-17

Adkins-Reagan, E. (1999). Foam produced by male *Coturnix* quail: what is its function? *Auk* **116**: 184-193

Alaghabri, A., Engel, H. N. and Froman, D. P. (1993). Modulation of the fertilizing ability of spermatozoa from roosters carrying the sd (sperm degeneration) allele. *Biology of Reproduction* **48**: 308-312

Alaghabri, A., Engel, H. N. and Froman, D. P. (1992). Analysis of seminal plasma from roosters carrying the sd (sperm degeneration) allele. *Biology of Reproduction* **47**: 1059-1063

Andersson, M. (1994). *Sexual Selection*. Princeton University Press, Princeton

Bahr, J. M. and Johnson, P. A. (1991). Reproduction in poultry. In: *Reproduction in Domestic Animals*. 4th edn. Edited by Cupps, P. T., Academic Press, San Diego, pp. 173-219

Bakst, M. R. and Vinyard, B. T. (2002). Oviducal sperm storate in turkeys: spatial distribution of sperm within the uterovaginal junction sperm-storage tubules. *Journal of Experimental Zoology* **292**: 206-209

Birkhead, T. R. (2000) Defining and demonstrating post-copulatory female choice – Again. *Evolution* **54**: 1057-1060

Birkhead, T. R. and Møller, A. P. (1992). *Sperm Competition in Birds: Evolutionary Causes and Consequences*. Academic Press, San Diego

Birkhead, T.R. and Fletcher, F. (1995). Male phenotype and ejaculate quality in the zebra finch, *Taeniopygia guttata*. *Proceedings of the Royal Society of London, series B* **262**: 329-334

Birkhead, T. R. and Parker, G. A. (1997). Sperm competition and mating systems. In: *Behavioural Ecology and Evolutionary Approach*. Edited by Krebs, J.R. and Davies, N.B., Oxford University Press, Oxford, pp. 323-343

Birkhead, T. R. and Biggins, J. D. (1998). Sperm competition mechanisms in birds: models and data. *Behavioural Ecology* **9**: 253-260

Birkhead, T. R. and Pizzari, T. (2002). Post-insemination sexual selection. *Nature Reviews. Genetics* **3**: 262-273

Birkhead, T. R., Wishart, G. J. and Biggins, J. D. (1995). Sperm precedence in the domestic fowl. *Proceedings of the Royal. Society of London, series B* **261**: 285-292

Birkhead, T. R., Martinez, J. G., Burke, T. and Froman, D. P. (1999). Sperm mobility determines the outcome of sperm competition in the domestic fowl. *Proceedings of the Royal. Society of London, series B* **266**: 1759-1764

Birkhead, T. R., Chaline, N., Biggins, J. D., Burke, T. A. and Pizzari, T. (2004). Nontransitivity of paternity in a bird. *Evolution* **58**: 416-420

Brillard, J. P. (1993). Sperm storage and transport following natural mating and artificial insemination. *Poultry Science* **72**: 923-928

Chapman, T. (2001). Seminal fluid-mediated fitness traits in Drosophila. *Heredity* **87**: 511-521

Chapman, T., Bangham, J., Vinti, G., Seifried, B., Lung, O., Wolfner, M. F., Smith, H. K. and Partridge, L. (2003). The sex peptide of *Drosophila melanogaster*: female post-mating responses analyzed by using RNA interference. *Proceedings of the National Academy of Sciences USA* **100**, 9923-9928

Charlesworth, D. and Charlesworth, B. (1987). Inbreeding depression and its evolutionary consequences. *Annual Review of Ecology and. Systematics* **18**, 237-268

Cheng, K. M., Burns, J. T. and Shoffner, R. N. (1985). Mating behaviour and fertility in domestic chickens. I. Inbreeding. *Applied Animal Behaviour Science* **13**: 371-381

Clark, A. G., Aguadé, M., Prout, T., Harshman, L. G. and Langley, C. H. 1995. Variation in sperm displacement and its association with accessory gland protein loci in *Drosophila melanogaster*. *Genetics* **139**:

189-201

Colegrave, N., Birkhead, T. R. and Lessells, C. M. (1995). Sperm precedence in zebra finches does not require special mechanisms of sperm competition. *Proceedings of the Royal. Society of London, series B* **259**: 223-228

Collias, N. E. and Collias, E. C. 1996. Social organization of a red jungle fowl, *Gallus gallus*, population related to evolution theory. *Animal Behaviour* **51**, 1337-1354

Compton, M. M., Van Krey, H. P. and Siegel, P. B. (1978). The filling and emptying of the uterovaginal sperm-host glands in the domestic hen. *Poultry Science* **57**: 1696-1700

Cornwallis, C. K. (2004). PhD Thesis. University of Sheffield

Cornwallis, C. K. and Birkhead, T. R. (2006). Social status and availability of females determine patterns of sperm allocation in the fowl. *Evolution* **60**: 1486-1493

Craig, J. V. and Baruth, R. A. 1965. Inbreeding and social dominance ability in chickens. *Animal Behaviour* **13**: 109-113

Cunningham, E. J. A. and Cheng, K. M. (1999) Biases in sperm use in the mallard: no evidence for selection by females based on sperm genotypes. *Proceedings of the Royal Society of London, series B* **266**: 905-910

Darwin, C. (1871). *The Descent of Man, and Selection in Relation to Sex*. John Murray, London

Denk, A. G., Holzmann, A., Peters, A., Vermeirssen, E. L. M. and Kempenaers, B. (2005). Paternity in mallards: effect of sperm quality and female sperm selection for inbreeding avoidance. *Behavioural Ecology* **16**: 825-833

Dewsbury, D. A. (1982). Ejaculate cost and male choice. *American Naturalist* **119**: 601-610

Dunn, P. O., Afton, A. D., Gloutney, M. L. and Alisauskas, R. T. (1999). Forced copulation results in few extrapair fertilizations in Ross's and lesser snow geese. *Animal Behaviour* **57**: 1071-1081

Dziuk, P. J. (1996). Factors that influence the proportion of offspring sired by a male following heterospermic insemination. *Animal Reproduction Science* **43**: 65-88

Eberhard, W.G. (1996). *Female Control. Sexual Selection by Cryptic Female Choice*. Princeton University Press, Princeton

Etches, R. J. (1996). *Reproduction in Poultry*. C.A.B. International, Oxon

Freedman, S. L., Akuffo, V. G. and Bakst, M. R. (2001). Evidence for the innervation of sperm storage tubules in the oviduct of the turkey (*Meleagris gallopavo*). *Reproduction* **121**: 809-814

Froman, D. P. (2003). Deduction of a model for sperm storage in the oviduct of the domestic fowl (*Gallus domesticus*). *Biology of Reproduction* **69**: 248-253

Froman, D. P. and Feltmann, A. J. (2000). Sperm mobility: phenotype in roosters (*Gallus domesticus*) determined by concentration of motile sperm and straight line velocity. *Biology of Reproduction* **62**: 303-309

Froman, D. P., Feltmann, A. J., Rhoads, M. L. and Kirby, J. D. (1999). Sperm mobility: a primary determinant of fertility in the domestic fowl (*Gallus domesticus*). *Biology of Reproduction* **61**: 400-405

Froman, D. P., Pizzari, T., Feltmann, A., Castillo-Juarez, H. and Birkhead, T. R. (2002) Sperm mobility: mechanisms of fertilising efficiency, heritability and phenotypic relationship with male social status in the fowl, *Gallus g. domesticus*. *Proceedings of the Royal Society of London, series B* **269**: 607-612

Fujihara, N. (1992). Accessory reproductive fluids and organs in male domestic birds. *World Poultry Science* **48**: 39-56

Gage, M. J. G. and Morrow, E. H. (2003). Experimental evidence for the evolution of numerous tiny sperm via sperm competition. *Current Biology* **13**: 754-757

Gage, M. J. G., McFarlane, C. P., Yeates, S., Ward, R. G., Searle, J. B. and Parker, G. A. (2006). Spermatozoal traits and sperm competition in Atlantic salmon: relative sperm velocity is the primary determinant of fertilization success. *Current Biology* **14**: 44-47

García-González, F. (2004). Infertile matings and sperm competition: the effect of 'nonsperm representation' on intraspecific variation in sperm precedence patterns. *American Naturalist* **164**: 457-472

Griffith, S. C., Owens, I. P. F. and Thuman, K. A. (2002) Extra-pair paternity in birds: a review of interspecific variation and adaptive function. *Molecular Ecology* **11**: 2195-2212

Harshman, L.G. and Prout, T. (1994). Sperm displacement without sperm transfer in *Drosophila melanogaster*. *Evolution* **48**: 758-766

Hedrick, P. W. (2002). Pathogen resistance and genetic variation at MHC loci. *Evolution* **56**: 1902-1908

Hill, G. E. (1993). Male mate choice and the evolution of female plumage coloration in the houce finch. *Evolution* **47**: 1515-1525

Hunter, F. M., Harcourt, R., Wright, M. and Davis, L. S. (2000). Strategic sperm allocation of ejaculates by male Adelie penguins. *Proceedings of the Royal Society of London, series B* **267**: 1541-1545

Jones, I. L. and Hunter, F. M. (1993). Mutual sexual selection in a monogamous seabird. *Nature* **362**: 238-239

Lake, P. E. (1971). In: *Physiology and Biochemistry of the Domestic Fowl*. Vol. 3 Edited by Bell, D.J. and Freeman, B.M., Academic Press, London, pp. 1411-1447

Lake, P. E., Ravie, O. and Waddington, D. (1985). Some effects of the composition of inseminated semen and the site of its deposition on fertility in *Gallus domesticus*. *Animal Reproduction Science* **9**: 273-284

Lessells, C. M. and Birkhead, T. R. (1990). Mechanisms of sperm competition in birds: mathematical models. *Behaviour Ecology and Sociobiology* **27**:325-337

Levitan, D. R. and Petersen, C. (1995). Sperm limitation in the sea. *Trends Ecology and Evolution* **10**: 228-231

Liu, H. and Kubli, E. (2003). Sex-peptide is the molecular basis of the sperm effect in *Drosophila melanogaster. Proceedings of the National Academy of Sciences USA* **100**: 9929-9933

Løvlie, H., Cornwallis, C. K. and Pizzari, T. (2005). Male mounting alone reduces female promiscuity in the fowl. *Current Biology* **15**: 1222-1227

Lynch, M. and Walsh, B. (1998) *Genetics and Analysis of Quantitative Traits*. Sinauer Associates, Inc., Sunderland

Marshall, R. C., Buchanan, K. L. and Catchpole, C. K. (2003). Sexual selection and individual genetic diversity in a songbird. *Proceedings of the Royal Society of London, series B Suppl.* **2**: 248-250

Martin, P. A., Reimers, T. J., Lodge, J. R. and Dziuk, P. J. (1974). The effect of ratios and numbers of spermatozoa mixed from two males on proportions of offspring. *Journal of Reproduction and Fertility* **39**: 251-258

Mazzoldi, C., Scaggiante, M., Ambrosin, E. and Rasotto, M. B. (2000). Mating system and alternative male mating tactics in the grass goby *Zosterisessor ophiocephalus* (Teleostei: Gobiidae). *Marine Biology* **137**: 1041-1048

Mohan, J., Saini, M. and Joshi, P. (1995). Isolation of a spermatozoa motility inhibiting factor from chicken seminal plasma with antibacterial property. *Biochemica et Biophysica Acta, General Subjects* **1245**: 407-413

Nakatsuru, K. and Kramer, D. L. (1982). Is sperm cheap? Limited male fertility and female choice in the lemon tetra (Pisces, Characidae). *Science* **216**: 753-755

Nicholls, E. H., Burke, T. A. and Birkhead, T. R. (2001). Ejaculate allocation by male sand martins, *Riparia riparia. Proceedings of the Royal Society of London, series B* **268**: 1265-1270

King, L. A., Brillard, J. P., Garrett, W. A., Bakst, A. R. and Donoghue, A. A. (2002). Segregation of spermatozoa within sperm storage tubules of fowl and turkey hens. *Reproduction* **123**: 79-86

Kirby, J.D. and Froman, D.P. (2000). Reproduction in male birds. In: *Sturkie's Avian Physiology*, 5th Edition. Edited by Whittow, G.C. Academic Press, San Diego, CA, pp. 597-615

Kokko, H. and Ots, I. (2006). When not to avoid inbreeding. *Evolution* **60**: 467-475

Persaud K.N. and Galef, B.G. Jr. (2004) Fertilized female quail avoid conspecific males: female tactics when potential benefits of sexual encounters are reduced. *Animal Behaviour* **68**: 1411-1416

Parker, G. A. (1979). Sexual selection and sexual conflict. In: *Sexual Selection and Reproductive Competition in Insects*. Edited by Blum, M.S. and Blum, N. A., Academic Press, New York, pp. 123-166

Parker, G. A. (1984). Sperm competition and the evolution of animal mating strategies. In: *Sperm Competition and the Evolution of Animal Mating Systems*. Edited by Smith, R. L., Academic Press, Orlando, pp. 1-60

Parker, G.A. (1998). Sperm competition and the evolution of ejaculates: towards a theory base. In: *Sperm Competition and Sexual Selection*. Edited by Birkhead, T. R. and Møller, A. P., Academic Press, London, pp. 3-54

Parker J.E., McKenzie, F. F. and Kempster, H. L. (1940). Observations on the sexual behaviour of New Hampshire males. *Poultry Science* **19**: 191-197

Parker, J. E., McKenzie, F. F. and Kempster, H. L. (1942). Fertility in male domestic fowl. *Missouri Agricultural Experimental Station Research Bulletin* **347**: 1-50

Pellatt, J. E. and Birkhead, T. R. (1994). Ejaculate size in zebra finches *Taenioppygia guttata* and a method for obtaining ejaculates from passerine birds. *Ibis* **136**: 97-101

Penn, D. J., Damjanovich, K. and Potts, W. K. (2002). MHC heterozygosity confers a selective advantage against multiple-strain infections. *Proceedings of the National Academy of Sciences USA* **99**: 11260-11264

Pitnick, S. and Brown, W. D. (2000) Criteria for demonstrating female sperm choice. *Evolution* **54**: 1052-1056

Pizzari, T. (2002) Sperm allocation, the Coolidge effect and female polyandry. *Trends in Ecology and Evolution* **17**: 456

Pizzari, T. and Birkhead, T. R. (2000). Female feral fowl eject sperm of subdominant males. *Nature* **405**: 787-789

Pizzari, T. and Birkhead, T. R. (2002). Sexually-selected sperm hypothesis: sex-biased inheritance and sexual antagonism. *Biological Review* **77**: 183-209

Pizzari, T., Froman, D. P. and Birkhead, T. R. (2002). Pre- and post-insemination episodes of sexual selection in the fowl. *Heredity* **89**: 112-116

Pizzari, T., Cornwallis, C. K., Løvlie, H., Jakobsson, S. and Birkhead, T. R. (2003). Sophisticated sperm allocation in male fowl. *Nature* **426**: 70-74

Pizzari, T., Løvlie, H. and Cornwallis, C. K. (2004). Sex-specific, counteracting responses to inbreeding in a bird. *Proceedings of the Royal Society of London, series B* **271**: 2115-2121

Potts, W. K. and Slev, P. R. (1995). Pathogen-based models favoring MHC genetic diversity. *Immunological Reviews* **143**: 181-197

Price, C. S. C., Kim, C. H., Posluszny, J. and Coyne, J. A. (2000). Mechanisms of conspecific sperm precedence in *Drosophila. Evolution* **54**: 2028-2037

Prout, T. and Clark, A. G. (1996). Polymorphism in genes that influence sperm displacement. *Genetics* **144**: 401-408

Qazi, M. C. B. and Wolfner, M. F. (2003). An early role for the *Drosophila melanogaster* male seminal protein Acp36DE in female sperm storage. *Journal of Experimental Biology* **206**: 3521-3528

Sæther, S. A., Fiske, P. and Kålås, J. A. (2001). Male mate choice, sexual conflict and strategic allocation of copulations in a lekking bird. *Proceedings of the Royal Society of London, series B* 268: 2097-2102

Shuster, S. M. and Wade, M. J. (2003). *Mating Systems and Strategies*. Princeton University Press, Princeton

Simmons, L. W. (2001). *Sperm Competition in the Insects and its Evolutionary Consequences*. Princeton

University Press, Princeton

Simmons, L. W. and Siva-Jothy, M. J. (1998). Sperm competition in insects: mechanisms and the potential for selection. In: *Sperm Competition and Sexual Selection*. Edited by Birkhead, T.R. and Moller, A. P., Academic Press, London, pp. 341-434

Tarín, J. J., Perez-Albala, S. and Cano, A. (2000). Consequences on offspring of abnormal function in ageing gametes. *Human Reproduction Update* **6**: 532-549

Tregenza, T and Wedell, N. (2000). Genetic compatibility, mate choice and patterns of parentage. *Molecular Ecology* **9:** 1013-1027

Tufvesson, M., Tufvesson, B., von Schantz, T., Johansson, K. and Wilhelmson, M. (1999). Selection for sexual male characters and their effects on other fitness related traits in white leghorn chickens. *Journal of Animal Breeding and Genetics* **116:** 127-138

Trivers, R. L. (1972). Parental investment and sexual selection. In: *Sexual Selection and the Descent of Man, 1872-1972*. Edited by Campbell, B., Aldine Publishing Co., Chicago, pp. 136–179

Van Krey, H. P., Balander, R J. and Compton, M. M. (1981). Storage and evacuation of spermatozoa from the uterovaginal sperm-host glands in the domestic fowl. *Poultry Science* **60**: 871-878

Wagner, R.H., Helfenstein, F. and Danchin, E. (2004). Female choice of young sperm in a genetically monogamous bird. *Proceedings of the Royal Society of London, series B, Suppl.* **271**: S134-S137

Wedell, N., Gage, M.J.W. and Parker, G.A. (2002). Sperm competition, male prudence and sperm-limited females. *Trends in Ecology and Evoution* **17**: 313–320

Westneat, D. F., Rambo, L. A., Fraterrigo, J. M., Birkhead, T. R. and Fletcher, F. (1998). Patterns of courtship behavior and ejaculate characteristics in male red-winged blackbirds. *Behavioural Ecology and Sociobiology* **43**: 161-171

Wilson, J. R., Kuehn, R. E. and Beach, F. A. (1963). Modification in the sexual behavior of male rats produced by changing the stimulus female. *Journal of Comparative Physiology and Psychology* **56:** 636-644

Wolfner, M. F. (2002). The gift that keeps on giving: physiological functions and evolutionary dynamics of male seminal proteins in *Drosophila*. *Heredity* **88**: 85-93

Zheng, W. M., Nishibori, M., Isobe, N. and Yoshimura, Y. (2001). An in situ hybridization study of the effects of artificial insemination on the localization of cells expressing MHC call II mRNA in the chicken oviduct. *Reproduction* **122**: 581-586

Spermatology. SRF Vol. 65. ERS Roldan and M Gomendio (eds) Nottingham University Press, Nottingham

Sperm fate and function in reproductive isolation in *Drosophila*

Therese Ann Markow[1], Laura K. Reed[2] and Erin S. Kelleher[1]

[1]*Department of Ecology and Evolutionary Biology, University of Arizona, Tucson, Arizona 85721, USA;* [2]*Current address: Department of Genetics, N.C. State University, Raleigh, NC 27695-7614, USA*

Central to the study of speciation is the identification of the isolating mechanisms that reduce gene flow. Of special interest are those isolating mechanisms that are the earliest to arise. Sperm fate and function play a major role in two distinct types of reproductive isolation in the genus *Drosophila*. First, postcopulatory-prezygotic (PCPZ) incompatibilities between the male ejaculate and a heterospecific female reproductive tract can result in a reduction in sperm fertilisation success. Although PCPZ incompatibilities have, until recently, received little attention, overwhelming diversity in reproductive morphology and biochemistry indicates that they may play an important role in speciation in this genus. The second type of isolating mechanism that affects sperm occurs in the testes of hybrid males produced by heterospecific matings. These individuals often suffer from dysfunction in spermatogenesis, resulting in the production of aberrant sperm. Hybrid male sterility and hybrid inviability are examples of postzygotic (PZ) reproductive isolation. The observation that hybrid sterility is pervasive among males of all taxonomic groups is known as Haldane's rule. Here we discuss both the evolutionary origins, and functional causes of both PCPZ incompatibilities, which affects sperm fate and function in females, and one type of PZ incompatibility, hybrid male sterility, or sperm dysfunction in hybrid males. Although these two mechanisms of isolation are quite distinct, they are similarly caused by breakdowns in epistatic interactions which occur in the encounter between two divergent genomes. Molecular, cytological, and empirical data are discussed, as is relevant evolutionary theory.

Introduction

Flies of the genus *Drosophila* provide an unusually attractive model system for studies of the reproductive isolating mechanisms that impede gene flow. The genus name *Drosophila* once was used interchangeably with *D. melanogaster,* but the genomes of 12 *Drosophila* species are now sequenced (http://rana.lbl.gov/drosophila) and genetic tools, once only available in *D. melanogaster,* are being developed for other species. Phylogenetic relationships of hundreds of *Drosophila* species are well defined (Markow and O'Grady, 2005a; Remsen and O'Grady, 2002) and the astounding variability among these species in their reproductive biology (Markow and O'Grady, 2005b; Markow, 1996, 2002) reveals that *D. melanogaster* is not representative

Corresponding author E-mail: tmarkow@arl.arizona.edu

of its many congeners. Furthermore, much of the interspecific and even intraspecific variability in reproductive characters first described by Throckmorton (1962) turns out to have important consequences for the fate and function of sperm once they are transferred to females. Hundreds of *Drosophila* species easily can be reared and used in the laboratory (Markow and O'Grady, 2005a), and many of these species are themselves actually in the early stages of speciation. Thus, *Drosophila* are invaluable resources with which to reveal the mechanisms underlying even the earliest events in reproductive isolation.

A framework widely utilised to explain the genetic basis of the evolution of reproductive isolation is the Dobzhansky-Muller model of incompatibilities (Dobzhansky, 1937; Muller, 1940, 1942) and the "snowballing" of these incompatibilities with increasing time in the divergence between two taxa (Orr, 1995; Orr and Turelli, 2001). Each of two diverging lineages becomes fixed for contrasting alleles at two different but interacting loci. Each substitution is neutral in its own genetic background, but when they are combined, an epistatic interaction between the two diverged alleles results in an incompatibility of some sort. Given the large number of interacting loci at which mutations can occur and become fixed, the number of potential incompatibilities "snowballs" with time (Fig. 1). Efforts to understand the very earliest stages of reproductive isolation should therefore focus upon taxa at the earliest stage of divergence.

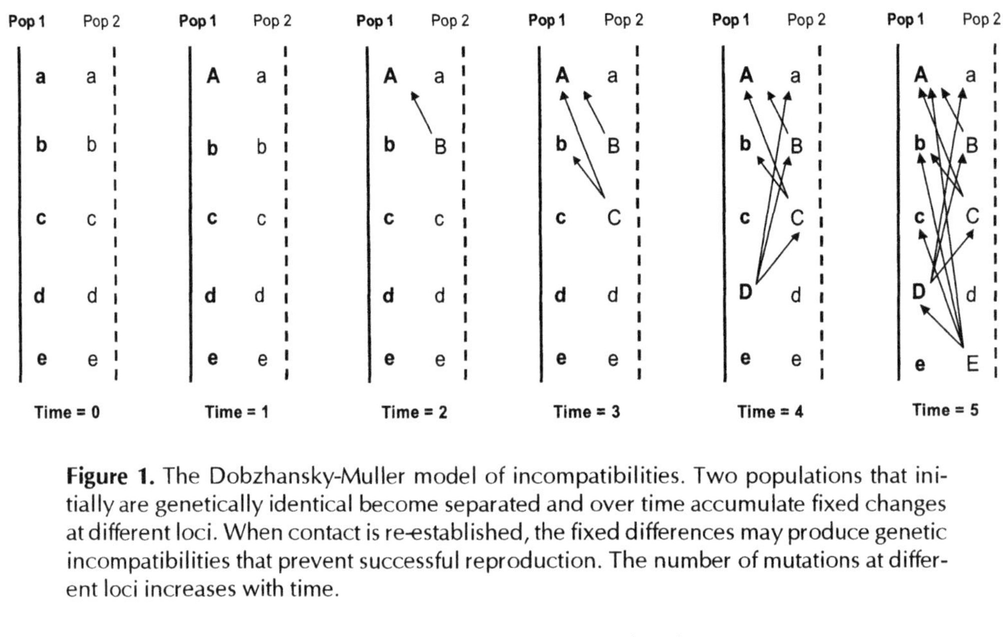

Figure 1. The Dobzhansky-Muller model of incompatibilities. Two populations that initially are genetically identical become separated and over time accumulate fixed changes at different loci. When contact is re-established, the fixed differences may produce genetic incompatibilities that prevent successful reproduction. The number of mutations at different loci increases with time.

Reproductive isolating mechanisms can act at three different levels: precopulatory (PC), postcopulatory but prezygotic (PC-PZ), and postzygotic (PZ). While the Dobzhansky-Muller model is typically applied to cases of postzygotic processes, its central concept, that of incompatibilities between divergent lineages, can be applied to other forms of reproductive isolation (Fig. 2). Sperm fate and function are important components of the last two of these isolating mechanisms: (1) In the absence of complete sexual isolation, interactions between the ejaculate and the female reproductive tract may result in nonrandom sperm storage and or utilization by mated females such that heterospecific or heterotypic sperm are less likely to achieve fertilisation, (2) Should fertilisation and the development of hybrids be successful, a common postzygotic problem is hybrid male sterility (HMS). These two aspects of sperm fate and function will be the focus of this paper.

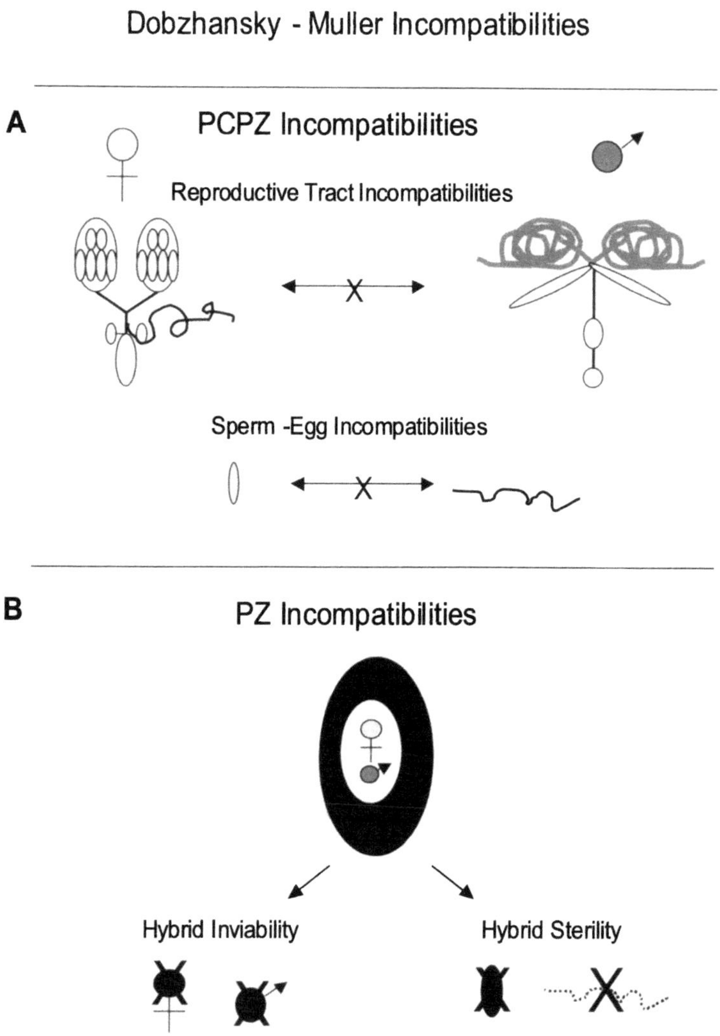

Figure 2. Reproductive isolation through Dobzhansky-Muller incompatibilities at the (**A**) postcopulatory-prezygotic (PCPZ) and (**B**) postzygotic (PZ) stages. PCPZ incompatibilities may occur between the male ejaculate and female reprodúctive tract or between the sperm and the egg. PZ incompatibilities cause reduced hybrid fitness, through either inviability or hybrid sterility.

PC-PZ processes: The fate and function of sperm in *Drosophila* females

In order to investigate the potential ways in which intersexual reproductive tract interactions can promote reproductive isolation, we first review the reproductive anatomy and biochemistry of both sexes as well as what is known of the processes of copulation, sperm storage and utilization.

Reproductive tract anatomy

Male *Drosophila* internal reproductive tracts consist of a pair of testes each connected to seminal vesicle, paired accessory glands, all connected to an ejaculatory duct and bulb and intromission organ, the aedeagus (Fig. 3). Spermatogenesis occurs in the testes and mature sperm are deposited in the seminal vesicle. Accessory glands produce most of the seminal proteins or accessory proteins (ACPs) passed to females during mating. Testes and accessory glands can vary widely in size and structure in association with ejaculate characteristics. Mature sperm and seminal proteins are combined in the ejaculatory duct, before they are passed through the vas deferens and out the intromission organ to the female.

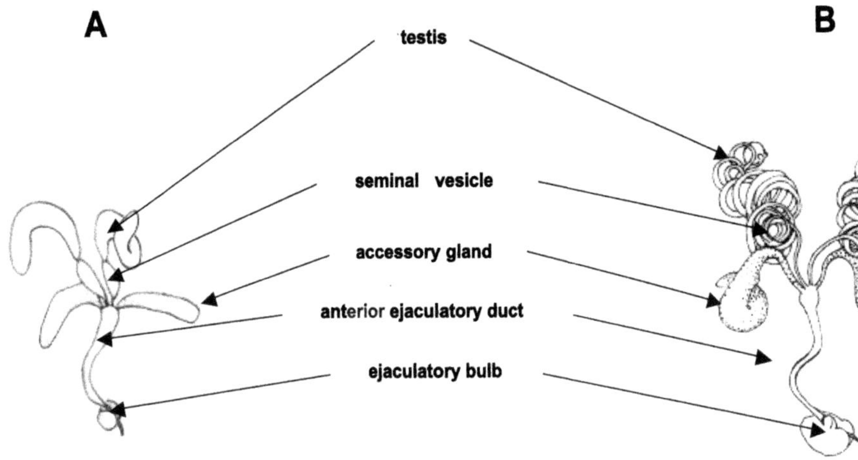

Figure 3. Components of the *Drosophila* male reproductive tract in (**A**) representative of those species producing sperm of a length similar to *D. melanogaster* and (**B**) representative of those species making giant sperm, such as *D. hydei*. Note the length of the testes required to produce giant sperm.

Drosophila female reproductive tracts have paired ovaries connected by lateral oviducts to a common oviduct and eventually the uterus. Ovaries contain ovarioles, long structures in which oocytes are matured from the distal to the proximal portions. Oocytes mature in chambers whose stages were originally classified by King (1970) as 1-14. A stage 14 oocyte is fully mature and ready to be released, fertilised and oviposited. Ovariole number differs markedly among *Drosophila* species, which in turn influences the relative number of oocytes females of the different species produce (Markow, 1996). As in many insects, *Drosophila* females possess sperm storage organs, a ventral seminal receptacle and a pair of spermathecae. In addition to its function in sperm storage, the spermathecal capsule may also generate a secretion important in the maintenance of healthy sperm (Anderson, 1945). The size and morphology of these storage organs are quite variable among species (Fig. 4). Although females of all *Drosophila* species possess both types of storage organs, species differ in whether they use the spermathecae, the seminal receptacle, or both for sperm storage (Pitnick, Markow and Spicer, 1999). Female reproductive tracts also possess parovaria, or accessory glands, whose function has received little attention, but may play a role in fertilisation (Gotteschewski, 1947; Anderson, 1945).

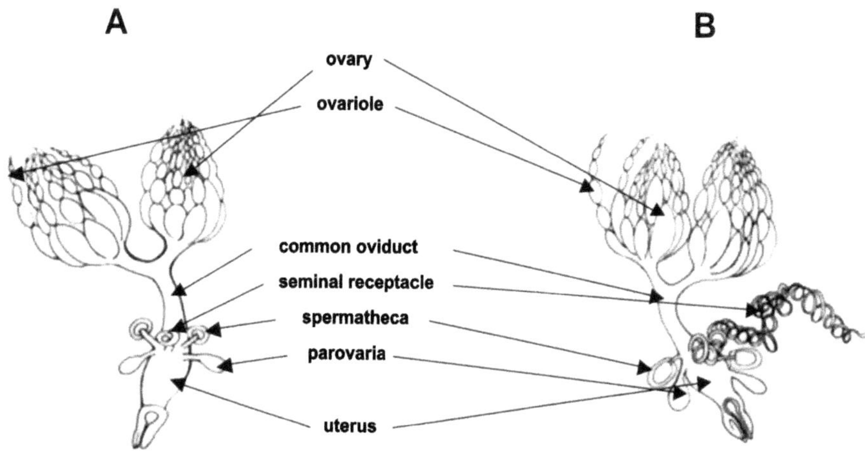

Figure 4. Components of the female reproductive tract in (**A**) representative of species in which males produce sperm of a length similar to *D. melanogaster* and (**B**) representative of species in which males produce giant sperm, as in *D. hydei.*

Reproductive tract biochemistry

Biochemical aspects of reproductive tissue have been more extensively characterised in males than in females, particularly in *D. melanogaster.* Attention has focused largely on the ~80 seminal proteins synthesised in the male accessory glands and subsequently transferred in the seminal fluid to females with the sperm (Wolfner, 1997, 2002, this volume; Kubli, 1992, 2003; Chapman and Davies, 2004). A driving factor in the attention the accessory gland proteins, or ACPs, have received is the discovery that, in *D. melanogaster,* these proteins influence female reproductive behaviour, sperm storage, sperm utilization, and even female longevity (Chapman and Davies, 2004). Accessory glands are assumed to be the primary source of seminal fluid components, but other structures such as the ejaculatory duct and ejaculatory bulb are known to produce at least one protein, glucose dehydrogenase, passed to females in seminal fluid of a number of species (Schiff, Feng, Quine, Krasney and Cavener, 1992; Ross, Fong and Cavener, 1994). Seminal fluid of several insect species has been shown also to contain elements such as sodium (Smedley and Eisner, 1995) and, in *Drosophila,* phosphorus (Markow, Copolla and Watts, 2001) that also are passed to females. For *Drosophila,* little attention has been paid, however, to other seminal fluid constituents, such as carbohydrates, or lipids with the exception of *cis* vaccenyl acetate (Butterworth, 1969; Brieger and Butterworth, 1970).

Biochemistry of *Drosophila* female reproductive tracts is even less well characterised. Lack of knowledge of the physiology and biochemistry of the tract is probably a reflection of the complexity of this organ system. Genes with enriched expression in the female reproductive tract, which may interact with sperm or accessory proteins, include genes involved in proteolysis, signal transduction, transport, and defense or immunity (Swanson, Wong, Wolfner and Aquadro, 2004). Biochemical characteristics of each organ have not been systematically characterised; however, differences in the expression pattern of one protein, glucose dehydrogenase, suggest that each tissue may have a unique composition (Schiff *et al.,* 1992).

Copulation, sperm storage and sperm utilization

In *D. melanogaster,* in the laboratory, copulations last approximately 20 minutes (Fowler, 1973) and sperm transfer is complete within the first 8 minutes. The additional 10-12 minutes of copulation do not appear to affect male fertilisation success but rather, the time to female remating (Gilchrist and Partridge, 2000). These results indicate that copulation serves a function above the simple transfer of sperm.

After sperm transfer, sperm must enter storage, remain viable, and be released from storage to fertilise eggs. Processes that underlie successful storage and retrieval of functional sperm and their use in fertilisation are highly complex and still not well understood (Nonidez, 1920; Bloch-Qazi, Heifetz and Wolfner, 2003). The fact that certain ACPs often become localised in particular regions of the female reproductive tract suggests that they interact with female tissues and may influence the process of sperm storage and recovery (Ravi-Ram, Ji and Wolfner, 2005). Additionally, changes in gene expression in mated females are indicative of physiological responses that likely play a critical role in mediating post-copulatory processes (Begun and Lawacnizack, 2004; McGraw, Gibson, Clark and Wolfner, 2004).

In *D. melanogaster*, sperm remain in the uterus for several minutes before entering first the seminal receptacle and then the two spermathecae (Nonidez, 1920). This process is known to depend on several components of the male ejaculate, including the lipase *esterase*-6 (Gilbert, 1981; Gilbert, Richmond and Sheehan, 1981; Richmond and Senior, 1981), glucose dehydrogenase (Iido and Cavener, 2004) and ACP36DE (Neubaum and Wolfner, 1999; Bloch-Qazi and Wolfner, 2003). Sperm must be nurtured within the female storage organ until they can be utilised to fertilise eggs. The presence of male-specific lipids in the ejaculate, and elevated lipase activity in both the male and mated female reproductive tracts suggest that fat digestion may be important for the maintenance of healthy spermatozoa (Butterworth, 1969; Smith, Rothwell, Wood, Yeaman and Bownes, 1994). Additionally, both the spermathecal capsule and the parovaria are known to produce secretions that are important for sperm longevity (Anderson, 1945).

Successful fertilisation requires coordination of sperm release from storage with the progression of a mature oocyte through the lateral and common oviducts to the fertilisation site. Seminal proteins, which trigger egg production and ovulation, may assist in this process (Heifetz, Lung, Frongillo and Wolfner, 2000; Chapman, Herndon, Heifetz, Partridge and Wolfner, 2001). Additionally the enzyme glucose dehydrogenase, produced in both the male ejaculatory bulb and the female reproductive tract, has been shown to influence the efficiency of sperm release (Iido and Cavener, 2004). Protein-carbohydrate interactions are important in gamete recognition in deuterosomes (reviewed in Mengerink and Vacquier, 2001), and the presence of glycoconjugates and glycosidases on sperm surfaces indicates this may also be true in *Drosophila* (Perotti and Pasini, 1995; Pasini, Cattaneo, Verni, Hackstein and Perotti, 1999; Cattaneo, Ogiso, Hoshi, Perotti and Pasini, 2002). Two ß-N-hexosaminidases found in sperm cell surfaces are known to be involved in primary sperm-egg interactions (Perotti, Cattaneo, Pasini, Verni and Hackstein, 2001). A second type of glycosidase found on the surface of *Drosophila* sperm a fucosidase, is specific to sugar residues found primarily on the micropyle of the egg where fertilisation occurs (Intra, Cenni and Perotti, 2006). How the function of these sperm-egg interactions depends on aspects of the surrounding chemical environment such as pH and ion concentration remains to be examined.

Potential PC-PZ incompatibilities and how they might work

In the preceding section we have described an array of biochemical and morphological interactions that are essential for reproduction in *D. melanogaster*. However, the genus is characterised by incredible diversity of morphological (Fig. 3) and biochemical reproductive traits. Gonad size, sperm size and sperm number produced by males of different *Drosophila* species show the greatest described variation in the animal kingdom (Pitnick, 1996). Sperm size may be particularly important in male reproductive success, and is under selection in the *obscura* species group (Snook, 1997). Across the genus, sperm size varies by two orders of magnitude, and sperm gigantism has several independent evolutionary origins (Pitnick, 1996; Pitnick *et al.*, 1999). Sperm heteromorphism, or the production of two sizes of sperm, the smaller of which are fertilisation incompetent (Snook, Markow and Karr, 1994), occurs only in *obscura* group species (Snook, 1997).

Male biochemistry also exhibits extreme diversity. Expression patterns of the gene glucose dehydrogenase, whose role in sperm storage and release is discussed in the previous section, differ between species (Schiff *et al.*, 1992; Ross *et al.*, 1994). The size and presence of an ejaculatory donation, or a protein component of the ejaculate that is incorporated into female somatic and ovarian tissue, is also species specific (Markow and Ankney, 1988; Pitnick, Spicer and Markow, 1997). At the molecular level, male accessory proteins are known to be extremely diverse and some have experienced strong selection (Begun, 2000; Swanson, Clark, Waldrip-Dail, Wolfner and Aquadro, 2001; Kern, Jones and Begun, 2004, Wagstaff and Begun, 2005; Schully and Hellberg, 2006). Additionally, rapid changes in the composition of the male ejaculate have resulted in the presence of many lineage specific accessory proteins (Begun and Lindfors, 2005; Mueller, Ravi-Ram, McGraw, Bloch-Qazi, Siggia, Clark, Aquadro and Wolfner, 2005).

More limited examination of female anatomy and biochemistry indicates they also exhibit divergence of reproductive function between species (Fig. 4). Size, morphology, and patterns of sperm storage organ use are extremely variable (Pitnick *et al.*, 1999). Variation in expression patterns of glucose dehydrogenase within the female tract indicates organ biochemistry and function may also differ between species (Schiff *et al.*, 1992; Ross *et al.*, 1994). The insemination reaction, or an opaque mass that forms in the uterus of certain species after mating, shows both interspecific and interpopulation differences (Paterson, 1946; Wheeler, 1947; Patterson and Stone, 1947; Knowles and Markow, 2001). Additionally, certain secreted proteins and cell surface receptors in the female reproductive tract show signatures of adaptive evolution (Swanson, 2004; Panhuis and Swanson, 2006).

This variation in the reproductive traits of both sexes is indicative of sexual selection that occurs in the reproductive tract of female *Drosophila*, the strength of which may depend on the level of female promiscuity, and therefore the number of competing ejaculates (Markow, 2002). Strong selection predicts divergence between closely related taxa and creates the possibility for morphological and biochemical incompatibilities, as predicted by the Dobzhansky-Muller model, when male ejaculates encounter an unfamiliar female reproductive tract.

Postcopulatory-prezygotic reproductive isolation originating from anatomical or morphological incompatibilities could arise at any of a number of steps following copulation. Optimal reproduction requires a full-length copulation in which both sperm and seminal fluids are transferred. Copulation duration is known to be extremely variable between species ranging from 30 seconds to over 2 hours (Markow, 1996). Interspecific differences in copulation duration, if accompanied by concomitant differences in the nature or number of seminal components transferred could impact the outcome of an interspecific mating. Because male genitalia often are extremely different between closely related *Drosophila* (Markow and O'Grady, 2005b), problems with sperm transfer may be caused by incompatibilities in genital morphology. For example, in crosses between *D. simulans* females and *D. sechellia* males copulations involve the transfer of relatively few sperm. Although most of these sperm later fertilise eggs, the productivity of the cross is significantly reduced when compared to homospecific matings (Price, Kim, Gronlund and Coyne, 2001). Furthermore, if genitalic differences affect the composition or quantity of transferred ejaculate they could also have implications for the fate of sperm once it is in the female. In heterospecific crosses in the *D. simulans* complex, aberrant copulation durations are associated with abnormal quantities of transferred sperm and/or reduced sperm storage efficiency (Price *et al.*, 2001).

Mismatches between the size or length of sperm and the spermathecae or receptacle could interfere with sperm entering, remaining, or exiting storage organs at the appropriate time. Following artificial selection that modified the relationship between sperm length and ventral receptacle length in different strains of *D. melanogaster,* Miller and Pitnick (2002) found that nonrandom fertilisation occurred. The likelihood of this sort of incompatibility occurring in crosses between natural populations of *Drosophila* is high, given the existence of intraspecific differences in sperm length and ventral receptacle length (Miller, Starmer and Pitnick, 2003). Because fertilisation requires that sperm pass through the egg micropyle, differences in sperm head morphology (Pasini, Redi, Caviglia and Perotti, 1996) may also affect fertilisation success. Though not systematically studied in a large number of species, interspecific variation has been reported for micropyle morphology (Counce, 1959).

The large number of seminal fluid proteins and their roles in sperm fate and the likely importance of their compatibility with the chemistry of the female reproductive tract, suggest that biochemical interactions hold numerous possibilities for reproductive failures. Given the evidence of strong selection on both male and female reproductive proteins, closely related species or even different populations of the same species may not function optimally together. Additionally, differences in the composition of the male ejaculate could mean an essential protein might be absent in an interspecific cross. The failure of just one interaction of the many that must occur between male ejaculate and receptors or enzymes in the female, could mitigate any of these normal postcopulatory processes. Such a failure could be considered as a passive incompatibility resulting, effectively, in a loss of function or signal. For example, Ross *et al.* (1994) demonstrate that while the parovaria of female *D. erecta* and *D. simulans* produce the enzyme glucose dehyrogenase, the other species in the *melanogaster* complex (*D. melanogaster, D. sechellia, D. mauritiana, D. teissieri, D. yakuba)* do not. Given that this protein is known to have an important role in sperm storage and release (Iido and Cavener, 2004), it is easy to envision problems following interspecifc matings with females from species where the parovaria lack this function.

There are several examples of incompatibilities that affect passage, storage, maintenance, and fertilisation success of heterospecific sperm in the genus, although the mechanistic basis of these incompatibilities be they morphological or biochemical remains unknown. A number of interspecific crosses have been examined for the efficacy of sperm storage following single matings. Problems with sperm storage were suggested in an early study of *D. mojavensis* females mated to males of their sister species *D. arizonae* (Baker, 1947). Additionally, reciprocal crosses between *D. mauritiana* and *D. simulans* show a high failure rate in sperm storage (Price *et al.,* 2001). Finally, *D. melanogaster* females fail to store *D. tessieri* and *D. orena* sperm transferred by developmentally modified, mosaic *D. melanogaster* males (Sanchez and Santamaria, 1997). This case is particularly intriguing because the foreign sperm was transferred with *D. melanogaster* seminal proteins, indicative of either a sperm-ACP incompatibility, or a sperm-female incompatibility. Although sperm viability in interspecific crosses has not been explicitly examined in any system, sperm mortality in storage has been reported in many insects including *D. melanogaster* (Snook and Hosken, 2004). Additionally, sperm loss from storage in crosses involving *D. mauritiana* females and *D. simulans* males, severely affects the productivity of this cross (Price *et al.*, 2001).

Stored sperm must also successfully fertilise eggs, which requires temporal coordination of sperm-egg release as well as the appropriate biochemical environment for fertilisation. While no clear examples are known of perturbation of this process in heterospecific crosses, at least two *Drosophila* species exhibit a type of self-sterility or self-incompatibility mechanism in which singly mated females do not utilise sperm from closely related males (Markow, 1997). While the responsible mechanism was not identified, the fact that the sperm of the related male can be "rescued" by seminal fluid of an unrelated male, suggests that the responsible factor is a seminal fluid component (rather than a feature of the sperm) necessary for either oviposition or sperm release and that the responsible factor or factors exhibit genetic variation. This is consistent with the observation that in *D. melanogaster* oogenesis and ovulation are stimulated in mated females by two different seminal proteins acp70A (Chen, Stumm-Zollinger, Aigaki, Balmer, Bienz and Bohlen, 1988; Aigaki, Fleischman, Chen and Kubli, 1997; Nakayama, Kaiser and Aigaki, 1997), and acp26Aa (Herndon and Wolfner, 1995; Heifetz *et al.*, 2000; Heifetz, Vandenberg, Cohn and Wolfner, 2005). An inter- or even an intraspecific difference in the functional properties of such proteins could lead to the inability of a male signal to trigger these processes in mated females. Given that both of these proteins exhibit signatures of adaptive evolution (Aguade, 1999; Begun *et al.*, 2000), these types of incompatibilities seem probable.

Other incompatibilities could conceivably be more "active", blocking sperm movement into or out of storage or the progression of oocytes to the fertilisation site and out the ovipositor. The insemination reaction mass found in certain *Drosophila* species, particularly in the *repleta* group, is known to be more severe in interspecific and interpopulation crosses (Baker, 1947; Patterson, 1947; Knowles and Markow, 2001), affecting oviposition and possibly sperm survival and progression to storage (Baker, 1947; Patterson, 1947). Additionally, the mating plug, a coagulation of seminal proteins which forms in the distal uterus of *Drosophila melanogaster* and *D. hibisci* mated females (Bairati, 1967; Bairati and Perrotti, 1970; Polak, Starmer and Barker, 1998), could also be perturbed in divergent crosses. This possibility, however, has never been examined experimentally.

The biochemical environment of fertilisation and the surface chemistry of the sperm acrosome and the egg micropyle could also have a role in reproductive isolation. Decreased fertilisation success in heterospecific crosses is common in marine invertebrates (Glabe, 1979; Vacquier, Swanson, Metz and Stout, 1999), and depends on molecular interactions between proteins in the sperm acrosome and egg surface (reviewed in Vacquier, 1998). Evidence for similar protein-carbohydrate mediated fertilisation (Perotti and Pasini, 1995; Pasini *et al.*, 1999, Perotti *et al.*, 2001, Cattaneo *et al.*, 2002; Cattaneo, Casini, Intra, Matsumoto, Briani, Hoshi and Perotti, 2006; Intra *et al.*, 2006), and a rapidly evolving seminal protein which interacts directly with sperm (Peng, Chen, Busser, Liu, Honegger and Kubli, 2005) raises the potential for similar fertilisation incompatibilities in *Drosophila*. However, conservation of sperm surface proteins (Intra *et al.*, 2006), and a lack of evidence that sperm-binding seminal proteins interact directly with the egg, provide little support for species-specific gamete recognition in *Drosophila*. The only putative example of a sperm-egg incompatibility occurs when *D. melanogaster* females of the Zimbabwe strain mate with males of a cosmopolitan strain. These heterotypically mated females lay an unusually large proportion of unfertilised or aberrantly fertilised eggs (Alipaz, Wu and Karr, 2001). Whether this is due to problems in gamete recognition, fertilisation environment, or sperm release remains unclear.

When homospecific and heterospecific ejaculates compete in the reproductive tract of a single female, a final class of isolating mechanisms appears. Conspecific sperm precedence (CSP) occurs when the conspecific sperm fertilises a greater proportion of the brood than expected and Is well known in many taxa (Howard, 1999). In *Drosophila,* CSP has been reported for the species pairs *D. mauritiana* and *D. simulans* (Price, 1997; Price, Kim, Posluszni and Coyne, 2000), *D. sechellia* and *D simulans* (Price, 1997), *D. yakuba* and *D. santomea* (Chang,

2004) and the subspecies pair *D. pseudoobscura pseudoobscura* and *D. p. bogotana* (Dixon, Coyne and Noor, 2003). However, because sperm viability can have implications for the outcome of sperm competition (Garcia-Gonzalez and Simmons, 2005), CSP may frequently be an artifact of male x female interactions rather than antagonistic interactions between the ejaculates. Only in crosses between *D. simulans* females and *D. mauritiana* males is there evidence that the presence of homospecific semen decreases the fertilisation success of heterospecific sperm (Price et *al.*, 2000).

PC-PZ interactions with the potential to cause isolation actually represent a collection of processes or mechanisms, either anatomical or biochemical, that can occur at one or more of many of the critical steps that lie between copulation and fertilisation. Whether certain classes of interactions, for example anatomical versus biochemical, or storage versus fertilisation, have a greater potential to produce isolation than others, especially early in the speciation process, remains a mystery. Reproductive characters exhibiting relatively greater variability within and between species, such as sperm length, may have a greater potential to cause incompatibilities than less variable traits. As a more complete understanding of all of relevant process underlying the successful transfer, storage, maintenance, retrieval and use of sperm within a species, is generated, our ability to infer how PCPZ interactions can impair normal sperm fate and/or function in interspecific matings will expand.

PZ reproductive isolation: Sperm and hybrid male sterility in *Drosophila*

Haldane (1922) observed that decreases in hybrid fitness are more common in the heterogametic sex. In other words, reduced hybrid fitness will first be seen in the sex with differentiated sex chromosomes (XY or WZ). This early observation has been verified in a wide range of taxa, including *Drosophila* and mammals where males are XY, but also in birds and Lepidoptera, where females are WZ (Coyne and Orr, 2004, pp. 284-286). Because of the robustness of this pattern, it has become known as Haldane's Rule. Hybrid fitness decrements covered by Haldane's Rule include both sterility and inviability. There is evidence from studies in *Drosophila* that HMS develops before hybrid inviability (Wu, 1992; Coyne and Orr, 1997).

Genetics of spermatogenesis

What causes HMS and what is the nature of the defect or defects observed in sterile hybrid *Drosophila* males? In order to address these questions, we first need to understand normal *Drosophila* spermatogenesis. Fortunately, a good deal is known about spermatogenesis in *D. melanogaster* (Lindsley and Tokuyasu, 1980; Fuller, 1993, 1998). Each testis is a long tube-like structure in which spermatogenesis proceeds from the proximal apex to the distal end, and is joined to the seminal vesicle, where mature sperm wait prior to copulation (Fig. 3). Thus, at the proximal tip of the testis are found the earliest stages of spermatogenesis while mature sperm are found distally. In *D. melanogaster,* it takes 260 hours to go from a mitotically dividing stem cell to a mature, coiled sperm. The process of spermatogenesis is ongoing, continuously, in each testis, or its primordium, from the time of the first larval instar.

For the purposes of genetic studies of male sterility, spermatogenesis can be viewed as occurring in two phases. The first phase includes the premeiotic and the meiotic stages. Spermiogenesis, the second phase, includes nebenkern formation, elongation, individualization, and coiling. A cytological study of 2216 EMS-induced autosomal male sterile mutations in *D. melanogaster* (Wakimoto, Lindsley and Herrera, 2004) revealed that the majority (73%) of the defects could be localised to spermiogenesis, primarily in elongation, individualization, or coiling. Only 20% exhibited problems in meiosis or earlier, suggesting that later stages of spermatogenesis may be more sensitive to disruption by genetic factors.

Genetics of hybrid male sterility

From studies of male sterility in *D. melanogaster*, we know that a large number of genes, approximately 500, are involved in spermatogenesis and thus have the potential to cause male sterility (Wakimoto *et al.*, 2004). It also appears that some loci affecting male fertility have very few alleles, while others are highly polymorphic. Whether these same genes are those most likely to contribute to HMS is not known. Considering the large number of loci capable of influencing male fertility (Wakimoto *et al.*, 2004), the genetic mechanisms underlying HMS are likely to be complex. Furthermore, evolutionary biologists have tended to utilise rather different approaches than developmental biologists in studying male sterility and its causes.

A critical component in studies of HMS is the way in which it is scored. *Drosophila* developmental biologists have been able to utilise a range of genetic tools, in addition to standard cytological techniques, to fully characterise the defects caused by male sterile mutations. Developmental studies, however, have been restricted primarily to *D. melanogaster* and its relatives. Genetically marked stocks with balancers, important tools of developmental biology, are unavailable in most other *Drosophila* species. *Drosophila* evolutionary biologists have utilised several, more general phenotypic measures of infertility: (1) sperm number (MacDonald and Goldstein, 1999), (2) sperm motility (Coyne, 1984; Vigneault and Zouros, 1986; Coyne and Charlesworth, 1989; Orr, 1989; Orr and Coyne, 1989; Coyne, Rux and David, 1991; Davis and Wu, 1996; MacDonald and Goldstein, 1999; Orr and Irving, 2001; Reed and Markow, 2004; Kopp and Frank, 2005), (3) morphological abnormalities of the testes (Dobzhansky, 1935; Sturtevant and Dobzhansky, 1936; Marin, Pla, Ruiz and Fontedevila, 1993; Joly, Bazin and Singh, 1997; MacDonald and Goldstein, 1999), and (4) the production of offspring (Coyne and Charlesworth, 1989; Marin *et al.*, 1993; Davis and Wu, 1996; True, Weir and Laurie,1996; Tao, Chen, Hartl and Laurie, 2003a, Reed and Markow, 2004). Some studies have examined more than one of these phenotypes simultaneously. While it may seem at first glance to be a discrete or dichotomous trait, sperm motility has been scored in different ways making comparisons among studies difficult. In most cases hybrid males are dissected and sperm scored as motile or non-motile, where the observation of only one motile sperm among many results in a score of motility. Other studies have used a scale or range of motilities, resulting in a more quantitative set of observations (Davis and Wu, 1996; MacDonald and Goldstein, 1999; Reed, LaFlamme and Markow, 2006). Regardless of the scale employed to score motility, the fact that sperm can be motile but functionally sterile (Wakimoto *et al.*, 2004; Reed and Markow, 2004) means that motility cannot be taken as an absolute criterion of male fertility. Reed *et al.* (2006) also found that depending upon whether sperm motility was scored as a binary or graded scale, immotility showed either dominant or additive inheritance, respectively. In addition, studies scoring HMS solely on the basis of offspring production by mated hybrid males may be confounded by reproductive tract interactions of the nature described above.

Observing HMS, regardless of how it is scored, typically has been the endpoint for phenotypic characterisations. The primary question of interest has been assessing the genetic architecture. With few exceptions the actual defects in sterile hybrid males have not been cytologically examined, making it difficult to assess similarities, either phenotypically or genetically, with male sterility found in mutagenic screens of *D. melanogaster*. Kulathinal and Singh (1998) however, in crosses among three species of the *simulans* clade of the *melanogaster* group, examined the cytological basis of male sterility. Sterility phenotypes fell into two classes: premeiotic-meiotic and spermiogenic or postmeiotic, with the majority of crosses showing defects of the latter type, much like the proportion of the mutations affecting male fertility in *D. melanogaster* (Wakimoto *et al.*, 2004).

What is known about the genetic architecture of HMS in *Drosophila*? Studies that identify the number of loci contributing to HMS range in their estimates from 3 to 60 (Coyne and Orr, 2004). This range does not appear to correlate with divergence time. For example, three different studies of the species pair *D. simulans* and *D. mauritiana* estimate the number of HMS factors to be 5, 15, and 60 (Coyne, 1984; True *et al.*, 1996; Tao and Hartl, 2003; Tao *et al.*, 2003a; Tao, Zeng, Li, Hartl and Laurie, 2003b). Also, species pairs with different degrees of divergence show the same number of estimated incompatibilities such as *D. virilis/D. texana* (D = 0.580, Throckmorton, 1982) and *D. willistoni/D. quechua* (D = 0.214, Ayala and Tracey, 1974) which each have an estimated three loci contributing to the male sterility seen in their hybrids (Dobzhansky, 1975; Lamnissou, Loukas and Zouros, 1996). It is generally accepted that HMS is a trait involving multiple genes as has been shown in all studies that have estimated its genetic basis, but it remains unclear how many genes are really involved at the very earliest stages of speciation. One study attempted to estimate the number of loci in two recently diverged subspecies of *D. pseudoobscura* (Orr and Irving, 2001). They found, using a backcross design, that at least 5 genomic regions are involved in HMS at this early stage of divergence. However, because it is impossible to use sterile males in breeding, most investigators resort to introgression and backcrosses in order to dissect the sterility phenotype, breaking up in the process, the very coadapted gene complexes that may be of interest. Genotypes studied above are thus not identical to those of the sterile F_1 males.

An additional complication in most studies of the genetic basis of HMS is the use of mass matings, *i.e.*, placing together many flies from each of two separate populations or species, instead of single pairings from isofemale lines. Use of mass matings precludes detection of early polymorphisms since there is no way to partition any variance in the hybrid male sterile phenotype. Three studies have overcome this problem, however, discovering surprising levels of within-species polymorphism for factors giving interspecific HMS (Reed and Markow, 2004; Kopp and Frank, 2005; Reed, unpublished). From these studies it is clear that in at least three *Drosophila* species, the potential to produce HMS is maintained as a polymorphic trait at varying frequencies in different populations of the same species. In addition the large amount of within population variation observed in consistent with a polygenic basis to the trait.

Another investigation actually sought to identify the nature and number of genetic loci affecting male sterility directly in the F_1 hybrid using within species polymorphism (Reed *et al.*, 2006). They mapped the factors giving HMS within the species *D. mojavensis* in crosses with its sister species, *D. arizonae*. Using quantitative trait locus (QTL) mapping, they found that HMS arises with a complex genetic basis. Since multiple loci were polymorphic simultaneously, it suggests that the loci controlling HMS are likely to accumulate as neutral variants within these species, allowing them to be "caught" in a polymorphic state. Examining species pairs that are early in their divergence trajectories are thus more likely to be successful at capturing the loci where the initial changes arise.

Hybrid male sterility and Haldane's rule

Despite the robustness of Haldane's rule its underlying genetic mechanisms remain elusive. Several well supported theories attempt to explain Haldane's rule but all lack the necessary ubiquity to explain all of its manifestations. The sex chromosomes play an important role in many of these theories since it is the heterogametic sex that is disproportionally affected in hybrids. Turelli and Begun (1997) considered whether the size of the X-chromosome influenced the rate of evolution of Haldane's rule across a variety of *Drosophila* species pairs. They found that the strength of Haldane's rule at early stages of divergence was positively correlated with the size of the X-chromosome but the effect was only relevant to HMS. Hybrid inviability and female sterility showed no correlation with X-chromosome size. In addition, the "large X-effect" has been shown in a variety of introgression studies and hybrid zone studies, where the X-chromosome shows greater influence on apparent hybrid incompatibilities than do the autosomes (Coyne and Orr, 2004, p. 290).

There are two prominent explanations for the large role of X-chromosome in Haldane's rule. First, dominance theory (Muller, 1940, 1942) predicts that the X-chromosome carries recessive alleles that only express their Dobzhansky-Muller incompatibilities when in a hemizygous state. This theory has been shown to be consistent with the pattern of hybrid inviability in *Drosophila* when hybrid females are made homozygous for the X-chromosome, thus allowing the expression of the recessive alleles (Orr, 1993; Coyne and Orr, 2004, p. 289). But, this theory does not explain hybrid sterility because hybrid females with the homozygous X do not necessarily show a loss of fertility. This implies that genes causing hybrid inviability affect both sexes while genes involved in sterility are sex specific. Surprisingly, this finding is not predicted by Wakimoto *et al.* (2004) where nearly 40% of the EMS-induced sterile male mutants also showed female sterility.

Another theory for Haldane's rule is called the "faster-X theory" which hypothesises that genes on the X-chromosome are evolving at a faster rate than those in the rest of the genome. The evidence for this hypothesis remains inconclusive. First, when the number of HMS factors on the X-chromosome in the *simulans* clade is estimated, different studies find dramatically different results. Hollocher and Wu (1996) find no difference between the X and autosomes for the proportion of HMS factors, while others find there are more on the X (50%, True *et al.*, 1996; 250%, Tao *et al.*, 2003a). In addition, there is weak evidence for an accelerated rate of evolution of X-linked genes in general (Betancourt, Presgraves and Swanson, 2002; Counterman, Ortiz-Barrientos and Noor, 2004). Finally, despite the evidence for a large X-effect and more HMS factors on the X-chromosome, there is a paucity of sex-linked male somatic and testes specific genes (Parisi, Nuttall, Naiman, Bouffard, Malley, Andrews, Eastman and Oliver, 2003). These results present a quandary, why is the X-chromosome so important if it presents so few targets for mutations causing HMS? Little work has expanded this analysis to other taxa. In one *Drosophila* species pair, however, *D. arizonae*/*D. mojavensis*, no evidence for a disproportionate effect of the X-chromosome has been found (Zouros, 1981; Zouros, Lofdahl and Martin, 1988; Zouros, 1991; Reed *et al.*, 2006). Caution should therefore be exercised when making generalizations about the genetic basis of Haldane's rule and about HMS specifically until the patterns reviewed here are substantiated in other systems.

The "faster male" theory is another, non-mutually exclusive, explanation for Haldane's rule but it is only relevant to heterogametic hybrid males. The idea here is that genes underlying HMS may simply evolve faster, developing incompatibilities more quickly than those for hybrid female sterility. True *et al.* (1996) found that in crosses with *D. mauritiana* and *D. simulans*, far more genetic regions caused HMS than hybrid female sterility. Additional studies have confirmed this pattern in the other members of the *melanogaster* species group (Hollocher and Wu, 1996; Tao *et al.*, 2003; Tao and Hartl, 2003).

In *Drosophila melanogaster* group species male-specific genes have an elevated substitution rate (Zhang, Hambuch and Parsch, 2004) and male specific genes diverge more quickly in expression patterns between *D. melanogaster* and *D. simulans* (Ranz, Castillo-Davis, Meiklejohn and Hartl, 2003). Further evidence for the faster male theory is the patterns observed in aberrant gene expression in hybrids. In hybrids between two *Drosophila* species pairs (*D. simulans*/*D. mauritiana* and *D. persimilis*/*D. pseudoobscura*), male specific genes show greater perturbation relative to their parental species than either female specific or non-specific genes (Michalak and Noor, 2003; Noor, Michalak and Donze, 2003; Michalak and Noor, 2004; Noor, 2005; Ranz, Namgyal, Gibson and Hartl, 2004). Some of those genes showing aberrant expression patterns in hybrids also reflect increased rates of protein evolution (Noor, 2005). One piece of conflicting evidence for this pattern is that when three genes specific to sperm development were tested for rates of evolution in *D. melanogaster* and *D. simulans*, only one showed evidence for adaptive evolution (Civetta, Rajakumar, Brouwers and Bacik, 2006). Thus, based on this limited sample it is not clear that sperm development is a specific target for the evolution of HMS factors.

Issues of experimental design, especially scoring of male sterility, use of introgression, mass versus single pair mating, and degree of divergence between populations or species being crossed all have substantial impacts on the outcome and interpretation of HMS studies. Furthermore, a preponderance of *Drosophila* investigations have been confined to a small number of species and their relatives, largely those of the *melanogaster* group no doubt owing to the availability of genetic markers. While these studies point to a major effect of the X chromosome, studies of other *Drosophila* species are not characterised by significant X-chromosome effects leaving this generalization in need of additional experiments across a wider range of *Drosophila* species.

Conclusions

Sperm are critical to reproductive isolation at two levels: postcopulatory but prezygotic and postzygotic. We have shown that both of these types of reproductive isolation are complex and depend upon multiple pathways and processes. Incompatibilities may stem from changes either in protein coding genes, as exemplified by substitutions in accessory gland proteins, or by divergence in regulatory mechanisms, as reflected in the differences in tissue specific expression of the enzyme glucose dehydrogenase. While the Dobzhansky-Muller model has been strictly applied to postzygotic isolation, because of epistatic interactions, its central concept, one of incompatibilities accumulating through time, can be applied to different categories of prezygotic interactions between the sexes. Hybrid male sterility is likely to be more tractable for genetic studies as the gross phenotype is largely rooted in spermatogenesis, a process that has been intensively characterised genetically and phenotypically in *D. melanogaster,* and one in which the pathways are likely to be conserved. Postcopulatory-prezygotic impairments of sperm fate and function can occur at a range of steps, none of which are well understood even in homotypic matings. Given the complexity of postcopulatory-prezygotic processes, however, this category of reproductive isolation may turn out to be of major importance in speciation.

References

Aguade, M. (1999). Positive selection drives the evolution of the Acp29AB accessory gland protein in Drosophila. *Genetics* **152:** 543-551

Aigaki, T., Fleischmann, I., Chen, P.S. and Kubli, E. (1997). Ectopic expression of sex peptide alters reproductive behavior of female *D. melanogaster*. *Neuron* **7:** 557-563

Alipaz, J.A., Wu, C.I. and Karr, T.L. (2001). Gametic incompatibilities between races of *D. melanogaster*. *Proceedings of the Royal Society, series B* **268:** 789-795

Anderson, R.C. (1945). A study of the factors affecting fertility of lozenge females of *Drosophila melanogaster*. *Genetics* **30:** 280-296

Ayala, F.J. and Tracey, M.L. (1974). Enzyme variability in the *Drosophila willistoni* group. VIII. Genetic differentiation and reproductive isolation between two subspecies. *Journal of Heredity* **64:** 120-124

Bairati, A. (1967). Structure and ultrastructure of the male reproductive system of *D. melanogaster* Meig 2. The genital duct and accessory glands. *Monitore Zoologico Italiano* **2**:105-182

Bairati, A. and Perrotti, M.E. (1970). The formation of a compact plug in the genital duct of *Drosophila* females after mating. *Drosophila Information Service* **45**: 67-68

Baker, W.R (1947). A study of the isolating mechanisms found in *Drosophila arizonensis* and *Drosophila mojavensis*. University of Texas Publications **4720:** 126-136

Begun, D.J. and Lindfors, H.A. (2005). Rapid evolution of genomic Acp complement in the melanogaster subgroup of Drosophila. *Molecular Biology and Evolution* **22:** 2010-2021

Begun, D.J., Whitley, P., Todd, B.L., Waldrip-Dail, H.M. and Clark, A.G. (2000). Molecular population genetics of male accessory gland proteins in *Drosophila*. *Genetics* **156:** 1879-1888

Betancourt, A.B., Presgraves, D.C. and Swanson W.J. (2002). A test for faster X evolution in Drosophila. *Molecular Biology and Evolution* **19:** 1816-1819

Bloch-Qazi, M.C. and Wolfner, M.F. (2003). An early role for the *Drosophila melanogaster* male seminal protein Acp36DE in female sperm storage. *Journal of Experimental Biology* **206:** 3521-3528

Bloch-Qazi, M.C., Heifetz, Y. and Wolfner, M.F. (2003). The developments between gametogenesis and fertilization: ovulation and sperm storage in *Drosophila melanogaster*. *Development* **256:** 195-211

Brieger, G. and Butterworth, F.M. (1970). *Drosophila melanogaster*: identity of male lipid in reproductive system. *Science* **167:** 1262

Butterworth, F.M. (1969). Lipids of *Drosophila*: a newly detected lipid in the male. *Science* **163:** 1356-1357

Campbell, R.V. and Noor, M.A.F. (2001). Assessing hybrid male fertility in *Drosophila* species:correlation between sperm motility and production of offspring. *Drosophila Information Service* **84:** 6-9

Cattaneo, F., Pasini, M.E., Intra, J., Matsumoto, M., Briani, F., Hoshi, M. and Perotti, M.E. (2006). Identification and expression analysis of *Drosophila melanogaster* genes encoding ß-hexosaminidases of the sperm plasma membrane. *Glycobiology* **16:** 786-800

Cattaneo, F., Ogiso, M., Hoshi, M., Perotti, M.E. and Pasini, M.E. (2002). Purification and characterization of the plasma membrane glycosidases of *Drosophila melanogaster* spermatozoa. *Insect Biochemistry and Molecular Biology* **32:** 929-941

Cavener, D.R. (1980). Genetics of male-specific glucose oxidase and the identification of other unusual hexose enzymes in *Drosophila melanogaster*. *Biochemical Genetics* **18:** 929-937

Chang, A.S. (2004). Conspecific sperm precedence in sister species of *Drosophila* with overlapping ranges. *Evolution* **58:** 781-789

Chapman, T. and Davies, S.J. (2004). Functions and analysis of the seminal fluid proteins of male *Drosophila melanogaster* fruit flies. *Peptides* **25:**1477-1490

Chapman, T., Herndon, L.A., Heifetz, T., Partridge, L. and Wolfner, M.F. (2001). The Acp26Aa seminal fluid protein is a modulator of early egg hatchability in *Drosophila melanogaster*. *Proceedings of the Royal Society, series B* **268:** 1647-1654

Chen, P.S., Stumm-Zollinger, E., Aigaki, T., Balmer, J., Bienz, M. and Bohlen, P. (1998). A male accessory gland peptide that regulates reproductive behavior of female *D. melanogaster*. *Cell* **54:** 291-298

Civetta, A., Rajakumar, S.A., Brouwers, B. and Bacik, J.P. (2006). Rapid evolution and gene-specific patterns of selection for three genes of spermatogenesis in Drosophila. *Molecular Biology and Evolution* **23:** 655-662

Counce, S.J. (1959). Micropyle structure in the eggs of several Drosophila species. *Drosophila Information Service* **33:** 126-127

Counterman, B.A., Ortiz-Barrientos, D. and Noor, M.A.F. (2004). Using comparative genomic data to test for faster-X evolution. *Evolution* **58:** 656-660

Coyne, J.A. (1984). Genetic basis of male sterility in hybrids between two closely related species of *Drosophila*. *Proceedings of the National Academy of Sciences USA* **81:** 4444-4447

Coyne, J.A. and Charlesworth, B. (1989). Genetic analysis of X-linked sterility in hybrids between three sibling species of *Drosophila*. *Heredity* **62:** 97-106

Coyne, J.A. and Orr, H.A. (1997). "Patterns of speciation in Drosophila" revisited. *Evolution* **51:** 295-303

Coyne, J. A. and Orr, H. A. (2004). *Speciation*. Sinauer Associates Inc., Sunderland, MA.

Coyne, J.A., Rux, J. and David, J.R. (1991). Genetics of morphological differences and hybrid sterility between *Drosophila sechellia* and its relatives. *Genetical Research* **57:** 113-122

Davis, A. W. and Wu, C.I. (1996). The broom of the sorcerer's apprentice: the fine structure of a chromosomal region causing reproductive isolation between two sibling species of *Drosophila*. *Genetics* **143:** 1287-1298

Dixon, S.M., Coyne, J.A. and Noor, M.F. (2003). The evolution of conspecific sperm precedence in *Drosophila*. *Molecular Ecology* **12:** 1179-1184

Dobzhansky, T. (1935). *Drosophila miranda*, a new species. *Genetics* **20:** 377-391

Dobzhansky, T. (1937). *Genetics and the Origin of Species*. Columbia University Press, New York.

Dobzhansky, T. (1975). Analysis of incipient reproductive isolation within a species of *Drosophila*. *Proceedings of the National Academy of Sciences USA* **72:** 3638-3641

Fowler, G.L. (1973). Some aspects of the reproductive biology of *Drosophila:* sperm transfer, sperm storage, and sperm utilization. *Advances in Genetics* **17:** 293-360

Fuller, M.T. (1993). Spermatogenesis. In: *The Development of Drosophila*. Edited by Bates, M., Cold Spring Harbor Laboratory Press, New York, pp. 71-147

Fuller, M.T. (1998). Genetic control of cell proliferation and differentiation in *Drosophila* spermatogenesis. *Seminars in Cell and Developmental Biology* **9:** 433-444

Garcia-Gonzalez, F. and Simmons, L.W. (2005). Sperm viability matters in insect sperm competition. *Current Biology* **15:** 271-275

Gilbert, D.G. (1981). Ejaculate esterase 6 and initial sperm use by female *Drosophila melanogaster. Journal of Insect Physiology* **27:** 641-650

Gilbert, D.G., Richmond, R.C. and Sheehan, K.B. (1981). Studies of esterase 6 in *Drosophila melanogaster. Evolution* **35:** 21-37

Gilchrist, A.S. and Partridge, L. (2000). Why it is difficult to model sperm displacement in *Drosophila melanogaster*: the relation between sperm transfer and copulation duration. *Evolution* **54:** 534-542

Glabe, C.G. and Lennarz, W.J. (1979). Species-specific sperm adhesion in sea urchins: a quantitative investigation of bindin-mediated egg agglutination. *Journal of Cell Biology* **83:** 595–604

Gottescheski, G. (1937). Künstliche Befructrung bei *Drosophila. Naturewissenschaften* **25:** 650

Haldane, J.B.S. (1922). Sex ratio and unisexual sterility in hybrid animals. *Journal of Genetics* **12:** 101-109

Heifetz, Y., Lung, O., Frongillo Jr, E.A. and Wolfner M.F. (2000). The *Drosophila* seminal fluid protein Acp26Aa stimulates release of oocytes by the ovary. *Current Biology* **10:** 99-102

Heifetz, Y., Vandenberg, L.N., Cohn, H.I. and Wolfner, M.F. (2005). Two cleavage products of the *Drosophila* accessory gland protein ovulin can independently induce ovulation. *Proceedings of the National Academy of Sciences USA* **102:** 743-748

Herndon, L.A. and Wolfner, M.F. (1995). A *Drosophila* seminal fluid protein Acp26Aa, stimulates egg-laying in females for one day following mating. *Proceedings of the National Academy of Sciences USA* **92:** 10114-10118

Hollocher, H. and Wu, C.I. (1996). The genetics of reproductive isolation in the *Drosophila simulans* clade: X vs. autosomal effects and male vs. female effects. *Genetics* **143:** 1243-1255

Howard, D.J. (1999). Conspecific sperm and pollen precedence and speciation. *Annual Review of Ecology and Systematics* **30:** 109-132

Iido , K. and Cavener, D.R. (2004). Glucose dehydrogenase is required for normal sperm storage and utilization in female *Drosophila melanogaster. Journal of Experimental Biology* **207:** 675-681

Intra, J., Cenni, F. and Perotti, M.E. (2006). An alpha-L-fucosidase potentially involved in fertilization is present on *Drosophila* spermatozoa surface. *Molecular Reproduction and Development* **73:** 1149-1158

Joly, D., Bazin, C., Zeng, L.W. and Singh, R.S. (1997). Genetic basis of sperm and testis length differences and epistatic effect on hybrid inviability and sperm motility between *Drosophila simulans* and *D. sechellia. Heredity* **78:** 354-362

Kern, A.D., Jones, C.D. and Begun, D.J. (2004). Molecular population genetics of male accessory gland proteins in the *Drosophila simulans* complex. *Genetics* **167:** 725-735

King, R.C. (1970). *Ovarian Development in* Drosophila melanogaster. Academic Press, New York

Knowles, L.L. and Markow, T.A. (2001). Sexually antagonistic coevolution of a postmating prezygotic reproductive character in desert *Drosophila. Proceedings of the National Academy of Sciences USA* **98:** 8692-8696

Kopp, A. and Frank, A.K. (2005). Speciation in progress? A continuum of reproductive isolation in *Drosophila bipectinata. Genetica* **125:** 55-68

Kubli, E. (1992). The sex-peptide. *Bioessays* **14:** 779-784

Kubli, E. (2003). Sex-peptides: seminal peptides of the Drosophila male. *Cellular and Molecular Life Sciences* **60:** 1689-1704

Kulathinal, R.J. and Singh, R.S. (1998). Cytological characterization of premeiotic vs. postmeiotic defects producing hybrid male sterility among sibling species of the *Drosophila melanogaster* complex. *Evolution* **52:** 1067-1079

Lamnissou, K. Loukas, M. and Zouros, E. (1996). Incompatibilities between Y chromosome and autosomes are responsible for male hybrid sterility in crosses between *Drosophila virilis* and *Drosophila texana. Heredity* **76:** 603-609

Lawniczak, M.K. and Begun, D.J. (2004). A genome-wide analysis of courting and mating responses in *Drosophila melanogaster* females. *Genome* **47:** 900-910

Lindsley, D. L. and Tokuyasu, K.T. (1980). Spermatogenesis. In: *The Genetics and Biology of Drosophila*, vol. 2. Edited by Wright, T.F.R. Academic Press, New York, NY, pp. 225-294

MacDonald, S.J. and Goldstein, D.B. (1999). A quantitative genetic analysis of male sexual traits distinguishing the sibling species *Drosophila simulans and D. sechellia. Genetics* **153:**1683-1699

Marin, I., Ruiz, A., Pla, C. and Fontdevila, A. (1993). Reproductive relationships among ten species of the *Drosophila* repleta group from South America and the West Indies. *Evolution* **47:** 1616-1620

Markow, T.A. (1996). Evolution of *Drosophila* mating systems. *Evolutionary Biology* **29:** 73-106

Markow, T.A. (1997). Assortative fertilization in *Drosophila. Proceedings National Academy of Sciences USA* **94:** 7756-7760

Markow, T.A. (2002). Female remating, operational sex ratio, and the arena of sexual selection in *Drosophila. Evolution* **59:** 1725-1734

Markow, T.A. and Ankney, P. (1988). Insemination reaction in *Drosophila*: A copulatory plug in species showing male contribution to offspring. *Evolution* **42:** 1097-1100

Markow, T.A., Coppola, A. and Watts, T. (2001). How Drosophila males make eggs: it's elemental. *Pro-*

ceedings of the Royal Society, series B **268:** 1527-1532

Markow, T.A. and O.Grady, P.M. (2005a). *Drosophila: A Guide to Species Identification and Use*. Academic Press, London.

Markow, T.A. and O.Grady, P.M. (2005b). Evolutionary genetics of reproductive behavior in *Drosophila*. *Annual Review of Genetics* **39:** 263-291

McGraw, L.A., Gibson, G., Clark, A.G. and Wolfner, M.F. (2004) Genes regulated by mating, sperm, or seminal proteins in mated female *Drosophila melanogaster*. *Current Biology* **14:** 1509-1514

Mengerink, K.J. and Vacquier, V.D. (2001). Glycobiology of sperm-egg interactions in deuterostomes. *Glycobiology* **11:** 37R-43R

Michalak, P. and Noor, M.A.F. (2003). Genome-wide patterns of expression in adult *Drosophila* pure-species and hybrid males. *Molecular Biology and Evolution* **20:** 1070-1076

Michalak, P. and Noor, M.A.F. (2004). Association of misexpression with sterility in hybrids of *Drosophila simulans* and *D. mauritiana*. *Journal of Molecular Evolution* **59:** 277-282

Miller, G. T. and Pitnick, S. (2002). Sperm-female coevolution in *Drosophila*. *Science* **269:** 1230-1233

Miller, G.T., Starmer, W.T. and Pitnick, S. (2003). Quantitative genetic analysis of among-population variation in sperm and female sperm-storage organ length in *Drosophila mojavensis*. *Genetical Research* **81:** 213-220

Mueller J.L., Ravi-Ram K., McGraw L.A., Bloch-Qazi M.C., Siggia E.D., Clark A.G., Aquadro, C.F. and Wolfner, M.F. (2005). Cross-species comparison of Drosophila male accessory gland protein genes. *Genetics* **171:** 131-143

Muller, H. J. (1940). Bearing of the Drosophila work on systematics. In: *The New Systematics*. Edited by Huxley, J.S., Clarendon Press, Oxford, pp.185-268

Muller, H. J. (1942). Isolating mechanisms, evolution, and temperature. *Biological Symposia* **6:** 71-125

Nakayama, S., Kaiser, K. and Aigaki, T. (1997). Ectopic expression of sex peptide in a variety of tissues in *Drosophila* females using the P[GAL4] enhancer-trap system. *Molecular and General Genetics* **254:** 449-455

Neubaum, D.M. and Wolfner, M.F. (1999). Mated *Drosophila melanogaster* females require a seminal fluid protein, Acp36DE, to store sperm efficiently. *Genetics* **153:** 845–857

Nonidez, J.F. (1920). The internal phenomena of reproduction in *Drosophila*. *Biological Bulletin* **39:** 207-230

Noor, M.A.F. (2005). Patterns of evolution of genes disrupted in expression in *Drosophila* species hybrids. *Genetical Research* **85:** 119-125

Noor, M.A.F., Michalak, P. and Donze, D. (2003). Characterization of a male-predominant antisense transcript underexpressed in hybrids of *Drosophila pseudoobscura* and *D. persimilis*. *Genetics* **165:** 1823-1830

Orr, H.A. (1989). Genetics of sterility in hybrids between two subspecies of *Drosophila*. *Evolution* **43:** 180-189

Orr, H.A. (1993). Haldane's rule has multiple genetic causes. *Nature* **361:** 532-533

Orr, H.A. (1995). The population-genetics of speciation - the evolution of hybrid incompatibilities. *Genetics* **139:** 1805-1813

Orr, H.A. and Coyne, J.A. (1989). The genetics of postzygotic isolation in the *Drosophila virilis* group. *Genetics* **121:** 527-537

Orr, H.A. and Irving, S. (2001). Complex epistasis and the genetic basis of hybrid sterility in the *Drosophila pseudoobscura* Bogota-USA hybridization. *Genetics* **158:** 1089-1100

Orr, H.A. and Turelli, M. (2001). The evolution of postzygotic isolation: accumulating Dobzhansky-Muller incompatibilities. *Evolution* **55:** 1085-1094

Panhuis, T. and Swanson, W.J. (2006). Molecular evolution and population genetics of candidate female reproductive genes in *Drosophila*. *Genetics* **173:** 2039-2047

Parisi, M., Nuttall, R., Naiman, D., Bouffard, G., Malley, J., Andrews, J., Eastman, S. and Oliver, B. (2003). Paucity of genes on the *Drosophila* X chromosome showing male-biased expression. *Science* **299:** 697-700

Pasini, M.E., Cattaneo, F., Verni, F., Hackstein, J.H.P. and Perotti, M.E. (1999). ß-N-acetylglucosaminidase of sperm plasma membrane is crucial for egg fertilization in *Drosophila melanogaster*. *Molecular Biology of the Cell* **10 (Suppl.):** 360a

Pasini, M.E., Redi, C.A., Caviglia, O. and Perotti, M.E. (1996). Ultrastructural and cytochemical analysis of sperm dimorphism in *Drosophila subobscura*. *Tissue and Cell* **28:** 165-175

Patterson, J.T. (1946). A new type of isolating mechanism in *Drosophila*. *Proceedings of the National Academy of Sciences USA* **32:** 202-208

Patterson, J.T. (1947). The insemination reaction and its bearing on the problem of speciation in the mulleri subgroup. *University of Texas Publications* **4752:** 78-115

Patterson, J.T. and Stone W.S. (1952) The insemination reaction and other isolating mechanisms. In: *Evolution in the Genus* Drosophila. The MacMillan Company, Toronto, pp. 359-383

Peng, J., Chen, S., Busser, S., Liu, H., Honegger, T. and Kubli, E. (2005). Gradual release of sperm bound sex-peptide controls female postmating behavior in Drosophila. *Current Biology* **15:** 207-213

Perotti, M.E., Cattaneo, F., Pasini, M.E., Verni, F. and Hackstein, J.H.P. (2001). Male sterile mutant casanova gives clues to mechanisms of sperm-egg interactions in *Drosophila melanogaster*. *Molecular Reproduction and Development* **60:** 248-259

Perotti, M.E. and Pasini, M.E. (1995). Glycoconjugates of the surface of the spermatozoa of *Drosophila melanogaster*: a qualitative and quantitative study. *Journal of Experimental Zoology* **272:** 311 318

Pitnick, S. (1996). Investment in testes and the cost of making long sperm in Drosophila. *The American*

Naturalist **148:** 57-80

Pitnick, S., Markow T.A. and Spicer G.S. (1999). Evolution of multiple kinds of female sperm-storage organs in *Drosophila*. *Evolution* **53:** 1804-1822

Pitnick, S., Spicer, G.S. and Markow, T.A. (1997). Phylogenetic examination of male ejaculatory donations in *Drosophila*. *Evolution* **51:** 833-845

Polak, M., Starmer, W.T. and Barker, J.S.F. (1998). A mating plug and male mate choice in *Drosophila hibisci*. *Animal Behaviour* **56:** 919-926

Price, C.S., Kim, C.H., Posluszny, J. and Coyne, J.A. (2000). Mechanisms of conspecific sperm precedence in *Drosophila*. *Evolution* **54:** 2028-2037

Price, C.S. (1997). Conspecific sperm precedence in *Drosophila*. *Nature* **388:** 663-666

Price, S.C., Kim, C.H., Gronlundm C.J. and Coyne, J.A. (2001). Cryptic reproductive isolation in the *Drosophila simulans* species complex. *Evolution* **55:**81-92

Ranz, J.M., Castillo-Davis, C.I., Meiklejohn, C.D. and Hartl, D.L. (2003). Sex-dependent gene expression and evolution of the *Drosophila* transcriptome. *Science* **300:** 1742-1745

Ranz, J. M., Namgyal, K., Gibson, G. and Hartl, D.L. (2004) Anomalies in the expression profile of interspecific hybrids of *Drosophila melanogaster* and *Drosophila simulans*. *Genome Research* **14:** 373-379

Ravi Ram K., Ji S. and Wolfner, M.F. (2005). Fates and targets of male accessory gland proteins in mated female *Drosophila melanogaster*. *Insect Biochemistry and Molecular Biology* **35**:1059-1071

Reed, L.K., LaFlamme, B.A. and Markow, T.A. (2006). The genetic architecture of male sterility within *Drosophila mojavensis* in F_1 hybrids between *D. mojavensis* and *D. arizonae*. In review

Reed, L. K. and Markow, T.A. (2004). Early events in speciation: Polymorphism for hybrid male sterility in *Drosophila*. *Proceedings of the National Academy of Sciences USA* **101:** 9009-9012

Remsen J and O'Grady, P. (2002). Phylogeny of Drosophilinae (Diptera: Drosophilidae), with comments on combined analysis and character support. *Molecular Phylogenetics and Evolution* **24:** 249-264

Richmond, R.C. and Senior, A. (1981). Esterase 6 of *Drosophila melanogaster*: kinetics of transfer to females, decay in females and male recovery. *Journal of Insect Physiology* **27:** 849-853

Ross, J.L., Fong, P.P. and Cavener, D.R. (1994). Correlated evolution of the cis-acting regulatory elements and developmental expression of the *Drosophila* Gld gene in seven species from the subgroup *melanogaster*. *Developmental Genetics* **15:** 38-50

Sanchez L and Santamaria P. (1997). Reproductive isolation and morphogenetic evolution in *Drosophila* analyzed by breakage of ethological barriers. *Genetics* **147:** 231-242

Schiff, N.M., Feng, Y., Quine, J.A., Krasney, P.A. and Cavener, D.R. (1992). Evolution of the expression of the *Gld* gene in reproductive tract of *Drosophila*. *Molecular Biology and Evolution* **9:** 1029-1049

Schully, S.D. and Hellberg, M.E. (2006). Positive selection on nucleotide substitutions and indels in accessory gland proteins of the *Drosophila pseudoobscura* subgroup. *Journal of Molecular Evolution* **62:** 793-802

Smedley, S.R. and Eisner, T. (1995). Sodium uptake by puddling in a moth. *Science* **270:** 1816-1818

Smith, G.M., Rothwell, K., Wood, S.L., Yeaman, S.J. and Bownes, M. (1994). Specificity and localization of lipolytic activity in adult *Drosophila melanogaster*. *Biochemical Journal* **304:** 775-779

Snook, R.R. (1997). Is the production of multiple sperm types adaptive? *Evolution* **51:** 797-808

Snook, R.R. and Hosken, D.J. (2004). Sperm death and dumping in *Drosophila*. *Nature* **428:** 939-941

Snook, R.R., Markow, T.A. and Karr, T.L. (1994). Functional non-equivalence of sperm in *Drosophila pseudoobscura*. *Proceedings of the National Academy of Sciences USA* **91:** 1-5

Sturtevant, A. H. and Dobzhansky, T. (1936) Observations on species related to *Drosophila affinis*, with descriptions of seven new forms. *American Naturalist* **70**: 574-584

Swanson, W.J., Clark, A.G., Waldrip-Dail, H.M., Wolfner, M.F. and Aquadro, C.F. (2001). Evolutionary EST analysis identifies rapidly evolving male reproductive proteins in *Drosophila*. *Proceedings of the National Academy of Sciences USA* **98:** 7375-7379

Swanson, W.J., Wong, A., Wolfner, M.F. and Aquadro, C.F. (2004). Evolutionary expressed sequence tag analysis of *Drosophila* female reproductive tracts identifies genes subjected to positive selection. *Genetics*. **168:** 1457-1465

Tao, Y. and Hartl, D. L. (2003). Genetic dissection of hybrid incompatibilities between *Drosophila simulans* and *D.mauritiana*. III. Heterogeneous accumulation of hybrid incompatibilities, degree of dominance, and implications for Haldane's rule. *Evolution* **57**: 2580-2598

Tao, Y., ChenS. N., Hartl, D. L. and Laurie, C. C. (2003a.) Genetic dissection of hybrid incompatibilities between *Drosophila simulans* and *D. mauritiana*. I. Differential accumulation of hybrid male sterility effects on the X and autosomes. *Genetics* **164**: 1383-1397

Tao, Y., Zeng, Z. B., Li, J., Hartl, D. L. and Laurie, C. C. (2003b). Genetic dissection of hybrid incompatibilities between *Drosophila simulans* and *D. mauritiana*. II. Mapping hybrid male sterility loci on the third chromosome. *Genetics* **164**: 1399-1418

Throckmorton, L.H. (1962). The problem of phylogeny in the genus *Drosophila*. *University of Texas Publications* **6205:** 207-345

Throckmorton, L.H. (1982). The *virilis* species group. In: *The Genetics and Biology of* Drosophila, Vol. 3b. Edited by Ashburner M., Carson, H. L. and Thompson Jr., J.N., Academic Press, London, pp .227-296

True, J.R., Weir, B.S. and Laurie, C. C. (1996). A genome-wide survey of hybrid incompatibility factors by the introgression of marked segments of *Drosophila mauritiana* chromosomes into *Drosophila simulans*. *Genetics* **142**: 819-837

Turelli, M. and Begun, D.J. (1997). Haldane's rule and X-chromosome size in Drosophila. *Genetics* **147**: 1799-1815

Vacquier, V.D. (1998). Evolution of gamete recognition proteins. *Science* **281**: 1995-1998

Vacquier, V.D., Swanson, W.J., Metz, E. C. and Stout, C. D. (1999). Acrosomal proteins of abalone spermatozoa. *Advances in Developmental Biochemistry* **5**: 49–81

Vigneault, G. and Zouros, E. (1986). The genetics of asymmetrical male sterility in *Drosophila mojavensis* and *Drosophila arizonensis* hybrids: Interactions between the Y chromosome and the autosomes. *Evolution* **40**: 1160-1170

Wagstaff, B.J. and Begun, D.J. (2005). Molecular population genetics of accessory gland protein genes and testis-expressed genes in *Drosophila mojavensis* and *D. arizonae*. *Genetics* **171:** 1083-1101

Wakimoto, B.T., Lindsley, D.L. and Herrera, C. (2004). Toward a comprehensive genetic analysis of male fertility in *Drosophila melanogaster*. *Genetics* **167:** 207-216

Wheeler M.R. (1947). The insemination reaction in intraspecific matings of *Drosophila*. *University of Texas Publications* **4720:** 78-115

Wolfner, M.F. (1997) Tokens of love: functions and regulation of Drosophila male accessory gland products. *Insect Biochemistry and Molecular Biology* **27:** 179-92

Wolfner, M.F. (2002). The gifts that keep on giving: physiological functions and evolutionary dynamics of male seminal proteins in *Drosophila*. *Heredity* **88:** 85-93

Wu, C.I. (1992). A note on Haldane's rule: Hybrid inviability versus hybrid sterility. *Evolution* **46:** 1584-1587

Yeh, S.D., Chen, Y.J., Chang, A.C.Y., Ray, R., She, B.R., Lee, W.S., Chiang, H.S., Cohen, S.N. and Lin-Chao, S. (2002). Isolation and properties of Gas8, a growth arrest-specific gene regulated during male gametogenesis to produce a protein associated with the sperm motility apparatus. *Journal of Biological Chemistry* **277:** 6311—6317

Zhang, Z., Hambuch, T.M. and Parsch, J. (2004). Molecular evolution of sex-biased genes in *Drosophila*. *Molecular Biology and Evolution* **21:** 2130-2139

Zouros, E. (1981). An autosome-Y chromosome combination that causes sterility in *D. mojavensis* and *D. arizonensis* hybrids. *Drosophila Information Service* **56:** 167-168

Zouros, E. (1991). Searching for speciation genes in the species pair *Drosophila mojavenesis* and *D. arizonae*. In: *Molecular Techniques in Taxonomy*. Edited by Young, J.P.W., Springer-Verlag, Berlin, pp. 233-247

Zouros, E., Lofdahl, K. and Martin P.A. (1988). Male hybrid sterility in *Drosophila* - interactions between autosomes and sex-chromosomes in crosses of *Drosophila mojavensis* and *Drosophila arizonensis*. *Evolution* **42:** 1321-1331

Spermatology. SRF Vol. 65. ERS Roldan and M Gomendio (eds) Nottingham University Press, Nottingham

Making sperm: design, quality control and sperm competition

Tim R. Birkhead and Simone Immler

Department of Animal & Plant Sciences, University of Sheffield, Sheffield, S10 2TN, UK

Spermatozoa vary remarkably in design at several different levels: phyla, orders, families, species, individuals, within individuals and within ejaculates of the same individual. Three factors are thought to account for some of this variation: (i) fertilisation mode; (ii) phylogeny, and (iii) post-copulatory sexual selection (i.e. sperm competition and cryptic female choice). We focus here on the hypothesis that post-copulatory sexual selection shapes sperm design and how this hypothesis can be tested. We discuss the importance of controlling for fertilisation mode and phylogeny. In addition, we consider the way the intensity of post-copulatory sexual selection is measured and how this may influence the way we test this hypothesis. So far the evidence that post-copulatory sexual selection influences inter-specific differences in sperm is mixed. We also consider how post-copulatory sexual selection influences *variation* in sperm design between and within males. In two phyla (mammals and birds) we show that reduced selection via sperm competition results not only in reduced sperm numbers but also in unusual (immature but functional) sperm phenotypes, high inter-male variability and high levels of sperm pleiomorphy within ejaculates. We propose that these are all energy-saving adaptations.

Introduction

One of the major mysteries in reproductive biology is why spermatozoa exhibit such extraordinary morphological diversity. Interspecific variation in spermatozoa is greater than in any other animal cell type (Cohen, 1977). The underlying causes of this variation remain poorly known, but three sets of factors are thought to be important: (i) fertilisation mode (internal *versus* external), (ii) phylogeny, and (iii) post-copulatory sexual selection, mediated by sperm competition and/or cryptic female choice (Pitnick *et al.*, 1999). Externally fertilising species generally have less elaborate sperm morphology than internally fertilising species (Franzén, 1956), the latter presumably reflecting the more complex environment of the female reproductive tract than external media. Phylogeny is clearly of major importance in sperm design, not only between major phylogenetic groups (e.g. insects and birds, or fish and mammals), but also between species in the same family (Jamieson, 1991; Jamieson *et al.*, 1987, 1999). The importance of post-copulatory sexual selection on sperm design is less clear (Snook, 2005), partly because such effects may be masked or confounded by phylogenetic effects. However, since post-copulatory sexual selection has such an important effect on other reproductive traits (Pizzari and Birkhead, 2002) it seems likely that it also affects sperm design.

Corresponding author E-mail: T. R. Birkhead@sheffield.ac.uk

In order to establish the relative importance of post-copulatory sexual selection in sperm design we need to (a) define what we mean by 'design', (b) control for the possible confounding effects of fertilisation mode and phylogeny, and (c) identify where variation in sperm design exists.

Sperm design

We define sperm design as the gross morphology of the sperm, that is its shape, size (e.g. total length) and the relative and absolute size of different components (e.g. head, midpiece and flagellum). Although ultrastructure is an important element of sperm morphology, since it is much more evolutionarily conserved than gross morphology and hence more strongly influenced by phylogeny (Jamieson, 1991; Jamieson *et al.*, 1987, 1999), we do not consider it here.

Fertilisation mode and phylogeny

Controlling for fertilisation mode is easily accomplished by focussing on groups with only internal or only external fertilisation, although comparisons between the two in those taxa whether both modes of fertilisation occur (e.g. fish and amphibia) is potentially revealing (e.g. Jamieson *et al.*, 1993). Controlling for and assessing the relative importance of phylogeny in sperm design is also possible, using particular statistical techniques (e.g. Freckleton *et al.*, 2002; Birkhead *et al.*, 2006).

Levels at which variation in sperm design exists

We recognise four major levels at which variation exists: (1) inter-taxon variation, (2) inter-specific variation, (3) inter-male variation, and (4) intra-male variation.

Inter-taxon variation

The diversity in sperm phenotypes between higher taxa (phyla, orders, families), is considerable (Jamieson, 1999). Within birds for example, the difference between the two major orders, the passerine and non-passerine birds in gross morphology and sperm ultrastructure is pronounced. The sperm of non-passerines are relatively simple; similar to those of reptiles and characterised by being relatively short with a smooth, tapering head. In contrast, passerine sperm exhibit a more complex design, with a helical form and elongate midpiece twisted around the flagellum and are typically longer than those of non-passerines. This basic sperm phenotype appears to be highly conserved in passerines (Asa and Philips, 1987).

Inter-specific variation

At the next phylogenetic level the variation in the gross morphology of sperm between species is usually sufficiently pronounced that it can be used to distinguish between species (Jamieson, 1991, 1999; Jamieson *et al.*, 1987). In passerine birds for example, our observations show that with one or two notable exceptions most of the inter-specific variation in sperm design at this level consists of differences in the relative size of different sperm components (e.g. Birkhead *et al.*, 2006).

Inter-male variation

Although variation between individuals of the same species is typically less than that between species (Morrow and Gage, 2001; Malo *et al.*, 2006), in some cases it can be considerable. In the zebra finch *Taeniopygia guttata* for example, the coefficient of variation (CV) in different sperm traits varied between 6 and 13% between males (based on means of 5 sperm from each of 913 males) (Birkhead *et al.*, 2005). The amount of inter-male variation will therefore determine the sample size required to obtain a 'species mean' for any particular sperm trait.

Intra-male variation

In most species studied so far individual males typically show a high degree of consistency in their sperm phenotype across ejaculates (Morrow and Gage, 2001). In the zebra finch for example, the within-male repeatability of sperm traits is relatively high (head: 0.52; midpiece: 0.75; flagellum: 0.84; Birkhead *et al.*, 2005). Although there is no evidence that male sperm morphology differs between ejaculates, there is evidence that sperm velocity differs between ejaculates of the same male, in relation to the risk of sperm competition (C. K. Cornwallis and T. R. Birkhead, in preparation). The high consistency of sperm design within males between ejaculates is consistent with the observation that sperm morphology is largely under genetic control (Woolley, 1971; Beatty, 1972; Birkhead *et al.*, 2005).

The marked influence of genetic factors is also reflected in the fact that intra-male variation in sperm morphology is often high in species or populations that have experienced a genetic bottleneck and/or are subject to inbreeding (e.g. Gage *et al.*, 2006).

Sperm phenotypes and post-copulatory sexual selection

To date, most studies testing the hypothesis that sperm phenotypes are influenced by post-copulatory sexual selection have focussed on comparisons at the first two levels, inter-taxon and inter-specific variation and sought correlations between the level of sperm competition (estimated in various ways) and sperm traits (such as total length) using species mean values (e.g. Gomendio and Roldan, 1991; Briskie and Montgomerie, 1992). The results of these analyses have been mixed in terms of assessing the relative importance of post-copulatory sexual selection in sperm design (Snook, 2005). Researchers have rarely considered whether variation in sperm design at lower levels, between and within males of the same species for example, may also be explicable in terms of sperm competition.

One of the main ways the intensity of post-copulatory sexual selection is assessed is through relative (or residual) testes size, since in species with high levels of female promiscuity and sperm competition, males typically have relatively large testes and where sperm competition is reduced or absent relative testes size is small (Harcourt *et al.* 1981; Møller, 1988a,b; Birkhead and Møller, 1998). The functional significance of this is that larger testes produce more sperm than smaller testes, allowing males to either deliver larger or more frequent ejaculates (Parker, 1998). Since there may also be a trade-off between sperm numbers and sperm design (e.g. Bjork and Pitnick, 2006), there is a potential problem of using testes size as an index of the intensity of post-copulatory sexual selection to examine its effect on sperm design. In other words, researchers need to take this potentially confounding effect into account.

Sperm are assumed to be costly to make and relatively large testes are also energetically costly – especially in taxa like birds, where they are likely to add substantially to flight costs. For example, comparing two similarly sized passerine birds: in the socially monogamous

Eurasian bullfinch *Pyrrhula pyrrhula* the paired testes constitute just 0.3% of male body mass (Birkhead *et al.*, 2006), whereas in the promiscuous Alpine accentor *Prunella collaris* they represent 6.5% (Nakamura, 1990). Theory predicts that in the absence of sperm competition males should minimise their investment in both testes and in the design of sperm themselves (Parker, 1998).

Examples from two different taxa are consistent with the hypothesis that relaxed selection in terms of post-copulatory processes, may result in an evolutionary change in sperm design. In rodents, sperm exhibit a complex design and are generally very long with an extensive midpiece and a falciform shaped head (Roldan *et al.*, 1992). However, several rodent lineages with extremely small testes for their body mass (indicating very low levels of sperm competition; Breed, 2002, 2005) exhibit unusual sperm design. Their sperm are characterised by: atypical sperm head shape (oval, flattened or round and lacking an apical hook: see Fig. 1); poor condensation of chromatin in the nucleus; an unusual structure of the acrosome (Breed, 1995) and an overall shorter length (Breed, 2002).

A similar phenomenon appears to exist in birds. The Eurasian bullfinch has the smallest testes for its body size of almost any species (Birkhead *et al.*, 2006), suggesting that sperm competition is extremely rare or absent. There have been no studies of parentage in this species, but the close pair bond between partners is consistent with a socially and sexually monogamous mating system. In addition to its relatively small testes, the gross morphology of the bullfinch's sperm is extremely unusual compared with all other passerine birds examined so far (Fig.1). Further, a study of the sperm ultrastructure of this species revealed that like the rodents in Breed's studies the sperm exhibited several traits typical of immature sperm. These included: a simple rounded acrosome; poorly condensed chromatin in the nucleus; the presence of two centrioles, and multiple mitochondria (rather than a single large fused mitochondrion as in other passerines) clustered tightly round the anterior portion of the flagellum (rather than forming an elongate helix along the length of the flagellum, as in other passerines; Birkhead *et al.*, in press). These traits are typical of immature sperm (i.e. spermatids) in other passerines, and we interpret these as an energy-saving device in the response to a lack of sperm competition. When selection via sperm competition is relaxed, any design - as long as it works - will be sufficient for fertilisation.

Not only does relaxed post-copulatory sexual selection result in a change in mean sperm design, it may also result in a reduction in sperm-quality control, and hence an increase in variation at levels 3 and 4, that is, between and within males. In both Breed's rodents and the bullfinch sperm pleiomorphy - large within-male variation in sperm design – is pronounced. In the Eurasian bullfinch the coefficient of variation (CV) of different sperm traits was 8-18% compared with 4-5% for the dunnock *Prunella modularis* – a species with intense sperm competition (S. Calhim, personal communication). Again, we consider this to be consistent with an energy saving adaptation in response to a lack of sperm competition. Sperm pleiomorphy is effectively a quality-control issue. In the absence of sperm competition, there is little benefit from expending energy refining quality control. Since the female tract of both birds and mammals is highly selective, poor sperm designs are likely to be prevented from entering the 'fertilising set'. On the other hand, when sperm competition is intense, every sperm counts and selection will favour tight quality control.

Conclusions

Sperm competition appears to result in the evolution of suites of male traits, including for example, genital morphology, testes size, ejaculate size, copulation behaviour, and sperm design. The problem is deciding which of these traits has been subject to direct selection, and which are a consequence of selection on other traits. In the present context, low levels of sperm competition are associated with relatively small testes and low sperm numbers and (a)

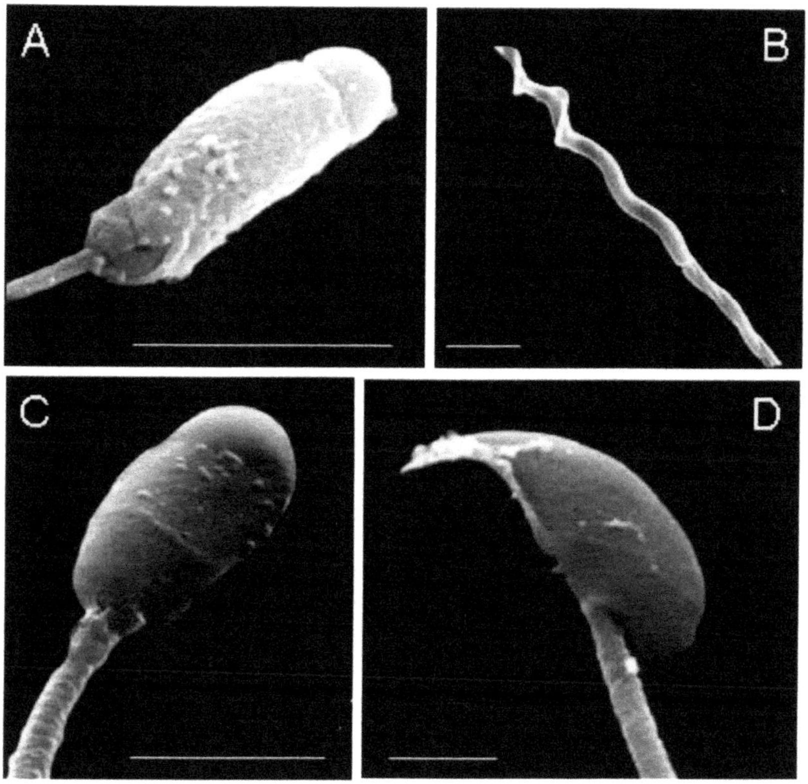

Figure 1. Sperm heads of two passerine birds and two rodents. (**A**) The atypical Eurasian bullfinch *Pyrrhula pyrrhula*; (**B**) A typical passerine, the House sparrow *Passer domesticus*; (**C**) the atypical Heath mouse *Pseudomys shortridgei*; and (**D**) the typical rodent House mouse. *Mus musculus* (see text). Scale bar = 2 µm. Photos by R. Dallai (**A** and **B**), and W.G. Breed (**C** and **D**).

simple (immature) sperm design, (b) high sperm pleiomorphy. To what extent are these features the result of direct selection, and to what extent might they simply be the consequence of small relative testis size? It seems unlikely that sperm design and pleiomorphy are the result of small testis size. Several insects have extremely small testes, yet produce high quality ejaculates and very long sperm (Pitnick *et al.*, 1995; Pitnick, 1996). Comparisons across phyla might not be appropriate however. But within birds or mammals, there are species whose testes are absolutely smaller than those in Breed's rodents, or the Eurasian bullfinch that produce more elaborate and consistent sperm designs with low levels of pleiomorphy (e.g. rodents: house mouse *Mus musculus,* New Holland mouse *Pseudomys novaehollandiae*; passerines: great reed warbler *Acrocephalus arundinaceus*, collared flycatcher *Ficedula albicolla*).

Finally, we need more information on the amounts of variation in sperm design at different levels and better tests of the hypothesis that much of this variation can be accounted for by post-copulatory sexual selection. Specifically, we need independent verification (via paternity studies) that sperm competition is infrequent in Breed's rodents (e. g. Tete veld rat *Aethomys ineptus*, greater Bandicoot rat *Bandicota indica*, harvest mouse *Micromys minutus*, Sulawesi soft-furred rat *Eropeplus canus*, Ryukyu spiny rat *Tokudaia osimensis*, heath mouse *Pseudomys shortridgei*) and the Eurasian bullfinch. It is also clear that phylogeny accounts for quite a lot of the variation in sperm design between species, which means that in order to estimate the relative importance of post-copulatory sexual selection we need to take phylogeny into account. We also need to consider possible inbreeding effects on sperm design. Finally, an outstanding problem in understanding the evolution of sperm design and numbers is the issue of how costly sperm are to produce.

Acknowledgements

We are grateful to the Leverhulme Trust for financial support. We thank Romano Dallai and Bill Breed for allowing us to use their photographs, and Montse Gomendio and Eduardo Roldan for constructive comments on the manuscript.

References

Asa, C.S. and Phillips, D.M. (1987). Ultrastructure of avian spermatozoa: a short review. In: *New Horizons in Sperm Cell Research*. Edited by Mohri, H., Japan Scientific Society Press, Tokyo, pp. 365-373

Beatty, R.A. (1972). The genetics of size and shape of spermatozoan organelles. In: *The Genetics of the Spermatozoa*. Edited by Gluecksohn-Waelsch, S., Edinburgh University Press, Edinburgh, p. 97-115

Birkhead, T.R. and Møller, A.P. (1998). *Sperm Competition and Sexual Selection*. Academic Press, London

Birkhead, T.R., Pellatt, E.J., Brekke, P., Yeates, R. and Castillo-Juarez, H. (2005). Genetic effects on sperm design in the zebra finch. *Nature* **434**: 383-387

Birkhead, T.R., Immler, S., Pellatt, E.J. and Freckleton, R.P. (2006). The unusual sperm morphology of the Northern bullfinch *Pyrrhula pyrrhula*. *Auk* **123**: 383-392

Birkhead, T.R., Giusti, F., Immler, S. and Jamieson, B.G.M. (in press). Ultrastructure of the unusual spermatozoon of the Eurasian bullfinch (*Pyrrhula pyrrhula*). *Acta Zoologica* in press

Bjork, A. and Pitnick, S. (2006). Intensity of sexual selection along the anisogamy-isogamy continuum. *Nature* **441**: 742-745

Breed, W. (1995). Spermatozoa of murid rodents from Africa: morphological diversity and evolutionary trends. *Journal of Zoology, London* **237**: 625-651

Breed, W. (2002). Variation in sperm form in rats and mice (subfamily Murinae). In: *Proceedings of 9th International Symposium on Spermatology*. Cape Town, South Africa, Edited by van der Horst, G., Franken, D., Bornman, R., Bornman, T. and Dyer, S., Monduzzi Editore, Bologna, pp. 269-274

Breed, W. (2005). Evolution of the spermatozoon in muroid rodents. *Journal of Morphology* **265**: 271-290.

Briskie, J.V. and Montgomerie, R. (1992). Sperm size and sperm competition in birds. *Proceedings of the Royal Society of London Series B*, **247**: 89-95

Cohen, J. (1977). *Reproduction*. Butterworths, London and Boston.

Franzén, A. (1956). On spermiogenesis, morphology of the spermatozoan, and biology of fertilization among invertebrates. *Zool. Bidr. Uppsala* **31**, 355-480

Freckleton, R.P., Harvey, P.H. and Pagel, M. (2002). Phylogenetic analysis and comparative data: a test and review of evidence. *American Naturalist* **160**: 712-725.

Gage, M.J.G. and Freckleton, R.P. (2003). Relative testis size and sperm morphometry across mammals: no evidence for an association between sperm competition and sperm length. *Proceedings of the Royal Society of London Series B* **270**: 625-632

Gage, M.J.G., Surridge, A.K., Tomkins, J.L., Green, E., Wiskin, L., Bell, D.J. and Hewitt, G.M. (2006). Reduced neterozygosity depresses sperm quality in wild rabbits *Oryctolagus cuniculus*. *Current Biology* **16**: 612-617

Gomendio, M. and Roldan, E.R.S. (1991). Sperm competition influences sperm size in mammals. *Proceedings of the Royal Society of London Series B* **243**: 181-185

Harcourt, A.H., Harvey, P.H., Larson, S.G. and Short, R.V. (1981). Testis weight, body weight and breeding system in primates. *Nature* **293**: 55-57

Jamieson, B.G.M. (1991). *Fish Evolution and Systematics: Evidence from Spermatozoa*. Cambridge University Press, Cambridge

Jamieson, B.G.M. (1999). Spermatozoal phylogeny of the vertebrata. In: *The Male Gamete: From Basic Science to Clinical Applications*. Edited by Gagnon, C. Cache River Press, Vienna, IL, pp. 304-331

Jamieson, B.G.M., Afzelius, B. and Dallai, R. (1987). *The Ultrastructure and Phylogeny of Insect Spermatozoa*. Cambridge University Press, Cambridge

Jamieson, B.G.M, Lee, M.S.Y. and Long, K. (1993). Ultrastructure of the spermatozoon of the internally fertilizing frog *Ascaphus truei* (Ascaphidae: Anura, Amphibia) with phylogenetic considerations. *Herpetologica* **49:** 52-65

Malo, A.F., Gomendio, M., Garde, J.J., Lang-Lenton, B., Soler, A.J. and Roldan, E.R.S. (2006). Sperm design and sperm function. *Biology Letters* **2:** 246-249

McFarlane, R.W. (1963). The taxonomic significance of avian sperm. *Proceedings of 13th International Ornithology Congress* **1:** 91-102

Møller, A.P. (1988a). Testes size, ejaculate quality and sperm competition in birds. *Biological Journal of the Linnean Society* **33:** 273-283

Møller, A.P. (1988b) Ejaculate quality, testes size and sperm competition in primates. *Journal of Human Evolution* **17:** 479-488

Morrow, E.H. and Gage, M.J.G. (2001). Consistent significant variation between individual males in spermatozoal morphometry. *Journal of Zoology, London* **254:** 147-153

Nakamura, M. (1990). Cloacal protuberance and copulatory behaviour of the Alpine accentor (*Prunella collaris*). *Auk* **107:** 284-295

Parker, G. A. (1998). Sperm competition and the evolution of ejaculates: Towards a theory base. In: *Sperm Competition and Sexual Selection*. Edited by Birkhead, T.R. and Møller, A. P., Academic Press, London, pp. 3-54

Pitnick, S. (1996). Investment in testes and the cost of making long sperm in *Drosophila*. *American Naturalist* **148:** 57-80

Pitnick, S., Spicer, G.S. and Markow, T.A. (1995). How long is a giant sperm? *Nature* **375:** 109

Pitnick, S., Markow, T. and Spicer, G.S. (1999). Evolution of multiple kinds of female sperm-storage organs in *Drosophila*. *Evolution* **53:** 1804-1822

Pizzari, T. and Birkhead, T.R. (2002). The sexually-selected sperm hypothesis: sex-biased inheritance and sexual antagonism. *Biological Review* **77:** 183-209

Roldan, E.R.S., Gomendio, M. and Vitullo, A.D. (1992). The evolution of eutherian spermatozoa and underlying selective forces: female selection and sperm competition. *Biological Review* **67:** 551-593

Snook, R.R. (2005). Sperm in competition: not playing by the numbers. *Trends in Ecology and Evolution* **20:** 46-53

Woolley, D.M. (1971). Selection for the length of spermatozoan midpiece in the mouse. *Genetical Research Cambridge* **16:** 261-275

Spermatology. SRF Vol. 65. ERS Roldan and M Gomendio (eds) Nottingham University Press, Nottingham

"S.P.E.R.M." (Seminal Proteins (are) Essential Reproductive Modulators): the view from *Drosophila*

M. F. Wolfner

Department of Molecular Biology and Genetics, Cornell University, Ithaca NY 14853, USA

The seminal fluid that females receive from their mates contains a suite of proteins that have important effects on sperm, as well as on reproduction in general. Seminal proteins are vital for the fertility of mating animals in several diverse taxonomic groups. For example, in *Drosophila melanogaster,* the ~70-106 accessory gland proteins (Acps) that are a major part of the seminal fluid are essential for the storage and utilization of sperm, as well as for increasing egg production and laying by the female. In addition, Acps have been implicated in modifying the female's eating behaviour, her receptivity to re-mating and her longevity. This review will first summarise the molecular nature and reproductive function of *Drosophila* Acps in general, as elucidated by genetic/ transgenesis, biochemical, and physiological experiments. The article will then focus on Acps that affect, or interact with, sperm. Sperm storage is a stepwise process in *Drosophila* and Acps facilitate at least some of these steps. For example, Acps promote sperm entry into storage, apparently by modulating muscle contractions in the female's reproductive tract. One Acp is known to be essential for the entry of sperm into storage. This Acp, which is cleaved after entering females, binds to sperm and enters the sperm-storage organs. Egg production, which is also modulated by Acps, can affect the transition between the steps in sperm storage, although not the rate of release of sperm from storage. Results on additional roles of Acp-sperm interaction in *Drosophila* will be reviewed.

Introduction

Internal fertilisation requires more than just sperm and eggs. Sperm are transferred in the company of secretions from male reproductive glands. These seminal secretions, which include proteins, carbohydrates, lipids and hormones, are not simply a passive delivery (or nutritive) medium for sperm. Instead, seminal secretions play critical roles in enhancing the fertility of organisms ranging from insects to mammals. These roles include stimulating the reproductive capacity of the female (e.g. humans: Bellinge, Copeland, Thomas, Mazzucchelli, O'Neil and Cohen, 1986; Coulam and Stern, 1995; Tremellen, Valbuena, Landeras, Ballesteros, Martinez, Mendoza, Norman, Robertson and Simon, 2000; Robertson, Bromfield and Tremellen, 2003;

Corresponding author E-mail: mfw5@cornell.edu

Drosophila: Heifetz and Wolfner, 2004), mediating the efficient movement and/or retention of sperm to storage or fertilisation sites and assisting sperm stability at these sites (e.g. cattle: Ignotz, Lo, Perez, Gwathmey and Suarez, 2001; *Drosophila*: Neubaum and Wolfner, 1999; Tram and Wolfner, 1999), and facilitating efficient and productive meeting of gametes (e.g. induced ovulation in some mammals (camels): Chen, Yuen and Pan, 1985; Pan, Chen, Liu, Li, Xie, Ling and Fang, 2001; Li and Zhao, 2004; *Drosophila*: Heifetz, Lung, Frongillo and Wolfner, 2000). The role of seminal proteins in fertility is illustrated, for example, by the sterility of male mice that lack the seminal protease inhibitor protease nexin-1 (Murer, Spetz, Hengst, Altrogge, deAgostini and Monard, 2001), or of men with non-functional protein C inhibitor (PCI; He, Lin and Liu, 1999) or abnormal levels of prostate-specific antigen (PSA) (Elzanaty, Richthoff, Malm and Giwercman, 2002). Recent studies suggest that semen may increase the chance of a positive pregnancy outcome during assisted reproductive technology procedures (Bellinge *et al.*, 1986; Tremellen *et al.*, 2000; Robertson *et al.*, 2003).

Drosophila is an excellent model system with which to identify and dissect the roles of seminal proteins, because of the availability of powerful genetic, molecular and genomic techniques (reviewed in Hall, 1994; Wolfner, 1997; Wolfner, 2002), and the well-described, stereotyped mating behaviour in this organism. Comprehensive dissection of the roles of *Drosophila* seminal proteins in fertility, behaviour, chemical communication and sperm storage/competition can inform applied uses in reproductive medicine and agricultural productivity in terms of enhancing or, alternatively, controlling fertility. Results with *Drosophila* can also be valuable to consider for regulating insect reproduction, either to hinder reproduction by pest insects or to improve reproduction by beneficial insects such as pollinators. This article will review results on seminal proteins in *Drosophila*, with a primary focus on the aspects of their function that regulate or depend upon sperm.

Roles of seminal proteins in *Drosophila* reproduction

Mating induces physiological and behavioural changes in Drosophila *females*

Drosophila females undergo a characteristic set of changes upon mating (reviewed in Wolfner, 1997; Chapman, 2001; Wolfner, 2002; Kubli, 2003; Chapman and Davies, 2004; Wolfner, Heifetz and Applebaum, 2005). Mated *Drosophila* females, like animals of many other internally-fertilising species, retain the sperm they receive during mating. *Drosophila* sperm are stored in two types of storage organs in females (the spermathecae and the seminal receptacle). Storage can last ~2 weeks (Kaufmann and Demerec, 1942), which is over a generation time, and allows for prolonged production of progeny from a single mating and for sperm competition if a female remates. Mated *Drosophila* females also elevate their egg production up to 25-fold (Kalb, DiBenedetto and Wolfner, 1993; Soller, Bownes and Kubli, 1999) and increase their rates of ovulation (Heifetz *et al.*, 2000) and oviposition. Mated *Drosophila* females also show altered behaviour. No longer willing to mate, they actively reject the advances of courting males, for as long as females contain sperm in their reproductive tracts (Manning, 1962; Chen, Stumm-Zollinger, Aigaki, Balmer, Bienz and Bohlen, 1988; Aigaki, Fleischmann, Chen and Kubli, 1991; Chapman, Bangham, Vinti, Seifried, Lung, Wolfner, Smith and Partridge, 2003; Liu and Kubli, 2003). Finally, multiply-mated *Drosophila* females have a decreased lifespan relative to unmated controls (Partridge, Green and Fowler, 1987; Fowler and Partridge, 1989; Chapman, Liddle, Kalb, Wolfner and Partridge, 1995).

Seminal secretions trigger the post-mating changes in females

Two types of experiment have indicated that seminal secretions, and in particular the "Acp" proteins produced by males' accessory glands, are crucial for the post-mating changes in *Drosophila* females. Injection of male accessory glands or their extracts into unmated female *Drosophila* (Garcia-Bellido, 1964; Merle, 1968; Chen *et al.*, 1988) caused those females to reject mating attempts and to increase egg production, behaviours that are normally only seen in mated females. The importance of accessory gland secretions in post-mating effects on egg production and receptivity was shown *in vivo* in experiments using males without (or with non-functional) accessory glands (Kalb *et al.*, 1993; Xue and Noll, 2000). Acp-deficient males were also used to show that secretions from male accessory glands are essential for sperm storage (Tram and Wolfner, 1999), and play roles in sperm competition (Harshman and Prout, 1994), appetite regulation (Carvalho, Kapahi, Anderson and Benzer, 2006) and in reducing the lifespan of the mated female (Chapman *et al.*, 1995). In addition, an *in vitro* study has shown that some Acps have anti-microbial activity (Lung, Kuo and Wolfner, 2001), suggesting that they protect the female, her reproductive tract, or the sperm she contains from microbes introduced during mating. Most Acps persist in female *Drosophila* for only a few hours after mating (Monsma, Harada and Wolfner, 1990; Bertram, Neubaum and Wolfner, 1996; Ravi Ram, Ji and Wolfner, 2005) and their effects are transient (< 1 day) in the absence of sperm; sperm need to be present in mated females for the effects of Acps to persist (Manning, 1962; Peng, Chen, Busser, Liu, Honegger and Kubli, 2005a, and see below).

Several approaches have been taken to determine how Acps mediate their effects

Acps' sequences predict some of their functions. Sixty-two individual Acps have been identified through cDNA hybridization, EST and proteomics screens (Schäfer, 1986; DiBenedetto, Lakich, Kruger, Belote, Baker and Wolfner, 1987; Simmerl, Schäfer and Schäfer, 1995; Wolfner, Harada, Bertram, Stelick, Kraus, Kalb, Lung, Neubaum, Park and Tram, 1997; Swanson, Clark, Waldrip-Dail, Wolfner and Aquadro, 2001; Holloway and Begun, 2004; Mueller, Ravi Ram, McGraw, Bloch Qazi, Siggia, Clark, Aquadro and Wolfner, 2005; Walker, Rylett, Keen, Audsley, Sajid, Shirras and Isaac, 2006). There are predicted to be about 70-106 Acps in *Drosophila melanogaster*; these are secreted proteins with predominant expression in male accessory glands (Mueller *et al.*, 2005). The putative biochemical functions of these molecules has been inferred from sequence predictions, structural modeling and, in some cases, *in vitro* studies (Swanson *et al.*, 2001; Lung, Tram, Finnerty, Eipper-Mains, Kalb and Wolfner, 2002; Mueller, Ripoll, Aquadro and Wolfner, 2004; Mueller *et al.*, 2005; Walker *et al.*, 2006). *Drosophila* Acps fall into several protein classes found in seminal fluid across a diverse array of other organisms (e.g. Mueller *et al.*, 2004). For example, *Drosophila* seminal fluid is rich in proteolysis regulators (proteases and protease inhibitors): 1/4 of the Acps identified in an EST screen fall into this class. As mentioned above, seminal fluid proteolysis regulators are found in other animals and are known to play important roles in, for example, mammal fertility. Acps also include lipid-modifying enzymes, and proteins that are likely to exert protective functions (anti-microbial proteins, redox-damage inhibitors) – again classes that are conserved across organisms. In addition, ~40% of *Drosophila* Acps are peptides, or known or predicted peptide precursors, which are typically not conserved in non-insects, and often not beyond Drosophilids (Lung *et al.*, 2002; Mueller, *et al.*, 2004), and are thus likely modulate physiological or behavioural responses unique to *Drosophila* (or insects). Other molecules, such as large glycoproteins, and members of the CRISP family of sperm-binding proteins found in mammalian seminal fluid, are also found in *Drosophila* seminal fluid (Bertram *et al.*, 1996; Mueller *et al.*, 2004). It will be intriguing to see if the functions of these conserved protein classes are similar from insects to mammals, especially in the face of the strong selection pressures placed on reproductive proteins.

The targets of Acps in mated females also suggest functions. Once transferred to females, Acps target to a variety of tissues. The sites in the mated female to which a specific Acp localises can give clues as to its function. For example, an Acp that regulates ovulation might be expected to target the ovary base (where eggs are released) or to enter the circulation (from which it can access neural or endocrine targets); just such a pattern is seen for the ovulation hormone ovulin (Acp26Aa) (Monsma *et al.*, 1990; Herndon and Wolfner, 1995; Heifetz *et al.*, 2000). A protein involved in sperm storage might be seen to associate with sperm and/or accompany them into storage, as is the case for the sperm-storage protein Acp36DE (Bertram *et al.*, 1996; Neubaum and Wolfner, 1999). To identify the targets of Acps, and thus gain insights about their function, sites to which Acps localise in mated females were identified for 13 additional Acps (Ottiger, Soller, Stocker and Kubli, 2000; Ravi Ram *et al.*, 2005). Each Acp has a characteristic targeting pattern. Ten Acps cross a special, permeable, part of the reproductive tract wall and enter the circulation, from where they might be expected to modulate behaviours, lifespan or physiological processes (Lung and Wolfner, 1999; Lung *et al.*, 2002; Ravi Ram *et al.*, 2005); an eleventh is hypothesised to enter the circulation (Aigaki *et al.*, 1991), and can bind to targets in the nervous system in an *in vitro* experiment (Ottiger *et al.*, 2000). Other Acps target the ovary or oviducts (6 Acps), sperm-storage organs (7 Acps) or associate with the mating plug (14 Acps) (Bertram *et al.*, 1996; Bloch Qazi and Wolfner, 2003; Ravi Ram *et al.*, 2005). These targeting patterns can help suggest specific post-mating processes to investigate for the roles of these specific Acps.

Association studies have suggested functions for some Acps. Flies collected from natural populations can carry an assortment of alleles at a given locus. Lines carrying chromosomes from natural populations were tested for strength of effect on sperm competition. Alleles at four Acp loci (Acp26Aa/Ab, Acp29B, Acp36DE and Acp53E) were found to correlate with the ability of sperm from one male to resist displacement by sperm from a subsequent male ("sperm defense" or P1). Further, alleles of a fifth Acp locus (Acp33A) were associated with both sperm defense and the ability of sperm from one male to displace the sperm of a previous male ("sperm offense" or P2). (Clark, Aguadé, Prout, Harshman and Langley, 1995; Fiumera, Dumont and Clark, 2005). Although the nature of association mapping precludes definitive identification of the gene responsible for the sperm competition effect (it could be the Acp, or something tightly linked to it), these results suggest that Acps in general and these five in particular, are excellent candidates to test directly for further roles in sperm competition.

Genetic/transgenic analysis has identified functions for several Acps. The most conclusive way to determine the function of a given Acp involves experiments that control the levels of individual Acps in females. One way to accomplish this is by modulating the amount of particular Acps produced by the male, either by genetically knocking out the Acp or using RNAi to knock down its levels. Alternatively, one also can control the amount of an individual Acp in females by injecting synthesised peptide or through ectopic expression of the Acp in unmated females. For example, the 36 amino acid "sex peptide" (Acp70A)(reviewed in Kubli, 2003; Swanson, 2003) was shown by several assays (injection, ectopic expression, knockout mutation, RNAi) to induce egg production and decrease receptivity in female flies (Chen *et al.*, 1988; Aigaki *et al.*, 1991; Soller *et al.*, 1999; Chapman *et al.*, 2003; Liu and Kubli, 2003). Further examination of sex peptide knockout flies have shown that this peptide decreases the longevity of the mated female, regulates her appetite and stimulates her production of antimicrobial peptides (Peng, Zipperlen and Kubli, 2005b; Wigby and Chapman, 2005; Carvalho *et al.*, 2006). *In vitro* studies suggest that sex peptide controls levels of juvenile hormone (Moshitzky, Fleischmann, Chaimov, Saudan, Klauser, Kubli and Applebaum, 1996), perhaps underlying the multiplicity of its effects on females (Schmidt, Choffat, Klauser and Kubli, 1993; Fan, Rafaeli, Gileadi, Kubli and Applebaum, 1999; Eliyahu, Nagalakshmi, Applebaum, Kubli, Choffat and Rafaeli, 2003).

Mutational and ectopic expression studies (Herndon and Wolfner, 1995; Heifetz *et al.*, 2000; Heifetz, Vandenberg, Cohn and Wolfner, 2005) have also identified ovulin as a regulator of ovulation in mated female *Drosophila*, an essential step in increasing egg production. Ovulin is a 264 amino acid self-interacting (Wong, Albright and Wolfner, submitted) protein, with a region similar to egg-laying hormones in the mollusk *Aplysia* (Monsma and Wolfner, 1988; Heifetz *et al.*, 2000). The prohormone is cleaved when it enters females (Monsma and Wolfner, 1988; Monsma *et al.*, 1990; Park and Wolfner, 1995), in a process that includes activation of an Acp-protease provided by the male (Ravi Ram, Sirot and Wolfner, 2006). At least two of its cleavage products can each, independently, trigger ovulation (Heifetz *et al.*, 2005).

Studies of effects of Acps upon ectopic expression in females have also identified four Acps that are toxic, suggesting that they might contribute to reducing the lifespan of the mated female (Lung *et al.*, 2002; Mueller, Page and Wolfner, 2006). At least two of the four are known to enter the female's circulatory system (Lung and Wolfner, 1999; Ravi Ram *et al.*, 2005), suggesting appropriate targeting for toxicity. Two of the three are known or predicted protease inhibitors (Lung *et al.*, 2002; Mueller *et al.*, 2004), so their toxicity might come about by inhibiting essential proteolytic cascades.

Finally, genetic studies have identified Acp36DE as essential for sperm storage, as discussed in more detail below (Neubaum and Wolfner, 1999).

Drosophila females are poised to switch to a mated state

The studies above have identified roles for specific Acps, and further studies of this type are expected to further elucidate the repertoire of Acp functions. One pressing question about Acp function is whether Acps elicit radical cellular responses that underlie the post-mating changes in females or whether the female is already poised to switch to a mated physiology and behaviour. Evidence exists for the latter, where Acps appear to serve as triggers that enable the transition from one physiological state to the other. The evidence for this is of two types.

Microarray studies have shown that expression levels of about 13% of the transcriptome changes by 1-3 hr post-mating, though almost all changes are less than two-fold (Lawniczak and Begun, 2004; McGraw, Gibson, Clark and Wolfner, 2004). Acps are responsible for about 10% of these changes (McGraw *et al.*, 2004), including induction (more than two-fold) of molecules that enhance immunity of mated females. Sperm are responsible for triggering changes in another 30% of the mating-regulated transcriptome (McGraw *et al.*, 2004). But, by and large the changes at this early post-mating time are small, even though major physiological changes (sperm storage, egg-production elevation) have occurred by then. This supports the hypothesis that mature females are "poised" to switch rapidly to a mated state, and that Acps are agents that trigger this change.

Consistent with this hypothesis, studies of vesicle release at nerve termini in the reproductive tract indicate that mature female flies already possess vesicles, whose subsequent release is modulated by mating (Heifetz and Wolfner, 2004). Some vesicles' release is regulated by receipt of Acps, release by others is regulated by receipt of sperm, and the behaviour of still others is regulated by the receipt of both Acps and sperm. Different parts of the reproductive tract respond to mating (or Acps, or sperm) differently, at different times. For example, shortly after mating, vesicles in the uterine wall release their contents; this release is triggered by the act of mating. A couple of hours after that, when sperm storage is maximal and egg deposition has begun, Acps received by the female modulate the release of vesicles in the seminal receptacle, and the presence of sperm suppresses release by some vesicles in the upper reproductive tract. Since neuromodulators released from vesicles mediate muscle contraction in insect, nematode, and mammalian reproductive tracts (Trent, Tsung and Horvitz, 1983; Orchard and Lange, 1985; Lange, Orchard and Adams, 1986; Lange, Orchard and Brugge, 1991; McCormick and Nichols, 1993; Bamji and Orchard, 1995; Lange and Nykamp, 1996; Monastirioti, Linn and White, 1996; Hesla, Preutthipan, Maguire, Chang, Wallach and Dharmarajan, 1997; Jacklet, 1997; Yamauchi, Miyazaki, Iwasaki, Kishi, Kuroshima, Tei and Yoshimura, 1997; Tshugguel *et al.*, 1998; Waggoner, Zhou, Schafer and Schafer, 1998; Lee, Seong, Kim, Davis and Han, 2003; Monastirioti, 2003), regulated vesicle release post-mating is likely to underlie some Acp effects, for example ovulation, egg deposition, and movement of sperm into storage.

Some Acps evolve rapidly

Although some of the effects of mating and Acps are common to many insects (Wolfner *et al.*, 2005), and although Acps belong to protein families that are conserved among seminal proteins of many animals (Mueller *et al.*, 2004), an unusually high proportion of Acps (~15%, at least 10-fold higher than non-Acp genes; Swanson *et al.*, 2001) shows rapid evolution at the primary sequence level. For example, the ovulation hormone ovulin is one of the most rapidly-evolving genes identified to date in *Drosophila* (Tsaur and Wu, 1997; Aguadé, 1998; Tsaur, Ting and Wu, 1998; Tsaur, Ting and Wu, 2001). Several other Acps have been found to be species-specific within *Drosophila* suggesting further rapid evolution (Aguadé, Miyashita and Langley, 1992; Tsaur *et al.*, 1998; Begun, Whitley, Todd, Waldrip-Dail and Clark, 2000; Swanson *et al.*, 2001; Holloway and Begun, 2004; Kern, Jones and Begun, 2004; Begun and Lindfors, 2005; Mueller *et al.*, 2005; Schully and Hellberg, 2006). This seemingly counterintuitive phenomenon, of rapid evolution of proteins that mediate essential and conserved reproductive processes, surprisingly occurs in many animals and plants (reviewed in Swanson and Vacquier, 2002). A number of explanations have been proposed for this phenomenon, including conflict between the reproductive "interests" of males and females (Eberhard, 1996; Cordero and Eberhard, 2003) or selection for highly species-specific reproductive molecules as species diverge (Swanson and Vacquier, 2002).

One outcome of sexual conflict can be degeneracy (the presence of totally different molecules serving the same function; Edelman and Gally, 2001), as males evolve multiple effectors to compensate for ones to which females have become resistant. Such degeneracy can be seen in the case of two Acps that each can stimulate egg production. Both sex peptide and ovulin are insufficient alone to cause the full elevation in egg production after mating (Herndon and Wolfner, 1995; Heifetz *et al.*, 2000; Chapman *et al.*, 2003; Liu and Kubli, 2003), and males provide additional molecules that can stimulate egg production, e.g. Dup99B in *D. melanogaster* (Saudan, Hauck, Soller, Choffat, Ottiger, Sporri, Ding, Hess, Gehrig, Klauser, Hunziker and Kubli, 2002), ED-OSS in *D. biarmipes* (Imamura, Haino-Fukushima, Aigaki and Fuyama, 1998), and a yet-unidentified peptide (OSS) in *D. suzukii* (Ohashi, Hamo-Fukushima and Fuyama, 1991). Similarly, males provide several proteases and protease inhibitors that, based on their sequence, would be predicted to have similar or redundant functions, and include some with very similar targets in mated females (Mueller *et al.*, 2004; Ravi Ram *et al.*, 2005). Functional redundancy can also come about by controlling different steps of a multi-step process such as egg production or sperm storage; having regulators specific for each step allows the overall process to be "tuned" at several levels (Heifetz *et al.*, 2000; Bloch Qazi, Heifetz and Wolfner, 2003). In fact, we see such redundancy in Acps that modulate different steps of the same biological process, such as oogenesis and ovulation within the process of elevated egg deposition (Herndon and Wolfner, 1995; Heifetz *et al.*, 2000; Chapman *et al.*, 2003; Liu and Kubli, 2003). Such redundancy could also guard against loss of effectiveness of a particular modulator due to rapid sequence evolution.

How do seminal proteins affect sperm or, alternatively, depend on sperm?

Although seminal proteins have important effects on their own, it is particularly illustrative (as well as relevant to this book) to consider their effects in relation to sperm. Acp-sperm interactions can be considered on several levels.

Acps are important for sperm entry into storage

Drosophila males transfer up to ~4000-6000 sperm to females (Kaufmann and Demerec, 1942; Kaplan, Tinderholt and Gugler, 1962) and 1/3 to 1/2 of these sperm get stored (Lefevre and Jonsson, 1962), though there is variation between strains. *Drosophila* sperm are very long, e.g. 1-2 mm in *D. melanogaster* (Pitnick, Spicer and Markow, 1995) and thus cannot swim to, or into, storage in the same way as the much smaller sperm of (for example) mammals. *Drosophila* sperm enter the female intertwined in a "sperm mass", though not in a "packet" (spermatophore) as in some other insects (e.g. mealworm beetles; Gadzama and Happ, 1974; Frenk and Happ, 1976). *Drosophila* sperm then move into two types of storage organs. Sperm are thought to enter the seminal receptacle first. This long, coiled, blind-ended tubule opens at the base of the lower oviduct. The length of the seminal receptacle appears to coevolve with sperm length in *Drosophila* species (Miller, Starmer and Pitnick, 2001; Miller and Pitnick, 2002; Miller and Pitnick, 2003; Miller, Starmer and Pitnick, 2003; Pitnick, Miller, Schneider and Markow, 2003). Sperm also enter the other sperm storage organ, the spermatheca (there are normally two in *D. melanogaster*; Filosi and Perotti, 1975). Spermathecae are chitinous ovoid capsules; each spermatheca is connected to the upper uterus/lower reproductive tract by a thin stalk. In some arthropods, spermathecae or their associated glands secrete molecules that support the sperm in storage (e.g. Anderson, 1945; Davey and Webster, 1967; Bhatnagar and Musgrave, 1971; Gillott, 1988; Kressin, Sommer and Schnorr, 1996; Michalik, Reiher, Tintelnot-Suhm, Coyle and Alberti, 2005); it is not yet known whether this is the case in *Drosophila*. The openings of the seminal receptacle and spermathecae are near to where the micropyle of the egg will come to rest, when the egg reaches the uterus after ovulation; thus sperm are released from storage near to the site of fertilisation.

That Acps are essential for sperm storage was shown by experiments that examined sperm storage in mates of *D. melanogaster* males that lacked Acps. The poor fertility of these mated females (Kalb *et al.*, 1993; Tram and Wolfner, 1999) suggested roles for Acps in sperm transfer or storage. This was confirmed by the finding that without Acps, sperm storage levels drop to ~10% of normal (Tram and Wolfner, 1999). Further examination of mates of Acp-deficient males showed that Acps are not necessary for the transfer of sperm from the male, but rather mediate the storage of sperm within mated females (Tram and Wolfner, 1999). As a first step in dissecting the roles of Acps in sperm storage, we determined that sperm storage occurs in a series of sequential steps (Bloch Qazi *et al.*, 2003), taking about 1.5 h to reach maximal levels.

Sperm are initially deposited in the lower reproductive tract, and must either move or be moved through the uterus towards the storage organs. Once in storage, they must be retained and remain viable; it is not known whether any sort of "capacitation" or modification of sperm occurs or is needed in *Drosophila*. Finally, sperm must be released from storage at a regulated rate, to coordinate with the availability of oocytes for fertilisation. These steps mentioned above can be modulated independently, though their coordination is necessary to maximise fertility.

One way that Acps could affect sperm entry into storage is through their participation in structures that form in the reproductive tract of the mated female. Seminal secretions congeal into a mating plug in the *Drosophila* uterus. This plug has been suggested to form a scaffold that assists sperm in moving, or orienting, towards the sperm storage organs (Bairati and Perotti, 1970). A major constituent of the mating plug is the seminal protein PEB (derived from the male's ejaculatory bulb; thus not an Acp), whose sequence suggests the ability to form homopolymers and filaments, which might provide structure to the mating plug (Ludwig, Uspensky, Ivanov, Kopantseva, Dianov, Tamarina and Korchin, 1991; Lung and Wolfner, 2001). Acps also contribute to the mating plug: 14 of the 15 tested Acps are found within it (Lung and Wolfner, 2001; Ravi Ram *et al.*, 2005), and thus might form part of a scaffold along which sperm could move to their storage sites. Assisting with sperm storage is not the only function ascribed to mating plugs. For example, in *D. hibisci*, the presence of a mating plug has been suggested to prevent a female from productively re-mating with a subsequent male (Polak, Starmer and Barker, 1998).

Experiments using Acp-deficient *Drosophila* males demonstrated that Acps also facilitate the movement of sperm towards storage organs (the first step in storage) (Adams and Wolfner, 2006). During mating, the lower reproductive tract of the *Drosophila* female undergoes a stereotyped set of contractions and shape changes as the sperm mass moves "up", towards the openings of the sperm storage organs (Adams and Wolfner, 2006). The shape changes and contractions also move aside a flap that had previously covered the openings of the sperm storage organs into the lower reproductive tract; this may permit sperm to enter these organs. To determine the triggers for these muscle contractions, we mated females to males that lacked sperm or Acps. We found that Acps were essential for the muscle contractions and shape changes (Adams and Wolfner, 2006). Perhaps Acps trigger these muscle contractions and conformational changes by regulating the release of neuromodulators at nerve termini (Heifetz and Wolfner, 2004), as discussed above. Interestingly, although Acps are essential for the conformational changes and contractions of the uterus that move sperm into storage, sperm themselves have no effect on these events (Heifetz and Wolfner, 2004). The conformational changes in the lower reproductive tract occur whether or not a male provides sperm to the female. It is intriguing to wonder whether the modulation of muscle contraction in the reproductive tract, possibly facilitating sperm storage, is a function of seminal proteins in many taxa. In other insects, seminal secretions are known to cause muscle contractions in the female's reproductive tract (e.g. bedbugs, tsetse flies; Davey, 1958; Gillott and Langley, 1981) and muscle contractions occur in the reproductive tracts of mated female mammals (e.g. Crane and Martin, 1991; Langendijk, Soede and Kemp, 2005) though if or how seminal proteins trigger them is not known.

The Acp(s) that trigger(s) conformational changes in the reproductive tracts of mated *Drosophila* females is unknown, although there is a candidate. Loss of the large glycoprotein Acp36DE (Bertram *et al.*, 1996; Wolfner *et al.*, 1997) causes a defect in sperm storage. Although the number of sperm transferred to females is normal in the absence of Acp36DE, only 10% of control numbers of sperm are stored (Neubaum and Wolfner, 1999). This is similar to what is seen in females that had mated to males lacking all Acps (Tram and Wolfner, 1999). By examining the location of sperm in females mated to males that lack Acp36DE (relative to females mated to control males) we found that lack of Acp36DE affected only an early step in sperm storage. There was no abnormality in the motility of the mutant male's sperm, or in the spatial arrangement, or the retention or viability of the few sperm that got into in storage in the absence of Acp36DE (Bloch Qazi and Wolfner, 2003). Without Acp36DE, the first sperm entered storage at the normal time but the efficiency with which subsequent sperm followed into storage was reduced. This suggested that the first sperm to enter storage uses a signal that is different from the Acp36DE-dependent mechanism that recruits later sperm into storage. One possible mechanism for Acp36DE action is that this Acp is needed for the muscle contractions that assist the entire sperm mass in reaching storage. If these muscle contractions are compromised, a sperm nearer to the sperm storage entry site might get into storage at the normal time, but the remaining sperm in the sperm mass would not. Another possible mechanism for Acp36DE's action stems from the finding (Bloch Qazi and Wolfner, 2003) that, although this protein is normally bound to sperm (Neubaum and Wolfner, 1999), Acp36DE can enter storage in the absence of sperm. This suggests a model in which Acp36DE acts from within the female sperm storage organs as a signal to trigger muscle contractions. Such contractions could "suck" sperm into storage, as has been proposed for another Dipteran (Linley, 1981). Consistent with this, Acp36DE that is present in the female from a prior mating assists the storage of sperm from a subsequent mating (Chapman, Neubaum, Wolfner and Partridge, 2000). It is also formally possible that Acp36DE has a steric function as discussed above for mating plugs. This is suggested by the presence of Acp36DE in the anterior part of the mating plug (Lung and Wolfner, 2001), and by its binding to a site in the oviduct just above the openings of the sperm storage organs (Bertram *et al.*, 1996), which might allow it to restrict the sperm from moving past the openings of the storage organs.

Acps are important for sperm utilization

Interspecific matings between *D. pulchrella* and *D. suzukii* first suggested that seminal secretions might play a role in sperm utilization (Fuyama, 1983). When males of one species mated with females of another, no progeny were produced. However, if the females were subsequently injected with an extract of reproductive tissues of conspecific males, a small number of interspecies progeny were produced. This experiment suggested that some sperm had been stored in the initial mating, but could not be utilised until a subsequent male provided an appropriate seminal molecule.

This hypothesis was supported further by "copulation complementation" experiments done in *D. melanogaster* by Xue and Noll (2000). These authors created transgenic males that lacked accessory glands, but still produced and transferred sperm. When these males mated, their females did not produce progeny until the females had re-mated with (spermless) *D. melanogaster* males, which provided Acps. This experiment indicated that Acps play an essential role in the utilization of sperm after they had entered storage, perhaps providing an activity that allows efficient fertilisation of eggs.

It is unclear how Acps might foster the utilization of sperm. Since two Acps are known to bind to sperm (sex peptide: Peng *et al.*, 2005a; Acp36DE: Bertram *et al.*, 1996) and/or to enter the sperm storage organs (several Acps that are predicted to regulate proteolysis, Lung *et al.*, 2002; Ravi Ram *et al.*, 2005), they might play protective roles, or potentially alter ("capacitate") sperm to allow their utilization. This is an area ripe for investigation. Finally, Acps are not the only male secretions needed to support sperm once they are stored. Genetic experiments indicate that GLD, a protein secreted by the male's ejaculatory duct and provided in the seminal fluid, also plays a role in sperm storage and utilization in *Drosophila* (Iida and Cavener, 2004).

Acps may play a role in sperm competition

In animals that mate multiply, there is the opportunity for sperm from different males to compete for storage or access to eggs (Birkhead and Hunter, 1990; Birkhead and Moller, 1993) or, alternatively (or in addition), for the female to exert a preference in terms of the sperm used to fertilise her eggs (Eberhard, 1996). Although *D. melanogaster* females are reluctant to re-mate when sperm are in storage, they do mate multiply, thus allowing for sperm competition in this species.

The first indication that Acps might play a role in sperm competition came from experiments in which mated females were subsequently mated with a spermless male that either lacked or did not lack Acps (Harshman and Prout, 1994). Females had more progeny sired by the first male's sperm if the second male lacked Acps. These data suggested that Acp functions include activities that enhanced a male's sperm-competitive ability (e.g., sperm displacement). How seminal fluids (or mating) enhance the sperm-competitive ability of the second male is not known, though several mechanisms have been suggested (e.g. Price, Dyer and Coyne, 1999; Snook and Hosken, 2004).

A second indication that Acps might be involved in sperm competition came from the association studies described above, that suggested roles in sperm competition for Acps 26Aa/Ab, 29AB, 33A, 36DE and 53Ea (Clark *et al.*, 1995). Subsequent studies using the Acp36DE knockout indicated that its effect on sperm ability to defend against competition was most likely a consequence of Acp36DE's role in sperm storage (Chapman *et al.*, 2000): differences in the number of sperm stored in the presence or absence of Acp36DE would affect the number of sperm available to be competed and hence their ability to resist competition. With the recent finding of a mutant in Acp29AB (Albright, 2003), it will be possible to test directly whether this Acp plays a role in sperm competition and/or storage. Interestingly allelic sequence differences at or near Acp26Aa (ovulin) correlated with sperm competition levels in the Clark *et al.* study (Clark *et al.*, 1995). Although a null mutant of ovulin does not show abnormalities in sperm defense (Herndon and Wolfner, 1995), recent results suggest that some natural alleles of this gene may affect the "offense" component of sperm competition (Fiumera *et al.*, 2005). Perhaps by accelerating or decelerating ovulation, ovulin alleles can affect the rate in which sperm from a given male are pushed out of the female, thereby impacting the number available to compete (or defend against) another male's sperm. Alternatively, the effect might be due to Acp26Ab, which is located within 1 kb of ovulin and thus whose alleles might be in linkage disequilibrium with ovulin's. Transgenic technologies should make it possible to conduct experiments to determine the role of specific alleles of a given Acp on sperm competition, and this should ultimately lead to a mechanistic understanding of how individual Acps, and their variants, affect sperm competition.

Acps' action can also affect sperm indirectly

In addition to regulating sperm storage, Acps' effects on reproductive phenomena like ovulation, receptivity, and longevity can indirectly impact the use of sperm. For example, decreasing females' receptivity to a second mating will give a male's sperm greater opportunity to fertilise eggs, by decreasing the chance of competition from the sperm or seminal fluids of another male. In addition, Acp-stimulated egg production may affect the rate of transition from one stage of sperm storage to another (Bloch Qazi and Wolfner, 2005). Mated, genetically identical female flies that contain or lack eggs have differences in the rate of depletion of sperm from storage – but only at early post-mating times. This finding suggests that in the presence of eggs, females transition more rapidly from accumulating sperm in storage to regulating retention vs. release of stored sperm. Perhaps the increased ovulation of eggs triggered by Acps (Heifetz *et al.*, 2000) pushes out sperm that have not yet been stored, thus terminating the entry of sperm into storage sooner than would occur in the absence of eggs (and thus of ovulation). However, after the early time-points, long-term rates of sperm depletion from storage are not affected by the presence of eggs.

Acps associate with sperm, and some of these associations are needed to prolong the post-mating changes

The sections above focused on how Acps' actions can affect sperm. But sperm also play an important role in modulating the responses initiated by Acps. As mentioned above, the effects of Acps alone on egg production and receptivity are short-term, if sperm are not provided during mating (Manning, 1962; Chen *et al.*, 1988; Kalb *et al.*, 1993). Moreover, changes in receptivity and egg-laying occasioned by mating only persist as long as sperm remain in storage; once the sperm storage organs no longer contain sperm, females revert to virgin-like behaviours (Manning, 1962).

Models have been suggested for how sperm could prolong effects triggered by Acps. In one model, sperm and Acps exert analogous effects but by different means, and with different timing of onset and duration. This model is based on the finding that effects of mating are transient in the absence of sperm in *Drosophila* (Manning, 1962; Kalb *et al.*, 1993), as are effects of injection of at least one Acp (the sex peptide, Chen *et al.*, 1988), that Acps cannot be detected biochemically in mated females by several hours post-mating (Monsma *et al.*, 1990; Bertram *et al.*, 1996; Ravi Ram *et al.*, 2005), and that filled sperm storage organs can trigger post-mating changes in *Anopheles gambiae* mosquitoes (Klowden, 2006), and are known to do this via neural signals in some moths (e.g. Ramaswamy, Qiu and Park, 1996). According to this model, *Drosophila* Acps would quickly trigger changes in females' physiology and behaviour. During the short time in which females' post-mating physiology and behaviour is driven by Acps, the females would also be storing sperm. Once the sperm storage organs were full, this would initiate the alternate, long-term pathway that maintained the behavioral and physiological changes, perhaps through neural signaling as in moths. In this model, Acps would cause rapid modification of the female's behaviour and physiology while the separate, slower but more permanent, signal was set up. This model also explains why the effects of mating on the female's behaviour and physiology persist as long as she has sperm in storage but then disappear (Manning, 1962).

Recent experiments from Eric Kubli's laboratory (University of Zürich) have suggested an alternative model for how sperm prolong the effects of Acps in *D. melanogaster*. Mutant males lacking sex peptide failed to induce prolonged changes in egg-laying and receptivity, even if they provided females with sperm (Liu and Kubli, 2003). This suggested that sex peptide itself might be needed for the later changes. Peng *et al.* (2005a) found that sex peptide binds to sperm and that sperm in storage contain bound sex peptide. Over time, the amount of the active region of sex peptide that was detected on stored sperm diminished, and experiments with a mutant form of sex peptide suggested that the active part of the molecule is normally cleaved from the stored sperm. Thus, sperm are suggested to prolong the effects of sex peptide by acting as "slow-release capsules", sequestering sex peptide by binding it, and then releasing it slowly into the circulation over time (Peng *et al.*, 2005). Therefore the long-term maintenance of post-mating effects is still due to Acps in this model, with sperm acting as the "carriers" that allow the effect to persist.

Conclusion

Studies in *Drosophila* have identified roles of a major class of seminal proteins called Acps in the induction and persistence of post-mating effects in females, including fascinating interactions between the actions of Acps and sperm.

On the one hand, Acps are essential for the proper storage and utilization of sperm, and mediate sperm competition. This raises interesting questions. For example, do Acps facilitate sperm storage by forming physical entities, like scaffolds along which sperm can move, or through the effects of the muscle contractions and conformational changes that they cause in the uterus (Bloch Qazi and Wolfner, 2003; Adams and Wolfner, 2006)? How do Acps exert these functions, and which Acps in particular do so? It will also be intriguing to determine how Acps mediate efficient utilization of sperm (Xue and Noll, 2000). Do they function to protect sperm from proteolysis or oxidative damage or to keep them viable in storage? Do they help to "activate" the stored sperm, or to control their release from storage so that they can be used to fertilise eggs? How do Acps participate in sperm competition?

On the other hand, the recent finding that sperm (Peng *et al.*, 2005a) may act as slow-release capsules for sex peptide also raises some interesting questions. Are other Acps also bound to and released from sperm? What causes released sex peptide to enter the circulation? Is sex peptide release coupled to the rate of fertilisation?

Finally, the interesting evolutionary dynamics of Acps also raises important questions. Is their rapid evolution at the sequence level related to Acps' role in sperm competition, sexual conflict or intersexual interaction (Swanson and Vacquier, 2002)? And what about the fact that many Acps belong to conserved protein classes (Mueller *et al.*, 2004): is there overall conservation of function across internally-fertilising animals, despite differences in reproductive strategies, perhaps extending to parallels in the roles of seminal proteins in "sperm management" (storage, utilization, release)? Might some of these functions even be conserved in organisms that fertilise externally?

Despite all these questions, one thing is clear. With what is already known about *Drosophila* Acps, the availability of tools for analysing protein function in this model system and the existence of genomic and proteomic tools that can extend the studies to other organisms, the next few years will be a fertile period for answering "seminal" questions in reproductive biology.

Acknowlegements

I thank Dr. E. Roldan for the opportunity to write this review. I thank the past and present members of my lab, and of the Acp field, for enjoyable and stimulating discussions and experiments. I am grateful to M. Bloch Qazi, L. McGraw, J. Mueller, K. Ravi Ram, L. Sirot, and A. Wong for helpful comments on the manuscript. I appreciate the support for our studies on *Drosophila* Acps from NIH grant HD38921.

References

Adams, E. A. and Wolfner, M. F. (2006). Seminal proteins but not sperm induce morphological changes in the *Drosophila melanogaster* female reproductive tract during sperm storage. *Journal of Insect Physiology* (in press)

Aguadé, M., Miyashita, N. and Langley, C. (1992). Polymorphism and divergence in the *Mst26A* male accessory gland gene region in *Drosophila*. *Genetics* **132**: 755-770

Aguadé, M. (1998). Different forces drive the evolution of the Acp26Aa and Acp26Ab accessory gland genes in the *Drosophila melanogaster* species complex *Genetics* **150**: 1079-1089

Aigaki, T., Fleischmann, I., Chen, P. S. and Kubli, E. (1991). Ectopic expression of sex peptide alters reproductive behavior of female *D. melanogaster*. *Neuron* **7**: 557-563

Albright, S. A. (2003). Molecular and genetic characterization of Acp29AB, and identification of Acp interactors in *Drosophila melanogaster*. Ph.D. thesis, Cornell University, Field of Genetics and Development

Anderson, R. C. (1945). A study of the factors affecting fertility of lozenge females of *Drosophila melanogaster*. *Genetics* **30**: 280-296

Bairati, A. and Perotti, M. E. (1970). Occurence of a compact plug in the genital tract of *Drosophila* females after mating. *Drosophila Information Service* **45**: 67-68

Bamji, S. X. and Orchard, I. (1995). Pharmacological profile of octopamine and 5HT receptor on the lateral oviducts of the cockroach, *Periplaneta americana*. *Archives of Insect Biochemistry and Physiology* **28**: 49-62

Begun, D. J., Whitley, P., Todd, B. L., Waldrip-Dail, H. M. and Clark, A. G. (2000). Molecular population genetics of male accessory gland proteins in *Drosophila*. *Genetics* **156**: 1879-1888.

Begun, D. J. and Lindfors, H. A. (2005). Rapid evolution of genomic Acp complement in the *melanogaster* subgroup of *Drosophila*. *Molecular Biology and Evolution* **22**: 2010-2021

Bellinge, B. S., Copeland, C. M., Thomas, T. D., Mazzucchelli, R. E., O'Neil, G. and Cohen, M. J. (1986). The influence of patient insemination on the implantation rate in an in vitro fertilization and embryo transfer program. *Fertility and Sterility* **46**: 252-256

Bertram, M. J., Neubaum, D. M. and Wolfner, M. F. (1996). Localization of the *Drosophila* accessory gland protein Acp36DE in the mated female suggests a role in sperm storage. *Insect Biochemistry and Molecular Biology* **26**: 971-980

Bhatnagar, R. D. and Musgrave, A. J. (1971). Aspects of histophysiology of the spermathecal gland of *Sitophilus granarius* (L.) (Coleoptera). *Canadian Journal of Zoology* **49**: 275-277

Birkhead, T. R. and Hunter, F. M. (1990). Mechanisms of sperm competition. *Trends in Ecology and Evolution* **5**: 48-52

Birkhead, T. R. and Moller, A. P. (1993). Sexual selection and the temporal separation of reproductive events: sperm storage data from reptiles, birds and mammals. *Biological Journal of the Linnean Society* **50**: 295-311

Bloch Qazi, M. C., Heifetz, Y. and Wolfner, M. F. (2003). The developments between gametogenesis and fertilization: ovulation and female sperm storage in *Drosophila melanogaster*. *Developmental Biology* **256**: 195-211

Bloch Qazi, M. C. and Wolfner, M. F. (2003). An early role for the *Drosophila melanogaster* male seminal protein Acp36DE in female sperm storage. *Journal of Experimental Biology* **206**: 3521-3528

Bloch Qazi, M. C. and Wolfner, M. F. (2005). Sperm release from female storage is effected by egg-independent factors in *Drosophila melanogaster*. *Biology Letters of Proceedings of the Royal Society London* **2**: 128-130

Carvalho, G. B., Kapahi, P., Anderson, D. J. and Benzer, S. (2006). Allocrine modulation of appetite by the sex peptide of *Drosophila*. *Current Biology* **16**: 692-696

Chapman, T., Liddle, L. F., Kalb, J. M., Wolfner, M. F. and Partridge, L. (1995). Cost of mating in *Drosophila melanogaster* females is mediated by male accessory gland products. *Nature* **373**: 241-244

Chapman, T., Neubaum, D. M., Wolfner, M. F. and Partridge, L. (2000). The role of male accessory gland protein Acp36DE in sperm competition in *Drosophila melanogaster*. *Proceedings of the Royal Society of London, series B* **267**: 1097-1105

Chapman, T. (2001). Seminal fluid-mediated fitness traits in *Drosophila*. *Heredity* **87**: 511-521

Chapman, T., Bangham, J., Vinti, G., Seifried, B., Lung, O., Wolfner, M. F., Smith, H. K. and Partridge, L. (2003). The sex peptide of *Drosophila melanogaster*: female post-mating responses analyzed by using RNA interference. *Proceedings of the National Academy of Sciences USA* **100**: 9923-9928

Chapman, T. and Davies, S. J. (2004). Functions and analysis of the seminal fluid proteins of male *Drosophila melanogaster* fruit flies. *Peptides* **25**: 1477-1490

Chen, B. X., Yuen, Z. X. and Pan, G. W. (1985). Semen-induced ovulation in the bactrian camel (*Camelus bactrianus*). *Journal of Reproduction and Fertility* **74**: 335-339

Chen, P. S., Stumm-Zollinger, E., Aigaki, T., Balmer, J., Bienz, M. and Bohlen, P. (1988). A male accessory gland peptide that regulates reproductive behavior of female *D. melanogaster*. *Cell* **54**: 291-298

Clark, A. G., Aguadé, M., Prout, T., Harshman, L. G. and Langley, C. H. (1995). Variation in sperm displacement and its association with accessory gland protein loci in *Drosophila melanogaster*. *Genetics*

139: 189-201

Cordero, C. and Eberhard, W. G. (2003). Female choice of sexually antagonistic male adaptations: a critical review of some current research. *Journal of Evolutionary Biology* **16**: 1-6

Coulam, C. B. and Stern, J. J. (1995). Effect of seminal plasma on implantation rates. *Early Pregnancy* **1**: 33-36

Crane, L. H. and Martin, L. (1991). Postcopulatory myometrial activity in the rat as seen by video-laparoscopy. *Reproduction, Fertility and Development* **3**: 685-98

Davey, K. G. (1958). The migration of spermatozoa in the female of *Rhodnius prolixus* Stal. *Journal of Experimental Biology* **35**: 694-701

Davey, K. G. and Webster, G. F. (1967). Structure and secretion of spermatheca of *Rhodnius prolixus* Stal - a histochemical study. *Canadian Journal of Zoology* **45**: 653

DiBenedetto, A. J., Lakich, D. M., Kruger, W. D., Belote, J. M., Baker, B. S. and Wolfner, M. F. (1987). Sequences expressed sex-specifically in *Drosophila melanogaster* adults. *Developmental Biology* **119**: 242-251

Eberhard, W. G. (1996). *Female Control: Sexual Selection by Cryptic Female Choice*. Princeton University Press, Princeton, 472 pp

Edelman, G. M. and Gally, J. A. (2001). Degeneracy and complexity in biological systems. *Proceedings of the National Academy of Sciences USA* **98**: 13763-8

Eliyahu, D., Nagalakshmi, V., Applebaum, S. W., Kubli, E., Choffat, Y. and Rafaeli, A. (2003). Inhibition of pheromone biosynthesis in *Helicoverpa armigera* by pheromonostatic peptides. *Journal of Insect Physiology* **49**: 569-574

Elzanaty, S., Richthoff, J., Malm, J. and Giwercman, A. (2002). The impact of epididymal and accessory sex gland function on sperm motility. *Human Reproduction* **17**: 2904-2911

Fan, Y., Rafaeli, A., Gileadi, C., Kubli, E. and Applebaum, S. W. (1999). *Drosophila melanogaster* sex peptide stimulates juvenile hormone synthesis and depresses sex pheromone production in *Helicoverpa armigera*. *Journal of Insect Physiology* **45**: 127-133

Filosi, M. and Perotti, M. E. (1975). Fine structure of the spermatheca of *Drosophila melanogaster Journal of Submicroscopic Cytology* **7**: 259-270

Fiumera, A. C., Dumont, B. L. and Clark, A. G. (2005). Sperm competitive ability in *Drosophila melanogaster* associated with variation in male reproductive proteins. *Genetics* **169**: 243-257

Fowler, K. and Partridge, L. (1989). A cost of mating in female fruitflies. *Nature* **338**: 760-761

Frenk, E. and Happ, G. M. (1976). Spermatophore of the mealworm beetle: immunochemical characteristics suggest affinities with male accessory gland. *Journal of Insect Physiology* **22**: 891-895

Fuyama, Y. (1983). Species specificity of paragonial substances as an isolating mechanism in *Drosophila*. *Experientia* **39**: 190-192

Gadzama, N. M. and Happ, G. M. (1974). The structure and evacuation of the spermatophore of *Tenebrio molitor* L. (Coleoptera: Tenebrionidae). *Tissue Cell* **6**: 95-108

Garcia-Bellido, A. (1964). Das Sekret der Paragonien als Stimulus der Fekundität bei Weibchen von *Drosophila melanogaster*. *Zeitschrift für Naturforschung* **19b**: 491-495

Gillott, C. and Langley, P. A. (1981). The control of receptivity and ovulation in the tsetse fly, *Glossina morsitans*. *Physiological Entomology* **6**: 269-281

Gillott, C. (1988). Arthropoda – Insecta. In: *Reproductive Biology of Invertebrates*. Volume III (Accessory Sex Glands). Edited by Adiyodi, K.G. and Adiyodi, R.G., Wiley Inc., Chichester, pp. 319-471

Hall, J. C. (1994). The mating of a fly. *Science* **264**: 1702-1714

Harshman, L. G. and Prout, T. (1994). Sperm displacement without sperm transfer in *Drosophila melanogaster*. *Evolution* **48**: 758-766

He, S., Lin, Y. L. and Liu, Y. X. (1999). Functionally inactive protein C inhibitor in seminal plasma may be associated with infertility. *Molecular Human Reproduction* **5**: 513-519.

Heifetz, Y., Lung, O., Frongillo, E. A. and Wolfner, M. F. (2000). The *Drosophila* seminal fluid protein Acp26Aa stimulates release of oocytes by the ovary. *Current Biology* **10**: 99-102

Heifetz, Y. and Wolfner, M. F. (2004). Seminal fluid and mating mediate changes in nerve termini innervating the *Drosophila* reproductive tract. *Proceedings of the National Academy of Sciences USA* **101**: 6261-6266

Heifetz, Y., Vandenberg, L. N., Cohn, H. I. and Wolfner, M. F. (2005). Two cleavage products of the *Drosophila* accessory gland protein ovulin can independently induce ovulation *Proceedings of the National Academy of Sciences USA* **102**: 743-748

Herndon, L. A. and Wolfner, M. F. (1995). A *Drosophila* seminal fluid protein, Acp26Aa, stimulates egg-laying in females for one day following mating. *Proceedings of the National Academy of Sciences USA* **92**: 10114-10118

Hesla, J. S., Preutthipan, S., Maguire, M. P., Chang, T. S. K., Wallach, E. E. and Dharmarajan, A. M. (1997). Nitric oxide modulate human chorionic gonadotropin-induced ovulation in the rabbit. *Fertility and Sterility* **67**: 548-552

Holloway, A. K. and Begun, D. J. (2004). Molecular evolution and population genetics of duplicated accessory gland protein genes in *Drosophila*. *Molecular Biology and Evolution* **21**: 1625-1628

Ignotz, G. G., Lo, M. C., Perez, C. L., Gwathmey, T. M. and Suarez, S. S. (2001). Characterization of a fucose-binding protein from bull sperm and seminal plasma that may be responsible for formation of the oviductal sperm reservoir. *Biology of Reproduction* **64**: 1806-1811

Iida, K. and Cavener, D. R. (2004). Glucose dehydrogenase is required for normal sperm storage and utilization in female *Drosophila melanogaster*. *Journal of Experimental Biology* **207**: 675-681

Imamura, M., Haino-Fukushima, K., Aigaki, T. and Fuyama, Y. (1998). Ovulation stimulating substances in *Drosophila biarmipes* males: their origin, genetic variation in the response of females, and molecular characterization. *Insect Biochemistry and Molecular Biology* **28**: 365-372

Jacklet, J. W. (1997). Nitric oxide signaling in invertebrates. *Invertebrate Neuroscience* **3**: 1-14

Kalb, J., DiBenedetto, A. J. and Wolfner, M. F. (1993). Probing the function of *Drosophila melanogaster* accessory glands by directed cell ablation. *Proceedings of the National Academy of Sciences USA* **90**: 8093-8097

Kaplan, W. D., Tinderholt, V. E. and Gugler, D. H. (1962). The number of sperm present in reproductive tracts of *D. melanogaster* females. *Drosophila Information Service* **36**: 82

Kaufmann, B. P. and Demerec, M. (1942). Utilization of sperm by the female *Drosophila melanogaster*. *American Naturalist* **76**: 445-469

Kern, A. D., Jones, C. D. and Begun, D. J. (2004). Molecular population genetics of male accessory gland proteins in the *Drosophila simulans* complex. *Genetics* **167**: 725-735

Klowden, M. J. (2006). Switchover to the mated state by spermathecal activation in female *Anopheles gambiae* mosquitoes. *Journal of Insect Physiology* **52**: 679-684

Kressin, M., Sommer, U. and Schnorr, B. (1996). The spermathecal epithelium of the queen bee (*Apis mellifera*): morphology, age-dependent changes and cell contacts. *Anatomical and Histological Embryology* **25**: 31-35

Kubli, E. (2003). Sex-peptides: seminal peptides of the *Drosophila* male. *Cell and Molecular Life Sciences* **60**: 1689-1704

Lange, A. B., Orchard, I. and Adams, M. E. (1986). Peptidergic innervation of insect reproductive tissue: The association of proctolin with oviduct visceral musculature. *Journal of Comparative Neurology* **254**: 279-286

Lange, A. B., Orchard, I. and Brugge, V. A. (1991). Evidence for the involvement of a SchistoFLRF-amide-like peptide in the neural control of locust oviduct. *Comparative Biochemistry and Physiology A* **168**: 383-391

Lange, A. B. and Nykamp, D. A. (1996). Signal transduction pathways regulating the contraction of an insect visceral muscle. *Archives of Insect Biochemistry and Physiology* **33**: 183-196

Langendijk, P., Soede, N. M. and Kemp, B. (2005). Uterine activity, sperm transport, and the role of boar stimuli around insemination in sows. *Theriogenology* **63**: 500-513

Lawniczak, M. K. and Begun, D. J. (2004). A genome-wide analysis of courting and mating responses in *Drosophila melanogaster* females. *Genome* **47**: 900-910

Lee, H. G., Seong, C. S., Kim, Y. C., Davis, R. L. and Han, K. A. (2003). Octopamine receptor OAMB is required for ovulation in *Drosophila melanogaster*. *Developmental Biology* **264**: 179-190

Lefevre, G. and Jonsson, V. B. (1962). Sperm transfer, storage, displacement and utilization in *Drosophila melanogaster*. *Genetics* **47**: 1719-1736

Li, X. and Zhao, X. (2004). Separation and purification of ovulation-inducing factors in the seminal plasma of the Bactrian camel (*Camelus bactrianus*). *Veterinary Research Communications* **28**: 235-245

Linley, J. R. (1981). Emptying of the spermatophore and spermathecal filling in Culicoides melleus (Coq.)(Diptera: Ceratopogonidae). *Canadian Journal of Zoology* **59**: 347-356

Liu, H. and Kubli, E. (2003). Sex-peptide is the molecular basis of the sperm effect in *Drosophila melanogaster*. *Proceedings of the National Academy of Sciences USA* **100**: 9929-9933

Ludwig, M. Z., Uspensky, I. I., Ivanov, A. I., Kopantseva, M. R., Dianov, C. M., Tamarina, N. A. and Korchin, L. I. (1991). Genetic control and expression of the major ejaculatory bulb protein PEB-me in *Drosophila melanogaster*. *Biochemical Genetics* **29**: 215-240

Lung, O. and Wolfner, M. F. (1999). *Drosophila* seminal fluid proteins enter the circulatory system through the walls of the posterior vagina. *Insect Biochemistry and Molecular Biology* **29**: 1043-1052

Lung, O., Kuo, L. and Wolfner, M. F. (2001). *Drosophila* males transfer antibacterial proteins from their accessory gland and ejaculatory duct to their mates. *Journal of Insect Physiology* **47**: 617-622.

Lung, O. and Wolfner, M. F. (2001). Identification of and characterization of the major *D. melanogaster* mating plug protein. *Insect Biochemistry and Molecular Biology* **31**: 543-551

Lung, O., Tram, U., Finnerty, C., Eipper-Mains, M., Kalb, J. M. and Wolfner, M. F. (2002). The *Drosophila melanogaster* seminal fluid protein Acp62F is a protease inhibitor that is toxic upon ectopic expression. *Genetics* **160**: 211-224

Manning, A. (1962). A sperm factor affecting the receptivity of *Drosophila melanogaster* females. *Nature* **194**: 252-253

McCormick, J. and Nichols, R. (1993). Spatial and temporal expression identify dromyosuppressin as a brain-gut peptide in *Drosophila melanogaster*. *Journal of Comparative Neurology* **338**: 279-288

McGraw, L. A., Gibson, G., Clark, A. G. and Wolfner, M. F. (2004). Genes regulated by mating, sperm or seminal proteins in mated female *Drosophila melanogaster*. *Current Biology* **14**: 1509-1514

Merle, J. (1968). Fonctionnement ovarien et réçeptivité sexuelle de *Drosophila melanogaster* après implantation de fragments de l'appareil génital male. *Journal of Insect Physiology* **14**: 1159-1168

Michalik, P., Reiher, W., Tintelnot-Suhm, M., Coyle, F. A. and Alberti, G. (2005). Female genital system of the folding trapdoor spider *Antrodiactus unicolor* (Hentz, 1842) (Antrodiaetidae, Araneae): ultrastructural study of form and function with notes on reproductive biology of spiders. *Journal of Morphology*

263: 284-309

Miller, G. T., Starmer, W. T. and Pitnick, S. (2001). Quantitative genetics of seminal receptacle length in *Drosophila melanogaster*. *Heredity* **87**: 25-32

Miller, G. T. and Pitnick, S. (2002). Sperm-female coevolution in *Drosophila*. *Science* **298**: 1230-1233

Miller, G. T. and Pitnick, S. (2003). Functional significance of seminal receptacle length in *Drosophila melanogaster*. *Journal of Evolutionary Biology* **16**: 114-126

Miller, G. T., Starmer, W. T. and Pitnick, S. (2003). Quantitative genetic analysis of among-population variation in sperm and female sperm-storage organ length in *Drosophila mojavensis*. *Genetical Research* **81**: 213-220

Monastirioti, M., Linn, C. E. and White, K. (1996). Characterization of Drosophila tyramine beta-hydroxylase gene and isolation of mutant flies lacking octopamine. *Journal of Neuroscience* **16**: 3900-3911

Monastirioti, M. (2003). Distinct octopamine cell population residing in the CNS abdominal ganglion controls ovulation in *Drosophila melanogaster*. *Developmental Biology* **264**: 38-49

Monsma, S. A. and Wolfner, M. F. (1988). Structure and expression of a *Drosophila* male accessory gland gene whose product resembles a peptide pheromone precursor. *Genes and Development* **2**: 1063-1073

Monsma, S. A., Harada, H. A. and Wolfner, M. F. (1990). Synthesis of two *Drosophila* male accessory gland proteins and their fate after transfer to the female during mating. *Developmental Biology* **142**: 465-475

Moshitzky, P., Fleischmann, I., Chaimov, N., Saudan, P., Klauser, S., Kubli, E. and Applebaum, S. (1996). Sex-peptide activates juvenile hormone biosynthesis in the *Drosophila melanogaster* corpus allatum. *Archives of Insect Biochemistry and Physiology* **32**: 363-374

Mueller, J. L., Page, J. and Wolfner, M. F. (2006). An ectopic expression screen reveals *Drosophila* seminal fluid proteins with protective and toxic effects. *Genetics* (in press)

Mueller, J. L., Ripoll, D. R., Aquadro, C. F. and Wolfner, M. F. (2004). Comparative structural modeling and inference of conserved protein classes in *Drosophila* seminal fluid. *Proceedings of the National Academy of Sciences USA* **101**: 13542-13547

Mueller, J. L., Ravi Ram, K., McGraw, L. A., Bloch Qazi, M. C., Siggia, E. D., Clark, A. G., Aquadro, C. F. and Wolfner, M. F. (2005). Cross-species comparison of *Drosophila* male accessory gland protein genes. *Genetics* **171**: 131-143

Murer, V., Spetz, J. F., Hengst, U., Altrogge, L. M., deAgostini, A. and Monard, D. (2001). Male fertility defects in mice lacking the serine protease inhibitor protease nexin-1. *Proceedings of the National Academy of Sciences USA* **98**: 3029-3033

Neubaum, D. M. and Wolfner, M. F. (1999). *Drosophila* females require a seminal fluid protein, Acp36DE, to store sperm efficiently. *Genetics* **153**: 845-857

Ohashi, Y. Y., Hamo-Fukushima, K. and Fuyama, Y. (1991). Purification and characterization of an ovulation stimulating substance from the male accessory glands of *Drosophila suzukii*. *Insect Biochemistry* **21**: 413-419

Orchard, I. and Lange, A. B. (1985). Evidence for octopaminergic modulation of an insect visceral muscle. *Journal of Neurobiology* **16**: 171-181

Ottiger, M., Soller, M., Stocker, R. F. and Kubli, E. (2000). Binding sites of *Drosophila melanogaster* sex peptide pheromones. *Journal of Neurobiology* **44**: 57-71.

Pan, G., Chen, Z., Liu, X., Li, D., Xie, Q., Ling, F. and Fang, L. (2001). Isolation and purification of the ovulation-inducing factor from seminal plasma in the bactrian camel (*Camelus bactrianus*). *Theriogenology* **55**: 1863-1879

Park, M. and Wolfner, M. F. (1995). Male and female cooperate in the processing of a *Drosophila melanogaster* seminal fluid protein. *Developmental Biology* **171**: 694-702

Partridge, L., Green, A. and Fowler, K. (1987). Effects of egg-production and of exposure to males on female survival in *Drosophila melanogaster*. *Journal of Insect Physiology* **10**: 745-749

Peng, J., Chen, S., Busser, S., Liu, H., Honegger, T. and Kubli, E. (2005a). Gradual release of sperm bound sex-peptide controls female postmating behavior in *Drosophila*. *Current Biology* **15**: 207-213

Peng, J., Zipperlen, P. and Kubli, E. (2005b). *Drosophila* sex-peptide stimulates female innate immune system after mating via the Toll and Imd pathways. *Current Biology* **15**: 1690-1694

Pitnick, S., Spicer, G. S. and Markow, T. A. (1995). How long is a giant sperm? *Nature* **375**: 109

Pitnick, S., Miller, G. T., Schneider, K. and Markow, T. A. (2003). Ejaculate-female coevolution in *Drosophila mojavensis*. *Proceedings of the Royal Society of London, series B* **270**: 1507-1512

Polak, M., Starmer, W. T. and Barker, J. S. F. (1998). A mating plug and male mate choice in *Drosophila hibisci* Bock. *Animal Behaviour* **56**: 919-926

Price, C. S., Dyer, K. A. and Coyne, J. A. (1999). Sperm competition between *Drosophila* males involves both displacement and incapacitation. *Nature* **400**: 449-452

Ramaswamy, S. B., Qiu, Y. and Park, Y. I. (1996). Neuronal control of post-coital pheromone production in the moth *Heliothis virescens*. *Journal of Experimental Zoology* **274**: 255-263

Ravi Ram, K., Sirot, L. K. and Wolfner, M. F. (2006). A predicted seminal astacin-like protease is required for the processing of reproductive proteins in *Drosophila melanogaster*. *Proceedings of the National Academy of Sciences USA* **103**: 18674-18679

Ravi Ram, K., Ji, S. and Wolfner, M. F. (2005). Fates and targets of male accessory gland proteins in mated female *Drosophila melanogaster*. *Insect Biochemistry and Molecular Biology* **35**: 1059-1071

Robertson, S. A., Bromfield, J. J. and Tremellen, K. P.

(2003). Seminal 'priming' for protection from pre-eclampsia-a unifying hypothesis. *Journal of Reproductive Immunology* **59**: 253-265

Saudan, P., Hauck, K., Soller, M., Choffat, Y., Ottiger, M., Sporri, M., Ding, Z., Hess, D., Gehrig, P. M., Klauser, S., Hunziker, P. and Kubli, E. (2002). Ductus ejaculatorius peptide 99B (DUP99B), a novel *Drosophila melanogaster* sex-peptide pheromone. *European Journal of Biochemistry* **269**: 989-997.

Schäfer, U. (1986). Genes for male-specific transcripts in *Drosophila melanogaster. Molecular and General Genetics* **202**: 219-225

Schmidt, T., Choffat, Y., Klauser, S. and Kubli, E. (1993). The *Drosophila melanogaster* sex-peptide: a molecular analysis of structure-function relationships. *Journal of Insect Physiology* **39**: 361-368

Schully, S. D. and Hellberg, M. E. (2006). Positive selection on nucleotide substitutions and indels in accessory gland proteins of the *Drosophila pseudoobscura* subgroup. *Journal of Molecular Evolution* **62**: 793-802

Simmerl, E., Schäfer, M. and Schäfer, U. (1995). Structure and regulation of a gene cluster for male accessory gland transcripts in *Drosophila melanogaster. Insect Biochemistry and Molecular Biology* **25**: 127-137

Snook, R. R. and Hosken, D. J. (2004). Sperm death and dumping in Drosophila. *Nature* **428**: 939-941

Soller, M., Bownes, M. and Kubli, E. (1999). Control of oocyte maturation in sexually mature *Drosophila* females *Developmental Biology* **208**: 337-351

Swanson, W. J., Clark, A. G., Waldrip-Dail, H. M., Wolfner, M. F. and Aquadro, C. F. (2001). Evolutionary EST analysis identifies rapidly evolving male reproductive proteins in *Drosophila. Proceedings of the National Academy of Sciences USA* **98**: 7375-7379.

Swanson, W. J. and Vacquier, V. D. (2002). The rapid evolution of reproductive proteins. *Nature Reviews Genetics* **3**: 137-144

Swanson, W. J. (2003). Sex peptide and the sperm effect in *Drosophila melanogaster. Proceedings of the National Academy of Sciences USA* **100**: 9643-9644

Tram, U. and Wolfner, M. F. (1999). Male seminal fluid proteins are essential for sperm storage in *Drosophila melanogaster. Genetics* **153**: 837-844

Tremellen, K. P., Valbuena, D., Landeras, J., Ballesteros, A., Martinez, J., Mendoza, S., Norman, R. J., Robertson, S. A. and Simon, C. (2000). The effect of intercourse on pregnancy rates during assisted human reproduction. *Human Reproduction* **15**: 2653-2658

Trent, C., Tsung, N. and Horvitz, R. (1983). Egg-laying defective mutants of the nematode *Caenorhabditis elegans. Genetics* **104**: 619-647

Tsaur, S. C. and Wu, C.-I. (1997). Positive selection and the molecular evolution of a gene of male reproduction, Acp26Aa of *Drosophila. Molecular Biology and Evolution* **14**: 544-549

Tsaur, S. C., Ting, C. T. and Wu, C.-I. (1998). Positive selection driving the evolution of a gene of male reproduction, Acp26Aa, of *Drosophila*: II. Divergence versus polymorphism. *Molecular Biology and Evolution* **15**: 1040-1046

Tsaur, S. C., Ting, C. T. and Wu, C. I. (2001). Sex in *Drosophila mauritiana*: a very high level of amino acid polymorphism in a male reproductive protein gene, Acp26Aa. *Molecular Biology and Evolution* **18**: 22-26

Tshugguel, W., Schneeberger, C., Unfried, G., Brautigam, G., Wieser, F., Czerwenka, K., Vytiska-Binstorfer, E., Kurz, C., Weninger, W., Mildner, M., Waselmayer, B., Bursch, W., Kaider, A., Waldhor, T., H., B., Ellinger, A. and Huber, J. C. (1998). The role of nitric oxide in reproduction. *Gynakologisch-Geburtshilfliche Rundschau* **38**: 44-46

Waggoner, L. E., Zhou, G. T., Schafer, R. W. and Schafer, W. R. (1998). Control of alternative behavioral states by serotonin in *Caenorhabditis elegans. Neuron* **21**: 203-214

Walker, M. J., Rylett, C. M., Keen, J. N., Audsley, N., Sajid, M., Shirras, A. D. and Isaac, R. E. (2006). Proteomic identification of *Drosophila melanogaster* male accessory gland proteins, including a pro-cathepsin and a soluble gamma-glutamyl transpeptidase *Proteome Science* **4**: 9

Wigby, S. and Chapman, T. (2005). Sex peptide causes mating costs in female *Drosophila melanogaster. Current Biology* **15**: 316-321

Wolfner, M. F. (1997). Tokens of love: functions and regulation of *Drosophila* male accessory gland products. *Insect Biochemistry and Molecular Biology* **27**: 179-192

Wolfner, M. F., Harada, H. A., Bertram, M. J., Stelick, T. J., Kraus, K. W., Kalb, J. M., Lung, Y. O., Neubaum, D. M., Park, M. and Tram, U. (1997). New genes for accessory gland proteins in *D. melanogaster. Insect Biochemistry and Molecular Biology* **27**: 825-834

Wolfner, M. F. (2002). The gifts that keep on giving: physiological functions and evolutionary dynamics of male seminal proteins in *Drosophila. Heredity* **88**: 85-93

Wolfner, M. F., Heifetz, Y. and Applebaum, S. W. (2005). Gonadal glands and their gene products. In: *Comprehensive Molecular Insect Science.* Volume 1. Edited by Gilbert, L.I., Iatrou, K. and Gil, S.S. , Elsevier Press, London, pp. 179-212

Wong, A., Albright, S. A. and Wolfner, M. F. (2006). Evidence for structural constraint on ovulin, a rapidly evolving *Drosophila melanogaster* seminal protein. *Proceedings of the National Academy of Sciences USA* **103**: 18644-18649

Xue, L. and Noll, M. (2000). *Drosophila* female sexual behavior induced by sterile males showing copulation complementation. *Proceedings of the National Academy of Sciences USA* **97**: 3272-3275

Yamauchi, J., Miyazaki, T., Iwasaki, S., Kishi, I., Kuroshima, M., Tei, C. and Yoshimura, Y. (1997). Effects of nitric oxide on ovulation and ovarian steroidogenesis and prostaglandin production in the rabbit. *Endocrinology* **138**: 3630-3637

Spermatology. SRF Vol. 65. ERS Roldan and M Gomendio (eds) Nottingham University Press, Nottingham

Insights into structure-function correlations of ungulate seminal plasma proteins

Juan J. Calvete and Libia Sanz

Instituto de Biomedicina de Valencia (CSIC), Jaime Roig 11, 46010-Valencia, Spain

The seminal plasmas of a variety of mammalian species contain factors that prevent inappropriate acrosome reactions ("decapacitating factors") and proteins that upon binding to the sperm surface enhance the fertilising potential of spermatozoa. The emerging view is that the membrane remodelling events that underlie capacitation of ejaculated spermatozoa in the female's genital tract is essentially conserved though it can be accomplished by different seminal plasma proteins in different species. The major seminal plasma proteins of ungulates (i.e., pig, bull, ram, stallion) belong to the spermadhesin family and to the Fn2-choline-binding proteins. Although the final effect of different seminal plasmas on homologous sperm capacitation may represent an example of functional convergence, individual proteins from the same family appear to function in a species-specific manner. Thus, differences in their structures, relative abundance, and expression patterns along the male genital tract may determine species-specific effects on sperm physiology of homologous proteins. Discussed here are a plethora of biochemical and biophysical studies on seminal plasma proteins from ungulates, including the determination of the crystal structures of porcine and bovine spermadhesins and bull PDC-109, which have provided over the past two decades insights into the molecular and structural basis of their role along the fertilisation process.

Introduction

Mammalian spermatozoa do not possess the ability to fertilise the egg immediately upon ejaculation, but acquire fertilisation competence following residence in the female tract for a finite period. This time-dependent maturational process, known as "capacitation", was first observed in 1951 independently by Chang and Austin. The seminal plasma (SP), constituted by secretions from the testes, epididymis and the accessory sex glands, contains a variety of factors (amino acids, lipids, fatty acids, osmolytes, peptides, and proteins) that influence the viability of ejaculated spermatozoa and the ability of acrosome-intact sperm to undergo the acrosome reaction in response to the zona pellucida (Mann and Lutwak-Mann, 1981; Shivaji, Scheit and Bhargava, 1990; Yanagimachi, 1994). Thus, the seminal plasmas of a variety of mammalian species contain both factors that prevent inappropriate acrosome reactions (decapacitating factors) and proteins that upon binding to the sperm surface enhance the fertilising potential of spermatozoa (Killian, Chapman and Rogowski, 1993; Thérien, Soubeyrand and Manjunath,

Corresponding author E-mail: jcalvete@ibv.csic.es

1997; Rodríguez-Martínez, Saravia, Wallgren, Tienthai, Johannisson, Vázquez, Martínez, Roca, Sanz and Calvete, 2005). However, the precise role of most of the seminal plasma proteins on sperm physiology remains obscure. In addition, the effect of seminal plasma on spermatozoa is variable among species, males of the same species, and ejaculates from a single male. The distinct effects exerted by different seminal plasmas on sperm functionality have been in part ascribed to variability of the composition and concentration of some proteins (Maxwell and Johnson, 1999; Centurión, Vázquez, Calvete, Roca, Sanz, Parrilla, Garcia and Martínez, 2003). Nevertheless, our current view is that the membrane remodelling events that underlie capacitation of ejaculated spermatozoa in the female's genital tract is essentially conserved (e.g. activation of transmembrane signal transduction pathways upon membrane fluidity changes induced by the removal of cholesterol from the sperm plasma membrane) (Thérien, Moreau and Manjunath, 1998; Visconti, Ning, Fornes, Álvarez, Stein, Connors and Kopf, 1999) though it can be accomplished by different seminal plasma proteins in different species. Supporting this concept, capacitation can be mimicked *in vitro* by incubating epididymal or ejaculated sperm in a defined medium containing appropriate concentrations of electrolytes, metabolic energy source, and a protein source that usually is the membrane lipid-sequestering protein bovine serum albumin. Hence, the effects of seminal plasma proteins on homologous sperm capacitation may represent an example of functional convergence of distinct molecular players. The major seminal plasma proteins involved in porcine and bovine sperm capacitation belong to different protein families and have been characterised in depth using biochemical, biophysical, and structural techniques. These studies are sumarised in this short review.

Spermadhesins

Boar AQN-1, AQN-3, AWN, PSP-I and PSP-II

Spermadhesins are male secretory proteins detected so far in ungulates (pig, cattle, horse and ram) (Calvete, Sanz, Dòstalovà and Töpfer-Petersen, 1995a; Bergeron, Villemure, Lazure and Manjunath, 2005; Haase, Schlötterer, Ekhlasi-Hundrieser, Kuiper, Distl and Töpfer-Petersen, 2005). In the pig, this family of proteins comprises five members, AQN-1, AQN-3, AWN, PSP-I and PSP-II, whose genes are clustered on SCC 14q28-q29 (Haase *et al.*, 2005). Together they represent over 90% of the total boar seminal plasma proteins (Dostàlovà, Calvete, Sanz and Töpfer-Petersen, 1994) and are thought to participate in different aspects of porcine fertilisation. Porcine spermadhesins, 110-133-residue polypeptides, are synthesised by the epididymis and accessory glands (Ekhlasi-Hundrieser, Sinowatz, Greiser de Wilke, Waberski and Töpfer-Petersen, 2002) and exhibit distinct sperm-coating and ligand-binding capabilities. Sequence variation, glycosylation, and the aggregation state of spermadhesins contribute to their specific pattern of biological activities (Calvete, Sanz, Dostàlovà and Töpfer-Petersen, 1993; Calvete, Solís, Sanz, Díaz-Mauriño, Schäfer, Mann and Töpfer-Petersen, 1993; Dostàlova, Calvete and Töpfer-Petersen, 1995; Calvete, Mann, Schäfer, Raida, Sanz and Töpfer-Petersen, 1995b). Thus, zinc ions induce the unfolding and self-association of boar spermadhesin PSP-I and promote its binding to heparin (Campanero-Rhodes, Menéndez, Sáiz, Sanz, Calvete and Solís, 2006). The dramatic effect of zinc on the stability and heparin-binding ability of PSP-I provides a rationale for the simultaneous occurrence in seminal plasma of PSP-I forms with different conformational status and ligand-binding capabilities. Further, only non-aggregated AWN-1 and AQN-3 are able to interact with phospholipid matrices, suggesting that this spermadhesin subpopulation can bind directly to the lipids of the sperm membrane (Dostàlovà, Calvete and Töpfer-Petersen, 1995). Aggregated spermadhesins could then become coated on top of this first layer, serving as stabilizing factors that protect the acrosome membrane from premature acrosome reaction. These spermadhesins are thought to stabilise the plasma membrane over the acrosomal vesicle and are mainly released from the spermatozoal surface during capacitation (Sanz, Calvete, Mann, Gabius and Töpfer-Petersen, 1993; Dostàlovà *et al.*, 1994; Calvete, Ensslin, Mburu, Iborra, Martínez, Adermann, Waberski, Sanz, Töpfer-Petersen, Weitze, Einarsson and Rodríguez-Martínez, 1997).

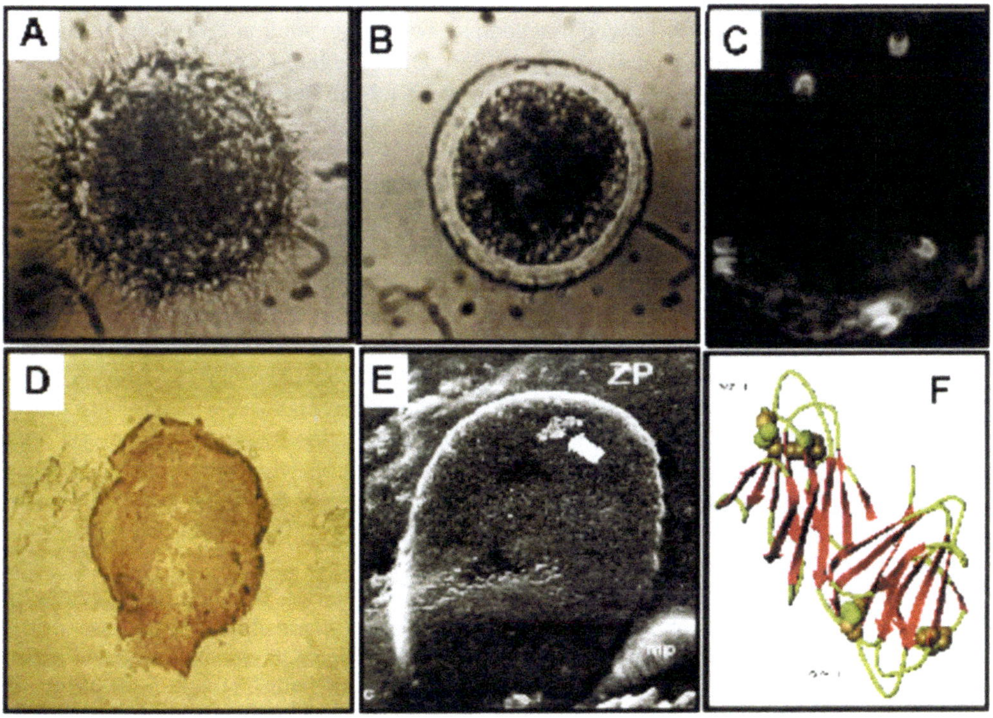

Figure 1. Spermadhesins are sperm-coating proteins involved in fertilisation. Binding of porcine spermatozoa to a zona pellucida-encased oocyte in the absence (**A**) or the presence (**B**) of purified boar spermadhesins AQN-1, AQN-3 or AWN. Indirect immunofluorescence detection of spermadhesin AWN on the acrosomal cap of boar spermatozoa (**C**) and on isolated zona pellucida (**D**). (**E**) Immunoelectronmicroscopic imaging of spermadhesin AWN epitopes on plasmalemmal remnants of a boar spermatozoon bound to the zona pellucida *in vivo*. (**F**) Crystal structure of the spermadhesin PSP-I/PSP-II heterodimer.

Purified spermadhesins AQN-1, AQN-3, and AWN inhibit the *in vitro* attachment of spermatozoa to zona pellucida-encased oocytes (Fig. 1A,B), coat the sperm acrosomal cap surface at ejaculation (Fig. 1C) and display zona pellucida glycoprotein-binding capability (Fig. 1D) and affinity for ß-galactosides and heparin (Sanz, Calvete, Mann, Gabius and Töpfer-Petersen, 1993; Dostàlovà, Calvete, Sanz and Töpfer-Petersen, 1995; Calvete, Dostàlovà, Sanz, Adermann, Thole and Töpfer-Petersen, 1996; Calvete, Carrera, Sanz and Töpfer-Petersen, 1996; Rodríguez-Martínez, Iborra, Martínez and Calvete, 1999). Moreover, among the heparin-binding spermadhesins, AQN-3 has been reported to form part of a sperm motility-inhibitor factor complex (Iwamoto, Hiroaki, Furuichi, Wada, Satoh, Satoh and Gagnon, 1995) and AQN-1 and AWN are sperm-associated acrosin inhibitor acceptor proteins (Sanz, Calvete, Jonàkovà and Töpfer-Petersen, 1992). These activities support a decapacitating function for the heparin-binding spermadhesins. Immunoelectronmicroscopic studies have revealed spermadhesin AWN epitopes on plasmalemmal remnants of boar spermatozoa bound *in vivo* to the zona pellucida of oocyte-sperm complexes recovered by surgery from naturally mated sows (Rodríguez-Martínez *et al.*, 1999) (Fig. 1E). This evidence clearly showed that at least spermadhesin AWN fulfils the requirements expected for a *bona-fide* zona pellucida-binding protein acting at the site of *in vivo* fertilisation. On the other hand, PSP-I and PSP-II are the most abundant boar seminal plasma proteins, and form a non-heparin-binding glycosylated heterodimer (Calvete *et al.*, 1995b; (Nimtz, Grabenhorst, Conradt, Sanz and Calvete, 1999). The crystal structure of the PSP-I/PSP-II complex showed that each spermadhesin subunit is built by a single CUB domain architecture (Romero, Romão, Varela, Kölln, Dias, Carvalho, Sanz, Töpfer-Petersen and Calvete, 1997) (Fig. 1F) and revealed the structural basis for the specific association of PSP-I and PSP-II (Varela, Romero, Sanz, Romão, Töpfer-Petersen and Calvete, 1997).

Accumulating evidence points to a role for the PSP-I/PSP-II spermadhesin complex as an exogenous modulator of the uterine immune activity thus ensuring reproductive success (Fig. 2). Hence, PSP-I and PSP-II are immunostimulatory for lymphocyte activity *in vitro* (Leshin, Raj, Smith, Kwok, Kraeling and Li, 1998; Yang, Kwok, Leshin, Bollo and Li, 1998), and binding of PSP-I to lymphocyte has been demonstrated (Yang *et al.*, 1998). Furthermore, the PSP-I/PSP-II heterodimer and its isolated subunits induce the recruitment of neutrophils into the peritoneal cavity of rats (Assreuy, Calvete, Alencar, Cavada, Rocha-Filho, Melo, Cunha and Ribeiro, 2002) and pigs (Rodríguez-Martínez *et al.*, 2005). The neutrophil migration-inducing activity of PSP-I/PSP-II, and possibly of the PSP-II subunit, is mediated by the stimulation of resident macrophages, which release a neutrophil chemotactic substance (Assreuy *et al.*, 2002), while PSP-I appears to act directly on neutrophils (Assreuy, Alencar, Cavada, Rocha-Filho, Feitosa, Cunha, Calvete and Ribeiro, 2003). The benefit of these proinflammatory and immunostimulatory activities would be to prevent possible infections of the lower reproductive tract and to provide a foreign cell-free uterine environment for the descending early embryos.

Dimerization markedly affects the binding abilites of the PSP-I/PSP-II subunits. Notably, the heparin and mannose-6-phosphate binding capabilities of PSP-II are abolished by its association with PSP-I, indicating that the corresponding binding sites may be located at (or near) the dimer interface (Solís, Romero, Jiménez, Díaz-Mauriño and Calvete, 1998). Hence, to be of biological importance these cryptic activities must be exposed in specific physiological environments. Relevant to this point, biophysical analyses of the stability of the spermadhesin PSP-I/PSP-II have shown that, although the integrity of the heterodimer in seminal plasma seems to be guaranteed by its high concentration, dissociation may be facilitated in the female genital tract because of dilution of the protein in the intraluminal fluids of the cervix and the uterus (Campanero-Rhodes, Menéndez, Sáiz, Sanz, Calvete and Solís, 2005). Such a mechanism might be relevant for regulating the uterine immune reactions triggered by PSP-I and PSP-II.

Besides their physiological functions in porcine reproduction, the heparin-binding and the non-heparin-binding spermadhesins exert opposite effects on spermatozoa which have been subjected to high dilution to mimic the conditions of sex sorting by flow cytometry. Thus, while the pooled heparin-binding spermadhesins (AQN-1, AQN-3, and AWN) cause a concentration-dependent sperm membrane damage, the purified non-heparin-binding PSP-I/PSP-II spermadhesin complex contributes to maintaining sperm with high viability, motility, and mitochondrial activity for at least 5 hours at physiological temperature (Centurión *et al.*, 2003). The beneficial effect of the native PSP-I/PSP-II on the functionality of highly diluted boar spermatozoa is largely preserved in its isolated PSP-II subunit and does not appear to require its glycan moiety (García, Vázquez, Calvete, Sanz, Caballero, Parrilla, Gil, Roca and Martínez, 2006). The beneficial effect of the PSP-I/PSP-II heterodimer points to this spermadhesin as a candidate for an additive to improve the viability of highly diluted porcine spermatozoa (i.e. for flow-cytometric sorting for chromosomal sex).

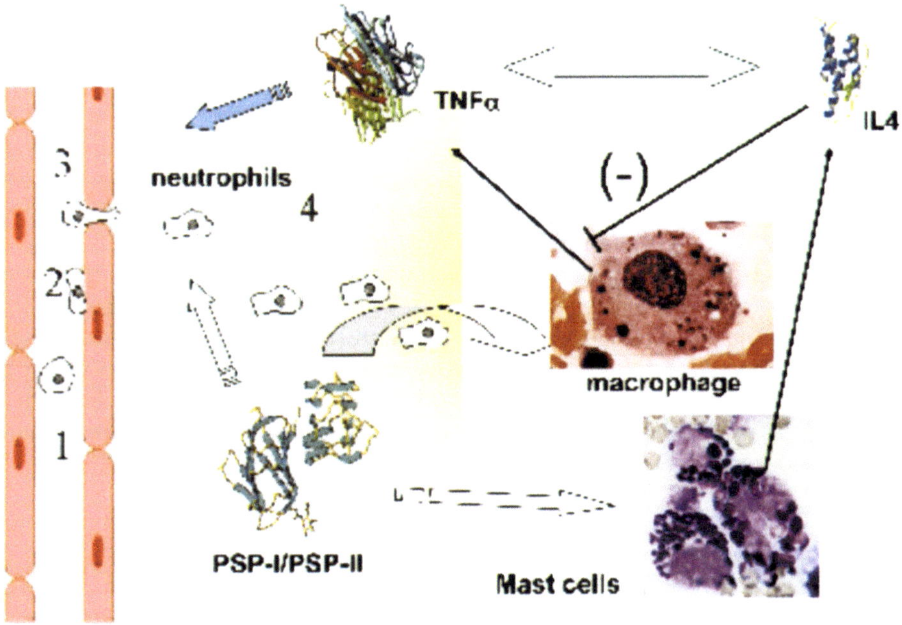

Figure 2. Neutrophil migration-inducing activity of PSP-I/PSP-II. Proposed mechanism for the action of PSP-I/PSP-II and its isolated subunits as exogenous modulators of the uterine immune activity. The emerging view is that the balance between cytokines released by macrophages (TNFα) and mast cells (IL-4) appears to control the onset of neutrophil migration *in response to the inflammatory stimuli of PSP-I/PSP-II, thereby avoiding excessive neutrophil recruitment. This regulation seems to be crucial during situations in which the excessive accumulation of neutrophils would lead to tissue damage (i.e., in postmating endometritis).*

Bull aSFP

Wempe, Einspanier and Scheit (1992) have cloned a bull seminal plasma 114-residue non-glycosylated polypeptide that is structurally related to boar spermadhesins. However, this protein, called acidic seminal fluid protein (aSFP), showed neither sperm-coating activity nor zona pellucida-binding ability (Dostàlovà *et al.*, 1994). This suggested that spermadhesin molecules in different species may play different functional roles. aSFP is the product of a single-copy gene, which is mainly expressed in seminal vesicle and ampulla tissues and weakly in the epididymis.

In reproductively active bulls, aSFP is a major component of seminal fluid, where its concentration ranges from 2 to 7 mg/ml (Dostàlovà *et al.*, 1994). aSFP displays around 50% amino acid sequence identity with the porcine members of the spermadhesin protein family. aSFP binds only loosely to the surface of ejaculated bovine sperm and is quantitatively released during *in vitro* capacitation (Dostàlovà *et al.*, 1994). Furthermore, aSFP possesses neither carbohydrate- nor zona pellucida-binding capabilities, strongly indicating that aSFP may not be involved in gamete interaction. On the other hand, at its physiological concentration, aSFP seems to protect spermatozoa from oxidative damage by diminishing lipid peroxidation (Schöneck, Einspanier, Schallenberger and Schams, 1993). In addition, aSFP appears to stimulate in a dose-dependent manner (in the ng/ml range) *in vitro* cell division of lymphocytes and progesterone secretion by bovine endometrium and ovarian granulosa cells (Einspanier, Einspanier, Wempe and Scheit, 1991). This suggests a possible sperm protecting function for aSFP and a stimulatory effect on ovulation upon insemination.

The crystal structure of aSFP revealed its CUB ß-sandwich topology (Romão, Kölln, Dias, Carvalho, Romero, Varela, Sanz, Töpfer-Petersen and Calvete, 1997) (Fig. 3). It is noteworthy that Cys54 forms a disulphide bond with Cys75, which adopts two conformations, indicating that cysteines 54 and 75 alternate between reduced and oxidized states (Fig. 3). This is in accordance with biochemical evidence showing that the two disulphide bonds of aSFP possess different reactivity yielding three molecular forms (completely oxidised, partly reduced, and fully reduced) which differ in their isoelectric points (Einspanier, Krause, Calvete, Töpfer-Petersen, Klostermeyer and Karg, 1994). The reactivity of Cys54 towards platinum confirmed and identified the reactive pair of cysteine residues. It has been suggested that the redox equilibrium exhibited by aSFP cysteine residues could potentially explain the fact that aSFP can prevent lipid peroxidation, thereby protecting sperm from oxidative damage (Einspanier *et al.*, 1994) by radicals such as those released from the electron transport chains of mitochondria (reviewed by Dean, Fu, Stocker and Davies, 1997).

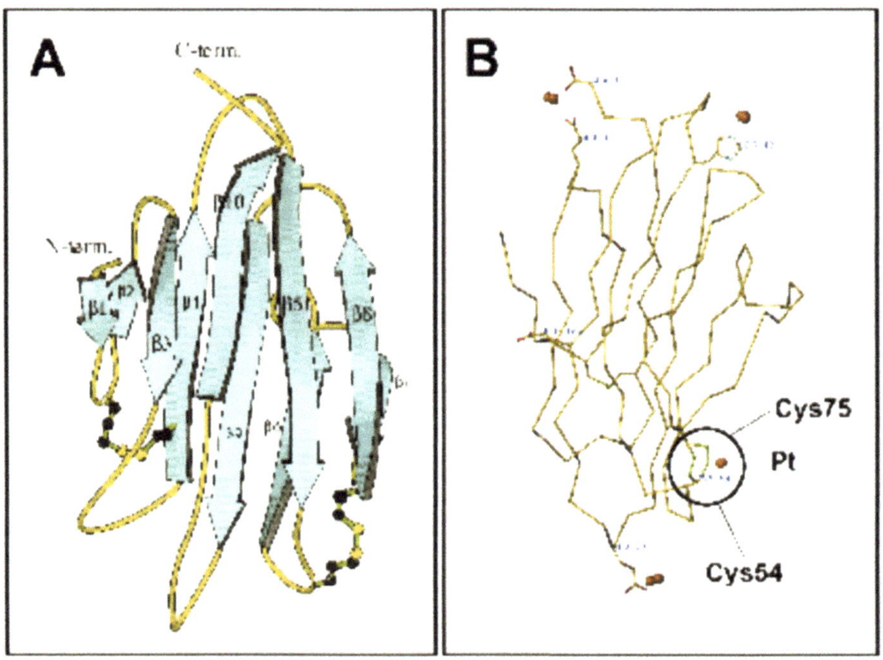

Figure 3. Structure of bull spermadhesin aSFP. (**A**) The molecular structure of aSFP is built by a single CUB domain architecture, a ß-sandwich topology organised into two 5-stranded ß-sheets, each of which contains two parallel and four antiparallel strands. The two disulphide bonds are displayed in the ball-and stick model. (**B**) Location of heavy atom-binding sites within a Cα plot of the aSFP molecule showing the binding of a platinum (Pt) atom to Cys54 and the two conformations of the disulphide bond Cys54-Cys75.

A second spermadhesin molecule has been purified from the seminal plasma of Holstein Friesian bulls (Tedeschi, Oungre, Mortarino, Negri, Maffeo and Ronchi, 2000). This non-glycosylated spermadhesin, termed Z13, is unusual in that it is made up of two disulphide-linked identical 13 kDa subunits. Its biological role remains to be disclosed.

Stallion HSP-7

A single spermadhesin molecule (HSP-7) has been found in stallion seminal plasma (Reinert, Calvete, Sanz, Mann and Töpfer-Petersen, 1996). The amino acid sequence of HSP-7 differs in only three amino acids from that of boar AWN. The large (97.7%) sequence identity between the equine and porcine proteins explains their strong immunological crossreactivity (Calvete, Nessau, Mann, Sanz, Sieme, Klug and Töpfer-Petersen, 1994). HSP-7 is synthesised in spermatogonia, rete testis, the ductus epididymidis and the seminal vesicle epithelial cells (Hoshiba and Sinowatz, 1998). HSP-7 exhibits zona pellucida-binding activity and its topography on the surface of stallion spermatozoa appears to be restricted to the equatorial segment (Reinert, Calvete, Sanz and Töpfer-Petersen, 1997) (Fig. 4). The different spermatozoal compartimentalization in the stallion and in the boar might suggest that homologous spermadhesin molecules could be involved in species-specific aspects of fertilisation.

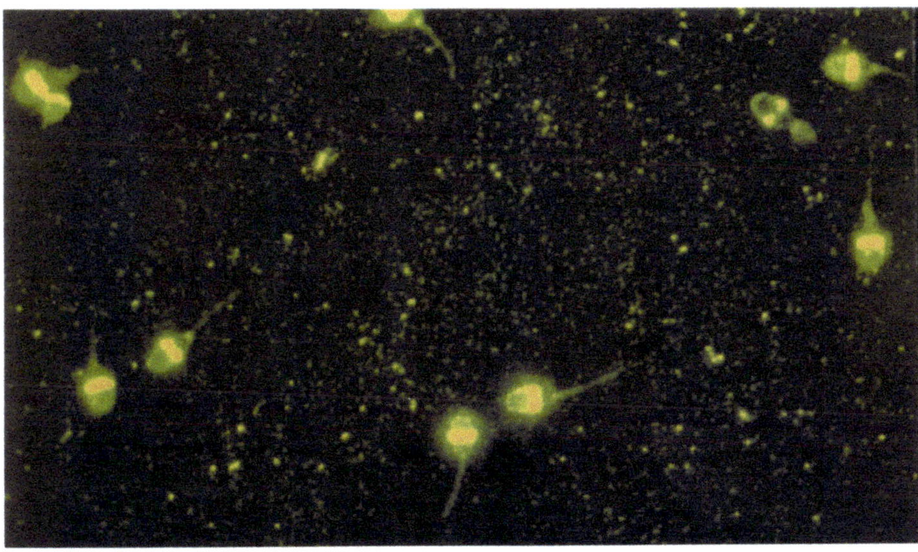

Figure 4. Topography of HSP-7 on stallion spermatozoa. The localization of HSP-7 epitopes on the surface of stallion spermatozoa recovered from cauda epididymis was studied by indirect immunofluorescence microscopy using a chicken anti-porcine-AWN polyclonal antibody. Notice that, in contrast to boar spermatozoa where AWN is localised on the acrosomal cap (Fig. 1C), stallion HSP-7 is restricted to the equatorial segment.

The primary structure similarity between equine HSP-7 and porcine AWN proteins is much greater than the sequence similarity between any other pair of porcine (or bovine) spermadhesin molecules (which ranges 40-60% similarity). Since perissodactyls (e.g., horse) and artiodactyls (e.g., pig) had a common ancestor more than 50 million years ago (Carroll, 1988), the unusually low mutational rate might suggests that the whole AWN structure is under strong selective pressure, pointing to a highly conserved conformation and, most probably, a common biological function for AWN and HSP-7 in both vertebrate species. Although the actual involvement of HSP-7 in horse fertilisation has not been established, the zona pellucida-binding activity displayed by the isolated protein supports its classification as a putative sperm-egg adhesion molecule. The hypothesis that the porcine and equine AWN molecules were diverging under functional constraints whereas the other polypeptides of the boar seminal plasma spermadhesin family were diverging more rapidly in their amino acid sequences would be in accordance with the proposal that, following gene duplication, one copy of the gene will divergently evolve under pressure dictated by the ancestral function, while the duplicate gene(s), unencumbered by a functional role, are free to search for new physiological roles. The reasons for, and consequences of, the presence of multiple spermadhesin molecules in porcine seminal plasma, but only one member of the family in stallion seminal plasma, are at present unclear.

PDC-109, a bull seminal plasma heparin- and choline-binding protein involved in capacitation

Considerable amount of work has been done during the last two decades on a group of bovine seminal plasma (BSP) proteins which bind to bovine sperm plasma membranes and appear to play a crucial role in the heparin-mediated sperm membrane lipid modification that occurs during the sperm capacitation process (Manjunath and Thérien, 2002; Swamy, 2004). Three proteins, BSP-A1/-A2 (also termed PDC-109 or Major Protein), BSP-A3, and BSP-30 (Calvete, Mann, Sanz, Raida and Töpfer-Petersen, 1996; Swamy, 2004) represent the major heparin-binding proteins of bull (*Bos taurus*) seminal plasma. Synthesis of these proteins is under androgen control and is restricted to the epithelia of the seminal vesicles and the *ampullae* (Scheit, Kemme, Aumüller, Seitz, Hagendorff and Zimmer, 1988). The BSP proteins alone cannot sufficiently increase the capacitation of epididymal sperm. They require other factors present in follicular fluid. Elegant investigations from P. Manjunath's laboratory have shown that the BSP proteins bind, in addition to heparin, apolipoprotein A-I associated with high-density lipoprotein (HDL) (Thérien *et al.*, 1997), the only class of lipoprotein present in bovine follicular and oviductal fluids, and to choline phospholipids of the sperm surface (Manjunath, Chandonnet, Leblond and Desnoyers, 1994). BSP proteins, isolated or complexed with HDL, stimulate the efflux of cholesterol and phospholipids from the plasma membrane that occurs during the early stages of sperm capacitation. Thus, at ejaculation the BSP proteins coat choline lipids increasing the heparin- and HDL-docking sites on the sperm surface, induce initial cholesterol and phospholipid efflux, and prime spermatozoa to undergo capacitation at the appropriate place, the caudal isthmus of the oviduct. BSP proteins accelerate the capacitation of bovine epididymal sperm triggered by HDL and potentiate capacitation induced by heparin *via* different, non-additive mechanisms (Thérien *et al.*, 1997, 1998; Lane, Thérien, Moreau and Manjunath, 1999). Heparin-induced bovine sperm capacitation is associated with stimulation of the uptake of Ca^{2+}, intracellular alkalinization, and an increase in cAMP-dependent protein phosphorylation (Visconti, Galantino-Homer, Moore, Bailey, Ning, Fornés and Kopf, 1998). On the other hand, HDL-induced sperm cholesterol removal may result in a further decrease in cholesterol:phospholipid molar ratio leading to fluidity changes and destabilization of the sperm membrane and to a reorganization of membrane bound components. The combined effects of these intracellular and extracellular events may complete capacitation, rendering spermatozoa responsive to inducers of the acrosome reaction.

Recent biochemical and biophysical studies have outlined the binding properties and sperm capacitation potential of PDC-109 (BSP-A1/A2), the most abundant protein of bovine seminal plasma. PDC-109 is present at about 15-20 mg/ml in the seminal fluid, and several million molecules coat the sperm surface at ejaculation (Calvete, Raida, Sanz, Wempe, Scheit, Romero and Töpfer-Petersen, 1994). PDC-109 is a 109-amino-acid modular protein composed of an N-terminal O-glycosylated and highly acidic extension followed by two tandemly-arranged fibronectin type II (Fn2) domains (Esch, Ling, Bohlen, Ying and Guillemin, 1983; Gerwig, Calvete, Töpfer-Petersen and Vliegenthart, 1996). The heparin-binding region of PDC-109 is conformational- and quaternary structure-dependent and involves basic residues of both Fn2 domains (Calvete, Campanero-Rhodes, Raida and Sanz, 1999). PDC-109 binds with high specificity the choline group of phosphorylcholine-containing phospholipids (Desnoyers and Manjunath, 1992; Anbazhagan and Swamy, 2005), which comprise over 70% of total bull sperm plasma membrane phospholipids (Watson, 1981). Equilibrium dialysis experiments demonstrated that each PDC-109 molecule binds 2 choline moieties with a low binding affinity of 0.95 mM (Desnoyers and Manjunath, 1993), and the combining sites have been mapped to the Fn2 domains by combination of limited proteolysis and affinity chromatography.

Purified PDC-109 displays affinity for solvent-exposed choline head group and interacts also with the apolar part of phosphatidylcholine vesicles (Müller, Erlemann, Müller, Calvete, Töpfer-Petersen, Marienfeld and Hermann, 1998) in a very rapid, biphasic process with half times of less than one second. The interaction of PDC-109 with dimyristoylphosphatidylcholine (DMPC) membranes is favoured by a strong entropic contribution, whereas negative entropic contribution is primarily responsible for the rather weak interaction of the seminal plasma protein with dimyristoylphosphatidic acid and dimyristoylphosphatidylglycerol (Thomas, Anbazhagan, Ramakrishnan, Sultan, Surolia and Swamy, 2003). The interaction of PDC-109 with DMPC membranes has been investigated by spin-label electron spin resonance spectroscopy employing phosphatidylcholine bearing the spin labels at different positions along the sn-2 acyl chain. The results indicated that the protein penetrates into the hydrophobic interior of the membrane and interacts with the lipid acyl chains up to the 14th C-atom (Ramakrishnan, Anbazhagan, Pratap, Marsh and Swamy, 2002). Maximal coating to small unilamellar vesicles was achieved with a stoichiometry of 10-12 phosphatidylcholine molecules/PDC-109 molecule (Müller *et al.*, 1998; Gasset, Magdaleno and Calvete, 2000). Upon binding to phosphatidylcholine vesicles, PDC-109 undergoes a conformational change characterised by 5-6% increase in the solvent-exposed loop content and causes rigidification of the membrane (Gasset, Saiz, Laynez, Sanz, Gentzel, Töpfer-Petersen and Calvete, 1997; Gasset *et al.*, 2000; Greube, Müller, Töpfer-Petersen, Herrmann and Müller, 2001) by preventing lipid molecules from participating in the gel-to-liquid phase transition due to enhancement of both acyl chain disorder and interfacial hydration (Gasset *et al.*, 2000) and leading to an enhanced fraction of immobilised cholesterol (Müller, Greube, Töpfer-Petersen and Herrmann, 2002). Furthermore, the interaction of PDC-109 with phosphatidylcholine vesicles provoked permeability changes and partial disruption of the lipid vesicles leading to the budding of 10-40 nm particles (Gasset *et al.*, 2000), which resembled the lipoprotein particle efflux observed upon incubation of bovine epididymal sperm with purified PDC-109 (Moreau and Manjunath, 1999). This effect was not modified by heparin but was inhibited in a concentration-dependent manner by cholesterol (Gasset *et al.*, 2000). PDC-109 does not interact directly with cholesterol, but the presence of relatively low concentrations of cholesterol leads to an increased association of PDC-109 to phospholipids, thus modulating the interaction of PDC-109 with membranes (Swamy, Marsh, Anbazhagan and Ramakrishnan, 2002).

PDC-109 exists in seminal plasma as a polydisperse multimeric, self-associated molecule whose aggregation state can be modulated by the ionic strength of the medium and ligand binding (Gasset *et al.*, 1997). The average size of the oligomers can be reduced by both o-phosphorylcholine interaction and by increasing the ionic strength of the solution, suggesting that both ionic and hydrophobic interactions are responsible for the aggregation tendency of PDC-109 monomers. On the other hand, unpublished results from our laboratory indicate that binding of heparins with increasing chain length (3-15 kDa) to PDC-109 induce the oligomerization of the bovine seminal plasma protein. The X ray crystal structure of PDC-109/o-phosphorylcholine complex clearly showed that specific recognition of the choline moiety by each Fn2 module of the protein is mediated by a cation-π interaction between the quaternary ammonium group and a core tryptophan, and several hydrogen bonds between the phosphate group of the ligand and the hydroxyl groups of a ring of exposed tyrosine residues (Wah, Fernández-Tornero, Sanz, Romero and Calvete, 2002). This mode of choline binding has been reported in pneumococcal surface virulence factors that mediate anchorage of the pathogen to the cell wall through their choline binding domains (Fernández-Tornero, López, García, Gimenez-Gallego and Romero, 2001; Hermoso, Lagartera, González, Stelter, García, Martínez-Ripoll, Gacía and Menéndez, 2005). Clearly, the choline-binding mechanism has evolved through evolutionary convergence in otherwise structurally unrelated protein scaffolds.

Compared with apo-PDC-109, a change in the conformation of the ß2-ß3 loop of the Fn2 domains upon phosphorylcholine binding leads to formation of PDC-109 dimers in which the four choline-binding sites are situated of the same protein face and at the corners of the PDC-109 dimer (Fig. 5). The relative orientations of the choline-binding pockets explain the biophysical results showing that PDC-109 interacts optimally with choline lipid-rich domains containing 12-15 lipid molecules/protein molecule. On the other hand, amino acids implicated in binding heparin form a large conformational patch of basic residues oriented towards the oposite face of the choline-binding surface of PDC-109 n-mer (Fig. 5). The non-contiguous cationic clusters are distant 20-30 Å from each other resembling the topology of glycosaminoglycan-binding sites of other proteins. This, in turn, explains why oligomeric, but not monomeric, PDC-109 possesses heparin-binding activity, and suggests a plausible mechanism for the docking of heparin-like glycosaminoglycans to clusters of PDC-109 molecules coated to the sperm surface during capacitation.

Figure 5. Proposed mechanism of bovine sperm capacitation by PDC-109. In this model, perturbation of the lipid order upon PDC-109 binding to phosphatidylcholine-lipids and the subsequent aggregation of membrane-bound PDC-109 by factors (e.g. heparin-like glycosaminoglycans) present in follicular or oviductal fluids leads to perturbation of the sperm plasma membrane and to the budding of cholesterol-enriched phospholipid-PDC-109 particles.

The combined biochemical and biophysical data provide further insights into our understanding of the process of membrane lipid modification by PDC-109, including the process of cholesterol efflux. Thus, although a full appreciation of the physiological consequences of the biophysical behaviour of PDC-109 summarised here clearly needs further experimentation carried out as close as possible to the *in vivo* situation, the available data suggest a possible mechanism for PDC-109-mediated bovine sperm capacitation by heparin. In this model, schematically depicted in Figure 5, PDC-109 molecules exist in seminal plasma, at physiological ionic strength and temperature, in a partially aggregated state. Binding of PDC-109 molecules to choline lipids of the sperm surface at ejaculation will drive the dissociation of protein oligomers which may partly integrate into the lipid bilayer. During residence in the cow genital tract, interaction of heparin-like glycosaminoglycans with these sperm surface-bound PDC-109-rich domains could promote the aggregation of PDC-109 molecules (and complexes of PDC-109 with other molecules, i.e. apoprotein AI, high-density lipoproteins) within the plane of the membrane, thereby perturbing the sperm plasma membrane and leading to the budding of cholesterol-enriched phospholipidic particles. This membrane perturbation induced by PDC-109 may underlie the initial and time-dependent lipoprotein particle efflux detected upon incubation of bull epididymal spermatozoa with PDC-109, rendering spermatozoa responsive to inducers of the acrosome reaction. Lowering the high cholesterol content in the membrane of ejaculated spermatozoa may increase the lipid bilayer fluidity and may facilitate the diffusion and interactions of receptors involved in transducing the sperm capacitated state across the membrane to the cell signalling machinery.

PDC-109, a fucose-binding protein?

Recent studies have suggested that PDC-109 binds L-fucose, the Lewis^a trisaccharide (α-L-Fuc[1,4]-ß-D-Gal[1,3]-D-GlcNAc), D-fructose, and D-mannose (Ignotz, Lo, Perez, Gwathmey and Suarez, 2001). Since bull sperm appears to bind to the oviductal epithelium through a Ca^{2+}-dependent lectin that recognises the Lewis^a structure (Suarez, Revah, Lo and Kölle, 1998), the authors concluded that PDC-109 could represent the fucose-binding protein responsible for enabling sperm to bind to oviductal epithelium, thus creating a sperm reservoir. To test this hypothesis, we have assessed the possible interaction of a large collection of mono- and oligosaccharides to isolated PDC-109 using both affinity chromatography and surface plasmon resonance. We have been unable to detect any interaction of the seminal plasma protein with sugars, including the fucosylated oligosaccharides Lewis^x and Lewis^a structures. Moreover, PDC-109 was crystallised in the presence of L-fucose but the crystal structure did not show L-fucose bound to the crystallised protein. Clearly, the putative fucose-binding activity of PDC-109 and its proposed functional relevance need a thorough re-evaluation.

Choline-binding proteins in pig and stallion

Heparin- and choline-binding proteins structurally similar to the BSP proteins have been reported in other ungulate species (e.g. pig, horse, and ram) (Calvete, Raida, Gentzel, Urbanke, Sanz and Töpfer-Petersen, 1997; Ekhlasi-Hundrieser, Schäfer, Kirchhoff, Hess, Bellair, Müller and Töpfer-Petersen, 2004; Bergeron *et al.*, 2005). The interaction of these Fn2-proteins with model membranes has been only investigated for the equine HSP-1/HSP-2 proteins (Greube, Müller, Töpfer-Petersen, Herrmann and Müller, 2004). The results indicated that the interaction of SP-1/2 with membranes is similar to that of PDC-109, i.e. SP-1/2 has a restricted specificity for the phosphocholine headgroup of phospholipids. Upon binding to lipid vesicles, the protein intercalates into the hydrophobic membrane core resulting in rigidification of the lipid phase and, at high preotein concentration, also in the perturbation of the membrane integrity. However, compared to PDC-109, the effect of the equine protein is less intense regarding the degree of protein-mediated immobilization of membrane lipids, suggesting that the equine proteins may exhibit conserved but also species-specific membrane-interacting properties and physiological consequences that would need to be assessed in a case-by-case manner.

Acknowledgements

Current work in the authors' laboratory is partly financed by grant BFU2004-01432 from the Ministerio de Educación y Ciencia, Madrid, Spain.

References

Anbazhagan, V. and Swamy, M.J. (2005). Thermodynamics of phosphorylcholine and lysophosphatidylcholine binding to the major protein of bovine seminal plasma, PDC-109. *FEBS Letters* **579**: 2933-2938

Assreuy, A.M., Calvete, J.J., Alencar, N.M.N., Cavada, B.S., Rocha-Filho, D., Melo, S., Cunha, F. and Ribeiro, R. (2002). Spermadhesin PSP-I/PSP-II heterodimer and its isolated subunits induced neutrophil migration into peritoneal cavity of rats. *Biology of Reproduction* **67**: 1796-1803

Assreuy, A.M.S., Alencar, N.M.N., Cavada, B.S., Rocha-Filho, D.R., Feitosa, R.F.G., Cunha, F.Q., Calvete, J.J. and Ribeiro, R.A. (2003). Porcine spermadhesin PSP-I/PSP-II stimulates macrophages to release a neutrophil chemotactic substance: modulation by mast cells. *Biology of Reproduction* **68**: 1836-1841

Austin, C.R. (1951). Observations on the penetration of the sperm into the mammalian egg. *Australian Journal of Scientific Research* **4**: 581-596

Bergeron, A., Villemure, M., Lazure, C. and Manjunath, P. (2005). Isolation and characterization of the major proteins of ram seminal plasma. *Molecular Reproduction and Development* **71**: 461-470

Calvete, J.J., Sanz, L., Dostàlovà, Z. and Töpfer-Petersen, E. (1993). Characterization of AWN-1 glycosylated isoforms help define the zona pellucida and serine proteinase inhibitor binding region on boar spermadhesins. *FEBS Letters* **334**: 37-40

Calvete, J.J., Solís, D., Sanz, L., Diaz-Mauriño, T., Schäfer, W., Mann, K., Töpfer-Petersen, E. (1993). Characterization of two glycosylated boar spermadhesins. *European Journal of Biochemistry* **218**: 719-725

Calvete, J.J., Nessau, S., Mann, K., Sanz, L., Sieme, H., Klug, E. and Töpfer-Petersen, E. (1994). Isolation and biochemical characterization of stallion seminal plasma proteins. *Reproduction in Domestic Animals* **29**: 411-426

Calvete, J.J., Raida, M., Sanz, L., Wempe, F., Scheit, K.H., Romero, A. and Töpfer-Petersen, E. (1994). Localization and structural characterization of an oligosaccharide O- linked to bovine PDC-109. Quantitation of the glycoprotein in seminal plasma and on the surface of ejaculated and capacitated spermatozoa. *FEBS Letters* **350**: 203-206

Calvete, J.J., Sanz, L., Dostàlovà, Z and Töpfer-Petersen, E. (1995a). Spermadhesins: sperm-coating proteins involved in capacitation and zona pellucida binding. *Fertilität* **11**: 35-40

Calvete, J.J., Mann, K., Schäfer, W., Raida, M., Sanz, L. and Töpfer-Petersen, E. (1995b). Boar spermadhesin PSP-II: localization of posttranlational modifications, heterodimer formation with PSP-I glycoforms and effect of dimerization on the ligand-binding capabilities of the subunits. *FEBS Letters* **365**: 179-182

Calvete, J.J., Mann, K., Schäfer, W., Sanz, L., Reinert, M., Nessau, S., Raida, M. and Töpfer-Petersen, E. (1995). Amino acid sequence of HSP-1, a major protein of stallion seminal plasma: effect of glycosylation on its heparin-and gelatin-binding capabilities. *Biochemical Journal* **310**: 615-622

Calvete, J.J., Dostálová, Z., Sanz, L., Adermann, K., Thole, H.H. and Töpfer-Petersen, E. (1996). Mapping the heparin-binding domain of boar spermadhesins. *FEBS Letters* **379**: 207-211

Calvete, J.J., Mann, K., Sanz, L., Raida, M. and Töpfer-Petersen, E. (1996). The primary structure of BSP-30K, a major lipid-, gelatin-, and heparin-binding glycoprotein of bovine seminal plasma. *FEBS Letters* **399**: 147-152

Calvete, J.J., Carrera, E., Sanz, L. and Töpfer-Petersen, E. (1996). Boar spermadhesins AQN-1 and AQN-3: oligosaccharide- and zona pellucida-binding characteristics. *Biological Chemistry* **377**: 521-527.

Calvete, J.J., Ensslin, M., Mburu, J., Iborra, A., Martínez, P., Adermann, K., Waberski, D., Sanz, L., Töpfer-Petersen, E., Weitze, K.F., Einarsson, S. and Rodríguez-Martínez, H. (1997). Monoclonal antibodies against boar sperm zona pellucida-binding protein AWN-1. Characterization of a continuous antigenic determinant and immunolocalization of AWN epitopes in inseminated sows. *Biology of Reproduction* **57**: 735-742

Calvete, J.J., Raida, M., Gentzel, M., Urbanke, C., Sanz,

L. and Töpfer-Petersen, E. (1997). Isolation and characterization of heparin- and phosphorylcholine-binding proteins of boar and stallion seminal plasma. Primary structure of porcine pB1. *FEBS Letters* **407**: 201-206

Calvete, J.J., Campanero-Rhodes, M.A., Raida, M. and Sanz, L. (1999). Characterisation of the conformational and quaternary structure-dependent heparin-binding region of bovine seminal plasma protein PDC-109. *FEBS Letters* **444**: 260-264

Campanero-Rhodes, M.A., Menéndez, M., Sáiz, J.L., Sanz, L., Calvete, J.J. and Solís, D. (2005). Analysis of the stability of the spermadhesin PSP-I/PSP-II heterodimer. Effects of Zn^{2+} and acidic pH. *FEBS Journal.* **272**: 5663-5670

Campanero-Rhodes, M.A., Menéndez, M., Sáiz, J.L., Sanz, L., Calvete, J.J. and Solís, D. (2006). Zinc ions induce the unfolding and self-association ofboar spermadhesin PSP-I, a protein with a single CUB domain architecture, and promote its binding to heparin. *Biochemistry* **45**: 8227-8235

Carroll, R.L. (1988) *Vertebrate Paleontology and Evolution.* Freeman, New York

Centurión, F., Vázquez, J.M., Calvete, J.J., Roca, J., Sanz, L., Parrilla, I., García, E.M. and Martínez, E.A. (2003). Influence of porcine spermadhesins on the susceptibility of boar spermatozoa to high dilution. *Biology of Reproduction* **69**: 640-646

Chang, M.C. (1951). Fertilizing capacity of spermatozoa deposited into the fallopian tubes. *Nature* **168**: 697-698

Dean, R.T., Fu, S., Stocker, R. and Davies, M.J. (1997). Biochemistry and pathology of radical-mediated protein oxidation. *Biochemical Journal* **324**: 1-18

Desnoyers, I. and Manjunath, P. (1992). Major proteins of bovine seminal plasma exhibit novel interactions with phospholipids. *Journal of Biological Chemistry* **267**: 10149-10155

Desnoyers, I. and Manjunath, P. (1993). Interaction of novel class of phospholipid-binding proteins of bovine seminal fluid with different affinity matrices. *Archives of Biochemistry and Biophysics* **305**: 341-349

Dostàlovà, Z., Calvete, J.J., Sanz, L. and Töpfer-Petersen, E. (1994). Quantitation of boar spermadhesins in accessory sex gland fluids and on the surface of epididymal, ejaculated and capacitated spermatozoa. *Biochimica et Biophysica Acta* **1200**: 48-54

Dostàlovà, Z., Calvete, J.J., Sanz, L. and Töpfer-Petersen, E. (1995). Boar spermadhesin AWN-1: oligosaccharide and zona pellucida binding characteristics. *European Journal of Biochemistry* **230**: 329-336

Dostàlovà, Z., Calvete, J.J. and Töpfer-Petersen, E. (1995a). Interaction of non-aggregated boar AWN-1 and AQN-3 with phospholipid matrices. A model for coating of spermadhesins to the sperm surface. *Biological Chemistry Hoppe-Seyler* **376**: 237-242

Einspanier, R., Einspanier, A., Wempe, F. and Scheit, K.H. (1991). Characterization of a new bioactive protein from bovine seminal fluid. *Biochemical and Biophysical Research Communications* **179**: 1006-1010

Einspanier, R., Krause, I., Calvete, J.J., Töpfer-Petersen, E., Klostermeyer, H. and Karg, H. (1994). Bovine seminal plasma aSFP: localization of disulfide bridges and detection of three different isoelectric forms. *FEBS Letters* **344**: 61-64

Ekhlasi-Hundrieser, M., Sinowatz, F., Greiser de Wilke, I., Waberski, D. and Töpfer-Petersen, E. (2002). Expression of spermadhesin genes in porcine male and female reproductive tracts. *Molecular Reproduction and Development* **61**:32-41

Ekhlasi-Hundrieser, M., Schäfer, B., Kirchhoff, C., Hess, O., Bellair, S., Müller,P. and Töpfer-Petersen, E. (2004). Structural and molecular characterization of equine sperm-binding fibronectin-II module proteins. *Molecular Reproduction and Development* **70**: 45-57

Esch, F.S., Ling, N.C., Bohlen, P., Ying, S.Y. and Guillemin, R. (1983). Primary structure of PDC-109, a major protein constituent of bovine seminal plasma. *Biochemical and Biophysical Research Communications* **113**: 861-867

Fernández-Tornero, C., López, R., García, E., Gimenez-Gallego, G. and Romero, A. (2001). A novel solenoid fold in the cell wall anchoring domain of the pneumococcal virulence factor LytA. *Nature Structural Biology* **8**: 1020-1024

García, E.M., Vázquez, J.M., Calvete, J.J., Sanz, L., Caballero, I., Parrilla, I., Gil, M.A., Roca, J. and Martínez, E.A. (2006). Dissecting the protective effect of the seminal plasma spermadhesin PSP-I/PSP-II on boar sperm functionality. *Journal of Andrology* **27**: 434-443

Gasset, M., Saiz, J.L., Laynez, J., Sanz, L., Gentzel, M., Töpfer-Petersen, E. and Calvete, J.J. (1997). Conformational features and thermal stability of bovine seminal plasma protein PDC-109 oligomers and phosphorylcholine-bound complexes. *European Journal of Biochemistry* **250**: 735-744

Gasset, M., Magdaleno, L. and Calvete, J.J. (2000). Biophysical study of the perturbation effect of model membrane structure caused by seminal plasma protein PDC-109. *Archives of Biochemistry and Biophysics* **374**: 241-247

Gerwig, G.J., Calvete, J.J., Töpfer-Petersen, E. and Vliegenthart, J.F.G. (1996). The structure of the O-linked carbohydrate chain of bovine seminal plasma protein PDC-109 revised by 1H-NMR spectroscopy. A correction. *FEBS Letters* **387**: 99-100

Greube, A., Müller, K., Töpfer-Petersen, E. Herrmann, A. and Müller, P. (2001). Influence of the bovine seminal plasma protein PDC-109 on the physical state of membranes. *Biochemistry* **40**: 8326-8334

Greube, A., Müller, K., Töpfer-Petersen, E. Herrmann, A. and Müller, P. (2004). Interaction of fibronectin type II proteins with membranes: the stallion seminal plasma protein SP-1/2. *Biochemistry* **43**: 464-472

Haase, B., Schlötterer, C., Ekhlasi-Hundrieser, M., Kuiper, H., Distl, O., Töpfer-Petersen, E. and Leeb, T. (2005). Evolution of the spermadhesin gene family. *Gene* **352**: 20-29

Hermoso, J.A., Lagartera, L., González, A., Stelter, M., García, P., Martínez-Ripoll, M., García J.L. and Menéndez, M. (2005). Insights into pneumococcal pathogenesis from the crystal structure of the modular teichoic acid phosphorylcholine esterase Pce. *Nature Structural and Molecular Biology* **12**: 533-538

Hoshiba, H. and Sinowatz, F. (1998). Immnuno-histochemical localization of the spermadhesin AWN-1 in the equine male genital tract. *Anatomy, Histology and Embryology* **27**: 351-353

Ignotz, G.G., Lo, M.C., Perez, C.L., Gwathemey, T.M. and Suarez, S.S. (2001). Characterization of a fucose-binding protein from bull sperm and seminal plasma that may be responsible for formation of the oviductal sperm reservoir. *Biology of Reproduction* **64**: 1806-1811

Iwamoto, T., Hiroaki, H., Furuichi, Y., Wada, K., Satoh, M., Satoh, M., Osada, T. and Gagnon, C. (1995). Cloning of boar SPMI gene which is expressed specifically in seminal vesicle and codes for a sperm motility inhibitor protein. *FEBS Letters* **368**: 420-424.

Killian, G.J., Chapman, D.A. and Rogowski, L.A. (1993). Fertility-associated proteins in Holstein bull seminal plasma. *Biology of Reproduction* **49**: 1202-1207

Lane, M-E., Thérien, I., Moreau, R. and Manjunath, P. (1999). Heparin and high-density lipoprotein mediate bovine sperm capacitation by different mechanisms. *Biology of. Reproduction* **60**: 169-175

Leshin, L.S., Raj, S.M., Smith, C.K., Kwok, S.C., Kraeling, R.R. and Li, W.I. (1998). Immunostimulatory effects of pig seminal proteins on pig lymphocytes. *Journal of Reproduction and Fertility* **114**: 77-84

Manjunath, P. Chandonnet, L., Leblond, E. and Desnoyers, L. (1994). Major proteins of bovine seminal plasma bind to spermatozoa. *Biology of Reproduction* **50**: 27-37

Manjunath, P. and Thérien, I. (2002). Role of seminal plasma phospholipid-binding proteins in sperm membrane lipid modification that occurs during capacitation. *Journal of Reproductive Immunology* **53**: 109-119

Mann, T. and Lutwak-Mann, C. (1981). *Male Reproductive Function and Semen*. Springer-Verlag, Berlin-Heidelberg-New York

Maxwell, W.M.C. and Johnson, L.A. (1999). Physiology of spermatozoa at high dilution rates: the influence of seminal plasma. *Theriogenology* **52**: 1353-1362

Moreau, R. and Manjunath, P. (1999). Characterization of lipid efflux particles generated by seminal phospholipid-binding proteins. *Biochimica et Biophysica Acta* **1438**: 175-184

Müller, P., Erlemann, K-R., Müller, K., Calvete, J.J., Töpfer-Petersen, E., Marienfeld, K. and Herrmann, A. (1998). Biophysical characterization of the interaction of bovine seminal plasma protein PDC-109 with phospholipid vesicles. *European Biophysical Journal* **27**: 33-41

Müller, P., Greube, A., Töpfer-Petersen, E. and Herrmann, A. (2002). Influence of the bovine seminal plasma protein PDC-109 on cholesterol in the presence of phospholipids. *European Biophysical Journal* **31**: 438-447

Nimtz, M., Grabenhorst, E., Conradt, H.S., Sanz, L. and Calvete, J.J. (1999). Structural characterization of the oligosaccharide chains of native and crystallized boar seminal plasma spermadhesin PSP-I and PSP-II glycoforms. *European Journal of Biochemistry* **265**: 703-718

Ramakdrishnan, M., Anbazhagan, V., Pratap, T.V., Marsh, D. and Swamy, M.J. (2002). Membrane insertion and lipid-protein interacttions of bovine seminal plasma protein PDC-109 investigated by spin-label electron spin resonance spectroscopy. *European Journal of Biophysics* **81**: 438-447

Reinert, M., Calvete, J.J., Sanz, L., Mann, K. and Töpfer-Petersen, E. (1996). Primary structure of stallion seminal plasma protein HSP-7, a zona pellucida-binding protein of the spermadhesin family. *European Journal of Biochemistry* **242**: 636-640

Reinert, M., Calvete, J.J., Sanz, L., and Töpfer-Petersen, E. (1997). Immunohistochemical localization in the stallion genital tract, and topography on spermatozoa of seminal plasma protein SSP-7, a member of the spermadhesin protein family. *Andrologia* **29**: 179-186

Rodríguez-Martínez, H., Iborra, A., Martínez, P. and Calvete, J.J. (1999). Immunoelectronmicroscopic imaging of spermadhesin AWN epitopes on boar spermatozoa bound *in vivo* to the zona pellucida. *Reproduction, Fertility and Development* **10**: 491-497

Rodríguez-Martínez, H., Saravia, F., Wallgren, M., Tienthai, P., Johannisson, A., Vázquez, J.M., Martínez, E., Roca, J., Sanz, L. and Calvete, J.J. (2005). Boar spermatozoa in the oviduct. *Theriogenology* **63**: 514-535

Romão, M.J., Kölln, I., Dias, J.M., Carvalho, A.L., Romero, A., Varela, P.F., Sanz, L., Töpfer-Petersen, E. and Calvete, J.J. (1997). Crystal structure of acidic seminal fluid protein (aSFP) at 1.9 Å resolution: a bovine polypeptide of the spermadhesin family. *Journal of Molecular Biology* **274**: 650-660

Romero, A., Romão, M.J., Varela, P.F., Kölln, I., Dias, J.M., Carvalho, A.L., Sanz, L., Töpfer-Petersen, E. and Calvete, J.J. (1997). The crystal structures of two spermadhesins reveal the CUB domain fold. *Nature Structural Biology* **4**: 783-788

Sanz, L., Calvete, J.J., Jonàkovà, V. and Töpfer-Petersen, E. (1992). Boar spermadhesins AQN-1 and AWN are sperm-associated acrosin inhibitor acceptor proteins. *FEBS Letters* **300**: 63-66

Sanz, L., Calvete, J.J., Mann, K., Gabius, H-J. and Töpfer-Petersen, E. (1993). Isolation and biochemical of heparin-binding proteins from boar seminal plasma: a dual role for spermadhesins in fertilization. *Molecular Reproduction and Development* **35**: 37-43

Scheit, K.H., Kemme, M., Aumüller, G., Seitz, J., Hagendorff, G. and Zimmer, M. (1988). The major protein of bull seminal plasma: biosynthesis and biological function. *Bioscience Reports* **8**: 589-608

Schöneck, C., Einspanier, R., Schallenberger, E. and Schams, D. (1993). Effects of the bovine seminal plasma protein aSFP: protection of spermatozoa and rapid uptake by the female mucosa. *Journal of Reproduction and Fertility (Abstract Series)* **12**: 21

Shivaji, S., Scheit, K-H. and Bhargava, P.M. (1990). *Proteins of Seminal Plasma*. Wiley and Sons, New York

Solís, D., Romero, A., Jiménez, M., Díaz-Mauriño, T. and Calvete, J.J. (1998). Binding of mannose-6-phosphate and heparin by boar seminal plasma PSP-II, a member of the spermadhesin protein family. *FEBS Letters* **431**: 273-278

Suarez, S.S., Revah, I., Lo, M. and Kölle, S. (1998). Bull sperm binding to oviductal epithelium is mediated by a Ca^{2+}-dependent lectin on sperm that recognizes Lewis-a trisaccharide. *Biology of Reproduction* **59**: 39-44

Swamy, M.J., Marsh, D., Anbazhagan, V. and Ramakrishnan, M. (2002). Effect of cholesterol on the interaction of seminal plasma protein, PDC-109 with phosphatidylcholine membranes. *FEBS Letters* **528**: 230-234

Swamy, M.J. (2004). Interaction of bovine seminal plasma proteins with model membranes and sperm plasma membranes. *Current Science* **87**: 203-211

Tedeschi, G., Oungre, E., Mortarino, M., Negri, A., Maffeo, G. and Ronchi, S. (2000). Purification and primary structure of a new bovine spermadhesin. *European Journal of Biochemistry* **267**: 6175-6179

Thérien, I., Soubeyrand, S. and Manjunath, P. (1997). Major proteins of bovine seminal plasma modulate sperm capacitation by high-density lipoprotein. *Biology of Reproduction* 57:1080-1088

Thérien, I., Moreau, R. and Manjunath, P. (1998). Major proteins of bovine seminal plasma and high-density lipoprotein induce cholesterol efflux from epididymal sperm. *Biology of Reproduction* **59**: 768-776

Thomas, C.J., Anbazhagan, V., Ramakrishnan, M., Sultan, N., Surolia, I. and Swamy, M.J. (2003). Mechanism of membrane binding by the bovine seminal plasma protein, PDC-109: a surface plasmon resonance study. *Biophysical Journal* **84**: 3037-3044

Varela, P.F., Romero, A., Sanz, L., Romão, M.J., Töpfer-Petersen, E. and Calvete, J.J. (1997). The 2.4 Å resolution crystal structure of boar seminal plasma PSP-I/PSP-II: a zona pellucida-binding glycoprotein heterodimer of the spermadhesin family built by a CUB domain architecture. *Journal of Molecular Biology* **274**: 635-649

Visconti, P.E., Galantino-Homer, H., Moore, G.D., Bailey, J.L., Ning, X., Fornes, M. and Kopf, G.S. (1998). The molecular basis of sperm capacitation. *Journal of Andrology* **19**: 242-248

Visconti, P.E., Ning, X., Fornés, M.W., Álvarez, J.G., Stein, P., Connors, S.A. and Kopf, G.S. (1999). Cholesterol efflux-mediated signal transduction in mammalian sperm: cholesterol release signals an increase in protein tyrosine phosphorylation during mouse capacitation. *Developmental Biology* **214**: 429-443

Wah, D.A., Fernández-Tornero, C., Sanz, L., Romero, A. and Calvete, J.J. (2002). Sperm coating mechanism from the 1.8 Å crystal structure of PDC-109-phosphorylcholine complex. *Structure* **10**: 505-514

Watson, P.F. (1981). The effects of cold shock on sperm cell membranes. In: *Effects of Low Temperature on Biological Membranes*. Edited by Morris, G.J. and Clarke, A., Academic Press, London, pp. 189-218

Wempe, F., Einspanier, R. and Scheit, K.H. (1992). Characterization by cDNA cloning of the mRNA of a new growth factor from bovine seminal plasma: acidic seminal fluid protein. *Biochemical and Biophysical Research Communications* **183**: 232-237

Yanagimachi, R. (1994). Mammalian fertilization. In: *The Physiology of Reproduction*, 2nd edition. Edited by Knobil, E. and Neill, J.D. Raven Press, New York, pp. 189-317

Yang, W.C., Kwok, S.C.M., Leshin, S., Bollo, E. and Li, W.I. (1998). Purified porcine seminal plasma protein enhances in vitro immune activities of porcine peripheral lymphocytes. *Biology of Reproduction* **59**: 202-207

Spermatology. SRF Vol. 65. ERS Roldan and M Gomendio (eds) Nottingham University Press, Nottingham

Seminal plasma proteins: functions and interaction with protective agents during semen preservation

Puttaswamy Manjunath[1,2,3], Annick Bergeron[3], Jasmine Lefebvre[2,3] and Jinjiang Fan[3]

Departments of [1]Medicine and [2]Biochemistry, University of Montreal and [3]Guy-Bernier Research Center, Maisonneuve-Rosemont Hospital, Montreal, Quebec, Canada, H1T 2M4

Seminal plasma contains factors that are beneficial and/or detrimental to sperm function and/or storage. However, the nature and characteristics of these factors are not well understood. The major protein fraction (50-70%) of **b**ovine **s**eminal **p**lasma is represented by a family of phospholipid-binding proteins collectively called BSP proteins. The BSP protein signature is characterised by two tandemly repeated fibronectin type 2 (Fn2) domains. It is now well established that BSP proteins and their relatives represent a new emerging superfamily of proteins in mammals. They bind to sperm membrane choline phospholipids at ejaculation. They also bind to capacitation factors, namely, high-density lipoproteins and glycosaminoglycans and promote sperm capacitation induced by these molecules, indicating their beneficial role in sperm function and fertility. In contrast, BSP proteins also induce changes in the sperm plasma membrane by stimulating cholesterol and phospholipid efflux. Thus, the continuous exposure of sperm to seminal plasma that contains BSP proteins is detrimental to the sperm membrane, which may render the membrane very sensitive to sperm storage in the liquid or frozen states. Interestingly, BSP proteins specifically bind low-density lipoproteins present in egg yolk, a compound commonly used in semen extenders. This interaction appears to abolish the detrimental effect of BSP proteins on the sperm membrane. Therefore, BSP proteins in seminal plasma act like a double-edged sword, being both beneficial and detrimental to sperm.

Introduction

Seminal plasma is a complex mixture of secretions from the testes, epididymides and accessory sex glands (seminal vesicles, ampulla, prostate, bulbourethral glands). Despite its physiological significance as the carrier of spermatozoa to the female reproductive tract, the biochemical characteristics and physiological roles of the various seminal plasma proteins are poorly understood. While some seminal plasma proteins appear to be beneficial to sperm functions such as motility, capacitation, acrosome reaction, viability and/or sperm-egg interaction, others seem to be detrimental, particularly to sperm preservation including cryopreservation (for a review see Manjunath and Therien, 2002; Bergeron and Manjunath, 2006). In this review, we describe

Corresponding author E-mail: puttaswamy.manjunath@umontreal.ca the characteristics and functions of a family of phospholipid-binding proteins widely distributed in mammalian seminal plasma. In addition, we describe the interaction of these proteins with egg yolk extender components and discuss the implication of this interaction in the context of sperm storage in the liquid and frozen states. We also explain why these lipid-binding seminal plasma proteins are beneficial to sperm function, yet detrimental to sperm preservation.

BSP proteins

In bovine, the major protein fraction of seminal plasma (30-50 mg/ml) constitutes a family of closely related proteins named BSP-A1, BSP-A2, BSP-A3 and BSP-30-kDa (BSP proteins) (Manjunath, 1984; Manjunath and Sairam, 1987). These proteins are the secretory products of the seminal vesicles. The biochemical properties and structure of these proteins have been studied in detail (Manjunath, 1984; Manjunath and Sairam, 1987; Manjunath *et al.*, 1987; Seidah, Manjunath, Rochemont, Sairam and Chretien, 1987; Manjunath, Baillargeon, Marcel, Seidah, Chretien and Chapdelaine, 1988; Leblond, Desnoyers and Manjunath, 1993; Desnoyers, Therien and Manjunath, 1994). They are small acidic proteins with apparent molecular weights of 15-16 kDa (BSP-A1, -A2 and -A3) and 28-30 kDa (BSP-30-kDa), which are all glycoproteins with the exception of BSP-A3. BSP-A1 and BSP-A2 (also called PDC-109) originate from the same gene but differ in the degree of glycosylation (glycoforms). All of these proteins contain two homologous type II domains similar to those found in the gelatin-binding domains of fibronectin (called Fn2 domains) and an amino-terminal extension that is variable between each protein (Fig. 1) (reviewed in Manjunath and Therien, 2002; Fan, Lefebvre and Manjunath, 2006). The crystal structure of BSP-A1/-A2 (PDC-109) has been determined (Wah, Fernandez-Tornero, Sanz, Romero and Calvete, 2002), allowing us to create homology models of BSP-A3 and BSP-30kDa (Fig. 2). Homologues of BSP proteins have been isolated and characterised in seminal plasma or seminal vesicle secretions from numerous other mammals, namely boar and stallion (Menard, Nauc, Lazure, Vaillancourt and Manjunath, 2003; Calvete, Mann, Schafer, Sanz, Reinert, Nessau, Raida and Topfer-Petersen, 1995; Calvete, Raida, Gentzel, Urbanke, Sanz and Topfer-Petersen, 1997; Sanz, Calvete, Mann, Gabius and Topfer-Petersen, 1993), goat (Villemure, Lazure and Manjunath, 2003), bison (Boisvert, Bergeron, Lazure and Manjunath, 2004), ram (Bergeron, Villemure, Lazure and Manjunath, 2005) and water buffalo (unpublished). All of these species contain 3-4 molecular forms, except boar, which contains one form (Table 1). BSP-like antigens are also present in rat, mouse and hamster seminal vesicle fluid as well as in human seminal plasma (Leblond *et al.*, 1993). The concentration of BSP homologues in seminal plasma varies from species to species (Table 2). At the genetic level, the cloning and sequencing of the cDNA corresponding to BSP proteins has been reported (Salois, Menard, Paquette and Manjunath, 1999; Kemme and Scheit, 1988). In a recent study, three new BSP related genes (*BSPH4, BSPH5* and *BSPH6*) were discovered in bovine (Fan *et al.*, 2006). In addition, an exhaustive genomic search indicated the presence of BSP-related DNA sequences in human, rat, mouse, chimpanzee, dog, horse and rabbit (Fan *et al.*, 2006). Molecular evolutionary analysis revealed that all BSP-related sequences could be grouped into three sub-families: BSPH4, which includes BSP-A1/-A2, BSP-A3 and BSP-30 kDa and is expressed in seminal vesicles, BSPH5 and BSPH6, which are expressed in the epididymis and testis (Fig. 3). It is becoming increasingly evident that BSP proteins and their homologous proteins represent a new emerging superfamily in mammals.

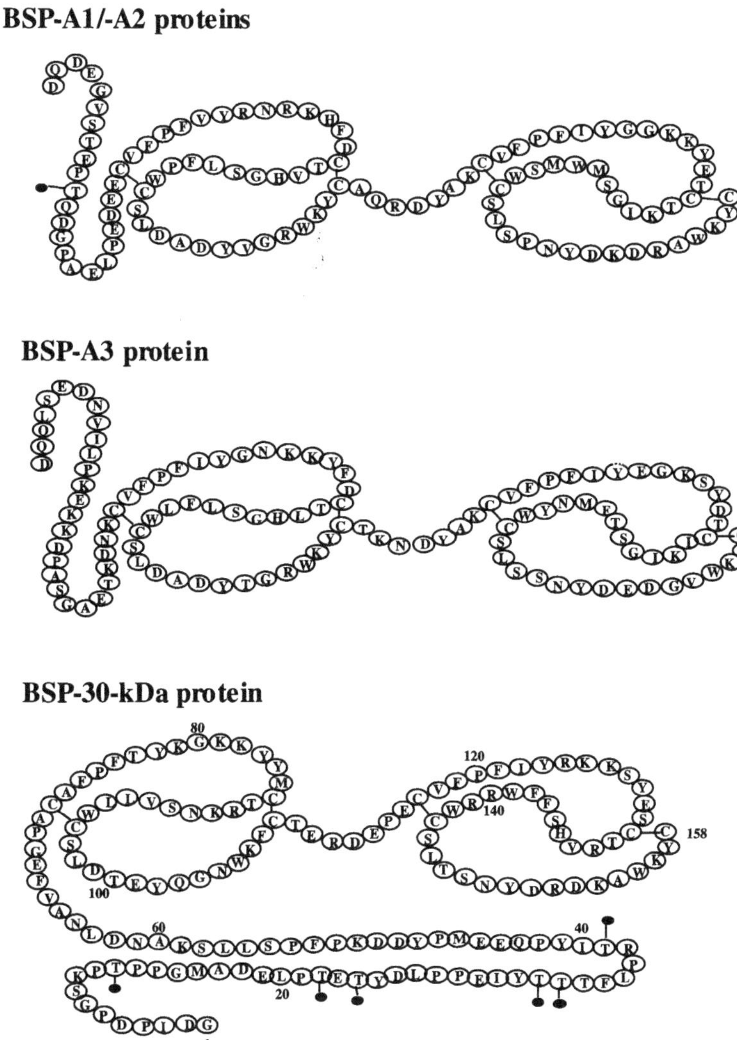

Figure 1. Primary structure of BSP proteins. The carbohydrate moieties are indicated by the symbol (●). Modified from Manjunath and Therien (2002).

BSP proteins play a key role in sperm capacitation

Studies aimed at elucidating the biological role of BSP proteins revealed that they are secreted by seminal vesicles (Manjunath *et al.*, 1987) and bind to the sperm surface at ejaculation (Manjunath, Chandonnet, Leblond and Desnoyers, 1994). Using various biochemical techniques, it was established that the binding sites for BSP proteins on the sperm membrane are choline phospholipids (Desnoyers and Manjunath, 1992), specifically phosphatidylcholine, phosphatidylcholine plasmalogen and sphingomyelin. Such choline phospholipids account for >60% of

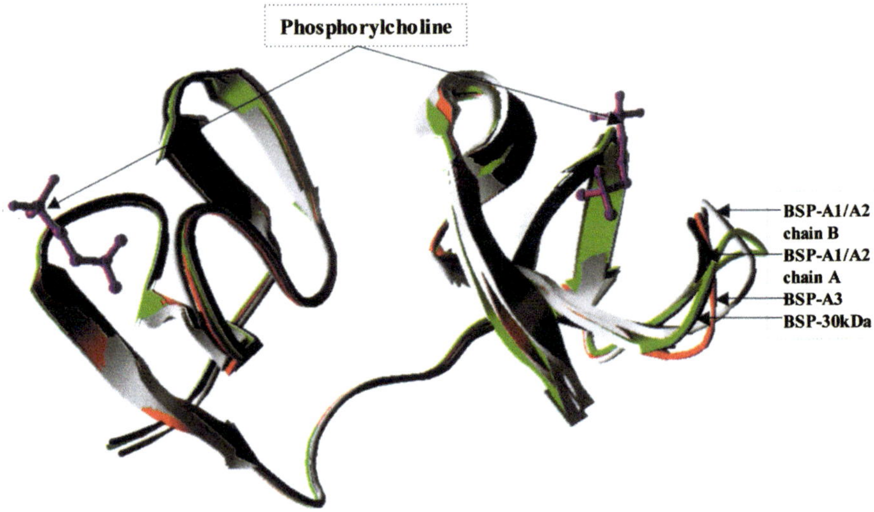

Figure 2. Ribbon representation of homology models of BSP proteins. The potential ligand, phosphorylcholine, is represented in ball and stick structure. The figure was prepared with Swiss-PbdViewer 3.7b2 (Guex and Peitsch, 1997) and MOLMOL 2K.2 (Koradi, Billeter and Wuthrich, 1996). The BSP-A1/-A2 (PDC-109) template is from the Research Collaboratory for Structural Bioinformatics Protein Data Bank (1H8P), in which chains A and B correspond to 2 monomers, because this protein forms dimers when associated with phosphorylcholine.

Table 1. BSP protein homologues isolated from seminal plasma of various species.

Species	BSP protein homologs
Bull	BSP-A1, BSP-A2, BSP-A3, BSP-30kDa
Boar	pB1
Stallion	HSP-1, HSP-2, HSP-12kDa
Goat	GSP-14kDa, GSP-15kDa, GSP-20kDa, GSP-22kDa
Ram	RSP-15kDa, RSP-16kDa, RSP-22kDa, RSP-24kDa
Bison*	BiSV-16kDa, BiSV-17kDa, BiSV-18kDa, BiSV-28kDa

*Isolated from seminal vesicle secretions.
From Bergeron and Manjunath (2006).

Table 2. Concentration of BSP protein homologues in the seminal plasma of different mammals.

	BSP proteins/homologues in seminal plasma (mg/ml)	(% of total proteins)
Bull	39.1	65
Bison*	3.0	25
Ram	15.0	30
Goat	5.6	20
Stallion	2.0	20
Boar	0.3	1.1

*BSP proteins in seminal vesicle secretions.
From Bergeron and Manjunath (2006).

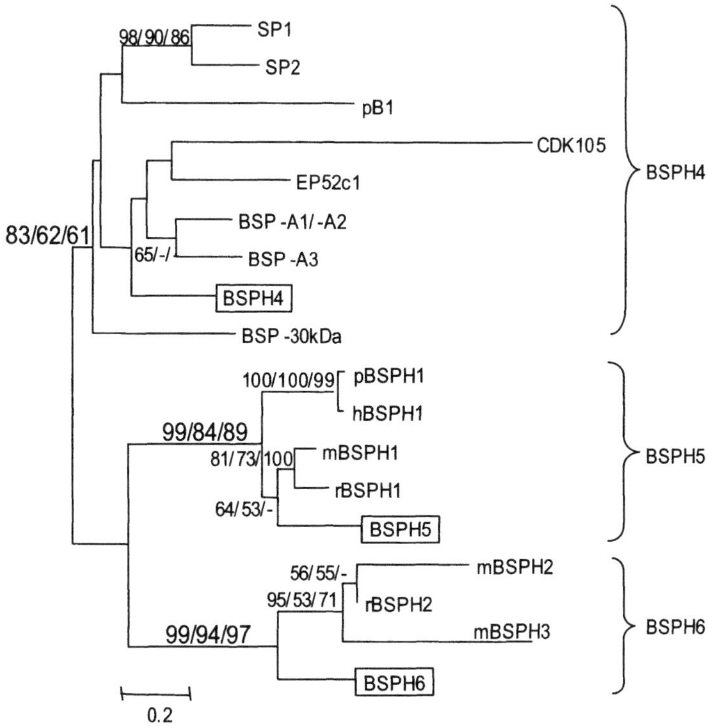

Figure 3. Neighbour-joining trees of two Fn2 domains of BSP-related proteins. BSP proteins and their relatives are separated into three subfamilies: BSPH4, BSPH5 and BSPH6. Bootstrap values from 1000 replicates for neighbour-joining (1st number), 1000 replicates for maximum parsimony (2nd number) and 100 replicates for maximum likelihood (3rd number) are indicated at the nodes and were used to assess the robustness of the trees. Only bootstrap values larger than 50% are shown and the bold numbers indicate the three main branches. Genetic distance is indicated as the number of substitutions per amino acid site. BSP protein, bovine seminal plasma protein; BSPH, BSP protein homologue; SP1/2, equine seminal plasma protein 1/2; pB1, boar seminal plasma protein pB1; CDK105, canine gene; EP52c1, epididymal protein 52 from rabbit. The small letter preceding each abbreviation represents the name of the species (h, *Homo sapiens*; p, *Pan troglodytes*; m, *Mus musculus*; r, *Rattus norvegicus*) from which the orthologous protein originates. Modified from Fan *et al.* (2006).

the total phospholipids of the sperm membrane (Parks, Arion and Foote, 1987). Further studies indicated that BSP proteins bind to high-density lipoproteins (HDL) (Manjunath *et al.*, 1988; Manjunath, Marcel, Uma, Seidah, Chretien and Chapdelaine, 1989; Therien, Soubeyrand and Manjunath, 1997; Therien, Bousquet and Manjunath, 2001), and glycosaminoglycans (GAGs) such as heparin (Chandonnet, Roberts, Chapdelaine and Manjunath, 1990), heparan sulfate and chondroitin sulfate (Therien, Bergeron, Bousquet and Manjunath, 2005). HDL and GAGs are components of follicular and oviductal fluids and are physiological inducers of sperm capacitation. Furthermore, BSP proteins potentiate capacitation induced by follicular fluid HDL and GAGs (Therien *et al.*, 1997, Therien *et al.*, 2001). It was shown that epididymal sperm exposed to BSP proteins followed by incubation with GAGs or HDL exhibit a higher incidence of capacitation compared to GAGs or HDL alone. In addition, BSP proteins mediate cholesterol efflux from sperm, which is also accompanied by choline phospholipid efflux (Therien, Moreau and Manjunath, 1998; Therien, Moreau and Manjunath, 1999). The removal of cholesterol from the sperm membrane is deemed to be essential during capacitation.

The studies described above led us to propose that BSP proteins participate in the sperm membrane lipid modification events that occur during capacitation in the female reproductive tract (Manjunath and Therien, 2002; Therien *et al.*, 2001, Therien *et al.*, 2005). Two distinct mechanisms, both mediated by BSP proteins but one induced by HDL and the other by GAGs, were deduced (Fig. 4). In brief, at ejaculation, sperm are mixed with accessory gland secretions containing BSP proteins, contributed by the seminal vesicles. During this brief exposure, BSP proteins remove some cholesterol (first cholesterol efflux) accompanied by the release of some phospholipids. This lipid efflux may slightly destabilise the sperm membrane (priming). At the same time, BSP proteins coat the sperm surface via their interaction with the membrane choline phospholipids. Within the next 10-20 min, sperm travel through the cervical mucus into the uterus, leaving behind most of the seminal plasma. During this time, bound BSP proteins may prevent the free movement of phospholipids, thereby stabilising the sperm membrane (arrested state). BSP protein-coated sperm then travel through the female genital tract and reach the oviduct, the site of fertilisation, where they encounter HDL and/or GAGs. With *HDL-induced capacitation*, oviductal- or follicular fluid-HDL remove BSP proteins from the sperm membrane. As a result, sperm membrane lipids are free to move. In addition, HDL may induce a second efflux of cholesterol. Since cholesterol is recognised to have a stabilising effect on membranes (Yeagle, 1985), its efflux would be expected to provoke further reorganization or destabilization of the membrane and trigger some unknown signal transduction pathways. Such events could regulate the surface expression of sperm zona pellucida receptors, and adhesion to the zona pellucida could trigger the acrosome reaction. With *GAG-induced capacitation*, sperm-bound BSP proteins may play a role, via their interaction with GAGs, in the uptake of Ca^{2+}, intracellular alkalinization and protein tyrosine phosphorylation (Therien *et al.*, 2001; Therien *et al.*, 2005). The role of tyrosine phosphorylation in capacitation in various species (human, bovine, murine, porcine) has been investigated in detail (Galantino-Homer, Visconti and Kopf, 1997; Leclerc, de Lamirande and Gagnon, 1996; Aitken, Harkiss, Knox, Paterson and Irvine, 1998; Tardif, Dube, Chevalier and Bailey, 2001; Osheroff, Visconti, Valenzuela, Travis, Alvarez and Kopf, 1999; de Lamirande, Leclerc and Gagnon, 1997; Visconti, Westbrook, Chertihin, Demarco, Sleight and Diekman, 2002; Naz and Rajesh, 2004; Baldi, Luconi, Bonaccorsi, Muratori and Forti, 2000). HDL-induced capacitation, in contrast to GAG-induced capacitation, is not associated with an increase in tyrosine phosphorylation (Lane, Therien, Moreau and Manjunath, 1999).

The first functional role identified for BSP proteins is that they promote the capacitation of bull sperm, which involves a complex series of events and is schematised in Figure 4 (Therien, Bleau and Manjunath, 1995; Therien *et al.*, 1997, 1998, 2005; Manjunath and Therien, 2002). Recently, two additional functions were proposed for BSP proteins. The group of Yu *et al.* (Yu, Zhao, Zhao, Chen, Liu, Zhang, Fu, Zong, Yu and Guan, 2003) demonstrated that, *in vitro*, the activity of protein kinase C (PKC) and of tyrosine protein kinase (TPK) are inhibited by PDC-109 (BSP-A1/-A2) and proposed that the inhibition of PKC may serve to prevent the premature acrosome reaction of sperm in the female reproductive tract. The Suarez group demonstrated that BSP proteins mediate the binding of sperm to the oviductal epithelium and proposed that they are involved in prolonging sperm survival during storage and in the maintaining sperm motility in the oviduct (Ignotz, Lo, Perez, Gwathmey and Suarez, 2001; Gwathmey, Ignotz and Suarez, 2003; Gwathmey, Ignotz, Mueller, Manjunath and Suarez, 2006). Therefore, BSP proteins play multiple roles in sperm functions and have a beneficial effect on sperm fertility.

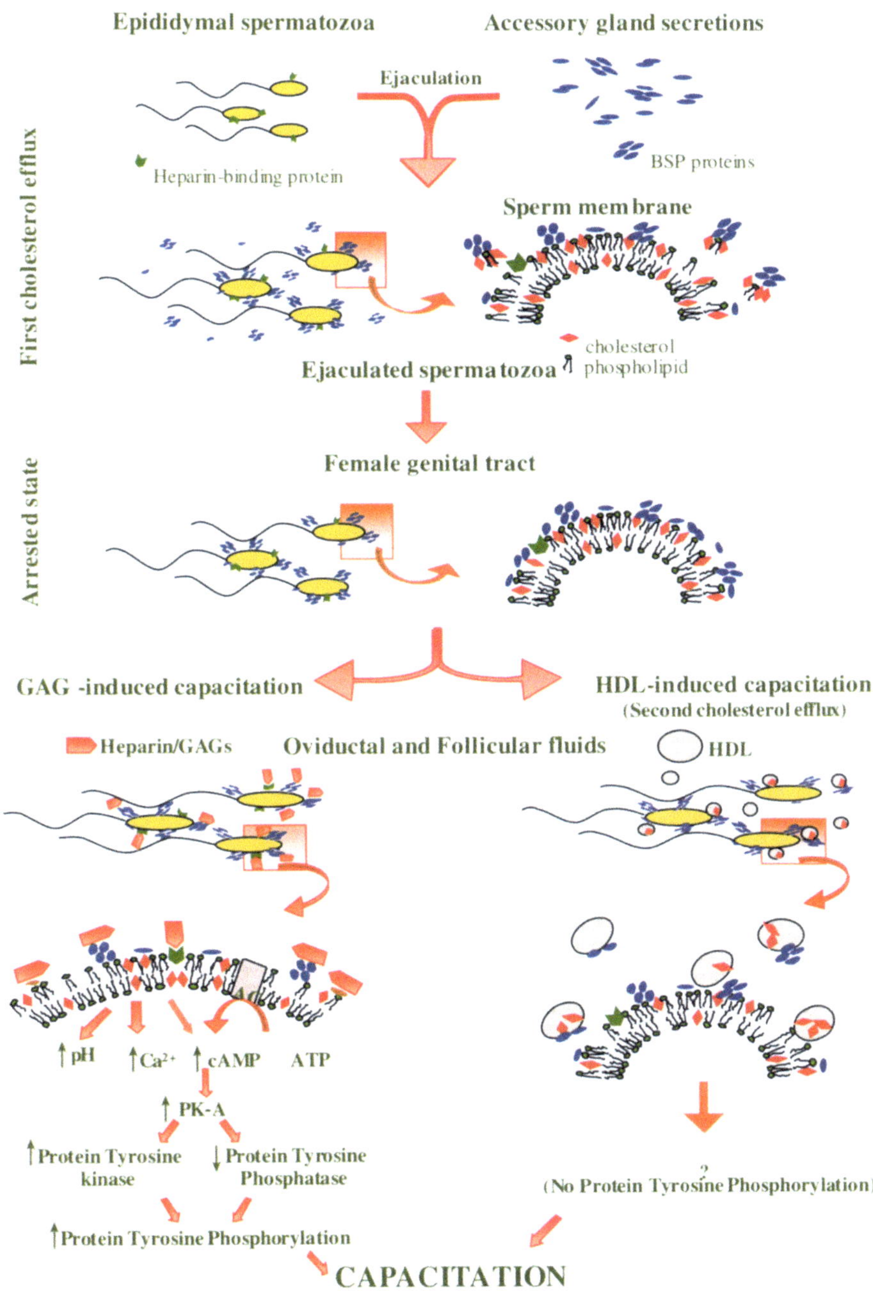

Figure 4. Mechanism of BSP-mediated sperm capacitation. Modified from Therien *et al.* (2001) and Therien *et al.* (2005).

BSP proteins are detrimental to sperm storage

In contrast to their beneficial role in sperm function, BSP proteins may be detrimental to sperm in the context of sperm storage. This is because BSP proteins stimulate cholesterol and phospholipid efflux from the sperm membrane in a time and concentration dependent manner (Therien *et al.*, 1998, 1999). Therefore, a continuous exposure of sperm to seminal plasma (as in the case of semen collected for *in vitro* storage), which contains BSP proteins, causes continuous cholesterol and phospholipid removal from the sperm membrane, which can render sperm very sensitive to storage in the liquid or the frozen states (for a review, see (Bergeron and Manjunath, 2006)). Thus, BSP proteins are beneficial or detrimental to sperm depending on the concentration and duration of exposure.

Egg yolk low-density lipoproteins interact with BSP proteins

Since BSP proteins are choline phospholipid-binding proteins and that low-density lipoproteins (LDL) contain these lipids, we investigated whether or not BSP proteins interact with LDL, the component responsible for sperm protection by egg yolk (Manjunath, Nauc, Bergeron and Menard, 2002). Interestingly, BSP proteins bind to LDL in a rapid and saturable manner. Moreover, the binding capacity of LDL is very high. In fact, one molecule of LDL (M_r 0.6-1.4 x 10^6 Da) can bind 240-555 molecules of BSP proteins. After freeze-thawing semen, BSP proteins are still associated with egg yolk-LDL. This interaction appears to have significant effects on the preservation of sperm functions (Bergeron, Crete, Brindle and Manjunath, 2004).

Novel mechanism of sperm protection by egg yolk-LDL

In view of the above study, a novel mechanism of sperm protection by LDL has been proposed (Fig. 5). Upon ejaculation, BSP proteins secreted by the seminal vesicles are added to sperm (Manjunath *et al.*, 1994). BSP proteins then bind to the sperm membrane (Desnoyers and Manjunath, 1992; Manjunath *et al.*, 1994) and induce cholesterol and phospholipid efflux (Therien *et al.*, 1998, 1999). If semen is not diluted, sperm are continuously exposed to a high concentration of BSP proteins and the lipid removal continues, resulting in decreased sperm resistance to cold shock and freezing. Since ejaculates are diluted with egg yolk extenders within minutes after collection, LDL sequestrates most of the BSP proteins present in semen. This could result in minimal modification of the sperm plasma membrane and allow better sperm storage. Thus, LDL may offer protection to sperm by reducing the deleterious effect of BSP proteins on the sperm membrane. In support of this novel mechanism, we observed that when semen is diluted with egg yolk- or LDL-containing extenders, 50-80% less BSP proteins associate with sperm and concomitantly, lipid (cholesterol and phospholipid) efflux from the sperm membrane is prevented during storage (Bergeron *et al.*, 2004). Consequently, sperm functions such as motility, acrosomal integrity and viability are protected. Since BSP proteins are involved in sperm capacitation and sperm binding to oviductal cells (Manjunath and Therien, 2002; Gwathmey *et al.*, 2003, 2006), the decrease in sperm-bound BSP proteins in the presence of egg yolk extender may be one of the reasons why cryopreserved sperm have impaired fertility compared to fresh semen.

It is remarkable that extender-containing egg yolk is also used to store semen from other mammals such as boar, ram, goat, stallion, buffalo, dog, monkey, gazelle and human (for a review, see Bergeron and Manjunath, 2006). Since BSP proteins are ubiquitous among mammals and the BSP homologues found in the seminal plasma of stallion, boar, goat, bison and

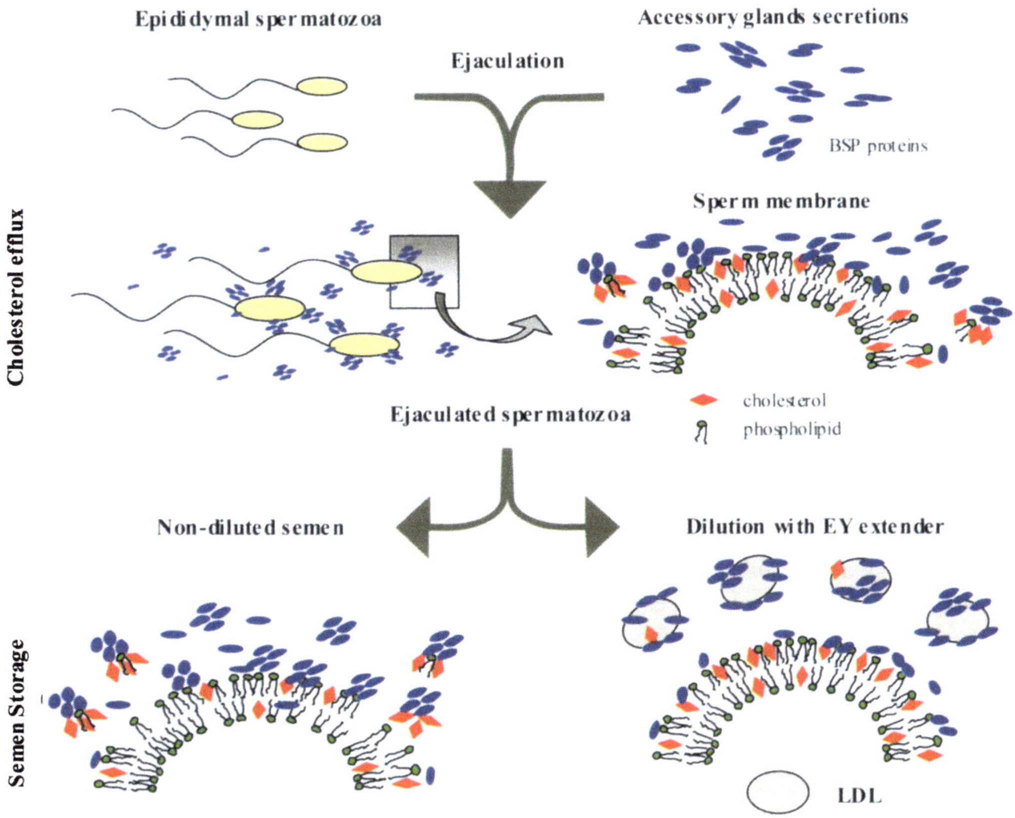

Figure 5. Mechanism of sperm protection by egg yolk (Manjunath *et al.*, 2002).

ram also bind to LDL (Menard *et al.*, 2003; Villemure *et al.*, 2003; Boisvert *et al.*, 2004; Bergeron *et al.*, 2005), the mechanism of sperm protection by egg yolk may be the same for all mammals.

Conclusions

Elucidating the mechanisms occurring during fertilisation is fundamental to resolve causes of infertility, develop contraceptives and sperm storage media, and optimise *in vitro* fertilisation. One of the key events regulating fertility is sperm capacitation, which is a complex and multi-step phenomenon. It occurs in the female reproductive tract and requires factors present in the oviduct. Despite more than 50 years of research and efforts by hundreds of scientists, the molecular events involved in sperm capacitation are still being investigated. Our laboratory has identified a family of lipid-binding proteins of seminal plasma (BSP) proteins that play a key role in this complex process. Based on our studies aimed at investigating the effects of purified BSP proteins, isolated follicular fluid-HDL and -GAGs on epididymal sperm, we have proposed the sequence of events shown in Figure 4. This investigation also unravelled another important mechanism that is sperm protection by egg yolk during storage (Fig. 5). While a brief exposure of sperm to BSP proteins (as in natural mating) is beneficial to sperm (promotes capacitation), a continuous exposure to BSP proteins (as in the case of semen collected for artificial insemination) is detrimental. However, the detrimental effect of seminal plasma BSP proteins is abolished or minimised when they associate with LDL, the protective agent present in egg yolk extender. This discovery may lead to improvements in the existing protocols for semen preservation and in developing novel protocols for the preservation of semen from domestic, farm, zoo and endangered animals as well as from human. The existence of beneficial and detrimental factors for sperm in seminal plasma has been known for many years. However, the fact that the same family of proteins acts like a double edged-sword that is both beneficial and detrimental to sperm depending on the context (capacitation or preservation) is surprising. We hope that this review will create renewed interest in the significance of seminal plasma proteins as well as their effect on sperm storage.

Acknowledgements

This review, which is based on the plenary talk by PM to the 10th International Symposium on Spermatology, was possible due to a "IHDCYH Lectureship in Child and Reproductive Health" awarded by the Institute of Human Development, Child and Youth Health of the Canadian Institutes of Health Research. This work was supported by grants from the Canadian Institutes of Health Research, Natural Sciences and Engineering Research Council of Canada, Lalor Foundation, World Health Organization, Semex Alliance, Cattle Breeding Research Council and L'Alliance Boviteq (Canada). We appreciate the collaboration of Mr Yves Brindle, Centre d'Insémination Artificielle du Québec. We thank all the graduate students, postdoctoral fellows and research assistants who contributed to the progress of this research since the project was initiated in 1984.

References

Aitken, R.J., Harkiss, D., Knox, W., Paterson, M. and Irvine, D. S. (1998). A novel signal transduction cascade in capacitating human spermatozoa characterised by a redox-regulated, cAMP-mediated induction of tyrosine phosphorylation. *Journal of Cell Science* **111:** 645-656

Baldi, E., Luconi, M., Bonaccorsi, L., Muratori, M. and Forti, G. (2000). Intracellular events and signaling pathways involved in sperm acquisition of fertilizing capacity and acrosome reaction. *Frontiers in Bioscience* **5:** 110-123

Bergeron, A., Crete, M.H., Brindle, Y. and Manjunath, P. (2004). Low-density lipoprotein fraction from hen's egg yolk decreases the binding of the major proteins of bovine seminal plasma to sperm and prevents lipid efflux from the sperm membrane. *Biology of Reproduction* **70:** 708-717

Bergeron, A. and Manjunath, P. (2006). New insights towards understanding the mechanism of sperm protection by egg yolk and milk. *Molecular Reproduction and Development* **73:** 1338-1344

Bergeron, A., Villemure, M., Lazure, C. and Manjunath, P. (2005). Isolation and characterization of the major proteins of ram seminal plasma. *Molecular Reproduction and Development* **71:** 461-470

Boisvert, M., Bergeron, A., Lazure, C. and Manjunath, P. (2004). Isolation and characterization of gelatin-binding bison seminal vesicle secretory proteins. *Biology of Reproduction* **70:** 656-661

Calvete, J.J., Mann, K., Schafer, W., Sanz, L., Reinert, M., Nessau, S., Raida, M. and Topfer-Petersen, E. (1995). Amino acid sequence of HSP-1, a major protein of stallion seminal plasma: effect of glycosylation on its heparin- and gelatin-binding capabilities. *Biochemical Journal* **310:** 615-622

Calvete, J.J., Raida, M., Gentzel, M., Urbanke, C., Sanz, L. and Topfer-Petersen, E. (1997). Isolation and characterization of heparin- and phosphorylcholine-binding proteins of boar and stallion seminal plasma. Primary structure of porcine pB1. *FEBS Letters* **407:** 201-206

Chandonnet, L., Roberts, K.D., Chapdelaine, A. and Manjunath, P. (1990). Identification of heparin-binding proteins in bovine seminal plasma. *Molecular Reproduction and Development* **26:** 313-318

de Lamirande, E., Leclerc, P. and Gagnon, C. (1997). Capacitation as a regulatory event that primes spermatozoa for the acrosome reaction and fertilization. *Molecular Human Reproduction* **3:** 175-194

Desnoyers, L. and Manjunath, P. (1992). Major proteins of bovine seminal plasma exhibit novel interactions with phospholipid. *Journal of Biological Chemistry* **267:** 10149-10155

Desnoyers, L., Therien, I. and Manjunath, P. (1994). Characterization of the major proteins of bovine seminal fluid by two-dimensional polyacrylamide gel electrophoresis. *Molecular Reproduction and Development* **37:** 425-435

Fan, J., Lefebvre, J. and Manjunath, P. (2006). Bovine seminal plasma proteins and their relatives: a new expanding superfamily in mammals. *Gene* **375:** 63-74

Galantino-Homer, H.L., Visconti, P.E. and Kopf, G.S. (1997). Regulation of protein tyrosine phosphorylation during bovine sperm capacitation by a cyclic adenosine 3'5'-monophosphate-dependent pathway. *Biology of Reproduction* **56:** 707-719

Guex, N. and Peitsch, M. C. (1997). SWISS-MODEL and the Swiss-PdbViewer: an environment for comparative protein modeling. *Electrophoresis* **18:** 2714-2723

Gwathmey, T.M., Ignotz, G.G., Mueller, J.L., Manjunath, P. and Suarez, S.S. (2006). Bovine seminal plasma proteins PDC-109, BSP-A3, and BSP-30-kDa share functional roles in storing sperm in the oviduct. *Biology of Reproduction* **75:** 501-507

Gwathmey, T.M., Ignotz, G.G. and Suarez, S.S. (2003). PDC-109 (BSP-A1/A2) promotes bull sperm binding to oviductal epithelium in vitro and may be involved in forming the oviductal sperm reservoir. *Biology of Reproduction* **69:** 809-815

Ignotz, G.G., Lo, M.C., Perez, C.L., Gwathmey, T.M. and Suarez, S.S. (2001). Characterization of a fucose-binding protein from bull sperm and seminal plasma that may be responsible for formation of the oviductal sperm reservoir. *Biology of Reproduction* **64:** 1806-1811

Kemme, M. and Scheit, K.H. (1988). Cloning and sequence analysis of a cDNA from seminal vesicle tissue encoding the precursor of the major protein of bull semen. *DNA* **7:** 595-599

Koradi, R., Billeter, M. and Wuthrich, K. (1996). MOLMOL: a program for display and analysis of macromolecular structures. *Journal of Molecular Graphics* **14:** 51-55, 29-32

Lane, M., Therien, I., Moreau, R. and Manjunath, P. (1999). Heparin and high-density lipoprotein mediate bovine sperm capacitation by different mechanisms. *Biology of Reproduction* **60:** 169-175

Leblond, E., Desnoyers, L. and Manjunath, P. (1993). Phosphorylcholine-binding proteins from the seminal fluids of different species share antigenic determinants with the major proteins of bovine seminal plasma. *Molecular Reproduction and Development* **34:** 443-449

Leclerc, P., de Lamirande, E. and Gagnon, C. (1996). Cyclic adenosine 3',5'monophosphate-dependent regulation of protein tyrosine phosphorylation in relation to human sperm capacitation and motility. *Biology of Reproduction* **55:** 684-692

Manjunath, P. (1984). Gonadotropin release stimulatory and inhibitory proteins in bull seminal plasma. In: *Gonadal Proteins and Peptides and their Biological Significance,* Edited by Sairam, M.R. and Atkinson, L.E., World Science Publishing, Singapore, pp. 49-61

Manjunath, P., Baillargeon, L., Marcel, Y.L., Seidah, N.G., Chretien, M. and Chapdelaine, A. (1988). Diversity of novel proteins in gonadal fluids. In: *Molecular Biology of Brain and Endocrine Peptidergic Systems,* Edited by McKerns, K.W. and Chretien. M., Plenum, New York, pp. 259-273

Manjunath, P., Chandonnet, L., Leblond, E. and Desnoyers, L. (1994). Major proteins of bovine seminal vesicles bind to spermatozoa. *Biology of Reproduction* **50:** 27-37

Manjunath, P., Marcel, Y.L., Uma, J., Seidah, N.G., Chretien, M. and Chapdelaine, A. (1989). Apolipoprotein A-I binds to a family of bovine seminal plasma proteins. *Journal of Biological Chemistry* **264:** 16853-16857

Manjunath, P., Nauc, V., Bergeron, A. and Menard, M. (2002). Major proteins of bovine seminal plasma bind to the low-density lipoprotein fraction of hen's egg yolk. *Biology of Reproduction* **67:** 1250-1258

Manjunath, P. and Sairam, M.R. (1987). Purification and biochemical characterization of three major acidic proteins (BSP-A1, BSP-A2 and BSP-A3) from bovine seminal plasma. *Biochemical Journal* **241:** 685-692

Manjunath, P., Sairam, M.R. and Uma, J. (1987). Purification of four gelatin-binding proteins from bovine seminal plasma by affinity chromatography. *Biosciences Reports* **7:** 231-238

Manjunath, P. and Therien, I. (2002). Role of seminal plasma phospholipid-binding proteins in sperm membrane lipid modification that occurs during capacitation. *Journal of Reproductive Immunology* **53:** 109-119

Menard, M., Nauc, V., Lazure, C., Vaillancourt, D. and Manjunath, P. (2003). Novel purification method for mammalian seminal plasma phospholipid-binding proteins reveals the presence of a novel member of this family of protein in stallion seminal fluid. *Molecular Reproduction and Development* **66:** 349-357

Naz, R.K. and Rajesh, P.B. (2004). Role of tyrosine phosphorylation in sperm capacitation/acrosome reaction. *Reproductive Biology and Endocrinology* **2:** 75

Osheroff, J.E., Visconti, P.E., Valenzuela, J.P., Travis, A.J., Alvarez, J. and Kopf, G.S. (1999). Regulation of human sperm capacitation by a cholesterol efflux-stimulated signal transduction pathway leading to protein kinase A-mediated up-regulation of protein tyrosine phosphorylation. *Molecular Human Reproduction* **5:** 1017-1026

Parks, J.E., Arion, J.W. and Foote, R.H. (1987). Lipids of plasma membrane and outer acrosomal membrane from bovine spermatozoa. *Biology of Reproduction* **37:** 1249-1258

Salois, D., Menard, M., Paquette, Y. and Manjunath, P. (1999). Complementary deoxyribonucleic acid cloning and tissue expression of BSP-A3 and BSP-30-kDa: phosphatidylcholine and heparin-binding proteins of bovine seminal plasma. *Biology of Reproduction* **61:** 288-97

Sanz, L., Calvete, J.J., Mann, K., Gabius, H.J. and Topfer-Petersen, E. (1993). Isolation and biochemical characterization of heparin-binding proteins from boar seminal plasma: a dual role for spermadhesins in fertilization. *Molecular Reproduction and Development* **35:** 37-43

Seidah, N.G., Manjunath, P., Rochemont, J., Sairam, M.R. and Chretien, M. (1987). Complete amino acid sequence of BSP-A3 from bovine seminal plasma. Homology to PDC-109 and to the collagen-binding domain of fibronectin. *Biochemical Journal* **243:** 195-203

Tardif, S., Dube, C., Chevalier, S. and Bailey, J. L. (2001). Capacitation is associated with tyrosine phosphorylation and tyrosine kinase-like activity of pig sperm proteins. *Biology of Reproduction* **65:** 784-792

Therien, I., Bergeron, A., Bousquet, D. and Manjunath, P. (2005). Isolation and characterization of glycosaminoglycans from bovine follicular fluid and their effect on sperm capacitation. *Molecular Reproduction and Development* **71:** 97-106

Therien, I., Bleau, G. and Manjunath, P. (1995). Phosphatidylcholine-binding proteins of bovine seminal plasma modulate capacitation of spermatozoa by heparin. *Biology of Reproduction* **52:** 1372-1379

Therien, I., Bousquet, D. and Manjunath, P. (2001). Effect of seminal phospholipid-binding proteins and follicular fluid on bovine sperm capacitation. *Biology of Reproduction* **65:** 41-51

Therien, I., Moreau, R. and Manjunath, P. (1998). Major proteins of bovine seminal plasma and high-density lipoprotein induce cholesterol efflux from epididymal sperm. *Biology of Reproduction* **59:** 768-776

Therien, I., Moreau, R. and Manjunath, P. (1999). Bovine seminal plasma phospholipid-binding proteins stimulate phospholipid efflux from epididymal sperm. *Biology of Reproduction* **61:** 590-598

Therien, I., Soubeyrand, S. and Manjunath, P. (1997). Major proteins of bovine seminal plasma modulate sperm capacitation by high-density lipoprotein. *Biology of Reproduction* **57:** 1080-1088

Villemure, M., Lazure, C. and Manjunath, P. (2003). Isolation and characterization of gelatin-binding proteins from goat seminal plasma. *Reproductive Biology and Endocrinology* **1:** 39

Visconti, P.E., Westbrook, V.A., Chertihin, O., Demarco, I., Sleight, S. and Dieckman, A.B. (2002). Novel signaling pathways involved in sperm acquisition of fertilizing capacity. *Journal of Reproductive Immunology* **53:** 133-150.

Wah, D.A., Fernandez-Tornero, C., Sanz, L., Romero, A. and Calvete, J.J. (2002). Sperm coating mechanism from the 1.8 A crystal structure of PDC-109-phosphorylcholine complex. *Structure (Camb)* **10:** 505-514

Yeagle, P.L. (1985). Cholesterol and the cell membrane. *Biochimica et Biophysica Acta* **822:** 267-287.

Yu, B., Zhao, Y., Zhao, W., Chen, F., Liu, Y., Zhang, J., Fu, W., Zong, Z., Yu, A. and Guan, Y. (2003). The inhibitory effect of BSP-A1/-A2 on protein kinase C and tyrosine protein kinase. *Cell Biochemistry and Function* **21:** 183-188

Spermatology. SRF Vol. 65. ERS Roldan and M Gomendio (eds) Nottingham University Press, Nottingham

Ion channels in sperm motility and capacitation

A. Darszon,[1] C.L. Treviño,[1] C. Wood,[1] B. Galindo,[1,2] E. Rodríguez-Miranda,[1,3] J.J. Acevedo,[4] E.O. Hernandez-González,[5] C. Beltrán,[1] P. Martínez-López[1] and T. Nishigaki[1]

[1]Departamento de Genética del Desarrollo y Fisiología Molecular, Instituto de Biotecnología, Universidad Nacional Autónoma de México (UNAM), Cuernavaca, Morelos, México; [2] Centro de Investigación y de Estudios Avanzados del Instituto Politécnico Nacional (CINVESTAV-IPN), Unidad Monterrey, Monterrey, Nuevo León, México; [3]Instituto Tecnológico Superior de Irapuato (ITESI). Carretera Irapuato, Guanajuato, México; [4]Departamento de Fisiología y Farmacología, Facultad de Medicina, Universidad Autónoma del Estado de Morelos (UAEM), Cuernavaca, México; [5]Departmento de Biología Celular, CINVESTAV-IPN, México D.F., México

Spermatozoa depend upon ion channels to rapidly exchange information with the outside world and to fertilise the egg. These efficient ion transporters participate in many of the most important sperm processes, such as motility and capacitation. It is well known that sperm swimming is regulated by $[Ca^{2+}]_i$. In the sea urchin sperm speract, a decapeptide isolated from egg outer envelope, induces changes in intracellular Ca^{2+} ($[Ca^{2+}]_i$), Na^+, cAMP and cGMP, membrane potential (E_m) and pH (pH_i). Photoactivation of a speract analogue induces Ca^{2+} fluctuations that generate turns that are followed by straighter swimming paths. A fast component of the $[Ca^{2+}]_i$ increase that most likely occurs through voltage dependent Ca^{2+} channels (Ca_vs) is essential for these turns. The Ca_vs involved are modulated by the E_m changes triggered by speract. On the other hand, mammalian sperm gain the ability to fertilise the egg after undergoing a series of physiological changes in the female tract. This maturational process, known as capacitation, encompasses increases in $[Ca^{2+}]_i$ and pH_i, as well as an E_m hyperpolarization in mouse sperm. Our electrophysiological, immunological and molecular-biological experiments indicate that inwardly rectifying K^+ channels regulated by ATP (KATP channels) and epithelial Na^+ channels (ENaCs) are functionally present in mouse spermatogenic cells and sperm. Notably, pharmacological experiments indicate that the opening of KATP channels and closure of ENaCs may contribute to the hyperpolarization that accompanies mouse sperm capacitation. Remarkably, both in the sea urchin sperm speract response and in the mouse sperm capacitation, E_m hyperpolarization seems necessary to remove inactivation from Ca_v channels so they can then open.

Corresponding author E-mail: darszon@ibt.unam.mx

Introduction

Fertilisation requires mature and competent male and female gametes. Sperm and egg must find each other and fuse to generate a new individual. The codes that govern how sperm gather information from the external world and translate signals coming from the egg are yet to be fully understood. Ion transport regulation lies at the heart of the signalling events that allow gametes to respond and interpret information from their vicinity (Darszon, Nishigaki, Wood, Trevino, Felix and Beltran, 2005; Quill, Wang and Garbers, 2006). This review will focus on the work we have carried out studying the role of ion channels in two of the most important sperm processes, motility and capacitation.

Sea urchins release billions of sperm, whose motility, powered by a microscopic flagellar engine, is exquisitely regulated by chemical signals from the environment and the egg that alter their ion permeability (Darszon *et al.*, 2005). On the other hand, sperm from internal fertilisers (reptiles, birds and mammals) develop the potential for motility, probably also undergoing ion transport modulation, as they travel across the vas deferens (a duct connecting the epididymis to the urethra) and the epididymis (Yanagimachi, 1994; Morisawa, 1994). Ejaculation into the female reproductive tract activates sperm that need to further mature through a process named capacitation which includes membrane permeability changes (Visconti, Westbrook, Chertihin, Demarco, Sleight and Diekman, 2002). In mammals only few sperm, from the millions released, reach their destination in the oviduct (Suarez and Ho, 2003). In recent years it has become clear that sperm chemotaxis is not only important for marine organisms but for mammals too (Eisenbach and Giojalas, 2006).

Motility

For many cells movement is fundamental and in many cases they are propelled by cilia and flagella that have a conserved structure. Therefore it is important to understand how flagellar beating is regulated. The axoneme, a complex machine lined by membrane and constituted by microtubules, dynein ATPases and hundreds of other proteins, is responsible for flagellar movement (Mitchell, 2004). Sperm are highly specialised flagellated cells evolved to deliver their genetic material to the female of their species. Inherent in this fundamental function is the need for sperm to be motile.

Though it is easy to observe sperm swimming, our understanding of the molecular mechanisms involved in this essential function is far from being complete. Only recently, developments in our technical ability to record ion permeability changes in individual sperm are revealing how ion channels participate in motility.

Early observations that the ionic composition of the outside media can have a determinate influence on how sperm swim indicated the possible relevance of ion channels in the regulation of this process. For example, sea urchin sperm and other marine organisms are immotile in the testes prior to their release. This is mainly due to the high extracellular K^+ concentration ($[K^+]_e$) (Darszon *et al.*, 2005). Upon spawning sperm activate as $[K^+]_e$ is reduced. On the other hand, it has been known for many years that exposing permeabilised or demembranated invertebrate or mammalian sperm to high $[Ca^{2+}]$ increases flagellar asymmetry and can even inhibit flagellar movement (Brokaw, 1979; Tash and Means, 1982). Chemotaxis, the reorientation of the sperm trajectory toward a chemoattractant gradient, such as may surround the egg, strictly depends on the presence of external Ca^{2+} (Eisenbach and Giojalas, 2006). This phenomenon is most widely observed and studied in sperm from invertebrate marine animals, but occurs also in mammalian and plant sperm. Because chemotaxis requires external Ca^{2+}, it is most likely that Ca^{2+}-permeant pathways are involved in this process.

Ion channels and sperm motility regulation in sea urchin sperm

Not much is known about the participation of ion channels in the regulation of invertebrate sperm motility. Sperm chemotaxis has been documented in many marine species and in all cases it depends on $[Ca^{2+}]_e$ (Miller, 1985; Morisawa, 1994; Darszon, Acevedo, Galindo, Hernandez-Gonzalez, Nishigaki, Trevino, Wood and Beltran, 2006a). A chemotactic response is often associated to a series of turns interspersed with periods of straighter swimming (Kaupp, Hildebrand and Weyand, 2006). When Ca^{2+} is removed from the extracellular medium sperm become unable to turn. Sea urchins are deuterostomes that share a common ancestor with the vertebrates, thus studies of their sperm chemotaxis and motility regulation have general implications.

Sea urchin sperm are particularly suitable models for investigating motility because: (a) their flagella beat in near-planar waves, confining them at water-glass and water-air interfaces (Cosson, Huitorel and Gagnon, 2003), (b) they swim in circles due to an inherent asymmetry in the flagellar planar waveform with respect to the long axis of the head. This circular swimming in a two-dimensional plane above the cover slip surface persists for many minutes, facilitating image analysis of their motile behaviour.

Small activating peptides (10-15 amino acids) (SAPs) that influence sperm motility have been isolated from the egg jelly of sea urchins (Ward, Brokaw, Garbers and Vacquier, 1985; Garbers, 1989). So far, resact, a SAP from *Arbacia punctulata*, is the only sea urchin egg peptide demonstrated to be a chemoattractant (Ward *et al.*, 1985). Photoactivation of 'caged' resact, induces a sequence of transient 'chemotactic' turns followed by longer periods of straighter swimming that allow sperm to swim up the resact gradient (Kaupp, Solzin, Hildebrand, Brown, Helbig, Hagen, Beyermann, Pampaloni and Weyand, 2003). Even though speract, a SAP isolated from *Strongylocentrotus purpuratus* (Garbers, 1989), was shown to alter sperm motility under certain conditions (Cook, Brokaw, Muller and Babcock, 1994), it has not been shown to be a chemoattractant.

Speract binds to the sperm flagellum where it transiently activates a guanylyl cyclase (Garbers, 1989; Cardullo, Herrick, Peterson and Dangott, 1994). Subsequently, according to the canonical model (reviewed in Darszon, Beltran, Felix, Nishigaki and Trevino, 2001; Darszon *et al.*, 2006a), the cGMP elevation opens a K^+-selective channel (Lee and Garbers, 1986, Babcock, Bosma, Battaglia and Darszon, 1992) that allows K^+ efflux and hyperpolarises sperm. Though it has been proposed that cGMP directly activates the K^+ channel (Galindo, Beltran, Cragoe and Darszon, 2000), this has not been fully demonstrated. The speract-induced hyperpolarization also activates an increase in pH_i (Lee and Garbers, 1986; Babcock *et al.*, 1992; Reynaud, De de La Torre, Zapata, Lievano and Darszon, 1993) and a transient $[Ca^{2+}]_i$ decrease (Nishigaki, Wood, Tatsu, Yumoto, Furuta, Elias, Shiba, Baba and Darszon, 2004), the hyperpolarization has been suggested to be required to remove inactivation from voltage-gated Ca^{2+} (Ca_v) channels (Lee and Garbers, 1986; Granados-Gonzalez, Mendoza-Lujambio, Rodriguez, Galindo, Beltran and Darszon, 2005). These Ca_v channels can then open as a consequence of a depolarization, such as the speract-induced repolarization and trigger the complex $[Ca^{2+}]_i$ changes essential for motility regulation (Schackmann and Chock, 1986; Babcock *et al.*, 1992; Darszon *et al.*, 2006a). Increases in cAMP and intracellular Na^+ also follow the hyperpolarization (Garbers, 1989; Kaupp *et al.*, 2003; Rodriguez and Darszon, 2003).

The canonical model is being questioned by recent findings. Fast stopped-flow population measurements using *A. punctulata* sperm reveal that low physiological resact concentrations (pM) increase cGMP 30-fold in ~50 ms, while cAMP increases more slowly and only four-fold (Kaupp *et al.*, 2003). The sperm $[Ca^{2+}]_i$ increase induced by pM resact is biphasic showing an early peak within 2 s, followed by a shoulder and, at the highest concentrations, a second peak at ~10-15 s. The early peak displays a ~0.2 s delay which the authors attribute to the time taken to synthesise cGMP. We had also recorded ~0.2 s delay for the $[Ca^{2+}]_i$ increase induced by high speract concentrations in stopped-flow population experiments in *S. purpuratus* sperm (Nishigaki, Zamudio, Possani and Darszon, 2001). Kaupp *et al.* (2003) did not see the early decrease in $[Ca^{2+}]_i$ postulated by Cook *et al.* (1994) even with a 10 ms resolution. Their results suggest that a rise in cGMP triggers the early, rapid $[Ca^{2+}]_i$ transient while a later rise in cAMP generates the slower, second $[Ca^{2+}]_i$ peak. In a following paper Kaupp's group reported that pM resact, in contrast to the canonical model, increases $[Ca^{2+}]_i$ before pH_i. These results and others suggest that the resact induced pH_i change is not necessary for the increases in $[Ca^{2+}]_i$ (Solzin, Helbig, Van, Brown, Hildebrand, Weyand and Kaupp, 2004), though they do not discard that it may influence their kinetics.

However, simultaneously we obtained time resolved pH_i and $[Ca^{2+}]_i$ changes induced by uncaging speract, cGMP and cAMP in *S. purpuratus* sperm populations that differ from those obtained by Kaupp's group. We find that the increase in pH_i precedes the $[Ca^{2+}]_i$ rise at all the speract concentrations we have tested, and that indeed a fast $[Ca^{2+}]_i$ decrease does occur (duration ~280 ms) when sperm are exposed to speract or cGMP but not to cAMP. This $[Ca^{2+}]_i$ decrease is inhibited by increasing external $[K^+]$ and stimulated by IBMX, which enhances sperm hyperpolarization (Nishigaki et *al.*, 2004). The alternative model thus far proposed by Kaupp's group fails to explain why high external K^+ inhibits all responses to speract but the increase in cGMP (Harumi, Hoshino and Suzuki, 1992; Darszon et *al.*, 2005). At this time the physiological relevance of the decrease in $[Ca^{2+}]_i$ and increase in pH_i are unknown, though they could modulate the delay to the $[Ca^{2+}]_i$ increase and in this manner influence the synchronization between sperm swimming and the responses to the SAP gradient.

The time-resolved measurements discussed above represent the average $[Ca^{2+}]_i$ changes of the sperm population. Independently and contemporaneously we have used high resolution Ca^{2+} imaging to measure $[Ca^{2+}]_i$ increases in single *S. purpuratus* sperm attached to coverslips (Wood, Darszon and Whittaker, 2003). We found that sperm responded to low, physiological concentrations of speract with periodic fluctuations in $[Ca^{2+}]_i$ that were particularly prominent in the sperm flagella. Noticeably, ~85% of the fluorescent dye signal from the sperm came from Ca^{2+} in the head. Thus $[Ca^{2+}]_i$ measurements in sperm suspensions, even at low peptide concentrations, are biased towards the sperm head $[Ca^{2+}]_i$. A full understanding of $[Ca^{2+}]_i$ signalling and motility regulation in this highly polarised cell requires analysis of individual swimming sperm.

$[Ca^{2+}]_i$ changes in swimming sea urchin sperm

Typically sea urchin sperm perform a "chemotactic turn" in 50-100 ms, so the ionic changes that regulate it occur during this time. We have been able for the first time to simultaneously measure $[Ca^{2+}]_i$, flagellar form and trajectory in individual swimming sea urchin sperm using a stroboscopic photodiode illumination in our fluorescence imaging system (Wood, Nishigaki, Furuta, Baba and Darszon, 2005; Nishigaki, Wood, Shiba, Baba and Darszon, 2006)(see Fig. 1). Uncaging cGMP induces Ca^{2+} entry via at least two distinct pathways: (1) a fast (ms) transitory nimodipine-sensitive pathway compartmentalised in the flagella that is a key regulator of flagellar bending and directed motility changes and (2) a sustained $[Ca^{2+}]_i$ elevation in the head and tail whose physiological role is yet to be defined. Contrary to the then-current models (Brokaw, 1979; Cook et *al.*, 1994), we found that flagellar bending does not vary in proportion to overall $[Ca^{2+}]_i$. Instead we proposed a new model whereby flagella bending is increased by Ca^{2+} flux through the nimodipine-sensitive pathway but unaffected by $[Ca^{2+}]_i$ elevation through alternative pathways. An important question, probably relevant to all ciliated cells, is how flagella distinguish between Ca^{2+} uptake pathways. Our findings revealed that cGMP uncaging changes the *S. purpuratus* sperm swimming behaviour in a similar manner to that induced by resact or cGMP in *A. punctulata* sperm (Kaupp *et al.*, 2003). Sperm redirect their trajectory in the direction orthogonal and centrifugal to the arc that they were describing when cGMP was uncaged (see Fig. 2), an observation consistent with a chemotactic-like response (Wood *et al.*, 2005).

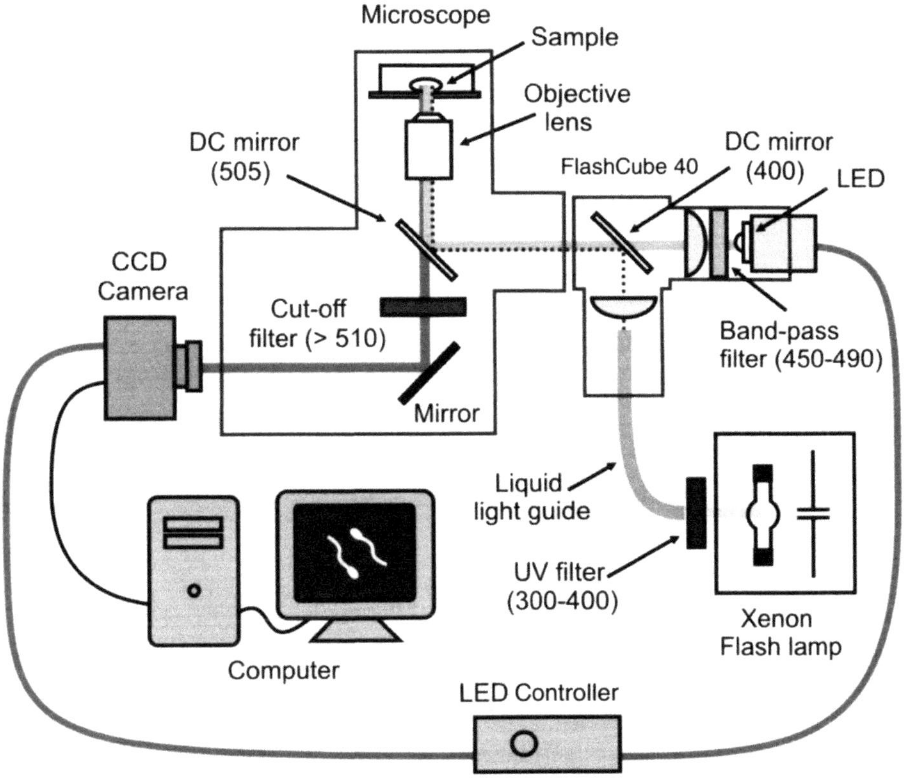

Figure 1. Schematic diagram of the imaging equipment. The LED (Luxeon V Star Lambertian Cyan LED, Lumileds Lighting LLC) is attached on a custom-built aluminum holder and mounted on a Flash Cube 40 assembly (Rapp Opto Electronic, Hamburg) connected to the rear epifluorescence illumination port of the microscope. The LED is controlled by custom-built stroboscopic power supply that provides 3 A current pulses of 1 ms duration, which is synchronously triggered by a CCD camera. Photolysis of the caged compounds is performed using a UV flash lamp system, JML-C2 (Rapp Opto Electronic, Hamburg) with a UV band-pass filter (300-400 nm). The flash light is introduced into the microscope by a liquid light guide (2 mm diameter, NA 0.45) connected to the FlashCube 40 with a dichroic mirror (400 nm).

Soon after, Kaupp's group reported $[Ca^{2+}]_i$ responses to the uncaging of cGMP and resact in swimming *A. punctulata* sperm (Bohmer, Van, Weyand, Hagen, Beyermann, Matsumoto, Hoshi, Hildebrand and Kaupp, 2005). In both cases they recorded $[Ca^{2+}]_i$ fluctuations that regulate sperm swimming, as we had proposed earlier from our experiments in immobilised *S. purpuratus* sperm (Wood *et al.*, 2003). Their responses to a resact gradient allowed them to correlate $[Ca^{2+}]_i$ fluctuations to a chemotactic response and to propose a model, a very significant contribution to the sperm motility field (Bohmer *et al.*, 2005).

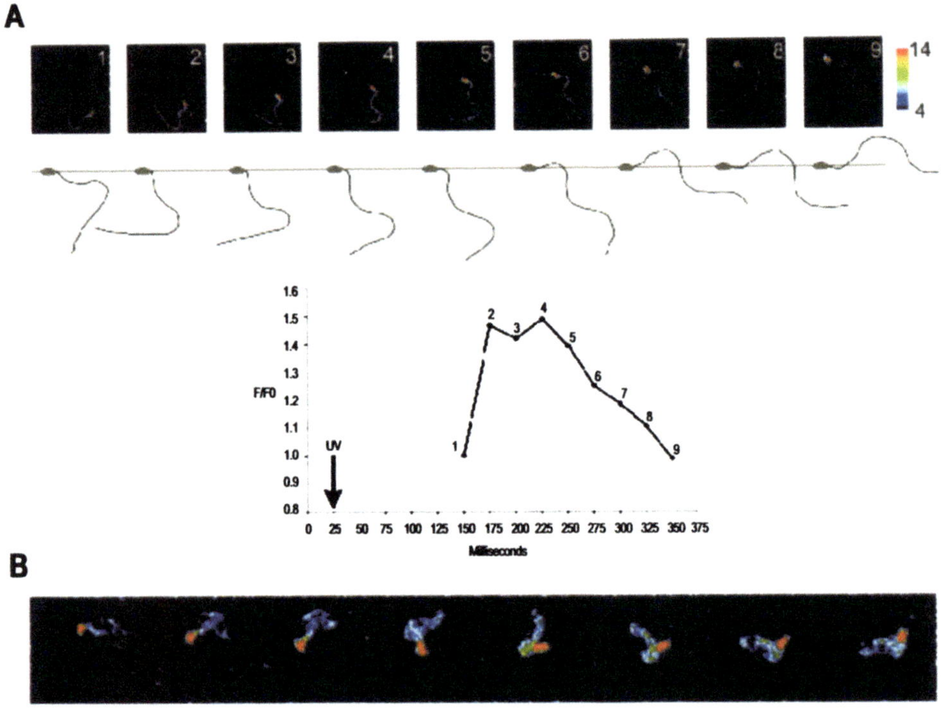

Figure 2. (**A**) Releasing caged cGMP by pulsing sperm with UV light increases both $[Ca^{2+}]_i$ and flagellar asymmetry in a swimming *S. purpuratus* sperm. The upper panels are images of fluo-4 (a Ca^{2+} indicator) fluorescence that show an increase in Ca^{2+} in the flagellum after exposure to a UV pulse. Camera exposure time is set equivalent to LED excitation light duration, in this case 1 ms. Total delay between successive images = 25 ms. Note that the short exposure time possible due to the stroboscopic nature of LED illumination clearly defines flagellar form. Below each fluorescence image the normalised position of the sperm shows that the flagellar asymmetry (in relation to the long axis of the head) increases in synchrony with the rise in $[Ca^{2+}]_i$, and that the flagellum subsequently returns to a more symmetric form while $[Ca^{2+}]_i$ remains relatively elevated. The graph shows the increase in fluo-4 fluorescence as an average along the length of the flagellum, with the numbers corresponding to the images above. F/F0 is the normalised flagellar fluorescence, where F0 is the flagellar fluorescence in the first image in which the flagellum is visible post-UV flash. (**B**) The increase in $[Ca^{2+}]_i$ induced by a UV pulse in a sperm under mercury-lamp epifluorescent illumination. Due to the necessary increase in exposure time to 18 ms the flagellar form is poorly defined and of lesser temporal resolution (total time between successive images 40 ms).

Our new unpublished results uncaging speract show that indeed the $[Ca^{2+}]_i$ fluctuations that we observed in tethered *S. purpuratus* sperm occur in swimming sperm. Each $[Ca^{2+}]_i$ fluctuation in the flagella induces a rapid increase in flagellar asymmetry that results in a sharp turn in the sperm swimming trajectory (Fig. 2). Each sperm turn was most frequently superimposed on a background trajectory of decreased curvature, resulting in a series of turns interspersed with periods of straighter swimming. This motility pattern is not exclusive to sea urchin sperm but occurs in sperm from other marine species undergoing chemotaxis, and has been termed the 'turn-and-run' model (Bohmer *et al.*, 2005). In three dimensions this circular motion would translate into a helical trajectory which upon encounter with a SAP, would display turn-and run responses that would deviate sperm from their original path and, in principle, increase their probability of finding an egg.

cGMP, the first signalling intermediate generated upon binding of egg-derived SAP, induces chemotaxis-like motility changes that involve Ca^{2+} entry into the flagellum through a specific pathway sensitive to Ca_v channel inhibitors (Wood *et al.*, 2005). Recently we immunolocalised Ca_v1.2 and 2.3 to the sea urchin sperm flagella (Granados-Gonzalez *et al.*, 2005), hinting at a role for these channels in regulating chemotaxis. In this work we also demonstrated functionally that Ca_v channels are present in *S. purpuratus* sperm. The inositol 1,4,5-trisphosphate (IP_3) receptor, another Ca^{2+} channel, has been localised to the flagella of sea urchin sperm (Zapata, Ralston, Beltran, Parys, Chen, Longo and Darszon, 1997). This receptor is usually associated with liberation of Ca^{2+} from internal Ca^{2+} stores, of which none have been described in the flagella of sea urchin sperm. Two hyperpolarization-activated and cyclic nucleotide-gated channels (HCN) are also present in *S. purpuratus* sperm flagella, SpHCN1 (Gauss, Seifert and Kaupp, 1998) and SpHCN2 (Galindo, Neill and Vacquier, 2005). These channels could be activated by the hyperpolarization and/or increases in cyclic nucleotides induced by SAP. An inhibitor of HCN channels alters the kinetics of the early $[Ca^{2+}]_i$ changes induced by speract (Nishigaki *et al.*, 2004). Figure 3 presents our presently preferred model.

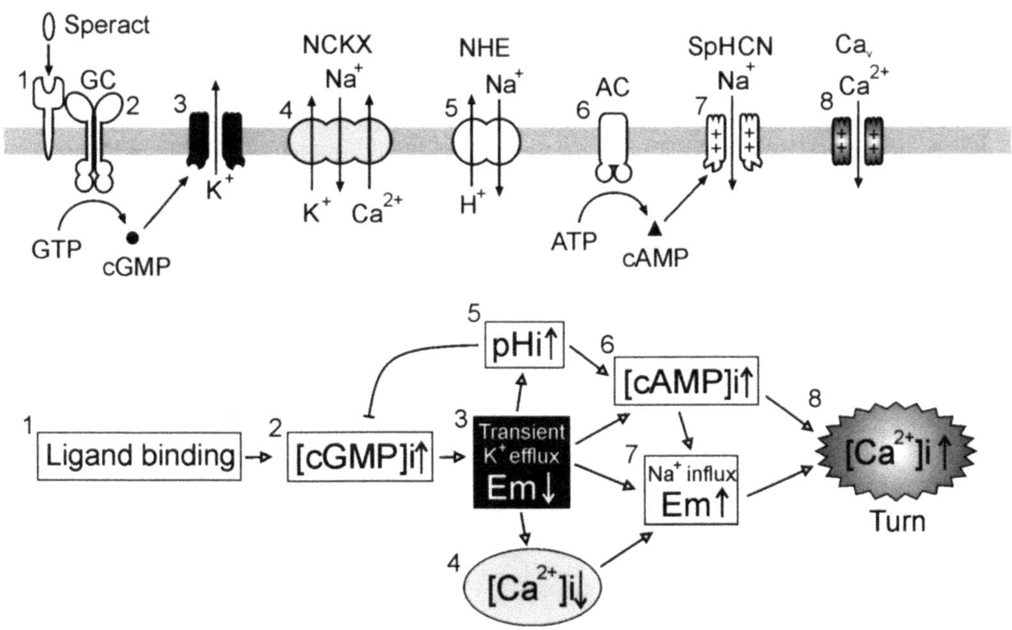

Figure 3. Speract signalling model. Proteins involved in speract signalling and their relationship are shown: 1, speract receptor; 2, guanylate cyclase (GC); 3, cGMP-regulated K^+ channel; 4, K^+-dependent Na^+/Ca^{2+} exchanger (NCKX); 5, Na^+/H^+ exchanger (NHE); 6, adenylate cyclase (AC); 7, sperm HCN channel (SpHCN); 8, voltage-gated Ca^{2+} (Ca_v) channel. Since GC is inactivated by pH_i-dependent dephosphorylation and cGMP is rapidly hydrolysed by phosphodiesterases, the opening of the cGMP-regulated K^+ channel is transient. E_m hyperpolarization facilitates Ca^{2+} extrusion activity of NCKX and removes inactivation from Ca_v channels which can then open upon repolarization and depolarization. Depolarization and Ca^{2+} influx may activate Cl^- and/or K^+ channels which can re-hyperpolarise sperm E_m and sustain $[Ca^{2+}]_i$ oscillations.

Ion channels and sperm motility regulation in mammalian sperm

Mammalian sperm can display two main motility modes, activated and hyperactivated. Freshly ejaculated sperm swim with a relatively low-amplitude flagellar beat that propels them across the entrance of the female reproductive tract. This activated motility is stimulated by Ser/Thr and Tyr phosphorylation of flagellar proteins (reviewed in Turner, 2006). The phosphorylation state of these proteins is regulated by cAMP, protein kinase A (PKA) and a soluble adenylyl cyclase (sAC) (Litvin, Kamenetsky, Zarifyan, Buck and Levin, 2003; Esposito, Jaiswal, Xie, Krajnc-Franken, Robben, Strik, Kuil, Philipsen, van Duin, Conti and Gossen, 2004). The activity of sAC depends on HCO_3^- and Ca^{2+} (Litvin *et al.*, 2003), therefore Ca^{2+} transport influences its operation.

Hyperactivated sperm swim with exaggerated, large-amplitude flagellar beats that result in characteristic "figure-of-eight" trajectories in normal medium, while in high-viscosity medium, they are more progressive (Suarez and Ho, 2003). This motility mode often coincides with the initiation of capacitation, although the two processes are now considered independent by some authors (Ho and Suarez, 2003). Sperm isolated from the upper regions of the female reproductive tract display hyperactivated motility. Advancement of sperm through the viscous environment of the oviduct, and possibly penetration of the egg cumulus may rely on hyperactivated motility.

Sperm hyperactivation requires Ca^{2+} to regulate components of the axoneme. $[Ca^{2+}]_i$ fluctuations occurring in hyperactivated hamster sperm correlate with their flagellar beat frequency (Suarez, Varosi and Dai, 1993). The Ca^{2+} required for this process seems to originate from a reticular structure at the flagellum neck, the redundant nuclear envelope (RNE) which contains IP_3 receptors (IP3Rs) (Ho and Suarez, 2003). It is interesting that a progesterone gradient induces long-lasting $[Ca^{2+}]_i$ oscillations in human sperm originating from the RNE that are comparable to those occurring during hyperactivation and correlate with increases in flagellar activity. Apparently, ryanodine receptors (RyRs) and not IP3Rs are involved in these oscillations (Harper, Barratt and Publicover, 2004). In contrast, in bull sperm RyRs have not been detected (Suarez and Ho, 2003; Jimenez-Gonzalez , Michelangeli, Harper, Barratt and Publicover, 2006). We have detected both IP3Rs and RyRs in mouse sperm (Trevino, Santi, Beltran, Hernandez-Cruz, Darszon and Lomeli, 1998). Very recent evidence indicates that, at pM concentrations, progesterone is chemotactic (Teves, Barbano, Guidobaldi, Sanchez, Miska and Giojalas, 2006).

New findings related to a novel class of Ca^{2+} channels, named CatSpers, some of which are expressed mainly in testes, have emphasised the important participation of ion channels in sperm motility. Of the four members of this family that have been described so far, CatSpers 1 and 2 localise to the sperm flagella (Ren, Navarro, Perez, Jackson, Hsu, Shi, Tilly and Clapham, 2001; Quill, Ren, Clapham and Garbers, 2001; Lobley, Pierron, Reynolds, Allen and Michalovich, 2003; Jin, O'Doherty, Wang, Zheng, Sanders and Yan, 2005). Interestingly, sperm from *catsper1* and *catsper2* null mice are unable to undergo hyperactivation and these animals are infertile (Ren *et al.*, 2001; Carlson, Westenbroek, Quill, Ren, Clapham, Hille, Garbers and Babcock, 2003). The phenotypes of these null mice are identical since neither protein alone localises to the flagellum without co-expression of the other.

Unfortunately it has been impossible to functionally express CatSpers in heterologous systems. On the other hand, studying sperm ion channels by traditional patch-clamping has been hampered by their small size and plasma membrane rigidity (reviewed in Darszon *et al.*, 2006a). Though spermatogenic cells have been used to study sperm ion channels electrophysiologically and using the tools of molecular biology (Hagiwara and Kawa, 1984; Arnoult, Cardullo, Lemos and Florman, 1996; Lievano, Santi, Serrano, Trevino, Bellve, Hernandez-Cruz and Darszon, 1996), these cells can re-distribute and dispose of their channels during the last stages of spermatogenesis (Serrano, Trevino, Felix and Darszon, 1999). Two new strategies have recently emerged that allow ion channel recordings directly on sperm: (1) the "smart patch-clamp" which, in addition to gigaseal formation and electrophysiological characterization, permits determination of the position and regulation of sperm ion channels (Gorelik, Gu, Spohr, Shevchuk, Lab, Harding, Edwards, Whitaker, Moss, Benton, Sanchez, Darszon, Vodyanoy, Klenerman and Korchev, 2002; Gu, Kirkman-Brown, Korchev, Barratt and Publicover, 2004; Shevchuk, Frolenkov, Sanchez, James, Freedman, Lab, Jones, Klenerman and Korchev, 2006), and (2) establishing a gigaseal on the sperm cytoplasmic sperm droplet, a remnant of the precursor germ cell cytoplasm located along the midpiece of maturing sperm that is shed around the time of ejaculation, to record whole sperm patch-clamp currents (Kirichok, Navarro and Clapham, 2006). Using this approach Kirichok *et al.* recorded a constitutively active, weakly voltage-dependent Ca^{2+} current which is strongly potentiated by intracellular alkalinization. *catsper1*-null mice lack this current indicating that CatSper1 is a component of a flagellar Ca^{2+} channel. The channel is permeable to Na^{+} in the absence of external Ca^{2+}.

Various Ca_v channel subunits have been found in mammalian sperm (reviewed in Darszon, López-Martínez, Acevedo, Hernández-Cruz and Trevino, 2006b). In this section we will discuss those found in the flagella (Fig. 4). $Ca_v1.2$, 2.1 and 2.3 subunits were localised to the proximal piece of mouse sperm flagella (Westenbroek and Babcock, 1999), as was the auxiliary ß3 subunit (Serrano *et al.*, 1999). Though fertile, sperm from $Ca_v2.3$ null mice swim with increased linearity (Sakata, Saegusa, Zong, Osanai, Murakoshi, Shimizu, Noda, Aso and Tanabe, 2002). The three T-type Ca_v3 channel isoforms are also present in the mouse and human sperm flagellum (Trevino, Felix, Castellano, Gutierrez, Rodriguez, Pacheco, Lopez-Gonzalez, Gomora, Tsutsumi, Hernandez-Cruz, Fiordelisio, Scaling and Darszon, 2004). Very recently CatSper1 and 2 have been shown to interact and down-regulate $Ca_v3.3$ in the principal piece of human sperm, a notable finding which could have functional implications related to motility (Zhang , Chen, Saraf, Cassar, Han, Rogers, Brioni, Sullivan and Gopalakrishnan, 2006).

A

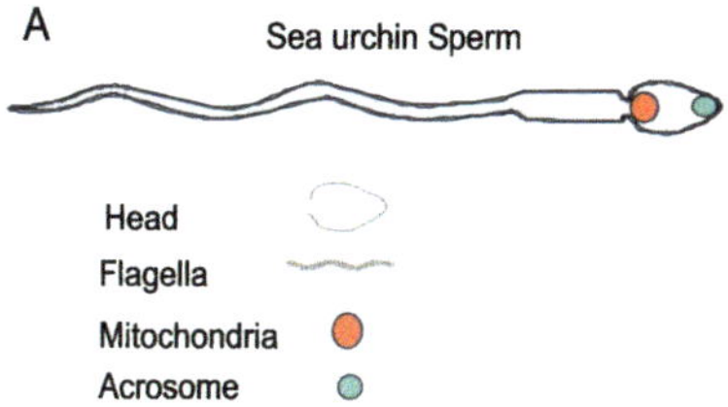

C Channels possibly involved in Chemotaxis/Motility

Sea urchin sperm	
Localization	Channels detected
Flagella	$Ca_v1.2$, $Ca_v2.3$, IP3R spHCN1, spHCN2
Mammalian sperm	
Localization	Channels detected
Flagelar principal piece	RyR3, Catsper1, Catsper2, $Ca_v2.2$, ß3, $Ca_v3.1$, $Ca_v3.2$, $Ca_v3.3$, CNGA1, CNGB1
Flagellar midpiece	$Ca_v1.2$, $Ca_v2.3$, Cav3.1, $Ca_v3.2$, Ca_v 3.3, TRPC1, CNGA3
Nuclear redundant envelope	IP3R, TRPC6

B

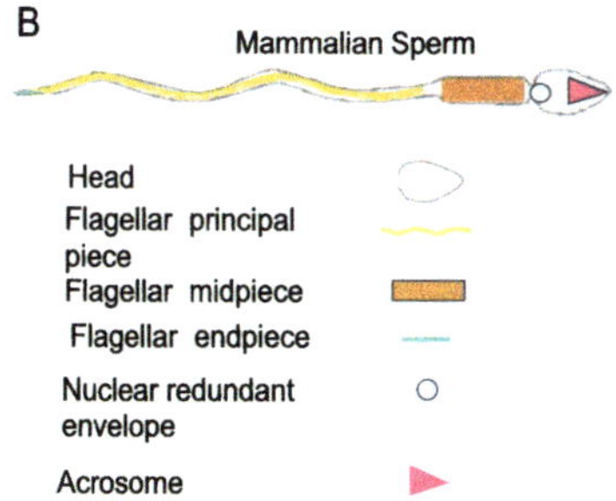

D Channels possibly involved in Capacitation

Localization	Channels detected
Head	$Ca_v1.2$, $Ca_v2.1$, $Ca_v2.2$, $Ca_v2.3$, $Ca_v3.1$, $Ca_v3.2$ TRPC1,TRPC2, TRPC3, TRPC4, TRPC5, TRPC6 Kv1.1, Kv1.2, Kir6.2, SUR1
Acrosome	ENaC
Flagellar midpiece	$Ca_v1.2$, $Ca_v2.1$, $Ca_v2.3$, $Ca_v3.1$, $Ca_v3.2$, $Ca_v3.3$, TRPC1, TRPC3, TRPC4, TRPC5, Kir6.1, Kir6.2, Kv1.1, Kv1.5, SUR1, ENaC
Flagellar principal piece	$Ca_v1.2$, $Ca_v2.2$, $Ca_v2.3$, $Ca_v3.1$, $Ca_v3.2$, $Ca_v3.3$, TRPC1, TRPC3, TRPC4, TRPC6, Kv1.2, Kv3.1, K2p5.1, SUR2

Figure 4. Ion channels present in sperm cells. Diagram of sea urchin (**A**) and mammalian (**B**) sperm illustrating the principal cell components. Summary of the channels that have been immunolocalised to distinct sperm section and may participate during the process of chemotaxis and/or motility (**C**) and capacitation (**D**).

Using the new strategy to obtain a whole-cell configuration on the citoplasmic droplet, we have reproducibly obtained patch-clamp recordings from individual mouse sperm isolated from testicle and caput epididymus. Under conditions to record Ca_V channels (Lopez-Gonzalez, De La Vega-Beltran, Santi, Florman, Felix and Darszon, 2001), we have corroborated, for the first time, that T-type Ca^{2+} currents are present in mouse sperm. As anticipated, these currents are blocked by micromolar concentrations of mibefradil and nifedipine (Fig. 5). As in mouse spermatogenic cells, the T-type Ca^{2+} currents are also blocked by 50 μm Ni^{2+} (unpublished results). These findings are consistent with early work indicating the participation of Ca_V channels in mammalian sperm physiology, with the presence of their mRNAs both in spermatogenic cells and sperm and with their immunolocalization (reviewed in Jimenez-Gonzalez, Michelangeli, Harper, Barratt and Publicover, 2006; Darszon *et al.*, 2006a) (see Fig. 4).

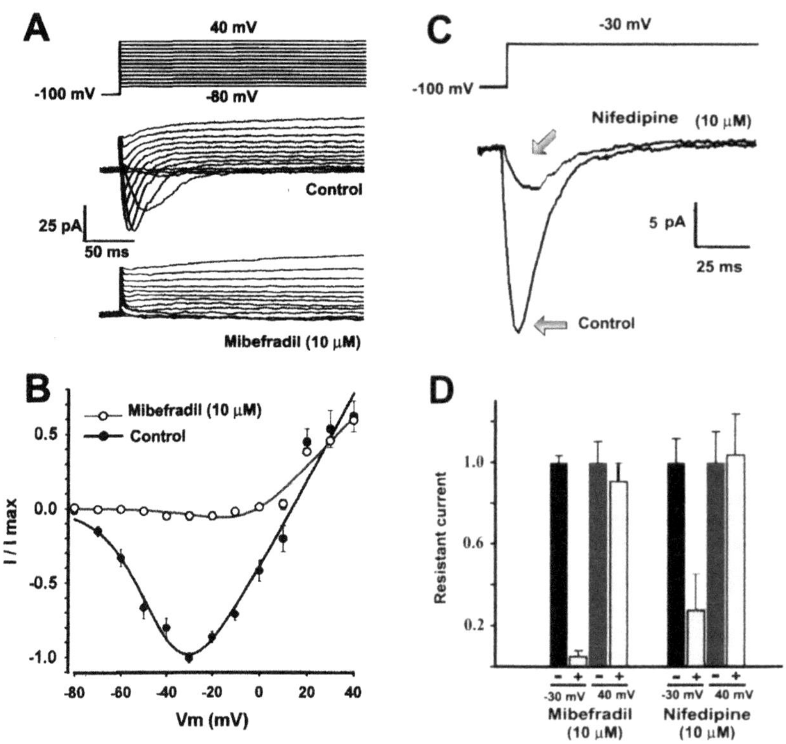

Figure 5. Whole cell patch-clamp recordings on testicular mouse sperm. (**A**) Representative family of total ionic currents from testicular sperm sealing on the cytoplasmic droplet. Ionic currents were evoked by 300 ms depolarising steps from -80 up to 40 mV from a holding potential of -70 mV. Before the test pulses a hyperpolarising pre-pulse was applied for 1 s. An inward and an outward ion current are apparent under control conditions in these representative upper traces. Inward current amplitude was determined at the peak (0-50 ms after activation), while outward current amplitude was measured as the mean of the whole trace. Mibefradil (10 µM), a T-type channel blocker, inhibits the inward component of the sperm ion currents (lower traces) but does not alter the outward component. (**B**) I-V plot of the ion currents shown in panel A in the absence (control, closed circles; n = 20) and presence of mibefradil (10 µM, open circles; n = 4). (**C**) Inhibition of sperm inward ion current by 10 µM nifedipine, another blocker of T-type currents in mouse spermatogenic cells. A 300 ms test pulse from a HP of -70 mV to -30 mV was used to evoke the currents. Around 80% of the inward current is inhibited by this nifedipine concentration

Figure 5 (contd). suggesting that indeed T-type Ca^{2+} channels are present in mouse sperm.

(**D**) Summary of the results with the two T-type channel blockers. Both mibefradil (10 μM; $n = 4$) and nifedipine (10 μM; $n = 4$) inhibit 94% and 80% of the inward component, respectively. Neither mibefradil nor nifedipine affect the outward component of the sperm ion currents. In all plots, symbols represent the mean ± S.E.M.

Our perseverant interest in the sperm acrosome reaction led us to also study the role of TRP channel family members in mouse and human sperm physiology. TRPC1, C3 and C6 were found in the flagella of human sperm while TRPC1, C3, C4 and C6 were detected in mouse sperm tail hinting for a role in motility. Interestingly, human sperm motility was decreased by some TRP channel inhibitors (Castellano, Trevino, Rodriguez, Serrano, Pacheco, Tsutsumi, Felix and Darszon, 2003).

Cyclic nucleotide-gated (CNG) channels are expressed in the mammalian sperm flagella. Actually CNGA3, a member of this family was the first sperm ion channel cloned (Weyand, Godde, Frings, Weiner, Muller, Altenhofen, Hatt and Kaupp, 1994). It is distributed along the flagellum. CNGA3 knockout mice are fertile (Turner, 2006); however their motility has not been characterised in detail. On the other hand, CNGB1 is confined to the principal piece in bovine sperm (Wiesner, Weiner, Middendorff, Hagen, Kaupp and Weyan, 1998). Major efforts are still required to fully understand sperm motility.

Mammalian sperm capacitation

While in transit through the female reproductive tract, mature epididymal sperm become competent to fertilise. This maturational process is named capacitation and involves complex sperm changes which include plasma membrane reorganization, cholesterol removal, protein tyrosine phosphorylation, E_m hyperpolarization (in mouse sperm) and increases in pH_i and $[Ca^{2+}]_i$ (reviewed in Visconti *et al.*, 2002; Darszon *et al.*, 2005). As indicated earlier, capacitation is associated with the emergence of hyperactivation (Suarez and Ho, 2003), and encompasses changes that prime sperm to respond to the zona pellucida (ZP), the extracellular egg coat that triggers the acrosome reaction. The molecular mechanisms involved are not fully understood and include Tyr-phosphorylation regulated by a cAMP-dependent pathway involving PKA (Visconti *et al.*, 2002). sAC appears to be the principal cyclase responsible for the cAMP changes that occur during capacitation (Esposito *et al.*, 2004; Hess, Jones, Marquez, Chen, Ord, Kamenetsky, Miyamoto, Zippin, Kopf, Suarez, Levin, Williams, Buck and Moss, 2005), although transmembrane ACs may also contribute (Fraser, Adeoya-Osiguwa, Baxendale, Mededovic and Osiguwa, 2005).

Capacitation *in vitro* depends on the presence of Ca^{2+}, K^+, HCO_3^- and Na^+ in the medium (Visconti *et al.*, 2002; Darszon *et al.*, 2005). In addition, bovine serum albumin (BSA) is also required. How and which of the ion channels and transporters present in sperm regulate E_m, pH_i and $[Ca^{2+}]_i$ to achieve capacitation is not well established (Darszon *et al.*, 2005). Non-capacitated mouse sperm are relatively depolarised (E_m ~ -35 to -45 mV) and hyperpolarise to ~ -80 mV during capacitation (Arnoult, Kazam, Visconti, Kopf, Villaz and Florman, 1999; Munoz-Garay, De la Vega-Beltran, Delgado, Labarca, Felix and Darszon, 2001). It is thought that the hyperpolarization is needed to remove inactivation from T-type Ca_v channels, recruiting them to a closed state from which they can open in response to ZP to induce the acrosome reaction (Arnoult *et al.*, 1999). Sperm Ca_v channels other than T-type, as well as several members of the TRP family (see Fig. 4), may influence capacitation.

It was proposed that in mouse sperm the K^+ permeability participates in the capacitation process since external K^+ and K^+-channel blockers affect the hyperpolarization that accompanies this process (Arnoult *et al.*, 1999; Munoz-Garay *et al.*, 2001; Acevedo, Mendoza-Lujambio, de la

Vega-Beltran, Trevino, Felix and Darszon, 2006). Voltage-gated K^+ channels (Salvatore, D'Adamo, Polishchuk, Salmona and Pessia, 1999), Ca^{2+}-activated K^+ channels (Chan, Wu, Sun, Leung, Wong, Chung, So, Zhou and Yan, 1998) and inwardly rectifying K^+ (Kir) channels (Munoz-Garay *et al.*, 2001; Felix, Serrano, Trevino, Munoz-Garay, Bravo, Navarro, Pacheco, Tsutsumi and Darszon, 2002; Acevedo *et al.*, 2006) have been reported to be present in spermatogenic and sperm cells. Sperm pH_i is relatively acidic before capacitation and may down-regulate sperm Kir channels. This could maintain E_m depolarised keeping Ca_v channels inactivated and unable to cause unregulated Ca^{2+} entry and the acrosome reaction (Zeng, Oberdorf and Florman, 1996).

Using whole-cell patch clamp recordings in mouse spermatogenic cells, we found Kir currents sensitive to Ba^{2+}, glucose, tolbutamide and glibenclamide (Acevedo *et al.*, 2006). These last two compounds are sulfonylureas known to block Kir channels regulated by ATP (KATP channels). Furthermore, we were able to detect transcripts for the KATP channel subunits SUR1, SUR2, Kir6.1 and Kir6.2 in total RNA from elongated spermatids in RT-PCR assays, and immunolocalised them in mature mouse sperm (Fig. 4). Incubation of sperm under capacitating conditions with tolbutamide abolishes the hyperpolarization and significantly inhibits the acrosome reaction. Since mouse sperm capacitation is accompanied by a pH_i increase and an ATP decrease (Baker and Aitken, 2004), it is feasible that KATP channels contribute to the capacitation-associated hyperpolarization in mouse sperm.

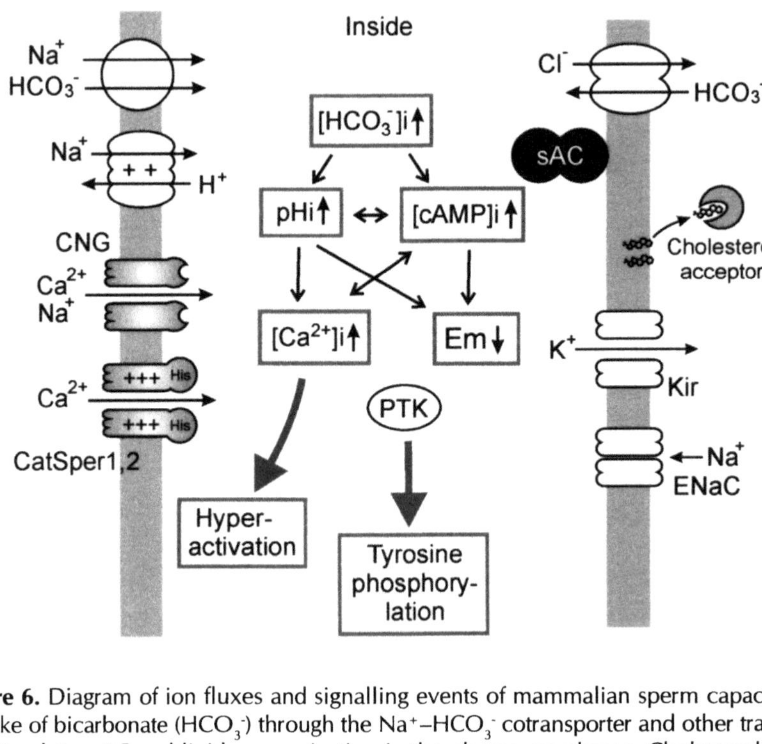

Figure 6. Diagram of ion fluxes and signalling events of mammalian sperm capacitation. Uptake of bicarbonate (HCO_3^-) through the Na^+–HCO_3^- cotransporter and other transporters stimulates sAC and lipid reorganization in the plasma membrane. Cholesterol acceptors such as albumin also promote the process of lipid reorganization. cAMP has several targets such as CNG channels, PKA and ENaC possibly indirectly. Also, an increase in intracellular Ca^{2+} may directly activate sAC and an increase in pH_i can promote Ca^{2+} influx through CatSper1 and 2 channels. Therefore, there is a cAMP, Ca^{2+} and pH_i synergism. Furthermore, the changes in these intracellular parameters lead to tyrosine phosphorylation involving PKA activation, hyperactivation of sperm motility and the hyperpolarization by opening K^+ channels (Kir) and closing ENaC. PTK, protein tyrosine kinase.

Recently we found that an electrogenic Na^+ transporter, possibly involving an amiloride sensitive Na^+ channel, contributes to depolarise the sperm resting E_m before capacitation. Indeed, the α and δ subunits of epithelial Na^+ channels (ENaCs) are present in the sperm flagellar midpiece and acrosome region, respectively (Hernandez-Gonzalez, Sosnik, Edwards, Acevedo, Mendoza-Lujambio, Lopez-Gonzalez, Demarco, Wertheimer, Darszon and Visconti, 2005). These channels are amiloride-sensitive and contribute to the resting E_m in cells by displacing it towards the Na^+ equilibrium potential (Awayda, Boudreaux, Reger and Hamm, 2000). Furthermore, spermatogenic cells display amiloride-sensitive inward Na^+ currents whose characteristics match those of ENaCs. Addition of cell-permeant cAMP analogues which lead to *in vitro* capacitation, decrease the sperm depolarization induced by addition of external Na^+. These findings indicate that the increases in cAMP that accompany capacitation may inhibit ENaCs and contribute to achieve the hyperpolarization required for this process to occur. Figure 6 illustrates our working model for the involvement of ion channels and transporters in mouse sperm capacitation.

Acknowledgements

The authors would like to acknowledge José Luis de la Vega for technical assistance. The work was supported by grants DGAPA-UNAM (to AD, CB, TN and CT), CONACYT (to AD and CB), the Wellcome Trust (to AD) and FIRCA R03 TW 006121 (to AD).

References

Acevedo, J.J., Mendoza-Lujambio, I., de la Vega-Beltran, J.L., Trevino, C.L., Felix, R. and Darszon, A. (2006). KATP channels in mouse spermatogenic cells and sperm, and their role in capacitation. *Developmental Biology* **289:** 395-405

Arnoult, C., Cardullo, R.A., Lemos, J.R. and Florman, H.M. (1996). Activation of mouse sperm T-type Ca^{2+} channels by adhesion to the egg zona pellucida. *Proceedings of the National Academy of Sciences USA* **93:** 13004-13009

Arnoult, C., Kazam, I.G., Visconti, P.E., Kopf, G.S., Villaz, M. and Florman, H.M. (1999). Control of the low-voltage-activated calcium channel of mouse sperm by egg ZP3 and by membrane hyperpolarization during capacitation. *Proceedings of the National Academy of Sciences USA* **96:** 6757-6762

Awayda, M.S., Boudreaux, M.J., Reger, R.L. and Hamm, L.L. (2000). Regulation of the epitelial Na^+ channel by extracelular acidification. *American Journal of Physiology, Cell Physiology* **279:** C1896-905.

Babcock, D.F., Bosma, M.M., Battaglia, D.E. and Darszon, A. (1992). Early persistent activation of sperm K^+ channels by the egg peptide speract. *Proceedings of the National Academy of Sciences USA* **89:** 6001-6005

Baker, M.A. and Aitken, R.J. (2004). The importance of redox regulated pathways in sperm cell biology. *Molecular and Cellular Endocrinology* **216:** 47-54

Bohmer, M., Van, Q., Weyand, I., Hagen, V., Beyermann, M., Matsumoto, M., Hoshi, M., Hildebrand, E. and Kaupp, U.B. (2005). Ca^{2+} spikes in the flagellum control chemotactic behavior of sperm. *EMBO Journal* **24:** 2741-2752

Brokaw, C.J. (1979). Calcium-induced asymmetrical beating of triton-demembranated sea urchin sperm flagella. *Journal of Cell Biology* **82:** 401-411

Cardullo, R.A., Herrick, S.B., Peterson, M.J. and Dangott, L.J. (1994). Speract receptors are localized on sea urchin sperm flagella using a fluorescent peptide analog. *Developmental Biology* **162:** 600-607

Carlson, A.E., Westenbroek, R.E., Quill, T., Ren, D., Clapham, D.E., Hille, B., Garbers, D.L. and Babcock, D.F. (2003). CatSper1 required for evoked Ca^{2+} entry and control of flagellar funcion in sperm. *Proceedings of the National Academy of Sciences USA* **100:** 14864-14868

Castellano, L.E., Trevino, C.L., Rodriguez, D., Serrano, C.J., Pacheco, J., Tsutsumi, V., Felix, R. and Darszon, A. (2003). Transient receptor potential (TRPC) channels in human sperm: expression, cellular localization and involvement in the regulation of flagellar motility. *FEBS Letters* **541:** 69-74

Chan, H.C., Wu, W. L., Sun, Y.P., Leung, P.S., Wong, T.P., Chung, Y.W., So, S.C., Zhou, T.S. and Yan, Y.C. (1998). Expression of sperm Ca^{2+}-activated K^+ channels in *Xenopus* oocytes and their modulation by extracellular ATP. *FEBS Letters* **438:** 177-182

Cook, S.P., Brokaw, C.J., Muller, C.H. and Babcock, D.F. (1994). Sperm chemotaxis: egg peptides control cytosolic calcium to regulate flagellar responses. *Developmental Biology* **165:** 10-19

Cosson, J., Huitorel, P. and Gagnon, C. (2003). How spermatozoa come to be confined to surfaces. *Cell Motility and the Cytoskeleton* **54:** 56-63

Darszon, A., Acevedo, J.J., Galindo, B.E., Hernandez-Gonzalez, E.O., Nishigaki, T., Trevino, C.L., Wood, C. and Beltran, C. (2006a). Sperm channel diversity and functional multiplicity. *Reproduction* **131:** 977-988

Darszon, A., Beltran, C., Felix, R., Nishigaki, T. and Trevino, C.L. (2001). Ion transport in sperm signaling. *Developmental Biology* **240:** 1-14

Darszon, A., López-Martínez, P., Acevedo, J.J., Hernández-Cruz, A. and Trevino, C.L. (2006b). T-type Ca^{2+} channels in sperm function. *Cell Calcium* **40:** 241-252

Darszon, A., Nishigaki, T., Wood, C., Trevino, C.L., Felix, R. and Beltran, C. (2005). Calcium channels and Ca^{2+} fluctuations in sperm physiology. *International Review of Cytology* **243:** 79-172

Eisenbach, M. and Giojalas, L.C. (2006). Sperm guidance in mammals - an unpaved road to the egg. *Nature Reviews. Molecular Cell Biology* **7:** 276-285

Esposito, G., Jaiswal, B.S., Xie, F., Krajnc-Franken, M.A., Robben, T.J., Strik, A.M., Kuil, C., Philipsen, R.L., van Duin, M., Conti, M. and Gossen, J. A. (2004). Mice deficient for soluble adenylyl cyclase are infertile because of a severe sperm-motility defect. *Proceedings of the National Academy of Sciences USA* **101:** 2993-2998

Felix, R., Serrano, C.J., Trevino, C.L., Munoz-Garay, C., Bravo, A., Navarro, A., Pacheco, J., Tsutsumi, V. and Darszon, A. (2002). Identification of distinct K^+ channels in mouse spermatogenic cells and sperm. *Zygote,* **10,** 183-188.

Fraser, L.R., Adeoya-Osiguwa, S., Baxendale, R.W., Mededovic, S. and Osiguwa, O.O. (2005). First messenger regulation of mammalian sperm function via adenylyl cyclase/cAMP. *Journal of Reproduction and Development* **51:** 37-46

Galindo, B.E., Beltran, C., Cragoe, E.J., Jr. and Darszon, A. (2000). Participation of a K^+ channel modulated directly by cGMP in the speract-induced signaling cascade of *Strongylocentrotus purpuratus* sea urchin sperm. *Developmental Biology* **221:** 285-294

Galindo, B.E., Neill, A.T. and Vacquier, V.D. (2005). A new hyperpolarization-activated, cyclid nucleotide-gated channel from sea urchin sperm flagella. *Biochemical and Biophysical Research Communications* **334:** 96-101

Garbers, D.L. (1989). Molecular basis of fertilization. *Annual Review of Biochemistry* **58:** 719-742

Gauss, R., Seifert, R. and Kaupp, U.B. (1998). Molecular identification of a hyperpolarization-activated channel in sea urchin sperm. *Nature* **393:** 583-587

Gorelik, J., Gu, Y., Spohr, H.A., Shevchuk, A.I., Lab, M.J., Harding, S.E., Edwards, C.R., Whitaker, M., Moss, G.W., Benton, D.C., Sanchez, D., Darszon, A., Vodyanoy, I., Klenerman, D. and Korchev, Y.E. (2002). Ion channels in small cells and subcellular structures can be studied with a smart patch-clamp system. *Biophysical Journal* **83:** 3296-3303.

Granados-Gonzalez, G., Mendoza-Lujambio, I., Rodriguez, E., Galindo, B.E., Beltran, C. and Darszon, A. (2005). Identification of voltage-depencent Ca^{2+} channels in sea urchin sperm. *FEBS Letters* **579:** 6667-6672

Gu, Y., Kirkman-Brown, J.C., Korchev, Y., Barratt, C.L.R. and Publicover, S.J. (2004). Multi-state, 4-aminopyridine-sensitive ion channels in human spermatozoa. *Developmental Biology* **274:** 308-317

Hagiwara, S. and Kawa, K. (1984). Calcium and potassium currents in spermatogenic cells dissociated from rat seminiferous tubules. *Journal of Physiology* **356** 135-149

Harper, C.V., Barratt, C.L.R. and Publicover, S.J. (2004). Stimulation of human spermatozoa with progesterone gradients to simulate approach to the oocyte. Induction of $[Ca^{2+}]_i$ oscillations and cyclical transitions in flagellar beating. *Journal of Biological Chemistry* **279:** 46315-46325

Harumi, T., Hoshino, K. and Suzuki, N. (1992). Effects of sperm-activating peptide I on *Hemicentrotus pulcherrimus* spermatozoa in high potassium sea water. *Development Growth and Differentiation* **34:** 163-172

Hernandez-Gonzalez, E.O., Sosnik, J., Edwards, J., Acevedo, J.J., Mendoza-Lujambio, I., Lopez-Gonzalez, I., Demarco, I., Wertheimer, E., Darszon, A. and Visconti, P.E. (2005). Sodium and epithelial sodium channels participate in the regulation of the capacitation-associated hyperpolarization in mouse sperm. *Journal of Biological Chemistry* **281:** 5623-5633.

Hess, K.C., Jones, B.H., Marquez, B., Chen, Y., Ord, T.S., Kamenetsky, M., Miyamoto, C., Zippin, J.H., Kopf, G.S., Suarez, S.S., Levin, L.R., Williams, C.J., Buck, J. and Moss, S.B. (2005). The "soluble" adenylyl cyclase in sperm mediated multiple signaling events re quired for fertilization. *Developmental Cell* **9:** 249-259

Ho, H.C. and Suarez, S.S. (2003). Characterization of the intracellular calcium store at the base of the sperm flagellum that regulates hyperactivated motility. *Biology of Reproduction* **68:** 1590-1596

Jimenez-Gonzalez, C., Michelangeli, F., Harper, C.V., Barratt, C.L.R. and Publicover, S. J. (2006). Calcium signaling in human spermatozoa: a specialized 'toolkit' of channels, transporters and stores. *Human Reproduction Update* **12:** 253-267

Jin, J.L., O'Doherty, A.M., Wang, S., Zheng, H., Sanders, K.M. and Yan, W. (2005). Catsper3 and catsper4 encode two cation channel-like proteins exclusively expressed in the testis. *Biology of Reproduction* **73:** 1235-1242

Kaupp, U.B., Hildebrand, E. and Weyand, I. (2006). Sperm chemotaxis in marine invertebrates—molecules and mechanisms. *Journal of Cellular Physiology* **208:** 487-494

Kaupp, U.B., Solzin, J., Hildebrand, E., Brown, J.E., Helbig, A., Hagen, V., Beyermann, M., Pampaloni, F. and Weyand, I. (2003). The signal flow and motor response controling chemotaxis of sea urchin sperm. *Nature Cell Biology* **5:** 109-117

Kirichok, Y., Navarro, B. and Clapham, D.E. (2006). Whole-cell patch-clamp measurements of spermatozoa reveal an alkaline-activated Ca^{2+} channel. *Nature* **439:** 737-740

Lee, H.C. and Garbers, D.L. (1986). Modulation of the voltage-sensitive Na^{+}/H^{+} exchange in sea urchin spermatozoa through membrane potential changes induced by the egg peptide speract. *Journal of Biological Chemistry* **261:** 16026-16032

Lievano, A., Santi, C.M., Serrano, C.J., Trevino, C.L., Bellve, A.R., Hernandez-Cruz, A. and Darszon, A. (1996). T-type Ca^{2+} channels and alpha1E expression in spermatogenic cells, and their possible relevance to the sperm acrosome reaction. *FEBS Letters* **388:** 150-154

Litvin, T.N., Kamenetsky, M., Zarifyan, A., Buck, J. and Levin, L.R. (2003). Kinetic properties of "soluble" adenylyl cyclase. Synergism between calcium and bicarbonate. *Journal of Biological Chemistry* **278:** 15922-15926

Lobley, A., Pierron, V., Reynolds, L., Allen, L. and Michalovich, D. (2003). Identification of human and mouse CatSper3 and CatSper4 genes: characterisation of a common interaction domain and evidence for expression in testis. *Reproductive Biology and Endocrinology* **1:** 53

Lopez-Gonzalez, I., De La Vega-Beltran, J.L., Santi, C.M., Florman, H.M., Felix, R. and Darszon, A. (2001). Calmodulin antagonists inhibit T-type Ca^{2+} currents in mouse spermatogenic cells and the zona pellucida-induced sperm acrosome reaction. *Developmental Biology* **236:** 210-219

Miller, R.L. (1985). Sperm chemo-orientation in the Metazoa. In: *Biology of Fertilization*. Edited by Metz, C.B. and Monroy, A., Academic Press, New York, pp. 275-337

Mitchell, D.R. (2004). Speculations on the evolution of 9+2 organelles and the role of central pair microtubules. *Biology of the Cell* **96:** 691-696

Morisawa, M. (1994). Cell signaling mechanisms for sperm motility. *Zoological Science* **11:** 647-662

Munoz-Garay, C., De la Vega-Beltran, J.L., Delgado, R., Labarca, P., Felix, R. and Darszon, A. (2001). Inwardly rectifying K^{+} channels in psermatogenic cells: functional expression and implication in sperm capacitation. *Developmental Biology* **234:** 261-274

Nishigaki, T., Wood, C.D., Tatsu, Y., Yumoto, N., Furuta, T., Elias, D., Shiba, K., Baba, S.A. and Darszon, A. (2004). A sea urchin egg jelly peptide induces a cGMP-mediated decrease in sperm intracellular Ca^{2+} before its increase. *Developmental Biology* **272:** 376-388

Nishigaki, T., Zamudio, F.Z., Possani, L.D. and Darszon, A. (2001). Time-resolved sperm responses to an egg peptide measured by stopped-flow fluorometry. *Biochemical and Biophysical Research Communications* **284:** 531-535

Nishigaki, T.W., Wood, C.D., Shiba, K., Baba, S.A. and Darszon, A. (2006). Stroboscopic illumination with light-emitting diode (LED) has significant advantages for fluorescence live cell imaging: a demonstration by fluorescence imaging of motile sperm. *BioTechniques* (in the press)

Quill, T.A., Ren, D., Clapham, D.E. and Garbers, D.L. (2001). A voltage-gated ion channel expressed specifically in spermatozoa. *Proceedings of the National Academy of Sciences USA* **98:** 12527-12531

Quill, T.A., Wang, D. and Garbers, D.L. (2006). Insights into sperm cell motility signaling through sNHE and the CatSpers. *Molecular and Cellular Endocrinology* **250:** 84-92.

Ren, D., Navarro, B., Perez, G., Jackson, A.C., Hsu, S., Shi, Q., Tilly, J.L. and Clapham, D.E. (2001). A sperm ion channel required for sperm motility and male fertility. *Nature* **413:** 603-609

Reynaud, E., De de La Torre, L., Zapata, O., Lievano, A. and Darszon, A. (1993). Ionic bases of the membrane potential and intracelullar pH changes induced by speract in swollen sea urchin sperm. *FEBS Letters* **329:** 210-214

Rodriguez, E. and Darszon, A. (2003). Intracellular sodium changes during the speract response and the acrosome reaction in sea urchin sperm. *Journal of Physiology* **546:** 89-100

Sakata, Y., Saegusa, H., Zong, S., Osanai, M., Murakoshi, T., Shimizu, Y., Noda, T., Aso, T. and Tanabe, T. (2002). $Ca_v2.3$ alpha1E Ca^{2+} channel participates in the control of sperm function. *FEBS Letters* **516:** 229-233

Salvatore, L., D'Adamo, M.C., Polishchuk, R., Salmona, M. and Pessia, M. (1999). Localization and age-dependent expression of the inward rectifier K^{+} channel subunit Kir 5.1 in a mammalian reproductive system. *FEBS Letters* **449:** 146-152

Schackmann, R.W. and Chock, P.B. (1986). Alteration of intracellular Ca^{2+} in sea urchin sperm by the egg peptide speract. Evidence that increased intracellulara Ca^{2+} is coupled to a Na^{+} entry and increased intracellular pH. *Journal of Biological Chemistry* **261:** 8719-8728

Serrano, C.J., Trevino, C.L., Felix, R. and Darszon, A. (1999). Voltage-dependent Ca^{2+} channel subunit expression and immunolocalization in mouse spermatogenic cells and sperm. *FEBS Letters* **462:** 171-176

Shevchuk, A.I., Frolenkov, G.I., Sanchez, D., James, P.S., Freedman, N., Lab, M.J., Jones, R., Klenerman, D. and Korchev, Y.E. (2006). Imaging proteins in membranes of living cells by high-resolution scanning ion conductance microscopy. *Angewandte Chemie- International Edition in English* **45:** 2212-2216

Solzin, J., Helbig, A., Van, Q., Brown, J.E., Hildebrand, E., Weyand, I. and Kaupp, U.B. (2004). Revisiting the role of H+ in chemotactic signaling of sperm. *Journal of General Physiology* **124:** 115-124

Suarez, S.S. and Ho, H.C. (2003). Hyperactivation of mammalian sperm. *Cellular and Molecular Biology (Noisy-le-grand)* **49:** 351-356

Suarez, S.S., Varosi, S.M. and Dai, X. (1993). Intracellular calcium increases with hyperactivation in intact, moving hamster sperm and oscillates with the flagellar beat cycle. *Proceedings of the National Academy of Sciences USA* **90:** 4660-4664

Tash, J.S. and Means, A.R. (1982). Regulation of protein phosphorylation and motility of sperm by cyclic adenosine monophosphate and calcium. *Biology of Reproduction* **26:** 745-763

Teves, M.E., Barbano, F., Guidobaldi, H.A., Sanchez, R., Miska, W. and Giojalas, L.C. (2006). Progesterone at the picomolar range is a chemoattractant for mammalian spermatozoa. *Fertility and Sterility* **86:** 745-749

Trevino, C.L., Felix, R., Castellano, L.E., Gutierrez, C., Rodriguez, D., Pacheco, J., Lopez-Gonzalez, I., Gomora, J.C., Tsutsumi, V., Hernandez-Cruz, A., Fiordelisio, T., Scaling, A.L. and Darszon, A. (2004). Expression and differential cell distribution of low-threshold Ca^{2+} channels in mammalian male germ cells and sperm. *FEBS Letters* **563:** 87-92

Trevino, C.L., Santi, C.M., Beltran, C., Hernandez-Cruz, A., Darszon, A. and Lomeli, H. (1998). Localisation of inositol trisphosphate and ryanodine receptors during mouse spermatogenesis: possible functional implications. *Zygote* **6:** 159-172

Turner, R. M. (2006). Moving to the beat: A review of mammalian sperm motility regulation. *Reproduction Fertility and Development* **18:** 25-38

Visconti, P.E., Westbrook, V.A., Chertihin, O., Demarco, I., Sleight, S. and Diekman, A. B. (2002). Novel signaling pathways involved in sperm acquisition of fertilizing capacity. *Journal of Reproductive Immunology* **53:** 133-150

Ward, G.E., Brokaw, C.J., Garbers, D.L. and Vacquier, V.D. (1985). Chemotaxis of *Arbacia punctulata* spermatozoa to resact, a peptide from the egg jelly layer. *Journal of Cell Biology* **101:** 2324-2329

Westenbroek, R.E. and Babcock, D.F. (1999). Discrete regional distributions suggest diverse functional roles of calcium channel alpha1 subunits in sperm. *Developmental Biology* **207:** 457-469

Weyand, I., Godde, M., Frings, S., Weiner, J., Muller, F., Altenhofen, W., Hatt, H. and Kaupp, U.B. (1994). Cloning and functional expression of a cyclic-nucleotide-gated channel from mammalian sperm. *Nature* **368:** 859-863

Wiesner, B., Weiner, J., Middendorff, R., Hagen, V., Kaupp, U.B. and Weyand, I. (1998). Cyclic nucleotide-gated channels on the flagellum control Ca^{2+} entry into sperm. *Journal of Cell Biology* **142:** 473-484

Wood, C.D., Darszon, A. and Whitaker, M. (2003). Speract induces calcium oscillations in the sperm tail. *Journal of Cell Biology* **161:** 89-101

Wood, C.D., Nishigaki, T., Furuta, T., Baba, S.A. and Darszon, A. (2005). Real-time analysis of the role of Ca^{2+} in flagellar movement and motility in single sea urchin sperm. *Journal of Cell Biology* **169:** 725-731

Yanagimachi, R. (1994). Mammalian fertilization. In *The Physiology of Reproduction*, Vol. 1. Edited by Knobil, E. and Neill, J.D., Raven Press, New York, pp. 189-317

Zapata, O., Ralston, J., Beltran, C., Parys, J.B., Chen, J.L., Longo, F.J. and Darszon, A. (1997). Inositol triphosphate receptors in sea urchin sperm. *Zygote* **5:** 355-364

Zeng, Y., Oberdorf, J.A. and Florman, H.M. (1996). pH regulation in mouse sperm: identification of Na^+-, Cl^--, and HCO_3^--dependent and arylaminobenzoate-dependent regulatory mechanisms and characterization of their roles in sperm capacitation. *Developmental Biology* **173:** 510-520

Zhang, D., Chen, J., Saraf, A., Cassar, S., Han, P., Rogers, J.C., Brioni, J.D., Sullivan, J. P. and Gopalakrishnan, M. (2006). Association of CatSper1 or 2 with $Ca_v3.3$ leads to suppression of T-type calcium channel activity. *Journal of Biological Chemistry* **281:** 22332-22341

Spermatology. SRF Vol. 65. ERS Roldan and M Gomendio (eds) Nottingham University Press, Nottingham

Signalling pathways involved in sperm capacitation

Ana M. Salicioni[1], Mark D. Platt[2], Eva V. Wertheimer[1], Enid Arcelay[1], Alicia Allaire[1], Julian Sosnik[1] and Pablo E. Visconti[1]

[1]*Department of Veterinary and Animal Sciences, University of Massachusetts, Amherst, MA 01003, USA;* [2]*Rensselaer Polytechnic Institute, Biotech Building, Troy, NY, USA*

After ejaculation, mammalian sperm have not yet acquired full fertilising capacity. They will require a finite period of residence in the female reproductive tract before they become fertilisation competent. The molecular, biochemical, and physiological changes that occur to sperm while in the female tract are collectively referred to as capacitation. During capacitation, changes in membrane properties, enzyme activities, and motility render spermatozoa responsive to stimuli that induce the acrosome reaction and prepare spermatozoa for penetration of the egg investments prior to fertilisation. These changes are facilitated by the activation of cell signalling cascades in the female reproductive tract *in vivo* or in defined media *in vitro*. The purposes of this review are to consider some recent contributions towards our understanding of capacitation, to summarise open questions in this field, and to discuss future avenues of research.

Introduction

Mammalian sperm are not able to fertilise eggs immediately after ejaculation. Chang (1951) and Austin (1951) independently demonstrated that sperm acquire fertilisation capacity after residing in the female tract for a finite period of time. Following these original observations, many studies confirmed that the environment of the female tract induces a series of physiological changes in the sperm; these changes are collectively called "capacitation". Taking these initial investigations into account capacitation became defined using fertilisation as end-point. However, various pieces of evidence suggest that the functional changes occurring in spermatozoa during capacitation are not one event, but a combination of sequential processes. Thus, capacitation can be divided in: (a) fast events such as the initiation of sperm motility, occurring as soon as the sperm are released from the epididymis, and (b) slow events such as changes in the motility pattern (e.g. hyperactivation) and the acquisition of the sperm capacity to undergo agonist-stimulated acrosome reaction, which are activated only after a certain period of time in conditions that support the sperm ability to fertilise the egg. In addition to these sequential events, capacitation is also the result of concomitant processes involving changes at the molecular level occurring both in the head (i.e., preparation for the acrosome reaction) and in the tail (i.e., motility changes). Although more than 50 years have passed since sperm capacitation was first reported, it is noteworthy that the molecular basis of this process is still today not well understood.

Corresponding author E-mail: pvisconti@vasci.umass.edu

Molecular basis of capacitation

Fast processes

The molecular basis of fast occurring processes in sperm has been recently the focal point of several studies (Fig. 1).

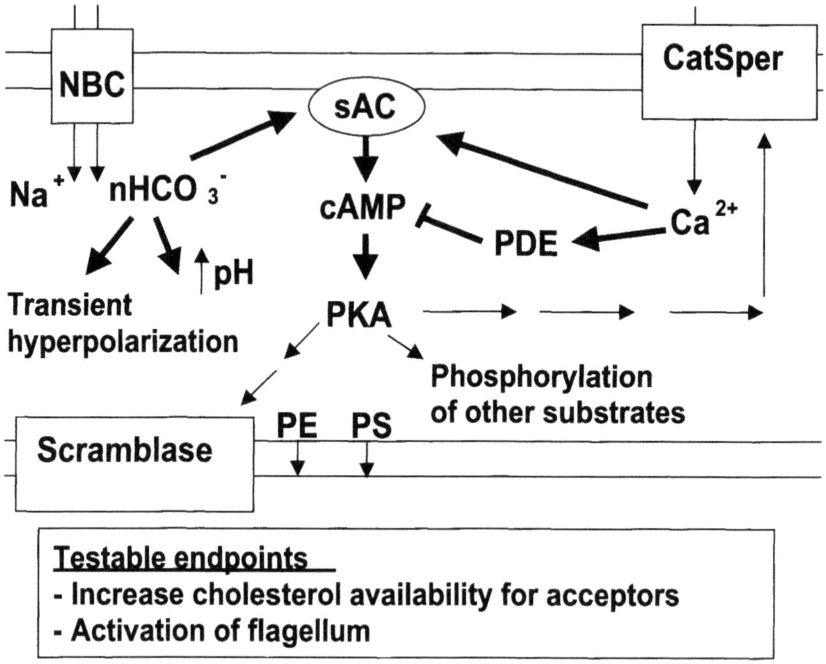

Figure 1. Fast capacitation-associated processes. Although it is controversial whether events that occur immediately after ejaculation are part of capacitation, it is important to notice that, physiologically, these processes take place in the female tract and that are likely to be necessary for fertilisation. This figure is based in work from several laboratories. (1) After ejaculation, the HCO_3^- concentration in the sperm surrounding milieu increased significantly (Setchell, Maddocks and Brooks, 1994). (2) HCO_3^- enters the sperm through a Na^+/HCO_3^- cotransporter (Demarco *et al.*, 2003). (3) Increased HCO_3^- concentration activates sAC and consequently PKA (Chen, Cann, Litvin, Iourgenko, Sinclair, Levin and Buck, 2000). (4) Activation of PKA modulates the response of CatSper to changes in membrane potential (Wennemuth *et al.*, 2003), activates a phospholipids scramblase (Harrison *et al.*, 1996; Gadella and Harrison, 2002) and increases the availability of cholesterol for external acceptors (Flesch, Brouwers, Nievelstein, Verkleij, van Golde, Colenbrander and Gadella, 2001b).

Among those, it is important to highlight work from Babcock and collaborators, and from Harrison, Gadella and collaborators. While Babcock's group has concentrated on mechanisms regulating flagellar movement, Harrison and Gadella have worked in the regulation of sperm membrane dynamics. From these studies it is possible to conclude that: (1) Exposure of sperm to HCO_3^- immediately triggers activation of their flagella and increases the depolarization-evoked rate of rise of intracellular Ca^{2+} concentration ($[Ca^{2+}]_i$)(Wennemuth, Carlson, Harper and Babcock, 2003). (2) Physiological levels of HCO_3^- induce a rapid collapse of the sperm phospholipid asymmetry mediated by scramblases (Harrison, Ashworth and Miller, 1996; Gadella and Harrison, 2002). (3) In both cases, HCO_3^- actions appear to be mediated by a cAMP pathway through activation of the soluble adenylyl cyclase (sAC, see below). (4) As a consequence of the increase in cAMP levels, protein kinase A (PKA) is activated resulting in the fast phosphorylation of a subset of proteins (Harrison, 2004). (5) A new family of Ca^{2+} channels, named CatSper, appears to mediate the depolarization-induced increase in $[Ca^{2+}]_i$ (Carlson, Westenbroek, Quill, Ren, Clapham, Hille, Garbers and Babcock, 2003). (6) Interestingly, contrary to other slower capacitation-associated events, these fast processes do not require the presence of cholesterol-acceptors such as bovine serum albumin (BSA). Although this review will be focused mainly on processes that occur after a significant period of capacitation, it is important to consider these rapidly occurring events; first, initial fast signalling in sperm capacitation is likely to be essential for the slower processes to take place; second, both fast and slow processes appear to be regulated by similar molecules (e.g. HCO_3^-, sAC, cAMP).

Slow processes

Despite the importance of fast occurring events in capacitation, until recently, most researchers have considered only the slow processes (Fig. 2) as part of sperm capacitation.

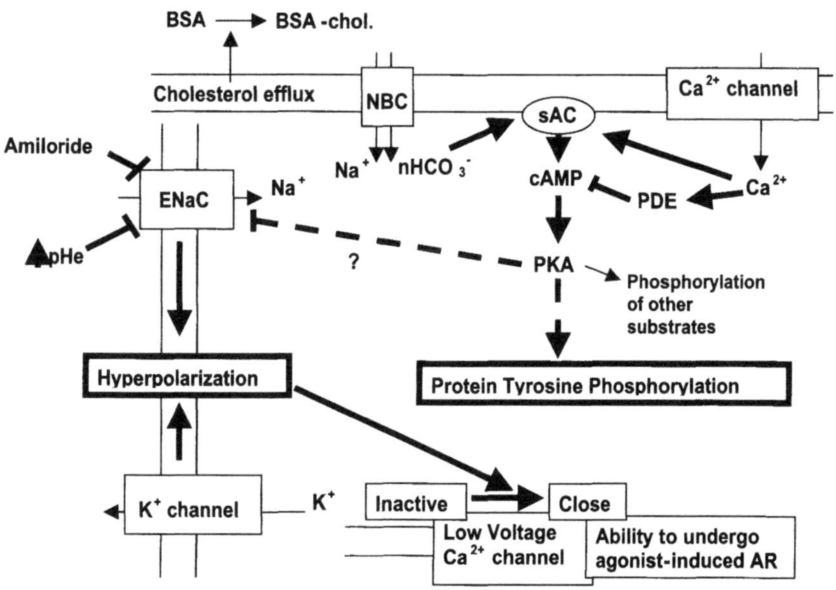

Figure 2. Slow capacitation-associated processes. Capacitation is a complex series of molecular events that occurs in sperm after epididymal maturation and confers on sperm the ability to fertilise an egg (Yanagimachi, 1994). In most cases, capacitation media contain energy substrates, such as pyruvate, lactate and glucose, a cholesterol acceptor (usually serum albumin), $NaHCO_3$, Ca^{2+}, low K^+, and physiological Na^+ concentrations. The mechanism of action by which these compounds promote capacitation is poorly understood at the molecular level; however, some molecular events significant to the initiation of capacitation have been identified and are represented in this model.

What follows is a list of endpoints that have been associated to capacitation over the years: (1) Ability of the sperm to fertilise the egg: this is the initial definition of capacitation and still, the most important evidence that sperm are capacitated. (2) Sperm hyperactivation: this change in the motility pattern is considered necessary for fertilisation; therefore, in this review we will consider hyperactivation as part of capacitation. (3) Preparation of the sperm to undergo an agonist-induced acrosome reaction: despite some controversy about the exact nature of the physiological inducer of the sperm acrosome reaction, two of the more likely agonists are the zona pellucida (ZP) of the egg and progesterone. In both cases, only capacitated sperm will react in the presence of these reagents. (4) Sperm chemotaxis: although chemotaxis in mammalian sperm is still controversial, it has been proposed that only capacitated sperm undergo chemotactic behaviour (Eisenbach, 1999). At the molecular level, during the last ten years, a series of studies have established that some signalling pathways are activated during sperm capacitation. These signalling pathways are summarised in Figure 2. This figure is based on work from many laboratories (Langlais and Roberts, 1985; Espinosa and Darszon, 1995; Visconti, Bailey, Moore, Pan, Olds-Clarke and Kopf, 1995a; Visconti, Moore, Bailey, Leclerc, Connors, Pan, Olds-Clarke and Kopf, 1995b; Zeng, Clark and Florman, 1995; Hernandez-Gonzalez, Sosnik, Edwards, Acevedo, Mendoza-Lujambio, Lopez-Gonzalez, Demarco, Wertheimer, Darszon and Visconti, 2006). Different aspects of this model have been the focus of our investigations and are summarised below.

HCO_3^- and the cAMP pathway in mammalian sperm

Capacitation is a HCO_3^--dependent process (Lee and Storey, 1986; Boatman and Robbins, 1991; Shi and Roldan, 1995; Gadella and Harrison, 2000; Visconti, Westbrook, Chertihin, Demarco, Sleight and Diekman, 2002). Recent evidence strongly suggests that HCO_3^- transport in these cells is mediated at least in part by a member of the Na^+/HCO_3^- cotransporter family (Romero and Boron, 1999). This conclusion is based on findings that HCO_3^- transport in sperm has the following properties (Demarco, Espinosa, Edwards, Sosnik, De La Vega-Beltran, Hockensmith, Kopf, Darszon and Visconti, 2003): (1) it is electrogenic, (2) it is Na^+-dependent, (3) it increases pH_i, and (4) it is blocked by stilbenes, such as DIDS. The transmembrane movement of HCO_3^- has been associated with the increase in intracellular pH (pH_i) observed during capacitation (Parrish, Susko-Parrish and First, 1989; Zeng, Oberdorf and Florman, 1996). However, another more likely target for HCO_3^- action in sperm is the regulation of cAMP metabolism (Garbers, Tubb and Hyne, 1982) through stimulation of a unique type of adenylyl cyclase (Okamura, Tajima, Soejima, Masuda and Sugita, 1985; Garty, Galiani, Aharonheim, Ho, Phillips, Dekel and Salomon, 1988; Visconti, Muschietti, Flawia and Tezon, 1990). Two types of adenylyl cyclases are responsible for cAMP synthesis in eukaryotes: transmembrane adenylyl cyclases (tmAC), and the recently isolated sAC (Buck, Sinclair, Schapal, Cann and Levin, 1999). sAC and tmACs are regulated by different pathways; sAC is insensitive to G-protein or forskolin regulation and is more active in the presence of Mn^{2+} than Mg^{2+}. Although it is still controversial how many tmACs are present in sperm (Baxendale and Fraser, 2003), multiple evidence suggests that sAC is activated during capacitation. This conclusion is supported by recent studies (Esposito, Jaiswal, Xie, Krajnc-Franken, Robben, Strik, Kuil, Philipsen, van Duin, Conti and Gossen, 2004; Hess, Jones, Marquez, Chen, Ord, Kamenetsky, Miyamoto, Zippin, Kopf, Suarez, Levin, Williams, Buck and Moss, 2005) indicating that sperm from sAC null mutant mice are not able to capacitate and consequently are infertile. In addition, a specific sAC inhibitor is able to block capacitation (Hess *et al.*, 2005). One of the targets for cAMP action is protein kinase A (PKA). Once activated, PKA phosphorylates various target proteins which are presumed to initiate several signalling pathways (Harrison and Miller, 2000; Harrison, 2004). In pig sperm exposed to HCO_3^-, cAMP rises to a maximum within 60 sec (Harrison and Miller, 2000), and the increase in PKA-dependent phosphorylation begins within 90 sec (Harrison, 2004). It is interesting that cAMP levels fall after their initial rise and then, after 7 min, begin to rise again; PKA-catalysed protein phosphorylation follows a similar time course. This second rise in cAMP levels appears to be a sustained response to HCO_3^-.

Changes that occur at the level of the plasma membrane

Capacitation is correlated with changes in the sperm plasma membrane architecture. These changes can be rapid (Harrison, 1996; Harrison and Miller, 2000; Gadella and Harrison, 2002) or slow (Visconti, Ning, Fornes, Alvarez, Stein, Connors and Kopf, 1999; Flesch, Wijnand, van de Lest, Colenbrander, van Golde and Gadella, 2001a). Interestingly, rapid changes in the sperm plasma membrane appear to be mediated by a fast activation of the HCO_3^-/sAC/PKA pathway. On the other hand, the late effects are related to the presence of cholesterol acceptors in the *in vitro* incubation medium. Bovine serum albumin (BSA) is a critical component of *in vitro* capacitation media; it is believed to function as a sink for cholesterol by removing it from the sperm plasma membrane (Davis, Byrne and Hungund, 1979; Langlais and Roberts, 1985; Cross, 1996; Visconti *et al.*, 1999). How cholesterol efflux couples to the regulation of signal transduction pathways intrinsic to capacitation is not clear at present. One possibility is that before capacitation, cholesterol concentrates in specialised plasma membrane microdomains or lipid rafts. Current concepts attribute important signalling properties to the existence of these rafts, acting to bring protein assemblies together. Taking this into consideration, depletion or supplementation of cholesterol within the plasma membrane will have profound effects on the behaviour of the raft complexes (Zajchowski and Robbins, 2002). In somatic cells, cholesterol removal is thought to disrupt lipid rafts, thus activating signalling events involving tyrosine kinases, G proteins, and/or other molecules (Kabouridis, Magee and Ley, 1997; Brown and London, 1998; Roy, Luetterforst, Harding, Apolloni, Etheridge, Stang, Rolls, Hancock and Parton, 1999). Because the activation of similar signalling events during capacitation correlates with the removal of cholesterol from the plasma membrane, it can be hypothesised that in the sperm, cholesterol may likewise be concentrated in lipid rafts and its efflux is related to changes in sperm lipid rafts. Supporting this hypothesis, caveolin has been detected in the plasma membrane overlying the acrosomal region and the flagellum of mouse and guinea pig sperm (Travis, Foster, Rosenbaum, Visconti, Gerton, Kopf and Moss, 1998; Treviño, Serrano, Beltran, Felix and Darszon, 2001) suggesting the presence of a special type of lipid raft, called caveolae, in these cells. More recent studies have shown that a 2-h incubation in HCO_3^- and BSA-containing capacitation medium induces lateral redistribution of raft marker proteins (Cross, 2004; Shadan, James, Howes and Jones, 2004) as well as disruption of lipid rafts (Sleight, Miranda, Plaskett, Maier, Lysiak, Scrable, Herr and Visconti, 2005). Among others, a possible result of cholesterol loss could be related to alterations in the steady-state intracellular ion concentrations with the resultant modification of the sperm resting membrane potential.

Sperm membrane potential

Mouse sperm capacitation is accompanied by the hyperpolarization of its plasma membrane potential (Zeng, Clark and Florman, 1995). This change results from a combination of electrogenic ion permeability changes that shift the membrane potential towards the K^+ equilibrium potential. Although the functional role of the capacitation-associated hyperpolarization is not clear, Florman's group (Florman, Arnoult, Kazam, Li and O'Toole, 1998) has proposed that since capacitation prepares the sperm for the acrosome reaction, hyperpolarization may regulate the ability of sperm to generate transient intracellular Ca^{2+} elevations during the acrosome reaction induced by physiological agonists (e.g. ZP). The major function of the voltage sensitive Ca^{2+} (Ca_v) channels is to convert changes in membrane potential into a Ca^{2+} signal. The Ca_v channel permeation pathway is formed by its α_1 subunit, encoded by a family of at least 10 genes. Ca_v channels fall into two major functional classes: high voltage-activated (HVA) and low voltage activated (LVA). LVA channels open following weak depolarizations and are encoded by the Ca_v3 subfamily of genes ($Ca_v3.1$ to $Ca_v3.3$) (Perez-Reyes, 2003). A property of Ca_v3 channels is that they are inactive at a membrane potential equivalent to those of non-capacitated sperm (~-30 mV) (Lievano, Santi, Serrano, Trevino, Bellve, Hernandez-Cruz and Darszon, 1996). Thus, if Ca_v3 channels are involved in the regulation of the acrosome reaction, the sperm membrane potential should hyperpolarise before becoming able to undergo this exocytotic event (Florman *et al.*, 1998). Interestingly, single sperm studies (Arnoult, Kazam, Visconti, Kopf, Villaz and Florman, 1999) indicated that after capacitation, sperm can be divided in two fractions. Approximately 50% of the sperm remain at a membrane potential close to the uncapacitated population while the rest hyperpolarise to -80 mV, a potential that can remove inactivation from Ca_v3 channels. Only this last population was able to undergo the acrosome reaction when exposed to solubilised ZP.

Sperm maintain an internal ion concentration markedly different from that in the extracellular medium. This difference is determined by the relative permeability of the plasma membrane to the ions found in the media, to their gradients and to the metabolic state of the cell. It is likely that the capacitation-associated hyperpolarization results from changes in the activity of ion-selective channels and transporters that control the extent of ion flow. Consistent with this hypothesis, different components of the capacitation media play important roles in the regulation of the sperm membrane potential. In mouse sperm, it has been shown that in the absence of BSA or HCO_3^- the changes in membrane potential do not occur (Demarco *et al.*, 2003). These data suggest that HCO_3^- as well as cholesterol efflux have a role in controlling events leading to hyperpolarization. Additionally, Muñoz-Garay, De la Vega-Beltran, Delgado, Labarca, Felix and Darszon (2001) demonstrated with patch clamp techniques that inward rectifying K^+ (Kir) channels are expressed in mouse spermatogenic cells and proposed that these channels may be responsible for the capacitation-associated hyperpolarization. Supporting this hypothesis, pharmacological experiments suggested that the K_{ATP} channels contribute to the capacitation-associated hyperpolarization (Acevedo, Mendoza-Lujambio, de la Vega-Beltran, Trevino, Felix and Darszon, 2006). In addition to these findings, our studies have revealed that amiloride sensitive Na^+ channels are present in mouse sperm (Hernandez-Gonzalez *et al.*, 2006). These channels constitute a new class of proteins known as epithelial Na^+ channels (ENaCs) (Kellenberger and Schild, 2002) that are expressed in many tissues of invertebrate and vertebrate organisms. Interestingly, ENaC family members may be regulated by pH, Ca^{2+}, Na^+, Cl^-, and their state of phosphorylation or that of proteins that regulate them (Kellenberger and Schild, 2002). Some of these parameters change during capacitation. ENaCs have been implicated in reproductive and early development processes in *Drosophila* (Darboux, Lingueglia, Champigny, Coscoy, Barbry and Lazdunski, 1998). How HCO_3^-, BSA and other ions integrate to regulate the changes in the sperm membrane potential is not known.

Phosphorylation events in mammalian sperm

Protein phosphorylation plays a role in the regulation of many intracellular events such as transduction of extracellular signals, intracellular transport, and cell cycle progression. Conditions favouring capacitation of mouse sperm promote tyrosine phosphorylation of a subset of proteins of Mr 40 – 120 Kda. In the absence of BSA, Ca^{2+} or $NaHCO_3$ neither capacitation nor the increase in tyrosine phosphorylation are observed (Visconti *et al.*, 1995a). Interestingly, the increase in tyrosine phosphorylation is regulated by a cAMP-dependent pathway in mouse sperm (Visconti *et al.*, 1995b) and other species (Leclerc, de Lamirande and Gagnon, 1996; Galantino-Homer, Visconti and Kopf, 1997; Kalab, Peknicova, Geussova and Moos, 1998). Because PKA is a serine/threonine kinase and not a tyrosine kinase, these experiments strongly suggest that capacitation is regulated by a protein kinase cascade (see Fig. 2). Despite this knowledge, except for PKA, the identity and functions of other kinases and their targets in sperm are not well defined. An initial approach to investigate the role of phosphorylation in capacitation is to identify proteins phosphorylated during this process and to characterise the kinases involved in their phosphorylation. In this respect, the use of two dimensional (2-D) polyacrylamide gel electrophoresis (PAGE) followed by tandem mass spectrometry (MS/MS) provides a comprehensive approach to the analysis of proteins involved in cell signalling (Blomberg, 1997; Alms, Sanz, Carlson and Haystead, 1999; Soskic, Gorlach, Poznanovic, Boehmer and Godovac-Zimmermann, 1999; Lewis, Hunt, Aveline, Jonscher, Louie, Yeh, Nahreini, Resing and Ahn, 2000). Specifically, changes in tyrosine phosphorylation can be monitored using 2-D gel electrophoresis (O'Farrell, 1975; Gorg, Postel and Gunther, 1988) followed by Western blot analysis with anti phosphotyrosine (α PY) antibodies. Proteins that undergo changes in tyrosine phosphorylation during cellular processes can then be isolated from a duplicated gel stained with silver and sequenced by MS/MS. We have used this strategy to identify several sperm proteins that undergo tyrosine phosphorylation during capacitation (Ficarro, Chertihin,Westbrook, White, Jayes, Kalab, Marto, Shabanowitz, Herr, Hunt and Visconti, 2003).

Identification of phosphorylation substrates using 2-D PAGE is a powerful approach. However, this methodology has some limitations. (1) It is not completely appropriate to identify proteins that undergo phosphorylation in serine or threonine residues since antibodies against those phosphorylated residues have not achieved sufficient quality and sensitivity. (2) MS/MS on proteins isolated by PAGE has detection limits several orders of magnitude lower than MS/MS performed on proteins not embedded in gels. (3) Although in some cases it is possible to get the exact site of phosphorylation of a candidate protein, more often the phosphorylation site remains elusive due to the presence of more abundant peptides that do not present the phosphorylation site.

Recently, several methods for the selective detection and enrichment of phosphopeptides have been developed (Porath, 1992; Cao and Stults, 1999; Posewitz and Tempst, 1999; Cao and Stults, 2000; Annan, Huddleston, Verma, Deshaies and Carr, 2001). However, most of them have been applied only on a protein-by-protein basis. Another method for enrichment of phosphoproteins is the use of Fe^{3+}-immobilised metal affinity chromatography (IMAC) prior to MS/MS to enrich digests for peptides containing phosphorylated amino acids. This technique was used by a number of investigators (Cao and Stults, 1999; Cao and Stults, 2000; Zarling, Ficarro, White, Shabanowitz, Hunt and Engelhard, 2000) but proved to generate false positives as acidic residues (i.e. glutamic and aspartic acid) will readily bind to IMAC (Muszynska, Dobrowolska, Medin, Ekman and Porath, 1992). To increase the selectivity of the IMAC column for phosphopeptides, we have used a modification of this technique (Ficarro, McCleland, Stukenberg, Burke, Ross, Shabanowitz, Hunt and White, 2002) in which acidic residues are converted to methyl esters to block their binding to iron before IMAC is employed. A flow diagram of the method is summarised in Fig. 3. Using this methodology, we have characterised 5 sites of tyrosine, 56 sites of serine and 2 sites of threonine phosphorylation in capacitated human sperm (Ficarro *et al.*, 2003).

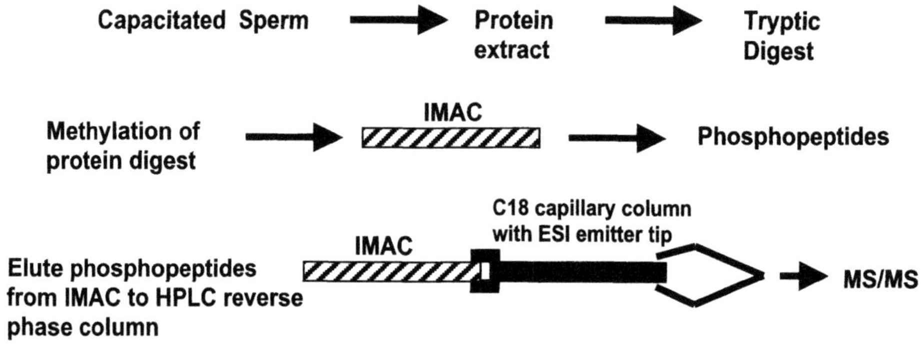

Figure 3. Flow chart of the IMAC/RP-HPLC/ESI/MS/MS analysis of capacitated sperm. Carboxy-methylated phosphopeptides were enriched using IMAC and then identified using nano-flow reverse phase (RP)-HPLC micro-electrospray ionization (nHPLC-mESI) mass spectrometry (MS) on a hybrid linear quadrupole ion trap/Fourier transform (LTQ/FT) ion cyclotron resonance mass spectrometer.

Although the combination of IMAC and MS/MS is ideally suited for the characterization of phosphorylation sites on proteins in complex mixtures, it is also important to determine which sites are phosphorylated in response to a particular signalling pathway. Common methods toward this goal include comparison of proteins labelled *in vitro* with [^{32}P]ATP, use of α PY antibodies or other phospho-specific antibodies, and *in vivo* labelling using inorganic [^{32}P] followed by immunoprecipitation and/or 2D gel analysis. We have developed an alternative method to compare phosphorylation of defined sequences in two different cell populations by differential isotopic labelling (Ficarro *et al.*, 2003). Details on this method are given below. The evaluation of differential phosphorylation added to the knowledge of the exact phosphorylated sequence goes beyond the sperm capacitation field and could be used to understand signalling mechanisms in multiple biological systems.

Differential MS/MS analysis of phosphopeptides during the capacitation process

Similarly to other high throughput methodologies such as microarrays, the advantage of a global MS/MS approach to identify phosphopeptides is that this methodology is able to generate large amount of data in a relatively short time. On the minus side, these high throughput techniques are often regarded as a descriptive analysis of a particular problem. More information can be achieved when these techniques are used to analyse functional changes occurring in a biological process. One interesting approach used differential isotopic labelling to compare phosphorylated sequences from capacitated and non-capacitated sperm populations (Ficarro *et al.*, 2003) (Fig. 4). This analysis is based in mass spectrometry; therefore, differential labelling is achieved by the use of isotopes of different mass for each sample. In our case, to compare non-capacitated with capacitated sperm populations by differential labelling, the aforementioned carboxymethylation reaction of acidic residues was used. Briefly, tryptic peptides from each cell population were converted to peptide methyl esters with deuterated (CD_3OH) (d3) and nondeuterated (CH_3OH) (d0) methanol, respectively. Both samples are then mixed in equal proportions and the mixture purified by IMAC to ensure that only phosphopeptides are retained. Signals for phosphopeptides present in both samples appear as doublets separated by a mass calculated as: n(3Dalton)/z (where n is the number of carboxylic acid groups in the peptide and z is the charge on the peptide). The ratio of the two signals in the doublet changes as a function of the phosphorylation or dephosphorylation that occurs during capacitation. Peptides of interest are then targeted for sequence analysis subsequently performed on the ion trap instrument. Moreover, comparison of doublets will give a quantitative estimate of the level of phosphorylation of each sequence. The differential isotopic labelling of phosphopeptides gives for the first time the possibility to analyse any phosphorylation site in a single experiment. In addition, this technique completes the information given by Western blot analysis since it identifies the exact sequence in a particular protein that becomes phosphorylated after a specific stimulus.

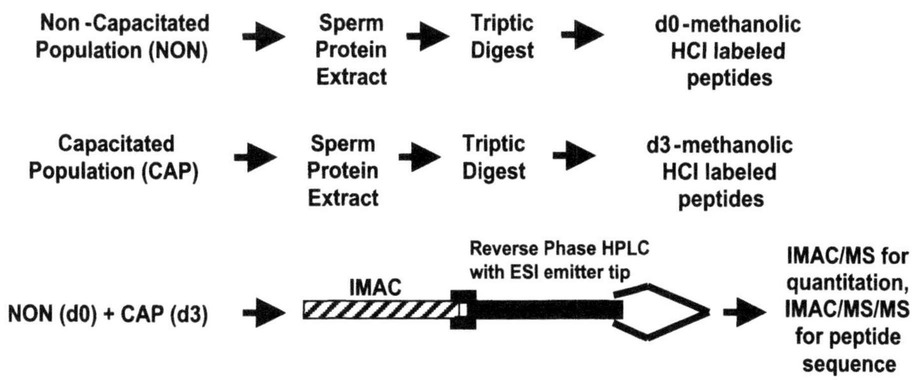

Figure 4. Schematic representation of the procedure for phosphopeptide comparison between two samples. Differential labeling is achieved by the use of a different isotope of hydrogen for each sample. Tryptic peptides from two samples of cells (i.e. non-capacitated *vs* capacitated sperm) are converted to peptide methyl esters with deuterated (CD_3OH) (d3) and nondeuterated (CH_3OH) (d0) methanol, respectively. Both samples are then mixed in equal proportions and the mixture purified by IMAC to ensure that only phosphopeptides are retained. Signals for phosphopeptides present in both samples appear as doublets separated by n(3 Da)/z (where n is the number of carboxylic acid groups in the peptide and z is the charge on the peptide). The ratio of the two signals in the doublet changes as a function of the phosphorylation or dephosphorylation that occurs during capacitation. Peptides of interest are then targeted for sequence analysis subsequently performed on the ion trap instrument.

Kinases in mammalian sperm

Protein kinases play essential roles in the regulation of cellular processes. Therefore, it is not surprising that several protein kinases have been shown to be involved in spermatogenesis (Sassone-Corsi, 1997). However, it is not known which of these kinases remains in the mature sperm and has a function during capacitation. Few kinases have been described in mature mammalian sperm using antibodies and, apart from PKA, their functional role in sperm have not been established. Some of these kinases are protein kinase C (PKC) (Rotem, Paz, Homonnai, Kalina and Naor, 1990), GSK 3 (Vijayaraghavan, Mohan, Gray, Khatra and Carr, 2000), casein kinase II (Chaudhry, Nanez and Casillas, 1991a; Chaudhry, Newcomer and Casillas, 1991b), MAP kinase (Luconi, Barni, Vannelli, Krausz, Marra, Benedetti, Evangelista, Francavilla, Properzi, Forti and Baldi, 1998) and at least one member of the testis specific serine kinase (Tssk) family (Hao, Jha, Kim, Vemuganti, Westbrook, Chertihin, Markgraf, Flickinger, Coppola, Herr and

Visconti, 2004). Less is known about the identity of tyrosine kinases in sperm. Although a tyrosine kinase activity has been partially purified from boar sperm (Berruti, 1994), the identity of the kinase responsible for this activity is still not known. There is some evidence for the presence of yes (Leclerc and Goupil, 2002) and src (Baker, Hetherington and Aitken, 2006) and also of a tyrosine kinase receptor in human sperm (ZRK) (Burks, Carballada, Moore and Saling, 1995); these kinases could mediate the effect the capacitation-associated increase in protein tyrosine phosphorylation. In particular, it is interesting the presence of csk in mouse sperm (Baker *et al.*, 2006). This kinase negatively regulates src kinase activity and it is at the same time directly inhibited by PKA, opening the possibility that this pathway is involved in the regulation of the capacitation-associated increase in tyrosine phosphorylation (Baker *et al.*, 2006). Undoubtedly, the presence of this kinase cascade in sperm warrants further investigation. Because protein kinases have become major targets for the development of novel drugs, identification of those that regulate capacitation could offer new opportunities towards an alternative approach for contraception.

Capacitation and the acrosome reaction

The acrosome is a membrane-limited organelle which overlies the sperm nucleus. In response to either physiological or pharmacological stimuli, the outer acrosomal membrane and the overlying plasma membrane undergo fusion and vesiculation leading to the exposure of the acrosomal contents to the extracellular environment (Yanagimachi, 1994). This exocytotic process is called acrosome reaction and its completion is an absolute prerequisite for fertilisation. It is noteworthy that a physiologically induced acrosome reaction cannot occur in sperm that have not undergone capacitation. Recent work has shown that N-ethyl maleimide-sensitive fusion protein (NSF) and soluble NSF attachment protein receptors (SNAREs) (Jahn and Sudhof, 1999) are present in sea urchin (Schulz, Wessel and Vacquier, 1997; Schulz, Sasaki and Vacquier, 1998) and mammalian sperm (Michaut, Tomes, De Blas, Yunes and Mayorga, 2000; Ramalho-Santos, Moreno, Sutovsky, Chan, Hewitson, Wessel, Simerly and Schatten, 2000; Yunes, Michaut, Tomes and Mayorga, 2000). These observations support the idea that the sperm acrosome reaction might be regulated in similar ways as exocytotic processes in somatic cells. However, the acrosome reaction also presents differences with other known exocytotic events. Some of these differences are: (1) The acrosome is a single secretory vesicle. (2) There are multiple fusion points between the outer acrosomal membrane and the plasma membrane. (3) Both these membranes form mixed vesicles that are shed during the acrosome reaction, resulting in membrane loss. (4) The acrosome reaction is a one-shot fusion event; thus there is no membrane recycling.

One may postulate that components of the sperm exocytotic machinery are modified during capacitation. Some of these alterations may involve changes in the phosphorylation status of certain proteins, changes in protein localization, and/or modification of protein-protein interactions. Experiments leading to the identification and characterization of these effector molecules will further increase our understanding of capacitation. Among the proteins that undergo tyrosine phosphorylation during capacitation, we identified valosin containing protein (VCP). The 97-kDa VCP is a member of the type II AAA (ATPases associated with a variety of activities) ATPases, which are characterised by the presence of two conserved ATPase domains (Wang, Song and Li, 2004). As other AAA proteins, VCP is an enzymatic machine. It catalyses ATP hydrolysis to generate energy and uses the energy to perform mechanical work in cells. VCP, also known as p97, is highly evolutionarily conserved and homologues of this protein can be found in archaebacteria (VAT), in yeast (CDC48) and in *Drosophila* (TER94) (Woodman, 2003).

In the last few years, emerging biochemical and genetic evidence has associated VCP with a range of functions. Among them, VCP is related to retranslocation of unfolded protein from the endoplasmic reticulum and to fusion events during homotypic fusion of smooth endoplasmic reticulum membranes and in the reformation of Golgi cisternae that occurs as the cell exits mitosis.

Since capacitation prepares sperm to undergo a regulated exocytosis (e.g. acrosome reaction), phosphorylation of proteins involved in fusion events may regulate this process and are of particular interest. VCP undergoes tyrosine phosphorylation during capacitation. This conclusion is based on two independent lines of evidence. First, using 2-D gel electrophoresis, a protein spot that undergoes tyrosine phosphorylation during human sperm capacitation was identified as VCP (Ficarro *et al.*, 2003). Second, α PY immunoprecipitates of capacitated human sperm contain more VCP than the equivalent immunoprecipitates from a non-capacitated population (Ficarro *et al.*, 2003). Tyrosine phosphorylation of VCP during sperm capacitation has also been reported in boar sperm (Geussova, Kalab and Peknicova, 2002). In addition, in human sperm VCP changes its immunofluorescence pattern during capacitation (Ficarro *et al.*, 2003). It can be hypothesised that VCP function in sperm is regulated by phosphorylation during capacitation and that changes in localization of VCP during capacitation are involved in the regulation of the acrosome reaction. More work will be necessary to evaluate this hypothesis.

Conclusions

This review is an attempt to summarise the current knowledge on sperm capacitation. Although this process was discovered more than 50 years ago, its molecular basis is still not well defined. Technical advances in the last years as well as knock out studies on sperm proteins warrant new discoveries in the mechanisms of capacitation and in the regulation of sperm function in general. In addition, it is important to notice that most discoveries in sperm capacitation come from *in vitro* experiments. Despite the importance of these observations, it should be taken into account for future research the ability of the female tract to control the speed of capacitation and the delivery of capacitated sperm to the site of fertilisation.

Acknowledgements

This review was supported by the National Institutes of Health, Grants HD38082 and HD44044 (to PEV).

References

Acevedo, J.J., Mendoza-Lujambio, I., de la Vega-Beltran, J.L., Trevino, C.L., Felix, R. and Darszon, A. (2006). K_{ATP} channels in mouse spermatogenic cells and sperm, and their role in capacitation. *Developmental Biology* **289**: 395-405

Alms, G.R., Sanz, P., Carlson, M. and Haystead, T.A. (1999). Reg1p targets protein phosphatase 1 to dephosphorylate hexokinase II in Saccharomyces cerevisiae: characterizing the effects of a phosphatase subunit on the yeast proteome. *EMBO Journal* **18**: 4157-4160

Annan, R.S., Huddleston, M.J., Verma, R., Deshaies, R.J. and Carr, S.A. (2001). A multidimensional electrospray MS-based approach to phosphopeptide mapping. *Analytical Chemistry* **73**: 393-404

Arnoult, C., Kazam, I.G., Visconti, P.E., Kopf, G.S., Villaz, M. and Florman, H.M. (1999). Control of the low voltage-activated calcium channel of mouse sperm by egg ZP3 and by membrane hyperpolarization during capacitation. *Proceeding National Academy Science USA* **96**: 6757-6762

Austin, C.R. (1951). Observations on the penetration of the sperm in the mammalian egg. *Australian Journal of Scientific Research (B)* **4**: 581-596

Baker, M.A., Hetherington, L. and Aitken, R.J. (2006). Identification of SRC as a key PKA-stimulated tyrosine kinase involved in the capacitation-associated hyperactivation of murine spermatozoa. *Journal of Cell Science* **119**: 3182-3192

Baxendale, R.W. and Fraser, L.R. (2003). Evidence for multiple distinctly localized adenylyl cyclase isoforms in mammalian spermatozoa. *Molecular Reproduction and Development* **66**: 181-189

Berruti, G. (1994). Biochemical characterization of the boar sperm 42 kilodalton protein tyrosine kinase: its potential for tyrosine as well as serine phosphorylation towards microtubule-associated protein 2 and histone H 2B. *Molecular Reproduction and Development* **38**: 386-392

Blomberg, A. (1997). Osmoresponsive proteins and functional assessment strategies in *Saccharomyces cerevisiae*. *Electrophoresis* **18**: 1429-1440

Boatman, D.E. and Robbins, R.S. (1991). Bicarbonate: carbon-dioxide regulation of sperm capacitation, hyperactivated motility, and acrosome reactions. *Biology of Reproduction* **44**: 806-813

Brown, D.A. and London, E. (1998). Functions of lipid rafts in biological membranes. *Annual Reviews in Cell and Developemental Biology* **14**: 111-136

Buck, J., Sinclair, M.L., Schapal, L., Cann, M.J. and Levin, L.R. (1999). Cytosolic adenylyl cyclase defines a unique signaling molecule in mammals. *Proceedings of the National Academy of Sciences USA* **96**: 79-84

Burks, D.J., Carballada, R., Moore, H.D. and Saling, P.M. (1995). Interaction of a tyrosine kinase from human sperm with the zona pellucida at fertilization. *Science* **269**: 83-86

Cao, P. and Stults, J.T. (1999). Phosphopeptide analysis by on-line immobilized metal-ion affinity chromatography-capillary electrophoresis-electrospray ionization mass spectrometry. *Journal of Chromatography A* **853**: 225-235

Cao, P. and Stults, J.T. (2000). Mapping the phosphorylation sites of proteins using on-line immobilized metal affinity chromatography/capillary electrophoresis/electrospray ionization multiple stage tandem mass spectrometry. *Rapid Communications in Mass Spectrometry* **14**: 1600-1606

Carlson, A.E., Westenbroek, R.E., Quill, T., Ren, D., Clapham, D.E., Hille, B., Garbers, D.L. and Babcock, D.F. (2003). CatSper1 required for evoked Ca^{2+} entry and control of flagellar function in sperm. *Proceedings of the National Academy of Sciences USA* **100**: 14864-14868

Chang, M.C. (1951). Fertilizing capacity of spermatozoa deposited into the fallopian tubes. *Nature* **168**: 697-698

Chaudhry, P.S., Nanez, R. and Casillas, E.R. (1991a). Purification and characterization of polyamine-stimulated protein kinase (casein kinase II) from bovine spermatozoa. *Archives in Biochemistry and Biophysics* **288**: 337-342

Chaudhry, P.S., Newcomer, P.A. and Casillas, E.R. (1991b). Casein kinase I in bovine sperm: purification and characterization. *Biochemical Biophysical Research Communications* **179**: 592-598

Chen, Y., Cann, M.J., Litvin, T.N., Iourgenko, V., Sinclair, M.L., Levin, L.R. and Buck, J. (2000). Soluble adenylyl cyclase as an evolutionarily conserved bicarbonate sensor [and comments]. *Science* **289**: 625-628

Cross, N.L. (1996). Effect of cholesterol and other sterols on human sperm acrosomal responsiveness. *Molecular Reproduction and Development* **45**: 212-217

Cross, N.L. (2004). Reorganization of lipid rafts during capacitation of human sperm. *Biology of Reproduction* **71**: 1367-1373

Darboux, I., Lingueglia, E., Champigny, G., Coscoy, S., Barbry, P. and Lazdunski, M. (1998). dGNaC1, a gonad-specific amiloride-sensitive Na^+ channel. *Journal of Biological Chemistry* **273**: 9424-9429

Davis, B.K., Byrne, R. and Hungund, B. (1979). Studies on the mechanism of capacitation. II. Evidence for lipid transfer between plasma membrane of rat sperm and serum albumin during capacitation in vitro. *Biochimica et Biophysica Acta* **558**: 257-266

Demarco, I.A., Espinosa, F., Edwards, J., Sosnik, J., De La Vega-Beltran, J.L., Hockensmith, J.W., Kopf, G.S., Darszon, A. and Visconti, P.E. (2003). Involvement of a Na^+/HCO_3^- cotransporter in mouse sperm capacitation. *Journal of Biological Chemistry* **278**: 7001-7009

Eisenbach, M. (1999). Mammalian sperm chemotaxis and its association with capacitation. *Developmental Genetics* **25**: 87-94

Espinosa, F. and Darszon, A. (1995). Mouse sperm membrane potential: changes induced by Ca^{2+}. *FEBS Letters* **372**: 119-125

Esposito, G., Jaiswal, B.S., Xie, F., Krajnc-Franken, M.A., Robben, T.J., Strik, A.M., Kuil, C., Philipsen, R.L., van Duin, M., Conti, M. and Gossen, J.A. (2004). Mice deficient for soluble adenylyl cyclase are infertile because of a severe sperm-motility defect. *Proceedings of the National Academy of Sciences USA* **101**: 2993-2998

Ficarro, S., Chertihin, O., Westbrook, V.A., White, F., Jayes, F., Kalab, P., Marto, J.A., Shabanowitz, J., Herr, J.C., Hunt, D. and Visconti, P.E. (2003). Phosphoproteome analysis of human sperm. Evidence of tyrosine phosphorylation of AKAP 3 and valosin containing protein/P97 during capacitation. *Journal of Biological Chemistry* **278**: 11579-11589

Ficarro, S.B., McCleland, M.L., Stukenberg, P.T., Burke, D.J., Ross, M.M., Shabanowitz, J., Hunt, D.F. and White, F.M. (2002). Phosphoproteome analysis by mass spectrometry and its application to Saccharomyces cerevisiae. *Nature Biotechnology* **20**: 301-306

Flesch, F.M., Wijnand, E., van de Lest, C.H., Colenbrander, B., van Golde, L.M. and Gadella, B.M. (2001a). Capacitation dependent activation of tyrosine phosphorylation generates two sperm head plasma membrane proteins with high primary binding affinity for the zona pellucida. *Molecular Reproduction and Development* **60**: 107-115

Flesch, F.M., Brouwers, J.F., Nievelstein, P.F., Verkleij, A.J., van Golde, L.M., Colenbrander, B. and Gadella, B.M. (2001b). Bicarbonate stimulated phospholipid scrambling induces cholesterol redistribution and enables cholesterol depletion in the sperm plasma membrane. *Journal of Cell Science* **114**: 3543-3555

Florman, H.M., Arnoult, C., Kazam, I.G., Li, C. and O'Toole, C.M. (1998). A perspective on the control of mammalian fertilization by egg-activated ion channels in sperm: a tale of two channels. *Biology of Reproduction* **59**: 12-16

Gadella, B.M. and Harrison, R.A. (2000). The capacitating agent bicarbonate induces protein kinase A-dependent changes in phospholipid transbilayer behavior in the sperm plasma membrane. *Development* **127**: 2407-2420

Gadella, B.M. and Harrison, R.A. (2002). Capacitation induces cyclic adenosine 3',5'-monophosphate-dependent, but apoptosis-unrelated, exposure of aminophospholipids at the apical head plasma membrane of boar sperm cells. *Biology of Reproduction* **67**: 340-350

Galantino-Homer, H.L., Visconti, P.E. and Kopf, G.S. (1997). Regulation of protein tyrosine phosphorylation during bovine sperm capacitation by a cyclic adenosine 3'5'-monophosphate-dependent pathway. *Biology of Reproduction* **56**: 707-719

Garbers, D.L., Tubb, D.J. and Hyne, R.V. (1982). A requirement of bicarbonate for Ca^{2+}-induced elevations of cyclic AMP in guinea pig spermatozoa. *Journal of Biological Chemistry* **257**: 8980-8984

Garty, N.B., Galiani, D., Aharonheim, A., Ho, Y.K., Phillips, D.M., Dekel, N. and Salomon, Y. (1988). G-proteins in mammalian gametes: an immunocytochemical study. *Journal of Cell Science* **91**: 21-31

Geussova, G., Kalab, P. and Peknicova, J. (2002). Valosine containing protein is a substrate of cAMP-activated boar sperm tyrosine kinase. *Molecular Reproduction and Development* **63**: 366-375

Gorg, A., Postel, W. and Gunther, S. (1988). The current state of two-dimensional electrophoresis with immobilized pH gradients. *Electrophoresis* **9**: 531-546

Hao, Z., Jha, K.N., Kim, Y.H., Vemuganti, S., Westbrook, V.A., Chertihin, O., Markgraf, K., Flickinger, C.J., Coppola, M., Herr, J.C. and Visconti, P.E. (2004). Expression analysis of the human testis-specific serine/threonine kinase (TSSK) homologues. A TSSK member is present in the equatorial segment of human sperm. *Molecular Human Reproduction* **10**: 433-444

Harrison, R.A. (1996). Capacitation mechanisms, and the role of capacitation as seen in eutherian mammals. *Reproduction Fertility and Development* **8**: 581-594

Harrison, R.A. (2004). Rapid PKA-catalysed phosphorylation of boar sperm proteins induced by the capacitating agent bicarbonate. *Molecular Reproduction and Development* **67**: 337-352

Harrison, R.A., Ashworth, P.J. and Miller, N.G. (1996). Bicarbonate/CO_2, an effector of capacitation, induces a rapid and reversible change in the lipid architecture of boar sperm plasma membranes. *Molecular Reproduction and Development* **45**: 378-391

Harrison, R.A. and Miller, N.G. (2000). cAMP-dependent protein kinase control of plasma membrane lipid architecture in boar sperm. *Molecular Reproduction and Development* **55**: 220-228

Hernandez-Gonzalez, E.O., Sosnik, J., Edwards, J., Acevedo, J.J., Mendoza-Lujambio, I., Lopez-Gonzalez, I., Demarco, I., Wertheimer, E., Darszon, A. and Visconti, P.E. (2006). Sodium and epithelial sodium channels participate in the regulation of the capacitation-associated hyperpolarization in mouse sperm. *Journal of Biological Chemistry* **281**: 5623-5633

Hess, K.C., Jones, B.H., Marquez, B., Chen, Y., Ord, T.S., Kamenetsky, M., Miyamoto, C., Zippin, J.H., Kopf, G.S., Suarez, S.S., Levin, L.R., Williams, C.J., Buck, J. and Moss, S.B. (2005). The "soluble" adenylyl cyclase in sperm mediates multiple signaling events required for fertilization. *Developmental Cell* **9**: 249-259

Jahn, R. and Sudhof, T.C. (1999). Membrane fusion and exocytosis. *Annual Reviews of Biochemistry* **68**: 863-911

Kabouridis, P.S., Magee, A.I. and Ley, S.C. (1997). S-acylation of LCK protein tyrosine kinase is essential for its signalling function in T lymphocytes. *EMBO Journal* **16**: 4983-4998

Kalab, P., Peknicova, J., Geussova, G. and Moos, J. (1998). Regulation of protein tyrosine phosphorylation in boar sperm through a cAMP-dependent pathway. *Molecular Reproduction and Development* **51**: 304-314

Kellenberger, S. and Schild, L. (2002). Epithelial sodium channel/degenerin family of ion channels: a variety of functions for a shared structure. *Physiological Reviews* **82**: 735-767

Langlais, J. and Roberts, J.D. (1985). A molecular membrane model of sperm capacitation and the acrosome reaction of mammalian spermatozoa. *Gamete Research* **12**: 183-224

Leclerc, P., de Lamirande, E. and Gagnon, C. (1996). Cyclic adenosine 3',5'monophosphate-dependent regulation of protein tyrosine phosphorylation in relation to human sperm capacitation and motility. *Biology of Reproduction* **55**: 684-692

Leclerc, P. and Goupil, S. (2002). Regulation of the human sperm tyrosine kinase c-yes. Activation by cyclic adenosine 3',5'-monophosphate and inhibition by Ca^{2+}. *Biology of Reproduction* **67**: 301-307

Lee, M.A. and Storey, B.T. (1986). Bicarbonate is essential for fertilization of mouse eggs: mouse sperm require it to undergo the acrosome reaction. *Biology of Reproduction* **34**: 349-356

Lewis, T.S., Hunt, J.B., Aveline, L.D., Jonscher, K.R., Louie, D.F., Yeh, J.M., Nahreini, T.S., Resing, K.A. and Ahn, N.G. (2000). Identification of novel MAP kinase pathway signaling targets by functional proteomics and mass spectrometry. *Molecular Cell* **6**: 1343-1354

Lievano, A., Santi, C.M., Serrano, C.J., Trevino, C.L., Bellve, A.R., Hernandez-Cruz, A. and Darszon, A. (1996). T-type Ca^{2+} channels and alpha1E expression in spermatogenic cells and their possible relevance to the sperm acrosome reaction. *FEBS Letters* **388**: 150-154

Luconi, M., Barni, T., Vannelli, G.B., Krausz, C., Marra, F., Benedetti, P.A., Evangelista, V., Francavilla, S., Properzi, G., Forti, G. and Baldi, E. (1998). Extracellular signal-regulated kinases modulate capacitation of human spermatozoa. *Biology of Reproduction* **58**: 1476-1489

Michaut, M., Tomes, C.N., De Blas, G., Yunes, R. and Mayorga, L.S. (2000). Calcium-triggered acrosomal exocytosis in human spermatozoa requires the coordinated activation of Rab3A and N-ethylmaleimide-sensitive factor. *Proceedings of the National Academy of Sciences USA* **97**: 9996-10001

Muñoz-Garay, C., De la Vega-Beltran, J.L., Delgado, R., Labarca, P., Felix, R. and Darszon, A. (2001). Inwardly rectifying K^+ channels in spermatogenic cells: functional expression and implication in sperm capacitation. *Developmental Biology* **234**: 261-274

Muszynska, G., Dobrowolska, G., Medin, A., Ekman, P. and Porath, J.O. (1992). Model studies on iron(III) ion affinity chromatography. II. Interaction of immobilized iron(III) ions with phosphorylated amino acids, peptides and proteins. *Journal of Chromatography* **604**: 19-28

O'Farrell, P.H. (1975). High resolution two-dimensional electrophoresis of proteins. *Journal of Biological Chemistry* **250**: 4007-4021

Okamura, N., Tajima, Y., Soejima, A., Masuda, H. and Sugita, Y. (1985). Sodium bicarbonate in seminal plasma stimulates the motility of mammalian spermatozoa through direct activation of adenylate cyclase. *Journal of Biological Chemistry* **260**: 9699-9705

Parrish, J.J., Susko-Parrish, J.L. and First, N.L. (1989). Capacitation of bovine sperm by heparin: inhibitory effect of glucose and role of intracellular pH. *Biology of Reproduction* **41**: 683-699

Perez-Reyes, E. (2003). Molecular physiology of low-voltage-activated T-type calcium channels. *Physiological Reviews* **83**: 117-161

Porath, J. (1992). Immobilized metal ion affinity chromatography. *Protein Expression and Purification* **3**: 263-281

Posewitz, M.C. and Tempst, P. (1999). Immobilized gallium(III) affinity chromatography of phosphopeptides. *Analytical Chemistry* **71**: 2883-2892

Ramalho-Santos, J., Moreno, R.D., Sutovsky, P., Chan, A.W., Hewitson, L., Wessel, G.M., Simerly, C.R. and Schatten, G. (2000). SNAREs in mammalian sperm: possible implications for fertilization. *Developmental Biology* **223**: 54-69

Romero, M.F. and Boron, W.F. (1999). Electrogenic Na^+/HCO_3^- cotransporters: cloning and physiology. *Annual Review of Physiology* **61**: 699-723

Rotem, R., Paz, G.F., Homonnai, Z.T., Kalina, M. and Naor, Z. (1990). Protein kinase C is present in human sperm: possible role in flagellar motility. *Proceedings of the National Academy of Sciences USA* **87**: 7305-7308

Roy, S., Luetterforst, R., Harding, A., Apolloni, A., Etheridge, M., Stang, E., Rolls, B., Hancock, J.F. and Parton, R.G. (1999). Dominant-negative caveolin inhibits H-Ras function by disrupting cholesterol-rich plasma membrane domains [and comments]. *Nature Cell Biology* **1**: 98-105

Sassone-Corsi, P. (1997). Transcriptional checkpoints determining the fate of male germ cells. *Cell* **88**: 163-166

Schulz, J.R., Sasaki, J.D. and Vacquier, V.D. (1998). Increased association of synaptosome-associated protein of 25 kDa with syntaxin and vesicle-associated membrane protein following acrosomal exocytosis of sea urchin sperm. *Journal of Biological Chemistry* **273**: 24355-24359

Schulz, J.R., Wessel, G.M. and Vacquier, V.D. (1997). The exocytosis regulatory proteins syntaxin and VAMP are shed from sea urchin sperm during the acrosome reaction. *Developmental Biology* **191**: 80-87

Setchell, B.P., Maddocks, S. and Brooks, D.E. (1994). Anatomy, vasculature, innervation, and fluids of the male reproductive tract. In: *The Physiology of Reproduction,* Volume 1. Edited by Knobil, E. and Neill, J.D., Raven Press, New York, pp. 1063-1175

Shadan, S., James, P.S., Howes, E.A. and Jones, R. (2004). Cholesterol efflux alters lipid raft stability and distribution during capacitation of boar spermatozoa. *Biology of Reproduction* **71**: 253-265

Shi, Q.X. and Roldan, E.R. (1995). Bicarbonate/CO_2 is not required for zona pellucida- or progesterone-induced acrosomal exocytosis of mouse spermatozoa but is essential for capacitation. *Biology of Reproduction* **52**: 540-546

Sleight, S.B., Miranda, P.V., Plaskett, N.W., Maier, B., Lysiak, J., Scrable, H., Herr, J.C. and Visconti, P.E. (2005). Isolation and proteomic analysis of mouse sperm detergent-resistant membrane fractions: evidence for dissociation of lipid rafts during capacitation. *Biology of Reproduction* **73**: 721-729

Soskic, V., Gorlach, M., Poznanovic, S., Boehmer, F.D. and Godovac-Zimmermann, J. (1999). Functional proteomics analysis of signal transduction pathways of the platelet-derived growth factor beta receptor. *Biochemistry* **38**: 1757-1764

Travis, A.J., Foster, J.A., Rosenbaum, N.A., Visconti, P.E., Gerton, G.L., Kopf, G.S. and Moss, S.B. (1998). Targeting of a germ cell-specific type 1 hexokinase lacking a porin- binding domain to the mitochondria as well as to the head and fibrous sheath of murine spermatozoa. *Molecular Biology of the Cell* **9**: 263-276

Treviño, C.L., Serrano, C.J., Beltran, C., Felix, R. and Darszon, A. (2001). Identification of mouse trp homologs and lipid rafts from spermatogenic cells and sperm. *FEBS Letters* **509**: 119-125

Vijayaraghavan, S., Mohan, J., Gray, H., Khatra, B. and Carr, D.W. (2000). A role for phosphorylation of glycogen synthase kinase-3alpha in bovine sperm motility regulation. *Biology of Reproduction* **62**: 1647-1654

Visconti, P.E., Bailey, J.L., Moore, G.D., Pan, D., Olds-Clarke, P. and Kopf, G.S. (1995a). Capacitation of mouse spermatozoa. I. Correlation between the capacitation state and protein tyrosine phosphorylation. *Development* **121**: 1129-1137

Visconti, P.E., Moore, G.D., Bailey, J.L., Leclerc, P., Connors, S.A., Pan, D., Olds-Clarke, P. and Kopf, G.S. (1995b). Capacitation of mouse spermatozoa. II. Protein tyrosine phosphorylation and capacitation are regulated by a cAMP-dependent pathway. *Development* **121**: 1139-1150

Visconti, P.E., Muschietti, J.P., Flawia, M.M. and Tezon, J.G. (1990). Bicarbonate dependence of cAMP accumulation induced by phorbol esters in hamster spermatozoa. *Biochimica et Biophysica Acta* **1054**: 231-236

Visconti, P.E., Ning, X., Fornes, M.W., Alvarez, J.G., Stein, P., Connors, S.A. and Kopf, G.S. (1999). Cholesterol efflux-mediated signal transduction in mammalian sperm: cholesterol release signals an increase in protein tyrosine phosphorylation during mouse sperm capacitation. *Developmental Biology* **214**: 429-443

Visconti, P.E., Westbrook, V.A., Chertihin, O., Demarco, I., Sleight, S. and Diekman, A.B. (2002). Novel signaling pathways involved in sperm acquisition of fertilizing capacity. *Journal of Reproductive Immunology* **53**: 133-150

Wang, Q., Song, C. and Li, C.C. (2004). Molecular perspectives on p97-VCP: progress in understanding its structure and diverse biological functions. *Journal of Structural Biology* **146**: 44-57

Wennemuth, G., Carlson, A.E., Harper, A.J. and Babcock, D.F. (2003). Bicarbonate actions on flagellar and Ca^{2+}-channel responses: initial events in sperm activation. *Development* **130**: 1317-1326

Woodman, P.G. (2003). p97, a protein coping with multiple identities. *Journal of Cell Science* **116**: 4283-4290

Yanagimachi, R. (1994). Mammalian fertilization. In: *The Physiology of Reproduction,* Volume 1. Edited by Knobil, E. and Neill, J.D., Raven Press, New York, pp. 189-317

Yunes, R., Michaut, M., Tomes, C. and Mayorga, L.S. (2000). Rab3A triggers the acrosome reaction in permeabilized human spermatozoa. *Biology of Reproduction* **62**: 1084-1089

Zajchowski, L.D. and Robbins, S.M. (2002). Lipid rafts and little caves. Compartmentalized signalling in membrane microdomains. *European Journal of Biochemistry* **269**: 737-752

Zarling, A.L., Ficarro, S.B., White, F.M., Shabanowitz, J., Hunt, D.F. and Engelhard, V.H. (2000). Phosphorylated peptides are naturally processed and presented by major histocompatibility complex class I molecules in vivo. *Journal of Experimental Medicine* **192**: 1755-1762

Zeng, Y., Clark, E.N. and Florman, H.M. (1995). Sperm membrane potential: hyperpolarization during capacitation regulates zona pellucida-dependent acrosomal secretion. *Developmental Biology* **171**: 554-563

Zeng, Y., Oberdorf, J.A. and Florman, H.M. (1996). pH regulation in mouse sperm: identification of Na^{+}-, Cl^{-}-, and HCO_3^{-}-dependent and arylaminobenzoate-dependent regulatory mechanisms and characterization of their roles in sperm capacitation. *Developmental Biology* **173**: 510-520.

Spermatology. SRF Vol. 65. ERS Roldan and M Gomendio (eds) Nottingham University Press, Nottingham

Cysteine rich secretory proteins in reproduction and venom

Gerard M. Gibbs[1] and Moira K. O'Bryan[1,2]

[1]*Monash Institute of Medical Research and the* [2]*The ARC Centre of Excellence in Biotechnology and Development, Monash University, Melbourne, Australia*

The cysteine rich secretory proteins (Crisp) are predominantly found in the mammalian male reproductive tract and in the venom of reptiles. Crisps are two domain proteins with a structurally similar yet evolutionarily diverse N-terminal domain and a characteristic cysteine rich C-terminal domain which we refer to as the Crisp domain. Since their identification 30 years ago Crisp research in mammals has focused on the characterisation of their expression localization to infer function. While no doubt important observations, these have not substantially led to an understanding of the biochemical activity of the Crisps and their role in sperm function or fertilisation. Recently, we demonstrated that the Crisp-2 Crisp domain has a structure similar to ion channel toxins ShK and BgK and was itself able to regulate Ca^{2+} flux through ryanodine receptors. These data build upon the previous characterizations of reptile venom Crisps as regulators of several types of ion channels and permits for the first time a dissection of the biochemical activity of mammalian Crisps.

Introduction

Cysteine rich secretory proteins (Crisps) are a group of vertebrate proteins which show a strong expression bias to the venom of reptiles, the mammalian male reproductive tract and to a lesser extent to non-reproductive tract secretory epithelia. They form a subgroup of a much more widely expressed superfamily of proteins initially characterised by the presence of a cysteine rich secretory protein, antigen 5 and pathogenesis related proteins domain (CAP domain). The defining feature of the Crisp sub-group is the possession of an unique C-terminal domain containing 10 absolutely conserved cysteines. It is the function of the C-terminal Crisp domain that is the primary focus of this review.

Crisp expression in mammals

Based on mRNA and protein expression data, it has been proposed that Crisps are involved in all stages of haploid male germ cell maturation, namely spermiogenesis, epididymal maturation, and sperm capacitation and fertilisation in the female reproductive tract. Crisp-2 (previously called Tpx-1) is a testis-specific protein which is incorporated into the developing acrosome (Foster and Gerton, 1996; O'Bryan *et al.*, 2001; Busso *et al.*, 2005) and sperm tail (specifically the outer dense fibres and connecting piece) (O'Bryan *et al.*, 1998; O'Bryan *et al.*, 2001).

Corresponding author E-mail: moira.obryan@med.monash.edu.au

Crisp-2 also facilitates adhesion of transfected Jurkat cells to cultured Sertoli cells, thus raising the possibility of a role in Sertoli-germ cell adhesion (Maeda *et al.*, 1998; Maeda *et al.*, 1999).

During epididymal transport and maturation, sperm incorporate all three additional Crisps. Crisp-1 is predominantly expressed by the principal cells of the corpus and cauda epididymis where two distinct forms of Crisp-1 (D and E), which differ by a glycosylation, coat the sperm surface in discrete but overlapping regions (Kohane *et al.*, 1980; Roberts *et al.*, 2002). Crisp-3 is also expressed and secreted into the epididymal lumen where it coats sperm (Kratzschmar *et al.*, 1996). Crisp-3 is the most widely expressed Crisp and is also found in tissues including the salivary glands, pancreas, B lymphocytes and the prostate (Kosari *et al.*, 2002; Udby *et al.*, 2002, 2005; Liao *et al.*, 2003). Lastly and only recently identified, Crisp-4 is produced by the principal cells of caput and corpus epididymis and again binds to the sperm surface (Jalkanen *et al.*, 2005; Nolan *et al.*, 2006).

Following ejaculation the fate and role of mammalian Crisps in general remains poorly defined, although there is compelling evidence that the dissociation of the D form of Crisp-1 from the sperm surface by diffusion is involved in the initiation of the global tyrosine phosphorylation associated with capacitation (Roberts *et al.*, 2002; Nixon *et al.*, 2006). Crisp-1, most likely the E form which remains tightly bound to the membrane of the rat sperm head (Roberts *et al.*, 2002), has also been implicated in sperm-egg fusion (Cohen *et al.*, 2000).

Crisp function in reptiles

Reptile venoms contain a complex mix of proteins which when injected into mammals disrupt homeostasis through a number of mechanisms, including activating or inhibiting blood coagulation factors and blocking ion channels. A number of reptile venom Crisps have been isolated and demonstrated to affect ion channel function and in a number of cases specific ion channels have been identified (Table 1). Helothermine is a Crisp from the venom of the Mexican beaded lizard, which when injected into mice resulted in the blockage of voltage gated Ca^{2+} receptors, voltage gated K^+ receptors and ryanodine receptors (RyR) leading to lethargy, partial hind limb paralysis, an inability to regulate body temperature and ultimately death (Mochca-Morales *et al.*, 1990).

Table 1. Snake venom Crisp activity on specific ion channels.

Crisp	Ion channel or current type	Ki	Reference
Helothermine	Voltage gated K^+		Nobile *et al.*, 1994
	- I_A-type transient current	0.52 μM	
	- Delayed-rectifier current	0.86 μM	
	Voltage gated Ca^{2+}	0.25 μM	Nobile *et al.*, 1996
	RyR1 (skeletal)	0.2 μM	Morrissette *et al.*, 1995
	RyR2 (cardiac)	7.0 μM	
PsTx	CNGA1	70 nM	Yamazaki *et al.*, 2002a
	CNGA2	15 nM	
	CNGA3	> 0.5 μM	Brown *et al.*, 1999
Pseudocin	CNGA1	1000 nM	Yamazaki *et al.*, 2002a
	CNGA2	460 nM	
Natrin	High conductance Ca^{2+} activated K^+ channel (BK_{Ca})	34.4 nM	Wang *et al.*, 2005
Ablomin	L-type Ca^{2+} channel	0.21 μM	Yamazaki *et al.*, 2002b

The range of ion channels affected by Crisps is diverse and specific activities vary considerably despite often very high protein identity (Table 1). For example, for the snake venom Crisps

PsTx and pseudecin have a 30 fold difference in potency on CNCA2 ion channels despite them sharing 97% identity at an amino acid level (Yamazaki *et al.*, 2002a). Detailed analyses of the venom Crisps PsTx and natrin has shown that they interact with ion channels in a 1:1 molar ratio, i.e. one molecule of Crisp interacts with the target ion channel receptor to cause the effect (Brown *et al.*, 1999; Wang *et al.*, 2005). PsTx was also shown to interact with and occlude the ion channel pore thereby preventing ion flux (Brown *et al.*, 2003).

Crisp structure

Crisps have a highly conserved structure which is characterised by a ~21 kDa N-terminal domain and a unique ~6 kDa C-terminal domain, the Crisp domain. The N-terminal domain has homology to the CAP superfamily. The CAP domain is also referred to as the Pr-1 domain or the SCP domain (from Sperm Coating Glycoprotein). As of July 2006 there were 955 CAP domain containing proteins in the PFam database and they are present across the three major phylic divisions of Archea, Eukarya and Bacteria. There is very little known about CAP domain function.

At a primary sequence level, Crisps contain 16 absolutely conserved cysteine residues which have been definitively shown by both enzymatic and structural methods to be involved in intramolecular disulphide bonding (Eberspaecher *et al.*, 1995; Guo *et al.*, 2005; Shikamoto *et al.*, 2005; Wang *et al.*, 2005). Recently the structure of several reptile Crisps have been solved and we have reported the NMR structure of mouse Crisp-2 Crisp domain (Guo *et al.*, 2005; Shikamoto *et al.*, 2005; Wang *et al.*, 2005; Gibbs *et al.*, 2006). The crystal structure of the venom Crisps, stecrisp, natrin and triflin conclusively showed that Crisps are 2 domain proteins, containing the CAP domain and the Crisp domain. The Crisp domain of all reported Crisps, however, may be further structurally divided into two sub-regions, which we referred to as the hinge region and the ion channel regulator (ICR) region based upon functional data and structural similarities to known channel toxins from sea anemones (described below). The crystal structures of venom Crisps showed that the hinge was tightly associated with the CAP domain by a number of hydrogen bonds and acted as a linker region between the CAP domain and the ICR.

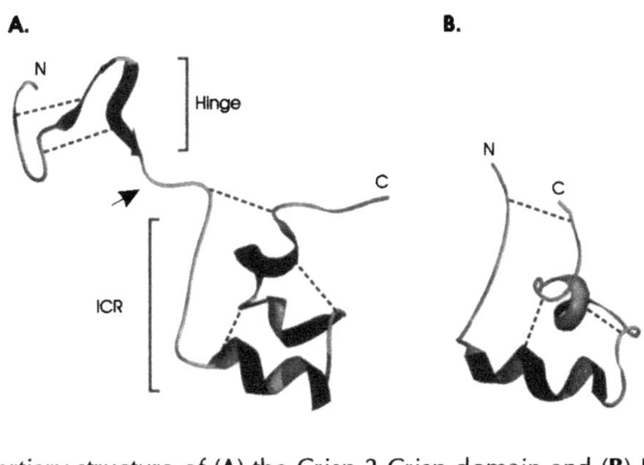

Figure 1. The tertiary structure of (**A**) the Crisp-2 Crisp domain and (**B**) BgK. Disulphide bonds are indicated by the dashed lines. The hinge and ICR domains are indicated on Crisp-2. The arrow marks the region of significant flexibility in the NMR structure of Crisp-2. The N and C terminus are marked where appropriate.

The mouse Crisp-2 Crisp domain structure also contained a hinge, Cys_{189}-Asn_{203}, and an ICR, Ser_{204}-His_{243} (Fig. 1). The hinge contained two short ß-strands that were stabilised by crossed disulphide bonds that formed a ß-hairpin. The ICR contained three disulphide bonds that locked together each of the three short helices into a very compact and rigid structure. Of particular note was that there were no stabilising interactions between the hinge and the ICR and they were essentially two discrete regions of the Crisp domain. With the hinge being tightly associated to the CAP domain, the ICR exists essentially as a separate domain and is likely to display significant rotational freedom (Gibbs *et al.*, 2006). It should be noted that other CAP proteins contain hinge-like sequences, but only Crisps contain an ICR-like region. This, and the relatively late evolutionary acquisition of the ICR encoding exon (exon 10 in Crisp-2), suggest that the two structural entities (CAP domain and the ICR) have separate biological activities. This hypothesis can only be tested upon determination of the biochemical activity of the CAP domain within Crisps, but is supported by the identification of unique CAP domain and Crisp domain interacting proteins (Jamsai, Bianco and O'Bryan, unpublished).

In determining the functions of highly disulphide bonded recombinant proteins, like the Crisps, it is critical to consider and check their disulphide connectivity. Some Crisps, such as those found in reptile venom and Crisp-1 in seminal plasma, are secreted in sufficient quantities so as to allow the purification of native protein. For our NMR studies we generated large amounts of correctly folded and disulphide bonded recombinant Crisp-2 Crisp domain using an *E. coli* expression system that possessed an oxidative cytoplasmic environment. It should be remembered that the production of recombinant Crisps in *E. coli* in the absence of such an environment will lead to inactive protein, artefacts and biologically misleading data.

The function of Crisps in mammals

Our structural analysis showed that the ICR of mouse Crisp-2 have homology to the ion channel toxins ShK (Tudor *et al.*, 1996) and BgK (Dauplais *et al.*, 1997) from the sea anemone. While the primary sequence of these proteins shared no homology, the disulphide bonding patterns were similar, resulting in similar overall topology (Fig. 1). Based upon these data and emerging reports describing the ion channel regulatory activity of reptile venom Crisps (Morrissette *et al.*, 1995; Brown *et al.*, 1999; Yamazaki and Morita, 2004), we hypothesised that the Crisp-2 ICR domain would also act on ion channels.

As Crisp-2 had been localised to intracellular compartments of sperm, we tested this hypothesis using recombinant Crisp-2 Crisp domain and RyR1 and RyR2 in lipid bilayer experiments. RyR are usually localised to intracellular membranes and are Ca^{2+} activated Ca^{2+} channels. We showed that the Crisp-2 Crisp domain acted as a differential regulator of RyR Ca^{2+} gating and that it was functional on both the cytoplasmic side and the luminal side of the receptor (Gibbs *et al.*, 2006). When added to the cytoplasmic side of RyR preparations, the Crisp-2 Crisp domain inhibited RyR2 opening at positive and negative applied bilayer potentials, however, it stimulated RyR1 channel opening at positive and negative applied bilayer potentials with a significant increase in mean open time. When added to the luminal side, the Crisp-2 Crisp domain activated both RyR1 and RyR2 opening at negative applied potentials only. This data was in partial contrast to RyR data from helothermine, which blocked both RyR1 and RyR2 gating activity when added to the cytoplasmic domain (Morrissette *et al.*, 1995). This highly selective ion channel regulatory activity and the ability of highly homologous venom Crisps to discriminate between receptor types underscores the exquisite specificity of Crisps on ion channels and may in fact be the key to the apparent over-abundance of Crisps in sperm biology. Specifically, we hypothesise different Crisps regulate different ion channels involved in sperm maturation and function and that specificity is dictated by the subtle amino acid differences and expression contexts. Given the absolute dependence of sperm Ca^{2+} signalling for the initiation of beat frequency (possibly involving Crisp-2), capacitation (clearly involving Crisp-1; Roberts *et al.*, 2003), hyperactivation, the acrosome reaction and potentially the early events of fertilisation, Crisps may be key regulators of male fertility.

While our RyR studies have definitively shown that the Crisp domain of mouse Crisp-2 can regulate RyR Ca^{2+} gating, it is presently unclear whether RyRs are the *in vivo* target for Crisp-2 activity. Immunochemical data on RyR in sperm is mixed, however, recent data (Harper *et al.*, 2004) provides compelling evidence that a RyR is found in human sperm in the connecting piece and possibly the acrosome. Both are consistent with the expression of Crisp-2 in mouse, rat and human sperm (O'Bryan *et al.*, 2001, and unpublished data). Both regions are also known to be intracellular Ca^{2+} storage vesicles and known to contain several ion channel types (Ho and Suarez, 2003; Harper *et al.*, 2004; Herrick *et al.*, 2005). Ca^{2+} fluxes emanating from the acrosome and the flagellum are involved in the initiation of the acrosome reaction and the control of the flagellar beat frequency and wave form respectively. As outlined above, we believe it is equally likely that the bioactivity of the Crisp domain of Crisp-2 (and other Crisps) will be limited by expression context.

The function of the CAP domain

There is little direct data on the function of CAP domains. Screening of proteins from the cone sail *Conus textile* venom duct has, however, identified the CAP Tex31 as a site specific protease with the ability to cleave pro-peptide forms of neurotoxins (Milne *et al.*, 2003). Further, XCrisp which is a Crisp expressed exclusively in the hatching gland of *Xenopus laevis* and accumulates on the surface of the presumptive hatching gland cells, was shown to be sufficient to induce the degradation of vitelline envelope (Schambony *et al.*, 2003), a function consistent with the involvement of a protease. The potential role for CAP domains as protease is also supported by several CAP domains structures (Fernandez *et al.*, 1997; Szyperski *et al.*, 1998; Henriksen *et al.*, 2001), each of which identify a putative active site pocket that has some similarity to known proteases. Sequence variation in this putative active site is hypothesised to determine the selectivity of the protease activity and may explain the difficulties in identifying cleavage targets. There is no evidence of the activity of mammalian CAP domains although several (including Crisp-3) are up-regulated under conditions of pathology (Murphy *et al.*, 1995; Kosari *et al.*, 2002).

Conclusions

The Crisps are a remarkable group of proteins whose expression bias toward venoms and cells of the reproductive tract initially appears confounding. With the function of venom being to disrupt cellular homeostasis, in part through disruption of Ca^{2+} signalling and with Ca^{2+} signalling being essential for normal sperm function, the two apparently opposing physiological end points are achieved through a single class of protein. How remarkable evolution can be!

References

Brown, R.L., Haley, T.L., West, K.A. and Crabb, J.W. (1999). Pseudechetoxin: a peptide blocker of cyclic nucleotide-gated ion channels. *Proceedings of the National Academy of Sciences USA* **96**: 754-759

Brown, R.L., Lynch, L.L., Haley, T.L. and Arsanjani, R. (2003). Pseudechetoxin binds to the pore turret of cyclic nucleotide-gated ion channels. *Journal of General Physiology* **122**: 749-760

Busso, D., Cohen, D.J., Hayashi, M., Kasahara, M. and Cuasnicu, P.S. (2005). Human testicular protein TPX1/

CRISP-2: localization in spermatozoa, fate after capacitation and relevance for gamete interaction. *Molecular Human Reproduction* **11**: 299-305

Cohen, D.J., Ellerman, D.A. and Cuasnicu, P.S. (2000). Mammalian sperm-egg fusion: evidence that epididymal protein DE plays a role in mouse gamete fusion. *Biology of Reproduction* **63**: 462-468

Dauplais, M., Lecoq, A., Song, J., Cotton, J., Jamin, N., Gilquin, B., Roumestand, C., Vita, C., de Medeiros, C.L., Rowan, E.G., Harvey, A.L. and Menez, A. (1997). On the convergent evolution of animal toxins. Conservation of a diad of functional residues in potassium channel-blocking toxins with unrelated structures. *Journal of Biological Chemistry* **272**: 4302-4309

Eberspaecher, U., Roosterman, D., Kratzschmar, J., Haendler, B., Habenicht, U.F., Becker, A., Quensel, C., Petri, T., Schleuning, W.D. and Donner, P. (1995). Mouse androgen-dependent epididymal glycoprotein CRISP-1 (DE/AEG): isolation, biochemical characterization, and expression in recombinant form. *Molecular Reproduction and Development* **42**: 157-172

Fernandez, C., Szyperski, T., Bruyere, T., Ramage, P., Mosinger, E. and Wuthrich, K. (1997). NMR solution structure of the pathogenesis-related protein P14a. *Journal of Molecular Biology* **266**: 576-593

Foster, J.A. and Gerton, G.L. (1996). Autoantigen 1 of the guinea pig sperm acrosome is the homologue of mouse Tpx-1 and human TPX1 and is a member of the cysteine-rich secretory protein (CRISP) family. *Molecular Reproduction and Development* **44**: 221-229

Gibbs, G.M., Scanlon, M.J., Swarbrick, J., Curtis, S., Gallant, E., Dulhunty, A.F. and O'Bryan, M.K. (2006). The cysteine-rich secretory protein domain of Tpx-1 is related to ion channel toxins and regulates ryanodine receptor Ca^{2+} signaling. *Journal of Biological Chemistry* **281**: 4156-4163

Guo, M., Teng, M., Niu, L., Liu, Q., Huang, Q. and Hao, Q. (2005). Crystal structure of cysteine-rich secretory protein stecrisp reveals the cysteine-rich domain has a K^+-channel inhibitor-like fold. *Journal of Biological Chemistry* **280**: 12405-12412

Harper, C.V., Barratt, C.L. and Publicover, S.J. (2004). Stimulation of human spermatozoa with progesterone gradients to simulate approach to the oocyte. Induction of $[Ca^{2+}]_i$ oscillations and cyclical transitions in flagellar beating. *Journal of Biological Chemistry* **279**: 46315-46325

Henriksen, A., King, T.P., Mirza, O., Monsalve, R.I., Meno, K., Ipsen, H., Larsen, J.N., Gajhede, M. and Spangfort, M.D. (2001). Major venom allergen of yellow jackets, Ves v 5: structural characterization of a pathogenesis-related protein superfamily. *Proteins* **45**: 438-448

Herrick, S.B., Schweissinger, D.L., Kim, S.W., Bayan, K.R., Mann, S. and Cardullo, R.A. (2005). The acrosomal vesicle of mouse sperm is a calcium store. *Journal of Cellular Physiology* **202**: 663-671

Ho, H.C. and Suarez, S.S. (2003). Characterization of the intracellular calcium store at the base of the sperm flagellum that regulates hyperactivated motility. *Biology of Reproduction* **68**: 1590-1596

Jalkanen, J., Huhtaniemi, I. and Poutanen, M. (2005). Mouse cysteine-rich secretory protein 4 (CRISP4): a member of the crisp family exclusively expressed in the epididymis in an androgen-dependent manner. *Biology of Reproduction* **72**: 1268-1274

Kohane, A.C., Cameo, M.S., Pineiro, L., Garberi, J.C. and Blaquier, J.A. (1980). Distribution and site of production of specific proteins in the rat epididymis. *Biology of Reproduction* **23**: 181-187

Kosari, F., Asmann, Y.W., Cheville, J.C. and Vasmatzis, G. (2002). Cysteine-rich secretory protein-3: a potential biomarker for prostate cancer. *Cancer Epidemiology Biomarkers and Prevention* **11**: 1419-1426

Kratzschmar, J., Haendler, B., Eberspaecher, U., Roosterman, D., Donner, P. and Schleuning, W.D. (1996). The human cysteine-rich secretory protein (CRISP) family. Primary structure and tissue distribution of CRISP-1, CRISP-2 and CRISP-3. *European Journal of Biochemistry* **236**: 827-836

Liao, Q., Kleeff, J., Xiao, Y., Guweidhi, A., Schambony, A., Topfer-Petersen, E., Zimmermann, A., Buchler, M.W. and Friess, H. (2003). Preferential expression of cystein-rich secretory protein-3 (CRISP-3) in chronic pancreatitis. *Histology and Histopathology* **18**: 425-433

Maeda, T., Nishida, J. and Nakanishi, Y. (1999). Expression pattern, subcellular localization and structure—function relationship of rat Tpx-1, a spermatogenic cell adhesion molecule responsible for association with Sertoli cells. *Development Growth and Differentiation* **41**: 715-722

Maeda, T., Sakashita, M., Ohba, Y. and Nakanishi, Y. (1998). Molecular cloning of the rat Tpx-1 responsible for the interaction between spermatogenic and Sertoli cells. *Biochemical and Biophysical Research Communications* **248**: 140-146

Milne, T.J., Abbenante, G., Tyndall, J.D., Halliday, J. and Lewis, R.J. (2003). Isolation and characterization of a cone snail protease with homology to CRISP proteins of the pathogenesis-related protein superfamily. *Journal of Biological Chemistry* **278**: 31105-31110

Mochca-Morales, J., Martin, B.M. and Possani, L.D. (1990). Isolation and characterization of helothermine, a novel toxin from Heloderma horridum horridum (Mexican beaded lizard) venom. *Toxicon* **28**: 299-309

Morrissette, J., Kratzschmar, J., Haendler, B., el-Hayek, R., Mochca-Morales, J., Martin, B.M., Patel, J.R., Moss, R.L., Schleuning, W.D., Coronado, R. and et al. (1995). Primary structure and properties of helothermine, a peptide toxin that blocks ryanodine receptors. *Biophysical Journal* **68**: 2280-2288

Murphy, E.V., Zhang, Y., Zhu, W. and Biggs, J. (1995). The human glioma pathogenesis-related protein is structurally related to plant pathogenesis-related proteins and its gene is expressed specifically in brain tumors. *Gene* **159**: 131-135

Nixon, B., MacIntyre, D.A., Mitchell, L.A., Gibbs, G.M., O'Bryan, M. and Aitken, R.J. (2006). The identification of mouse sperm-surface-associated proteins and characterization of their ability to act as decapacitation factors. *Biology of Reproduction* **74**: 275-287

Nobile, M., Magnelli, V., Lagostena, L., Mochca-Morales, J., Possani, L.D. and Prestipino, G. (1994). The toxin helothermine affects potassium currents in newborn rat cerebellar granule cells. *Journal of Membrane Biology* **139**: 49-55

Nobile, M., Noceti, F., Prestipino, G. and Possani, L.D. (1996). Helothermine, a lizard venom toxin, inhibits calcium current in cerebellar granules. *Experimental Brain Research* **110**: 15-20

Nolan, M.A., Wu, L., Bang, H.J., Jelinsky, S.A., Roberts, K.P., Turner, T.T., Kopf, G.S. and Johnston, D.S. (2006). Identification of rat cysteine-rich secretory protein 4 (Crisp4) as the ortholog to human CRISP1 and mouse Crisp4. *Biology of Reproduction* **74**: 984-991

O'Bryan, M.K., Loveland, K.L., Herszfeld, D., McFarlane, J.R., Hearn, M.T. and de Kretser, D.M. (1998). Identification of a rat testis-specific gene encoding a potential rat outer dense fibre protein. *Molecular Reproduction and Development* **50**: 313-322

O'Bryan, M.K., Sebire, K., Meinhardt, A., Edgar, K., Keah, H.H., Hearn, M.T. and De Kretser, D.M. (2001). Tpx-1 is a component of the outer dense fibers and acrosome of rat spermatozoa. *Molecular Reproduction and Development* **58**: 116-125

Roberts, K.P., Ensrud, K.M. and Hamilton, D.W. (2002). A comparative analysis of expression and processing of the rat epididymal fluid and sperm-bound forms of proteins D and E. *Biology of Reproduction* **67**: 525-533

Roberts, K.P., Wamstad, J.A., Ensrud, K.M. and Hamilton, D.W. (2003). Inhibition of capacitation-associated tyrosine phosphorylation signaling in rat sperm by epididymal protein Crisp-1. *Biology of Reproduction* **69**: 572-581

Schambony, A., Hefele, J.A., Gentzel, M., Wilm, M. and Wedlich, D. (2003). A homologue of cysteine-rich secretory proteins induces premature degradation of vitelline envelopes and hatching of *Xenopus laevis* embryos. *Mechanisms of Development* **120**: 937-948

Shikamoto, Y., Suto, K., Yamazaki, Y., Morita, T. and Mizuno, H. (2005). Crystal structure of a CRISP family Ca^{2+}-channel blocker derived from snake venom. *Journal of Molecular Biology* **350**: 735-743

Szyperski, T., Fernandez, C., Mumenthaler, C. and Wuthrich, K. (1998). Structure comparison of human glioma pathogenesis-related protein GliPR and the plant pathogenesis-related protein P14a indicates a functional link between the human immune system and a plant defense system. *Proceedings of the National Academy of Sciences USA* **95**: 2262-2266

Tudor, J.E., Pallaghy, P.K., Pennington, M.W. and Norton, R.S. (1996). Solution structure of ShK toxin, a novel potassium channel inhibitor from a sea anemone. *Nature Structural Biology* **3**: 317-320

Udby, L., Bjartell, A., Malm, J., Egesten, A., Lundwall, A., Cowland, J.B., Borregaard, N. and Kjeldsen, L. (2005). Characterization and localization of cysteine-rich secretory protein 3 (CRISP-3) in the human male reproductive tract. *Journal of Andrology* **26**: 333-342

Udby, L., Calafat, J., Sorensen, O.E., Borregaard, N. and Kjeldsen, L. (2002). Identification of human cysteine-rich secretory protein 3 (CRISP-3) as a matrix protein in a subset of peroxidase-negative granules of neutrophils and in the granules of eosinophils. *Journal of Leukocyte Biology* **72**: 462-469

Wang, J., Shen, B., Guo, M., Lou, X., Duan, Y., Cheng, X.P., Teng, M., Niu, L., Liu, Q., Huang, Q. and Hao, Q. (2005). Blocking effect and crystal structure of natrin toxin, a cysteine-rich secretory protein from Naja atra venom that targets the BKCa channel. *Biochemistry* **44**: 10145-10152

Yamazaki, Y., Brown, R.L. and Morita, T. (2002a). Purification and cloning of toxins from elapid venoms that target cyclic nucleotide-gated ion channels. *Biochemistry* **41**: 11331-11337

Yamazaki, Y., Koike, H., Sugiyama, Y., Motoyoshi, K., Wada, T., Hishinuma, S., Mita, M. and Morita, T. (2002b). Cloning and characterization of novel snake venom proteins that block smooth muscle contraction. *European Journal of Biochemistry* **269**: 2708-2715

Yamazaki, Y. and Morita, T. (2004). Structure and function of snake venom cysteine-rich secretory proteins. *Toxicon* **44**: 227-231

Spermatology. SRF Vol. 65. ERS Roldan and M Gomendio (eds) Nottingham University Press, Nottingham

Proteasome activity and its relationship with protein phosphorylation during capacitation and acrosome reaction in human spermatozoa

Patricio Morales, Emilce S. Díaz and Milene Kong

Unit of Reproductive Biology, Faculty of Health Sciences, University of Antofagasta, Antofagasta, Chile

We have shown that the proteasome is present in mammalian sperm and plays a role during fertilisation. In this work we studied the relationship between protein phosphorylation and proteasomal activity in human sperm. Aliquots of motile sperm were incubated for 0, 5 and 18 h at 37°C, 5% CO_2, with different concentration of the kinase inhibitors genistein, H89 or tamoxifen. Control aliquots were treated with the inhibitor solvent. The chymotrypsin-like activity of the proteasome was assayed using a fluorogenic substrate. Aliquots of spermatozoa capacitated during 18 h were incubated for 30 min with kinase inhibitors and then with 7 μM progesterone (P). The percentage of viable acrosome-reacted sperm was evaluated using FITC-labeled *Pisum sativum* agglutinin. The results indicate that spermatozoa treated with different concentrations of genistein and tamoxifen did not modify the chymotrypsin-like activity of the proteasome during capacitation. On the other hand, proteasome activity was significantly decreased by incubation with H89. Sperm treatment with genistein, H89 and tamoxifen significantly inhibited the P-induced acrosome reaction. Western blot analysis indicated that the proteasome inhibitor, epoxomicin, reduced serine protein phosphorylation. These results suggest that the enzymatic activity of the proteasome is modulated by protein kinase A, and that both enzymes are involved in the P-induced acrosome reaction.

Introduction

The phosphorylation/dephosphorylation of proteins is a universal postransductional biochemical mechanism that involves the reversible introduction of a phosphate group into an organic molecule. It is controlled by protein kinases and phosphatases, and takes place predominantly on serine, threonine and tyrosine residues (Hunter, 2000). The phosphorylation of proteins has demonstrated to play an important role in a variety of physiological processes including growth, differentiation, and cellular function.

There are numerous studies which show that before and during mammalian fertilisation, the sperm undergo a series of changes that are regulated by the activation of intracellular signalling pathways, which control the phosphorylation status of many proteins (Baldi *et al.*, 2002; Urner

Corresponding author E-mail: pmorales@uantof.cl

and Sakkas, 2003; Visconti *et al.*, 2002). Thus, several processes regulated by protein phosphorylation during sperm development and acquisition of fertilising capacity have been described, including sperm maturation, capacitation, hyperactivated motility, acrosome reaction, and gamete membrane fusion (Lewis and Aitken, 2001; Naz and Rajesh, 2004).

Most of the changes that have been studied during sperm protein phosphorylation take place in tyrosine residues. Thus, it has described that human sperm experience a progression in the phosphorylation status of its proteins during capacitation and zona pellucida binding (Sakkas *et al.*, 2003). The main proteins that are phosphorylated in tyrosine residues during capacitation correspond to the AKAPs and the CABYR, both families associated to the hyperactivated motility of the sperm (Vijayaraghavan *et al.*, 1997).

In contrast to the numerous studies related to sperm protein phosphorylation in tyrosine residues, little is known about the phosphorylation in serine/threonine residues. There are only a few studies that suggest an important role for the phosphorylation in ser/thr residues during the events of capacitation, hyperactivation and acrosome reaction (Jha and Shivaji, 2002; Naz and Rajesh, 2004). In addition, it has been shown that inhibition of the cAMP-dependent protein kinase (PKA) inhibit both capacitation and the acrosome reaction (reviewed in Baldi *et al.*, 2002). This kinase modifies other proteins by phosphorylation in ser/thr residues. In addition, PKA participates in the cross-talk between different intracellular signalling cascades like tyrosine kinases, MAPKs, and ser/thr phosphatases.

Sperm proteases play important roles during fertilisation. Recently, it was shown that spermatozoa from several mammalian species, including human, possess a multi-enzymatic protease complex or proteasome, which has trypsin-like, chymotrypsin-like and peptidylglutamyl peptide-hydrolyzing activities (Tipler *et al.*, 1997; Pizarro *et al.*, 2004). The proteasome degrades most nuclear and cytosolic proteins, after they have been covalently labelled to ubiquitin molecules, and the ubiquitin-proteasome pathway is responsible for most of the cell proteolysis (Ciechanover, 1998). Proteasome is a multi-subunit protease composed of two 19S regulatory complexes, capping both sides of a barrel-shaped 20 S proteasomal core. The 19S particle is composed of at least 17 proteasomal regulatory complex subunits, including ATP-dependent and ATP-independent subunits. The 20S core is composed of 14 subunits, 7 of α-type and 7 of ß-type. The α-subunits possess regulatory functions and the ß-subunits catalytic functions (Coux *et al.*, 1996).

In marine invertebrate sperm, the proteasome is involved in multiple steps of the fertilisation process, from the acrosome reaction induced by the egg jelly to penetration of the vitelline coat and fusion with the plasma membrane of the oocyte (Sawada *et al.*, 2002). In human sperm, we demonstrated that the proteasome is involved in the zona pellucida- and progesterone (P)-induced acrosome reaction. The aim of this work was to examine the relationship between proteasomal activity and protein phosphorylation in human sperm during capacitation and the P-induced acrosome reaction.

Protein kinase and proteasome activity

During capacitation, the chymotrypsin-like activity of the proteasome was inhibited by treating human spermatozoa with the PKA inhibitor H89 (Fig. 1). This is in agreement with the work of Marambaud *et al.* (1996), which showed that PKA effectors (like forskolin) but not protein kinase C agonists increased the activity and phosphorylation state of the proteasome in HK293 cells. This observation suggests that the enzymatic activity of the proteasome depends on PKA activation and protein phosphorylation in ser/thr residues. In addition, treatment with H89 significantly reduced the P-induced acrosome reaction (7 μM P = 54 ± 3.9% vs. 50 μM H89 +

7 μM P = 30.5±3%). On the other hand, none of the kinase inhibitors affected proteasome activity during induction of the acrosome reaction with P. This suggests that phosphorylation of the proteasome is not required during the occurrence of the acrosome reaction stimulated by P.

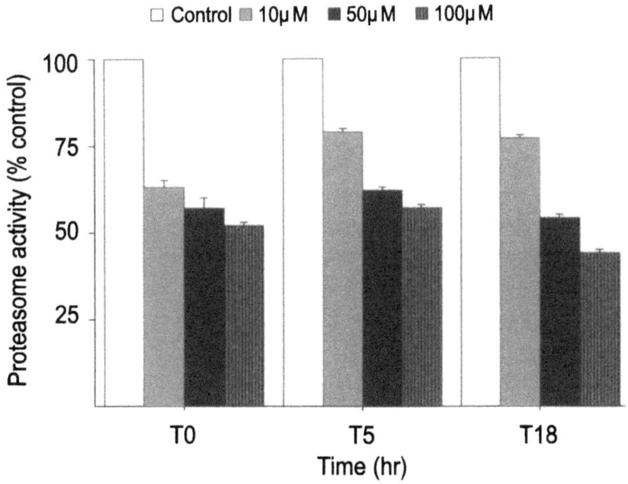

Figure 1. Effect of different concentrations of the protein kinase A inhibitor H89 on the chymotrypsin-like activity of human sperm proteasome. The control group is significantly higher than all the other groups.

Protein phosphorylation and proteasome activity

To determine the relationship between proteasome activity and protein phosphorylation, we proceeded to evaluate the pattern of protein phosphorylation during sperm capacitation and P-induced acrosome reaction in the presence of the proteasome inhibitor epoxomicin (Fig. 2). Aliquots of highly motile sperm were divided in six groups: non-capacitated (0); capacitated for 5 h (5); capacitated for 18 h (18); capacitated for 18 h in the presence of 10 μM epoxomicin (18+E); capacitated for 18 h and incubated with 7 μM P for 15 min (P); and capacitated for 18 h and then incubated with 10 μM epoxomicin for 30 min followed by incubation with P for 15 min (P+E). Then, a SDS-PAGE followed by a Western blot was performed. The bands were revealed with an anti-phosphoserine antibody (Fig. 2).

The densitometric analysis of the bands revealed an increase in protein phosphorylation in serine residues during capacitation, that is between the 0 and the 18 h of incubation. This increase was detected in seven bands, whose molecular weights were 155, 107, 86, 70, 28, 26 and 10 kDa, approximately (lanes 0, 5, and 18). In addition, sperm treated with epoxomicin during capacitation exhibited a diminution in the degree of phosphorylation in serine residues of those bands (lane 18+E). On the other hand, a band of 26 kDa increased its degree of phosphorylation in serine when the sperm were treated with P (lane P). This increase was not observed when the sperm were previously treated with epoxomicin (lane P+E).

When the phosphorylation pattern in threonine residues was analysed during capacitation, a gradual increase in phosphorylation of bands of molecular weight in the range from 38.5 to 53 kDa was observed (data not shown). This increase was not reverted by incubating the sperm with epoxomicin. Sperm treated with P did not show a phosphorylation pattern different from sperm capacitated for 18 h, with the exception of a band of approximately 23 kDa. There was a small increase in the degree of phosphorylation of this band after P treatment. There was a further increase in the phosphorylation of this band after the sperm were treated with epoxomicin and P.

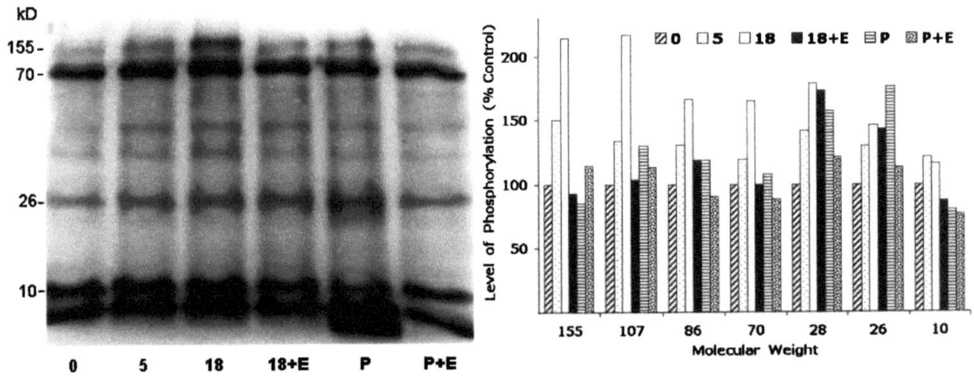

Figure 2. Western blot analysis of proteins phosphorylated in serine residues during human sperm capacitation (0, 5, and 18 h) and progesterone-induced acrosome reaction (P). Other sperm aliquots, capacitated for 18 h, were incubated with the proteasome inhibitor epoxomicin (18+E) or with progesterone and epoxomicin (P+E). The panel on the right shows the densitometric analysis of the phosphorylated bands.

An increase in the pattern of protein phosphorylation in tyrosine residues was observed during capacitation in bands of an approximate molecular weight from 16 to 111 kDa. When the sperm were incubated in the presence of epoxomicin, a slight increase in the phosphorylation degree of the band corresponding to an approximate molecular weight of 26 kDa was observed.

Conclusions

On the basis of the results obtained, we suggest that the activity of the sperm proteasome is directly or indirectly modulated by PKA. In turn, the proteasome is able to modulate not only the activity of protein kinases but also of phosphatases.

Acknowledgements

This work was financed by FONDECYT 1040295 and DIRINV 1322-06 and 1316-06

References

Baldi, E., Luconi, M., Bonaccorsi, L. and Forti, G. (2002). Signal transduction pathways in human spermatozoa. *Journal of Reproductive Immunology* **53**: 121-131

Ciechanover, A. (1998). The ubiquitin-proteasome pathway: on protein death and cell life. *The EMBO Journal* **17**: 7151-7160

Coux, O., Tanaka, K. and Goldberg, A.L. (1996). Structure and function of the 20S and 26S proteasomes. *Annual Review of Biochemistry* **65**: 801-847

Hunter, T. (2000). Signaling—2000 and beyond. *Cell* **100**: 113-127

Jha, K. and Shivaji, S. (2002). Protein serine and threonine phosphorylation, hyperactivation and acrosome reaction in in vitro capacitated hamster spermatozoa. *Molecular Reproduction and Development* **63**: 119-130

Lewis, B. and Aitken, R. (2001). Impact of epididymal maturation on the tyrosine phosphorylation patterns exhibited by rat spermatozoa. *Biology of Reproduction* **64**: 1545-1556

Marambaud, P., Sherwin, W. and Checler, F. (1996). Protein kinase A phosphorylation of the proteasome: A contribution to the alpha-secretase pathway in human cells. *Journal of Neurochemistry* **87**: 2616-2619

Naz, R.K. and Rajesh, P.B. (2004). Role of tyrosine phosphorylation in sperm capacitation/acrosome reaction. *Reproductive Biology and Endocrinology* **2**: 75-87

Pizarro, E., Pastén, C., Kong, M. and Morales, P. (2004). Proteasomal activity in mammalian spermatozoa. *Molecular Reproduction and Development* **69**: 87-93

Sakkas, D., Leppens-Luisier, G., Lucas, H., Chardonnens, D., Campana, A., Franken, D. and Urner, F. (2003). Localization of tyrosine phosphorylated protein in human sperm and relation to capacitation and zona pellucida binding. *Biology of Reproduction* **68**: 1463-1469

Sawada, H., Takahashi, Y., Fujino, J., Flores, S.Y. and Yokosawa, H. (2002). Localization and roles in fertilization of sperm proteasomes in the ascidian Halocynthia roretzi. *Molecular Reproduction and Development* **62**: 271-276

Tipler, C.P., Hutchon, S.P., Hendil, K., Tanaka, K., Fishel, S. and Mayer, R.J. (1997). Purification and characterization of 26S proteasomes from human and mouse spermatozoa. *Molecular Human Reproduction* **3**: 1053-1060

Urner, F. and Sakkas, D. (2003). Protein phosphorylation in mammalian spermatozoa. *Reproduction* **125**: 17-26

Vijayaraghavan, S., Trautman, K., Goueli, S. and Carr, D. (1997). A tyrosine-phosphorylated 55-kilodalton motility-associated bovine sperm protein is regulated by cyclic adenosine 3',5'-monophosphates and calcium. *Biology of Reproduction* **56**: 1450-1457

Visconti, P.E., Westbrook, V.A., Chertihin, O., Demarco, I., Sleight, S. and Diekman, A.B. (2002). Novel signaling pathways involved in sperm acquisition of fertilizing capacity. *Journal of Reproductive Immunology* **53**: 133-150

Spermatology. SRF Vol. 65. ERS Roldan and M Gomendio (eds) Nottingham University Press, Nottingham

Molecular mechanisms of membrane fusion during acrosomal exocytosis

Claudia Nora Tomes

Laboratorio de Biología Celular y Molecular, Instituto de Histología y Embriología (IHEM-CONICET), Facultad de Ciencias Médicas, CC 56, Universidad Nacional de Cuyo, 5500-Mendoza, Argentina

Sperm are attractive cells. Understanding their physiology has motivated researchers from all over the globe for decades. Initially came the description of sperm's overall shape and properties, together with their genesis and development in the testis. Later, the study of exocytosis took off owing to ultrastructural analysis that achieved exquisite levels of detail. Biochemical analysis ensued, identifying ligands and signalling pathways whose end point was exocytosis. Somehow, the unveiling of the molecular mechanisms involved in membrane fusion itself lagged behind all this progress. The picture changed dramatically in the last few years, due to an explosion in our knowledge of the many proteins required for exocytosis and its regulation, and the discovery that very similar versions of these proteins play the same roles in virtually all membrane fusion models. Luckily, sperm are not the exception to this rule. For instance, fusion of the outer acrosomal to the plasma membrane depends on Rab3 activation, α-SNAP/NSF, synaptotagmin, and SNAREs; it also requires an efflux of calcium from the acrosomal lumen. Convergence of Rab- and toxin-sensitive SNARE-dependent pathways is a hallmark of the acrosome reaction that makes it an attractive mammalian model to study the different phases of the membrane fusion cascade. Finally, because nature has endowed sperm with a cellular specialization that gives them a single, irreversible chance to fertilise an egg, the acrosome reaction is more straightforward to dissect than fusion in other cell types, where the same substances are secreted again and again, requiring the membranes and fusion machinery to recycle multiple times.

Membrane fusion

Fusion of a transport vesicle with its target membrane is a fundamental process essential to the organization and functions of all eukaryotic cells. Fusion results in the coalescence of two distinct biological membranes into a single bilayer and requires the targeting and docking of membrane-bound transport carriers to the appropriate destinations (Szule and Coorssen, 2003). Exocytosis is a highly regulated process consisting of multiple definable stages that culminate in the attachment of secretory vesicles to the plasma membrane, followed by the calcium-triggered opening of fusion pores connecting the interior of the granule to the extracellular medium (Burgoyne and Morgan, 1998a;Burgoyne and Morgan, 2003). During the last few

Corresponding author E-mail: ctomes@fcm.uncu.edu.ar

years, a number of molecules involved in this complex cascade have been identified and characterised by genetic approaches, reconstitution assays and biochemical means (Jahn and Sudhof, 1999; Bruns and Jahn, 2002; Jahn, Lang and Sudhof, 2003; Sollner, 2003). Several of them are conserved from yeast to human, and are shared between constitutive and regulated exocytosis as well as in intracellular membrane merger events. This has been interpreted as indicating that fusion is fundamentally conserved in different cell types, built on a common core machinery upon which accessory components regulate efficiency, sensitivity to trigger and speed (Szule and Coorssen, 2003). Hence, it is not surprising that, despite the fact that exocytosis of the acrosomal vesicle (acrosome reaction, AR) is somewhat unique, all of the involved molecules so far identified had been initially found in somatic cells and implicated in various membrane-fusion events.

The acrosome reaction

The acrosome is a large, Golgi-derived, membrane-limited granule which overlies the anteriormost two-thirds of the nucleus in the apical region of the sperm head. During the course of fertilisation, the acrosome undergoes the AR, a special type of calcium-dependent exocytosis, near the time that the sperm encounters the zona pellucida. Only sperm that have completed the AR can successfully penetrate the zona pellucida to reach the oolema. Morphologically, the AR in eutherian mammals proceeds through a series of steps that begin with the apposition of the sperm plasma membrane with the membrane directly adjacent to it, the outer acrosomal membrane, in the anterior region of the sperm head. The subsequent steps include swelling of the acrosomal contents, prior to or concomitant with fusion of the outer acrosomal membrane with the overlying plasma membrane at the points of apposition, followed by formation of multiple pores at these points. In most cells, the pores widen, the granule contents are discharged, and granule membrane remodelling eventually occurs. In sperm, however, the pores widen, but since the outer acrosomal membrane is as large as the area of plasma membrane it is fusing with, the result of the pore widening is the fenestration of the fusion membranes and joining of pores to produce a reticulum of tubules and hybrid plasma-outer acrosomal membrane vesicles. Release of acrosomal contents begins at this point, until the inner acrosomal membrane, the one closest to the nuclear envelope, becomes the leading edge of the sperm (Barros, Bedford, Franklin and Austin, 1967; Yudin Gottlieb and Meizel, 1988; Storey, 1995) (Fig. 1A).

Methods to study the molecular mechanisms underlying the acrosome reaction. Advantages of plasma membrane permeabilization

Until recently, immunodetection of proteins in the acrosomal region was accepted as circumstantial evidence of their role in exocytosis. This approach was and remains unsatisfactory, since presence and/or localization alone do not demonstrate physiological relevance. Alternatively, a few groups have inferred the role of fusion proteins in the AR by adding peptides to intact (i.e. non-permeabilised) ram (Garde and Roldan, 1996) and mouse (Iida, Yoshinaga, Tanaka, Toshimori and Mori, 1999) sperm, or antibodies to bovine (Ramalho-Santos, Moreno, Sutovsky, Chan, Hewitson, Wessel, Simerly and Schatten, 2000) sperm in the absence or presence of calcium ionophores and observing their effect on the AR. It was suggested that these macromolecules permeate into the cytosol thanks to an initial fusion of the outer acrosomal membrane and the adjacent plasma membrane that creates permeable pores. Once inside

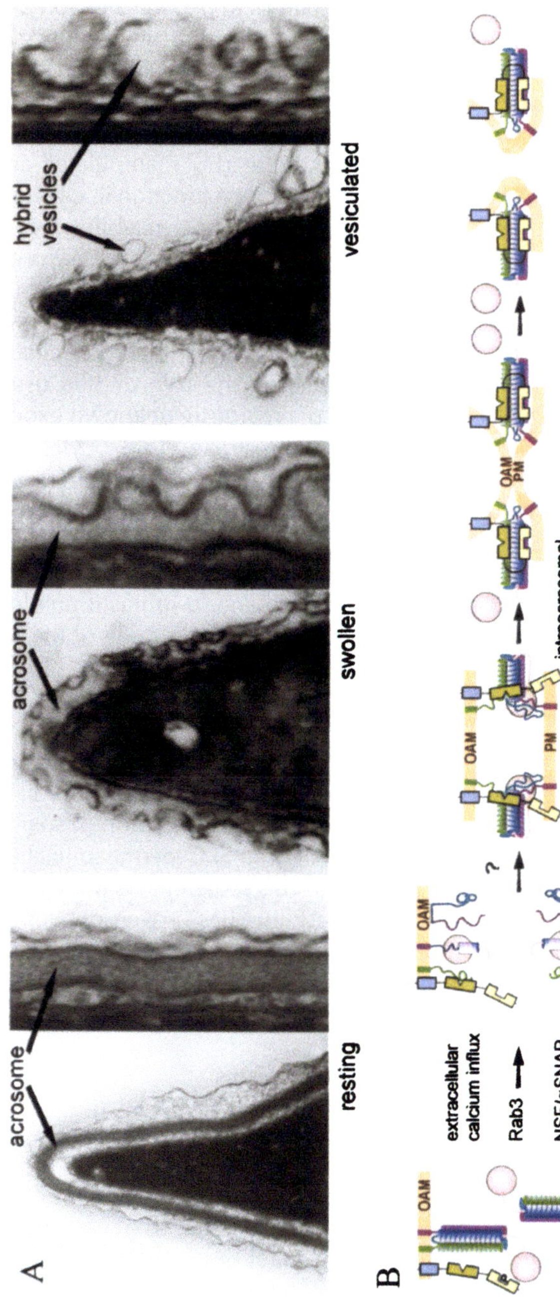

Figure 1. **(A)** Transmission electron micrographs of the progression of the acrosome reaction viewed through sagital sections of human sperm heads. Left: resting sperm; center: early AR stage, with swollen acrosomes; right: exocytosis in progress: membrane fusion has occured, generating hybrid plasma-outer acrosomal membrane vesicles. AR inducer: calcium ionophore A23187. Insets: higher magnification sections of the corresponding images (Photographs provided by Natalia Zanetti and Luis Mayorga, IHEM, Mendoza, Argentina). **(B)** Working model for the dynamics of SNAREs assembly and disassembly during the AR. Resting sperm (left): synaptotagmin is phosphorylated and SNARE proteins are engaged in *cis* complexes. All proteins are "inactive". Sperm stimulation dephosphorylates synaptotagmin and dissociates *cis* SNARE complexes (second from the left). As exocytosis progresses, SNAREs assemble in a loose *trans* configuration (third from the left) and remain thus until calcium is released from the acrosome. This calcium binds to synaptotagmin and causes the zippering up of the SNARE complexes into a tight *trans* configuration (fourth from the left). Last, the fusion pore opens, completing the sequence (right most drawing). Synaptobrevin is depicted in green, syntaxin in pink (amino terminal Habc domain) and purple (SNARE motif and transmembrane domain), SNAP-25 in blue, and synaptotagmin in blue (transmembrane domain), green (C2A) and yellow (C2B). OAM, outer acrosomal membrane; PM, plasma membrane. (Modified by Luis Mayorga, IHEM, Mendoza, Argentina, from De Blas *et al.*, 2005; this is an open access article. Copyright 2005 De Blas *et al.*).

the cells, they exert their influence on exocytosis before extensive content release takes place. A shortcoming of this method is its unsuitability to study regulatory steps that govern exocytosis before actual membrane fusion has begun.

Sperm's biggest drawback is their inability to transcribe genes and translate them into proteins, ruling out the use of standard powerful tools/techniques such as RNAi, transfection and overexpression. In addition, the possibility of studying the role of macromolecules or charged compounds is severely limited because such molecules are unable to cross an intact cell's membrane. To overcome these limitations, we have applied a widely used methodology in the study of exocytosis, that of controlled plasma membrane permeabilization with streptolysin O (SLO). The permeable sperm preparations we have established are readily manipulated and reflective of the *in vivo* organization of the cell, and allow us to measure a functional, calcium-regulated AR that is morphologically indistinguishable from that in non-permeabilised cells (Diaz, Dominguez, Fornes, Burgos and Mayorga, 1996; Yunes, Michaut, Tomes and Mayorga, 2000). The protocol allows penetration into the cell of exogenous factors - such as ions, neurotoxins, recombinant proteins, and antibodies - to assess the function of their specific targets. Permeabilization, like every system and every research approach, has its own inherent strengths and weaknesses. Among the former are the possibility of exploring the role of "the usual (membrane fusion) suspects" in the AR, as well as the discovery of previously unknown exocytosis players and/or their regulators. The latter may or may not be restricted to exocytosis in sperm, but most likely they will not be. Another strength of sperm permeabilization is the ability to reversibly halt exocytosis at a number of pre-fusion stages. This allows the unveiling of the sequence in which each player participates in the cascade. We have exploited these features and refined our functional assays, as well as the interpretations we derive from them, to create a molecular model of the calcium-triggered steps of membrane fusion as it unfolds. SLO-permeabilization has recently been adopted by other laboratories in the study of the AR (Hernandez-Gonzalez, Lecona-Valera, Escobar-Herrera and Mugica, 2000; Kitamura, Tanaka and Nishimune, 2003; Hutt, Baltz and Ngsee, 2005), and sperm motility (Johnson, Moss and Gerton, 1999).

An additional advantage of the use of SLO permeabilization is the possibility of uncovering relatively late calcium-related pathways (see below), otherwise masked in intact sperm. Last, several signalling pathways serve the purpose of regulating more than one function in sperm. This is particularly true when it comes to capacitation and the AR. The SLO-permeabilization approach bestows the possibility of isolating these two processes. This separation is mandatory to reveal the role of, for instance, protein phosphorylation, a task that has proven difficult to investigate in a heterogeneous population of intact sperm cells that are engaged in capacitation as well as exocytosis.

Exocytosis of the acrosome: molecular mechanisms

Calcium

Calcium is the most widely used intracellular messenger in cell signalling (Clapham, 1995; Berridge, Lipp and Bootman, 2000a,b; Nowycky and Thomas, 2002). Unsurprisingly, one common theme in the regulation of exocytosis is the role of intracellular calcium as a mediator of stimulus-secretion coupling. Thus, the elevation of calcium concentrations from resting (≤ 0.1 μM) to stimulated (~ 1-100 μM) levels is necessary and sufficient to initiate regulated exocytosis (Neher, 1998; Martin, 2003).

In sperm, calcium is involved in virtually all physiological functions, including capacitation, hyperactivation, and the AR. Two phases of solubilised zona pellucida- and progesterone-evoked calcium responses can be resolved with the use of ion-selective fluorescent probes. A first, transient phase is characterised by calcium influx from the extracellular milieu to the cytosol through voltage-gated channels. Subsequent activation of phospholipase Cs results in the production of inositol-1,4,5-trisphosphate (IP_3) and the opening of IP_3-sensitive calcium permeable channels in the outer acrosomal membrane, which promotes the release of calcium from an intra-acrosomal pool (see below). Mobilization of this store produces the gating signal that opens store-operated channels (SOCs) in the plasma membrane and produces the second, sustained phase of calcium entry into the cytosol (Florman, Arnoult, Kazam, Li and O'Toole, 1998; Trevino, Serrano, Beltran, Felix and Darszon, 2001; Kirkman-Brown, Punt, Barratt and Publicover, 2002b; Darszon, Nishigaki, Wood, Trevino, Felix and Beltran, 2005; Felix, 2005).

According to this model, the role of the intra-acrosomal calcium pool is to direct the opening of SOCs in the plasma membrane. Incubating permeabilised sperm with calcium resembles the physiological situation of calcium influx through open SOCs, because SLO permits free diffusion of ions. In consequence, permeabilization allows the examination of relatively late steps of the exocytic cascade, occurring after the sustained calcium influx, while bypassing earlier pathways whose end point is the opening of SOCs. With this in mind, we were surprised to discover a direct correlation between depletion of an acrosomal calcium pool and inhibition of the AR in permeabilised human sperm. In other words, exocytosis requires a second release of calcium from the acrosome at a step downstream of SOCs opening. This second phase drives exocytosis, utilizes IP_3-sensitive channels (De Blas, Michaut, Trevino, Tomes, Yunes, Darszon and Mayorga, 2002), and takes place once the biochemical machinery for fusion has been assembled. We have experimentally classified fusion factors as: (i) acting after SOCs opening but before the second intra-acrosomal calcium efflux [Rab3A, α-soluble NSF attachment protein (α-SNAP), and N-ethylmaleimide-sensitive factor (NSF)]; and (ii) acting during or after the second intra-acrosomal release [SNAP receptors (SNAREs) and synaptotagmin VI] (Table 1).

Table 1. Temporal action sites of a number of factors known to participate in sperm exocytosis. "Early" means they exhibit their role after SOC opening and before the second efflux of intra-acrosomal calcium. "Late" means they act concomitantly with or after the second efflux of intra-acrosomal calcium. The experimental method employed allows the identification of the latest step affected by the molecule under scrutiny. Thus, "late" does not exclude the possibility of earlier points of action. AR, acrosome action.

	Role in the AR cascade	Reference
Rab3	early	De Blas *et al.*, 2005
α-SNAP	early	Tomes *et al.*, 2005
NSF	early	Tomes *et al.*, 2005
Q-SNAREs, R-SNARE	late	De Blas *et al.*, 2005
synaptotagmin	late	De Blas *et al.*, 2005
cAMP/Epac	early	Branham et *al.*, 2006
cholesterol removal	early	Belmonte *et al.*, 2005
protein tyrosine kinases	late	Tomes *et al.*, 2004
Rap	early	Branham et *al.*, 2004
PTP1B	early	Zarelli *et al.*, 2005
calmodulin	late	Yunes, unpublished

The acrosome as a store of releasable calcium. Calcium efflux from membrane-bound vesicles is a prerequisite for transport in several systems such as the yeast vacuolar fusion model (Peters and Mayer, 1998; Merz and Wickner, 2004) and the secretion of exocytotic granules (Scheenen, Wollheim, Pozzan and Fasolato, 1998; Thirion, Troadec, Pivovarova, Pagnotta, Andrews, Leapman and Nicaise, 1999). The acrosomal granule had been postulated to be an internal reservoir of releasable calcium for several years. However, direct evidence for the existence of this pool was first acquired after developing a SLO plasma membrane permeabilization method. By using calcium sensitive probes, we could visualise directly the presence of a calcium pool inside the acrosome (Fig. 2A,B) and monitor the changes in this compartment under different experimental conditions that affect exocytosis (De Blas *et al.*, 2002). It is noteworthy that a clear and unequivocal observation of the acrosomal pool is hard to visualise due to the diffuse cytosolic staining observed in non-permeabilised sperm (Fig. 2C-E). By loading non-permeabilised mouse sperm with membrane permeant forms of calcium-sensitive fluorescent dyes and quenching fluorescence in the cytosol with high manganese concentrations, another laboratory has subsequently shown that the acrosome is a calcium store (Herrick, Schweissinger, Kim, Bayan, Mann and Cardullo, 2005).

The acrosome luminal calcium pool must be actively maintained for sperm to achieve exocytosis (De Blas *et al.*, 2002). Interestingly, AR depends on the second calcium efflux from the acrosome even in the presence of relatively high cytosolic levels of this ion (De Blas *et al.*, 2002; Tomes, Roggero, De Blas, Saling and Mayorga, 2004). The requirement stands regardless of whether exocytosis is initiated by calcium itself or by other inducers, such as Rab3A (De Blas *et al.*, 2002) or cyclic adenosine 3′, 5′-monophosphate (cAMP) (Branham, Mayorga and Tomes, 2006). The easiest explanation for such a prerequisite is that fusion reactions might only take place in a particular microenvironment characterised by highly localised, elevated calcium concentrations. It is conceivable that membranes maintained in close proximity during the fusion process create such partially isolated domains required by the fusion machinery.

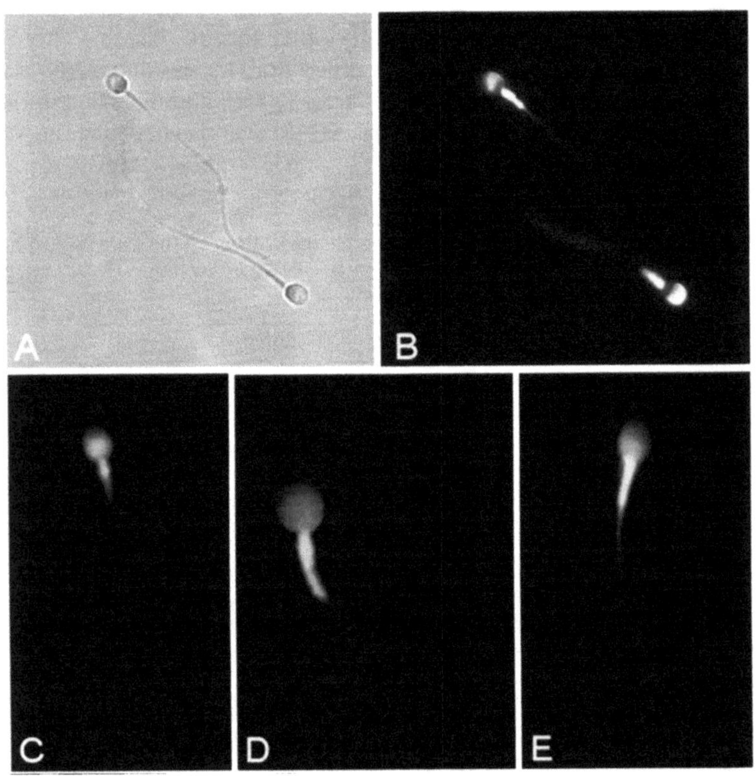

Figure 2. Visualization of intracellular calcium stores in human sperm. (**A**) Light micrograph corresponding to the field depicted in panel (**B**); (**B-E**) Fluo3-AM fluorescence images. A calcium accumulation is clearly observed in the acrosome and midpiece of permeabilised sperm (B), whereas a diffuse, whole head plus midpiece signal is detected in intact cells (**C-E**). Notice that the acrosomal store can only be visualised when sperm are permeabilised. (Photographs provided by Gerardo De Blas and Luis Mayorga, IHEM, Mendoza, Argentina).

Calcium sensing proteins in exocytosis. In the early stages of the secretory pathway, both in constitutive exocytosis and in the endocytic pathway, vesicle fusion follows immediately after docking. In regulated exocytosis a trigger is required, so that vesicle fusion begins only when an appropriate signal arrives (Burgoyne and Morgan, 1998b). The signal is most frequently an elevation of the intracellular calcium concentration. Because calcium must act on vesicles closely engaged with the plasma membrane and poised for immediate release, it is believed that a late-acting calcium sensor is likely to be associated with the fusion machinery. A number of calcium-binding proteins relevant to exocytosis have been reviewed elsewhere (Burgoyne and Morgan, 1998b). I will only cover here the two whose involvement in the AR have been demonstrated.

(a) *Synaptotagmin.* Synaptotagmins are sensor proteins that detect the concentration of calcium by reversibly binding it and thereafter signalling to other targets. They also bind phospholipids in a calcium-dependent manner. These properties are endowed by their two C2 domains, C2A and C2B, which are homologous to the C2 domain in protein kinase C (PKC). Preceding the tandem C2 domains is a variable linker region connecting to the transmembrane anchor and a short intravesicular or extracellular amino terminus (Mikoshiba, Fukuda, Ibata, Kabayama and Mizutani, 1999). In addition to calcium ions and phospholipids, synaptotagmins bind other synaptotagmins, calcium channels, isolated Q-SNAREs and those in the SNARE complex, α-SNAP, RIM and calmodulin (Sudhof and Rizo, 1996; Burgoyne and Morgan, 2003; Li and Chin, 2003; Sollner, 2003). How does synaptotagmin participate in membrane fusion? Synaptotagmin binds to SNAREs, contributing to their zippering into trans complexes (Bhalla, Chicka, Tucker and Chapman, 2006).

Synaptotagmins localise to the acrosomal region in mammalian sperm: isoform VI in human (Michaut, De Blas, Tomes, Yunes, Fukuda and Mayorga, 2001), VIII in mouse (Hutt, Cardullo, Baltz and Ngsee, 2002), I in stallion (Gamboa and Ramalho-Santos, 2005), and undefined isoforms in hamster, bull, rhesus monkey, mouse and human sperm (Ramalho-Santos *et al.*, 2000). Synaptotagmins I, VI (Michaut *et al.*, 2001; Roggero, Tomes, De Blas, Castillo, Michaut, Fukuda and Mayorga, 2005) and VIII (Hutt *et al.*, 2005) are implicated in the AR. Synaptotagmin VI displays a late role in the AR, simultaneous with or downstream of the second phase of calcium efflux from the acrosome (Table 1). This protein is a substrate for PKC in sperm and phosphorylation impairs its activity during the AR (Michaut *et al.*, 2001; Roggero *et al.*, 2005). Human sperm synaptotagmin is phosphorylated *in vivo* on threonine 419, within a polybasic region on the fourth beta strand of the C2B domain known as the effector domain. Phosphorylation diminishes upon triggering the AR, probably due to protein phosphatases that dephosphorylate synaptotagmin to make it functional in exocytosis. Dephosphorylation takes place prior to the second efflux of calcium from the acrosome (Roggero *et al.*, 2005).

(b) *Calmodulin.* Calmodulin binds calcium via its four EF-hand domains. Calcium-calmodulin has been invoked as a universal requirement for constitutive membrane fusion (Burgoyne and Clague, 2003; Burgoyne and Morgan, 2003), and for exocytosis in synapses through the activation of calmodulin-dependent protein kinase II (Burgoyne and Morgan, 1995). It exhibits a direct role in membrane fusion, through its binding to synaptobrevin 2 *in vitro* and in PC12 cells (Quetglas, Iborra, Sasakawa, De Haro, Kumakura, Sato, Leveque and Seagar, 2002; de Haro, Ferracci, Opi, Iborra, Quetglas, Miquelis, Leveque and Seagar, 2004), and behaves as the calcium sensor in the first post-docking phase of yeast vacuolar fusion (Peters and Mayer, 1998). V0, the membrane-integral sector of the vacuolar H^+-ATPase, is a binding partner for calcium-calmodulin. *Trans* V0 complexes assemble into channels that connect the fusing vacuoles, and have been proposed to mature under the influence of calcium-calmodulin into the fusion pore itself (Peters, Bayer, Buhler, Andersen, Mann and Mayer, 2001).

The presence of calmodulin in sperm has been known for many years and roles have been suggested in several functions (Marin-Briggiler, Jha, Chertihin, Buffone, Herr, Vazquez-Levin and Visconti, 2005). Calmodulin modulates more than one step during the AR. In an early phase, it positively regulates the opening of T-type calcium channels elicited by physiological inducers in sea urchin (Sano, 1983; Guerrero and Darszon, 1989), mouse (Bendahmane, Lynch and Tulsiani, 2001; Lopez-Gonzalez, Vega-Beltran, Santi, Florman, Felix and Darszon, 2001) and human (Yunes, Tomes, Michaut, De Blas, Rodriguez, Regazzi and Mayorga, 2002) sperm. Later on, calmodulin exhibits an inhibitory role in permeabilised human sperm (Yunes *et al.*, 2002). What its targets might be at this late phase is not yet known.

Vesicle-to plasma membrane fusion stages: an overview

I will briefly summarise here the definitions of pre-fusion stages that best fit our own concepts (Szule and Coorssen, 2003) and describe how we apply them to the AR. Next, I will cover the molecular machinery involved in each stage. The term "tethering" describes the initial recognition and stabilised contact between the membranes of the secretory vesicle and the cell. Priming refers to any molecular reaction(s) that contributes to optimization of the fusogenic potential of the vesicle, and to its fully docked state. Productive tethering is followed by membrane "docking", a stronger, and spatially closer, interaction of the two bilayers engaged in fusion. Docking is required for the efficient translation of triggering signals, and represents the optimal state of inter-membrane contact that can exist prior to actual fusion. In many systems, a proportion of vesicles are "release-ready", awaiting only the trigger (usually an elevation in local calcium concentrations), to initiate fusion. These vesicles are described as being functionally docked.

Because we have not yet developed a method to distinguish visually between tethered and non-tethered, docked and non-docked acrosomes, we employ the molecular definition of tethering as being SNARE-independent, triggered by Rabs and accomplished through the recruitment of tethering factors. Again, we use a molecular, rather than a morphological, definition of docking, achieved by *trans* SNARE pairing. We use the term "priming" to describe the activation of the fusion machinery. Priming includes - but is not limited to - the unpairing of *cis* SNARE complexes on both membranes by α-SNAP/NSF to make them available for *trans* pairing and subsequent fusion. It follows from these definitions that priming precedes docking. Because in the AR Rab3 functions prior to α-SNAP/NSF, we believe tethering precedes or is concomitant with priming. Tethering, priming and docking take place before the second efflux of calcium from the acrosomal store (Fig. 1B, Table 1).

Fusion molecules in the acrosome reaction

Rab3A. Rabs are small GTP-binding proteins that behave as molecular switches by cycling between inactive (GDP-bound, cytosolic) and active (GTP-bound, membrane-associated) states (Zerial and McBride, 2001). Rabs are responsible for the recognition and initial physical attachment of the membranes that are going to fuse. They accomplish this task through recruitment of tethering factors, large fibrous proteins or protein complexes containing coiled-coiled domains (Gonzalez Jr. and Scheller, 1999; Pfeffer, 1999; Waters and Hughson, 2000).

Rab3 participates in regulated exocytosis of neurotransmitters and hormones (Zerial and McBride, 2001; Coppola, Frantz, Perret-Menoud, Gattesco, Hirling and Regazzi, 2002). One of the four isoforms, Rab3A, is present in the acrosomal region of human (Yunes *et al.*, 2000), rat (Iida *et al.*, 1999), and mouse (Ward, Faundes and Foster, 1999) sperm. Rab3A has been reported to stimulate ram (Garde and Roldan, 1996) and human (Yunes *et al.*, 2000) sperm AR and to inhibit rat sperm AR (Iida *et al.*, 1999). Rab3A (the full length recombinant Rab3A, prenylated *in vitro*, and persistently activated with GTP-γ-S; "Rab3A" from now on) triggers sperm exocytosis to a magnitude similar to calcium in permeabilised sperm (Yunes *et al.*, 2000). Likewise, a permeant form of the protein induces the AR in intact human sperm (Lopez, Belmonte, De Blas and Mayorga, 2006). Rab3 is required for the AR triggered by calcium (Belmonte, Lopez, Roggero, De Blas, Tomes and Mayorga, 2005; De Blas, Roggero, Tomes, and Mayorga, 2005) and cAMP (Branham *et al.*, 2006). Capacitation leads to an enhanced association of Rab3A with sperm membranes, presumably correlating with its activation in preparation for the AR. Interestingly, acute removal of cholesterol from sperm membranes enhances both calcium-triggered exocytosis and Rab3A targeting to membranes (Belmonte *et al.*, 2005). Where in the fusion cascade involved in the AR does Rab3 play its part? When elicited by exogenous Rab3A, exocytosis depends on α-SNAP/NSF (Tomes, De Blas, Michaut, Farre, Cherhitin, Visconti and Mayorga, 2005), SNAREs (Tomes, Michaut, De Blas, Visconti, Matti and Mayorga, 2002), synaptotagmin VI (Michaut *et al.*, 2001), and an efflux of calcium from inside the acrosome (De Blas *et al.*, 2002), coinciding with its "post-SOC-pre-second acrosomal calcium release" role (Fig. 1B, Table 1). RIM, a Rab binding protein (Martin, 2002) is present in human sperm and appears to be required for exocytosis, opening the possibility of it being one of Rab3's tethering factors in sperm (Bello, Magadan and Mayorga, 2006).

SNAREs. SNAREs are a family of small membrane proteins that can form stable hetero-oligomeric complexes. It is believed that SNAREs anchored in the membrane of a transport vesicle pair with those in the target membrane through a region adjacent to their C-terminal anchor (the SNARE motif), forming molecular bridges that bring the fusing membranes into close apposition. It is assumed that these bridging interactions between SNARE proteins form a four-helix bundle similar to a complex in the *cis* configuration, when all proteins are on the same membrane. The bundle, or core complex, contains one helix from synaptobrevin, one from syntaxin and two from synaptosome-associated protein of 25 kDa (SNAP-25) (Sutton, Fasshauer, Jahn and Brunger, 1998; Bruns and Jahn, 2002; Fasshauer, 2003). Individual SNAREs assemble quasi irreversibly into ternary (*cis*) complexes that are highly resistant to denaturing reagents and can only undergo disassembly with the assistance of an ATPase (NSF) and its cofactor (α, ß or γ SNAPs) (May, Whiteheart and Weis, 2001). SNAREs are classified as R-SNAREs or Q-SNAREs, based on the identity of a highly conserved arginine or glutamine residue in an unusually hydrophilic layer in the middle of the hydrophobic bundle (Fasshauer, Sutton, Brunger and Jahn, 1998). Syntaxin and SNAP-25 are Q-SNAREs whereas synaptobrevin is an R-SNARE. Complex formation between membranes begins at the amino terminus portion of the molecules (Sorensen, Wiederhold, Muller, Milosevic, Nagy, de Groot, Brubmuller and Fasshauer, 2006). Their subsequent zippering into stable membrane-bridging (*trans*) complexes progresses from partly (loose) to fully (tight) assembled configurations. Elucidating the exact role of SNARE proteins in membrane fusion is the goal of much research effort. The topic has been reviewed extensively in the last few years (Bruns and Jahn, 2002; Szule and Coorssen, 2003).

Tetanus toxin (TeTx) and seven structurally related botulinum neurotoxins (BoNT/A, B, C1, D, E, F and G) are potent inhibitors of secretory vesicle release due to their highly specific, zinc-dependent, proteolytic cleavage of SNARE proteins. BoNT/A and E cleave SNAP-25, BoNT/C cleaves syntaxin and, with much lower efficiency, SNAP-25. The remaining BoNTs, as well as TeTx, are specific for synaptobrevin (Pellizzari, Rossetto, Schiavo and Montecucco., 1999). All SNAREs are only sensitive to cleavage by neurotoxins when not packed in tight heterotrimeric complexes (Hayashi, McMahon, Yamasaki, Binz, Hata, Sudhof and Niemann, 1994).

Members of the SNARE complex in sperm comprise all three families in sea urchin (Schulz, Wessel and Vacquier, 1997; Schulz, Sasaki and Vacquier, 1998; Schulz, De La Vega-Beltran, Beltran, Vacquier and Darszon., 2004), and mammals (Katafuchi, Mori, Toshimori and Iida, 2000; Ramalho-Santos *et al.*, 2000; Tomes *et al.*, 2002; De Blas *et al.*, 2005; Gamboa and Ramalho-Santos 2005; Hutt *et al.*, 2005). The onset of sperm's exocytosis relies on the productive assembly of toxin-sensitive members of all three SNARE families in ternary complexes regardless of the stimulus applied (Tomes *et al.*, 2002; De Blas et *al.*, 2005; Branham et *al.*, 2006).

α-SNAP/NSF. Defining when SNARE complexes disassemble has proven difficult and controversial. Dissociation of the *cis*-SNARE complex by α-SNAP/NSF has been reported to occur at a late, post-fusion step to recycle the fusion machinery accumulated in the plasma membrane as a result of vesicle fusion. Alternatively, an early, pre-fusion, priming role for α-SNAP/NSF is strongly supported by evidence from studies in a variety of models (Li and Chin, 2003; Sollner, 2003; Tomes, 2007).

NSF has been detected in human, mouse, rat, bovine, murine and rhesus macaque sperm. Indirect immunofluorescence experiments revealed that it localises to the acrosomal region (Michaut, Tomes, De Blas, Yunes and Mayorga., 2000; Ramalho-Santos and Schatten, 2004; Gamboa and Ramalho-Santos, 2005). Not surprisingly, α-SNAP exhibits a similar localization in human sperm (Tomes et *al.*, 2005) and mouse round spermatids (Ramalho-Santos, Moreno, Wessel, Chan and Schatten, 2001). α-SNAP and NSF are required for the AR when triggered by calcium, Rab3A (Michaut et *al.*, 2000; Tomes et *al.*, 2005), and cAMP (Branham et al., 2006). These proteins display a relatively early role, prior to the second efflux of calcium from the acrosome (Fig. 1B, Table 1). We believe the step catalysed by α-SNAP/NSF constitutes the priming of the fusion machinery, which is coupled to the tethering between the outer acrosomal and plasma membranes initiated by Rab3.

Molecular correlates of the fusion stages

The ratio of monomeric to assembled SNAREs depends on the type and physiological condition of the cell. Thus, while some studies indicate that most SNAREs are free in the plasma membrane (Darszon et *al.*, 2001), others suggest that they are engaged in complexes (Lang, Margittai, Holzler and Jahn, 2002; Rickman, Meunier, Binz and Davletov, 2004). Whatever the steady state configuration of SNAREs in neuroendocrine cells might be, exocytosis is blocked by neurotoxins, suggesting that SNAREs go through toxin-sensitive stages (Xu, Binz, Niemann and Neher, 1998; Chen, Scales, Patel, Doung and Scheller, 1999).

In sperm, α-SNAP/NSF are required for the AR. Since they only bind and dissociate *cis* SNARE complexes, this requirement suggests that sperm SNAREs might be engaged in such a configuration. Both R- and Q-SNAREs are protected from toxin cleavage in resting cells, instead of cycling through toxin-resistant and toxin-sensitive configurations as proposed in other systems. Sperm activation triggers SNARE disassembly in a step prior to the second efflux of calcium from the intra-acrosomal pool (De Blas et *al.*, 2005). Two SNARE configurations are compatible with toxin-sensitivity: monomeric and partially assembled in loose *trans* complexes. We have used differential sensitivity to BoNT/B and TeTx (Hua and Charlton, 1999) to establish that, following sperm stimulation, SNAREs are engaged in loose *trans* complexes awaiting the release of the second phase of intra-acrosomal calcium that will trigger the final steps of the AR (Fig. 1B). We refer to this intermediate, stable and, as far as we can tell, irreversible fusion stage as docking. Unlike what has been described in other exocytosis models, the assembly of *trans* SNARE complexes in sperm is calcium independent (De Blas et *al.*, 2005). In the yeast vacuolar fusion model, docking by *trans* SNARE pairing produces an unknown signal leading to the transient opening of calcium channels in the vacuolar membrane. The resulting calcium efflux activates targets that catalyse bilayer mixing (Peters and Mayer, 1998; Merz and Wickner, 2004). In sperm, on the contrary, productive *trans* SNARE pairing does not appear to be a prerequisite for intra-acrosomal calcium efflux, given that Rab3A is able to mobilise intravesicular calcium in cells treated with BoNT/E (De Blas *et al.*, 2002). We are currently devoting our efforts to elucidate what, other than mere docking, is the signal that opens the IP_3-sensitive channels in the outer acrosomal membrane to give rise to the second efflux of calcium.

Model

In resting sperm, SNAREs are locked in inactive *cis* complexes on plasma and outer acrosomal membranes. Synaptotagmin is phosphorylated in the polybasic C2B region and therefore maintained in an inactive conformation. Upon calcium entry into the cytoplasm, Rab3 is activated, initiating the tethering of the outer acrosomal to the plasma membrane (not shown for simplicity) and the disassembly of *cis* SNARE complexes by α-SNAP/NSF. Sperm stimulation with calcium also leads to dephosphorylation of synaptotagmin which de-represses this protein's ability to interact with its binding partners. By a not yet identified mechanism, monomeric SNAREs assemble in loose *trans* complexes, causing the irreversible docking of the acrosome to the plasma membrane. At this point, calcium is released from inside the acrosome through IP_3-sensitive channels to trigger the final steps of membrane fusion, which require SNAREs (presumably in tight *trans* complexes) and synaptotagmin (Fig. 1B).

Signalling during the acrosome reaction

There are a number of intracellular signalling molecules that carry the message from a given inducer bound to the sperm surface to the membranes that fuse. These include several signal-transducing components, second messengers, G proteins, phospholipases, lipids, protein kinases and phosphatases. The subject has been extensively covered in the literature (Yanagimachi, 1994; Breitbart and Spungin, 1997; Brewis and Moore, 1997; Roldan, 1998; Garde and Roldan, 2000; Darszon, Beltran, Felix, Nishigaki and Trevino, 2001; Gadella, Rathi, Brouwers, Stout and Colenbrander., 2001; Baldi, Luconi, Bonaccorsi and Forti, 2002; Breitbart, 2003; Evans and Florman, 2002; Herrero, de Lamirande and Gagnon, 2003; Neill and Vacquier, 2004; Felix, 2005; Fraser, Adeoya-Osiguwa, Baxendale, Mededovic, and Osiguwa, 2005; Wassarman, Jovine, Qi, Williams, Darie and Litscher, 2005). Since signalling pathways studies are, for the most part, not concerned with membrane fusion, I will not consider them further in this review, with the following two exceptions.

Protein kinases and phosphatases. Protein phosphorylation is a post-translational modification that allows the cell to control various cellular processes. The phosphorylation state of phosphoproteins is controlled by the activity of protein kinases and phosphatases. Mature spermatozoa have their own idiosyncrasies as highly specialised cells: they are highly compartmentalised, transcriptionally inactive, and unable to synthesise new proteins. Therefore, their reliance on protein phosphorylation as a means of modulating their function is greater than in other cells (Urner and Sakkas, 2003). Western blotting of sperm populations during capacitation and/or after exposure to AR inducers has been widely used in the reproductive biology field, giving rise to a catalogue of phosphorylated proteins that is, unfortunately, not very reproducible and, with a few exceptions, conveys little information about their identity and functions.

Protein tyrosine phosphorylation had been implicated in mammalian sperm AR by using pharmacological inhibitors (Leyton, LeGuen, Bunch and Saling., 1992; Tesarik, Moos and Mendoza., 1993; Luconi, Bonaccorsi, Krausz, Gervasi, Forti and Baldi, 1995; Meizel and Turner, 1996; Pukazhenthi, Wildt, Ottinger and Howard, 1998; Kirkman-Brown, Lefievre, Bray, Stewart, Barratt and Publicover., 2002a). The initiation of signalling cascades leading to the AR induces tyrosine kinase activity in human sperm in two waves, the first shortly after cytoplasmic calcium increase and the second in a later step, following the second intra-acrosomal calcium efflux (Tomes *et al.*, 2004). Calcium- and Rab3A-induced protein tyrosine phosphorylation are essential for the AR. Interestingly, there appear to be late acting kinases that presumably require a specific microenvironment - high and/or localised calcium concentrations achieved upon efflux of calcium from the acrosome - for optimal activity (Table 1). We are currently devoting some efforts to identify the targets of this kinase/s, but more importantly, to determine whether tyrosine phosphorylation is a signature of sperm's exocytosis or if represents a widespread and, as of yet, unexplored phenomenon in membrane fusion. The very recent discovery that Munc18 is tyrosine phosphorylated in response to stimuli and that its phosphorylation facilitates exocytosis (Oh and Thurmond, 2006) supports the latter.

As far as protein tyrosine dephosphorylation goes, the non-transmembrane protein tyrosine phosphatase1B (PTP1B) (Tonks, 2003) is the only enzyme described so far in sperm (Seligman, Zipser and Kosower, 2004; Tomes *et al.*, 2004). PTP1B is required for the AR downstream of Rab3-induced tethering and before the second efflux of intra-acrosomal calcium (Table 1). Interestingly, serine/threonine phosphatases PP1 and PP2A participate in membrane fusion in yeast (Peters, Andrews, Stark, Cesaro-Tadic, Glatz, Podtelejnikov, Mann and Mayer., 1999; Marash and Gerst, 2001; Bryant and James, 2003), and the tyrosine phosphatase PTP-MEG2 in mammalian cells (Huynh, Bottini, Williams, Cherepanov, Musumeci, Saito, *et al.*, 2004).

cAMP/Exchange protein directly activated by cAMP (Epac). In many cell types, an increase in intracellular cAMP concentration regulates calcium-triggered exocytosis (Holz, 2004; Seino and Shibasaki, 2005). In most cases, an elevation of cAMP alone in the absence of a calcium rise is not sufficient to elicit secretion (Burgoyne and Morgan, 2003). Yet, a limited range of cells use cAMP as a major trigger for exocytosis. The cAMP dependent pathways coexist with calcium-dependent ones for exocytosis in these cells, and it is likely that they use a common final SNARE-dependent mechanism. cAMP-triggered exocytosis in parotid gland, one of the best studied, requires cAMP dependent protein kinase (PKA) and synaptobrevin 2. The involvement of other components of the general fusion machinery in cAMP-triggered exocytosis has not, until now, been investigated (Burgoyne and Morgan, 2003), with the exception of the pathway cAMP-Epac2-Rim-Rab3-calcium sensor (Ozaki, Shibasaki, Kashima, Miki, Takahashi, Ueno, Sunaga, Yano, Matsuura, Iwanaga, Takai and Seino, 2000; Kashima, Miki, Shibasaki, Ozaki, Miyazaki, Yano and Seino., 2001; Fujimoto, Shibasaki, Yokoi, Kashima, Matsumoto, Sasaki, Tajima, Iwanaga and Seino, 2002; Shibasaki, Sunuga, Fujimoto, Kashima and Seino, 2004).

In sperm, various activation pathways required to achieve egg-fertilising ability depend on the intracellular rise of cAMP. Of particular interest here are those concerned with the AR (Lefievre, Jha, de Lamirande, Visconti and Gagnon, 2002; Breitbart, 2003). It has been known for some time that reagents that increase intracellular cAMP levels induce exocytosis in mammalian sperm (De Jonge, Han, Lawrie, Mack and Zaneveld, 1991; Garde and Roldan, 2000). One of the mechanisms proposed to explain these observations involves PKA activation by cAMP, leading to the opening of calcium channels (Breitbart, 2003). We have recently unveiled an alternative mechanism by which cAMP triggers exocytosis in human sperm. This cAMP-dependent step is located downstream of the influx of calcium into the cytosol through SOCs and is independent of PKA, suggesting that another cAMP target is involved. Epac, a small guanine nucleotide exchange factor for the small GTPase Rap (de Rooij, Zwartkruis, Verheijen, Cool, Nijman, Wittinghofer and Bos, 1998; Kawasaki, Springett, Mochizuki, Toki, Nakaya, Matsuda, Housman and Graybiel, 1998), is the cAMP target responsible for all its post-SOC opening effects in sperm exocytosis. Like calcium, cAMP - via Epac activation - has the ability to drive the whole cascade of events necessary to bring exocytosis to completion, including tethering and docking of the acrosome to the plasma membrane, priming of the fusion machinery, mobilization of intravesicular calcium, and ultimately, bilayer mixing and fusion. In fact, the AR elicited by cAMP proceeds in the virtual absence of cytosolic calcium, while all the effects of calcium on the AR appear to be mediated by cAMP/Epac (Braham et *al.*, 2006). These findings fit nicely into our molecular model of the AR (Fig. 1B). Upon calcium entry into the cytoplasm, a soluble adenylyl cyclase known as sAC (Buck, Sinclair, Schapal, Cann and Levin, 1999; Chen, Cann, Litvin, Iourgenko, Sinclair, Levin and Buck, 2000) is stimulated, leading to an increase in cAMP levels. In turn, cAMP binds to and activates Epac. Later on Rab3 is activated and the whole signalling cascade, alluded to before, unfolds.

Acknowledgements

Thanks to Luis Mayorga and all members of his and my laboratory for helpful discussions. Thanks also to Sean Patterson for critical reading of this review and valuable suggestions. Gerardo De Blas, Natalia Zanetti and Luis Mayorga's kind permission to use unpublished figures is acknowledged.

References

Baldi, E., Luconi, M., Bonaccorsi, L. and Forti, G. (2002). Signal transduction pathways in human spermatozoa. *Journal of Reproductive Immunology* **53**: 121-131

Barros, C., Bedford, J.M., Franklin, L.E. and Austin, C.R. (1967). Membrane vesiculation as a feature of the mammalian acrosome reaction. *Journal of Cell Biology* **34**: C1-C5

Bello O., Magadan J. and Mayorga, L. (2006). RIM participates in the acrosomal exocytosis of human sperm. *Biocell* **29** (Supplement): 99.

Belmonte, S.A., Lopez, C.I., Roggero, C.M., De Blas, G.A., Tomes, C.N. and Mayorga, L.S. (2005). Cholesterol content regulates acrosomal exocytosis by enhancing Rab3A plasma membrane association. *Developmental Biology* **285**: 393-408

Bendahmane, M., Lynch, C. and Tulsiani, D.R. (2001). Calmodulin signals capacitation and triggers the agonist-induced acrosome reaction in mouse spermatozoa. *Archives of Biochemistry and Biophysics* **390**: 1-8

Berridge, M.J., Lipp, P. and Bootman, M.D. (2000a). Signal transduction. The calcium entry pas de deux. *Science* **287**: 1604-1605

Berridge, M.J., Lipp, P. and Bootman, M.D. (2000b). The versatility and universality of calcium signalling. *Nature Review in Molecular Cell Biology* **1**: 11-21

Bhalla, A., Chicka, M.C., Tucker, W.C. and Chapman, E.R. (2006). Ca^{2+}-synaptotagmin directly regulates t-SNARE function during reconstituted membrane fusion. *Nature Structural Molecular Biology* **13**: 323-330

Branham, M.T., Mayorga,L.S. and Tomes,C.N. (2006). Calcium-induced acrosomal exocytosis requires cAMP acting through a protein kinase A-independent, Epac-mediated pathway. *Journal of Biological Chemistry*, **281**: 8656-8666

Breitbart, H. (2003). Signaling pathways in sperm capacitation and acrosome reaction. *Cellular and Molecular Biology (Noisy.-le-grand)* **49**: 321-327

Breitbart, H. and Spungin, B. (1997). The biochemistry of the acrosome reaction. *Molecular Human Reproduction* **3**: 195-202

Brewis, I.A. and Moore, H.D. (1997). Molecular mechanisms of gamete recognition and fusion at fertilization. *Human Reproduction* **12**: 156-165

Bruns, D. and Jahn, R. (2002). Molecular determinants of exocytosis. *Pflugers Archives* **443**: 333-338

Bryant, N.J. and James, D.E. (2003). The Sec1p/Munc18 (SM) protein, Vps45p, cycles on and off membranes during vesicle transport. *Journal of Cell Biology* **161**: 691-696

Buck, J., Sinclair, M.L., Schapal, L., Cann, M.J. and Levin, L.R. (1999). Cytosolic adenylyl cyclase defines a unique signaling molecule in mammals. *Proceedings of the National Academy of Sciences USA* **96**: 79-84

Burgoyne, R.D. and Clague, M.J. (2003). Calcium and calmodulin in membrane fusion. *Biochimica et Biophysica Acta* **1641**: 137-143

Burgoyne, R.D. and Morgan, A. (1995). Ca^{2+} and secretory-vesicle dynamics. *Trends in Neuroscience* **18**: 191-196

Burgoyne, R.D. and Morgan, A. (1998a). Analysis of regulated exocytosis in adrenal chromaffin cells: insights into NSF/SNAP/SNARE function. *Bioessays* **20**: 328-335

Burgoyne, R.D. and Morgan, A. (1998b). Calcium sensors in regulated exocytosis. *Cell Calcium* **24**: 367-376

Burgoyne, R.D. and Morgan, A. (2003). Secretory granule exocytosis. *Physiological Reviews* **83**: 581-632

Chen, Y., Cann, M.J., Litvin, T.N., Iourgenko, V., Sinclair, M.L., Levin, L.R. and Buck, J. (2000). Soluble adenylyl cyclase as an evolutionarily conserved bicarbonate sensor. *Science* **289**: 625-628

Chen, Y.A., Scales, S.J., Patel, S.M., Doung, Y.C. and Scheller, R.H. (1999). SNARE complex formation is triggered by Ca^{2+} and drives membrane fusion. *Cell* **97**: 165-174

Clapham, D.E. (1995). Calcium signaling. *Cell* **80**: 259-268

Coppola, T., Frantz, C., Perret-Menoud, V., Gattesco, S., Hirling, H. and Regazzi, R. (2002). Pancreatic b-cell protein granuphilin binds Rab3 and Munc-18 and controls exocytosis. *Molecular Biology of the Cell* **13**: 1906-1915

Darszon, A., Beltran, C., Felix, R., Nishigaki, T. and Trevino, C.L. (2001). Ion transport in sperm signaling. *Developmental Biology* **240**: 1-14

Darszon, A., Nishigaki, T., Wood, C., Trevino, C.L., Felix, R. and Beltran, C. (2005). Calcium channels and Ca^{2+} fluctuations in sperm physiology. *International Review of Cytology* **243**: 79-172

De Blas, G., Michaut, M., Trevino, C.L., Tomes, C.N., Yunes, R., Darszon, A. and Mayorga, L.S. (2002). The intraacrosomal calcium pool plays a direct role in acrosomal exocytosis. *Journal of Biological Chemistry* **277**: 49326-49331

De Blas, G.A., Roggero, C.M., Tomes, C.N. and Mayorga, L.S. (2005). Dynamics of SNARE assembly and disassembly during sperm acrosomal exocytosis. *Public Library of Science. Biology* **3**: 1801-1812

de Haro, L., Ferracci, G., Opi, S., Iborra, C., Quetglas, S., Miquelis, R., Leveque, C. and Seagar, M. (2004). Ca^{2+}/calmodulin transfers the membrane-proximal lipid-binding domain of the v-SNARE synaptobrevin from cis to trans bilayers. *Proceedings of the National Academy of Sciences USA* **101**: 1578-1583

De Jonge, C.J., Han, H.L., Lawrie, H., Mack, S.R. and Zaneveld, L.J. (1991). Modulation of the human sperm acrosome reaction by effectors of the adenylate cyclase/cyclic AMP second-messenger pathway. *Journal of Experimental Zoology* **258**: 113-125

de Rooij, J., Zwartkruis, F.J., Verheijen, M.H., Cool, R.H., Nijman, S.M., Wittinghofer, A. and Bos, J.L. (1998). Epac is a Rap1 guanine-nucleotide-exchange factor directly activated by cyclic AMP. *Nature* **396**: 474-477

Diaz, A., Dominguez, I., Fornes, M.W., Burgos, M.H. and Mayorga, L.S. (1996). Acrosome content release in streptolysin O permeabilized mouse spermatozoa. *Andrologia.* **28**: 21-26

Evans, J.P. and Florman, H.M. (2002). The state of the union: the cell biology of fertilization. *Nature Cell Biology* **4** Supplement: s57-s63

Fasshauer, D. (2003). Structural insights into the SNARE mechanism. *Biochimica et Biophysica Acta.* **1641**: 87-97

Fasshauer, D., Sutton, R.B., Brunger, A.T. and Jahn, R. (1998). Conserved structural features of the synaptic fusion complex: SNARE proteins reclassified as Q- and R-SNAREs. *Proceedings of the National Academy of Sciences USA* **95**: 15781-15786

Felix, R. (2005). Molecular physiology and pathology of Ca^{2+}-conducting channels in the plasma membrane of mammalian sperm. *Reproduction* **129**: 251-262

Florman, H.M., Arnoult, C., Kazam, I.G., Li, C. and O'Toole, C.M. (1998). A perspective on the control of mammalian fertilization by egg- activated ion channels in sperm: a tale of two channels. *Biology of Reproduction* **59**: 12-16

Fraser, L.R., Adeoya-Osiguwa, S., Baxendale, R.W., Mededovic, S. and Osiguwa, O.O. (2005). First messenger regulation of mammalian sperm function via adenylyl cyclase/cAMP. *Journal of Reproduction and Development* **51**: 37-46

Fujimoto, K., Shibasaki, T., Yokoi, N., Kashima, Y., Matsumoto, M., Sasaki, T., Tajima, N., Iwanaga, T. and Seino, S. (2002). Piccolo, a Ca^{2+} sensor in pancreatic b-cells. Involvement of cAMP-GEFII.Rim2.Piccolo complex in cAMP-dependent exocytosis. *Journal of Biological Chemistry* **277**: 50497-50502

Gadella, B.M., Rathi, R., Brouwers, J.F., Stout, T.A. and Colenbrander, B. (2001). Capacitation and the acrosome reaction in equine sperm. *Animal Reproduction Science* **68**: 249-265

Gamboa, S. and Ramalho-Santos, J. (2005). SNARE proteins and caveolin-1 in stallion spermatozoa: possible implications for fertility. *Theriogenology* **64**: 275-291

Garde, J. and Roldan, E.R.S. (1996). Rab3-peptide stimulates exocytosis of the ram sperm acrosome via interaction with cyclic AMP and phospholipase A_2 metabolites. *FEBS Letters* **391**: 263-268

Garde, J. and Roldan, E.R. (2000). Stimulation of Ca^{2+}-dependent exocytosis of the sperm acrosome by cAMP acting downstream of phospholipase A_2. *Journal of Reproduction and Fertility* **118**: 57-68

Gonzalez, L., Jr. and Scheller, R.H. (1999). Regulation of membrane trafficking: structural insights from a Rab/effector complex. *Cell* **96**: 755-758

Guerrero, A. and Darszon, A. (1989). Egg jelly triggers a calcium influx which inactivates and is inhibited by calmodulin antagonists in the sea urchin sperm. *Biochimica et Biophysica Acta* **980**: 109-116

Hayashi, T., McMahon, H., Yamasaki, S., Binz, T., Hata,

Y., Sudhof, T.C. and Niemann, H. (1994). Synaptic vesicle membrane fusion complex: action of clostridial neurotoxins on assembly. *EMBO Journal* **13**, 5051-5061

Hernandez-Gonzalez, E.O., Lecona-Valera, A.N., Escobar-Herrera, J. and Mujica, A. (2000). Involvement of an F-actin skeleton on the acrosome reaction in guinea pig spermatozoa. *Cell Motility and the Cytoskeleton* **46**, 43-58

Herrero, M.B., de Lamirande, E. and Gagnon, C. (2003). Nitric oxide is a signaling molecule in spermatozoa. *Current Pharmaceutical Design* **9**, 419-425

Herrick, S.B., Schweissinger, D.L., Kim, S.W., Bayan, K.R., Mann, S. and Cardullo, R.A. (2005). The acrosomal vesicle of mouse sperm is a calcium store. *Journal of Cell Physiology* **202**, 663-671

Holz, G.G. (2004). Epac: A new cAMP-binding protein in support of glucagon-like peptide-1 receptor-mediated signal transduction in the pancreatic beta-cell. *Diabetes* **53**, 5-13

Hua, S.Y. and Charlton, M.P. (1999). Activity-dependent changes in partial VAMP complexes during neurotransmitter release. *Nature. Neuroscience* **2**, 1078-1083

Hutt, D.M., Baltz, J.M. and Ngsee, J.K. (2005). Synaptotagmin VI and VIII and syntaxin 2 are essential for the mouse sperm acrosome reaction. *Journal of Biological Chemistry* **280**, 20197-20203

Hutt, D.M., Cardullo, R.A., Baltz, J.M. and Ngsee, J.K. (2002). Synaptotagmin VIII is localized to the mouse sperm head and may function in acrosomal exocytosis. *Biology of Reproduction* **66**, 50-56

Huynh, H., Bottini, N., Williams, S., Cherepanov, V., Musumeci, L., Saito, K., Bruckner, S., Vachon, E., Wang, X., Kruger, J., Chow, C.W., Pellecchia, M., Monosov, E., Greer, P.A., Trimble, W., Downey, G.P. and Mustelin, T. (2004). Control of vesicle fusion by a tyrosine phosphatase. *Nature Cell Biology* **6**, 831-839

Iida, H., Yoshinaga, Y., Tanaka, S., Toshimori, K. and Mori, T. (1999). Identification of Rab3A GTPase as an acrosome-associated small GTP-binding protein in rat sperm. *Developmental Biology* **211**, 144-155

Jahn, R., Lang, T. and Sudhof, T.C. (2003). Membrane fusion. *Cell* **112**, 519-533

Jahn, R. and Sudhof, T.C. (1999). Membrane fusion and exocytosis. *Annual Review of Biochemistry* **68**, 863-911

Johnson, L.R., Moss, S.B. and Gerton, G.L. (1999). Maintenance of motility in mouse sperm permeabilized with streptolysin O. *Biology of Reproduction* **60**, 683-690

Kashima, Y., Miki, T., Shibasaki, T., Ozaki, N., Miyazaki, M., Yano, H. and Seino, S. (2001). Critical role of cAMP-GEFII—Rim2 complex in incretin-potentiated insulin secretion. *Journal of Biological Chemistry* **276**, 46046-46053

Katafuchi, K., Mori, T., Toshimori, K. and Ilda, H. (2000). Localization of a syntaxin isoform, syntaxin 2, to the acrosomal region of rodent spermatozoa. *Molecular Reproduction and Development* **57**, 375-383

Kawasaki, H., Springett, G.M., Mochizuki, N., Toki, S., Nakaya, M., Matsuda, M., Housman, D.E. and Graybiel, A.M. (1998). A family of cAMP-binding proteins that directly activate Rap1. *Science* **282**, 2275-2279

Kirkman-Brown, J.C., Lefievre, L., Bray, C., Stewart, P.M., Barratt, C.L. and Publicover, S.J. (2002a). Inhibitors of receptor tyrosine kinases do not suppress progesterone-induced $[Ca^{2+}]_i$ signalling in human spermatozoa. *Molecular Human Reproduction* **8**, 326-332

Kirkman-Brown, J.C., Punt, E.L., Barratt, C.L. and Publicover, S.J. (2002b). Zona pellucida and progesterone-induced Ca^{2+} signaling and acrosome reaction in human spermatozoa. *Journal of Andrology* **23**, 306-315

Kitamura, K., Tanaka, H. and Nishimune, Y. (2003). Haprin, a novel haploid germ cell-specific RING finger protein involved in the acrosome reaction. *Journal of Biological Chemistry* **278**, 44417-44423

Lang, T., Margittai, M., Holzler, H. and Jahn, R. (2002). SNAREs in native plasma membranes are active and readily form core complexes with endogenous and exogenous SNAREs. *Journal of Cell Biology* **158**, 751-760

Lefievre, L., Jha, K.N., de Lamirande, E., Visconti, P.E. and Gagnon, C. (2002). Activation of protein kinase A during human sperm capacitation and acrosome reaction. *Journal of Andrology* **23**, 709-716

Leyton, L., LeGuen, P., Bunch, D. and Saling, P.M. (1992). Regulation of mouse gamete interaction by a sperm tyrosine kinase. *Proceedings of the National Academy of Sciences USA* **89**, 11692-11695

Li, L. and Chin, L.S. (2003). The molecular machinery of synaptic vesicle exocytosis. *Cellular and Molecular Life Sciences* **60**, 942-960

Lopez, C, Belmonte, S.A., De Blas, G. and Mayorga, L. (2006). Membrane permeant Rab3A triggers acrosomal exocytosis in the absence of extracellular calcium. *Biocell* **29** (Supplement), 66

Lopez-Gonzalez, I., Vega-Beltran, J.L., Santi, C.M., Florman, H.M., Felix, R. and Darszon, A. (2001). Calmodulin antagonists inhibit T-type Ca^{2+} currents in mouse spermatogenic cells and the zona pellucida-induced sperm acrosome reaction. *Developmental Biology* **236**, 210-219

Luconi, M., Bonaccorsi, L., Krausz, C., Gervasi, G., Forti, G. and Baldi, E. (1995). Stimulation of protein tyrosine phosphorylation by platelet- activating factor and progesterone in human spermatozoa. *Molecular and Cellular Endocrinology* **108**, 35-42

Marash, M. and Gerst, J.E. (2001). t-SNARE dephosphorylation promotes SNARE assembly and exocytosis in yeast. *EMBO Journal* **20**, 411-421

Marin-Briggiler, C.I., Jha, K.N., Chertihin, O., Buffone, M.G., Herr, J.C., Vazquez-Levin, M.H. and Visconti, P.E. (2005). Evidence of the presence of calcium/calmodulin-dependent protein kinase IV in human sperm and its involvement in motility regulation. *Journal of Cell Science* **118**, 2013-2022

Martin, T.F. (2002). Prime movers of synaptic vesicle

exocytosis. *Neuron.* **34**: 9-12

Martin, T.F. (2003). Tuning exocytosis for speed: fast and slow modes. *Biochimyca et Biophysica Acta* **1641**: 157-165

May, A.P., Whiteheart, S.W. and Weis, W.I. (2001). Unraveling the mechanism of the vesicle transport ATPase NSF, the N-Ethylmaleimide-sensitive factor. *Journal of Biological Chemistry* **276**: 21991-21994

Meizel, S. and Turner, K.O. (1996). Chloride efflux during the progesterone-initiated human sperm acrosome reaction is inhibited by lavendustin A, a tyrosine kinase inhibitor. *Journal of Andrology* **17**: 327-330

Merz, A.J. and Wickner, W.T. (2004). Trans-SNARE interactions elicit Ca^{2+} efflux from the yeast vacuole lumen. *Journal of Cell Biology* **164**: 195-206

Michaut, M., De Blas, G., Tomes, C.N., Yunes, R., Fukuda, M. and Mayorga, L.S. (2001). Synaptotagmin VI participates in the acrosome reaction of human spermatozoa. *Developmental Biology* **235**: 521-529

Michaut, M., Tomes, C.N., De Blas, G., Yunes, R. and Mayorga, L.S. (2000). Calcium-triggered acrosomal exocytosis in human spermatozoa requires the coordinated activation of Rab3A and N-ethylmaleimide-sensitive factor. *Proceedings of the National Academy of Sciences USA* **97**: 9996-10001

Mikoshiba, K., Fukuda, M., Ibata, K., Kabayama, H. and Mizutani, A. (1999). Role of synaptotagmin, a Ca^{2+} and inositol polyphosphate binding protein, in neurotransmitter release and neurite outgrowth. *Chemistry and Physics of Lipids* **98**: 59-67

Neher, E. (1998). Vesicle pools and Ca^{2+} microdomains: new tools for understanding their roles in neurotransmitter release. *Neuron* **20**: 389-399

Neill, A.T. and Vacquier, V.D. (2004). Ligands and receptors mediating signal transduction in sea urchin spermatozoa. *Reproduction* **127**: 141-149

Nowycky, M.C. and Thomas, A.P. (2002). Intracellular calcium signaling. *Journal of Cell Science* **115**: 3715-3716

Oh, E. and Thurmond, D.C. (2006). The stimulus-induced tyrosine-phosphorylation of Munc18c facilitates vesicle exocytosis. *Journal of Biological Chemistry* **281**: 17624-17634

Ozaki, N., Shibasaki, T., Kashima, Y., Miki, T., Takahashi, K., Ueno, H., Sunaga, Y., Yano, H., Matsuura, Y., Iwanaga, T., Takai, Y. and Seino, S. (2000). cAMP-GEFII is a direct target of cAMP in regulated exocytosis. *Nature Cell Biology* **2**: 805-811

Pellizzari, R., Rossetto, O., Schiavo, G. and Montecucco, C. (1999). Tetanus and botulinum neurotoxins: mechanism of action and therapeutic uses. *Philosophical Transactions of the Royal Society of London series B Biological Sciences* **354**: 259-268

Peters, C., Andrews, P.D., Stark, M.J., Cesaro-Tadic, S., Glatz, A., Podtelejnikov, A., Mann, M. and Mayer, A. (1999). Control of the terminal step of intracellular membrane fusion by protein phosphatase 1. *Science* **285**: 1084-1087

Peters, C., Bayer, M.J., Buhler, S., Andersen, J.S., Mann, M. and Mayer, A. (2001). Trans-complex formation by proteolipid channels in the terminal phase of membrane fusion. *Nature* **409**: 581-588

Peters, C. and Mayer, A. (1998). Ca^{2+}/calmodulin signals the completion of docking and triggers a late step of vacuole fusion. *Nature* **396**: 575-580

Pfeffer, S.R. (1999). Transport-vesicle targeting: tethers before SNAREs. *Nature Cell Biology* **1**: E17-E22

Pukazhenthi, B.S., Wildt, D.E., Ottinger, M.A. and Howard, J. (1998). Inhibition of domestic cat spermatozoa acrosome reaction and zona pellucida penetration by tyrosine kinase inhibitors. *Molecular Reproduction and Developement* **49**: 48-57

Quetglas, S., Iborra, C., Sasakawa, N., De Haro, L., Kumakura, K., Sato, K., Leveque, C. and Seagar, M. (2002). Calmodulin and lipid binding to synaptobrevin regulates calcium-dependent exocytosis. *EMBO Journal* **21**: 3970-3979

Ramalho-Santos, J., Moreno, R.D., Sutovsky, P., Chan, A.W., Hewitson, L., Wessel, G.M., Simerly, C.R. and Schatten, G. (2000). SNAREs in mammalian sperm: possible implications for fertilization. *Developmental Biology* **223**: 54-69

Ramalho-Santos, J., Moreno, R.D., Wessel, G.M., Chan, E.K. and Schatten, G. (2001). Membrane trafficking machinery components associated with the mammalian acrosome during spermiogenesis. *Experimental Cell Research* **267**: 45-60

Ramalho-Santos, J. and Schatten, G. (2004). Presence of N-ethyl maleimide sensitive factor (NSF) on the acrosome of mammalian sperm. *Archives of Andrology* **50**: 163-168

Rickman, C., Meunier, F.A., Binz, T. and Davletov, B. (2004). High affinity interaction of syntaxin and SNAP-25 on the plasma membrane is abolished by botulinum toxin E. *Journal of Biological Chemistry* **279**: 644-651

Roggero, C.M., Tomes, C.N., De Blas, G.A., Castillo, J., Michaut, M.A., Fukuda, M. and Mayorga, L.S. (2005). Protein kinase C-mediated phosphorylation of the two polybasic regions of synaptotagmin VI regulates their function in acrosomal exocytosis. *Developmental Biology* **285**: 422-435

Roldan, E.R.S. (1998). Role of phospholipases during sperm acrosomal exocytosis. *Frontiers of Bioscience* **3**: D1109-D1119

Sano, K. (1983). Inhibition of the acrosome reaction of sea urchin spermatozoa by a calmodulin antagonist, N-(6-aminohexyl)-5-chloro-1-naphthalenesulfonamide (W-7). *Journal of Experimental Zoology* **226**: 471-473

Scheenen, W.J., Wollheim, C.B., Pozzan, T. and Fasolato, C. (1998). Ca^{2+} depletion from granules inhibits exocytosis. A study with insulin-secreting cells. *Journal of Biological Chemistry* **273**: 19002-19008

Schulz, J.R., Sasaki, J.D. and Vacquier, V.D. (1998). Increased association of synaptosome-associated protein of 25 kDa with syntaxin and vesicle-associated membrane protein following acrosomal exocytosis of sea urchin sperm. *Journal of Biological Chemistry* **273**: 24355-24359

Schulz, J.R., Wessel, G.M. and Vacquier, V.D. (1997). The exocytosis regulatory proteins syntaxin and VAMP are shed from sea urchin sperm during the acrosome reaction. *Developmental Biology* **191**: 80-87

Schulz, J.R., De La Vega-Beltran, J., Beltran, C., Vacquier, V.D. and Darszon, A. (2004). Ion channel activity of membrane vesicles released from sea urchin sperm during the acrosome reaction. *Biochemical and Biophysical Research Communications* **321**: 88-93

Seino, S. and Shibasaki, T. (2005). PKA-dependent and PKA-independent pathways for cAMP-regulated exocytosis. *Physiological Reviews* **85**: 1303-1342

Seligman, J., Zipser, Y. and Kosower, N.S. (2004). Tyrosine phosphorylation, thiol status, and protein tyrosine phosphatase in rat epididymal spermatozoa. *Biology of Reproduction* **71**: 1009-1015

Shibasaki, T., Sunaga, Y., Fujimoto, K., Kashima, Y. and Seino, S. (2004). Interaction of ATP sensor, cAMP sensor, Ca^{2+} sensor, and voltage-dependent Ca^{2+} channel in insulin granule exocytosis. *Journal of Biological Chemistry* **279**: 7956-7961

Sollner, T.H. (2003). Regulated exocytosis and SNARE function. *Molecular Membrane Biology* **20**: 209-220

Sorensen, J.B., Wiederhold, K., Muller, E.M., Milosevic, I., Nagy, G., de Groot, B.L., Grubmuller, H. and Fasshauer, D. (2006). Sequential N- to C-terminal SNARE complex assembly drives priming and fusion of secretory vesicles. *EMBO Journal* **25**: 955-966

Storey, B.T. (1995). Interactions between gametes leading to fertilization: the sperm's eye view. *Reproduction Fertility and Development* **7**: 927-942

Sudhof, T.C. and Rizo, J. (1996). Synaptotagmins: C2-domain proteins that regulate membrane traffic. *Neuron* **17**: 379-388

Sutton, R.B., Fasshauer, D., Jahn, R. and Brunger, A.T. (1998). Crystal structure of a SNARE complex involved in synaptic exocytosis at 2.4 A resolution. *Nature* **395**: 347-353

Szule, J.A. and Coorssen, J.R. (2003). Revisiting the role of SNAREs in exocytosis and membrane fusion. *Biochimica et Biophysca Acta* **1641**: 121-135

Tesarik, J., Moos, J. and Mendoza, C. (1993). Stimulation of protein tyrosine phosphorylation by a progesterone receptor on the cell surface of human sperm. *Endocrinology* **133**: 328-335

Thirion, S., Troadec, J.D., Pivovarova, N.B., Pagnotta, S., Andrews, S.B., Leapman, R.D. and Nicaise, G. (1999). Stimulus-secretion coupling in neurohypophysial nerve endings: a role for intravesicular sodium? *Proceedings of the National Academy of Sciences USA* **96**: 3206-3210

Tomes, C.N. (2007). Acrosomal exocytosis. In: *Molecular Mechanisms of Exocytosis*. Edited by Regazzi, R., Landes Biosciences, Austin, TX, pp. 117-147

Tomes, C.N., De Blas, G.A., Michaut, M.A., Farre, E.V., Cherhitin, O., Visconti, P.E. and Mayorga, L.S. (2005). α-SNAP and NSF are required in a priming step during the human sperm acrosome reaction. *Molecular Human Reproduction* **11**: 43-51

Tomes, C.N., Michaut, M., De Blas, G., Visconti, P., Matti, U. and Mayorga, L.S. (2002). SNARE complex assembly is required for human sperm acrosome reaction. *Developmental Biology* **243**: 326-338

Tomes, C.N., Roggero, C.M., De Blas, G., Saling, P.M. and Mayorga, L.S. (2004). Requirement of protein tyrosine kinase and phosphatase activities for human sperm exocytosis. *Developmental Biology* **265**: 399-415

Tonks, N.K. (2003). PTP1B: from the sidelines to the front lines! *FEBS Letters* **546**: 140-148

Trevino, C.L., Serrano, C.J., Beltran, C., Felix, R. and Darszon, A. (2001). Identification of mouse trp homologs and lipid rafts from spermatogenic cells and sperm. *FEBS Letters* **509**: 119-125

Urner, F. and Sakkas, D. (2003). Protein phosphorylation in mammalian spermatozoa. *Reproduction* **125**: 17-26

Ward, C.R., Faundes, D. and Foster, J.A. (1999). The monomeric GTP binding protein, Rab3a, is associated with the acrosome in mouse sperm. *Molecular Reproduction and Development* **53**: 413-421

Wassarman, P.M., Jovine, L., Qi, H., Williams, Z., Darie, C. and Litscher, E.S. (2005). Recent aspects of mammalian fertilization research. *Molecular and Cellular Endocrinology* **234**: 95-103

Waters, M.G. and Hughson, F.M. (2000). Membrane tethering and fusion in the secretory and endocytic pathways. *Traffic* **1**: 588-597

Xu, T., Binz, T., Niemann, H. and Neher, E. (1998). Multiple kinetic components of exocytosis distinguished by neurotoxin sensitivity. *Nature Neuroscience* **1**: 192-200

Yanagimachi, R. (1994). Mammalian fertilization. In: *The Physiology of Reproduction*. Volume 1.Edited by Knobil, E. and Neill, J.D., Raven Press, New York, pp. 189-317

Yudin, A.I., Gottlieb, W. and Meizel, S. (1988). Ultrastructural studies of the early events of the human sperm acrosome reaction as initiated by human follicular fluid. *Gamete Research* **20**: 11-24

Yunes, R., Michaut, M., Tomes, C. and Mayorga, L.S. (2000). Rab3A triggers the acrosome reaction in permeabilized human spermatozoa. *Biology of Reproduction* **62**: 1084-1089

Yunes, R., Tomes, C., Michaut, M., De Blas, G., Rodriguez F., Regazzi, R. and Mayorga, L.S. (2002). Rab3A and calmodulin regulate acrosomal exocytosis by mechanisms that do not require a direct interaction. *FEBS Letters* **525**: 126-130

Zerial, M. and McBride, H. (2001). Rab proteins as membrane organizers. *Nature Review Molecular Cell Biology* **2**: 107-117

Spermatology. SRF Vol. 65. ERS Roldan and M Gomendio (eds) Nottingham University Press, Nottingham

Sperm motility in birds: insights from fowl sperm

David P. Froman

Animal Sciences Department, 112 Withycombe Hall, Oregon State University, Corvallis, Oregon 97331, USA

This review outlines how computer-assisted sperm motion analysis helped explain variation in fertility among roosters, investigate sperm cell energetics, and predict sperm allocation within the oviduct. The context for these experiments was the analysis of sperm mobility, a quantitative trait discovered in the mid-1990s. *Sperm mobility* denotes the movement of a sperm cell population against resistance at body temperature. Whereas mobile sperm must be motile, not all motile sperm are mobile. For example, straight-line velocity (VSL) must be > 30 μm/s for sperm to penetrate an Accudenz solution within a cuvette from an overlaid sperm suspension. Absorbance of the Accudenz solution is proportional to the number of mobile sperm within the semen sample. Phenotypic variation among roosters was explicable in terms of VSL distributions, which in turn reflected varying proportions of sperm containing dysfunctional mitochondria. In the case of mobile sperm, motility was enabled by calcium cycling, which drives phospholipase A_2 activity and thereby controls the release of endogenous fatty acids for ß-oxidation within mitochondria. In view of sperm storage tubule ultrastructure and sperm rheotaxis, experimental outcomes led to the realization that sperm residence within the hen's sperm storage tubules was most readily explained in terms of progressive motility and that decreased VSL may account for sperm egress. It is noteworthy that male fertility can be altered by genetic selection using *in vitro* sperm mobility as the selection criterion. Sperm immobility may arise from formation of the mitochondrial permeability transition pore. Ongoing work addresses underlying mechanisms and related genes.

Sperm mobility

Serendipity and invention enabled the discovery of this trait. Sperm mobility is measured *in vitro* by sperm penetration into Accudenz®, a nonionic, biologically inert cell separation medium (Accurate Chemical and Scientific Corp., Westbury, NY). By chance, Accudenz® was used to separate desialylated sperm from neuraminidase (Froman and Thursam, 1994). Our experimental goal was to perform competitive fertilisation using desialylated and intact sperm. Initially, a discrete interface was observed between the 12% (w/v) Accudenz® solution and the overlaid sperm suspension. However, sperm rapidly entered the Accudenz® solution under their own power, although the extent to which sperm penetrated the medium was limited in

Corresponding author E-mail: David.Froman@oregonstate.edu

depth. Two years later, the need arose for an inexpensive, objective, and reliable means of measuring fowl sperm motility under physiological conditions. In this case, the experimental goal was describing a mechanism underlying subfertility associated with homozygosity for the rose comb allele (McLean and Froman, 1996). The unequivocal distinction between *R/r* and *R/R* genotypes based upon sperm penetration into 6% (w/v) Accudenz® prompted two questions. First, could the assay distinguish individuals within a population of males with no genotypic distinction other than breed? Second, would the fertility of such individuals differ? Though validated as a sperm motility assay (Froman and McLean, 1996), a key distinction was made: the assay estimated the *mobility of a sperm population* rather than sperm motility per se. When the sperm mobility assay was applied to successive generations of random bred New Hampshire roosters, observed frequencies approximated normal distributions (Froman and McLean, 1996; Froman and Feltmann, 1998; Froman, Feltmann, Rhoads and Kirby, 1999). Thus, the designations *low*, *average*, and *high* sperm mobility had statistical designations because they denoted males whose measurements were less that one standard deviation from the mean, within a standard deviation of the mean, or greater than one standard deviation above the mean, respectively. Furthermore, phenotypic distinctions were independent of age (Froman and Feltmann, 1998). As anticipated, fertility differed among phenotypes as evidenced by non-competitive and competitive fertilisation (Froman *et al.*, 1997; Froman and Feltmann, 1998; Froman *et al.*, 1999; Birkhead, Martinez, Burke and Froman, 1999). In fact, fertility was shown to be a *function* of sperm mobility phenotype (Fig. 1). Subsequent experiments estimated the heritability of the trait and demonstrated that sperm mobility phenotype was subject to genetic selection (Froman, Pizzari, Feltmann, Castillo-Juarez and Birkhead, 2002). Thereafter, low and high sperm mobility lines were bred at the Oregon Agricultural Experiment Station.

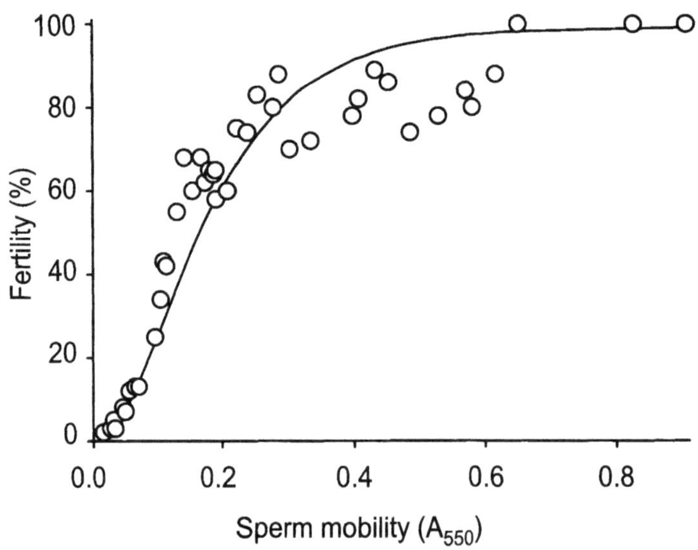

Figure 1. Fertility plotted as a function of sperm mobility, as measured by the absorbance in Accudenz® solution. Absorbance is proportional to the number of sperm that penetrated the Accudenz® solution from an overlaid sperm suspension containing a fixed number of sperm. Open circles denote observed fertility per male (n = 48). In each case, each of 8 to 12 hens was inseminated with 8 x 10^7 viable sperm. Eggs were collected for 7 days following artificial insemination and each egg examined after 4 days of incubation. Data approximated a skewed logistic function. The solid line represents the fitted curve. Reproduced with the permission of the Society for the Study of Reproduction.

Computer-assisted sperm motion analysis

A Hobson SpermTracker was procured (Hobson Tracking Systems, Ltd., Sheffield, UK) in order to explain phenotypic variation in sperm mobility among random bred roosters. This experimental goal seemed straightforward in view of the extent to which computer-assisted sperm motion analysis was used by the mid-1990s. However, a serious procedural setback was encountered: *in vitro* sperm motility was compromised at the dilution required for analysis. Whereas fowl semen typically contains 4 to 5 x 10^9 sperm/ml, concentrations near 1 x 10^6 sperm/ml were required for use in MicroCells with a chamber depth of 50 μm (Conception Technologies, San Diego, CA). Straight-line velocity (VSL) decreased as a function of time when such concentrations were used. As reviewed by Bakst, Wishart and Brillart (1994), very few sperm ascend the hen's vagina and enter the oviduct's sperm storage tubules following an insemination, typically ≤1%. Thus, loss of sperm cell function *in vitro* could have been a biological phenomenon. However, sperm cell hypoxia was a second possibility even though fowl sperm use anaerobic glycolysis to maintain intracellular ATP (Wishart, 1982) and the experimental medium contained glucose. Therefore, the following argument was made: if an erythrocyte monolayer within the chamber counters this effect, then hypoxia compromises sperm cell function. Upon loading a mixture of sperm and erythrocytes, the latter immediately settled to the bottom of the chamber and remained stationary while sperm moved over them. Thereafter, a steady state was verified as average VSL was proven to be independent of time (Fig. 2).

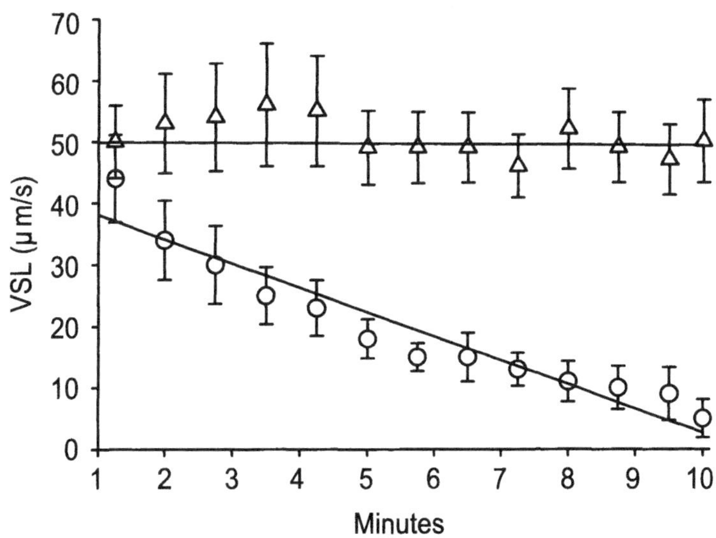

Figure 2. VSL of fowl sperm at body temperature in the presence and absence of an erythrocyte monolayer. Sperm from each semen donor (n = 7 high sperm mobility males) were diluted to 1.2 x 10^6 sperm/ml in either buffered saline (circles) or a 1:100 dilution of whole blood in buffered saline (triangles). Erythrocytes formed stationary monolayers once chamber filling was complete. Solid lines represent fitted curves. Reproduced with the permission of the Society for the Study of Reproduction.

Analysis of semen from low, average, and high sperm mobility phenotypes within a random bred population demonstrated that motile concentration and VSL were critical for phenotypic variation (Froman and Feltmann, 2000). For example, motile concentration and sperm mobility were correlated (r = 0.71), and average VSL of sperm from the low phenotype tended to be less than that of average and high phenotypes. However, the way in which these variables interacted was not clear. Subsequent computer-assisted sperm motion analysis conducted with commercial broiler breeders presented a conundrum: whereas average VSL typically was minimal for the low phenotype, many of these individuals were characterised with *high* motile concentration. The apparent inconsistency between low sperm mobility phenotypes within populations of New Hampshire and broiler breeder roosters was resolved by evaluating VSL distributions for individual males (Froman, Bowling and Wilson, 2003). Frequency distributions were constructed by assigning the VSL of individual sperm cells to 1 of 12 categories ranging from 1 to 120 micrometers per second in increments of ten. Each category was plotted as a proportion of motile concentration. Representative distributions are shown in Figure 3. Whereas distributions were uniformly skewed to the right, the area within a distribution's upper tail varied by phenotype. Consequently, semen donors were chosen to represent the phenotypic range. In each case, a suspension containing 5 x 10^8 sperm/ml was used for a sperm mobility assay and the construction of a VSL distribution. A linear relationship was observed when sperm mobility phenotype was plotted as a function of the number of sperm with a VSL > 30 µm/s (Fig. 4). This empirical approach provided a parsimonious explanation for the means by which motile concentration and VSL interacted to determine *in vitro* sperm mobility for any given male.

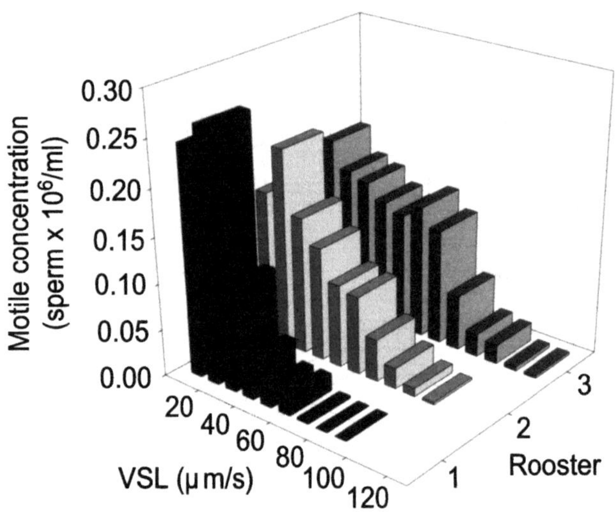

Figure 3. VSL distributions for males representing a range in sperm mobility phenotype. Roosters 1, 2, and 3 were categorised *a priori* by low, average, and high sperm mobility, respectively. Reproduced with permission of the Poultry Science Association.

Mitochondria

In addition to computer-assisted sperm motion analysis, the attempt to explain phenotypic variation in sperm mobility also entailed conventional biochemical tests. The reader should understand that the author is writing with hindsight. In other words, this narrative portrays common denominators that emerged through time using diverse types of experiments. In contrast, such clarity was not always a feature in the laboratory on a day-to-day basis. For example, whereas the Hobson SpermTracker was expected to provide an immediate, unequivocal explanation, the technology was useless until a steady state condition was envisioned and validated (Fig. 2). In short, the author had no *a priori* knowledge one decade ago of experimental methods that would prove to be useful.

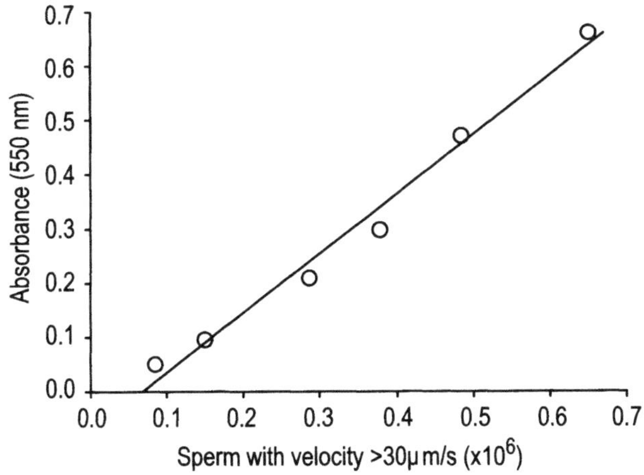

Figure 4. Sperm mobility, measured in absorbance units, as a function of the size of the upper tail of a VSL distribution. The solid line represents the fitted curve. Reproduced with the permission of the Poultry Science Association.

For example, the first possibilities to be tested and dismissed were sperm morphology (Froman and Feltmann, 1998) and sperm viability (Birkhead *et al.*, 1999). These outcomes supported inferences that phenotypic variation arose from *differences in degree* rather than kind. As such, correlation studies were performed using semen donors representing the phenotypic range within a random bred population. Sperm mobility phenotype neither varied with body weight nor testis mass as evidenced by product-moment correlation coefficients of -0.02 and 0.01, respectively (Froman and Feltmann, 1998). In contrast, phenotype did vary with sperm ATP content and oxygen consumption as evidenced by product-moment correlation coefficients of 0.80 and 0.83, respectively (Froman and Feltmann, 1998; Froman and Kirby, 2005). Such observations were consonant with the correlation between motile concentration and sperm mobility ($r = 0.71$; Froman and Feltmann, 2000) and the demonstration that mitochondrial oxidative phosphorylation was pivotal to phenotypic expression (Froman and Feltmann, 1998). Thus, collective experimental outcomes were explicable either in terms of mitochondrial number or quality.

Mitochondrial number was deemed equivalent to mitochondrial helix length. The relationship between sperm mobility phenotype and mitochondrial helix length was investigated based upon a knowledge of differences in this variable among species, the hunch that little was known about variation within a species, knowledge of a monoclonal antibody specific for poultry mitochondria (Fig. 5A), and the desire for avant-garde knowledge. This effort revealed two unequivocal answers (Froman and Kirby, 2005). First, whereas variation in mitochondrial helix length was observed among males within a species, such variation could not account for variation in sperm mobility phenotype (Table 1). Second, as evidenced by coefficients of variation, mitochondrial helix length varied little within males.

This outcome warranted the investigation of mitochondrial quality. Transmission electron microscopy was used for this purpose. Males characterised by average sperm mobility ($n = 10$)

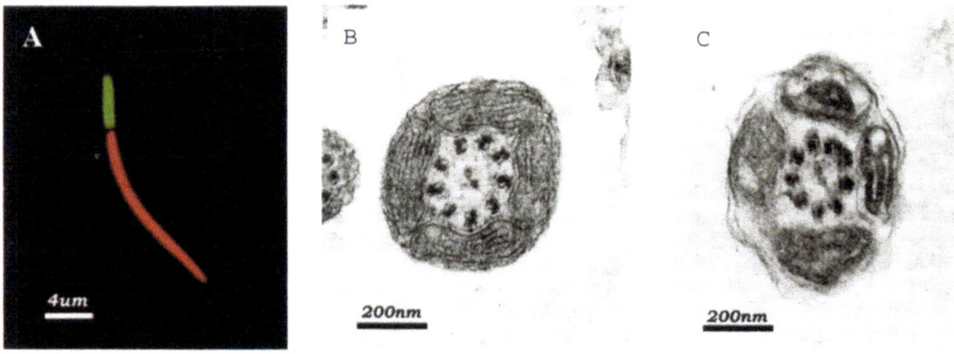

Figure 5. Fowl sperm mitochondria. (**A**) Epifluorescence micrograph of a fowl sperm cell. Mitochondria appear green due to staining with an FITC secondary antibody. The primary antibody was a mouse monoclonal antibody specific for poultry mitochondria (Korn *et al.*, 2002, *Poultry Science* **81:** 1077-1085). The nucleus is red due to counterstaining with ethidium bromide. The photograph was provided by Nancy Korn at Clemson University. (**B,C**) Transmission electron micrographs of fowl sperm midpieces. (**B**) Cross-section showing mitochondria arranged in a compact helix around the axoneme. Mitochondrial cristae are well-defined. (**C**) Cross-section showing aberrant mitochondria characterised by swelling, separation from the axoneme, and disorganised cristae. Reproduced with the permission of the Society for the Study of Reproduction.

Table 1. Estimates of mitochondrial helix length from roosters representing the range in sperm mobility phenotype within a base population of random bred roosters (n = 242). Helix length (mean ± SEM) was estimated using immunofluorescence and a sample size of 50 sperm/rooster. Reproduced with the permission of the Society for the Study of Reproduction.

Rooster	Sperm mobility (A_{550})	Helix length (mm)	CV (%)
1	0.130	3.85 ± 0.016	2.9
2	0.208	3.81 ± 0.018	3.2
3	0.415	3.74 ± 0.018*	3.3
4	0.496	3.86 ± 0.017	3.2
5	0.560	3.86 ± 0.014	2.5
6	0.737	3.84 ± 0.014	2.9
7	0.836	3.84 ± 0.014	3.7

*Different from all nonsupercripted means ($P < 0.05$).

were selected from a base population of random bred males (n = 242). By way of reminder, the sperm mobility assay enables mobile sperm to segregate from immobile sperm at body temperature. Consequently, sperm within the sperm suspension overlay were prepared for electron microscopy before and after the assay. Ultra thin sections located 125 μm apart were used so that distinct sperm cells were evaluated (n = 500 per treatment per male). The proportion of sperm with aberrant mitochondria (Fig. 5B,C) averaged 9 ± 2.2 % prior to the separation of mobile from immobile sperm. In contrast, 40 ± 6.7 % of the immobile sperm contained aberrant mitochondria. In subsequent work, nearly 50% of sperm from roosters selected for low sperm mobility contained aberrant mitochondria (Froman and Kirby, 2005). In contrast, only 4% of sperm from roosters selected for high sperm mobility contained aberrant mitochondria. Consequently, moribund mitochondria are deemed to be an attribute of immobile fowl sperm. As evidenced by the trajectory of motile sperm cells from either phenotype, the quality of sperm cell motion did not differ between phenotypes (Fig. 6). Thus, it was uncertain whether sperm immobility arose from an inability to become motile at ejaculation or an inability to remain motile after ejaculation. In either case, the systematic study of sperm motility was expected to provide a detailed account of a *process* that would afford a conceptual framework for interpreting sperm cell dysfunction.

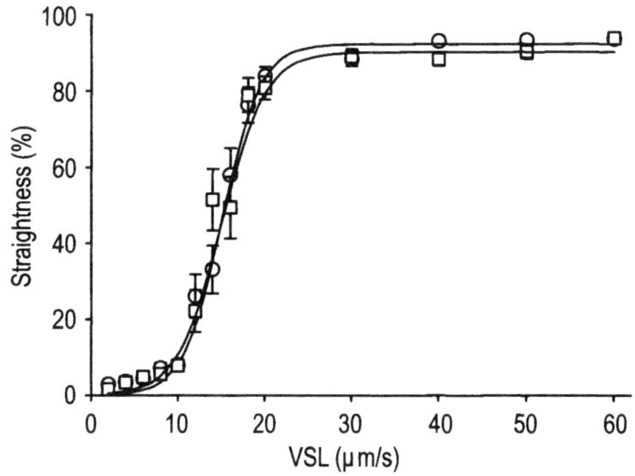

Figure 6. Effect of VSL on sperm cell trajectory. Straightness denotes the quotient between the straight-line distance between the first and last point on a sperm cell's smoothed path and the length of the smoothed path. Each data set approximated a logistic function. Each symbol denotes a mean ± SEM. Squares and circles denote low (n = 20 observations per mean) and high (n = 30 observations per mean) sperm mobility phenotypes. Solid lines represent fitted curves. Reproduced with the permission of the Society for the Study of Reproduction.

Maintenance of the motile state

The term *motile state* denotes that fowl sperm are immotile prior to ejaculation (Ashizawa and Sano, 1990), become motile at ejaculation, and most likely remain motile during their functional life span within the oviduct (Froman, 2003). Sperm pass through the excurrent ducts of the testis over the course of several days as evidenced by the relationship between daily sperm production and the size of the extragonadal sperm reserve (Lake, 1984). Sperm that ascend the vagina and enter sperm storage tubules can remain functional for several weeks as evidenced by duration of fertility following a single intravaginal insemination (Fig. 7). However, this fact must be tempered by the exponential egress of sperm from the hen's sperm storage tubules (Wishart, 1987). In any event, fowl sperm longevity is measured on a scale of days. Sperm become and then remain motile during this interval. As such, variables that maintain motility under physiological conditions are in effect determinants of male fecundity.

Maintenance of motility was studied using semen donors within a line of New Hampshire chickens selected for high sperm mobility. Selection of low and high sperm mobility lines began in 1998 using a single sire from a random bred population. The sire's mobility score was equivalent to the population mean. The sire was crossed with full sib sisters whose brothers were characterised with either low or high sperm mobility. This strategy was used following a heritability study. Whereas h^2 was estimated to be 0.30, a maternal additive genetic effect was observed (Froman *et al.*, 2002). In addition, phenotypic differences were enhanced by deleting glucose from media used to perform the sperm mobility assay (Froman *et al.*, 1999). Thus, selection was most stringent when *in vitro* sperm mobility depended upon mitochondrial oxidation of endogenous fatty acids. This conclusion was consonant with mitochondrial Ca^{2+} cycling (McCormack and Denton, 1999) and the relationship between extracellular Ca^{2+} and fowl sperm motility at body temperature (Ashizawa, Wishart and Tsuzuki, 2000). In review, high motile concentration afforded procedural sensitivity, and high motile concentration was an attribute of males selected for high sperm mobility.

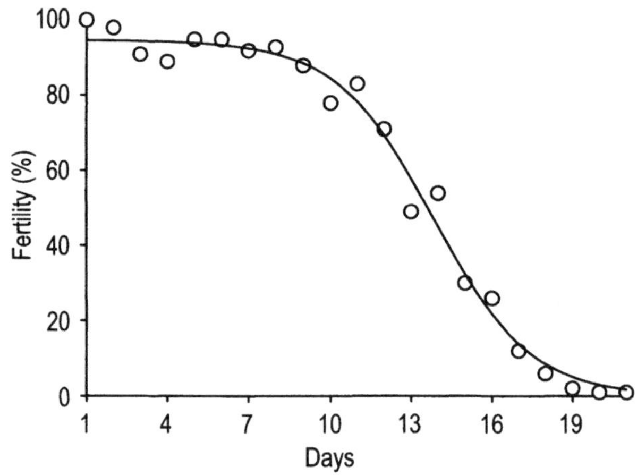

Figure 7. Fertility as a function of time following a single intravaginal insemination. Open circles denote percentages of fertilised eggs. In this case, each of 50 New Hampshire hens was inseminated with 5 x 10^7 sperm from semen pooled from donors characterised by high sperm mobility (n = 5). The solid line represents the fitted curve.

The experimental demand for oxidative phosphorylation at body temperature over a prolonged time course was met by incubating several hundred microliters of a mixture of sperm and erythrocytes in a polystyrene cuvette held within a heating block (Froman, 2003). This approach afforded repeated measures from a common sample and a steady state condition for a 60 min interval (Fig. 8A). The role of extracellular Ca^{2+} was evaluated by replacing 2 mM Ca^{2+} with 5 mM tetrasodium 1,2-*bis*-(o-amino-phenoxy) ethane-N,N,N′,N′-tetraacetic acid (BAPTA), a Ca^{+2} chelator. Thus, BAPTA served as an extracellular sink for the Ca^{2+} efflux known to occur from fowl sperm at body temperature (Ashizawa *et al.*, 2000). However, as shown in Figure 8B, a majority of sperm were not immotile upon dilution (Froman, 2003). Rather, they *became* immotile over the course of an hour. And once motility was inhibited, pretreatment motile concentration was observed when excess Ca^{2+} was added to overcome the effect of BAPTA (Fig. 8B). As evidenced by confocal microscopy and flow cytometry, Ca^{2+} was depleted from and restored to mitochondria (Froman, unpublished data). Likewise, an exponential decrease in motile concentration was observed when sperm were treated with 0, 30, 60, and 120 mM arachidonyltri-fluoromethyl ketone, a specific inhibitor of phospholipase A_2. Analysis of sperm acylcarnitine content confirmed the presence of long-chain fatty acids in motile sperm. Consequently, the phenomenon observed in Figure 9 was attributed to *Ca^{2+} cycling* driving phospholipase A_2 activity in order to generate long-chain fatty acids for ß-oxidation within mitochondria. However, the mechanism underlying Ca^{2+} cycling was another matter.

This issue had already been addressed inadvertently by asking whether extracellular Na^+ was required for fowl sperm motility in addition to extracellular Ca^{2+}. Exchanging either sucrose for NaCl or Li^+ for Na^+ decreased motile concentration (Fig. 9A). The sparing effect of lithium was attributed to the mitochondrial Na^+/Ca^{2+} exchanger, which can use Li^+ in lieu of Na^+

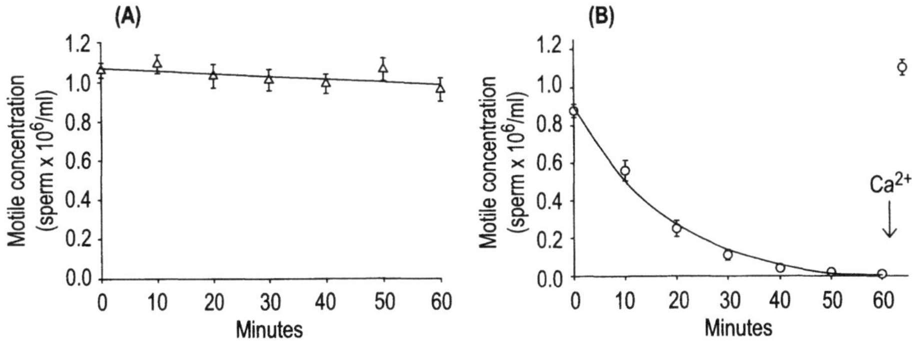

Figure 8. Maintenance of motility in fowl spermatozoa. (**A**) Effect of incubating fowl sperm at body temperature with erythrocytes in glucose-free isotonic buffered saline containing 2 mM Ca^{2+}. Each data point represents a mean ± SEM (n = 10). The solid line represents the fitted curve. Motile concentration was independent of time ($P > 0.05$). Based on a predicted concentration of 1.2 x 10^6 sperm/ml and a y-intercept of 1.07 x 10^6 sperm/ml, sperm motility was stable at 89%. Average VSL (data not shown) was also independent of time. The y-intercept for this variable was 35 µm/s. Thus, this data set represents a steady state. Reproduced with the permission of the Society for the Study of Reproduction. (**B**) Effect of 5 mM BAPTA on motile concentration at body temperature. The suspending medium was prepared with neither glucose nor Ca^{2+}. Each data point represents a mean ± SEM (n = 10). The solid line represents the fitted curve. Once all sperm were immotile, supplemental Ca^{2+} induced motile concentration equivalent to that of a pre-incubated control. Reproduced with the permission of the Society for the Study of Reproduction.

(Philipson, 1999). The role of this protein was tested using CGP 37157, a specific inhibitor of the mitochondrial Na^+/Ca^{2+} exchanger. As shown in Figure 9B, micromolar concentrations of CGP 37157 had an immediate and profound effect upon motile concentration. Inhibition of the mitochondrial uniporter was not tested because ruthenium red is typically applied to isolated mitochondria. Thus, this protein's role in Ca^{2+} cycling was assumed.

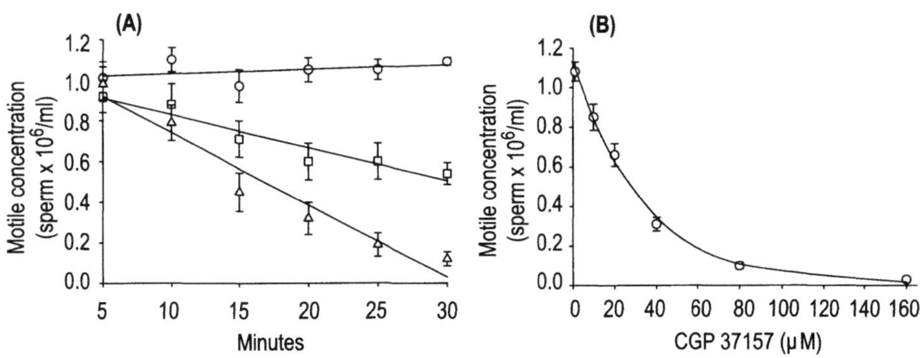

Figure 9. Ionic requirements for fowl sperm motility. (**A**) Effect of 2 mM Ca^{2+} in buffered (pH 7.4) isotonic NaCl (circles), LiCl (squares), or sucrose (triangles) on motile concentration at body temperature. Each symbol represents a mean ± SEM (n = 10). Solid lines represent fitted curves. Motile concentration of the NaCl control was independent of time. Whereas motile concentration was maintained by neither Li^+ nor sucrose, Li^+ had a sparing effect. Reproduced with the permission of the Society for the Study of Reproduction. (**B**) Effect of the mitochondrial Na^+/Ca^{2+} exchanger inhibitor CGP 37157 on motile concentration at body temperature. Each data point represents a mean ± SEM (n = 10). The solid line represents the fitted curve. Reproduced with the permission of the Society for the Study of Reproduction.

Cellular uptake of Ca^{2+} is mediated by six types of voltage-gated Ca^{2+} channels as well as NMDA glutamate channels. In contrast, Ca^{2+} efflux is mediated by Ca^{2+}-ATPase and a plasma membrane Na^+/Ca^{2+} exchanger. As a first approximation, nifedipine was used to test for the presence of L-type Ca^{2+} channels because T-, N-, P-, Q-, and R-type channels are insensitive to this channel antagonist. As shown in Figure 10A, a dose response was observed. Thus, voltage-gated Ca^{2+} channels appear to be one type of ion channel found within the fowl sperm's plasma membrane. Whereas Thomson and Wishart (1991) attributed Ca^{2+} efflux from fowl sperm to Ca^{2+}-ATPase, the ATPase inhibitor orthovanadate had no effect on either motile concentration or average VSL (Froman and Feltmann, 2005). Consequently, KB-R7943 mesylate was used to test for the presence of a plasma membrane Na^+/Ca^{2+} exchanger. However, sperm had to be rendered immotile with BAPTA in order to do this; for KB-R7943 mesylate inhibits the *reverse mode* of this exchange protein. In other words, Ca^{2+} influx and Na^+ efflux are inhibited rather than Ca^{2+} efflux and Na^+ influx. As shown in Figure 10B, a plasma membrane Na^+/Ca^{2+} exchanger was deemed to serve as the L-type Ca^{2+} channel's counterpart within the plasma membrane. Though extracellular Na^+ is essential for maintenance of motility (Fig. 9A), presumably at the level of the mitochondrial Na^+/Ca^{2+} exchanger (Fig. 9B), the means by which intracellular Na^+ is regulated remains a mystery because ouabain had no effect on either motile concentration or average VSL (Froman and Feltmann, 2005).

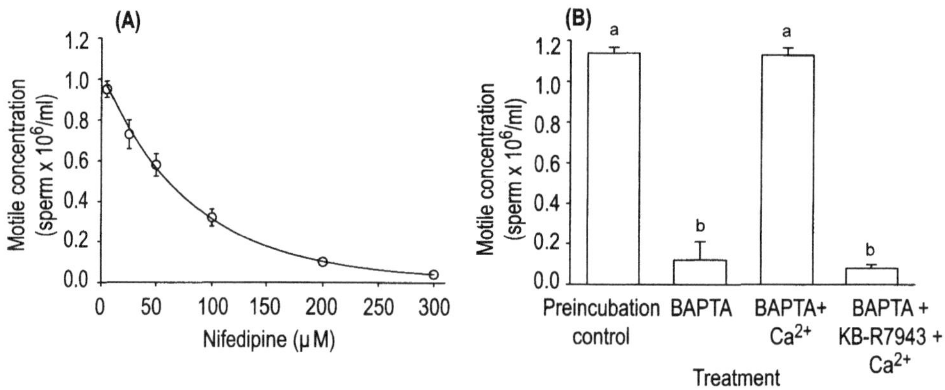

Figure 10. Ca^{2+} and Na^+ regulation in fowl spermatozoa. (**A**) Effect of a voltage-gated Ca^{2+} channel inhibitor on motile concentration. Each data point represents a mean ± SEM (n = 10). The solid line represents the fitted curve. Reproduced with the permission of the Society for the Study of Reproduction. (**B**) Inhibition of sperm reactivation with 100 μM KB-R7943 mesylate. Prior to treatment, sperm were incubated with BAPTA in order to render them immotile by Ca^{2+}-depletion. Whereas sperm motility was restored by excess extracellular Ca^{2+}, this effect was blocked when KB-R7943 mesylate was present. Means with different superscripts differed ($P < 0.05$). Reproduced with the permission of the Society for the Study of Reproduction.

In review, it appears that Ca^{2+} cycling is mediated by a millimolar concentration of extracellular Ca^{2+}, a voltage-gated Ca^{2+} channel, the mitochondrial uniporter, the mitochondrial Na^+/Ca^{2+} exchanger, and the plasma membrane Na^+/Ca^{2+} exchanger. Furthermore, Ca^{2+} cycling is dependent upon extracellular Na^+. If phospholipase A_2 is contained within the mitochondrion's intermembrane space and acts upon the outer mitochondrial membrane, then Ca^{2+} cycling would enable stable enzyme activity and thus stable ATP synthesis in the presence of steady state amounts of ADP and O_2. Though this account is based upon empiricism, it is not a nuanced explanation of the motile state; for it does not explain what may be the most important biological feature of a motile population of fowl sperm.

VSL distributions

The sperm mobility assay predicts male fertility (Fig. 1) because it affords a context in which the *consequences* of variation in sperm velocity become evident in time. Fertile males that were indistinguishable with conventional measures of semen quality were proven distinct by *in vitro* sperm mobility (Froman and Feltmann, 1998; Birkhead *et al.*, 1999). Experiments designed to explain phenotypic variation revealed that VSL distributions were skewed to the right (Fig. 3). Subsequent work demonstrated that the VSL distribution was malleable. For example, the upper tail was eliminated without affecting motile concentration when sperm were held at room temperature (Fig. 11A; Froman *et al.* 2003). Likewise, glutamate, a motility agonist, skewed the distribution to the left (Froman, 2003). This particular experiment demonstrated that most sperm are not at maximal VSL when held at body temperature. These results were not surprising. In contrast, VSL distributions skewed to the right were unexpected in two circumstances: when motile concentration decreased as a function of time due to sperm incubation in a Ca^{2+}-depleted medium (Froman, 2003), and when sperm were reactivated after treatment with BAPTA (Fig. 11B; Froman, unpublished data). Collectively, these experimental outcomes lend credence to the possibility that a VSL distribution skewed to the right is a *physiological attribute* of a sperm population and therefore a determinant of male fertility.

How? Only motile sperm ascend the vagina, presumably against a flux generated by ciliated epithelial cells (Bakst *et al.*, 1994). Thus, any sperm cell that cannot move against resistance at body temperature is eliminated from the oviduct. This may, in part, explain the difference between the number of sperm inseminated and the number of sperm that enter sperm storage tubules following an intravaginal insemination (Bakst *et al.*, 1994). In addition, sperm residence within the sperm storage tubules is explicable in terms of sperm motility (Froman, 2003). Upon entering the sperm storage tubules, it is probable that sperm maintain their position against a fluid current generated by secretion from non-ciliated epithelial cells. Likewise, it is probable that resident sperm oxidise exogenous fatty acids released from surrounding lipid-laden epithelial cells. Finally, it is probable that motile sperm emerge from the sperm storage tubules when VSL reaches a threshold at which retrograde movement ensues. In fact, the relationship between VSL and trajectory might expedite sperm egress (Fig. 6). As evidenced by error bars, sperm may experience greater drag at velocities between 10 and 20 μm/s. Emergent sperm with low VSL are consistent with antiperistalsis mediating sperm passage from the shell gland to the infundibulum, the site of fertilisation in the fowl (Bakst *et al.*, 1994).

Avian sperm motility in perspective

Since the mid-20th century, the objective study of sperm motility has been expedited by electron microscopy, computer-assisted sperm motion analysis, and confocal microscopy. These technologies provide detailed descriptions of the organelles that impart progressive motion, a wealth of information about the properties of individual motile sperm cells, and a spatial interpretation of cellular processes that enable progressive motility. For example, determining the location of intracellular Ca^{2+} was warranted once Ca^{2+} cycling was found to be responsible for progressive motility at body temperature (Fig. 8, 9, 10). Transmission electron microscopy and confocal microscopy enabled the deduction of a model explaining phenotypic variation in sperm mobility (Froman, Wardell and Feltmann, 2006). However, the localization of Ca^{2+}-

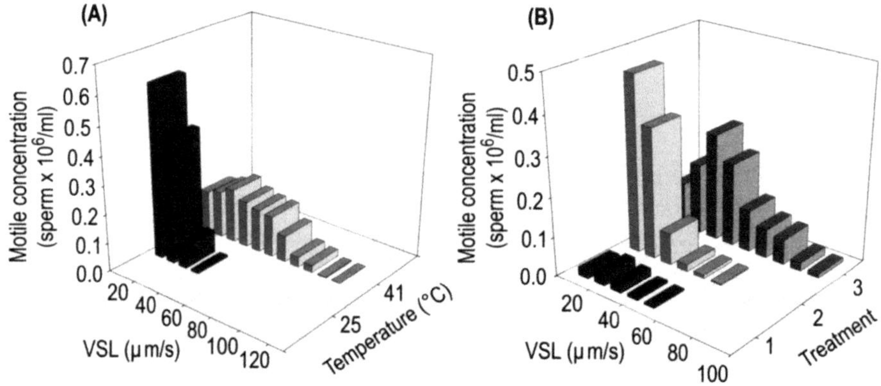

Figure 11. Factors affecting straight line velocity (VSL) distribution. (**A**) Effect of temperature on VSL distribution. Whereas distinct distributions were observed when sperm from a common source were held at either 25 or 41°C, motile concentrations were equivalent at 1.13 and 1.11 x 10^6 sperm/ml. Reproduced with the permission of the Poultry Science Association. (**B**) Effect of inactivation and reactivation on VSL distribution. Sperm were inactivated by washing through BAPTA, a Ca^{2+} chelator. The motility of such sperm was negligible (Treatment 1). Washed sperm resuspended in 2 mM Ca^{2+} were characterised by high motile concentration, but low VSL (Treatment 2). In contrast, washed sperm resuspended in 2 mM Ca^{2+} and 5 mM glucose yielded a VSL distribution skewed to the right (Treatment 3). This observation is consonant with that of Wishart (1984) who used exogenous glucose to reactivate fowl sperm after anaerobic storage.

specific dyes also presented an enigma. On one hand, rhod-2AM was expected to partition into mitochondria but was found primarily at the level of the nucleus, presumably within the nuclear envelope (Froman *et al.*, 2006). On the other hand, fluo-3AM was expected to behave like rhod-2AM but only stained mitochondria (Froman, unpublished data). Thus, integration of experimental methods affords conceptual advances and can prompt new questions.

The author acknowledges the community of scientists who have studied avian sperm motility for a variety of reasons. The author's emphasis on fowl sperm stems from his career as a poultry scientist. Objective measurements of fowl sperm motility have been made using spectrophotometry, videomicroscopy, and digital image analysis. Each method has advantages and limitations. However, as evidenced by Figure 11A, an investigator's method should reflect a well-defined and reasonable experimental goal. For example, Vernon and Woolley (1999) used videomicroscopy to conduct a biophysical analysis of sperm motility using passerine and non-passerine birds as semen donors. Most measurements were made at room temperature in order to expedite data acquisition. Fowl sperm, like the rest, were observed to roll around their progression axes. This observation provided a cause and effect explanation for the correlation between VSL and beat cross frequency observed with fowl sperm at body temperature ($r = 0.97$; Froman and Feltmann, 2000). As shown by Vernon and Woolley (1999), fowl sperm bore through space. In the case of an individual cell suspended in saline and approaching the observer, the sperm cell's head would rotate clockwise around the progression axis in response to a flagellar wave that is dextrally helical. Thus, videomicroscopy of a small number of sperm cells at room temperature provided exquisite information about motion within populations of sperm at body temperature.

It is the author's contention that sperm motility be viewed in the broadest sense possible.

The author's post-doctoral mentor, Rupert Amann, would often shock and amuse his students with a perspective he called the *sperm's eye view*. For example, were fowl sperm self-aware, the topic of cryopreservation might leave them morose in view of the probability of their survival. Whether or not Rupert got this idea from a Woody Allen movie is unknown. In any event, the lesson, as applied by the author, pertains to the distinction between human constructs - in particular those based upon a snapshot such as *percent motility* - and the life span of a cell in the midst of cohorts. Therefore, it follows that the concept of fowl sperm motility should include *sperm cell populations in the context of time*. This is the biological context for sperm within male and female reproductive tracts alike.

Future challenges

As reviewed by Kirby and Froman (2000), the flow of testicular fluid carries immotile sperm through the seminiferous tubules and into the excurrent ducts of the testis. Passage through the excurrent ducts confers the *potential* for progressive motility upon release from the male reproductive tract rather than fertilising ability. It must be noted that testicular sperm fertilise eggs if placed in the oviduct above the vaginal sphincter (Howarth, 1983). Based upon the work of Esponda and Bedford (1985), sperm maturation entails the adherence of secretory protein. However, the number and nature of such proteins remains unknown. The bulk of the extragonadal sperm reserve is contained within the deferent duct (Lake, 1984) and sperm within this duct are immotile (Ashizawa and Sano, 1990).

One curious attribute of deferent duct fluid is the presence of glutamate at a concentration of 80-90 mM (Freeman, 1984). Fowl sperm have ionotropic glutamate receptors as evidenced by the effect of D-homocysteinesulfinic acid, a potent agonist of the NMDA glutamate receptor. The channel portion of the NMDA glutamate receptor is permeable to Na^+, K^+, and Ca^{2+}. Both motile concentration and average VSL of ejaculated sperm increased in response to D-homocysteinesulfinic acid (Froman *et al.*, 2006). Nonetheless, sperm within the deferent duct are immotile even though they are suspended in a medium containing glutamate, 1 mM Ca^{2+} (Freeman, 1984), and presumably oxygen as evidenced by dense networks of capillaries within the wall of the deferent duct beneath the duct's inner epithelial surface (Nakai, Hashimoto, Kitagawa, Kon and Kudo, 1988).

How can this be? One-upmanship warrants invocation of the *mitochondrion's eye view*, which might be as follows. A mitochondrion's inner membrane potential is affected by two critical processes. The expulsion of H^+ from the mitochondrial matrix by action of the electron transport chain generates the proton motive force used by ATP synthase to regenerate ATP from ADP. In contrast, active transport of H^+ by the reverse mode of ATP synthase (the F_0F_1-ATPase) consumes ATP in order to maintain a membrane potential. If this occurs within the mitochondria of immotile sperm within the deferent duct, then one must consider the source of ATP. In motile sperm, mitochondrial creatine kinase makes ATP generated by oxidative phosphorylation interchangeable with phosphocreatine, which diffuses along the axoneme. Likewise, axonemal creatine kinase enables regeneration of ATP from ADP and phosphocreatine. Thus, the axoneme's pool of ATP is maintained by the phosphocreatine circuit in motile sperm (Walliman, Brdiczka, Nicolay and Eppenberger, 1992). It follows that ADP generated by active transport of H^+ in immotile sperm would be converted to ATP by the combined action of mitochondrial creatine kinase and phosphocreatine, at least for awhile. Ashizawa and Sano (1990) proposed that decreased temperature accompanying ejaculation initiates the motility of fowl sperm. However, the means by which such a stimulus might work is unknown.

Based upon experiments outlined in this manuscript as well as on-going research, it appears that sperm immobility following ejaculation may be a sperm cell version of cardiac reperfusion injury or neuronal excitotoxicity in which formation of the mitochondrial permeability pore terminates cell function. Perhaps glutamate serves to help Ca^{2+} enter immotile sperm and thereby synchronise them prior to ejaculation? Perhaps formation of the mitochondrial permeability transition pore is a deleterious side effect in any male? Perhaps the extent to which this happens depends on genotype? Perhaps even mobile sperm experience latent mitochondrial dysfunction? A preliminary test of this possibility was made by comparing patterns of sperm egress from the sperm storage tubules (Froman *et al.*, 2006). An insemination dose of 30 x 10^6 sperm per hen was used with semen pooled from males characterised by high sperm mobility. In contrast, an insemination dose of 90 x 10^6 sperm per hen was used with semen pooled from males characterised by low sperm mobility. The 3-fold difference in sperm number was used in the hope of maximising effective insemination dose at the level of the sperm storage tubules. As evidenced by the Y-intercepts in Figure 12, this goal was realised. Nonetheless, sperm from the low phenotype emerged more rapidly through time than did their counterparts. This effect was attributed to different rates at which mitochondrial function is lost within populations of *mobile* sperm. When mitochondrial function ceases within fowl sperm, they appear necrotic rather than apoptotic as evidenced by flow cytometry (Froman, unpublished data).

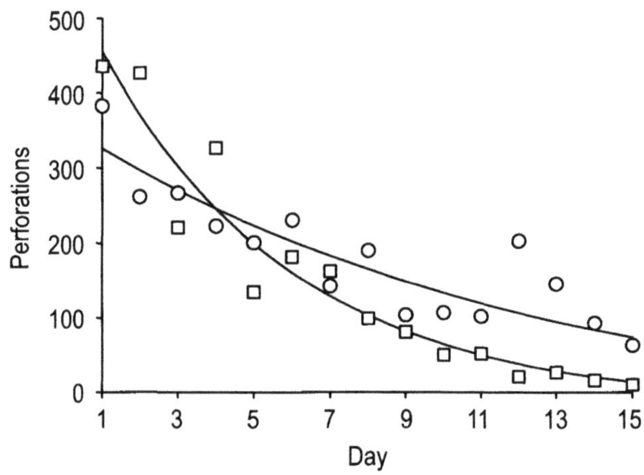

Figure 12. Patterns of sperm egress from sperm-storage tubules inferred from perforations through the inner perivitelline layer overlying the germinal disk. Each perforation was made by an individual, acrosome-reacted sperm cell. Each symbol respresents a mean from an average of 15 eggs per day. Open squares and circles represent data from low and high sperm mobility phenotypes, respectively. Solid lines represent predicted lines. Reproduced with the permission of the Society for the Study of Reproduction.

In conclusion, four additional important questions remain to be answered. Perhaps heterogeneity in VSL can be explained for sperm within ejaculates? Perhaps sperm mobility phenotype affects male competition at the level of the sperm storage tubule? Perhaps specific genes can be discovered that determine the efficacy of sperm cell populations and thereby male fitness? And finally, perhaps cryopreservation be improved by using mitochondrial function as an endpoint rather than post-thaw motility?

Acknowledgements

The author thanks his doctoral mentor, Ron Thurston, for teaching the discipline and power of the scientific method. With sadness, and yet fond memories, the late Peter Burrows is thanked for teaching inferential statistics based upon mathematical relationships. The author's post-doctoral mentor has been thanked surreptiously. A very special thanks is extended to Peter Lake, a renowned expert on male reproduction in poultry. Dr. Lake served as an external reviewer when the author was considered for indefinite tenure at Oregon State University, and Dr. Lake was not troubled to serve in a similar capacity many years later for post-tenure review. The author is indebted to Allen Feltmann, John Kirby, Nancy Korn, Tim Birkhead, and Tommaso Pizzari for their collaboration and encouragement. Finally, the research outlined in this manuscript would not have been possible without two awards from the USDA Cooperative State Research, Education, and Extension Service. The first award was National Research Initiative Competitive Grant number 97-35203-4807 entitled "Heritability & Basis of a New Reproductive Trait in Male Poultry." The second award was National Research Initiative Competitive Grant number 2003-35203-13343 entitled "Identification of Proteins Determining Sperm Mobility Phenotype."

References

Ashizawa, K. and Sano, R. (1990). Effects of temperature on the immobilization and the initiation of motility of spermatozoa in the male reproductive tract of the domestic fowl, *Gallus domesticus*. *Comparative Biochemistry and Physiology* **96A:** 297-301

Ashizawa, K., Wishart, G.J. and Tsuzuki, Y. (2000). Avian sperm motility: Environmental and intracellular regulation. *Avian and Poultry Biology Reviews* **11:** 161-172

Bakst, M.R., Wishart, G. and Brillard, J.-P. (1994). Oviducal sperm selection, transport, and storage in poultry. *Poultry Science Reviews* **5:** 117-143

Birkhead, T.R., Martinez, J.G., Burke, T. and Froman, D.P. (1999). Sperm mobility determines the outcome of sperm competition in the domestic fowl. *Proceedings of the Royal Society of London series B* **266:** 1759-1764

Esponda, P. and Bedford, J.M. (1985). Surface of the rooster spermatozoan changes in passing through the Wolffian duct. *Journal of Experimental Zoology* **234:** 441-449

Freeman, B.M. (1984). Appendix: Biochemical and physiological data. In: *Physiology and Biochemistry of the Domestic Fowl*, Volume 5. Edited by Freeman, B.M. Academic Press, London, pp. 407-424

Froman, D.P. (2003). Deduction of a model for sperm storage in the oviduct of the domestic fowl (*Gallus domesticus*). *Biology of Reproduction* **69:** 248-253

Froman, D.P. and Feltmann, A.J. (1998). Sperm mobility: A quantitative trait of the domestic fowl (*Gallus domesticus*). *Biology of Reproduction* **58:** 379-384

Froman, D.P. and Feltmann, A.J. (2000). Sperm mobility: Phenotype in roosters (*Gallus domesticus*) determined by concentration of motile sperm and straight line velocity. *Biology of Reproduction* **62:** 303-309

Froman, D.P. and Feltmann, A.J. (2005). Fowl (*Gallus domesticus*) sperm motility depends upon mitochondrial calcium cycling driven by extracellular sodium. *Biology of Reproduction* **72:** 97-101

Froman, D.P. and Kirby, J.D. (2005). Sperm mobility: Phenotype in roosters (*Gallus domesticus*) determined by mitochondrial function. *Biology of Reproduction* **72:** 562-567

Froman, D.P. and McLean, D.J. (1996). Objective measurement of sperm motility based upon sperm penetration of Accudenz®. *Poultry Science* **75:** 76-784

Froman, D.P. and Thursam, K.A. (1994). Desialylation of the rooster sperm's glycocalyx decreases sperm sequestration following intravaginal insemination of the hen. *Biology of Reproduction* **50:** 1094-1099

Froman, D.P., Bowling, E.R. and Wilson, J.L. (2003). Sperm mobility phenotype not determined by sperm quality index. *Poultry Science* **82:** 496-502

Froman, D.P., Feltmann, A.J. and McLean, D.J. (1997). Increased fecundity resulting from semen donor selection based upon in vitro sperm motility. *Poultry Science* **76:** 73-77

Froman, D.P., Wardell, J.C. and Feltmann, A.J. (2006). Sperm mobility: Deduction of a model explaining phenotypic variation in roosters (*Gallus domesticus*). *Biology of Reproduction* **74:** 487-491

Froman, D.P., Feltmann, A.J., Rhoads, M.L. and Kirby, J.D. (1999). Sperm mobility: A primary determinant of fertility in the domestic fowl (*Gallus domesticus*). *Biology of Reproduction* **61:** 400-405.

Froman, D.P., Pizzari, T., Feltmann, A.J., Castillo-Juarez, H. and Birkhead, T.R. (2002). Sperm mobility: Mechanisms of fertilizing efficency, genetic variation and phenotypic relationship with male status in the domestic fowl, *Gallus gallus domesticus*. *Proceedings of the Royal Society of London seires B* **269:** 607-612

Howarth, B., Jr. (1983). Fertilizing ability of cock spermatozoa from the testis, epididymis, and vas deferens following intramagnal insemination. *Biology of Reproduction* **28:** 586-590

Kirby, J.D. and Froman, D.P. (2000). Reproduction in male birds. In: *Sturkie's Avian Physiology,* 5th Edition. Edited by Whittow, G.C. Academic Press, San Diego, CA, pp. 597-615

Lake, P.E. (1984). The male in reproduction. In: *Physiology and Biochemistry of the Domestic Fowl*, Volume 5. Edited by Freeman, B.M. Academic Press, London, pp. 381-405

McCormack, J.G. and Denton, R.M. (1999). Calcium in the regulation of intramito-chondrial enzymes. In: *Calcium as a Cellular Regulator.* Edited by Carafoli, E. and Klee, C. Oxford University Press, New York, NY, pp. 529-544

McLean, D.J. and Froman, D.P. (1996). Identification of a sperm cell attribute responsible for subfertility of roosters homozygous for the rose comb allele. *Biology of Reproduction* **54:** 168-172

Nakai, M., Hashimoto, Y., Kitagawa, H., Kon, Y. and Kudo, N. (1988). Microvasculature of the epididymis and ductus deferens of domestic fowls. *Japanese Journal of Veterinary Medicine* **50:** 371-381

Philipson, K.D. (1999). Sodium-calcium exchange. In: *Calcium as a Cellular Regulator.* Edited by Carafoli, E. and Klee, C. Oxford University Press, New York, NY, pp. 279-294

Thompson, M.F. and Wishart, G.J. (1991). Temperature-mediated regulation of calcium flux and motility in fowl spermatozoa. *Journal of Reproduction and Fertility* **93:** 385-391

Vernon, G.G. and Woolley, D.M. (1999). Three-dimensional motion of avian spermatozoa. *Cell Motility and the Cytoskeleton* **43:** 149-161

Walliman, T., Wyss, M., Brdiczka, D., Nicolay, K. and Eppenberger, H.M. (1992). Intracellular compartmentalization, structure and function of creatine kinase isozymes in tissues with high and fluctuating energy demands: the 'phosphocreatine circuit' for cellular energy homeostasis. *Biochemical Journal* **281:** 21-40

Wishart, G.J. (1982). Maintenance of ATP concentrations in and of fertilizing ability of fowl and turkey spermatozoa in vitro. *Journal of Reproduction and Fertility* **66:** 457-462

Wishart, G.J. (1982). The effect of anoxia on fowl spermatozoa: Recovery of fertilizing ability, motility, and cellular concentrations of potassium and ATP. *Gamete Research* **10:** 83-89

Wishart, G.J. (1987). Regulation of the length of the fertile period in the domestic fowl by numbers of oviducal spermatozoa, as reflected by those trapped in laid eggs. *Journal of Reproduction and Fertility* **80:** 493-498

Spermatology. SRF Vol. 65. ERS Roldan and M Gomendio (eds) Nottingham University Press, Nottingham

Energy metabolism and sperm function

Kiyoshi Miki

Department of Cell and Developmental Biology, University of North Carolina, Chapel Hill, North Carolina 27599-7090, USA

Energy metabolism is a key factor supporting sperm function. Sustaining sperm motility and active protein modifications such as phosphorylation could be the reason why sperm require exceptionally more ATP than other cells. Many methods have been used to understand the relationship between energy metabolism and sperm function. These approaches have identified critical metabolic pathways that support specific processes during germ cell development and fertilisation. In round spermatids, lactate and pyruvate are the preferred substrates and the use of glucose is limited, however, during epididymal maturation sperm expand to use glycolysis. While the acrosome reaction requires lactate or pyruvate for ATP production by oxidative phosphorylation, gamete fusion requires glucose to produce NADPH by the pentose phosphate pathway. Sperm motility appears to be supported by relatively low ATP levels, but achievement of high ATP levels are essential for tyrosine phosphorylation linked to hyperactivation. Thus, each individual process and event requires a different substrate and metabolic pathway. Despite different preferences for energy substrates and metabolic pathways between species, analysis of knockout mice revealed that glycolysis is indispensable for mouse sperm function and that oxidative phosphorylation is not essential for male fertility. This suggests that glycolysis could compensate for the lack of oxidative phosphorylation and recover most sperm function. Spermatogenic cell-specific glycolytic enzymes may confer flexible use of substrates and adapt to unexpected conditions for substrates in the female reproductive tract.

Introduction

Spermatozoa go through many processes in order to fertilise an egg. Testicular sperm move to the epididymis, and here they are capable of moving actively during the process of maturation. Sperm reside for a period of time in the female reproductive tract, whereby they undergo a series of physiological changes termed capacitation. This process prepares the sperm head for the acrosome reaction while also changing the flagellar activity, resulting in a waveform with high amplitude and low beat-frequency. These changes cause a whiplash-like motion, termed hyperactivated motility (Yanagimachi, 1994). Sperm undergo the acrosome reaction and then bind and fuse to the egg, resulting in fertilisation. ATP is one of the basic components in a sperm cell and is used not only as an energy source but also for protein phosphorylation in cell signalling and as a cofactor regulating protein function. Glycolysis is the central metabolic

Corresponding author E-mail: miki@med.unc.edu

pathway producing ATP from a hexose molecule such as glucose. Metabolism of one glucose molecule produces 2 molecules each of pyruvate, ATP and NADH. Pyruvate then enters the TCA cycle in mitochondria for further oxidation. A combination of electron transfer and oxidative phosphorylation produces theoretically about 36 ATP in mitochondria. Sperm function is closely associated with ATP production, in terms of an ATP-producing pathway and a substrate. Sperm appear to select the metabolic pathway that they need. Although energy metabolism of sperm is affected by physiological conditions in the female reproductive tract, critical aspects of energy metabolism should be conserved beyond species. This review dissects the fertilisation process into individual steps and summarises the accumulated knowledge concerning energy metabolism of sperm. Also, the characteristics of spermatogenic cell-specific glycolytic isozymes will be summarised, while their potential functions in energy metabolism of sperm will be discussed.

Energy production in spermatogenic cells in the testis and epididymis

Energy production in round spermatids

In the context of energy metabolism, spermatids are at a transient stage where several spermatogenic cell-specific glycolytic enzymes begin to be expressed and replace somatic isozymes. Energy metabolism in rat round spermatids has been extensively analysed using *in vitro* culture systems. The comparison of CO_2 production from glucose and lactate revealed that lactate is effectively oxidised ~6 times more than glucose (Nakamura, Okinaga and Arai, 1984). Consistent with this result, glucose was not able to maintain intracellular ATP levels in round spermatids, but ~10 mM lactate was sufficient to support ATP levels (Mita and Hall, 1982). Pyruvate also supported most ATP production when it was added at around 1 mM, however, addition of higher concentrations (10 mM) of pyruvate resulted in reduced production of ATP (Mita and Hall, 1982; Nakamura, Okinaga and Arai, 1986). Pharmacological experiments also demonstrated the importance of lactate and pyruvate as a substrate in round spermatids. An uncoupler of oxidative phosphorylation or inhibitors of electron transport inhibited pyruvate and lactate consumption (Mita and Hall, 1982). Together this data confirmed that glucose is not efficiently utilised as an energy source, but that lactate and pyruvate are the preferred substrates. The unique metabolic approach of spermatids may be supported by Sertoli cells, which convert a large amount of glucose to lactate when cultured *in vitro* (Robinson and Fritz, 1981). Indeed, lactate is rich in the fluid of seminiferous tubules (Pande, Chowdhury, DasGupta, Chowdhury and Kar, 1966).

Why do round spermatids prefer lactate and pyruvate to glucose? Although pyruvate is synthesised via glycolysis, the production rate of pyruvate is too low to support full ATP generation via the TCA cycle and oxidative phosphorylation in mitochondria (Nakamura *et al.*, 1984). Round spermatids may have a unique regulation system to limit the use of glucose as a substrate. In general, phosphofructokinase (PFK) is a limiting enzyme for the glycolytic pathway. PFK is activated by AMP, ADP and inorganic phosphate, whereas it is inhibited by ATP, citrate, and fatty acids. Therefore, it is inactivated when energy supplies are abundant. Indeed, PFK was considered to be a limiting enzyme in sperm after taking into account enzyme activities, amount of intermediates and equilibrium constants (Harrison, 1971; Hoskins, Munsterman and Hall, 1975). However, round spermatids may be different. Mita and Hall (1982) postulated that utilization of glucose is limited by aldolase activity since round spermatids accumulate fructose 1,6-bisphosphate and dihydroxyacetone phosphate when they are cultured in medium containing 10 mM glucose. Nakamura, Fukiwara, Yasumasu, Okinaga and Arai (1982) suggested that glyceraldehyde-3-phosphate deydrogenase (GAPDH) could be a key enzyme regulating glycolysis in round spermatids, since the GAPDH activity in round spermatids is inhibited by adenine and its derivatives including ATP. However, it is still unknown how the activity of rate-limiting enzyme(s) is regulated in round spermatids.

Changes in metabolism during maturation

Cauda epididymal sperm have greater metabolic potential than testicular sperm. The amount of energy derived from endogenous substrates is constant for sperm in all regions of the epididymis. However, cauda sperm demonstrate a several fold increase of glucose and oxygen consumption and of lactate and CO_2 production compared to sperm from caput and corpus epididymidis (Inskeep and Hammerstedt, 1982). This is interesting in contrast to energy metabolism in round spermatids, which prefer lactate and pyruvate as substrates. Sperm appear to undergo a change in metabolism during transit from the distal corpus to the proximal cauda epididymidis, where the glycolytic pathway supplies enough amount of pyruvate required for the TCA cycle in mitochondria. Relative levels of fructose-1,6-bisphosphate and dihydoxyacetone phosphate were found to be dramatically reduced in cauda sperm as compared to caput sperm and round spermatids (Hoskins *et al.*, 1975; Mita and Hall, 1982). The maturation process that occurs in epididymis seems to change the regulation of glucose metabolism, resulting in activation of the glycolytic pathway required for sperm activity and survival in the female reproductive tract. This activation seems to result from modification of a glycolytic enzyme(s); however, the molecular mechanism underlying this regulation remains unknown. Increased glycolytic rates in cauda sperm could lead to an elevation in intracellular ATP levels. Indeed, ATP levels in cauda sperm are about two times higher than in caput sperm from bull (Hoskins *et al.*, 1975) and hamster sperm (White and Aitken, 1989). There are also some conflicting reports demonstrating no change or reduced ATP levels in cauda sperm. Contaminating blood cells and epididymal cells in sperm preparations make it difficult to prepare pure caput sperm, especially from small animals such as mice. ATP from these contaminants might affect the estimation of ATP levels in caput sperm preparations.

Energy production for supporting multiple cellular activities

Energy metabolism supporting hyperactivation

When sperm are incubated under capacitating conditions, sperm undergo changes in intracellular ion concentration, components of the plasma membrane, phosphorylation status of proteins and other protein modifications. These processes define capacitation and are essential for sperm to undergo hyperactivation and the acrosome reaction, followed by sperm-oocyte binding and fusion. *In vitro* experiments have shown that the fertilisation process requires glucose in addition to pyruvate or lactate (Hoppe, 1976; Niwa and Iritani, 1978). ATP is required for many biochemical events during fertilisation, therefore a supply of appropriate substrate(s) may be required for individual capacitation events and normal sperm function.

Glucose or fructose is required for hyperactivated motility of mouse (Fraser, 1977) and human sperm (Williams and Ford, 2001). Pharmacological experiments using inhibitors of glycolysis supported this observation. Ornidazole, an inhibitor of glycolysis, delivers metabolites including 3-chloroacetaldeyde and α-chlorohydrin which are known inhibitors of GAPDHS and triosephosphate isomerase (TPI) in human, rat and boar sperm (Jones, Stevenson, Hutton and Dawson, 1981; Jones and Stevenson, 1983). Bone, Jones, Kamp, Yeung and Cooper (2000) demonstrated that cauda epididymal sperm cultured under capacitating conditions from ornidazole-fed rats did not show hyperactivated motility when judged by computer-aided sperm analysis.

The cyclic AMP-dependent protein kinase (PKA) pathway leads to an increase in tyrosine phosphorylation of a subset of sperm proteins during capacitation. Protein tyrosine phosphorylation coincides with sperm hyperactivation (Visconti, Bailey, Moore, Pan, Olds-Carke and Kopf, 1995). Although it is not understood how tyrosine phosphorylation relates with hyperactivation, prototypical patterns of tyrosine phosphorylation represent the best molecularly defined hallmark of capacitation. The absence of glucose leads to a delay in the appearance of protein tyrosine phosphorylation when mouse sperm were cultured under capacitating conditions (Urner, Leppens-Luisier and Sakkas, 2001). However, omitting lactate and pyruvate from the capacitating medium did not have an affect on protein tyrosine phosphorylation in mouse sperm. Consistent with this observation, neither uncouplers of oxidative phosphorylation nor inhibitors of electron transfer affected patterns of protein tyrosine phosphorylation during capacitation (Travis, Jorgez, Merdiushev, Jones, Dess, Diaz-Cueto, Storey, Kopf and Moss, 2001). These data indicate that ATP specifically produced by the glycolytic pathway, as opposed to ATP generated by oxidative phosphorylation, is required for protein tyrosine phosphorylation during capacitation. Baker, Hetherington, Ecroyd, Roman and Aitker (2004) showed that the presence of Ca^{2+} in the medium decreased tyrosine phophorylation in both human and mouse sperm. Sperm have many Ca^{2+}-dependent ATP-utilising enzymes that reduce intracellular ATP levels. Therefore, extracellular Ca^{2+} suppresses tyrosine phosphorylation by decreasing the availability of intracellular ATP. Supporting this idea, the absence of glucose in the culture medium caused a decrease in sperm ATP levels, however, the addition of an inhibitor of electron transport in mitochondria did not have an affect on ATP levels (Baker *et al.*, 2004). In light of this finding, tyrosine phosphorylation seems to link to intracellular ATP levels. I speculate that high ATP levels may be required for some kinase(s) in the signalling pathway leading to tyrosine phosphorylation because of the kinase's high K_m value for ATP. Glycolysis, but not oxidative phosphorylation, may be able to meet this condition.

Energy metabolism supporting the acrosome reaction and sperm-oocyte binding and fusion

Binding to the zona pellucida of an oocyte stimulates sperm to undergo the acrosome reaction, followed by binding and fusion to the oocyte. Only capacitated sperm can complete these steps successfully. Inhibitors of electron transfer and uncouplers of oxidative phosphorylation reduced the incidence of hamster sperm acrosome reaction and fertilisation, without markedly affecting sperm motility (Rogers, Ueno and Yanagimachi, 1977). On the contrary, the zona pellucida-induced acrosome reaction of mouse sperm occurred in glucose-free medium (Urner and Sakkas, 1996a). The presence of glucose in the medium slightly increased the number of calcium ionophore-stimulated acrosome reactions, but did not increase spontaneous acrosome reactions (Rogers *et al.*, 1977; Williams and Ford, 2001). It remains unknown how ATP is supplied to the sperm head; however, the acrosome reaction appears to require ATP produced by oxidative phosphorylation in mitochondria. Although the ATP production rate by oxidative phosphorylation is lower than by glycolysis (Mukai and Okuno, 2004), it may be sufficient for inducing the acrosome reaction. Alternatively, the sperm head has geometrically better access to ATP produced by mitochondria than by the principal piece where glycolysis occurs (see below). The location of the ATP production site may be a critical factor.

The failure of sperm to achieve fertilisation in media lacking glucose is attributed in part to the absence of sperm hyperactivated motility as described above. Sperm undergo multiple processes before successful fertilisation. Some of these steps require glucose but others do not.

Urner and Sakkas (1996b) dissected the fertilisation process and showed that glucose is specifically required for sperm penetration into a zona-free oocyte (gamete fusion) but not for sperm binding to the oolemma. Interestingly, they showed that preincubation of oocytes with an inhibitor of glucose transporter did not prevent sperm entry into a zona-free oocyte. This result suggests that glucose uptake and metabolism by oocytes is not necessary for successful gamete fusion. Although gamete fusion appears to be the result of cooperation between male and female gametes, only sperm require glucose.

The pentose phosphate pathway (PPP) is a detour of part of the glycolytic pathway, corresponding to a flux from glucose-6-phosphate to glyceraldehyde-3-phosphate. The major role of PPP is to generate NADPH and ribose 5-phosphate, which are required for reductive reactions and synthesis of nucleic acids, respectively. Capacitated sperm metabolise glucose through PPP, although the level of metabolism is considerably lower than that through the glycolytic pathway. As mentioned above, sperm require glucose for entry into zone-free oocytes; however, NADPH was found to substitute for glucose in mouse gamete fusion (Urner and Sakkas, 1999). Furthermore, inhibition of gamete fusion by a PPP inhibitor was relieved by NADPH (Urner and Sakkas, 2005). These data indicate that sperm need glucose for production of NADPH by PPP at the stage of gamete fusion. Hexokinase 1 (HK1) is localised to the principal piece, as are other glycolytic enzymes, but also localised to the plasma membrane of the sperm head (Travis, Foster, Rosenbaum, Visconti, Gerton, Kopf and Moss, 1998). HK1 catalyses the conversion of glucose to glucose-6-phosphate, which is the branching point of PPP and the residual glycolytic pathway. An unusual distribution of HK1 in the sperm head may reflect the utilisation of glucose-6-phosphate for PPP. NADPH is unlikely to be used for energy production like NADH, rather it is proposed that NADPH provides antioxidant defense (Williams and Ford, 2004) and/or regulates reactive oxygen species that may affect a particular step of fertilisation (Urner and Sakkas, 2005).

ATP production and sperm motility

Glucose metabolism plays a significant role in hyperactivation. The absence of glucose in the capacitating medium causes decreased ATP levels and failure of protein tyrosine phosphorylation. These changes appear to suppress hyperactivation. Whereas, pyruvate alone or glucose plus an uncoupler of oxidative phosphorylation can maintain about 90% of ATP levels in mouse sperm (Mukai and Okuno, 2004). Under these conditions sperm keep vigorous motility both in beat frequency and waveform for more than 30 min. This finding suggests that sperm utilise glucose or pyruvate/lactate as a sole substrate of energy metabolism that supports normal sperm motility.

The waveform of sperm with progressive motility shows high beat frequency and these sperm swim straight and fast in medium. Upon hyperactivation, sperm change their waveform, having lower beat frequency but higher beat amplitude. Sperm no longer swim straight but wave the flagellum actively, showing whiplash-like motility. The name "hyperactivated motility" may suggest that sperm require more energy to undergo such a change in motility. Dynein is the motor protein of the flagellum and gives motive force by using ATP as an energy source. If it could be possible to measure the dynein ATPase activity of living sperm that show progressive or hyperactivated motility, a comparison of the activity would tell us whether hyperactivation is really the active state of the flagellum. However, no experiments looking at this have been reported perhaps due to technical difficulties. Some experiments do give insight into energy requirement for maintaining hyperactivation. First, calculation of the hydrodynamic power output indicates no significant change in hyperactivated motility of guinea pig and hamster sperm (Katz, Yanagimachi and Dresdner, 1978). This suggests that ATP consumption by dynein motor proteins does not change, although the sperm waveform is substantially changed. Second, while both hyperactivated motility and protein tyrosine phosphorylation of mouse sperm are observed after ~90 minutes of incubation under capacitating conditions (Visconti *et al.*, 1995), ATP levels of whole sperm populations are not elevated for up to 3 hours (Williams and Ford, 2001; Miki, Qu, Goulding, Willis, Bunch, Strader, Perreault, Eddy and O'Brien, 2004). Sperm are likely to undergo hyperactivation without increasing ATP levels since the percentage of viable sperm does not decrease throughout the incubation period. Last, reactivation of demembranated sperm experiments showed that low ATP concentration is sufficient for sperm reactivation. Maximum reactivation of demembranated sperm is attained with 0.1 ~ 1 mM ATP (Morton, 1973; Ishijima and Mohri, 1985). These ATP concentrations are extremely lower than intracellular ATP levels, which are 15 mM in rat sperm (Jeulin and Sourir, 1992). Flagellar curvature is one of the best parameters used to describe hyperactivated motility. A change in the curvature of demembranated sperm can be induced by Ca^{2+} in the reactivating solution (Lindermann, Goltz and Kanous, 1987). Ho, Granish and Suarez (2002) showed that increasing ATP in the reactivating solution strengthens the action of Ca^{2+} on curvature changes in demembranated sperm. It is unlikely for sperm to require additional ATP to undergo hyperactivated motility, since this effect was observed at significantly lower ATP concentrations as compared to intracellular concentrations in sperm.

Sperm maintain exceptionally high ATP levels while bulk cytosolic ATP concentration is about 1 mM in various mammalian somatic cells (Wang, Tao, Trumbore and Berger, 1997; Kennedy, Pouli, Ainscow, Jouaville, Rizzuto and Rutter, 1999). Why do sperm require high ATP levels for motility? Dynein is the motor protein in the axoneme, which extends throughout the flagellum. This is a basic structure seen in all flagella and cilia across species. Experiments using purified dynein from *Tetrahymena* showed that progressive movement was observed *in vitro* only at low ATP concentrations, less than 20 μM (Hirakawa, Higuchi and Toyoshima, 2000). In addition, tracheal epithelial cells were found to regulate beat frequency of cilia with micromolar levels of ATP (Zhang and Sanderson, 2003). According to reactivation experiments, low levels of ATP (~1 mM) are sufficient to reactivate demembranated sperm motility, while ATP levels higher than 1 mM clearly suppress sperm reactivation (White and Voglmayr, 1986). Although the molecular mechanism behind this suppression remains unknown, these results suggest that ~1 mM ATP is an optimal concentration for normal function of the dynein motor protein in sperm. If this is true, why are sperm ATP levels maintained at least 10 times higher than concentrations necessary for dynein activity? Capacitation is associated with many ATP-consuming events resulting in activation of a signal transduction cascade. This cascade includes the PKA signalling pathway, which leads to protein tyrosine phosphorylation (Visconti *et al.*, 1995). Influx of Ca^{2+} is also associated with capacitation and stimulates Ca^{2+}-dependent ATP-utilising enzymes (Baker *et al.*, 2004). Therefore, elevation of intracellular ATP levels may be essential for modulation of sperm function and waveform, and excess for supporting sperm motility.

Defects of sperm functions in knockout mouse models

Physiological experiments have identified substrates for energy metabolism during spermiogenesis, maturation and capacitation. The use of specific inhibitors identified the essential metabolic pathway required for sperm activities. A combination of these experiments outlined the relation between energy metabolism and sperm function. However, metabolism of living cells is complicated because bypass reactions and feedback regulations could affect major fluxes of substrates. *In vitro* experiments might lead to a theory that does not represent sperm located in the reproductive tract. Knockout models are powerful tools to confirm the essential role of metabolic pathways in sperm function. In general, genes involved in the central metabolic pathway, such as genes for glycolytic enzymes or components of mitochondrial respiration, cannot be targeted by the knockout approach because the elimination of the gene will result in embryonic lethality. In testis we see the expression of an exceptionally large number of genes encoding sperm-specific isoforms. Due to the strictly regulated expression in testis, these genes are an ideal target for generating knockout animals and analysing the role of metabolic pathways in sperm function.

Cytochrome c is present in the mitochondria and transports electrons from complex III to complex IV in the respiratory chain. Mammalian testes express two types of cytochrome c, the somatic isoform, Cyt c_S, and a testis-specific isoform, Cyt c_T (Goldberg, Sberna, Wheat, Urbanski and Margoliash, 1977). Cyt c_T replaces Cyt c_S in spermatocytes and becomes the sole cytochrome c expressed in all postmeiotic male germ cells including mature sperm (Hess, Miller, Kirby, Margoliash and Goldberg, 1993). Narisawa, Hecht, Goldberg, Boatright, Reed and Millan, (2002) generated *Cyt* c_T mutant mice by using a gene targeting protocol and demonstrated that homozygous mutants for the *Cyt* c_T gene developed normally and healthily. Male mutant mice were fertile although mutant sperm showed reduced motility, ATP concentrations, and achievement of *in vitro* fertilisation. ATP levels were 60% of wild-type sperm. This may be explained, in part, by an increase in immotile sperm in the vas deferens (Narisawa *et al.*, 2002). Nevertheless, the ATP levels of mutant sperm are consistent with results from pharmacological experiments using inhibitors of electron transfer (Baker *et al.*, 2004; Mukai and Okuno, 2004). In addition, low success rates of *in vitro* fertilisation may be attributed to reduced acrosome reactions in mutant sperm, since the acrosome reaction requires ATP production by oxidative phosphorylation (Rogers *et al.*, 1977). Regardless of their inferior activity, these mutant sperm are still functional *in vivo*. This is a striking finding and suggests that other metabolic pathways compensate for the lack of metabolites and energy production when oxidative phosphorylation is blocked.

One glycolytic enzyme, GAPDH, is another candidate for the gene targeting approach to analyse energy metabolism. Glyceraldehyde-3-phosphate dehydrogenase-S (GAPDHS) is a sperm-specific isozyme of GAPDH and is expressed in late spermatids, replacing the somatic isozyme (Bunch, Welch, Magyar, Eddy and O'Brien, 1998). Therefore, GAPDHS becomes the sole isozyme present in sperm, similarly to what happens with Cyt c_T replacing Cyt c_S. *Gapdhs*$^{-/-}$ males were infertile and had profound defects in sperm motility, although the sperm counts from these animals were normal (Miki *et al.*, 2004). Mutant sperm had dramatically reduced ATP levels that were ~10% of those in wild-type sperm, and did not recover ATP levels even after incubation with medium containing pyruvate and lactate in addition to glucose. This result has several important implications. First, glycolysis is important for sperm maturation. During maturation in the epididymis, sperm begin to consume more glucose and oxygen to increase ATP concentrations (Hoskins *et al.*, 1975; Inskeep and Hammerstedt, 1982; White and Aitken, 1989). However, ATP concentrations do not increase in sperm lacking GAPDHS during transit of the epididymal tract. If pyruvate is primarily supplied by glycolysis and not by intake from epididymal fluid during maturation, mutant sperm are short of a substrate for the TCA cycle, resulting in reduced activity of oxidative phosphorylation. Second, sperm lacking GAPDHS might not undergo functional maturation during epididymal transit. Epididymal maturation is associated with biochemical changes that may require ATP in multiple events, including modification/maintenance of the phosporylation status of signalling molecules (Vijayaraghavan, Mohan, Gray, Khatra and Carr, 2000; Huang and Vijayaraghavan, 2004).

Unsuccessful maturation due to low ATP levels may explain sluggish motility and infertility of mutant sperm. Lastly, oxidative phosphorylation alone can not maintain ATP levels in sperm lacking GAPDHS. Although mitochondrial oxygen consumption was unchanged and ATP was supposed to be generated efficiently in the mitochondria, mutant sperm did not recover intracellular ATP levels at all (Miki *et al.*, 2004). It is noteworthy that Mukai and Okuno (2004) demonstrated that ATP levels in sperm incubated with pyruvate were ~90% of those in sperm incubated with glucose; however, 2-deoxyglucose, an inhibitor of glycolysis, dramatically decreased ATP levels in sperm incubated with pyruvate as the sole substrate. This result is consistent with the phenotype of sperm lacking GAPDHS. The authors hypothesised that respiratory substrates function as substrates for gluconeogenesis in the middle piece, resulting in glucose that can diffuse to other regions of the flagellum. Both *in vivo* (knockout) and *in vitro* (pharmacological) experiments demonstrated that the glycolytic pathway is essential for maintaining functional ATP levels in sperm.

Mutant mice that lack the *Cyt* c_T or *Gapdh* gene are ideal animal models to evaluate the contribution of glycolysis and oxidative phosphorylation to sperm function. Analysis of these mutants clearly showed that ATP production by glycolysis can recover the absence of oxidative phosphorylation. Glycolysis, but not oxidative phosphorylation, is indispensable for male fertility.

Glycolysis in sperm

Spermatogenic cell-specific isozymes of glycolytic enzymes

Although the glycolytic pathway and its components are highly conserved for energy production, several enzymes in this pathway have spermatogenic cell-specific isozymes that are products of a splice variant or are a paralogue of a somatic gene (Fig. 1). Among the 11 glycolytic enzymes, at least 4 of these enzymes in sperm are spermatogenic cell-specific isozymes that are encoded at a locus distinct from the somatic gene. This group includes phosphofructokinase-2 (PFK-2) type 4, GAPDHS/GAPDH2, phosphglycerate kinase 2 (PGK2) and LDH-C_4. HK1-s isozyme is an alternative splicing product of the common hexokinase gene. Sperm-specific isozymes of aldolase, enolase and glucosephosphate isomerase (GPI) were identified based on unique mobility by electrophoresis, although corresponding genes have not been found. No spermatogenic cell-specific isozyme was found for the remaining 3 glycolytic enzymes, but phosphoglycerate mutase (PGAM), pyruvate kinase (PK) and triphosphate isomerase (TPI) show unique post-transcriptional regulation in post-meiotic germ cells. Thus, for all 11 glycolytic enzymes, spermatogenic cells have unique characteristics that are distinct from their somatic counterparts. Some of these characteristics result in different solubility, localization and/or kinetic properties.

HK is one of the rate-limiting enzymes, catalysing the first step of glycolysis where glucose is phosphorylated to make glucose-6-phosphate. Mouse spermatogenic cells express HK1-sa, HK1-sb and HK1-sc, all containing a spermatogenic cell-specific region (SSR) at the N-terminus instead of the porin-binding domain (PBD). The PBD is believed to be responsible for the binding of HK1 to the mitochondrial membrane (Mori, Welch, Fulcher, O'Brien and Eddy, 1993). HK1-sb mRNA appears to be the most abundant and is detected only in post-meiotic germ cells (Mori *et al.*, 1993). HK1-s protein expression was first found in round spermatids and the amount increased dramatically as spermiogenesis proceeded, with the highest amount in epididymal sperm (Travis *et al.*, 1998). The somatic isozyme of hexokinase is found as a soluble protein that under certain conditions is bound to the outer mitochondrial membrane through its PBD. HK1-s is mainly localised to the principal piece and is found in the soluble

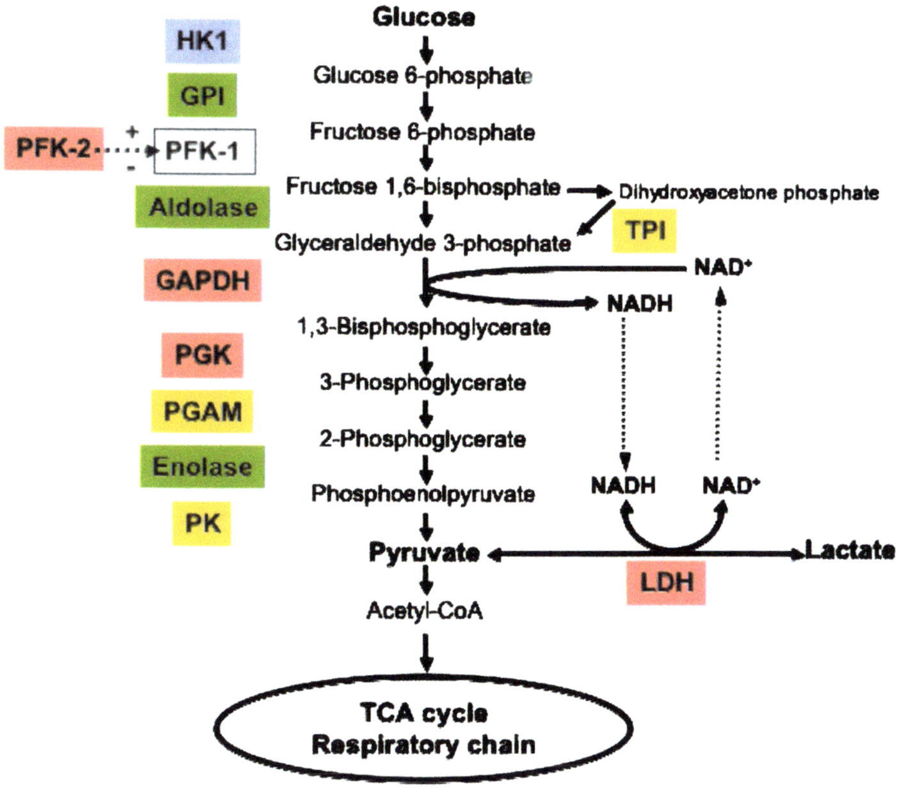

Figure 1. The glycolytic pathway. Glucose is metabolised to pyruvate by 10 glycolytic enzymes. Pyruvate is then used as a substrate for the TCA cycle in the mitochondria or released outside of the cell after conversion to lactate by LDH. In preference, extracellular lactate is taken into the cell and used as a substrate for the TCA cycle after conversion to pyruvate. Spermatozoa possess spermatogenic cell-specific glycolytic isozymes that are divided into 4 groups: isozymes encoded at a locus distinct from the somatic gene (red box), an isozyme derived from an alternative splice variant of the somatic isozyme (blue box), isozymes identified based on unique mobility by electrophoresis (green box), and no spermatogenic cell-specific glycolytic isozymes identified, but transcripts show unique post-transcriptional regulation in post-meiotic germ cells (yellow box). HK1; hexokinase 1, GPI; glucosephosphate isomerase, FPK; phosphofructokinase, TPI; triphosphate isomerase, GAPDH; glyceraldehyde-3-phosphate dehydrogenase, PGK; phosphoglycerate kinase, PGAM; phosphoglycerate mutase, PK; pyruvate kinase, LDH; lactate dehydrogenase. PFK-2 is not a glycolytic enzyme but catalyses the synthesis and breakdown of fructose-2,6-bisphosphate that is a regulator of PFK-1 in the glycolytic pathway.

fraction of sperm homogenate, suggesting that SSR substitution for PBD results in altered intracellular targeting of HK1-s (Visconti, Olds-Clarke, Moss, Kalab, Travis, de las Heras and Kopf, 1996; Mori, Nakamura, Welch, Gotoh, Goulding, Fujioka and Eddy, 1998; Travis *et al.*, 1998). In addition, HK1-s is also found in the sperm head, suggesting that HK1-s is involved in NADPH production by PPP described above (Mori *et al*, 1998; Travis *et al.*, 1998).

PFK-1 catalyses the conversion of fructose-6-phosphate to fructose-1,6-bisphospate, the early rate-limiting step of the glycolytic pathway. Three PFK-1 isozymes have been identified; however, no spermatogenic cell-specific isozyme has been reported. PFK-2 is a bifunctional enzyme responsible for the synthesis and breakdown of fructose-2,6-bisphosphate, which is a potent activator of PFK-1. Therefore, PFK-2 plays a central role in regulating the entire glycolytic pathway in somatic cells. Several isozymes of PFK-2 have been identified in mammalian tissues including liver, muscle, heart, fetus, brain, placenta and testis. At least four genes encode PFK-2 and with alternative splicing events this results in many diverse PFK-2 isozymes. The testis isozyme is encoded by the PFKFB4 gene (Manzano, Perez, Nadal, Estivill, Lange and Bartrons, 1999) and the deduced amino acid sequence shows 74% and 67% similarity with liver and heart isozymes in rat, respectively (Sakata, Abe and Uyeda, 1991). In addition to the testis isozyme, the placenta isozyme encoded by the PFKFB3 gene is also expressed in testis (Sakakibara, Okudaira, Fujiwara, Kato, Hirata, Yamanaka, Naito and Fukasawa, 1999); however, the tempo-spatial regulation of these isozymes in testis is unknown.

Aldolase catalyses the cleavage of fructose-1,6-bisphosphate to glyceraldehyde-3-phospate and dihydroxyacetone phosphate. There are three isozymes of aldolase; A, B and C. While aldolase B and C isozymes show a narrowed expression pattern, aldolase A is expressed in most tissues, with high concentrations found in skeletal muscle (Penhoet, Kochman and Rutter, 1969). Although muscle aldolase is found in the soluble fraction, sperm aldolase is associated with the fibrous sheath. Under optimised conditions soluble sperm aldolase represents only 10% of the total aldolase activity in bovine sperm (Gillis and Tamblyn, 1984). Furthermore, sperm aldolase was shown to have distinctive kinetic properties. Krisfalusi, Miki, Magyar and O'Brien (2006) isolated the fibrous sheath from mouse sperm and identified aldolase A in a 50 kDa band by mass spectrometry analysis, in addition to a 41 kDa band corresponding to the expected molecular weight of somatic aldolase A. These data suggest that sperm contain multiple isoforms of aldolase A.

GAPDH catalyses the conversion of glyceraldehyde-3-phosphate to 1,3-bisphosphoglycerate, generating the reducing power of NADH. The mouse spermatogenic cell-specific gene, *Gapdhs,* is a paralogue of the somatic isozyme *Gapdh.* GAPDHS is 71% identical to GAPDH, but contains 105 additional residues at the N-terminus (Welch, Schatte, O'Brien and Eddy, 1992). This region may play an important role in protein interactions and/or localization to the flagellum. While the amount of GAPDH decreases during spermiogenesis, GAPDHS first appears in condensing spermatids and becomes the only protein showing GAPDH/GAPDHS enzyme activity in sperm. This suggests that GAPDHS has a specific role in sperm function (Bunch *et al.*, 1998; Miki *et al.*, 2004). Immunogold labelling and immunofluorescence demonstrated that GAPDHS is exclusively localised to the fibrous sheath in the principal piece (Westhoff and Kamp, 1997; Bunch *et al.*, 1998). GAPDHS is tightly bound to the fibrous sheath and co-purified with PKA anchoring protein 4 (AKAP4), the major cytoskeletal component of the fibrous sheath (Krisfalusi *et al.*, 2006). In addition, sperm lacking AKAP4 fail to assemble GAPDHS into the flagellum, suggesting that GAPDHS is one of the cytoskeletal components of the fibrous sheath (Miki, Willis, Brown, Goulding, Fulcher and Eddy, 2002).

PGK catalyses one of the ATP-producing steps in the glycolytic pathway. *Pgk1* is a ubiquitously expressed X-linked gene, while *Pgk2* is an autosomal intronless gene that evolved by retroposition from the *Pgk1* gene (McCarrey and Thomas, 1987). The *Pgk2* gene is expressed at high levels in meiotic spermatocytes and post-meiotic spermatids (Kramer and Erickson, 1981). The PGK2 protein sequence is 85% homologous to the PGK1 protein sequence (McCarrey and Thomas, 1987). Although the precise localization of PGK2 is yet unknown, a part of PGK2 was found in the fibrous sheath fraction when sperm proteins were sequentially extracted (Krisfalusi *et al.*, 2006). It has been proposed that autosomal *Pgk2* gene expression in spermatocytes may be a compensatory response to the inactivation of the X-linked *Pgk1* gene in spermatogenic cells before meiosis.

LDH catalyses the interconversion of pyruvate and lactate. This step can work as a "drain" of the glycolytic pathway under anaerobic conditions or as a "suction pipe" of the TCA cycle under aerobic conditions. LDH is a tetramer composed of three subunits, A, B and C. Different combinations of subunits give rise to six LDH isozymes; A_4, A_3B_1, A_2B_2, A_1B_3, B_4 and C_4. Somatic cells contain these isozymes in various proportions. LDH-C_4 is present only in mature testis and sperm (Goldberg, 1965) and has been recognised to be the predominant LDH isozyme in spermatogenic cells (Zinkham, Blanco and Kupchyk, 1963; Li, O'Brien, Hou, Versola, Rockett and Eddy, 1989), whereas the LDHA subunit was also found in sperm (Krisfalusi *et al*., 2006). LDH-C_4 is localised in the principal piece of mouse sperm (Duan and Goldberg, 2003) and approximately 10% of its activity is associated with the plasma membrane (Alvarez and Storey, 1984). Krisfalusi *et al*. (2006) showed that the sperm LDHC subunit was solubilised by potassium thiocyanate solution while the LDHA subunit was tightly associated to the fibrous sheath. These isozymes may take part in different metabolic activities in different intracellular compartment.

Unique structures or characteristics have been reported for other glycolytic enzymes in spermatogenic cells, although a spermatogenic cell-specific gene or splicing variant has not been identified. Functional PGAM is a dimer of two distinct subunits, M and B (Bartrons and Carreras, 1982). The M subunit is present mainly in muscle and testis, whereas the B subunit is expressed ubiquitously (Zhang, Yu, Fu, Gao, Xie, Chen, Zhang, Liu and Zhao, 2001). In testis, *Pgam-m* is expressed in post-meiotic spermatogenic cells, with a longer poly (A) tail compared to *Pgam-m* mRNA expressed in muscle cells, suggesting that *Pgam-m* expression is developmentally regulated (Fundele, Winking, Illmensee and Jagerbauer1987; Broceno, Ruiz, Reina, Vilaro and Pons, 1995). PK shows a similar relationship between poly (A) tail extension and increase of gene expression during development. There are four isozymes for PK; R, L, M1 and M2 (reviewed in Ibsen, 1977). PK-M1 isozyme is most common in skeletal muscle, heart and brain, but is also found in testis. M1 mRNA is most abundant in condensing spermatids, and poly (A) tail extension occurs during spermiogenesis (de Luis and del Mazo, 1998). Another example of the use of alternative polyadenylation signal is TPI. Somatic cells express a 1.4 kb transcript of TPI; however, haploid spermatids specifically express a 1.5 kb transcript that is a result of the use of an alternative polyadenylation signal, suggesting that TPI expression is post-transcriptionally regulated in spermatids (Russell and Kim, 1996). Zymographs have identified isozymes of several glycolytic enzymes. Enolase (ENO) is found as a dimer formed from three subunits; α, ß and γ, encoded by distinct genes. While the α subunit is found in most cell types, the ß and γ subunits are expressed in the nervous system and muscle, respectively (Fletcher, Rider, Taylor, Adamson, Luke and Graham, 1978). In addition to ENO-$\alpha\alpha$, sperm have a unique isozyme, ENO-S, detected by zymography (Edwards and Grootegoed, 1983). ENO-S activity increases during sperm maturation in the epididymis while ENO-$\alpha\alpha$ activity decreases, suggesting that sperm enolase isozymes undergo post-translational regulation (Force, Viallard, Saez, Grizard and Boucher, 2004). Sperm-specific isozymes of GPI have also been detected by zymography using cellulose acetate electrophoresis (Buehr and McLaren, 1981).

The principal piece as a compartment producing ATP by glycolysis

The flagellum is divided into the middle piece, principal piece and end piece. The middle piece is adjacent to the head and is distinguished by the presence of the mitochondrial sheath. The principal piece occupies approximately three-fourths of the length of the flagellum. It is defined by the presence of the fibrous sheath, which is a cytoskeletal structure. The end piece is a short segment at the tip of the flagellum. The fibrous sheath is believed to stiffen the flagellum and to modulate its bending. AKAP4 is the major component of the fibrous sheath and is the product of an X-linked gene expressed only in spermatids (Carrera, Gerton and Moss, 1994; Fulcher, Mori, Welch, O'Brien, Klapper and Eddy, 1995). Since AKAP4 contains PKA-binding domains, it is assumed that AKAP4 recruits PKA to the fibrous sheath to facilitate phosphorylation of neighbouring proteins that regulate flagellar function (Visconti, Johnson, Oyaski, Fornes, Moss, Gerton and Kopf, 1997; Miki and Eddy, 1998, 1999). Indeed, AKAP4 is a scaffold protein that interacts with many sperm-specific proteins including GAPDHS, AKAP3 and Fibrous Sheath Interacting Protein 1 and 2 (Brown, Miki, Harper and Eddy, 2003). The definitive fibrous sheath did not develop in sperm lacking AKAP4, and GAPDHS failed to become associated with the fibrous sheath when AKAP4 was absent (Miki *et al.*, 2002). There is no strong evidence that glycolytic enzymes are directly associated with AKAP4 and are being phosphorylated by PKA. However, Krisfalusi *et al.* (2006) showed that glycolytic enzymes, including aldolase A, GAPDHS, LDH-A, PK, and PGK2, are found in the fibrous sheath fraction that is also rich in AKAP4. Immunofluorescence or immunogold staining has revealed that HK1-s (Visconti *et al.*, 1996; Mori *et al.*, 1998), GAPDHS (Westhoff and Kamp, 1997; Bunch *et al.*, 1998) and LDH-C_4 (Duan and Goldberg, 2003) are localised to the principal piece. In all, currently seven isozymes from six glycolytic enzymes are found to be localised to the principal piece, supporting the proposed idea that sperm possess compartmentalised metabolic pathways in specific regions: the middle piece generates mitochondrial respiration and the principal piece generates glycolysis (Westhoff and Kamp, 1997).

Co-localization of glycolytic enzymes is also observed in erythrocytes. Band 3 is a major membrane protein in erythrocytes and tethers GAPDH, aldolase and PFK through its N-terminus (Higashi, Richards and Uyeda, 1979). The association of this enzyme complex with the membrane is regulated by the oxygenation state of the cell (Campanella, Chu and Low, 2005). It is interesting that band 3 shows a similar role to AKAP4 in the principal piece of sperm. The physiological significance still needs to be investigated, but co-localization of enzymes must be beneficial for efficient production of ATP by glycolysis.

Conclusions

Glycolysis or mitochondrial respiration? This is the primary question when trying to understand how sperm obtain energy that is required for fertilisation. Despite the different preference for energy substrates between species, it is clear that operation of both energy production pathways is essential for optimal sperm functions. However, knockout studies of *Gapdhs* and *Cyt* c_T genes revealed that glycolysis is indispensable for mouse sperm functions, but oxidative phosphorylation is not essential for male fertility (Narisawa *et al.*, 2002; Miki *et al.*, 2004). This suggests that glycolysis is the primary pathway generating energy for fertility, and that oxidative phosphorylation is the backup system. Analysis of *Gapdhs* knockout mice highlighted the importance of glycolysis for sperm function. While the essential function of glycolysis in sperm maturation was suggested, this study alone did not identify the critical steps where sperm require glycolysis. Taken together with results from other experiments, glycolysis is essential for cell signalling during capacitation, especially hyperactivation. Sperm motility, however, appears to be supported by either glycolysis or oxidative phosphorylation. Although sperm maintain high ATP levels as compared to somatic cells, dynein ATPase seems to require a relatively low level of ATP for flagellar beating. Instead, some of other ATPases or protein kinases seem to require high levels of ATP generated by glycolysis in order to regulate sperm motility.

Although the glycolytic pathway is essential for mature sperm function, substrate use seems to be flexible. Sperm appear to adapt to their microenvironment and adjust the ratio of glycolysis and oxidative phosphorylation depending on the availability of substrate and oxygen. If the medium is poor in hexose, such as glucose or fructose, more lactate or pyruvate will be incorporated into sperm to compensate for the shortage of hexose. ATP generated by the mitochondria in the middle piece may be distributed along the flagellum by adenylate kinase and other shuttles (Ford, 2006). Some of mitochondrial ATP may be also used for glyconeogenesis, which provides substrates for re-extraction of ATP by the glycolytic pathway in the principal piece as observed in dog sperm (Albarracin, Fernandez-Novell, Ballester, Rauch, Quintero-Moreno, Pena, Mogas, Rigau, Yanez, Guinovart, Slebe, Concha and Rodriguez-Gil, 2004; Mukai and Okuno, 2004). It is unlikely that hexose-free medium can support all steps of fertilisation in any mammal. However, lactate and/or pyruvate seems to compensate for a shortage of hexose and supports normal sperm function.

Sperm contain a large number of isoproteins including spermatogenic cell-specific glycolytic isozymes. There are two possible functions of these isozymes. First, they may have distinct kinetic properties from the somatic isozyme. The activity of spermatogenic cell-specific isozymes may be optimised for the microenvironment in reproductive tracts and controlled by a different type of feedback regulation that is not observed in somatic cells. Furthermore, spermatogenic cell-specific isozymes possess unique kinetic properties. It is interesting to note that two LDH subunits, LDHA and LDHC, are present in the principal piece. LDH-A_4, the muscle-type isozyme, is not inhibited much by an excess of pyruvate and is consequently well adapted to anaerobic glycolysis and the accumulation of lactate (Cahn, Kaplan, Levine and Zwilling, 1962). LDH-A_4 may form an enzyme complex with other glycolytic enzymes in the fibrous sheath (Krisfaluse *et al.*, 2006) and facilitate active glycolysis by pumping out pyruvate after its conversion to lactate (Fig. 1). On the other hand, LDH-C_4 shows closer kinetic properties to LDH-B_4, a heart-type isozyme which requires aerobic conditions (Wilkinson and Withycombe, 1965). LDH-C_4 may be present in a different compartment from LDH-A_4 (Krisfalusi *et al.*, 2006) and primarily work on intake of lactate for mitochondrial respiration. The second possible function of spermatogenic cell-specific glycolytic isozymes is localization to a specific cellular site. These isozymes may contain a unique signal sequence that defines their localization. Also, a unique domain in this sequence may be responsible for protein interactions in the principal piece. A significant amount of sperm proteins associate with a cytoskeletal structure and are difficult to solubilise under mild conditions. Nevertheless, analysis of the protein complex containing spermatogenic cell-specific isozymes may give us insight into unique functions of spermatogenic cell-specific isozymes and lead to a better understanding about the energy metabolism of sperm.

Acknowledgements

The author thanks Ms. Soumya Vemuganti for editing the manuscript and other members of Dr. Deborah O'Brien's laboratory for fruitful discussion. This work was supported by NICHD/NIH through cooperative agreement U54 HD35041 as part of the Specialized Cooperative Centers Program in Reproductive Research.

References

Albarracin, J.L., Fernandez-Novell, J.M., Ballester, J., Rauch, M.C., Quintero-Moreno, A., Pena, A., Mogas, T., Rigau, T., Yanez, A., Guinovart, J.J., Slebe, J.C., Concha, I.I. and Rodriguez-Gil, J.E. (2004). Gluconeogenesis-linked glycogen metabolism is important in the achievement of in vitro capacitation of dog spermatozoa in a medium without glucose. *Biology of Reproduction* **71**: 1437-1445

Alvarez, J.G. and Storey, B.T. (1984). Assessment of cell damage caused by spontaneous lipid peroxidation in rabbit spermatozoa. *Biology of Reproduction* **30**: 323-331

Baker, M.A., Hetherington, L., Ecroyd, H., Roman, S.D. and Aitken, R.J. (2004). Analysis of the mechanism by which calcium negatively regulates the tyrosine phosphorylation cascade associated with sperm capacitation. *Journal of Cell Science* **117**: 211-222

Bartrons, R. and Carreras, J. (1982). Purification and characterization of phosphoglycerate mutase isozymes from pig heart. *Biochimica et Biophysica Acta* **708**: 167-177

Bone, W., Jones, N.G., Kamp, G., Yeung, C.H. and Cooper, T.G. (2000). Effect of ornidazole on fertility of male rats: inhibition of a glycolysis-related motility pattern and zona binding required for fertilization in vitro. *Journal of Reproduction and Fertility* **18**: 127-135

Broceno, C., Ruiz, P., Reina, M., Vilaro, S. and Pons, G. (1995). The muscle-specific phosphoglycerate mutase gene is specifically expressed in testis during spermatogenesis. *European Journal of Biochemistry* **227**: 629-635

Brown, P.R., Miki, K., Harper, D.B. and Eddy, E.M. (2003). A-kinase anchoring protein 4 binding proteins in the fibrous sheath of the sperm flagellum. *Biology of Reproduction* **68**: 2241-2248

Buehr, M. and McLaren, A. (1981). An electrophoretically detectable modification of glucosephosphate isomerase in mouse spermatozoa. *Journal of Reproduction and Fertility* **63**: 169-173

Bunch, D.O., Welch, J.E., Magyar, P.L., Eddy, E.M. and O'Brien, D.A. (1998). Glyceraldehyde 3-phosphate dehydrogenase-S protein distribution during mouse spermatogenesis. *Biology of Reproduction* **58**: 834-841

Cahn, R.S., Kaplan, N.O., Levine, L. and Zwilling, E. (1962). Nature and development of lactic dehydrogenases. *Science* **136**: 962-969

Campanella, M.E., Chu, H. and Low, P.S. (2005). Assembly and regulation of a glycolytic enzyme complex on the human erythrocyte membrane. *Proceedings of the National Academy of Sciences USA* **102**: 2402-2407

Carrera, A., Gerton, G.L. and Moss, S.B. (1994). The major fibrous sheath polypeptide of mouse sperm: structural and functional similarities to the A-kinase anchoring proteins. *Developmental Biology* **165**: 272-284

de Luis, O. and del Mazo, J. (1998). Gene expression of mouse M1 and M2 pyruvate kinase isoenzymes correlates with differential poly[A] tract extension of their mRNAs during the development of spermatogenesis. *Biochimica et Biophysica Acta* **1396**: 294-305

Duan, C. and Goldberg, E. (2003). Inhibition of lactate dehydrogenase C4 (LDH-C4) blocks capacitation of mouse sperm in vitro. *Cytogenetics and Genome Research* **103**: 352-359

Edwards, Y.H. and Grootegoed, J.A. (1983). A sperm-specific enolase. *Journal of Reproduction and Fertility* **68**: 305-310

Fletcher, L., Rider, C.C., Taylor, C.B., Adamson, E.D., Luke, B.M. and Graham, C.F. (1978). Enolase isoenzymes as markers of differentiation in teratocarcinoma cells and normal tissues of mouse. *Developmental Biology* **65**: 462-475

Force, A., Viallard, J.L., Saez, F., Grizard, G. and Boucher, D. (2004). Electrophoretic characterization of the human sperm-specific enolase at different stages of maturation. *Journal of Andrology* **25**: 824-829

Ford, W.C. (2006). Glycolysis and sperm motility: does a spoonful of sugar help the flagellum go round? *Human Reproduction Update* **12**: 269-274

Fraser, L.R. (1977). Motility patterns in mouse spermatozoa before and after capacitation. *Journal of Experimental Zoology* **202**: 439-444

Fulcher, K.D., Mori, C., Welch, J.E., O'Brien, D.A., Klapper, D.G. and Eddy, E.M. (1995). Characterization of Fsc1 cDNA for a mouse sperm fibrous sheath component. *Biology of Reproduction* **52**: 41-49

Fundele, R., Winking, H., Illmensee, K. and Jagerbauer, E.M. (1987). Developmental activation of phosphoglycerate mutase-2 in the testis of the mouse. *Developmental Biology* **124**: 562-566

Gillis, B.A. and Tamblyn, T.M. (1984). Association of bovine sperm aldolase with sperm subcellular components. *Biology of Reproduction* **31**: 25-35

Goldberg, E. (1965). Lactate dehydrogenase in trout: Evidence for a third subunit. *Science* **148**: 391-392

Goldberg, E., Sberna, D., Wheat, T.E., Urbanski, G.J. and Margoliash, E. (1977). Cytochrome c: immunofluorescent localization of the testis-specific form. *Science* **196**: 1010-1012

Harrison, R.A. (1971). Glycolytic enzymes in mammalian spermatozoa. Activities and stabilities of hexokinase and phosphofructokinase in various fractions from sperm homogenates. *Biochemical Journal* **124**: 741-750

Hess, R.A., Miller, L.A., Kirby, J.D., Margoliash, E. and Goldberg, E. (1993). Immunoelectron microscopic localization of testicular and somatic cytochromes c in the seminiferous epithelium of the rat. *Biology of Reproduction* **48**: 1299-1308

Higashi, T., Richards, C.S. and Uyeda, K. (1979). The interaction of phosphofructokinase with erythrocyte membranes. *Journal of Biological Chemistry* **254**: 9542-9550

Hirakawa, E., Higuchi, H. and Toyoshima, Y.Y. (2000). Processive movement of single 22S dynein molecules occurs only at low ATP concentrations. *Proceedings of the National Academy of Sciences USA* **97**: 2533-2537

Ho, H.C., Granish, K.A. and Suarez, S.S. (2002). Hyperactivated motility of bull sperm is triggered at the axoneme by Ca^{2+} and not cAMP. *Developmental Biology* **250**: 208-217

Hoppe, P.C. (1976). Glucose requirement for mouse sperm capacitation in vitro. *Biology of Reproduction* **15**: 39-45

Hoskins, D.D., Munsterman, D. and Hall, M.L. (1975). The control of bovine sperm glycolysis during epididymal transit. *Biology of Reproduction* **12**: 566-572

Huang, Z. and Vijayaraghavan, S. (2004). Increased phosphorylation of a distinct subcellular pool of protein phosphatase, PP1γ2, during epididymal sperm maturation. *Biology of Reproduction* **70**: 439-447

Ibsen, K.H. (1977). Interrelationships and functions of the pyruvate kinase isozymes and their variant forms: a review. *Cancer Research* **37**: 341-353

Inskeep, P.B. and Hammerstedt, R.H. (1982). Changes in metabolism of ram sperm associated with epididymal transit or induced by exogenous carnitine. *Biology of Reproduction* **27**: 735-743

Ishijima, S and Mohri, H. (1985). A quantitative description of flagellar movement in golden hamster spermatozoa. *Journal of Ex.perimental Biology* **114**: 463-475

Jeulin, C. and Soufir, J.C. (1992). Reversible intracellular ATP changes in intact rat spermatozoa and effects on flagellar sperm movement. *Cell Motility and the Cytoskeleton* **21**: 210-222

Jones, A.R. and Stevenson, D. (1983). Formation of the active antifertility metabolite of (S)-α-chlorohydrin in boar sperm. *Experientia* **39**: 784-785

Jones, A.R., Stevenson, D., Hutton, P. and Dawson, A.G. (1981). The antifertility action of α-chlorohydrin: metabolism by rat and boar sperm. *Experientia* **37**: 340-341

Katz, D.F., Yanagimachi, R. and Dresdner, R.D. (1978). Movement characteristics and power output of guinea-pig and hamster spermatozoa in relation to activation. *Journal of Reproduction and Fertility* **52**: 167-172

Kennedy, H.J., Pouli, A.E., Ainscow, E.K., Jouaville, L.S., Rizzuto, R. and Rutter, G.A. (1999). Glucose generates sub-plasma membrane ATP microdomains in single islet ß-cells. Potential role for strategically located mitochondria. *Journal of Biological Chemistry* **274**: 13281-13291

Kramer, J.M. and Erickson, R.P. (1981). Developmental program of PGK-1 and PGK-2 isozymes in spermatogenic cells of the mouse: specific activities and rates of synthesis. *Developmental Biology* **87**: 37-45

Krisfalusi, M., Miki, K., Magyar, P.L. and O'Brien, D.A. (2006). Multiple glycolytic enzymes are tightly bound to the fibrous sheath of mouse spermatozoa. *Biology of Reproduction* **75**: 270-278

Li, S.S., O'Brien, D.A., Hou, E.W., Versola, J., Rockett, D.L. and Eddy, E.M. (1989). Differential activity and synthesis of lactate dehydrogenase isozymes A (muscle), B (heart), and C (testis) in mouse spermatogenic cells. *Biology of Reproduction* **40**: 173-180

Lindemann, C.B., Goltz, J.S. and Kanous, K.S. (1987). Regulation of activation state and flagellar wave form in epididymal rat sperm: evidence for the involvement of both Ca^{2+} and cAMP. *Cell Motility and the Cytoskeleton* **8**: 324-332

Manzano, A., Perez, J.X., Nadal, M., Estivill, X., Lange, A. and Bartrons, R. (1999). Cloning, expression and chromosomal localization of a human testis 6-phosphofructo-2-kinase/fructose-2,6-bisphosphatase gene. *Gene* **229**: 83-89

McCarrey, J.R. and Thomas, K. (1987). Human testis-specific PGK gene lacks introns and possesses characteristics of a processed gene. *Nature* **326**: 501-505

Miki, K. and Eddy, E.M. (1998). Identification of tethering domains for protein kinase A type Ia regulatory subunits on sperm fibrous sheath protein FSC1. *Journal of Biological Chemistry* **273**: 34384-34390

Miki, K. and Eddy, E.M. (1999). Single amino acids determine specificity of binding of protein kinase A regulatory subunits by protein kinase A anchoring proteins. *Journal of Biological Chemistry* **274**: 29057-29062

Miki, K., Willis, W.D., Brown, P.R., Goulding, E.H., Fulcher, K.D. and Eddy, E.M. (2002). Targeted disruption of the *Akap4* gene causes defects in sperm flagellum and motility. *Developmental Biology* **248**: 331-342

Miki, K., Qu, W., Goulding, E.H., Willis, W.D., Bunch, D.O., Strader, L.F., Perreault, S.D., Eddy, E.M. and O'Brien, D.A. (2004). Glyceraldehyde 3-phosphate dehydrogenase-S, a sperm-specific glycolytic enzyme, is required for sperm motility and male fertility. *Proceedings of the National Academy of Sciences USA* **101**: 16501-16506

Mita, M. and Hall, P.F. (1982). Metabolism of round spermatids from rats: lactate as the preferred substrate. *Biology of Reproduction* **26**: 445-455

Mori, C., Nakamura, N., Welch, J.E., Gotoh, H., Goulding, E.H., Fujioka, M. and Eddy, E.M. (1998). Mouse spermatogenic cell-specific type 1 hexokinase (mHk1-s) transcripts are expressed by alternative splicing from the mHk1 gene and the HK1-S protein is localized mainly in the sperm tail. *Molecular Reproduction and Development* **49**: 374-385

Mori, C., Welch, J.E., Fulcher, K.D., O'Brien, D.A. and Eddy, E.M. (1993). Unique hexokinase messenger ribonucleic acids lacking the porin-binding domain are developmentally expressed in mouse spermatogenic cells. *Biology of Reproduction* **49**: 191-203.

Morton, B. (1973). Reactivation of progressive motility in hamster sperm modified by Triton X-100. *Experimental Cell Research* **79**: 106-110

Mukai, C. and Okuno, M. (2004). Glycolysis plays a

major role for adenosine triphosphate supplementation in mouse sperm flagellar movement. *Biology of Reproduction* **71**: 540-754

Nakamura, M., Fujiwara, A., Yasumasu, I., Okinaga, S. and Arai, K. (1982). Regulation of glucose metabolism by adenine nucleotides in round spermatids from rat testes. *Journal of Biological Chemistry* **257**: 13945-13950

Nakamura, M., Okinaga, S. and Arai, K. (1984). Metabolism of round spermatids: evidence that lactate is preferred substrate. *American Journal of Physiology* **247**: E234-242

Nakamura, M., Okinaga, S. and Arai, K. (1986). Studies of metabolism of round spermatids: glucose as unfavorable substrate. *Biology of Reproduction* **35**: 927-935

Narisawa, S., Hecht, N.B., Goldberg, E., Boatright, K.M., Reed, J.C. and Millan, J.L. (2002). Testis-specific cytochrome c-null mice produce functional sperm but undergo early testicular atrophy. *Molecular and Cellular Biology* **22**: 5554-5562

Niwa, K. and Iritani, A. (1978). Effect of various hexoses on sperm capacitation and penetration of rat eggs in vitro. *Journal of Reproduction and Fertility* **53**: 267-271

Pande, J.K., Chowdhury, S.R., DasGupta, P.R., Chowdhury, A.R. and Kar, A.B. (1966). Biochemical composition of the rat testis fluid. *Proceedings of the Society for Experimental Biology and Medicine* **121**: 899-902

Penhoet, E.E., Kochman, M. and Rutter, W.J. (1969). Ioslation of fructose diphosphate aldolases A, B, and C. *Biochemistry* **8**: 4391-4395

Robinson, R. and Fritz, I.B. (1981). Metabolism of glucose by Sertoli cells in culture. *Biology of Reproduction* **24**: 1032-1041

Rogers, B.J., Ueno, M. and Yanagimachi, R. (1977). Inhibition of hamster sperm acrosome reaction and fertilization by oligomycin, antimycin A, and rotenone. *Journal of Experimental Zoology* **199**: 129-136

Russell, D.L. and Kim, K.H. (1996). Expression of triosephosphate isomerase transcripts in rat testis: evidence for retinol regulation and a novel germ cell transcript. *Biology of Reproduction* **55**: 11-18

Sakakibara, R., Okudaira, T., Fujiwara, K., Kato, M., Hirata, T., Yamanaka, S., Naito, M. and Fukasawa, M. (1999). Tissue distribution of placenta-type 6-phosphofructo- 2-kinase/fructose-2,6-bisphosphatase. *Biochemical and Biophysical Research Communications* **257**: 177-181

Sakata, J., Abe, Y. and Uyeda, K. (1991). Molecular cloning of the DNA and expression and characterization of rat testes fructose-6-phosphate,2-kinase:fructose-2,6-bisphosphatase. *Journal of Biological Chemistry* **266**: 15764-15770

Travis, A.J., Foster, J.A., Rosenbaum, N.A., Visconti, P.E., Gerton, G.L., Kopf, G.S. and Moss, S.B. (1998). Targeting of a germ cell-specific type 1 hexokinase lacking a porin-binding domain to the mitochondria as well as to the head and fibrous sheath of murine spermatozoa. *Molecular Biology of the Cell* **9**: 263-276

Travis, A.J., Jorgez, C.J., Merdiushev, T., Jones, B.H., Dess, D.M., Diaz-Cueto, L., Storey, B.T., Kopf, G.S. and Moss, S.B. (2001). Functional relationships between capacitation-dependent cell signaling and compartmentalized metabolic pathways in murine spermatozoa. *Journal of Biological Chemistry* **276**: 7630-7636

Urner, F., Leppens-Luisier, G. and Sakkas, D. (2001). Protein tyrosine phosphorylation in sperm during gamete interaction in the mouse: the influence of glucose. *Biology of Reproduction* **64**: 1350-1357

Urner, F. and Sakkas, D. (1996a). Glucose is not essential for the occurrence of sperm binding and zona pellucida-induced acrosome reaction in the mouse. *International Journal of Andrology* **19**: 91-96

Urner, F. and Sakkas, D. (1996b). Glucose participates in sperm-oocyte fusion in the mouse. *Biology of Reproduction* **55**: 917-922

Urner, F. and Sakkas, D. (1999). A possible role for the pentose phosphate pathway of spermatozoa in gamete fusion in the mouse. *Biology of Reproduction* **60**: 733-739

Urner, F. and Sakkas, D. (2005). Involvement of the pentose phosphate pathway and redox regulation in fertilization in the mouse. *Molecular Reproduction and Development* **70**: 494-503

Vijayaraghavan, S., Mohan, J., Gray, H., Khatra, B. and Carr, D.W. (2000). A role for phosphorylation of glycogen synthase kinase-3α in bovine sperm motility regulation. *Biology of Reproduction* **62**: 1647-1654

Visconti, P.E., Bailey, J.L., Moore, G.D., Pan, D., Olds-Clarke, P. and Kopf, G.S. (1995). Capacitation of mouse spermatozoa. I. Correlation between the capacitation state and protein tyrosine phosphorylation. *Development* **121**: 1129-1137

Visconti, P.E., Johnson, L.R., Oyaski, M., Fornes, M., Moss, S.B., Gerton, G.L. and Kopf, G.S. (1997). Regulation, localization, and anchoring of protein kinase A subunits during mouse sperm capacitation. *Developmental Biology* **192**: 351-363

Visconti, P.E., Olds-Clarke, P., Moss, S.B., Kalab, P., Travis, A.J., de las Heras, M. and Kopf, G.S. (1996). Properties and localization of a tyrosine phosphorylated form of hexokinase in mouse sperm. *Molecular Reproduction and Development* **43**: 82-93

Wang, R.H., Tao, L., Trumbore, M.W. and Berger, S.L. (1997). Turnover of the acyl phosphates of human and murine prothymosin a in vivo. *Journal of Biological Chemistry* **272**: 26405-26412

Welch, J.E., Schatte, E.C., O'Brien, D.A. and Eddy, E.M. (1992). Expression of a glyceraldehyde 3-phosphate dehydrogenase gene specific to mouse spermatogenic cells. *Biology of Reproduction* **46**: 869-878

Westhoff, D. and Kamp, G. (1997). Glyceraldehyde 3-phosphate dehydrogenase is bound to the fibrous sheath of mammalian spermatozoa. *Journal of Cell Science* **110**: 1821-1829

Wilkinson, J.H. and Withycombe, W.A. (1965). Organ specificity and lactate-dehydrogenase activity: Some properties of human spermatozoal lactate dehydrogenase. *Biochemical Journal* **97**: 663-668

White, D.R. and Aitken, R.J. (1989). Relationship between calcium, cyclic AMP, ATP, and intracellular pH and the capacity of hamster spermatozoa to express hyperactivated motility. *Gamete Research* **22**: 163-177

Williams, A.C. and Ford, W.C. (2001). The role of glucose in supporting motility and capacitation in human spermatozoa. *Journal of Andrology* **22**: 680-695

Williams, A.C. and Ford, W.C. (2004). Functional significance of the pentose phosphate pathway and glutathione reductase in the antioxidant defenses of human sperm. *Biology of Reproduction* **71**: 1309-1316

Yanagimachi, R. (1994). Mammalian fertilization. In: *The Physiology of Reproduction*. 2nd edition. Edited by Knobil E. and Neill J.D, Raven Press, New York, pp. 189-317

Zhang, J., Yu, L., Fu, Q., Gao, J., Xie, Y., Chen, J., Zhang, P., Liu, Q. and Zhao, S. (2001). Mouse phosphoglycerate mutase M and B isozymes: cDNA cloning, enzyme activity assay and mapping. *Gene* **264**: 273-279

Zhang, L. and Sanderson, M.J. (2003). Oscillations in ciliary beat frequency and intracellular calcium concentration in rabbit tracheal epithelial cells induced by ATP. *Journal of Physiology* **546**: 733-749

Zinkham, W.H., Blanco, A. and Kupchyk, L. (1963). Lactate dehydrogenase in testis: Dissociation and recombination of subunits. *Science* **142**: 1303-1304

Spermatology. SRF Vol. 65. ERS Roldan and M Gomendio (eds) Nottingham University Press, Nottingham

Digital image analysis of flagellar beating and microtubule sliding of activated and hyperactivated sperm flagella

Sumio Ishijima

Department of Bioengineering, Graduate School of Bioscience and Biotechnology, Tokyo Institute of Technology, 2-12-1 W3-41, O-okayama, Meguro-ku, Tokyo 152-8551, Japan

Flagellar beatings of *Suncus*, golden hamster, and monkey spermatozoa before and after hyperactivation were analysed using high-speed video microscopy and digital image processing in order to examine the sliding mechanism of the flagellar beating and the function of accessory fibres of the mammalian spermatozoa. Although these spermatozoa have different morphology and movement characteristics, the flagellar beatings of hyperactivated spermatozoa had a few common features; i.e., sharp bends at the base of the flagellum and a low beat frequency. While nonhyperactivated (activated) spermatozoa exhibited nearly constant-curvature beating, the hyperactivated spermatozoa displayed a constant-frequency beating. A detailed analysis of the microtubule sliding of the activated and hyperactivated sperm flagella revealed that the sharp bends at the base of the flagella were induced by an increase in the total length of the microtubule sliding at the base of the flagella and that the sliding velocity of the activated and hyperactivated sperm flagella was consistent within each species. A comparison of the sliding velocity of the flagellar beating of *Suncus*, golden hamster, and monkey spermatozoa with the moment of inertia of the cross section of the flagellar base suggests that the sliding velocity is involved in the hardness of a sperm flagellum.

Introduction

Mammalian spermatozoa are a suitable material for studying the regulatory mechanism of flagellar beating because they alter their beating in both the male and female reproductive tracts (Ishijima and Mohri, 1985, 1990; Yanagimachi, 1994). The most dramatic change in the motility of mammalian spermatozoa is observed during the capacitation and is called hyperactivation (Yanagimachi, 1994). The characteristic beating of the hyperactivated spermatozoa is effective for sperm transport through the female reproductive tract (Suarez, 1987; DeMott and Suarez, 1992) and useful for sperm penetration through the zona pellucida (Stauss *et al.*, 1995). Various kinds of experiments show that this change in sperm motility is brought about by the increase in the intracellular concentrations of the calcium ion and cAMP (Yanagimachi, 1994). Using spermatozoa deprived of its plasma membrane using a non-ionic detergent and then reactivated with ATP, the factor inducing the sperm motility similar to

Corresponding author E-mail: sishijim@bio.titech.ac.jp

hyperactivation was identified; namely, the asymmetrical flagellar beating observed frequently in the early stage of hyperactivation was caused by calcium ion, whereas the later figure-of-eight motility of the hyperactivation resulted from low levels of calcium ion and high levels of cAMP (Ishijima *et al.*, 2006). Hyperactivated flagellar beating varies to a large extent among species (Yanagimachi, 1994), perhaps mainly due to the mechanical property of the sperm flagella. Therefore, it is important for understanding the regulatory mechanism of hyperactivation to examine the flagellar beating of different types of spermatozoa. The flagellar beating of hyperactivated spermatozoa of *Suncus*, golden hamster, and monkey was quantitatively analysed using high-speed video microscopy and digital image processing, and was compared with that of nonhyperactivated (activated) spermatozoa to clarify the change in the flagellar beating between the activated and hyperactivated spermatozoa and the sliding mechanism of the microtubule in these sperm flagella.

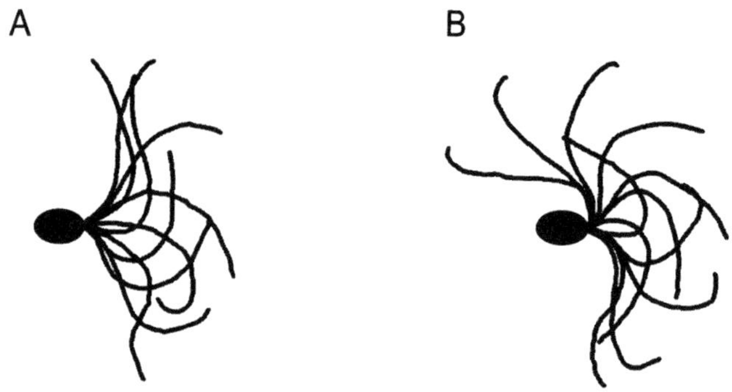

Figure 1. Change in flagellar beatings of *Suncus* spermatozoa before and after hyperactivation. Nonhyperactivated (activated) spermatozoa (**A**) and hyperactivated spermatozoa (**B**). Time interval between successive images is 30 ms in (**A**) and 70 ms in (**B**), during which period the sperm flagellum beats one complete cycle. The spermatozoa are firmly attached to the slide surface by their heads, so that the head of each tracing is exactly aligned.

Change in flagellar beating of spermatozoa before and after hyperactivation

Tracings of the flagellar beatings of activated and hyperactivated *Suncus* spermatozoa are shown in Figure 1. The most characteristic change in the flagellar beating between the activated and hyperactivated spermatozoa was the sharp bends at the base of the flagellum and a low beat frequency. On the other hand, flagellar bends in other regions of the flagellum appear to have a limited involvement in hyperactivation. These fundamental features of the flagellar beating in the hyperactivated spermatozoa were obtained from the monkey, golden hamster and *Suncus* spermatozoa, although these spermatozoa have a different morphology and different movement characteristics (Ishijima *et al.*, 2002, 2006; Ohmuro and Ishijima, 2006; Kaneko *et al.*, 2006). To characterise the flagellar beating in detail, flagellar bends were described in terms of curvature using digital image processing (Ishijima *et al.*, 2002). Examination of the relationship between the curvature and beat frequency of the activated and hyperactivated sperm flagella revealed that the activated flagellar beating was carried out at an almost constant curvature, whereas the hyperactivated flagellar beating exhibited beating with a constant frequency beat, suggesting that the hyperactivation is a mode conversion from constant-curvature beating to constant-frequency beating (Ishijima *et al.*, 2006; Ohmuro and Ishijima, 2006; Kaneko *et al.*, 2006).

Microtubule sliding of sperm flagella before and after hyperactivation

It is now well established that the basis for flagellar beating is an ATP-dependent sliding between the adjacent outer doublet microtubule (Gibbons, 1981). Thus, it is essential to understand the sliding-filament mechanism underlying the specific beating of the sperm flagella. To this end, the shear angle (Fig. 2) was measured since it is known that the degree and the rate of the shear angle are proportional to the amount of microtubule sliding and sliding velocity, respectively (Gibbons, 1982).

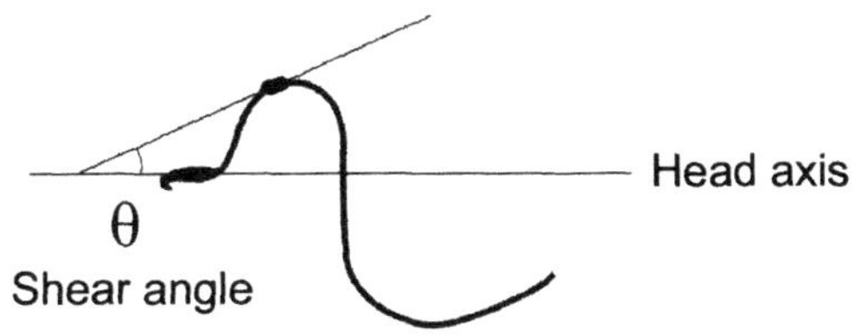

Figure 2. Shear angle is the angle of the tangent to the flagellar shaft with reference to the sperm head axis. The shear angle is proportional to the amount of microtubule sliding, the derivative of the shear angle with respect to the length is curvature, and the derivative of the shear angle with time is proportional to the sliding velocity.

A comparison of the shear angle of the flagellar beating of hyperactivated spermatozoa with that of activated spermatozoa revealed that the sharp bends at the base of the flagella of the hyperactivated spermatozoa were induced by the increase in the total length of the microtubule sliding at the flagellar base. Furthermore, the shear angle was inversely proportional to the beat frequency. This result suggests that the shear angle and beat frequency, in other words, the waveform and frequency of the flagellar beating, are dependently regulated, although they behave almost independently over a fairly wide range. This feature of flagellar beating of mammalian spermatozoa seems to be different from that of echinoderm spermatozoa because the flagellar beating of sea urchin spermatozoa may be regulated by two independent mechanisms (Gibbons, 1975). Since the product of the shear angle and beat frequency are proportional to the sliding velocity of the microtubule, the sliding velocity of the microtubule did not change before and after hyperactivation within each species, although the values of the sliding velocity were entirely different among monkey, golden hamster, and *Suncus* spermatozoa. To examine the effect of load on the sliding velocity, the relationship between the sliding velocity and the moment of inertia was investigated. The moment of inertia, which is proportional to the stiffness of the sperm flagella if Young's modulus of the axoneme and the accessory fibres is not significantly different in different species, was calculated using images of the cross-section of sperm flagella at the base and image analysis software. The relationship between them was described by an exponential equation, similar to the classical load-velocity relationship of muscle contraction (Fenn and Marsh, 1935; Hill, 1938). Data from sea urchin spermatozoa also lay on the line extrapolated from the mammalian spermatozoa data, suggesting that the same microtubule sliding system functions in both the mammalian and echinoderm spermatozoa.

References

DeMott, R.P. and Suarez, S.S. (1992). Hyperactivated sperm progress in the mouse oviduct. *Biology of Reproduction* **46:** 779-785

Fenn, W.O. and Marsh, B.S. (1935). Muscular force at different speeds of shortening. *Journal of Physiology (London)* **85:** 277-297

Gibbons, I.R. (1975). Mechanisms of flagellar motility. In: *Functional Anatomy of The Spermatozoon*. Edited by Afzelius, B.A., Pergamon Press, Oxford, pp. 127-140

Gibbons, I.R. (1981). Cilia and flagella of eukaryotes. *Journal of Cell Biology* **91:** 107s-124s

Gibbons, I.R. (1982). Sliding and bending in sea urchin sperm flagella. *Symposia of The Society for Experimental Biology* **35:** 225-287

Hill, A.V. (1938). The heat of shortening and the dynamic constants of muscle. *Proceedings of the Royal Society of London, series B* **126:** 136-195

Ishijima, S. and Mohri, H. (1985). A quantitative description of flagellar movement in golden hamster spermatozoa. *Journal of Experimental Biology* **114:** 463-475

Ishijima, S. and Mohri, H. (1990). Beating patterns of mammalian spermatozoa. In: *Controls of Sperm Motility: Biological and Clinical Aspects*. Edited by Gagnon, C., CRC Press, Boca Raton, pp. 29-42

Ishijima, S., Baba, S.A., Mohri, H. and Suarez, S.S. (2002). Quantitative analysis of flagellar movement in hyperactivated and acrosome-reacted golden hamster spermatozoa. *Molecular Reproduction and Development* **61:** 376-384

Ishijima, S., Mohri, H., Overstreet, J.W. and Yudin, A.I. (2006). Hyperactivation of monkey spermatozoa is triggered by Ca^{2+} and completed by cAMP. *Molecular Reproduction and Development* **73:** 1129-1139

Kaneko, T., Mori, T. and Ishijima, S. (2006). Digital image analysis of the flagellar beat of activated and hyperactivated Suncus spermatozoa. *Molecular Reproduction and Development* **74:** 478-485

Ohmuro, J. and Ishijima, S. (2006). Hyperactivation is the mode conversion from constant-curvature beating to constant-frequency beating under a constant rate of microtubule sliding. *Molecular Reproduction and Development* **73:** 1412-1421

Stauss, C.R., Votta, T.J. and Suarez, S.S. (1995). Sperm motility hyperactivation facilitates penetration of the hamster zona pellucida. *Biology of Reproduction* **52:** 1280-1285

Suarez, S.S. (1987). Sperm transport and motility in the mouse oviduct: observations in situ. *Biology of Reproduction* **36:** 203-210

Yanagimachi, R. (1994). Mammalian fertilization. In: *The Physiology of Reproduction* Volume 1, Second edition, Edited by Knobil, E. and Neill, J., Raven Press, New York, pp. 189-317

Spermatology. SRF Vol. 65. ERS Roldan and M Gomendio (eds) Nottingham University Press, Nottingham

Different regulatory systems operate in the midpiece and principal piece of the mammalian sperm flagellum

Susan S. Suarez, Becky Marquez, Tanya P. Harris and John C. Schimenti

Department of Biomedical Sciences, Cornell University, Ithaca, New York 14853, USA

In mammalian sperm, the flagellar midpiece and principal piece contain different signalling molecules and ion channels. For example, the soluble adenylyl cyclase (sAC), which is required for activation of motility, is restricted to the midpiece, while the plasma membrane calcium channels CatSper1 and CatSper2, which are required for hyperactivation of motility, are restricted to the principal piece. The midpiece and principal piece are partially separated by a barrier called the annulus, yet despite this and the differential distribution of signalling molecules, they normally appear to work together to produce a wave that propagates smoothly down the tail. Under certain circumstances, however, the two segments can act independently. This is the case for sperm of *Ste5Jcs1* mutant mice, which swim with stiff midpieces but vigorously-beating principal pieces, due apparently to a defect in the mitochondria.

In mammals, sperm motility is required for male fertility; therefore, an understanding of the signalling mechanisms that regulate flagellar movement is important. Regulation of flagellar movement includes the activation of motility when sperm leave the epididymis and the hyperactivation of motility in the oviduct. Activation is primarily stimulated by cAMP signalling, whereas hyperactivation is primarily stimulated by calcium signalling (Ho *et al.*, 2002).

Flagellar movement is dependent on the flagellar axoneme, especially its microtubule skeleton and dynein ATPase motor. The walking of the dynein arms along the microtubules, constrained by elastic skeletal elements, produces waves that propagate smoothly along the flagellum from the base to the tip. Surprisingly, the axoneme is subjected to different regulatory systems in the midpiece and principal piece of the flagellum. This is evidenced by the restriction of various ion channels and signalling molecules to one or the other segment of the flagellum. For example, the soluble adenylyl cyclase (sAC), which is required for activation of motility, is restricted to the midpiece (Hess *et al.*, 2005), while the plasma membrane calcium channels CatSper1 and CatSper2, which are required for hyperactivation of motility, are restricted to the principal piece (Carlson *et al.*, 2005). There is an anatomical barrier at the inner face of the plasma membrane at the junction of the midpiece and principal piece, visible by transmission electron microscopy as a ring of electron dense material (Fig. 1). This material, called the annulus, may form a protein fence between the plasma membrane domains of the midpiece and principal piece (Cesario and Bartles, 1994) and may also restrict diffusion of molecules in the near-membrane cytoplasm. The annulus, however, does not appear to seal off

Corresponding author E-mail: sss7@cornell.edu

the cytoplasm within and immediately surrounding the axoneme of the midpiece from that within the core of the principal piece.

In the midpiece of mammalian sperm, the axoneme is surrounded by a tight helix of mitochondria (Fig. 1) that creates a barrier between the plasma membrane and axoneme. Thus, ions entering through the plasma membrane are filtered by the mitochondria before reaching the axoneme.

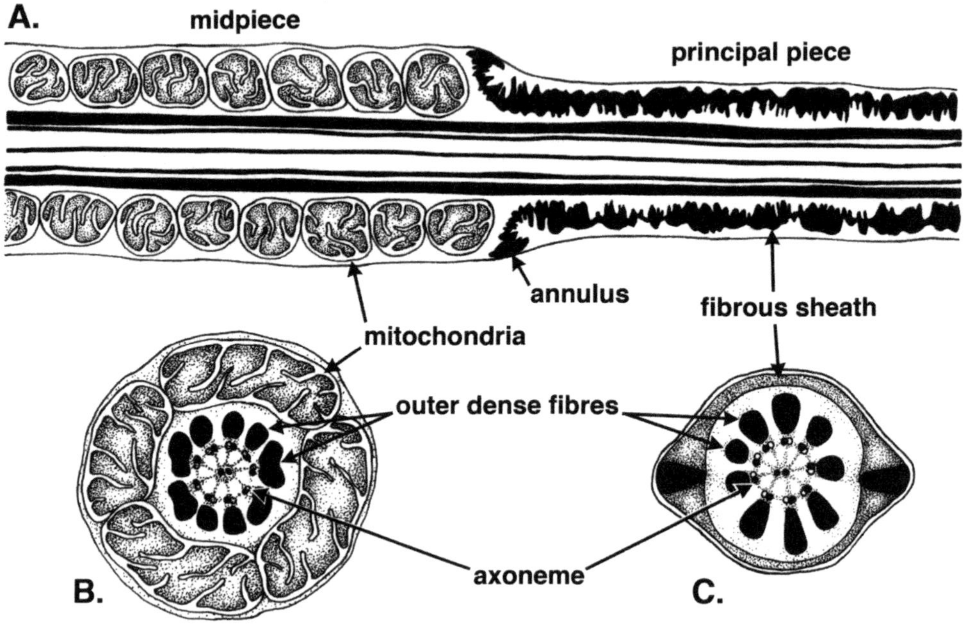

Figure 1. Diagram of the ultrastructure of the midpiece and principal piece of a mouse sperm flagellum. **(A)** Longitudinal section of part of the midpiece and principal piece. **(B)** Cross section of the midpiece. **(C)** Cross section of the principal piece. Drawn by C. Rose Gottlieb.

In the principal piece, the axoneme is surrounded by outer dense fibers and the fibrous sheath. These do not form a continuous barrier between the plasma membrane and axoneme as does the mitochondrial helix in the midpiece; therefore, it can be assumed that ions passing through the plasma membrane channels have some direct access to the axoneme. Thus, calcium ions (Ca^{2+}) entering through the CatSper channels can reach the axoneme directly and initiate hyperactivation. Furthermore, the plasma membrane Ca^{2+}-ATPase PMCA4, which is restricted to the principal piece in the flagellum (Okunade *et al.*, 2004), can regulate the levels of Ca^{2+} at the axoneme of the principal piece.

Glycolytic enzymes are present in the midpiece and are also anchored to the fibrous sheath in the principal piece to provide the flagellar axoneme with ATP (Krisfalusi *et al.*, 2006). In the midpiece, the mitochondria can provide ATP for flagellar movement via oxidative phosphorylation. Nevertheless, spermatozoa from some species continue to swim with normal progressive motility when oxidative phosphorylation is blocked by the uncoupler CCCP or the respiratory chain inhibitor antimycin A, as long as glycolytic substrates are available (Mukai and Okuno, 2004).

Because oxidative production of ATP for motility is not an essential function of the midpiece mitochondria, a more important role for them may be to regulate the ionic environment of the axoneme, particularly calcium ions and protons. In this respect it is curious that the midpiece plasma membrane lacks CatSper1 and CatSper2, as well as some other voltage-gated Ca^{2+} channels (Carlson *et al.*, 2005) and PMCA4 (Okunade *et al.*, 2004). On the other hand, there is evidence for the existence of an intracellular Ca^{2+} store in the redundant nuclear envelope in the base of the flagellum that could supplement the Ca^{2+} supply to the midpiece axoneme. The redundant nuclear envelope was so named because it represents the excess nuclear envelope left over during spermiogenesis when the nucleus condenses (Franklin, 1968). The evidence that these membranes contain a Ca^{2+} store includes the localization there of inositol trisphosphate receptors and the demonstration that pharmacological agents which release Ca^{2+} from stores produce a wave of Ca^{2+} initiating from the base of the flagellum (Ho and Suarez, 2003). The redundant nuclear envelope lies in close association with the base of the mitochondrial helix. Because mitochondria have been demonstrated to take up Ca^{2+} released from reticular stores (Bathori *et al.*, 2006), the sperm mitochondria may act to distribute Ca^{2+} ions down the midpiece. There are also alpha subunits of cyclic-nucleotide-gated plasma membrane Ca^{2+} channel proteins in the midpiece, although the principal piece contains both alpha and beta subunits of the channels (Wiesner *et al.*, 1998).

Despite the differences between Ca^{2+} regulation systems in the midpiece and principal piece, there are signs that the two regions interact in this regard. For example, CatSper1 and CatSper2 null mutant mouse sperm, both of which lack functional CatSper plasma membrane Ca^{2+} channels in the principal piece, show no signs of hyperactivation in either the principal piece or the midpiece. Since there are no CatSper1 or CatSper2 channels in the midpiece, one might have expected to see the start of a hyperactivated flagellar bend in the midpiece that dampened in the principal piece in the null mutants; however, this is not the case. Similarly, mutant mouse sperm that lack PMCA4 in the principal piece cannot hyperactivate at all; in fact, they become completely immotile when incubated *in vitro* under capacitating conditions that should lead to hyperactivation (Okunade *et al.*, 2004).

On the other hand, there is evidence that the midpiece and principal piece can be functionally isolated. Mouse sperm carrying a mutation (*Ste5Jcs1*) have reduced fertility and abnormal motility. The flagella of these mutants beat vigorously in the principal piece but are rigid in the midpiece. Transmission electron micrographs of the midpieces of mutant sperm reveal normal axonemes, dense outer fibers, and fibrous sheaths, but apparently abnormal electron-lucent areas in the mitochondrial matrix. Interestingly, the electron micrographs of the mitochondria resemble those of the PMCA4 null mutant sperm that had been incubated under capacitating conditions (Okunade *et al.*, 2004). The cause of this abnormality in the *Ste5Jcs1* mutants is as yet unknown; however, the behaviour of the mutant sperm demonstrates that regulation of motility can be compartmentalised to the midpiece and that the flagellum can continue to beat in the principal piece when it is paralysed in the midpiece.

In conclusion, in mammalian sperm, the flagellar midpiece and principal piece contain different signalling molecules and ion channels and are partially separated by a barrier, yet they normally appear to work together to produce a wave that propagates smoothly down the tail. Under certain circumstances, however, the two segments can act independently.

Acknowledgements

Supported by NSF MCB0421855 (SSS), NIH 5 F31 HD43693 (BM) and NIH HD35984 (JS).

References

Bathori, G., Csordas, G., Garcia-Perez, C., Davies, E. and Hajnoczky, G. (2006). Ca^{2+}-dependent control of the permeability properties of the mitochondrial outer membrane and VDAC. *Journal of Biological Chemistry* **281:** 17347-17358

Carlson, A.E., Quill, T.A., Westenbroek, R.E., Schuh, S.M., Hille, B. and Babcock, D.F. (2005). Identical phenotypes of CatSper1 and CatSper2 null sperm. *Journal of Biological Chemistry* **280:** 32238-32244

Cesario, M.M. and Bartles, J.R. (1994). Compartmentalization, processing and redistribution of the plasma membrane protein CE9 on rodent spermatozoa. Relationship of the annulus to domain boundaries in the plasma membrane of the tail. *Journal of Cell Science* **107:** 561-570

Franklin, L.E. (1968) Formation of the redundant nuclear envelope in monkey spermatids. *Anatomical Record* **161:** 149-161

Hess, K.C., Jones, B.H., Marquez, B., Chen, Y., Ord, T.S., Kamenetsky, M., Miyamoto, C., Zippin, J.H., Kopf, G.S., Suarez, S.S., Levin, L.R., Williams, C.J., Buck, J. and Moss, S.B. (2005). The "soluble" adenylyl cyclase in sperm mediates multiple signaling events required for fertilization. *Developmental Cell* **9:** 249-259

Ho, H.C. and Suarez, S.S. (2003). Characterization of the intracellular calcium store at the base of the sperm flagellum that regulates hyperactivated motility. *Biology of Reproduction* **68:** 1590-1596

Ho, H.C., Granish, K.A. and Suarez, S.S. (2002). Hyperactivated motility of bull sperm is triggered at the axoneme by Ca^{2+} and not cAMP. *Developmental Biology* **250:** 208-217

Krisfalusi, M., Miki, K., Magyar, P.L. and O'Brien, D.A. (2006). Multiple glycolytic enzymes are tightly bound to the fibrous sheath of mouse spermatozoa. *Biology of Reproduction* **75:** 270-278

Mukai, C. and Okuno, M. (2004). Glycolysis plays a major role for adenosine triphosphate supplementation in mouse sperm flagellar movement. *Biology of Reproduction* **71:** 540-547

Okunade, G.W., Miller, M.L., Pyne, G.J., Sutliff, R.L., O'Connor, K.T., Neumann, J.C., Andringa, A., Miller, D.A., Prasad, V., Doetschman, T., Paul, R.J. and Shull, G.E. (2004). Targeted ablation of plasma membrane Ca^{2+}-ATPase (PMCA) 1 and 4 indicates a major housekeeping function for PMCA1 and a critical role in hyperactivated sperm motility and male fertility for PMCA4. *Journal of Biological Chemistry* **279:** 33742-33750

Wiesner, B., Weiner, J., Middendorff, R., Hagen, V., Kaupp, U.B. and Weyand, I. (1998). Cyclic nucleotide-gated channels on the flagellum control Ca^{2+} entry into sperm. *Journal of Cell Biology* **142:** 473-484

Spermatology. SRF Vol. 65. ERS Roldan and M Gomendio (eds) Nottingham University Press, Nottingham

Sperm-egg binding requires a multiplicity of receptor-ligand interactions: new insights into the nature of gamete receptors derived from reproductive tract secretions

Robert Lyng and Barry D. Shur

Department of Cell Biology, Emory University School of Medicine, Atlanta, GA 30322, USA

Although the broad concepts of fertilisation are well defined, our understanding of the biochemical mechanisms underlying sperm-egg binding is limited. Early studies of fertilisation in the mouse implicated ß-1,4-galactosyltransferase-1 (GalT) as a sperm receptor for oligosaccharide ligands on the zona pellucida glycoprotein, ZP3. Binding of multiple ZP3 oligosaccharides induces GalT aggregation, leading to G-protein activation and initiation of the acrosome reaction. Consistent with this, GalT-null sperm show greatly reduced binding to ZP3 and are unable to undergo a zona-induced acrosome reaction. However, unexpectedly, GalT-null sperm are still able to bind to the intact zona pellucida. These and other studies indicate that sperm-egg binding requires at least two distinct molecular events: a GalT/ZP3-independent interaction that mediates initial gamete adhesion, followed by ZP3-dependent aggregation of GalT leading to acrosomal exocytosis. In this regard, recent studies have identified novel components of both sperm and eggs that are derived from reproductive tract secretions and which appear to participate in initial sperm-egg adhesion. SED1, also known as MFG-E8 and lactadherin, is secreted by the initial segment of the epididymis where it intercalates into the sperm plasma membrane and plays a critical role during sperm-zona binding. Similarly, an oviduct-derived glycoprotein has been identified that coats the zona pellucida of ovulated eggs and which has sperm-binding activity. The realization that successful sperm-egg binding results from the sequential action of molecularly distinct events may account for the large number of components that have been implicated in sperm-egg interactions, as well as why it has been difficult to reach a consensus regarding the underlying molecular mechanisms.

Introduction

Fertilisation is a highly dynamic and complex process that is dependent upon the coordination of complex cellular and molecular mechanisms. The cellular events leading to fertilisation are fairly well understood; however, the underlying molecular mechanisms are still a subject of

Corresponding author E-mail: barry@cellbio.emory.edu

intense investigation and debate. One factor that contributes to our current limited understanding of fertilisation is the realization that sperm-egg binding is dependent upon a multiplicity of receptors and their cognate ligands, rather than resulting from a single "lock and key" type of interaction. Some of the components now known to participate in sperm-egg binding are synthesised by the developing gametes, whereas others are added post-meiotically to sperm and eggs via secretions from the male and female reproductive tracts. In this article, we discuss some of the components that have been proposed to facilitate sperm binding to the zona pellucida, using the mouse as the principle model system, as well as the difficulties intrinsic to deciphering the molecular basis of this complex cellular interaction.

An early model for sperm-egg binding in mouse

The zona pellucida glycoprotein ZP3 behaves as a sperm-binding ligand and induces acrosomal exocytosis

Pioneering studies by Wassarman and colleagues showed that the murine zona pellucida (ZP) is secreted by growing oocytes and forms an extracellular matrix composed of three glycoproteins, referred to as ZP1, ZP2 and ZP3. Competitive binding assays using individually purified ZP glycoproteins demonstrated that ZP3 isolated from unfertilised eggs is able to competitively inhibit sperm binding to the ZP, whereas ZP1 and ZP2 cannot. Furthermore, ZP3 isolated from fertilised eggs, which no longer support sperm binding, fails to competitively inhibit sperm-egg binding (Bleil and Wassarman, 1980a,b).

The sperm-binding activity of ZP3 appears to reside in its oligosaccharide chains rather than the polypeptide backbone (Florman and Wassarman, 1985). Consistent with this, alkali release of *O*-linked chains destroys ZP3 binding activity and the released *O*-linked oligosaccharides bind to sperm and competitively inhibit sperm-zona binding. In contrast, removal of *N*-linked oligosaccharides by endo-H glycosidase digestion does not affect ZP3 sperm-binding activity. Although these studies suggest that *O*-linked oligosaccharides on ZP3 act as one of the primary binding epitopes for sperm, there remains contention surrounding the molecular composition of the sperm-binding oligosaccharides (discussed below).

Binding of ZP3 activates a range of intracellular signal cascades within sperm that culminate in fusion of the plasma membrane and underlying outer acrosomal membrane, i.e., the acrosome reaction (Bleil and Wassarman, 1983). The best studied among these are (i) a pertussis toxin (PTx)-sensitive heterotrimeric G-protein (Ward and Kopf, 1993), and (ii) a voltage-independent cation channel that regulates membrane potential, which in turn regulates a voltage-dependent cation channel (Florman, Arnoult, Kazam, Li and O'Toole, 1999). These cascades, among others, result in elevated intracellular calcium and pH that are required for acrosomal exocytosis. Zona binding also activates a sperm glycine receptor/chloride channel that may contribute to the control of voltage-sensitive calcium influx (Meizel, 1997).

The nature of the sperm receptor for ZP3

A wide variety of sperm surface components have been implicated in binding to the ZP. They have been identified by a range of approaches including (i) the biochemical analysis of sperm mutations thought to impact fertilising ability (GalT) (Shur and Bennett, 1979), (ii) development of inhibitory monoclonal antibodies (PH-20) (Primakoff, Hyatt and Myles, 1985), (iii) identification of sperm proteins that bind ZP affinity columns (zonadhesin, p47) (Hardy and Garbers, 1995; Ensslin, Vogel, Calvete, Thole, Schmidtke and Töpfer-Petersen, 1998), (iv) or that can be photoaffinity cross-linked to the ZP (sp56) (Bleil and Wassarman, 1990), (v) use of radiolabelled zona glycoproteins to overlay blots of sperm lysates (p95) (Leyton and Saling, 1989), (vi) by competitive glycoside substrates (mannosidase) (Cornwall, Tulsiani and Orgebin-Crist, 1991), (vii) by an indirect implication that they require the chaperone, calmegin, for proper folding and expression (Ikawa, Wada, Kominami, Watanabe, Toshimori, Nishimune and Okabe, 1997), among others approaches [galactosyl receptor (Rivkin, Tres, Kaplan-Kraicer, Shalgi, and Kierszenbaum, 2000); SGG (White, Weerachatyanukul, Gadella, Kamolvarin, Attar, and Tanphaichitr, 2000)].

Some of these sperm proteins have specifically been implicated as a/the primary receptor for ZP3, namely GalT, sp56 and p95. However, sp56 and p95 and are no longer thought to mediate initial sperm binding to ZP3 (Bork, 1996; Kim, Cha and Gerton, 2001), although sp56 (among others) remains an attractive candidate for mediating sperm-egg binding during and/or after the acrosome reaction. Thus far, among the sperm surface proteins thought to bind the egg coat, and ZP3 oligosaccharides in particular, sperm surface GalT satisfies many of the criteria expected of a ZP3 receptor. However, more recent observations discussed below indicate that GalT participates in secondary aspects of sperm-egg interactions and is not involved in initial gamete adhesion.

Some glycosyltransferases function as cell surface receptors for glycoside substrates

Glycosyltransferases are generally considered to be intracellular biosynthetic enzymes that catalyze the synthesis of complex oligosaccharides by the sequential transfer of sugar residues from nucleotide donors to growing polysaccharide chains (Joziasse, 1992). However, it is now known that GalT, as well as small number of other glycosyltransferases, are localised on the plasma membrane of various cells where they function in cell-cell or cell-matrix interactions (Shur, Evans and Lu, 1998). Since the sugar donor substrates required for catalysis are not found in extracellular fluids, it is hypothesised that the glycosyltransferase extracellular domain is able to form a stable complex upon binding to its glycoside substrate on adjacent cells or within the extracellular matrix.

The basis whereby glycosyltransferases may function as surface receptors for extracellular glycosides is best understood for GalT. Mammalian tissues synthesise two different GalT isoforms that have identical catalytic and transmembrane domains, but differ in the length of their cytoplasmic domains. The shorter protein has a cytoplasmic tail of 11 amino acids (aa), whereas the longer species contains an additional 13 aa sequence to produce a 24 aa cytoplasmic tail (Russo, Shaper and Shaper, 1990). The functions of the two isoforms remained an area of active debate until the creation of mice deficient for either one or both GalT isoforms (Rodeheffer and Shur, 2002). Collectively, the data indicate that the short GalT isoform is normally confined to the Golgi complex, where it serves a purely biosynthetic function; this is the isoform that shows extreme upregulation during lactation to accommodate lactose synthesis. In contrast, the long GalT isoform can function both biosynthetically in the Golgi, and due to its additional 13 aa cytoplasmic sequence can also function as a signal transducing receptor on the cell surface (Shur *et al.*, 1998); the extreme example being sperm that only produce the long isoform, all of which is localised on the dorsal, anterior aspect of the sperm head (Shur and Neely, 1988). Recent studies show that the ability of the long isoform to function as a cell surface receptor, in both somatic cells and sperm, is regulated, at least partly, by the phosphorylation status of a serine residue unique to the long cytoplasmic domain (Hathaway, Evans, Dubois, Foote, Elder and Shur, 2003).

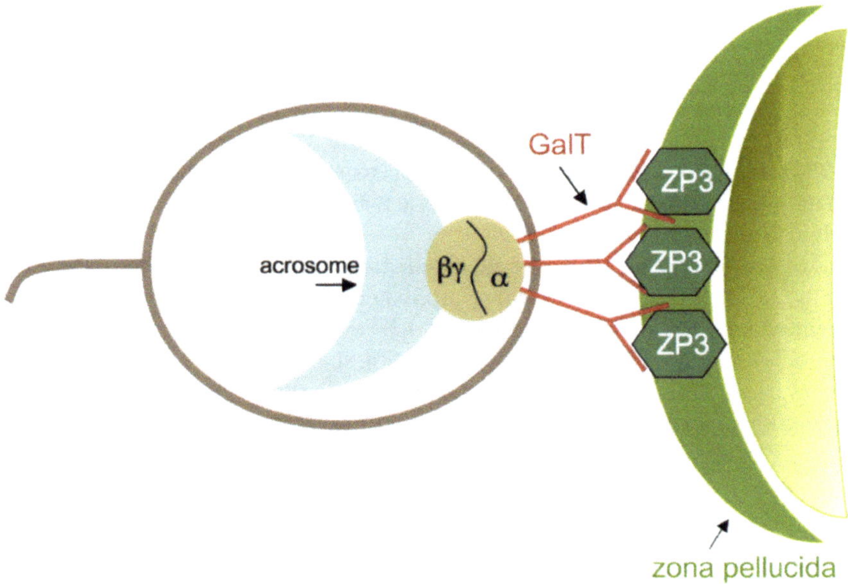

Figure 1. Early studies suggested that mouse sperm binding to the zona pellucida is mediated by a sperm surface ß1,4-galactosyltransferase (GalT) that selectively binds to a subset of oligosaccharides on the ZP3 glycoprotein. Aggregation of GalT by multiple ZP3 oligosaccharides, leads, directly or indirectly, to activation of a pertussis toxin-sensitive heterotrimeric G-protein that eventually culminates in acrosomal exocytosis.

Sperm surface GalT functions as a receptor for sperm-binding oligosaccharides on ZP3

A large number of observations are consistent with a role for sperm surface GalT in mediating adhesion to the ZP by selectively binding to specific oligosaccharides on ZP3 (Fig. 1). GalT is produced by pachytene spermatocytes and is eventually redistributed to the dorsal, anterior aspect of the sperm plasma membrane, where it behaves as a traditional integral membrane protein (Shur and Neely, 1988). In the epididymis, GalT is masked by epididymal glycosides possessing a polylactosamine core structure, which are subsequently shed from the sperm surface during capacitation (Shur and Hall, 1982a). The exposed GalT is now able to interact with its *N*-acetylglucosamine (GlcNAc)-terminating ligand on ZP3 (Shur and Hall, 1982b; Miller, Macek and Shur, 1992). In this regard, the sperm-binding activity of purified ZP3 can be abolished by either masking its GalT-binding site (by pregalactosylation) or by removing it (with GlcNAc'dase digestion). Binding of ZP3 oligosaccharides leads to GalT aggregation and subsequent activation of a heterotrimeric G-protein cascade, contributing to the acrosome reaction (Gong, Dubois, Miller and Shur, 1995). The acrosome releases hydrolytic enzymes, including *N*-acetylglucosaminidase (GlcNAc'dase), that facilitate sperm penetration through the zona (Miller, Gong and Shur, 1993). After egg activation, cortical granules release very high levels of GlcNAc'dase (~ 15-times higher than other glycosidases), which destroys the GalT recognition site on ZP3, thus contributing to the block to polyspermic binding (Miller, Gong, Decker and Shur, 1993). The participation of GalT during mouse sperm-egg binding has been confirmed by others (Benau, McGuire and Storey, 1990).

Altering the expression of sperm GalT impacts the efficacy of sperm-egg binding

The use of molecular genetics has enabled a more rigorous test of GalT function during fertility, ZP3 binding, G-protein activation, and acrosomal exocytosis. As briefly reviewed below, not only do these studies confirm the role of GalT as a receptor for ZP3 and as an inducer of the acrosome reaction, but they also lead to the realization that initial gamete adhesion relies upon GalT/ZP3-independent mechanisms.

Ectopic expression of GalT leads to ZP3-dependent binding and G-protein activation. Injection of constructs encoding the long isoform of murine GalT into *Xenopus* oocytes leads to GalT expression on the oocyte plasma membrane, whereas expression of the short isoform is confined to intracellular compartments. Oocytes that express surface-localised GalT bind ZP3, but do not bind other ZP glycoproteins. After oocyte maturation, ZP3 or anti-GalT IgG induces G-protein activation and triggers cortical granule exocytosis and egg activation. Furthermore, mutagenesis of a putative G-protein activation motif in the long GalT cytoplasmic domain, and present in both heptahelical and single-pass receptors (Wade, Lim, Lan, Chung, Nanamori and Neubig, 1999), eliminates G-protein activation in response to ZP3 or anti-GalT IgG. These results clearly show that GalT functions as a ZP3 receptor, and that, following aggregation, activates PTx-sensitive G-proteins leading to exocytosis (Shi, Amindari, Paruchuru, Skalla, Shur, and Miller, 2001).

Overexpressing GalT on sperm leads to increased ZP3 binding and accelerated acrosomal exocytosis. Transgenic mice that overexpress GalT on the sperm head bind more radiolabelled ZP3 than do wild-type sperm. Membranes from transgenic sperm show increased rates of both basal and ZP-dependent G-protein activation, relative to wild-type sperm. The increased G-protein activation correlates with greatly accelerated rates of acrosomal exocytosis. Thus, increasing GalT expression on sperm makes them hypersensitive to their ZP3 ligand (Youakim, Hathaway, Miller, Gong and Shur, 1994; Gong *et al.*, 1995).

Eliminating GalT makes sperm refractory to ZP3. In contrast to sperm that overexpress GalT, sperm from GalT-null males are unable to bind radiolabelled ZP3 as well as wild-type sperm, and fail to undergo ZP3-induced acrosomal exocytosis. However, GalT-null sperm respond normally to calcium ionophore, which induces acrosomal exocytosis by bypassing the requirement for ZP3 binding. The inability of GalT-null sperm to undergo a ZP3-induced acrosome reaction renders them physiologically inferior to wild-type sperm, since their ability to penetrate the egg coat and fertilise the oocyte *in vitro* is reduced to only 7% that of wild-type sperm (Lu and Shur, 1997). Importantly, control experiments eliminate the possibility that defects in sperm binding to ZP3 and in acrosomal induction are not the secondary consequences of defective intracellular galactosylation during spermatogenesis. Furthermore, the cytoplasmic domain unique to the long GalT isoform that is responsible for coupling to the G-protein complex appears to be absolutely required for ZP3-dependent signal transduction (Gong *et al.*, 1995; Shi *et al.*, 2001).

Sperm from GalT-null males still bind to the ZP and are fertile

Surprisingly, GalT-null sperm still bind to the intact ZP, even though they show low levels of binding to soluble ZP3 and cannot undergo a ZP3-induced acrosome reaction (Lu and Shur, 1997). The acrosome reaction has traditionally been viewed as a prerequisite for sperm penetration through the ZP; however, GalT-null sperm are fertile. Similarly, sperm null for the acrosomal protease, acrosin, are also fertile (Baba, Azuma, Kashiwabara and Toyoda, 1994). Although both of these observations suggest that the entire ZP3/GalT/G-protein-activated release of acrosomal contents is dispensable for fertilisation, it is noteworthy that the efficacy of fertilisation is greatly diminished; GalT-null sperm penetrate and fertilise the oocyte only 7% of wild-type levels, and acrosin-null sperm show reduced rates of zona penetration *in vitro*. Thus, in a noncompetitive environment where all sperm are phenotypically similar, as in the GalT-null male, a few sperm eventually find their way through the ZP (by either spontaneous acrosome reactions or other scenarios) and fertilise the egg. However, when compared to wild-type sperm, this process is extremely inefficient, suggesting that in a competitive environment the integrity of the GalT/ZP3/acrosome induction complex may be advantageous for successful fertilisation. This has been shown to be the case for acrosin-null sperm, since when they are mixed with wild-type sperm, either by mixed inseminations or in chimeric testes, the wild-type sperm compete out the ability of acrosin-null sperm to fertilise eggs (Adham, Nayernia and Engel, 1997).

In any event, the ability of GalT-null sperm to bind the ZP implies that sperm-egg binding likely involves at least two distinct events: a GalT/ZP3-independent mechanism that mediates initial gamete adhesion, followed by ZP3-dependent aggregation of GalT leading to acrosomal exocytosis. Conceptually, this paradigm is similar to the concerted action of selectins and integrins during lymphocyte adhesion to the endothelium (Lasky, 1992). It is therefore, not unreasonable to expect that successful binding of sperm and eggs may be dependent upon the sequential action of independent receptor-ligand interactions that stabilise the gametes as they encounter one another in a dynamic fluid environment.

Glycoproteins secreted from the female and male reproductive tracts participate in initial sperm-egg adhesion

Ovulated egg coats have a peripherally-associated ZP3-independent ligand

The ability of GalT-null sperm to bind to the ovulated ZP, but not soluble ZP3, suggested the presence of ZP3-independent sperm-binding ligands on the ovulated egg. Therefore, the sperm-binding activity of egg coats isolated from ovarian and ovulated oocytes was compared by assaying their ability to competitively inhibit sperm binding to ovulated oocytes. As expected, *ovarian* ZP glycoproteins inhibit the binding of wild-type sperm to ovulated eggs in a concentration-dependent manner. In contrast, *ovarian* ZP glycoproteins fail to competitively inhibit GalT-null sperm binding to ovulated eggs (Rodeheffer and Shur, 2004). This result is consistent with previous evidence demonstrating that GalT-null sperm do not bind ZP3 (Lu and Shur, 1997) and indicates the absence of any other binding ligand in the ovarian ZP for GalT-null sperm.

Nevertheless, GalT-null sperm still bind to ovulated eggs, which predicts that the *ovulated*, but not the *ovarian*, ZP might contain a ligand for GalT-null sperm. To test this hypothesis, wild-type or GalT-null sperm were incubated with ovulated eggs in the presence of solubilised *ovulated* ZP glycoproteins. Unlike that seen with *ovarian* ZP glycoproteins, zona glycoproteins from *ovulated* oocytes inhibit both wild-type and GalT-null sperm binding. Thus, as predicted, the *ovulated* ZP contains a ligand to which both wild-type and GalT-null sperm bind and which is absent from the *ovarian* ZP (Rodeheffer and Shur, 2004).

Stringent washing of ovulated ZP generated a soluble peripheral fraction and an insoluble zona matrix. The peripheral fraction possesses sperm-binding activity for both GalT-null and wild-type sperm; whereas, the solubilised zona matrix, containing ZP3, possesses sperm-binding activity for only wild-type sperm (Fig. 2). Further characterization of the sperm-binding ligand on the ovulated ZP shows that it does not result from contaminating ZP3, nor does it result from the action of enzymes released from egg cortical granules during preparation. The sperm-binding ligand behaves as a ~250 kDa, WGA-reactive glycoprotein with a basic isoelectric point, distinguishing it from the acidic glycoproteins that form the insoluble matrix of the ZP.

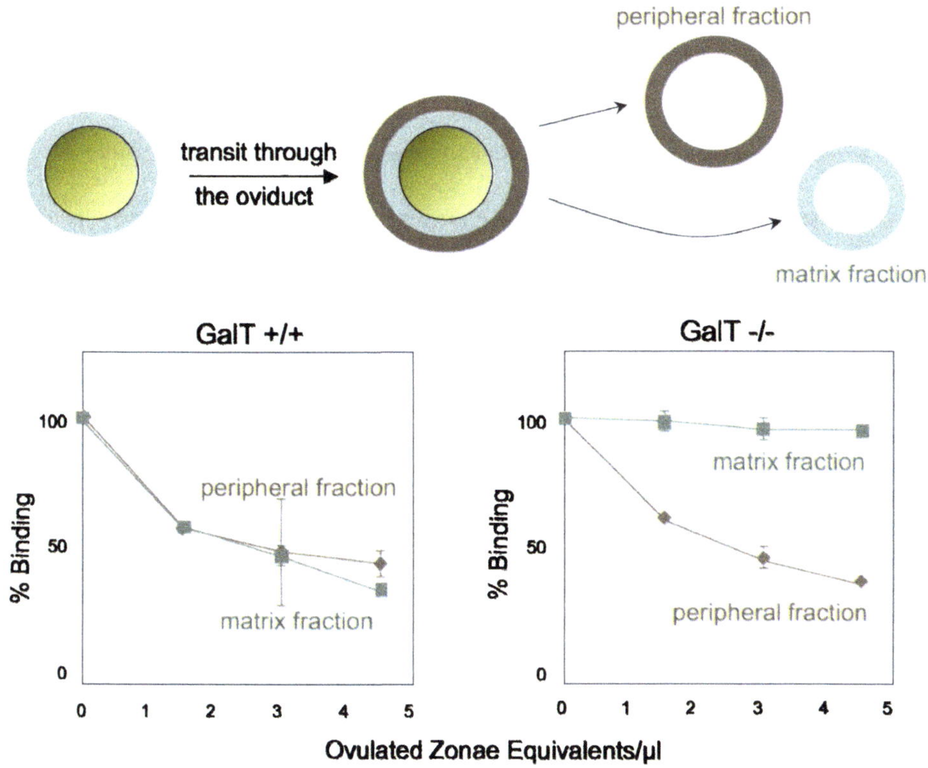

Figure 2. The zona pellucida of ovulated oocytes contains two distinct sperm-binding ligands. One is found in the insoluble zona matrix as ZP3, one of the three glycoproteins that form the developing zona pellucida. ZP3 is able to competitively inhibit the binding of wild-type ($GalT^{+/+}$) sperm to eggs, but has no effect on the binding of GalT-null ($GalT^{-/-}$) sperm to ovulated eggs. A second sperm-binding ligand is peripherally associated with the ovulated egg coat, and has sperm-binding activity for both wild-type and GalT-null sperm, as evidenced by its ability to competitively inhibit their binding to eggs (adapted from Rodeheffer and Shur, 2004).

Not surprisingly, subsequent studies indicate that the ZP3-independent ligand found on ovulated oocytes is likely derived from oviduct secretions, and is under hormonal control. In this regard, there is extensive literature regarding oviduct-secreted glycoproteins that bind to the zona matrix at ovulation (Malette, Paquette, Merlen and Bleau, 1995; Verhage, Mavrogianis, O'Day-Bowman, Schmidt, Arias, Donnelly, Boomsma, Thibodeaux, Fazleabas and Jaffe, 1998), the best characterised being Oviduct Secreted Glycoprotein (OGP) (Sendai, Komiya, Suzuki, Onuma, Kikuchi, Hoshi and Araki, 1995). However, OGP-null females are fertile and retain the ZP3-independent ligand, indicating they are distinct gene products. In any event, the GalT-null male is able to resolve gamete recognition into at least two distinct binding events, leading to the characterization of a novel, peripherally-associated, sperm-binding ligand on the ovulated ZP.

Identification of sperm SED1 as an adhesin for the ZP

Among the many sperm proteins implicated as receptors for the egg coat, as opposed to receptors specifically for ZP3 glycosides, two were of particular interest: zonadhesin and p47. Originally identified in the pig, zonadhesin is a sperm protein that binds to the ZP in a species-specific manner (Hardy and Garbers, 1995). Subsequent studies indicated that zonadhesin is localised in the acrosomal matrix, rather than on the sperm plasma membrane (Hardy, 2000). Thus, zonadhesin may mediate the binding of sperm to the egg coat during early stages of acrosomal exocytosis, rather than during initial sperm-egg binding. This possibility awaits testing.

p47, a peripherally-associated sperm surface protein, was identified by affinity chromatography of solubilised boar sperm plasma membranes applied to immobilised porcine ZP glycoproteins (Ensslin *et al.*, 1998). p47 displays a Notch-like mosaic structure organised into two amino-terminal EGF-like repeats followed by two domains with high similarity to the C2 domains found in blood clotting factors V and VIII. The second EGF-like domain contains an RGD sequence, a motif often found in integrin ligands. The biochemical evidence indicates that p47 has high affinity for unfractionated zona glycoproteins, but its function during fertilisation required more direct testing. Consequently, the mouse homologue of p47 was cloned and renamed SED1 to indicate that it is a *S*ecreted protein that contains Notch-like *E*GF repeats and *D*iscoidin/F5/8 type C (discoidin/C) domains.

Immunocytochemical analysis localises SED1 to the Golgi complex of spermatogenic cells, from which it is presumably secreted. However, the predominant source of SED1 appears to be the initial segment of the caput epididymis, where it is secreted and coats the sperm head overlying the acrosome by intercalation of the discoidin/C domains into the plasma membrane bilayer. SED1-derivatised beads bind selectively to the ZP of unfertilised oocytes by virtue of the discoidin/C domains, but do not bind to zona of fertilised eggs. Recombinant SED1 and anti-SED1 antibodies competitively inhibit sperm-egg binding, as do truncated SED1 proteins containing a discoidin/C domain, all suggestive of a role for SED1 in sperm adhesion to the ZP.

SED1 function during sperm-egg binding was confirmed by analysis of SED1-null male mice, which are subfertile and produce litter sizes ranging from no pups (sterility) to near normal, with an average litter size of about 1/3 that of control littermates. *In vitro* assays demonstrate a near complete inability of SED1-null sperm to bind to ZP of ovulated eggs. Importantly, SED1-null males show normal mating behaviour and normal sperm morphology, number, motility and ability to undergo acrosomal exocytosis. These studies clearly demonstrate a role for SED1 during initial sperm-egg adhesion (Ensslin and Shur, 2003). Preliminary blot overlay assays suggests that SED1 binds to ZP3 and to a lesser extent to ZP2, although a more detailed analysis of the ZP ligands for SED1 is required. Presumably, SED1 is docking to complex carbohydrate residues in ZP3 (and possibly ZP2), since the binding specificity of discoidin/C domains is dictated by hypervariable loops that dock with negatively charged phospholipids and/or carbohydrates (Fuentes-Prior, Fujikawa and Pratt, 2002).

Revised model for sperm-egg binding in mouse

The GalT-null mouse was instrumental in dissecting sperm-egg binding into two distinct molecular events: a GalT/ZP3-independent interaction that contributes to initial gamete adhesion, and ZP3-dependent aggregation of GalT that facilitates acrosomal exocytosis. This lead to the identification of two novel gamete receptors that appear to contribute to initial sperm-egg adhesion: sperm SED1 and the ZP3-independent ligand. These findings have enabled us to revise our earlier, and overly simplistic, model for sperm-egg interactions. We now hypothesize that SED1 and the ZP3-independent ligand, as well as still unidentified components, facilitate the initial stable adhesion between the sperm and the ZP. This initial gamete adhesion enables ZP3 oligosaccharides to interact with and aggregate GalT, which contributes to G-protein activation and acrosomal exocytosis (Fig. 3). We suspect these components are just some of those that form a macromolecular adhesion and signalling complex, not unlike the T-cell receptor. Clearly, this is simply a working model and no doubt will require revision as more detailed studies of SED1 and the ZP3-independent ligand are conducted.

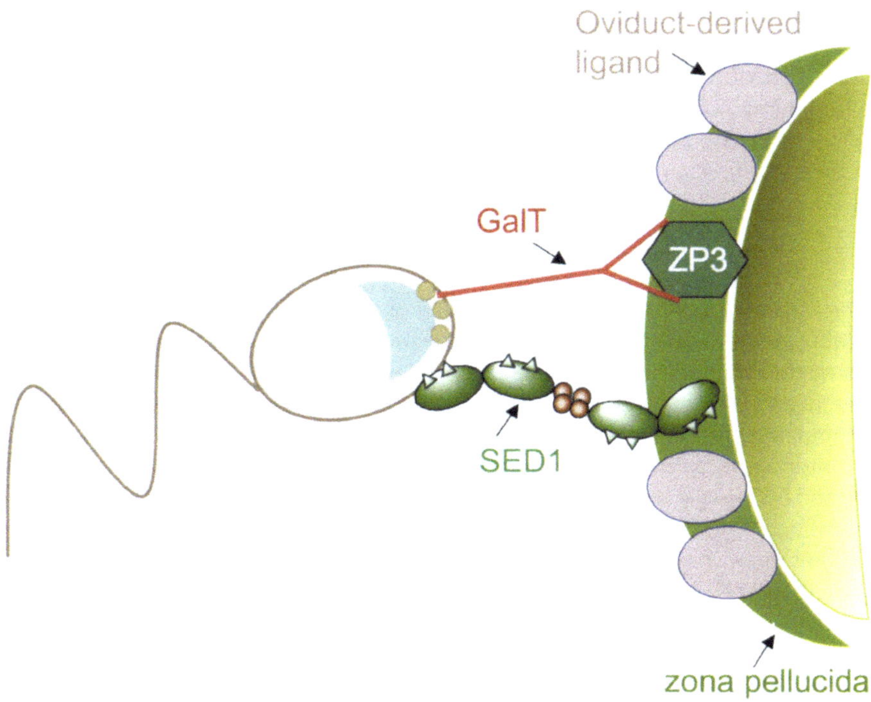

Figure 3. More recent studies have shown that sperm-egg interactions require at least two distinct binding mechanisms. Initial sperm-egg adhesion must rely upon mechanisms independent of GalT and ZP3, after which ZP3-dependent aggregation of GalT facilitates acrosomal exocytosis. Although a GlcNAc-terminating trisaccharide has been identified on ZP3 using radiolabelled tracers, its presence was not verified in a structural characterization of the ZP3 sugars. Nevertheless, the sperm-binding properties of purified ZP3 are dependent upon a terminal GlcNAc residue, as shown by the loss of sperm-binding activity following GlcNAc'dase treatment or by pre-galactosylation with GlcNAc-specific ß1,4-galactosyltransferase. Subsequent studies lead to the identification of SED1 as a sperm-associated protein that is required for binding to the ZP, as well as an oviduct-derived sperm-binding ligand that coats the zona pellucida upon ovulation. The nature of the SED1 binding sites within the ZP remain unknown, as is the mechanism of sperm binding to the oviduct-derived ZP3-independent ligand.

Conflicting issues and perspectives

The nature of the egg coat oligosaccharide recognised by sperm remains undefined

It is generally accepted by most investigators (Tulsiani, Yoshida-Komiya and Araki, 1997; Clark and Dell, 2006), but not all (Hoodbhoy and Dean, 2004), that gamete recognition occurs between defined carbohydrate structures of the ZP and their corresponding receptors on the sperm plasma membrane. However, there is considerable disagreement as to the composition of the sperm-binding oligosaccharide. Among the terminal glycosides suggested to serve as sperm-binding epitopes, are α-galactose (Bleil and Wassarman, 1988), ß-*N*-acetylglucosamine (GlcNAc) (Miller *et al.*, 1992), fucose (Johnston, Wright, Shaper, Hokke, Van den Eijinden and Joziasse, 1998), mannose (Tulsiani *et al.*, 1992), the disaccharide *N*-acetyllactosamine (Galß1,4GlcNAc) (Clark and Dell, 2006) and its α1,3-fucosylated trisaccharide, the Lewis X epitope (Johnston *et al.*, 1998; Kerr, Hanna, Shaper and Wright, 2004). Further complicating this issue, is the suggestion that *N*-linked oligosaccharides are involved in mouse sperm binding, in addition to *O*-linked oligosaccharides, since treatment of the intact ZP (as opposed to solubilized ZP3) with *N*-glycanase inhibits subsequent sperm binding (Yamagata, 1985). In this regard, *N*-linked oligosaccharides have also been implicated in the binding of porcine and bovine sperm to the egg ZP (Yonezawa, Aoki, Hatanaka and Nakano, 1995).

The use of competitive glycosides to inhibit sperm-egg binding must be interpreted with caution

Many of these studies have relied upon the ability of defined oligosaccharides to competitively interfere with sperm-zona binding. However, the interpretation of these competitive inhibition assays is brought into question by three recent observations.

First, the degree and type of oligosaccharide branching affects the ability of the terminal sugar to competitively inhibit sperm-egg binding. This raises the possibility that negative results may simply result from an inappropriate presentation of the terminal sugar, such that it is not *recognis*ed by the sperm receptor (Litscher, Suntunen, Seppo, Penttila, Niemela, Renkonen and Wassarman, 1995; Clark and Dell, 2006).

Second, immunocytochemical analysis at the ultrastructural level reveals that the sugar composition of the ZP is heterogeneous (Avilés, Jaber, Castells, Ballesta and Kan, 1997; Avilés, Okinaga, Shur and Ballesta, 2000). For example, α-galactosyl and *N*-acetylgalactosaminyl residues are confined to the inner portions of the mature ZP, whereas other sugars, such as GlcNAc, are dispersed uniformly throughout the zona. The spatial heterogeneity of zona glycosides makes it impossible to know whether soluble glycosides added as competitors of sperm-zona binding necessarily mimic oligosaccharides available to sperm at initial binding, or rather, during later aspects of zona penetration. In this regard, since α-galactosyl residues are not accessible to sperm on the zona surface, they cannot be involved in initial gamete recognition. Consistent with this, eggs from females bearing null mutations in α-galactosyltransferase are fertilised normally (Thall, Maly and Lowe, 1995), at least *in vivo*. On the other hand, the selective localization of α-galactosyl residues to the inner portions of the ZP suggest they may function during later stages of zona penetration. Similarly, Lewis X fucosyl residues have been implicated in initial sperm-egg adhesion (Johnston *et al.*, 1998); however, these structures are not present anywhere in the ovarian or ovulated mouse ZP, making it unlikely they participate in sperm-egg interactions (Avilés *et al.*, 2000). Nevertheless, the inhibitory activity of Lewis X glycosides suggests they may mimic a structurally similar epitope in the ZP.

Terminal GlcNAc residues are present throughout the zona matrix based upon both ultrastructural (Avilés *et al.*, 1997, 2000) and light microscopic evidence (Mori, Mori and Takasaki, 1997) and have also been implicated in initial sperm-zona binding. Either removing or masking terminal GlcNAc residues on ZP3 destroys its sperm-binding activity (Miller *et al.*, 1992). More importantly, masking terminal GlcNAc residues in the intact ZP by pregalactosylation (i.e., covalently adding a terminal galactose residue) destroys the sperm-binding activity of the intact ZP (Lopez, Bayna, Litoff, Shaper, Shaper and Shur, 1985). While the use of the intact ZP to probe the biochemical nature of initial sperm-egg interactions is subject to its own difficulties and limitations, it does allow one to examine initial aspects of gamete recognition without disruption of the zona matrix. Interestingly, the predominant *O*-linked oligosaccharide of ZP3 has been characterised as a small trisaccharide with a terminal GlcNAc (Nagdas, Araki, Chayko, Orgebin-Crist and Tulsiani, 1994), but it is not yet known if this specific trisaccharide has sperm-binding activity. This trisaccharide is most likely present in very small quantities relative to the principle glycoside chains, since structural analysis of the major ZP3 oligosaccharides failed to detect GlcNAc-terminating *O*-linked sugars. Terminal GlcNAc residues are present, however, on some *N*-linked chains of ZP3 (Easton, Patankar, Lattanzio, Leaven, Morris, Clark and Dell, 2000; Dell, Chalabi, Easton, Haslam, Sutton-Smith, Patankar, Lattanzio, Panico, Morris and Clark, 2003).

A third recent observation that impacts our identification of sperm-binding structures on the egg coat, is that it now appears there are at least two distinct binding events during the early stages of sperm-egg interaction: a ZP3-dependent and a ZP3-independent event (Rodeheffer and Shur, 2004). As discussed above, the newly ovulated oocyte is coated with glycoproteins derived from the oviduct epithelium, and which have carbohydrate-dependent sperm-binding activity. Thus, it is unknown which, if any, sperm-binding ligand is being inhibited by the addition of soluble competitive glycosides.

Structural characterization of the oligosaccharide chains on murine ZP glycoproteins

Dell, Clark and colleagues have recently reported the structural characterization of the glycoside chains of the murine ZP glycoproteins (Easton *et al.*, 2000; Dell *et al.*, 2003). This tour-de-force of structural glycobiology should ideally represent a catalogue of the potential sperm-binding oligosaccharides, as well as a means to eliminate sugars previously implicated in sperm binding, but which prove to be undetectable in the mouse ZP. However, this rich data set must be interpreted cautiously, since the ZP used for this analysis was isolated from ovarian homogenates that contain all stages of developing oocytes. Since immunocytochemical studies indicate differential glycosylation of the ZP glycoproteins during maturation, the relative contribution of the various sugar chains identified in the ovarian ZP, composed of all follicular stages, is likely to be distinct from those expressed on the surface of the mature ZP where sperm bind. Another consideration is whether the preparation of the ovarian homogenate enabled the release of cortical granule glycosidases, among other enzymes, that are known to inactivate the sperm-binding properties of the ZP. The most abundant cortical granule glycosidase is GlcNAc'dase, which is able to destroy the sperm-binding activity of ZP3 (Miller *et al.*, 1993). This raises the possibility that the inability to detect *O*-linked GlcNAc on ZP3 isolated from ovarian homogenates may be the result of GlcNAc'dase hydrolysis. Nevertheless, this database of glycoside chains is an excellent start at cataloguing the sugar modifications of the developing ZP, and hopefully the technology will soon enable the identification of sugars isolated directly from ZP of ovulated eggs.

Replacement of murine ZP glycoproteins with their human homologues has no effect on mouse-specific sperm binding

An exhaustive series of studies by Dean, Rankin and their colleagues has shown that mouse sperm still bind to eggs in which the mouse ZP3 (mZP3) polypeptide has been replaced by human ZP3 (huZP3), suggesting that ZP3 is not the primary receptor for species-specific sperm binding (Rankin, Tong, Castle, Lee, Gore-Langton, Nelson and Dean, 1998). Similar results are found when all three mouse ZP genes are replaced with their human homologues (Hoodbhoy and Dean, 2004). However, since the glycoside chains of ZP3 are elaborated by mouse oocyte glycosyltransferases, it would be expected that huZP3 and mZP3 synthesised in mouse oocytes would contain similar, if not identical, glycoside chains. This has recently been confirmed (Dell *et al.*, 2003). Thus, one simple interpretation for these results is that mouse sperm are still binding to their cognate glycoside epitopes whether they are presented on the human or murine ZP3 polypeptide backbone.

Interestingly, huZP2-containing eggs continue to support sperm binding after fertilisation, unlike normal eggs that no longer support binding after egg activation. This block to polyspermy is associated with, among other changes, cleavage of ZP2 by a protease released from the cortical granules. It is therefore noteworthy that the ability of hu-ZP2 eggs to support sperm binding after fertilisation is correlated with a failure of human ZP2 to undergo cleavage even though the cortical reaction occurs normally. Dean and colleagues argue these results speak against carbohydrate-mediated sperm binding, but rather suggest that sperm *recognise* a supramolecular scaffold composed of both ZP2 and ZP3 (Hoodbhoy and Dean, 2004). After fertilisation, they hypothesise that cleavage of ZP2 results in loss of the macromolecular structure and a concomitant loss of sperm binding. Similarly, they argue that the continued binding of sperm to hu-ZP2 fertilised eggs further speaks against carbohydrate-mediated sperm binding, since the released cortical granule enzymes should have destroyed the sperm binding glycosides. However, the same logic that these authors have used to suggest the cortical granule protease is incapable of cleaving huZP2, although it cleaves mZP2 even when both are present in the same egg, also applies to the ability of cortical granule glycosidases to destroy the sperm-binding epitope. Many carbohydrate-modifying enzymes *recognise* binding sites within the polypeptide backbone as well as the sugar chain, and therefore, the cortical granule glycosidases may not *recognise* their oligosaccharide substrates in the context of the human polypeptide backbone.

Finally, a significant complication of these "gene swapping" experiments, is that sperm binding is assayed to ovulated eggs, which possesses a ZP3-independent ligand that may obscure any differences due to the replacement of the ZP genes (Rodeheffer and Shur, 2004). Therefore, in order for the humanised mouse zona to be the useful resource that it could be, the biological activities (e.g., sperm binding) of the individually isolated huZP glycoproteins needs to be determined as was done for the original identification and characterization of ZP3 as well as the ZP3-independent ligand.

Lipid rafts likely influence the organization of multiple sperm receptors for the ZP

Lipid rafts, also known as DRMs (detergent resistant membranes), are highly specialised membrane domains capable of organizing cell adhesion and signalling molecules into macromolecular complexes. A characteristic feature of lipid rafts is their high cholesterol content that maintains the rigid lipid platform in an otherwise fluid bilayer. One of the hallmarks of sperm capacitation is a loss of cholesterol, resulting in increased membrane fluidity that is thought to be required for subsequent fusion of the outer acrosomal and overlying plasma membranes. It is therefore not surprising that capacitation leads to a reorganization of both the macromolecular components and spatial distribution of lipid rafts on the sperm head. This is well illustrated by a comparative analysis of proteins localised in rafts before and after capacitation (Sleight, Miranda,

Plaskett, Maier, Lysiak, Scrable, Herr and Visconti, 2005). Cholesterol efflux during capacitation results in the (transient?) dissolution of lipid rafts, leading to the release, or dispersion, of sperm components that may be required for subsequent interactions with the ZP and/or oolema.

One constituent of sperm rafts is the sulfoglycolipid, sulfogalactosylglycerolipid (SGG), which has been proposed to mediate binding to the ZP via electrostatic interactions between the sulfated galactosyl residues of SGG and glycoside moieties of zona glycoproteins. These weak electrostatic interactions may be facilitated by arylsulfatase A, a sperm associated enzyme that is thought to aid in the initial binding of SGG to zona glycosides (Khalil, Chakrabandhu, Xu, Weerachatyanukul, Buhr, Berger, Carmona, Vuong, Kumarathasan, Wong, Carrier and Tanphaichitr, 2006). Interestingly, in porcine sperm, the concentration of SGG increases in rafts as a function of capacitation, which correlates with an increased ability of rafts isolated from capacitated sperm to bind the ZP. These authors suggest that cholesterol loss during capacitation leads to a reorganization of components localised within lipid rafts, rather than to their complete dissolution. During this raft "remodeling" SGG is proposed to facilitate the coalescence of the zona-binding machinery into DRMs that are now competent to bind the ZP. In any event, the fact that sperm contain lipid rafts, or DRMs, the composition of which appears to change as a result of capacitation, sheds new light and complexity on the analysis of components that mediate sperm-ZP binding.

Molecular chaperones may influence the zona-binding machinery

Molecular chaperones are another class of membrane-associated components that has recently been suggested to facilitate the proper organization and presentation of the zona-binding machinery on sperm. Chaperones have long been known to be critical for the proper folding, transport and turnover of proteins, and several chaperones are found in mitotically-active spermatogenic cells. One chaperone that is expressed specifically during post-meiotic spermatogenic differentiation is calmegin, a testis-specific homologue of the endoplasmic reticulum chaperone, calnexin. Calmegin-dependent protein processing is required during spermatid differentiation for sperm to function normally, as shown by the inability of calmegin-null sperm to transport through the oviduct or bind the ZP (Ikawa, Wada, Kominami, Watanabe, Toshimori, Nishimune and Okabe, 1997).

Surprisingly, some chaperones are localised to the sperm plasma membrane, including heat-shock protein 1 (HSPD1) and tumor rejection antigen gp96 (TRA1), also known as endoplasmin. Although these chaperones are present in early spermatogenic cells, it is thought that the sperm-associated chaperones are derived from epididymal sections. Regardless, the expression of these two chaperones on the sperm surface increases significantly during capacitation, and is lost following the acrosome reaction. Their capacitation-dependent expression on the sperm surface suggest to Aitken and colleagues that they may be involved in the recruitment, organization and/or stabilization of sperm surface components required for subsequent interactions with the oviduct and/or binding to the ZP (Asquith, Harman, McLaughlin, Nixon and Aitken, 2005; Nixon, Asquith and Aitken, 2005).

Concluding thoughts

It is not surprising that the continued molecular and genetic dissection of mammalian sperm-egg binding has introduced an ever-increasing list of players (Fig. 4). This has resulted, in part, from technological advances that enabled us to develop more rigorous and physiologically-relevant assays. As more candidate receptors and their ligands are subjected to homologous

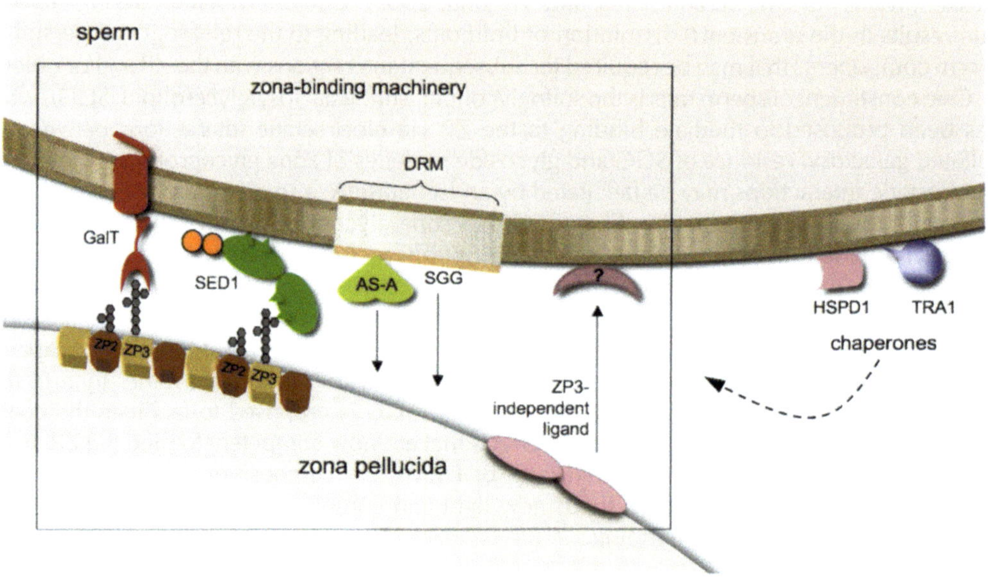

Figure 4. Diagram of some of the molecular components discussed in this review that have been implicated in sperm recognition and binding to the zona pellucida. SED1 is thought to mediate initial sperm adhesion to the zona pellucida by virtue of the hypervariable loops (triangles) in the discoidin/C domains that dictate binding specificity to either the sperm membrane and/or the zona matrix. Gamete adhesion may be further aided the binding of sperm to a ZP3-independent ligand on the egg coat that appears to be derived from oviduct secretions. Electrostatic interactions between sulfated galactosyl residues of SGG and glycoside chains of the zona glycoproteins are also thought to aid sperm-zona adhesion, which may be facilitated by a peripherally-associated arylsulfatase-A (AS-A). Eventually, the relevant ZP3 oligosaccharides bind to GalT on sperm, causing its aggregation and consequent activation of heterotrimeric G-protein cascades that contribute to acrosomal exocytosis. The presence of molecular chaperones on the sperm surface, such as HSPD1 and TRA1, suggest that they may modulate the organization or assembly of the sperm zona-binding machinery. Whether the function of these components is modulated by their presence in lipid rafts (DRM) or how they may be reorganised in the sperm membrane as a consequence of capacitation, must also be considered.

recombination, our understanding of their mode of action becomes more refined. Nevertheless, the complex array of cellular and molecular events that underlie fertilisation has yet to be recapitulated in any one *in vitro* system, thus forcing investigators to study molecules in isolation of their normal molecular and cellular environment. How these components are expressed and organised within the membrane; whether their functions are modulated by other membrane-associated components, and how these molecular events are affected by the physiological changes that occur during capacitation, are just a few of the obvious issues that complicate our understanding of their precise roles during this complex and critically important event.

References

Adham, I.M., Nayernia, K. and Engel, W. (1997). Spermatozoa lacking acrosin protein show delayed fertilization. *Molecular Reproduction and Development* **46**: 370-376

Asquith KL, Harman AJ, McLaughlin EA, Nixon B and Aitken RJ. (2005). Localization and significance of molecular chaperones, heat shock protein 1, and tumor rejection antigen gp96 in the male reproductive tract and during capacitation and acrosome reaction. *Biology of Reproduction* **72**: 328-337

Avilés, M., Jaber, L., Castells, M.T., Ballesta, J. and Kan, F.W.K. (1997). Modifications of carbohydrate residues and ZP2 and ZP3 glycoproteins in the mouse zona pellucida after fertilization. *Biology of Reproduction* **57**: 1155-1163

Avilés, M., Okinaga, T., Shur, B.D. and Ballesta, J. (2000). Differential expression of glycoside residues in the mammalian zona pellucida. *Molecular Reproduction and Development* **57**: 296-308

Baba, T., Azuma, S., Kashiwabara, S. and Toyoda, Y. (1994). Sperm from mice carrying a targeted mutation of the acrosin gene can penetrate the oocyte zona pellucida and effect fertilization. *Journal of Biological Chemistry* **269**: 31845-31849

Benau, D.A., McGuire, E.J. and Storey, B,T. (1990). Further characterization of the mouse sperm surface zona-binding site with galactosyltransferase activity. *Molecular Reproduction and Development* **25**: 393-399

Bleil, J.D. and Wassarman, P.M. (1980a). Mammalian sperm-egg interaction: identification of a glycoprotein in mouse egg zonae pellucidae possessing receptor activity for sperm. *Cell* **20**: 873-882

Bleil, J.D. and Wassarman, P.M. (1980b). Structure and function of the zona pellucida: identification and characterization of the proteins of the mouse oocytes zona pellucida. *Developmental Biology* **76**: 185-202

Bleil, J.D. and Wassarman, P.M. (1983). Sperm-egg interactions in the mouse: sequence of events and induction of the acrosome reaction by a zona pellucida glycoprotein. *Developmental Biology* **95**: 317-324

Bleil, J.D. and Wassarman, P.M. (1988). Galactose at the nonreducing terminus of O-linked oligosaccharides of mouse egg zona pellucida glycoprotein ZP3 is essential for the glycoprotein's sperm receptor activity. *Proceedings of the National Academy of Sciences USA* **85**: 6778-6782

Bleil, J.D. and Wassarman, P.M. (1990). Identification of a ZP3-binding protein on acrosome-intact mouse sperm by photoaffinity crosslinking. *Proceedings of the National Academy of Sciences USA* **87**: 5563-5567

Bork, P. (1996). Sperm-egg binding protein or proto-oncogene? *Science* **271**: 1431-1432

Clark, G.F. and Dell, A. (2006). Molecular models for murine sperm-egg binding. *Journal of Biological Chemistry* **281**: 13853-13856

Cornwall, G.A., Tulsiani, D.R.P. and Orgebin-Crist, M.C. (1991). Inhibition of the mouse sperm surface a-D-mannosidase inhibits sperm-egg binding in vitro. *Biology of Reproduction* **44**: 913-921

Dell, A., Chalabi, S., Easton, R.L., Haslam, S.M., Sutton-Smith, M., Patankar, M.S., Lattanzio, F., Panico, M., Morris, H.R. and Clark, G.F. (2003). Murine and human zona pellucida 3 derived from mouse eggs express identical *O*-glycans. *Proceedings of the National Academy of Sciences USA* **100**: 15631-15636

Easton, R.L., Patankar, M.S., Lattanzio, F.A., Leaven, T.H., Morris, H.R., Clark, G.F. and Dell, A. (2000). Structural analysis of murine zona pellucida glycans. Evidence for the expression of core 2-type *O*-glycans and the Sda antigen. *Journal of Biological Chemistry* **275**: 7731-7742

Ensslin, M., Vogel, T., Calvete, J.J., Thole, H.H., Schmidtke, J. and Töpfer-Petersen, E. (1998). Molecular cloning and characterization of P47, a novel boar sperm-associated zona pellucida-binding protein homologous to a family of mammalian secretory proteins. *Biology of Reproduction* **58**: 1057-1068

Ensslin, M.A. and Shur, B.D. (2003). Identification of mouse sperm SED1, a bi-motif EGF repeat and discoidin domain protein involved in sperm-egg binding. *Cell* **114**: 405-417

Florman, H.M. and Wassarman, P.M. (1985). *O*-linked oligosaccharides of mouse egg ZP3 account for its sperm receptor activity. *Cell* **41**: 313-324

Florman, H.M., Arnoult, C., Kazam, I.G., Li, C. and O'Toole, C.M.B. (1999). An intimate biochemistry: egg-regulated acrosome reactions of mammalian sperm. *Advances in Developmental Biochemistry* **5**: 199-233

Fuentes-Prior, P., Fujikawa, K. and Pratt, K.P. (2002). New insights into binding interfaces of coagulation factors V and VIII and their homologues – lessons from high resolution crystal structures. *Current Protein and Peptide Science* **3**: 313-339

Gong, X., Dubois, D.H., Miller, D.J. and Shur, B.D. (1995) Activation of a G-protein complex by aggregation of ß1,4-galactosyltransferase on the surface of sperm. *Science* **269**: 1718-1721

Hardy, D.M. (2000). Structure, activity, and species-diversity of zonadhesin. *The Third International Symposium on the Molecular and Cellular Biology of the Egg and Embryo Extracellular Matrix*. Abstract II1

Hardy, D.M. and Garbers, D.L. (1995). A sperm membrane protein that binds in a species-specific manner to the egg extracellular matrix is homologous to von Willebrand factor. *Journal of Biological Chemistry* **270**: 26025-26028

Hathaway, H.J., Evans, S.C., Dubois, D.H., Foote, C.I, Elder, B.H. and Shur, B.D. (2003). Mutational analysis of the cytoplasmic domain of ß1,4-galactosyltransferase I: influence of phosphorylation on cell surface expression. *Journal of Cell Science*

116: 4319-4330

Hoodbhoy, T. and Dean, J. (2004). Insights into the molecular basis of sperm – egg recognition in mammals. *Reproduction* **127**: 417-422

Ikawa, M., Wada, I., Kominami, K., Watanabe, D., Toshimori, K., Nishimune, Y. and Okabe, M. (1997). The putative chaperone calmegin is required for sperm fertility. *Nature* **387**: 607-611

Johnston, D.S., Wright, W.W., Shaper, J.H., Hokke, C.H., Van den Eijinden, D.H. and Joziasse, D.H. (1998). Murine sperm-zona binding, a fucosyl residue is required for a high affinity sperm-binding ligand. A second site on sperm binds a nonfucosylated, ß-galactosyl-capped oligosaccharide. *Journal of Biological Chemistry* **278**: 1888-1895.

Joziasse, D.H. (1992). Mammalian glycosyltransferases: genomic organization and protein structure. *Glycobiology* **2**: 271-277

Kerr, C.L., Hanna, W.F., Shaper, J.H. and Wright, W.W. (2004). Lewis X-containing glycans are specific and potent competitive inhibitors of the binding of ZP3 to complementary sites on capacitated, acrosome-intact mouse sperm. *Biology of Reproduction* **71**: 770-777

Khalil, M.B., Chakrabandhu, K., Xu, H., Weerachatyanukul, W., Buhr, M., Berger, T., Carmona, E., Vuong, N., Kumarathasan, P., Wong, P.T.T., Carrier, D., and Tanphaichitr, N. (2006). Sperm capacitation induces an increase in lipid rafts having zona pellucida binding ability and containing sulfogalactosylglycerolipid. *Developmental Biology* **290**: 220-235

Kim, K.-S., Cha, M.C. and Gerton, G.L. (2001). Mouse sperm protein sp56 is a component of the acrosomal matrix. *Biology of Reproduction* **64**: 36-43

Lasky, L.A. (1992). Selectins: interpreters of cell-specific carbohydrate information during inflammation. *Science* **258**: 964-969

Leyton, L. and Saling, P. (1989). 95 kd sperm proteins bind ZP3 and serve as tyrosine kinase substrates in response to zona binding. *Cell* **57**: 1123-113

Litscher, E.S., Suntunen, K., Seppo, A., Penttila, L., Niemela, R., Renkonen, O. and Wassarman, P.M. (1995). Oligosaccharide constructs with defined structures that inhibit binding of mouse sperm to unfertilized eggs in vitro. *Biochemistry* **34**: 4662-4669

Lopez, L.C., Bayna, E.M., Litoff, D., Shaper, N.L., Shaper, J.H. and Shur, B.D. (1985). Receptor function of mouse sperm surface galactosyltransferase during fertilization. *Journal of Cell Biology* **101**: 1501-1510

Lu, Q. and Shur, B.D. (1997). Sperm from b1,4-galactosyltransferase null mice are refractory to ZP3-induced acrosome reactions and penetrate the zona pellucida poorly. *Development* **124**: 4121-4131

Malette, B., Paquette, Y., Merlen, Y. and Bleau, G. (1995). Oviductins possess chitinase- and mucin-like domains: a lead in the search for the biological function of these oviduct-specific ZP-associating glycoproteins. *Molecular Reproduction and Development* **41**: 384-397

Meizel, S. (1997). Amino acid neurotransmitter receptor/chloride channels of mammalian sperm and the acrosome reaction. *Biology of Reproduction* **56**: 569-574

Miller, D.J., Gong, X. and Shur, B.D. (1993a). Sperm require ß-*N*-acetylglucosaminidase to penetrate through the egg zona pellucida. *Development* **118**: 1279-1289

Miller, D.J., Gong, X., Decker., G. and Shur, B.D. (1993b). Egg cortical granule *N*-acetylglucosaminidase is required for the mouse zona block to polyspermy. *Journal of Cell Biology* **123**: 1431-1440

Miller, D.J., Macek, M.B. and Shur, B.D. (1992). Complementarity between sperm surface ß1,4galactosyltransferase and egg coat ZP3 mediates sperm-egg binding. *Nature* **357**: 589-593

Mori, E., Mori, T. and Takasaki, S. (1997) Binding of mouse sperm to beta-galactose residues on egg zona pellucida and asialofetuin-coupled beads. *Biochemical and Biophysical Research Communications* **238**: 95-99

Nagdas, S.K., Araki, Y., Chayko, C.A., Orgebin-Crist, M.C. and Tulsiani, D.R.P. (1994). *O*-linked trisaccharide and *N*-linked poly-*N*-acetyllactosaminyl glycans are present on mouse ZP2 and ZP3. *Biology of Reproduction* **51**: 262-272

Nixon, B., Asquith, K.L., and Aitken, R.J. (2005). The role of molecular chaperones in mouse sperm-egg interactions. *Molecular and Cellular Endocrinology* **240**: 1-10

Primakoff, P., Hyatt, H. and Myles, D.G. (1985). A role for the migrating sperm surface antigen PH-20 in guinea pig sperm binding to the egg zona pellucida. *Journal of Cell Biology* **101**: 2239-2244

Rankin, T.L., Tong, Z.-B., Castle, P.E., Lee, E., Gore-Langton, R., Nelson, L.M. and Dean, J. (1998). Human ZP3 restores fertility in Zp3 null mice without affecting order-specific sperm binding. *Development* **125**: 2415-2424

Rivkin, E., Tres, L.L., Kaplan-Kraicer, R., Shalgi, R. and Kierszenbaum, A.L. (2000) Molecular cloning of rat sperm galactosyl receptor, a C-type lectin with in vitro egg binding activity. *Molecular Reproduction and Development* **56**: 401-411

Rodeheffer, C. and Shur, B.D. (2002). Targeted mutations in ß1,4 galactosyltransferase I reveal its multiple cellular functions. *Biochimica et Biophysica Acta Reviews* **1573**: 258-270

Rodeheffer, C. and Shur, B.D. (2004). Identification of a novel, ZP3-independent egg coat ligand that facilitates initial sperm-egg adhesion. *Development* **131**: 503-512

Russo, R.N., Shaper, N.L. and Shaper, J.H. (1990). Bovine ß1-4-galactosyltransferase: two sets of mRNA transcripts encode two forms of the protein with different amino-terminal domains. In vitro translation experiments demonstrate that both the short and the long forms of the enzyme are type II membrane-bound glycoproteins. *Journal of Biological Chemistry* **265**: 3324-3331

Sendai, Y., Komiya, H., Suzuki, K., Onuma, T., Kikuchi, M., Hoshi, H. and Araki, Y. (1995). Molecular cloning and characterization of a mouse oviduct-specific glycoprotein. *Biology of Reproduction* **53**: 285-294

Shi, X., Amindari, S., Paruchuru, K., Skalla, D., Shur, B.D. and Miller, D.J. (2001) Cell surface ß1,4-galactosyltransferase-I activates G protein-dependent exocytotic signaling. *Development* **128**: 645-654

Shur, B.D. and Bennett, D. (1979). A specific defect in galactosyltransferase regulation on sperm bearing mutant alleles of the T/t-locus. *Developmental Biology* **71**: 243-259

Shur, B.D. and Hall, N.G. (1982a). Sperm surface galactosyltransferase activities during in vitro capacitation. *Journal of Cell Biology.* **95**: 567-573

Shur, B.D. and Hall, N.G. (1982b). A role for mouse sperm surface galactosyltransferase in sperm binding to the egg zona pellucida. *Journal of Cell Biology.* **95**: 574-579

Shur, B.D. and Neely, C.A. (1988). Plasma membrane association, purification and characterization of mouse sperm ß1,4 galactosyltransferase. *Journal of Biological Chemistry* **263**: 17706-17714

Shur, BD, Evans, S and Lu, Q (1998). Cell surface galactosyltransferase: current issues. *Glycoconjugate Journal* **15**: 537-548

Sleight, S.B., Miranda, P.V., Plaskett, N.-W., Maier, B., Lysiak, J., Scrable, H., Herr, J.C. and Visconti, P.E. (2005). Isolation and proteomic analysis of mouse sperm detergent-resistant membrane fractions: evidence for dissociation of lipid rafts during capacitation. *Biology of Reproduction* **73**: 721-729

Thall, A.D., Maly, P. and Lowe, J.B. (1995). Oocyte Galα1,3Gal epitopes implicated in sperm adhesion to the zona pellucida glycoprotein ZP3 are not required for fertilization in the mouse. *Journal of Biological Chemistry* **270**: 21437-21440

Tulsiani, D.R.P., Nagdas, S.K., Cornwall, G.A. and Orgebin-Crist, M.C. (1992). Evidence for the presence of high mannose/hybrid oligosaccharide chain(s) on the mouse ZP2 and ZP3. *Biology of Reproduction* **46**: 93-100.

Tulsiani, D.R.P., Yoshida-Komiya, H. and Araki, Y. (1997). Mammalian fertilization: a carbohydrate-mediated event *Biology of Reproduction* **57**: 487-494

Verhage, H.G., Mavrogianis, P.A., O'Day-Bowman, M.B., Schmidt, A., Arias, E.B., Donnelly, K.M., Boomsma, R.A., Thibodeaux, J.K., Fazleabas, A.T. and Jaffe, R.C. (1998). Characteristics of an oviductal glycoprotein and its potential role in the fertilization process. *Biology of Reproduction* **58**: 1098-1101

Wade, S.M., Lim, W.K., Lan, K.-L., Chung, D.A., Nanamori, M. and Neubig, R.R. (1999). G_i activator region of α_{2A}-adrenergic receptors: distinct basic residues mediate G_i versus G_s activation. *Molecular Pharmacology* **56**: 1005-1013

Ward, C.R. and Kopf, G.S. (1993). Molecular events mediating sperm activation. *Developmental Biology* **158**: 9-34

White, D., Weerachatyanukul, W., Gadella, B., Kamolvarin, N., Attar, M. and Tanphaichitr, N. (2000). Role of sperm sulfogalactosylglycerolipid in mouse sperm-zona pellucida binding. *Biology of Reproduction* **63**: 147-155

Yamagata, T. (1985). The role of saccharides in fertilization of the mouse. *Development, Growth and Differentiation* **27**: 176-177

Yonezawa, N., Aoki, H., Hatanaka, Y. and Nakano, M. (1995). Involvement of *N*-linked carbohydrate chains of pig zona pellucida in sperm-egg binding. *European Journal of Biochemistry* **233**: 35-41

Youakim, A., Hathaway, H.J., Miller, D.J., Gong, X. and Shur, B.D. (1994). Overexpressing sperm surface ß1,4 galactosyltransferase in transgenic mice affects multiple aspects of sperm-egg interactions. *Journal of Cell Biology* **126**: 1573-1584

Spermatology. SRF Vol. 65. ERS Roldan and M Gomendio (eds) Nottingham University Press, Nottingham

Molecular mechanisms involved in gamete interaction: Evidence for the participation of cysteine-rich secretory proteins (CRISP) in sperm-egg fusion

V. Da Ros, D. Busso, D.J. Cohen, J. Maldera, N. Goldweic and P.S. Cuasnicu

Instituto de Biología y Medicina Experimental (IBYME-CONICET), Vuelta de Obligado 2490, (1428) Buenos Aires, Argentina

Epididymal protein DE and testicular protein Tpx-1 are two cysteine-rich secretory proteins also known as CRISP-1 and CRISP-2, respectively. DE/CRISP-1 is localised on the equatorial segment of acrosome-reacted sperm and participates in rat gamete fusion through its binding to egg-complementary sites. Recent results using bacterially-expressed recombinant fragments of DE as well as synthetic peptides revealed that the ability of DE to bind to the egg surface and inhibit gamete fusion resides in a region of 12 amino acids corresponding to an evolutionary conserved motif of the CRISP family (Signature 2). Given the high degree of homology between DE/CRISP-1 and Tpx-1/CRISP-2, we also explored the potential participation of the testicular intra-acrosomal protein in gamete fusion. Results showing the ability of recombinant Tpx-1 to bind to the surface of rat eggs (evaluated by indirect immunofluorescence) and to significantly inhibit zona-free egg penetration, support the participation of this protein in gamete fusion through its interaction with egg-binding sites. Interestingly, rat Tpx-1 exhibits only two substitutions in Signature 2 when compared to this region in DE. Together, these results provide evidence for the involvement of both epididymal DE/CRISP-1 and testicular Tpx-1/CRISP-2 in gamete fusion suggesting the existence of a functional cooperation between homologue molecules as a mechanism to ensure the success of fertilisation.

CRISP proteins

The CRISP (Cysteine-RIch Secretory Protein) family is a large group of secreted proteins with molecular weights of about 20-30 kDa, characterised by the presence of sixteen conserved cysteine residues, ten of which are clustered in the C-terminal portion of the molecule. The first member of this family was originally described by our laboratory in the rat epididymis (Cameo and Blaquier, 1976), and it is known as DE or CRISP-1. Since then, other members of the family have been identified in different mammalian tissues: CRISP-2, also known as Tpx-1, which is expressed in the testis and is synthetised exclusively in the developing spermatids (Hardy *et al.*, 1988; Kasahara *et al.*, 1989), CRISP-3, with a wider tissue distribution than the other CRISPs including reproductive (prostate and ovary) and non-reproductive (salivary gland, pancreas, thymus and co-

Corresponding author E-mail: cuasnicu@dna.uba.ar

lon) organs (Haendler *et al.*, 1993; Kjeldsen *et al.*, 1996; Udby *et al.*, 2005), and the recently described CRISP-4 which is exclusively expressed in the epididymis (Jalkanen et *al.*, 2005). Other members of the family are present in salivary secretions of certain snakes and lizards, and several proteins with significant homology to the N-terminal domain of CRISPs are present in plants, insects and fungi. Although CRISPs are found across a broad variety of living forms and exhibit diverse biological functions, the molecular mechanisms underlying these functions remain unknown for most of the CRISP family members.

Protein DE/CRISP-1

Evidence obtained in our laboratory indicates that, after its secretion into the epididymal lumen, DE associates with the sperm surface during epididymal maturation, and migrates from the dorsal to the equatorial segment of the acrosome concomitantly with the occurrence of the acrosome reaction (Rochwerger and Cuasnicu, 1992). The relocation of DE to the equatorial segment, the fusogenic region of the sperm head, opened the possibility for a role of this protein in gamete fusion. Results showing that exposure of zona-free rat eggs to purified DE produced a significant reduction in the percentage of egg penetration without affecting the first step of sperm-egg binding, indicated that this protein participates in an event subsequent to binding and leading to fusion, through complementary sites localised to the fusogenic area of the egg surface (Rochwerger *et al.*, 1992).

The molecular mechanism by which this epididymal protein mediates sperm-egg fusion is still unknown. Structure-function studies performed by our laboratory using both native and bacterially-expressed recombinant DE showed that the activity of this protein does not involve carbohydrates, and resides in the polypeptidic region of the molecule (Ellerman *et al.*, 2002). However, the analysis of the amino acid sequence of the protein indicated the lack of functional domains that could explain its involvement in gamete fusion. To elucidate the molecular mechanisms underlying DE function, a series of recombinant fragments of DE were expressed in a prokaryotic system and evaluated for their ability to both bind to the egg surface and interfere with gamete fusion. Indirect immunofluorescence (IIF) and sperm-egg fusion experiments using a first series of fragments revealed that the egg binding ability of DE would be contained within the N-terminal domain of the molecule. Subsequent experiments using a new series of recombinant fragments circumscribed this activity to a region of 45-amino acids (114-158) within this domain. Interestingly, the analysis of this region revealed that it contains the two feature motifs of the CRISP family named Signature 1 and Signature 2. To investigate whether these motifs were involved in the egg binding ability of DE, two synthetic peptides corresponding to each of these motifs were produced: Peptide 1 (P1): GHYTQVVWNST and Peptide 2 (P2): FYVCHYCPGGNY. The use of these peptides in IIF and sperm-egg fusion assays indicated that P2 but not P1 was capable of binding to the egg and interfering with gamete fusion. The lack of both egg labelling and fusion inhibition observed with a peptide containing the same amino acids than P2 but in a different order, confirmed the relevance of the Signature 2 region for the binding of DE to the egg. These results indicate that the egg-binding site of DE resides in an evolutionarily conserved region of the molecule, and provide novel information on the molecular mechanisms underlying CRISP-1 function in gamete fusion.

Tpx-1/CRISP-2

Tpx-1 was originally described as one of the primary autoantigens abundantly present within the guinea pig acrosome (Hardy *et al.*, 1988). Since then, several groups have studied different aspects of this protein. Although there is consensus on the specific expression of Tpx-1 in rodent male germ cells and its presence in mature sperm, the possible function of this sperm protein has not yet been clearly established. In rat sperm, in addition to its presence in the acrosome, Tpx-1 was found to exist as a component of the outer dense fibers of the tail (O'Bryan *et al.*, 2001), and as a surface molecule responsible for the specific interaction between spermatogenic and Sertoli cells (Maeda *et al.*, 1999). More recently, it has been shown that the C-terminal domain of recombinant Tpx-1 is able to regulate ryanodine Ca^{2+} channels (Gibbs *et al.*, 2006).

Results form our laboratory indicate that human Tpx-1 (TPX1) would be an intra-acrosomal protein which remains associated with capacitated and acrosome-reacted sperm and is relevant for gamete interaction (Busso *et al.*, 2005). In view of these observations, and considering the high degree of homology between DE/CRISP-1 and rat Tpx-1/CRISP2, we specifically explored the participation of Tpx-1 in gamete fusion. For this purpose, zona-free rat eggs were incubated with different concentrations of bacterially-expressed recombinant Tpx-1 (recTpx-1) coupled to maltose binding protein (MBP), and then inseminated with capacitated sperm. Results showed that the presence of recTpx-1 during gamete co-incubation produced a significant and dose-dependent decrease in the percentage of penetrated eggs compared to controls. These results support the participation of Tpx-1 in gamete fusion and suggest the existence of complementary sites for this protein on the egg surface. To study the localization of these binding sites, zona-free eggs were incubated with recTpx-1 or MBP, fixed and subjected to IIF using anti-MBP as primary antibody. Results showed that while eggs incubated with MBP were unlabelled, as previously described (Ellerman *et al.*, 2002), those incubated with recTpx-1 presented a bright fluorescent labelling similar to that found in DE-incubated eggs (Fig. 1).

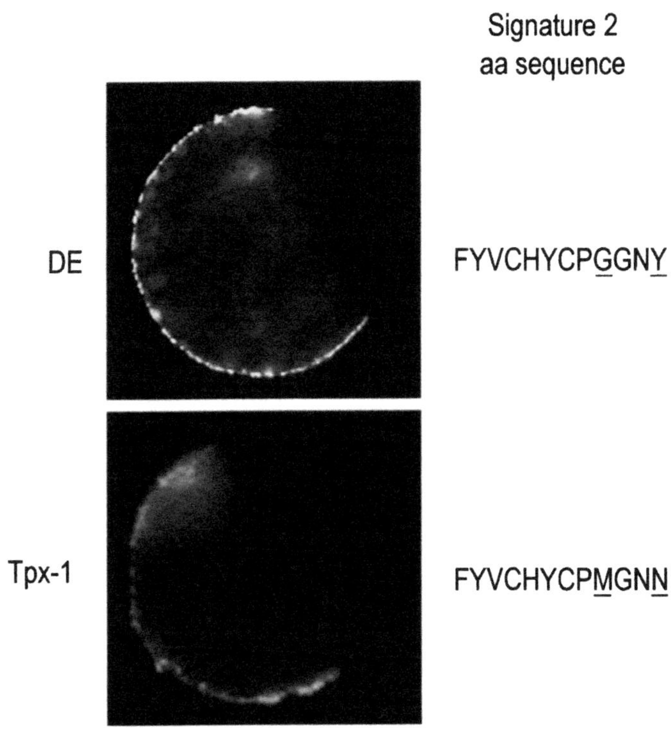

Figure 1. Immunolocalization of binding sites for DE and Tpx-1 on the egg surface. Zona-free rat eggs were incubated with 6 μM recombinant DE or recombinant Tpx-1, fixed and subjected to IIF using anti-MBP as primary antibody. Note the presence of the negative areas. The amino acid sequences on the right correspond to the Signature 2 region in each protein. The two distinct residues between DE and Tpx-1 are underlined.

Interestingly, the comparison of the amino acidic sequences corresponding to the Signature 2 motifs of DE and Tpx-1 revealed that 10 out of the 12 amino acids of this region are identical (Fig. 1), suggesting that both proteins might participate in gamete fusion through a similar molecular mechanism. The participation of two members of a family in the same event of fertilisation has already been reported. In this regard, fertilin ß and cyritestin, two members of the ADAM family, have both been involved in sperm-ZP interaction (Nishimura *et al.*, 2001). Similarly, CD81, an egg protein belonging to the tetraspanin family, has been shown to replace the function of its homologous protein CD9 during sperm-egg fusion (Kaji *et al.*, 2002).

Together, our results support the involvement of both epididymal DE/CRISP-1 and testicular Tpx-1/CRISP 2 in gamete fusion through their binding to the egg surface. These results contribute to a better understanding of the molecular mechanisms underlying gamete fusion and support the idea of a functional cooperation between homologue molecules as a mechanism to ensure the success of fertilisation.

References

Busso, D., Cohen, D.J., Hayashi, M., Kasahara, M. and Cuasnicu, P.S. (2005). Human testicular protein TPX1/ CRISP-2: localization in spermatozoa, fate after capacitation and relevance for gamete interaction. *Molecular Human Reproduction* **11**: 299-305

Cameo, M.S. and Blaquier, J.A. (1976). Androgen-controlled specific proteins in rat epididymis. *Journal of Endocrinology* **69**: 317-324

Ellerman, D.A., Da Ros, V.G., Cohen, D.J., Busso, D., Morgenfeld, M.M. and Cuasnicu, P.S. (2002). Expression and structure-function analysis of DE, a sperm cysteine-rich secretory protein that mediates gamete fusion. *Biology of Reproduction* **67**: 1225-1231

Gibbs, G.M., Scanlon, M.J., Swarbrick, J., Curtis, S., Gallant, E., Dulhunty, A.F. and O'Bryan, M.K. (2006). The cysteine-rich secretory protein domain of Tpx-1 is related to ion channel toxins and regulates ryanodine receptor Ca^{2+} signaling. *Journal of Biological Chemistry* **281**: 4156-4163

Haendler, B., Kratzschmar, J., Theuring, F. and Schleuning, W.D. (1993). Transcripts for cysteine-rich secretory protein-1 (CRISP-1; DE/AEG) and the novel related CRISP-3 are expressed under androgen control in the mouse salivary gland. *Endocrinology* **133**: 192-198

Hardy, D.M., Huang, T.T.F., Driscoll, W.J., Tung, K.S.K. and Wild, G.C. (1988). Purification and characterization of the primary acrosomal autoantigen of guinea pig epididymal spermatozoa. *Biology of Reproduction* **38**: 423-437

Jalkanen, J., Huhtaniemi, I. and Poutanen, M. (2005). Mouse cysteine-rich secretory protein 4 (CRISP4): a member of the crisp family exclusively expressed in the epididymis in an androgen-dependent manner. *Biology of Reproduction* **72**: 1268-1274

Kaji, K., Oda, S., Miyazaki, S. and Kudo, A. (2002). Infertility of CD9-deficient mouse eggs is reversed by mouse CD9, human CD9, or mouse CD81; polyadenylated mRNA injection developed for molecular analysis of sperm-egg fusion. *Developmental Biology* **247**: 327-334

Kasahara, M., Gutknecht, J., Brew, K., Spurr, N. and Goodfellow, P.N. (1989). Cloning and mapping of a testis-specific gene with sequence similarity to a sperm-coating glycoprotein gene. *Genomics* **5**: 527-534

Kjeldsen, L., Cowland, J.B., Johnsen, A.H. and Borregaard, N. (1996). SGP28, a novel matrix glycoprotein in specific granules of human neutrophils with similarity to a human testis-specific gene product and a rodent sperm-coating glycoprotein. *FEBS Letters* **380**: 246-250

Maeda, T., Nishida, J. and Nakanishi, Y. (1999). Expression pattern, subcellular localization and structure-function relationship of rat Tpx-1, a spermatogenic cell adhesion molecule responsible for association with Sertoli cells. *Development Growth and Differentiation* **41**: 715-722

Nishimura, H., Cho, C., Branciforte, D.R., Myles, D.G. and Primakoff, P. (2001). Analysis of loss of adhesive function in sperm lacking cyritestin or fertilin beta. *Developmental Biology* **233**: 204-213

O'Bryan, M.K., Sebire, K., Meinhardt, A., Edgar, K., Keah, H., Hearn, M.T. and De Kretser, D.M. (2001). Tpx-1 is a component of the outer dense fibers and acrosome of rat spermatozoa. *Molecular Reproduction and Development* **58**: 116-125

Rochwerger, L., Cohen, D.J. and Cuasnicú, P.S. (1992). Mammalian sperm-egg fusion: The rat egg has complementary sites for a sperm protein that mediates gamete fusion. *Developmental Biology* **153**: 83-90

Rochwerger, L. and Cuasnicu, P.S. (1992). Redistribution of a rat sperm epididymal glycoprotein after in vivo and in vitro capacitation. *Molecular Reproduction and Development* **31**: 34-41

Udby, L., Bjartell, A., Malm, J., Egesten, A., Lundwall, A., Cowland, J.B., Borregaard, N. and Kjeldsen, L. (2005). Characterization and localization of cysteine-rich secretory protein 3 (CRISP-3) in the human male reproductive tract. *Journal of Andrology* **26**: 333-342

Spermatology. SRF Vol. 65. ERS Roldan and M Gomendio (eds) Nottingham University Press, Nottingham

Comparison of human sperm morphometry assessment models based on zona pellucida selectivity

C. Garrett, D.Y. Liu and H.W.G. Baker

Department of Obstetrics and Gynaecology, and Reproductive Services, University of Melbourne, Royal Women's Hospital, Carlton 3053, Australia; Melbourne IVF, East Melbourne 3002, Australia

Computer image analysis improves the objectivity and reproducibility of conventional assessments of sperm morphology. We have extended this objectivity by using the physiological process of sperm-zona pellucida (ZP) binding to classify 'normal' sperm morphometry. Linear regression analysis of the morphometric differences between sperm in semen, swim-up and ZP bound samples (n=53) was used to establish 12 "zona preferred" morphometric parameters from which two different morphometric evaluations of an individual sperm's potential for motility and ability to bind to the ZP of human oocytes were calculated. The two models, %Z and %ZB, were then tested, together with conventional morphometry assessments of percent "normal" and other semen variables, against experimental sperm-ZP binding rates (n=64) and natural pregnancy rates in sub-fertile couples (n=1191). Although %ZB was the most significantly related semen variable to sperm-ZP binding rate, the simpler %Z was more significant in the Cox regression model for natural pregnancy rates in sub-fertile couples. In multivariate analysis, the only additional significant covariates were sperm straight line velocity and female age. %Z is therefore the most appropriate and also easiest of our morphometry assessments to be applied to commercial CASA systems which offer a morphometry module with stain density analysis.

Introduction

Despite well documented deficiencies in the manual assessment of human sperm morphology it is an important part of the assessment of semen quality (Ombelet, 1995; WHO, 1999). Computer image analysis has provided major advances in quantification, objectivity and reproducibility for sperm morphology. The question remains, how do we best apply this reliable source of raw morphometric information to provide a good prognostic tool for evaluation of male infertility?

To date, automated sperm morphometry systems have concentrated on 'perfection' of the conventional dichotomous classification of sperm based on historical, largely unsubstantiated criteria for normality 'Normal' sperm reflect the morphology of typical sperm in the semen of fertile men, rather than that of specific sperm capable of fertilisation. Inadequacies of the

Corresponding author E-mail: garrettc@unimelb.edu.au

classification are indicated by notable departures from the general positive relationship between normal morphology and fertilisation rates with standard *in vitro* fertilisation (IVF), particularly where good IVF results occur with semen classified as morphologically poor (Liu *et al.*, 1989).

ESHRE guidelines recommend that the selection of CASA variables should be based on a relevant functional end-point (ESHRE, 1998). Unfortunately it is not possible to ascertain the fertilising potential of the individual spermatozoon for direct comparison with its morphometry. Nor is examination of the relationship between IVF rates and semen morphometry of much value now because it has become standard practice to use intracytoplasmic sperm injection (ICSI) in the majority of patients presenting with severe to moderate teratozoospermia. Consequently, physiological endpoints other than fertilisation must be explored in order to establish criteria to assess morphology. Since the sperm-zona pellucida (ZP) binding process is an essential physiological precedent to fertilisation, both *in vivo* and in standard IVF procedures, it is expected to provide such criteria. The results of sperm-ZP binding tests have been shown to relate to fertilisation rates in standard IVF (Liu *et al.*, 1989) and, using either manual or computer automated assessments of morphology, previous studies have also shown that the sperm-ZP binding process is indeed morphologically selective (Menkveld *et al.*, 1991, Liu and Baker, 1992, Garrett *et al.*, 1997). The image analysis results enable identification of this selectivity in terms of specific morphometric parameters, not just predetermined cell shapes, thus providing a situation analogous with broadening motility measurements with CASA to include explicit parameters such as straight line velocity (VSL).

The work in this paper describes a search for the most effective application of the raw morphometric parameter output of our Sperm Head Automated Morphometric Analysis System (SHAMAS) (Garrett and Baker, 1995) to the assessment of semen quality, based on the quantitative selectivity exhibited by the sperm-ZP binding process. Considering that the degree of selectivity varies significantly between morphometric parameters, several different assessments were modelled and evaluated by comparison with experimental sperm-ZP binding rates and natural pregnancy rates in a large group of sub-fertile couples. SHAMAS explicitly examines sperm head dimensions and densitometry and is sensitive to some salient neck features but does not assess the tail. Since several motility measures, including CASA evaluation of sperm velocities were included in the multiple regression analysis of the fertility data, this limitation is not expected to compromise the evaluation.

Morphometric selectivity of swim-up and sperm-ZP binding

SHAMAS was used to measure 200 Shorr-stained sperm from 53 men for each of washed semen, swim-up preparation and those recovered from the surface of the ZP after binding to human oocytes which failed to fertilise *in vitro*. As previously described, linear regression analysis of this data was used to establish that 12 of the 32 SHAMAS parameters exhibited significant selectivity for both the swim-up and sperm-zona binding processes (Garrett *et al.*, 1997). For these selected parameters, a "preferred" value for each of the processes was also deduced for use in the semen morphometry assessment models.

Swim-up specifically selected longitudinally symmetric sperm heads with no neck anomalies. In addition, the sperm-ZP binding process favoured sperm with a large acrosomal region defined by low optical density. Absolute head dimensions and ellipticity of the posterior region did not show significant selection, consistent with our previous findings (Garrett *et al.*, 1997).

Semen morphometry assessment models

%N: the conventional percent normal morphology, assessed manually to WHO criteria.

%C: the percent conformity, the SHAMAS equivalent of %N.

%Z: the percentage of sperm in a sample conforming to within 1.5 SD of the 'ZP preferred' reference value for 12 of the SHAMAS parameters.

%ZB: a more physiological and non dichotomous evaluation which allows for variation in the relative importance of the deviation of each of the 12 parameters from the swim-up or ZP 'preferred' values, with respect to the swimming and binding ability of an individual sperm. The weighting factors for each of the selected parameters are calculated iteratively using least squares fitting between the predicted and measured parameter means for each ZP bound sample. %ZB denotes the theoretical average sperm-ZP binding probability, expressed as a percentage.

Comparison of experimental sperm-ZP binding rates and assessments of semen quality

Differential fluorescence labelling was used to measure sperm-ZP binding rates for sperm from 64 patients relative to control sperm (Liu, 1988). The only standard semen analysis variable significantly related to the sperm-ZP binding rate was concentration ($r^2=0.08$, $p=0.03$). Both morphometry measures based on ZP selectivity were significantly related (%ZB: $r^2=0.13$, $p=0.006$ and %Z: $r^2=0.07$, $p=0.04$), however, neither %C nor the manual equivalent %N reached significance (Fig. 1). Figure 1 also illustrates that the %ZB assessment of a semen sample had no cases of false negatives for sperm-ZP binding results, unlike the conventional assessments of morphology or morphometry.

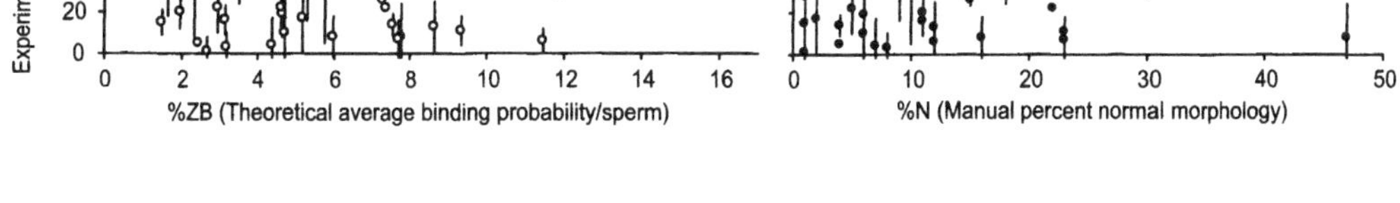

Figure 1. Linear regression of experimental sperm-ZP binding rates with respect to a ZP based morphometry assessment (%ZB) (left), compared with the conventional morphology assessment (%N) (right), for semen samples from 64 patients.

Prediction of natural pregnancy in a sub-fertile population

Natural pregnancy rate as determined by follow-up for up to 3 years, measured from the time of first semen analysis, was assessed by Cox regression analysis for 1191 couples with no known absolute barrier to conception (Fig. 2) (Garrett *et al.*, 2003). All morphometry assessments,

%N, sperm concentration, %progressive motility and VSL were significantly related to pregnancy rate by univariate analysis ($p < 0.05$). However, in multivariate analysis only three factors were found to be independently and significantly related, with %Z the most important, followed by VSL and then age of the female partner. The alternative measures of sperm morphology and morphometry were less significant than %Z in combination with VSL and age.

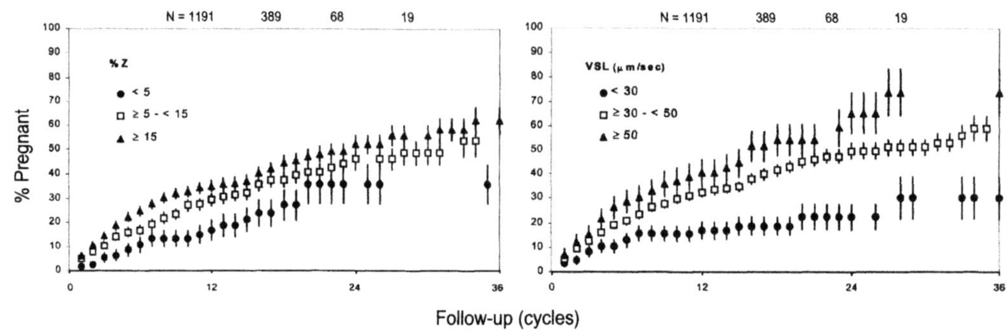

Figure 2. Effect of the percentage 'ZP preferred' morphometry (%Z) and straight line velocity (VSL) on natural pregnancy rates in 1191 subfertile couples, uncorrected for other covariates. The number of couples followed for 12, 24 and 36 cycles are denoted at the top of the graphs. Modified from Garrett *et al.* (2003), with permission from Oxford University Press.

Conclusions

Morphometry assessments based explicitly on sperm-ZP binding selectivity show an expected improved relationship with fertility compared with conventional morphology. Although the more sophisticated model %ZB best described the ZP binding ability of a semen sample, the simpler %Z was the more significant in the analysis of natural pregnancy in a large subfertile population. The %Z assessment has the added advantage of being more easily applied to other CASA systems which have a morphometry module which analyses stain density.

In summary, the results presented in this study illustrate the benefit of investigating alternative or supplementary assessments of sperm motility and morphology, such as VSL and %Z, made possible by the extensive parameterisation of information offered by image analysis.

References

ESHRE Andrology Special Interest Group (1998). Guidelines on application of CASA technology in the analysis of spermatozoa. *Human Reproduction* **13**: 142-145

Garrett, C. and Baker, H.W.G. (1995). A new fully automated system for the morphometric analysis of human sperm heads. *Fertility and Sterility* **63**: 1306-1317

Garrett, C., Liu, D.Y. and Baker, H.W.G. (1997). Selectivity of the human sperm-zona pellucida binding process to sperm head morphometry. *Fertility and Sterility* **67**: 362-371

Garrett, C., Liu, D.Y., Clarke, G.N., Rushford, D.D. and Baker, H.W.G. (2003). Automated semen analysis: 'zona pellucida preferred' sperm morphometry and straight line velocity are related to pregnancy rate in subfertile couples. *Human Reproduction* **18**: 1643-1649

Liu, D.Y., Lopata, A., Johnston, W.I.H. and Baker, H.W.G. (1988). A human sperm-zona pellucida binding test using oocytes that failed to fertilize *in vitro*. *Fertility and Sterility* **50**: 782-788

Liu, D.Y., Clarke, G.N., Lopata, A., Johnston, W.I.J. and Baker, H.W.G. (1989) A sperm-zona pellucida binding test and *in vitro* fertilization. *Fertility and Sterility* **52**: 281-287

Liu, D.Y. and Baker, H.W.G. (1992) Morphology of spermatozoa bound to the zona pellucida of human oocytes that failed to fertilise *in vitro*. *Journal of Reproduction and Fertility* **94**: 71-84

Menkveld, R., Franken, D.R., Kruger, T.F., Oehninger, S. and Hodgen, G.D. (1991). Sperm selection capacity of the human zona pellucida. *Molecular Reproduction and Development* **30**: 346-352

Ombelet, W., Menkveld, R., Kruger, T.F. and Steeno, O. (1995). Sperm morphology assessment: historical review in relation to infertility. *Human Reproduction Update* **1**: 543-557

World Health Organisation (1999). *WHO Laboratory Manual for the Examination of Human Semen and Sperm-Cervical Mucus Interaction*. 4th ed., Cambridge University Press, Cambridge

Spermatology. SRF Vol. 65. ERS Roldan and M Gomendio (eds) Nottingham University Press, Nottingham

Sperm-egg interaction and gene manipulated animals

Naokazu Inoue, Ryo Yamaguchi*, Masahito Ikawa and Masaru Okabe

*Genome Information Research Center and *Graduate School of Pharmaceutical Sciences, Osaka University, Yamadaoka 3-1, Suita, Osaka 565-0871, Japan*

It is intriguing that the disruption of a number of genes, e.g., Clgn, Ace, Adam1a, Adam2 and Adam3 results in a similar sperm phenotype, i.e., failure of sperm to bind to the zona pellucida (ZP). Because calmegin functions as a testis-specific molecular chaperone, there is a possibility of misfolding of ADAMs in sperm from Clgn$^{-/-}$ mice. In the first half of this review we describe the interaction of ADAMs with calmegin and try to elucidate the relationship of these proteins to establish the zona binding ability. In the second half of this review we describe other gene manipulated animals that lead to a defect in sperm-egg fusion. The first factor, CD9, was found serendipitously on the egg side, while the second factor, Izumo, on the sperm side, was discovered after 20 years of pursuit. Here we describe various gene manipulated animals that are useful to elucidate the mechanism of fertilisation.

Introduction

Mammalian fertilisation is accomplished through complex processes involving cell matrix and cell-cell interactions. The zona pellucida (ZP) is an extracellular matrix that serves as a barrier to protect the egg from physiological damage (Nichols and Gardner, 1989) and to block polyspermic fertilisation. Based on biochemical analyses, different sperm proteins have been postulated to play roles in ZP binding, penetration and fusion. However, when genes such as Acr (acrosin) (Baba, Azuma, Kashiwabara and Toyoda, 1994), B4galt1 (Lu and Shur, 1997), Spam1 (PH-20) (Baba, Kashiwabara, Honda, Yamagata, Wu, Ikawa, Okabe and Baba, 2002), Mfge8 (SED1) (Ensslin and Shur, 2003) and Cd46 (Inoue, Ikawa, Nakanishi, Matsumoto, Nomura, Seya and Okabe, 2003) were ablated, the predicted phenotypes were not demonstrated, suggesting that the various models explaining fertilisation need to be refined. The first example of a gene knockout experiment that leads to male infertility without impairing spermatogenesis was demonstrated in calmegin (Clgn). The calmegin (Clgn) gene was found to be essential for sperm to acquire the binding ability to egg ZP (Ikawa, Wada, Kominami, Watanabe, Toshimori, Nishimune and Okabe, 1997). After our finding, other genes, such as Adam1a, Adam2, Adam3, and the angiotensin-converting enzyme (Ace), also have emerged as essential genes for sperm-ZP binding (Krege, John, Langenbach, Hodgin, Hagaman, Bachman, Jennette, O'Brien and Smithies, 1995; Nishimura, Kim, Nakanishi and Baba, 2004; Shamsadin, Adham, Nayernia, Heinlein, Oberwinkler and Engel, 1999).

Corresponding author E-mail: okabe@gen-info.osaka-u.ac.jp

Why do calmegin knockout and ADAMs family knockout mice share the same phenotype?

We have previously shown that the disruption of calmegin expression results in defective sperm adhesion to egg ZP (Ikawa *et al.*, 1997). After our report, ADAM2$^{-/-}$ sperm were also reported to exhibit defective adhesion to ZP. Moreover, an impaired migration into the oviduct and a defective adhesion to egg plasma membrane were observed for ADAM2$^{-/-}$ sperm. As shown in Figure 1, calmegin$^{-/-}$ sperm were also defective in migration into the oviduct and in adhesion to the egg plasma membrane. To analyse egg-fertilising ability of calmegin$^{-/-}$ sperm further, we performed a partial zona dissection followed by *in vitro* fertilisation. Although the fertilisation rate, 30 ± 14 %, observed in calmegin$^{-/-}$ mice sperm was low (36% of wild type), the value was comparable to that reported in ADAM2$^{-/-}$ sperm (50% of wild type) (Cho, Bunch, Faure, Goulding, Eddy, Primakoff and Myles, 1998). From these experiments, calmegin was demonstrated to be required for sperm migration and zona pellucida adhesion but not essential for egg plasma membrane fusion as described for ADAM2$^{-/-}$ mice (Cho *et al.*, 1998; Cho, Ge, Branciforte, Primakoff and Myles, 2000; Yamagata, Nakanishi, Ikawa, Yamaguchi, Moss and Okabe, 2002). It is noteworthy that once the eggs were fertilised with calmegin$^{-/-}$ sperm, there were no significant differences in subsequent development.

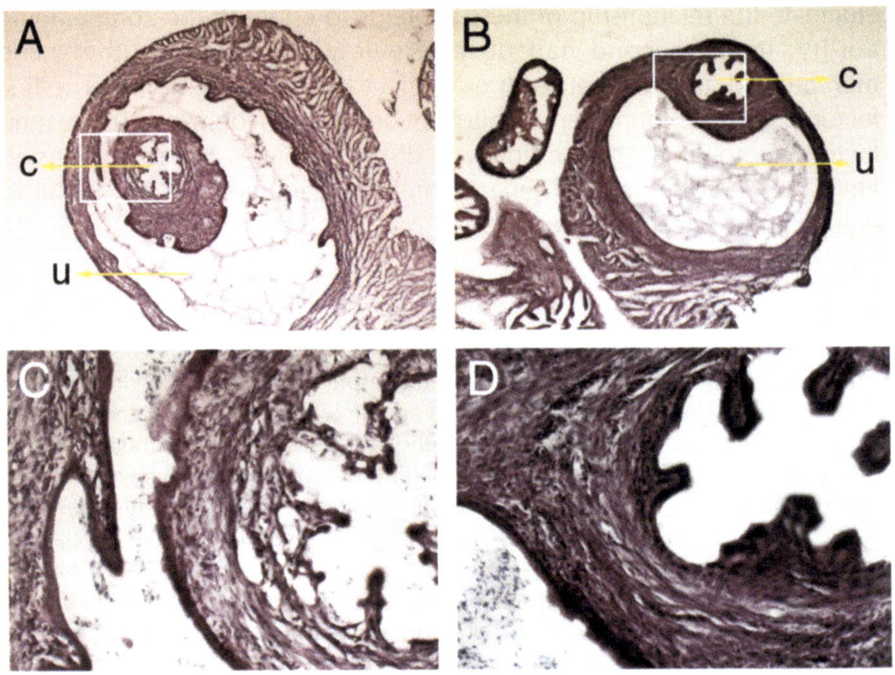

Figure 1. Failure of oviductal migration of calmegin$^{-/-}$ sperm. Sperm migration was observed by preparing a thin section of the utero-tubal junction, and staining with hematoxylin and eosin. (**A**,**C**) Wild type; (**B**,**D**) calmegin$^{-/-}$. Areas boxed in (**A**) and (**B**) were magnified in (**C**) and (**D**). Note that there are no spermatozoa inside the colliculus part of the oviduct when mating occurred with a calmegin$^{-/-}$ male, as shown in (**D**). c: colliculus, u: uterus. From Ikawa *et al.* (2001), reproduced with permission.

Since there are some cases in which apparent phenotypes of certain gene knockout mice were actually derived from the indirect disruption of other genes (Olson, Arnold, Rigby and Wold, 1996), we examined whether or not expression of calmegin from a transgene can restore the phenotype of calmegin$^{-/-}$ mice. We found that 2 of 4 transgenic mouse lines that express calmegin from the transgene during spermatogenesis regained their fecundity. These results clearly demonstrated that the deficiency of calmegin was the real cause of infertility of calmegin$^{-/-}$ mice.

Since calmegin was localised on endoplasmic reticulum (ER) membranes and disappeared from spermatogenic cells during the last stages of spermatogenesis (Watanabe, Sawada, Koshimizu, Kagawa and Nishimune, 1992; Yoshinaga, Tanii and Toshimori, 1999), calmegin could not participate in the sperm-egg adhesion event directly. In addition, since calmegin is considered to function like calnexin (Ikawa *et al.*, 1997; Watanabe, Yamada, Nishina, Tajima, Koshimizu, Nagata and Nishimune, 1994), we speculated that calmegin chaperones sperm membrane proteins required for sperm fertility. Therefore, the very similar phenotype of calmegin$^{-/-}$ and ADAM2$^{-/-}$ mouse lines led us to investigate the relationship between these two proteins. Immunoprecipitation followed by Western blot analysis revealed that both ADAM1 and ADAM2 specifically complexed with calmegin in the ER, but not with calnexin (Fig. 2). The disruption of calmegin was shown to cause impaired heterodimerisation of Adam1/Adam2 and finally resulted in the complete absence of ADAM2 from mature sperm (Fig. 3). Since Adam1 was not present in mature sperm when the ADAM2 gene was disrupted (Cho *et al.*, 1998), Adam1 was also predicted to be missing from mature calmegin$^{-/-}$ sperm. There is precedence for the disappearance of a membrane protein from the cell surface when the chaperone function of calnexin (a calmegin homologue) was disrupted. For example, in the absence of a functioning calnexin, the formation of the insulin receptor homodimer was repressed and the receptor failed to be presented on the cell surface (Bass, Chiu, Argon and Steiner, 1998). The present data not only indicated the importance of calmegin function for the sperm membrane protein maturation but also reinforced the concept of ER chaperone functioning to form dimerised proteins.

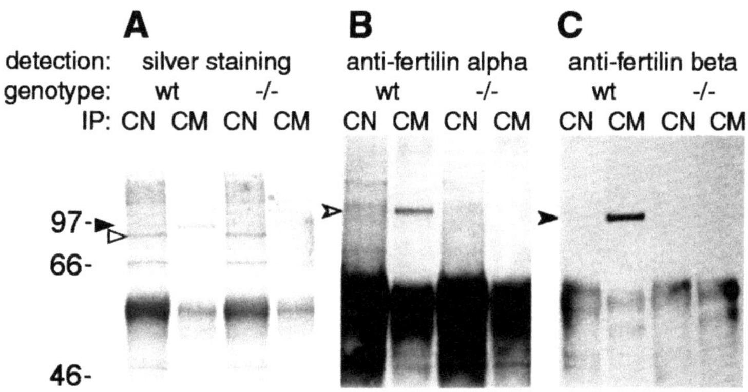

Figure 2. Specific binding of fertilin alpha (ADAM1) and beta (ADAM2) to calmegin. Testicular cell extracts prepared from wild-type and calmegin$^{-/-}$ mice were immunoprecipitated with anti-calnexin or anti-calmegin antibodies. (**A**) The immunoprecipitated calnexin (CN) and calmegin (CM) bands (90 and 93 kDa, respectively) were detectable even by silver staining after SDS-PAGE under reducing conditions. Comparable amounts of CN and CM immunoprecipitates were analysed in the following Western blotting experiments. (**B**,**C**) Proteins transferred to PVDF membrane were probed with anti-fertilin alpha (ADAM1) (**B**), or anti-fertilin beta (ADAM2) (**C**)-specific antibodies. From Ikawa *et al.* (2001), reproduced with permission.

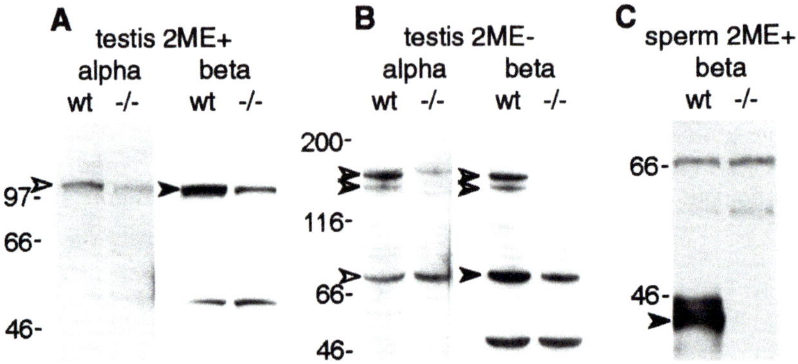

Figure 3. Impaired fertilin alpha/beta (ADAM1/2) assembly and disappearance of fertilin beta (ADAM2) from the sperm surface. (**A**,**B**) Testicular proteins were resolved by SDS-PAGE under reducing (**A**) and nonreducing conditions (**B**). (**C**) Sperm proteins were resolved by SDS-PAGE under reducing conditions. Proteins transferred to PVDF membrane were probed with anti-fertilin alpha (ADAM1)- or anti-fertilin beta (ADAM2)-specific antibodies. Open arrowhead, fertilin alpha; solid arrowhead, fertilin beta; half-solid arrowhead, fertilin alpha/beta heterodimer. From Ikawa *et al.* (2001), reproduced with permission.

These results indicate that the heterodimerisation is important for the subsequent presentation of ADAM1/ADAM2 heterodimer on sperm surface. The result could explain why calmegin- and ADAM2-disrupted mouse lines share the same phenotype (Fig. 4). However, the key factor that gives sperm a zona binding ability remains an unanswered question.

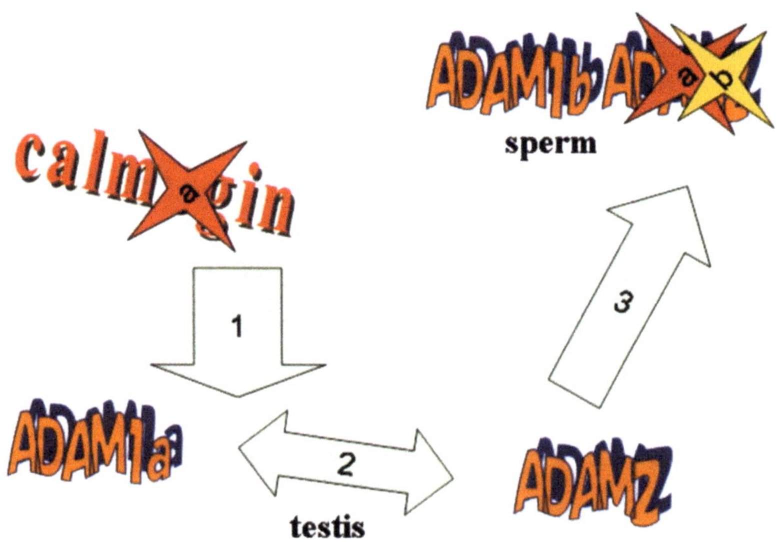

Figure 4. Schematic diagram of various factors involved in sperm-zona binding deduced from gene knockout experiment. The disruption of testis-specific molecular chaperone calmegin results in the failure of ADAM1a/2 formation in testis. The disappearance of ADAM1a/2 heterodimer (either by calmegin knockout or by ADAM1a knockout) leads to the disappearance of ADAM3 from the sperm surface. Thus all of the calmegin, ADAM1a, ADAM2 and ADAM3 knockout mouse lines share the same phenotype (Cho *et al.*, 1998, 2000; Ikawa *et al.*, 1997, 2001; Nishimura *et al.*, 2004).

Elucidation of sperm-egg fusion mechanism through gene manipulated animals

A number of genes have been disrupted using homologous recombination, and other disruptions are still under way. Since the entire knockout mouse lines have to be established by breeding, any gene having a role in reproduction can be found in this breeding process. Thus, the factors involved in mouse reproduction are emerging one by one, sometimes with a surprise.

The CD9-disrupted mouse line is one of the findings classified as serendipitous. The involvement of CD9 on eggs in sperm-egg fusion was not at all expected before the experiment. However, the phenotype was very drastically expressed in the fertilisation phenomena. The CD9-depleted mouse eggs could not fuse with sperm (Miyado, Yamada, Yamada, Hasuwa, Nakamura, Ryu, Suzuki, Kosai, Inoue, Ogura, Okabe and Mekada, 2000). CD9 was the first molecule proven to be involved in sperm-egg fusion. However, since this is the only factor, there should be more factors examined to elucidate the mechanism of sperm-egg fusion.

To identify factors involved in sperm-egg fusion, we utilised an anti-mouse sperm monoclonal antibody OBF13 that specifically inhibits the fusion process. The antigen was identified by separation of the crude extracts from mouse sperm by two-dimensional gel electrophoresis and subsequent immunoblotting with the monoclonal antibody. We termed the antigen "Izumo" after a Japanese shrine dedicated to marriage. The identified spot was analysed by liquid chromatography tandem mass spectrometry (LC-MS/MS), and 10 peptides, which were 100% identical to a part of the sequence listed in the RIKEN full-length database (NCBI accession number: XM_133424) were found. The registered DNA sequence was reconfirmed by sequencing following RT-PCR using total RNA prepared from the testis. A human homologue was found as an unverified gene on NCBI (accession number BC034769). The gene encodes a novel immunoglobulin superfamily (IgSF), type I membrane protein with one extracellular Ig domain that contains one putative glycosylation site.[1] Mouse Izumo was shown to be a testis (sperm)-specific 56.4 kDa antigen by Western blotting using a polyclonal antibody raised against recombinant mouse Izumo (Fig. 5). Coinciding with the fact that mammalian sperm are incapable of fertilising eggs when ejaculated and that fertilisation only occurs after the acrosome reaction, Izumo became detectable on sperm surface only after the acrosome reaction. This was probably because Izumo is not localised on plasma membrane of fresh spermatozoa, but is hidden under the plasma membrane and accessible after the acrosome reaction as in the case of CD46 on mouse sperm.

To address the physiological role of Izumo *in vivo*, we generated Izumo-deficient mice by homologous recombination. Izumo$^{-/-}$ mutant mice were healthy and showed no overt developmental abnormalities. Izumo$^{-/-}$ females demonstrated normal fecundity. Izumo$^{+/-}$ males also showed normal fertilising ability. However, Izumo$^{-/-}$ males were sterile despite normal mating behaviour and ejaculation with normal vaginal plug formations. After observing 28 plugs, 9 pairs of Izumo$^{-/-}$ male and wild-type female were kept for another 4 months but no pregnancies were observed. There was no reduction of sperm motility in Izumo$^{-/-}$ sperm; motility was measured 120 min after incubation by computer aided sperm analysis (CASA) (81.7% $\pm$ 7.7 in Izumo$^{+/-}$ sperm and 77% $\pm$ 8.9 in Izumo$^{-/-}$ sperm). The sterile nature of Izumo$^{-/-}$ sperm was also demonstrated in the *in vitro* fertilisation system (Fig. 6). The impaired fertilisation step undoubtedly followed zona-penetration, because sperm penetrated the zona pellucida and accumulated in the perivitelline space of the eggs.

Sperm and egg membrane interactions can be considered to occur in to two stages: binding of the sperm plasma membrane to that of the egg and actual membrane fusion. Izumo$^{-/-}$ sperm were capable of binding to the plasma membranes of eggs whose ZP had been mechanically removed. In this system, the Izumo$^{+/-}$ sperm incubated for 2 and 6 h fused to eggs at approximate ratios of 4.5 and 6 sperm/egg, respectively, but no Izumo$^{-/-}$ sperm fused with eggs.

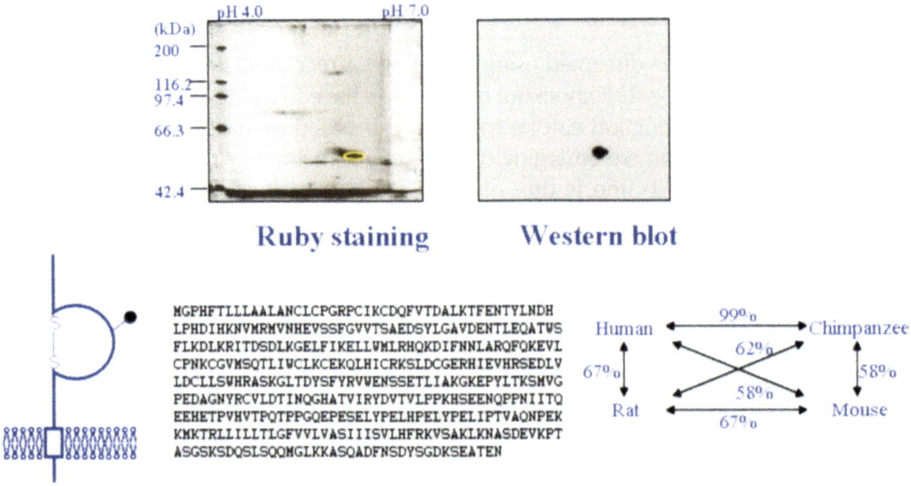

Figure 5. Identification and characterization of Izumo. After identifying the Izumo spot on a 2D-gel by Western blotting using OBF13 monoclonal antibody, the spot was subjected to LC/MS/MS analysis. The amino-acid sequence of the deduced antigen was as indicated. The antigen consisted of a signal peptide, one immunoglobulin-like domain, and a transmembrane domain, and was revealed as an immunoglobulin super family protein. The homologue of mouse Izumo was also found in rat, chimpanzee and human. Modified from Inoue *et al.* (2005), reproduced with permission.

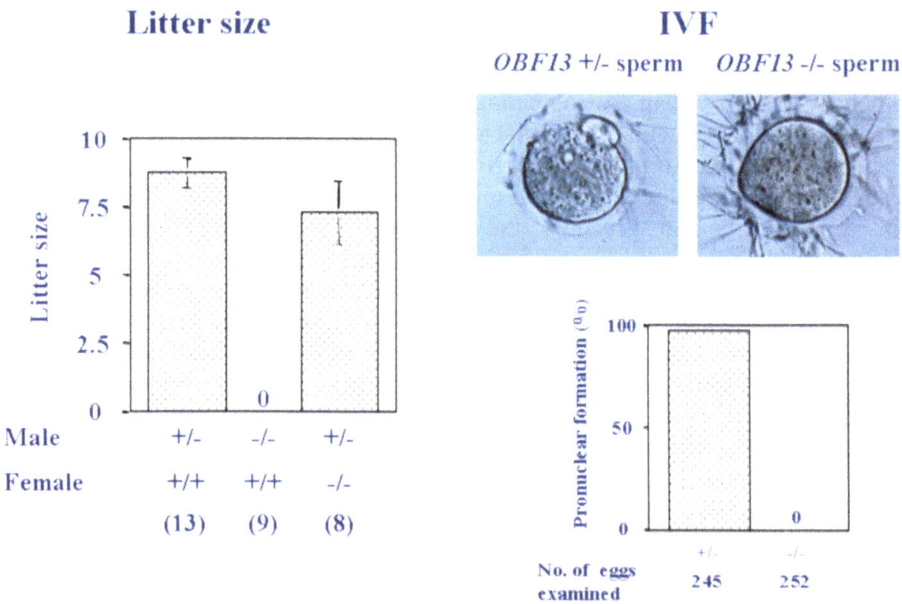

Figure 6. Male infertility caused by Izumo disruption. Left: fecundity of Izumo$^{+/-}$ and Izumo$^{-/-}$ males and Izumo$^{-/-}$ females. The numbers in parentheses indicate the numbers of mating pairs. Right: *in vitro* fertilisation (IVF) using spermatozoa from Izumo$^{+/-}$ and Izumo$^{-/-}$ mice. The eggs inseminated with Izumo$^{-/-}$ spermatozoa had many sperm cells on their zona pellucida, unlike the situation seen with Izumo$^{+/-}$ spermatozoa. This is due to the failure of sperm-egg fusion, probably leading to the absence of zona-reaction to lessen the sperm-binding ability of the zona pellucida. The error bars are not visible (n = 5). Modified from Inoue *et al.* (2005), reproduced with permission.

It is widely accepted that sperm cannot fuse with eggs unless sperm undergo an exocytotic process called the acrosome reaction. To verify the acrosomal status of Izumo$^{-/-}$ sperm, we stained the sperm accumulated in perivitelline spaces with the MN9 monoclonal antibody which immunoreacts only to the equatorial segment of acrosome-reacted sperm. The staining indicated that the Izumo$^{-/-}$ sperm had undergone the acrosome reaction but failed to fuse with eggs (Fig. 7). However, we could produce pups from Izumo$^{-/-}$ sperm by utilising intracytoplasmic sperm injection (ICSI). Eggs injected with Izumo$^{-/-}$ sperm were successfully activated and the fertilised eggs were transplanted into the oviducts of pseudo-pregnant females. The eggs implanted normally and the resulting embryos developed appropriately to term with rates similar to those of heterozygous mice (Inoue, Ikawa, Isotani and Okabe, 2005).

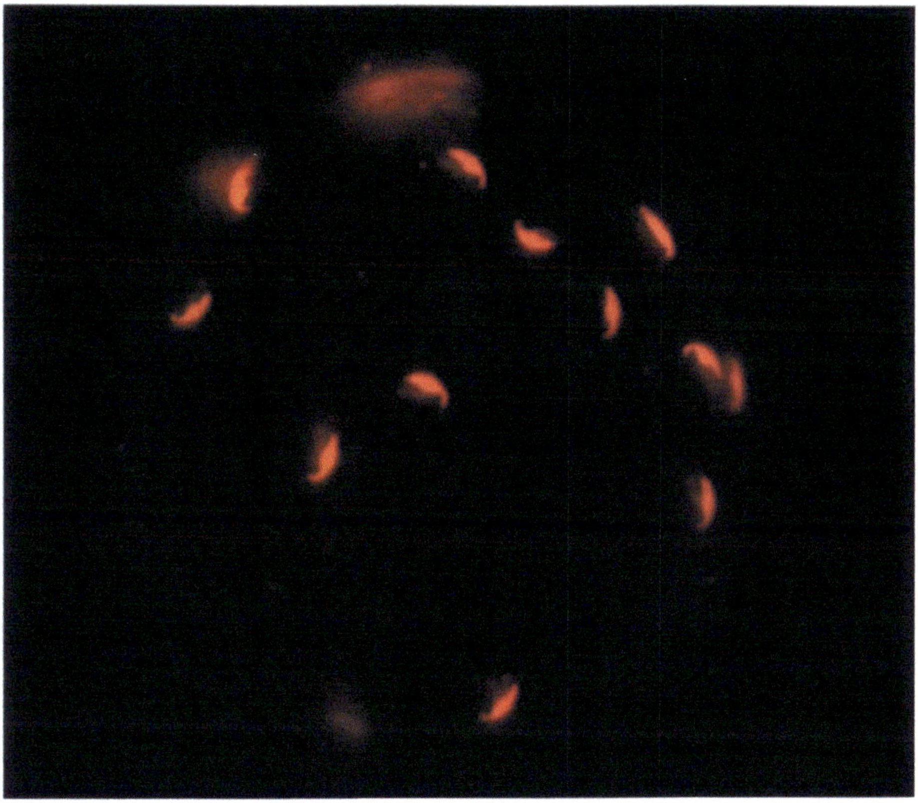

Figure 7. Accumulation of many sperm in the perivitelline space of eggs recovered from the females mated with Izumo$^{-/-}$ males. Izumo$^{-/-}$ spermatozoa failed to fuse with the egg, resulting in the absence of a zona reaction. This is why many sperm cells are seen in the perivitelline space when Izumo$^{-/-}$ sperm are used for *in vitro* fertilisation. All the spermatozoa that penetrated the zona pellucida were acrosome-reacted because they reacted to "acrosome-reacted sperm"-specific monoclonal antibody MN9 (Toshimori, Saxena, Tanii and Yoshinaga, 1998).

Sperm-egg fusion is known to be less species-specific than sperm-zona interaction. For example, human sperm cannot penetrate the hamster ZP but they can fuse with zona-free hamster eggs, and this system (zona-free hamster egg sperm penetration test. hamster test) has been used for assessment of human sperm fertility. We first examined the contribution of mouse

Izumo in a zona-free hamster egg sperm penetration assay. The mouse Izumo was essential not only in the homologous fusion system but also for xenogeneic fusion with hamster eggs. Likewise, when the anti-human Izumo polyclonal antibody was added to the incubation mixture, no fusion was observed, while the sperm treated with control IgG fused with eggs at an average of 5.9 ± 0.7 sperm/egg. The total numbers of eggs observed were 23 and 29, respectively (n = 3). These results suggested that human Izumo might be also involved in the fertilisation process in human sperm. However, further investigation will be required to elucidate the function of Izumo in human fertilisation because addition of the antibody inhibited not only fusion but also sperm binding to the egg plasma membrane in the xenogeneic sperm-egg fusion system (Inoue *et al.*, 2005).

As stated before, the phenotypes of gene knockout mice do not necessarily directly relate to the disrupted genes, but are sometimes caused by disruption of neighbouring genes. To examine whether or not the phenotype was directly derived from the lack of Izumo on sperm, we performed the rescue experiment by crossing $Izumo^{-/-}$ mice with transgenic mouse lines generated to express Izumo using the testis-specific calmegin promoter. The sterile phenotype was rescued with the transgenically expressed Izumo on mouse sperm (Inoue *et al.*, 2005).

Izumo is the first sperm membrane protein proven to be essential for fusion. It is not yet known whether sperm Izumo interacts with egg CD9, as in the case of placental IgSF protein PSG17. Nor do we know why the localization of Izumo after the acrosome reaction is not limited to the equatorial segment where fusion initially takes place. All we can say now is that continued study of this protein's function will undoubtedly lead to a fuller understanding of the cell-cell fusion process in fertilisation and perhaps in other somatic systems such as muscle cells or trophoblasts.

The finding not only provides insight into the enigmatic fusion mechanism but also promises future benefits for the clinical treatment of infertility and the potential development of novel contraceptive strategies.

Lastly, the "antigenically important" genes have not always presented the expected phenotypes by gene disruptions in the past. Were those genes not important for fertilisation? One may argue that the phenotypes did not turn out to be very severe because other factors compensated for the disrupted gene. Such a case might exist. However, sperm function in fertilisation is very different from, for example, liver cells, which can adjust their metabolic systems depending on the environment. In sperm, all the genes are shut down and they possess no organelle to synthesise proteins. Therefore, if the fertilisation process is made robust in anticipation of malfunctions, the factors must be redundantly prepared when sperm are made. If so, the addition of an antibody may not be able to show any inhibitory activity. We therefore believe that the mechanism of fertilisation should be primarily drawn by analysing many gene manipulated animals.

References

Baba, D., Kashiwabara, S., Honda, A., Yamagata, K., Wu, Q., Ikawa, M., Okabe, M. and Baba, T. (2002). Mouse sperm lacking cell surface hyaluronidase PH-20 can pass through the layer of cumulus cells and fertilize the egg. *Journal of Biological Chemistry* **277:** 30310-30314

Baba, T., Azuma, S., Kashiwabara, S. and Toyoda, Y. (1994). Sperm from mice carrying a targeted mutation of the acrosin gene can penetrate the oocyte zona pellucida and effect fertilization. *Journal of Biological Chemistry* **269:** 31845-31849

Bass, J., Chiu, G., Argon, Y. and Steiner, D.F. (1998). Folding of insulin receptor monomers is facilitated by the molecular chaperones calnexin and calreticulin and impaired by rapid dimerization. *Journal of Biological Chemistry* **141:** 637-646

Cho, C., Bunch, D.O., Faure, J.E., Goulding, E.H., Eddy, E.M., Primakoff, P. and Myles, D.G. (1998). Fertilization defects in sperm from mice lacking fertilin beta. *Science* **281:** 1857-1859

Cho, C., Ge, H., Branciforte, D., Primakoff, P. and Myles, D.G. (2000). Analysis of mouse fertilin in wild-type and fertilin beta$^{-/-}$ sperm: evidence for C-terminal modification, alpha/beta dimerization, and lack of essential role of fertilin alpha in sperm-egg fusion. *Developmental Biology* **222:** 289-295

Ensslin, M.A. and Shur, B.D. (2003). Identification of mouse sperm SED1, a bimotif EGF repeat and discoidin-domain protein involved in sperm-egg binding. *Cell* **114:** 405-417

Ikawa, M., Nakanishi, T., Yamada, S., Wada, I., Kominami, K., Tanaka, H., Nozaki, M., Nishimune, Y. and Okabe, M. (2001). Calmegin is required for fertilin alpha/beta heterodimerization and sperm fertility. *Developmental Biology* **240:** 254-261

Ikawa, M., Wada, I., Kominami, K., Watanabe, D., Toshimori, K., Nishimune, Y. and Okabe, M. (1997). The putative chaperone calmegin is required for sperm fertility. *Nature* **387:** 607-611

Inoue, N., Ikawa, M., Isotani, A. and Okabe, M. (2005). The immunoglobulin superfamily protein Izumo is required for sperm to fuse with eggs. *Nature* **434:** 234-238

Inoue, N., Ikawa, M., Nakanishi, T., Matsumoto, M., Nomura, M., Seya, T. and Okabe, M. (2003). Disruption of mouse CD46 causes an accelerated spontaneous acrosome reaction in sperm. *Molecular and Cell Biology* **23:** 2614-2622

Krege, J.H., John, S.W., Langenbach, L.L., Hodgin, J.B., Hagaman, J.R., Bachman, E.S., Jennette, J.C., O'Brien, D.A. and Smithies, O. (1995). Male-female differences in fertility and blood pressure in ACE-deficient mice. *Nature* **375:** 146-148

Lu, Q. and Shur, B.D. (1997). Sperm from beta 1,4-galactosyltransferase-null mice are refractory to ZP3-induced acrosome reactions and penetrate the zona pellucida poorly. *Development* **124:** 4121-4131

Miyado, K., Yamada, G., Yamada, S., Hasuwa, H., Nakamura, Y., Ryu, F., Suzuki, K., Kosai, K., Inoue, K., Ogura, A., Okabe, M. and Mekada, E. (2000). Requirement of CD9 on the egg plasma membrane for fertilization. *Science* **287:** 321-324

Nichols, J. and Gardner, R.L. (1989). Effect of damage to the zona pellucida on development of preimplantation embryos in the mouse. *Human Reproduction* **4:** 180-187

Nishimura, H., Kim, E., Nakanishi, T. and Baba, T. (2004). Possible function of the ADAM1a/ADAM2 Fertilin complex in the appearance of ADAM3 on the sperm surface. *Journal of Biological Chemistry* **279:** 34957-34962

Olson, E.N., Arnold, H.H., Rigby, P.W. and Wold, B.J. (1996). Know your neighbors: three phenotypes in null mutants of the myogenic bHLH gene MRF4. *Cell* **85:** 1-4

Shamsadin, R., Adham, I. M., Nayernia, K., Heinlein, U. A., Oberwinkler, H. and Engel, W. (1999). Male mice deficient for germ-cell cyritestin are infertile. *Biology of Reproduction* **61:** 1445-1451

Toshimori, K., Saxena, D.K., Tanii, I. and Yoshinaga K. (1998). An MN9 antigenic molecule, equatorin, is required for successful sperm-oocyte fusion in mice. *Biology of Reproduction* **59**: 22-29

Watanabe, D., Sawada, K., Koshimizu, U., Kagawa, T. and Nishimune, Y. (1992). Characterization of male meiotic germ cell-specific antigen (Meg 1) by monoclonal antibody TRA 369 in mice. *Molecular Reproduction and Development* **33:** 307-312

Watanabe, D., Yamada, K., Nishina, Y., Tajima, Y., Koshimizu, U., Nagata, A. and Nishimune, Y. (1994). Molecular cloning of a novel Ca^{2+}-binding protein (calmegin) specifically expressed during male meiotic germ cell development. *Journal of Biological Chemistry* **269:** 7744-7749

Yamagata, K., Nakanishi, T., Ikawa, M., Yamaguchi, R., Moss, S. B. and Okabe, M. (2002). Sperm from the calmegin-deficient mouse have normal abilities for binding and fusion to the egg plasma membrane. *Developmental Biology* **250:** 348-357

Yoshinaga, K., Tanii, I. and Toshimori, K. (1999). Molecular chaperone calmegin localization to the endoplasmic reticulum of meiotic and post-meiotic germ cells in the mouse testis. *Archives of Histology and Cytology* **62:** 283-293

[1]The GenBank accession numbers are AB195681 for mouse Izumo, AB195682 for human Izumo and AB195683 for rat Izumo.

Spermatology. SRF Vol. 65. ERS Roldan and M Gomendio (eds) Nottingham University Press, Nottingham

Ensuring continuity of the paternal genome: potential roles for spermatozoal RNA in mammalian embryogenesis

David Miller

Reproduction and Early Development Group, University Department of Obstetrics and Gynaecology, Level D, Clarendon Wing, Leeds General Infirmary, Belmont Grove, Leeds, Ll2 9NS, UK

The mammalian male gamete is traditionally considered to be a less complex cell than its female counterpart, primarily because all development (bar oocyte activation) is thought to be initiated by the egg. However, the spermatozoon is exquisitely specialised to deliver the paternal genome to the oocyte and, in this respect, is pared down to carry out that job as efficiently as possible. As such, it is transcriptionally silent as a consequence of the highly condensed architecture of its chromatin and yet some spermatozoal DNA may be organised into domains that resemble sites of transcriptional activity. There is also little or no cytoplasm capable of supporting translation; however, we now understand that these cells carry a cargo of translationally competent mRNAs. Moreover, a recent report has provided intriguing new evidence that, under certain conditions, spermatozoa can translate their mRNAs *de novo*. Indeed, delivery of an RNA cargo by the spermatozoon to the oocyte may be as fundamental a feature of this quiescent cell as delivery of the paternal genome itself. New evidence is available showing that spermatozoal RNA can potentially affect phenotypic traits in offspring. This epigenetic phenomenon may involve the transmission of extra-chromosomal episomal elements. Taken together, these data suggest that the spermatozoon is much more than the sum of its constituent parts and that the continued transmission of male-benefit genes and possibly selfish elements may be one reason for the male gamete's additional and unexpected investment in delivering the paternal genome safely and efficiently.

Introduction

Perhaps the most critical moment in any new sexually reproducing life is the coming together of egg and sperm and the successful formation of a zygote. Traditionally, syngamy was always considered to be under the full control of the oocyte; however the comparatively recent discoveries that the mammalian spermatozoon (with the notable exception of some rodent species) is responsible for introducing the centrosome and also an essential egg activating factor, has changed our preconception of the male gamete's role in fertilisation (Palermo, Colombero and

Corresponding author E-mail: d.miller@leeds.ac.uk

Rosenwaks, 1997; Saunders, Larman, Parrington, Cox, Royse, Blayney, Swann and Lai , 2002). Indeed, it is tempting to consider that these additional sperm-derived investments to the oocyte arose in part to ensure the successful transmission and continuity of the male genome. More recently, it has been confirmed that spermatozoa also deliver a cargo of RNA to the egg that has the potential to influence phenotype through an epigenetic alteration to gene expression (Rassoulzadegan, Grandjean, Gounon, Vincent, Gillot and Cuzin, 2006). Spermatozoal RNA can also be translated *de novo* by gametes that are undergoing capacitation and it is thought that this mechanism acts to replace or regenerate nascent proteins that are somehow denatured or degraded during transit in the female genital tract (Gur and Breitbart, 2006). Hence, we can now view the spermatozoon as far more than just a vessel for the safe delivery of the paternal genome, although that of course, is the ultimate and specialised role of this cell. Instead, the spermatozoon can be viewed as a dynamic cell that is optimised in a variety of interesting ways for fertilisation, including the delivery of extra-chromosomal components.

This review takes a closer look at how spermatozoal RNA has come back into focus recently, firstly as a record of spermatogenesis and as a proxy for the germ line in the testis; secondly as a reserve of information for the synthesis of new proteins and finally as a purveyor of epigenetic information, the function of which is the subject of informed speculation that I hope the reader will at least find thought provoking. The central theme of this essay is the proposition that the paternal genome is essentially 'redundant' (Kono, Obata, Wu, Niwa, Ono, Yamamoto, Park, Seo and Ogawa 2004) and has to make itself indispensable to the embryo in order to ensure its continuity and survival.

RNA carriage by the ejaculate spermatozoon

Historical perspective

The story of spermatozoal RNA goes back more than fifty years to early experiments aimed at understanding the flow of genetic information in these terminally differentiated cells (Krawetz, 2005; Miller, Ostermeier and Krawetz 2005). At the time, arguments arose over whether spermatozoa were capable of gene expression. This was due mainly to the understanding that the paternal genome was haploid and in a more highly condensed conformation than in somatic cells (Balhorn, Brewer, Corzett and Cosman, 1999). Moreover, during condensation and elongation of the round spermatid, most of the cytoplasm is extruded and phagocytosed by Sertoli cells leaving a remnant that was presumed to be unable to support translation of mRNA. Considering that spermatozoa are transcriptionally 'shut down' and translationally incompetent, it followed that they would not require mRNAs and should therefore not contain any.

Reports challenging those views arose in the late 1950s and early 1960s when spermatozoa were shown to be capable of incorporating radioactive tracers into both RNA and protein (Abraham and Bhargava, 1963; Bhargava, 1957). These data were criticised for not excluding the possibility that extra-spermatozoal elements in the semen (such as bacteria and contaminating somatic cells) were responsible for generating this activity (Martin and Brachet, 1959; Markewitz, Graff and Veenema, 1967). However in a series of landmark experiments, two independent laboratories concluded that spermatozoa were indeed transcriptionally active, but that this activity was restricted to the mitochondria (Premkumar and Bhargava, 1972; MacLaughlin and Terner, 1973). While additional reports suggested that spermatozoal RNA contained 28S and 18S subunits and hence components of the 80S cytoplasmic ribosome (Betlach and Erickson, 1976), a consensus emerged (that still prevails), accepting the notion that the spermatozoon nucleus is transcriptionally inactive. Indeed, a more recent experiment has confirmed these earlier conclusions although as we shall see, both transcriptional and translational repression may be lifted under exceptional circumstances (Grunewald, Paasch, Glander and Anderegg, 2005). Figure 1 shows a cartoon summary of some potential sites and uses for spermatozoal RNA. These range from establishing zones of demarcation between histone and protamine bound gene sequences (Fig. 1A), inducement of transcriptional silencing (Fig. 1B), translation on mitochondrial polysomes (Fig. 1C), epigenetically marking the female pronucleus or zygotic nucleus (Fig. 1D) and potential segregation points of RNAs based on association with the dividing centrosomes (Fig. 1E).

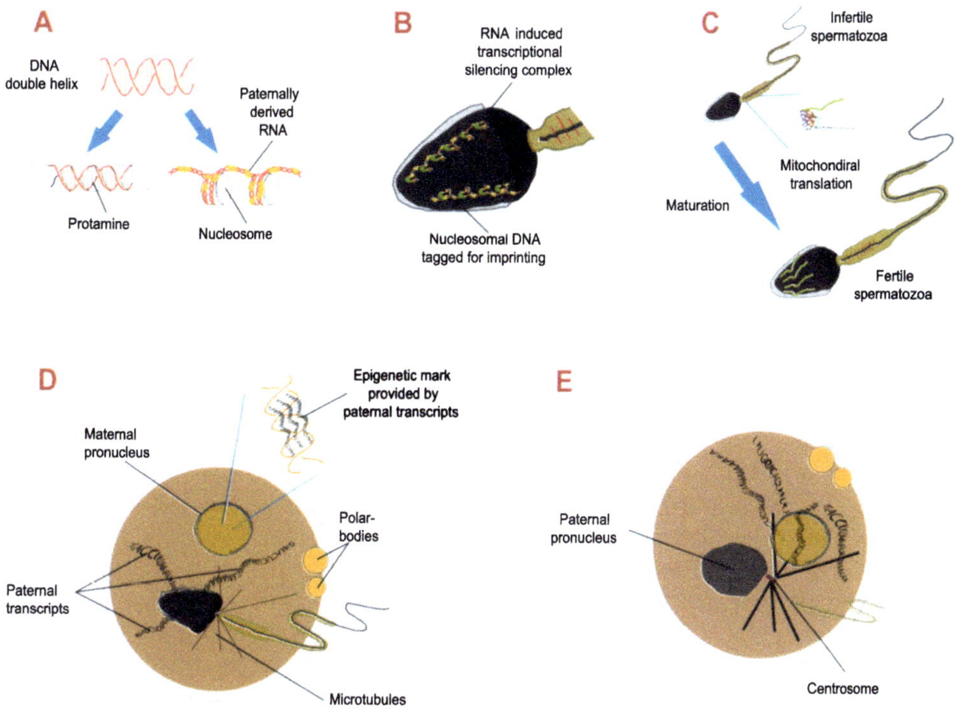

Figure 1. Predicted functionality of spermatozoal RNAs. The vast majority of DNA within the spermatozoan nucleus is packaged by highly charged protamines. However, a small proportion of the genome is packaged by histone-containing nucleosomes. It is proposed that these segments are associated with paternally derived RNA, as shown by yellow patches on the nucleosomal DNA (**A**). There is also mounting evidence that antisense RNAs play a critical role in establishing silenced chromatin domains. Perhaps recently discovered antisense spermatozoal RNAs could provide an epigenetic mark that is necessary for establishing and / or maintaining paternal imprints (**B**). Under certain conditions, some cytoplasmic mRNAs are apparently translated *de novo*, possibly on mitochondrial polysomes. When this translation is prevented, sperm maturational events associated with capacitation are not observed. This would suggest that the production of proteins from spermatozoal RNAs is necessary for the maintenance of male fertility (**C**). In addition to structural roles, it has been suggested that in common with oocyte RNA stores, spermatozoal transcripts may be functionally important to the zygote by actively promoting post-fertilisation development. Recent evidence also would suggest that paternal RNAs can provide epigenetic marks to the developing embryo that influence the phenotype of the offspring (**D**). Panel E illustrates possible RNA carriage by the paternally derived centrosome.

Clinical utility

Our laboratory's earliest foray into this controversial field came in 1992, when co-workers in my laboratory were discussing vasectomy and its reverse procedure. One of the routine assessments of these procedures is the visual observation of spermatozoa in the ejaculate. We reasoned that it should be possible to monitor clearance (or recovery) of spermatozoa from (or to) the ejaculate of men more conveniently and sensitively by looking for a specific and highly selectable molecular marker of spermatozoal mRNA with the relatively new reverse-transcription polymerase chain reaction (RT-PCR) assay. After close examination of the literature, we chose a short fragment of the protamine 2 (PRM2) gene that spanned an intron and looked for the presence of processed PRM2 mRNA in the (cells from) ejaculates of vasectomised and intact males (Miller, Tang, Skinner and Lilford, 1994). This work also used a number of 'housekeeping' markers as control mRNAs and suggested that spermatozoal RNA was a potentially useful resource for infertility studies. Earlier reports on the presence of c-MYC mRNA and protein in ejaculate human spermatozoa and of other RNAs residing in the sperm nucleus satisfied us of the reliability of our own findings (Rejon, Bajon, Blaize and Robert, 1988; Pessot, Brito, Figueroa, Concha, Yanez and Burzio, 1989; Brito, Concha, Pessot, Mezzano and Burzio, 1992; Chiang, Steuerwald, Lambert, Steinleitner and Main, 1993; Kumar, Patel and Naz, 1993). A later report localised PRM2, TNP2 and ß-actin mRNA to the human spermatozoal nucleus (Wykes, Visscher and Krawetz, 1997).

Many other reports have since appeared cataloguing a number of mRNAs in the spermatozoa of various species and the reader is directed to two recent reviews for additional information in this regard (Krawetz, 2005; Miller *et al.*, 2005). Suffice to say that human spermatozoa harbour a highly complex repertoire of mRNAs estimated at between 3,500 and 5,500 distinct species (Ostermeier, Dix, Miller, Khatri and Krawetz, 2002; David Dix, unpublished work). Ostermeier's original report also indicated that spermatozoal mRNAs were contained within a subset of testis mRNAs and suggested that spermatozoal RNAs might be a useful extra-testicular proxy for the germ line *in situ*. Figures for the number of distinct mRNA species in mature spermatozoa have been derived from both DNA array based studies and from a study using Serial Analysis of Gene Expression (SAGE) that provided additional information on the relative abundance of distinct mRNA species in these cells (Wang, Zhou, Xu, Li, Xiao, Xu and Sha, 2004; Zhao, Li, Yao, Wang, Zhou, Wang, Liu, Wang, Wang and Qiao, 2006). Indeed, the five most abundant mRNAs of human spermatozoa determined by SAGE analysis are GA 17 (most abundant) which encodes a fusogenic protein, spermatogenesis related protein 7 and the VAV2 oncogene (second and third most abundant, respectively) and COX5B encoding a (nuclear-encoded) subunit of cytochrome c oxidase (fourth most abundant).

Interestingly, ontological analysis of the 389 clustered genes with the most abundant SAGE tags revealed a highly selective grouping with one quarter (96) related to DNA dependent transcription and regulation of transcription (Table 1). We have unpublished data showing that genes involved in the regulation of transcription comprises the most significant ontological category of human spermatozoal DNA that is packaged into protamines (and hence deeply condensed). The second largest grouping (84) included genes related to protein biosynthesis and encoding ribosomal proteins. I shall return to this observation in a later section. Another notable finding from this study was the high number of genes encoding spermatozoal mRNA that are located on chromosome 1, a known 'hotspot' for a male fertility locus (Bache, Assche, Cingoz, Bugge, Tumer, Hjorth, Lundsteen, Lespinasse, Winther, Niebuhr *et al.*, 2004). Figure 2 shows the potential clinical impact of spermatozoal RNA on infertility studies where comparisons between normal and abnormal (Fig. 2A) testes can be made using their respective spermatozoal populations (Fig. 2B) in conjunction with suitable downstream analytical platforms such as microarrays (Fig. 2C).

Table 1. Top 12 ontological categories of gene transcripts identified by SAGE analysis of spermatozoal RNA. Modified from an original table in Zhao *et al.* (2006).

Number of gene sequences	Ontological description
96	Nucleus protein; related to transcription, transcription regulation (DNA-dependent)
84	Cytoplasm, ribosomal protein; related to protein biosynthesis
27	Proteolysis and peptidolysis; related to cellular protein catabolism
24	Protein kinase activity; related to protein amino acid phosphorylation, protein modification
22	RNA-binding; RNA processing and modification; RNA splicing
18	Cytoplasm protein; related to protein transportation and transmembrane localization
16	ANK repeat (many ankyrin repeat regions are known to function as protein–protein interaction domains)
10	Microtubule cytoskeleton; related to cytoskeleton organization and biogenesis
9	RNA-binding; RNA processing and modification; RNA splicing
9	ATPase activity; ATP binding, helicase; ATP-dependent RNA helicase activity; RNA processing and modification
9	Nuclear pore complex; related to protein–nucleus import
8	Calcium ion binding; EF-hand

Hypotheses explaining retention of RNA

Although there are currently no reports from avian studies, the presence of RNA in spermatozoa appears to be a general phenomenon having been reported in the male gametes of human, mouse, fern and maize (Engel, Chaboud, Dumas and McCormick, 2003). Some ideas exploring RNA retention are already illustrated in Fig. 1. However, the evidence for a complex and possibly selective population of mRNAs in these cells is at odds with the accepted dogma that they are (with the exception of the mictochondrial machinery), transcriptionally inactive. Of the numerous explanations that have been put forward to explain this retention, the notion that the RNA is residual to spermatogenesis is loosing credence (Miller, Briggs, Snowden, Hamlington, Rollinson, Lilford and Krawetz, 1999; Miller *et al.*, 2005). According to this hypothesis, spermatozoal RNA is a 'by product' of the gradual shut down of the nuclear transcriptional apparatus that precedes the switch from histone-based to protamine-based chromatin packaging via the transition proteins. Developing spermatids temporally uncouple the transcription of mRNA from its translation during spermiogenesis by many days in order to facilitate genomic repackaging (Kleene, 1996). The RNA or RNA containing particle (polysomal and non-polysomal) is also free to move between spermatids that remain connected to their neighbours (Caldwell and Handel, 1991). However, once their jobs are done, they are not degraded or removed and become 'trapped' as the cells' cytoplasm is extruded. This explanation was attractive because it complied with the notion of a cell that having used its RNA, relied on the oocyte to degrade it, much as it degrades other components of the spermatozoon following entry into the ooplasm (Sutovsky, 2003). Moreover, the SAGE tags identified by Zhao *et al.* (2006) included strong representation of messages that were likely to have arisen during spermiogenesis. However, in human spermatozoa at least, intact ribosomal 28S and 18S rRNAs are virtually absent although fragments derived from these subunits can be detected in total spermatozoal RNA extracts (Miller *et al.*, 1999; and unpublished observations). This indicates that there is either selective removal or degradation of 80S ribosomes in ejaculate spermatozoa but that mRNAs escape the removal mechanism because of additional requirements.

What might these additional requirements be? There are two alternative hypotheses for the retention or protection of spermatozoal mRNA that relies on a more active role for these messages. According to the second hypothesis, the RNA is required by the spermatozoon prior to its entry into the ooplasm. The third hypothesis suggests that the RNA performs some function beyond fertilisation. The second hypothesis is appealing because it fits well with the role of mRNA in every other cell type and indeed, new evidence in its support has recently been reported (Gur and

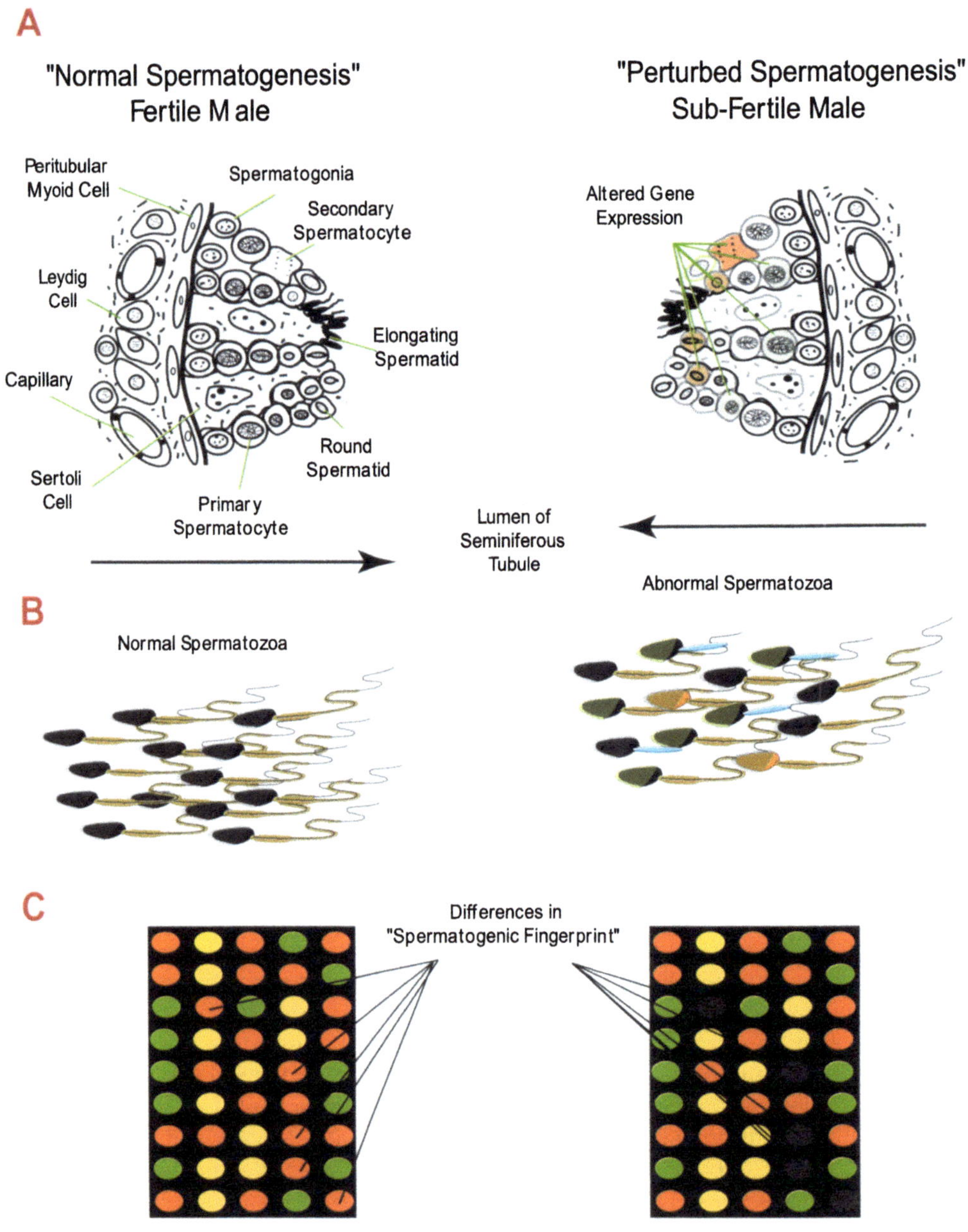

Figure 2. Clinical applications of spermatozoal RNA. Spermatozoal RNAs are transcribed in the testis during the production of spermatozoa. As indicated by spermatogenic and Leydig cells in highlighted colours (**A**), perturbations in spermatogenic gene expression may occur. This altered gene expression, can lead to the production of abnormal spermatozoa (**B**). Changes in spermatogenic gene expression leading to subfertility or infertility should be detected by variations in spermatozoal fingerprints utilising in this example, microarrays (**C**)

Breitbart, 2006). In this study, capacitating spermatozoa were shown to be capable of translating mRNA *de novo* as demonstrated by the uptake of labelled amino acids. While this essentially confirms the much earlier work described above, other aspects of the study were most intriguing. For example, translation was inhibited by antibiotics but not by cycloheximide. Moreover, compelling evidence supporting translation of cytoplasmic mRNAs on mitochondrial and not cytoplasmic polysomes was also presented. It is difficult to know how cytoplasmic RNA can be translated on mitochondrial machinery as they use quite distinct elongation initiation mechanisms. The suggestion that rather than being transported into the mitochondria, cytoplasmic mRNA is translated on mitochondrial ribosomes somehow transported into the nucleus is also interesting but equally challenging to explain. Another suggestion put forward in this study is that spermatozoal RNA is translated to supplement proteins that are degraded or denatured with newly synthesised replacements. These data are supported by contemporary reports indicating a gradual depletion of some mRNAs in motile spermatozoa, compared with non-motile cells suggesting their 'consumption' in healthy spermatozoa (Steger, Fink, Failing, Bohle, Kliesch, Weidner and Bergmann, 2003; Lambard, Galeraud-Denis, Martin, Levy, Chocat and Carreau, 2004; Aoki, Liu and Carrell, 2006). These reports infer that at least some spermatozoal mRNA acts as a reservoir for the replacement of degraded proteins, a notion that concords with maximising the potential of these cells to reach and fertilise the egg. The third hypothesis is perhaps the most challenging because it introduces a new factor into fertilisation that goes beyond delivery of the paternal genome and which I shall spend the remainder of this essay discussing.

That the sperm delivers more than just DNA to the egg is not disputed. The zygotic spindle cannot be organised without the duplication of a sperm-derived centrosome (with the notable exception of the mouse). We also know that a sperm-derived phosholipase is responsible for activating the egg at fertilisation. In 2004, however, Ostermeier and colleagues (2004) used the hamster oocyte human sperm penetration assay to provide proof of principle that the sperm also delivered its RNA to the egg where it could be detected by RT-PCR (Ostermeier, Miller, Huntriss, Diamond and Krawetz, 2004). This work provided the backdrop to a study showing that RNA introduced into the oocyte by the spermatozoon could alter the phenotype of a mouse without changing its genotype (Rassoulzadegan *et al.*, 2006). This study showed that the *c-Kit* derived heritable effect on tail tip and paw hair colour was strongly influenced by the presence of an aberrant spermatozoal RNA or RNAs derived from an affected individual. The original mutant giving rise to this phenotype was one in which one allele of the *c-Kit* gene was inactivated. Under normal circumstances, animals homozygous for the wild type alleles arising from heterozygous crosses should not have been affected. However it was soon noticed that they were far more frequently affected than standard Mendelian genetics predicted and that the effect could be stably transmitted to wild-type offspring. This non-Mendelian pattern of inheritance is reminiscent of epigenetic phenomena commonly observed in some plant species and referred to as paramutation. In this case, the effect is apparently caused by an RNA-mediated down regulation of one of the *c-Kit* wild-type alleles, possibly brought about by differential allele methylation. The closest analogous system to this phenomenon in mammals is the uniparental expression of imprinted genes such as *Igf2* and *h19* in mammals (Kerjean, Dupont, Vasseur, Le Tessier, Cuisset, Paldi, Jouannet and Jeanpierre, 2000).

Possible spermatozoal RNA mediated epigenetic effects on the zygote are illustrated in Fig. 3 where the spermatozoan's RNA and DNA separate following entry in the ooplasm (Fig. 3A; blue paternal nucleus surrounded by green RNA). In Fig. 3B, the two remain separate with the RNA probably converted to a more stable DNA form following egg activation. If the RNA is host derived, it may perform some 'invisible' function on the maternal pronucleus or later on in development. In Fig. 3C, the RNA directly alters the zygotic nucleus (change of colour) and induces a heritable change that may or may not produce an obvious phenotype (Rassoulzadegan *et al.*, 2006). In both scenarios, selfish elements (see below for more details) could be transferred from male to female gametes with or without noticeable effects.

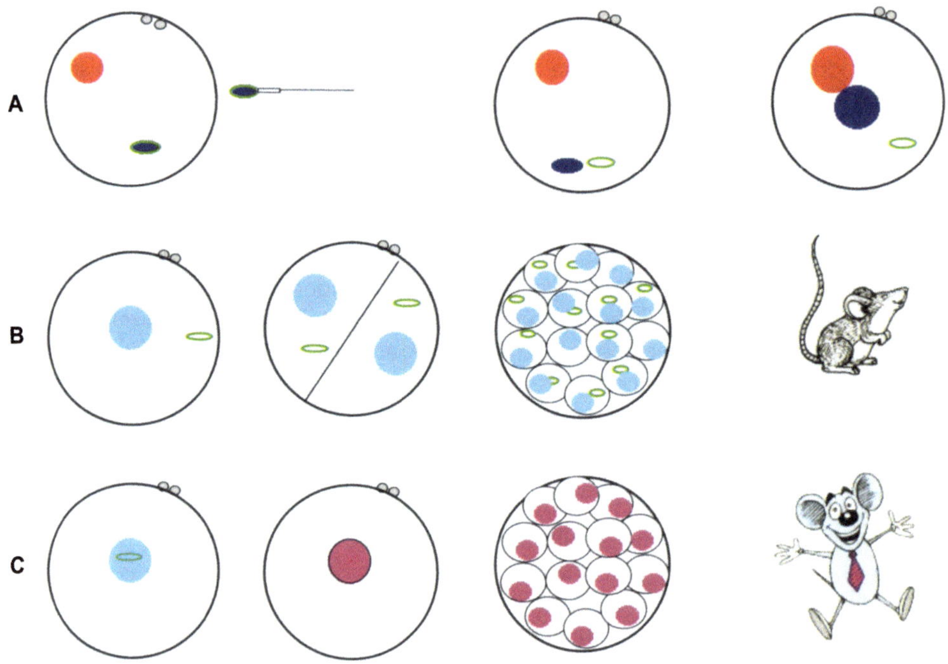

Figure 3. Proposed roles for spermatozoal RNA in the zygote. A spermatozoon carries with it an RNA signal (green nuclear periphery) that is introduced into the egg at fertilisation (**A**). The female and male pronuclei are in red and blue, respectively. One possibility (**B**) is that a sperm RNA acts independently of the genome from syngamy (turquoise shaded nucleus) onwards and autonomously replicates as embryonic development proceeds. This RNA (or an episomal DNA derived from it) does not interact with the genome and can be considered as a selfish element that uses reproductive cells to spread itself but has no other phenotypic effects on the offspring. In (**C**), a sperm derived RNA interacts with and alters the genome, epigenetically (shift from turquoise to purple shading) and the change is then passed on to all other cells of the developing embryo producing offspring with a changed phenotype.

To some extent, Rassoulzadegan's discovery was serendipitous in that their model animal displayed a phenotype that defied normal Mendelian rules. They concluded that the pattern of coat colour they had observed was most likely due to some non-genomic factor introduced by the spermatozoon (the paramutating RNA). This RNA is probably an aberrant c-*Kit* non-coding transcript that is certainly not sperm-specific because the same effect was induced by injecting RNA from affected animals' brains directly into oocytes as could the injection of micro RNA (miRNA) molecules known to target the c-*Kit* locus. These data, however, showed for the first time that spermatozoal RNA can affect phenotype via an epigenetic mechanism capable of modulating post-zygotic gene expression. The earlier work of Ostermeier *et al.* who examined the RNA composition of human spermatozoa and uncovered evidence for full length antisense RNAs therein takes on new significance in the light of Rassoulzadegan's report as several of the antisense RNAs reported were implicated in developmental pathways (Ostermeier, Goodrich, Moldenhauer, Diamond and Krawetz., 2005). Together, they show the potential of RNA delivered to the egg by the spermatozoon to epigenetically modify the zygotic genome.

Evidence for a more dynamic sperm nucleus

Endogenous nucleases of the spermatozoal nucleus

The recent work demonstrating *de novo* translation of spermatozoal RNA and the zygotic transmission of epigenetic traits reopens the old arguments over whether the spermatozoan nucleus is as 'switched off' as we have traditionally assumed. At stake is a possible reversal of the accepted compromise view that only the mitochondrial genome is transcriptionally active. In other words, given the recent evidence for the wider role of spermatozoal RNAs in translation and transmission of epigenetic traits, is there evidence that transcription might also occur? Certainly as we have seen, past evidence for this has been conflicting and one recent report failed to find any evidence for this in ejaculate human spermatozoa (Grunewald *et al.*, 2005). However, an unlikely source that is now challenging the prevailing view that transcription is not possible comes from work examining the mechanism of transgenic animal production by sperm-mediated gene transfer (SMGT). Originally a highly controversial field in itself, SMGT is now widely accepted. The reader is directed to recent reviews for a more comprehensive description of the field (Smith and Spadafora, 2005; Spadafora, this volume). In brief, spermatozoa washed free of seminal plasma and exposed to exogenous DNA are able to absorb and transmit it to offspring following intra-cytoplasmic sperm injection (ICSI). The DNA is rarely integrated into the genome; instead it normally exists in very low copy number in some episomal compartment that is wholly independent of chromosomal DNA and more reminiscent of autonomously replicating plasmids. It is, however, stably transmitted to offspring and hence may be analogous to the epigenetic transmission reported by Rassoulzadegan *et al.* (2006). Part of the process of DNA absorption and occasional random integration is the up-regulation of endogenous nuclear nucleases that can progressively degrade spermatozoal DNA, one effect of which is the liberation of nuclease-sensitive DNA from the spermatozoal nucleus (Pittoggi, Renzi, Zaccagnini, Cimini, Degrassi, Giordano, Magnano, Lorenzini, Lavia and Spadafora, 1999). This DNA has been partially characterised and shown to be derived primarily from the nucleohistone compartment of spermatozoal chromatin (see Fig. 1A for a cartoon representation). In murine spermatozoa, released DNA apparently contains high levels of repeat sequences that include Long Interspersed Nuclear Elements (LINEs) and Short Interspersed Nuclear Elements (SINEs) (Pittoggi, Zaccagnini, Giordano, Magnano, Baccetti, Lorenzini and Spadafora, 2000). The more abundant human nucleohistone compartment has been reported to be enriched in telomeric sequences and possibly in encoding sequences that may be important in early embryonic development (Zalenskaya, Bradbury and Zalensky, 2000; Miller *et al.*, unpublished observation). We also reported the presence of transcripts for the LINE 1 open reading frame 2 (ORF2) in ejaculate human spermatozoa (Miller *et al.*, 1999). ORF2 encodes a reverse transcriptase enzyme (Brouha, Schustak, Badge, Lutz-Prigge, Farley, Moran and Kazazian, 2003).

Evidence for RNA and DNA dependent polymerases

Spermatozoal RNA and DNA are possibly inter-convertible *de novo* as shown by elegant experiments demonstrating that RNA can be absorbed and reverse-transcribed into DNA and that absorbed DNA can be transcribed into fully processed RNA. DNA and RNA-dependent DNA polymerases (essentially DNA polymerase and reverse transcriptase) were described in human spermatozoa back in the 1970s (Witkin and Bendich, 1977; Witkin, Traganos and Bendich, 1977). The more recent work used SMGT with a hybrid murine leukemia virus/virus-like 30S (MLV/VL30) beta-galactosidase (beta-gal) gene-containing vector that was detected as DNA in murine spermatozoal DNA extracts, indicating the activity of an operational reverse transcriptase in mature spermatozoa (Sciamanna, Barberi, Martire, Pittoggi, Beraldi, Giordano, Magnano, Hogdson and Spadafora, 2003). More intriguing is the more recent observation that a DNA construct containing an intronic sequence can be processed by a spermatozoal machinery such that the intron is correctly spliced out (Pittoggi, Beraldi, Sciamanna, Barberi, Giordano, Magnano, Torosantucci, Pesarmona and Spadafora, 2006). To achieve this, the machinery would have to transcribe the DNA into complementary RNA using an RNA polymerase, correctly remove the intronic sequence, splice the RNA into a mature form and then reverse transcribe it back into a stable DNA (see Fig. 5A). Such a mechanism is strongly reminiscent of endogenous proviral and LINE replication and suggests that the episomal entity encoding the transgene in SMGT transfected animals is replicated in much the same way. These remarkable data demonstrate some unusual features of spermatozoa including the presence of a fully functional transcription apparatus that is probably derived from the nuclear machinery for appropriate intron splicing to be possible. While normally dormant, this molecular machinery may be capable of activation under certain conditions (Hecht and Williams, 1978).

What roles for sperm nucleases and polymerases?

What purposes could nucleases and reverse transcriptase serve the spermatozoon? One possibility is that the nucleases are there to disable any spermatozoon that comes into contact with exogenous nucleic acids, most likely in the form of viruses and perhaps DNA released from other cell types (including bacteria) encountered in the female genital tract (Sotolongo, Lino and Ward, 2003; Ward and Ward, 2004). Such an apoptosis-like mechanism would ensure that under normal circumstances, foreign DNA is unlikely to reach the egg [however, see Moreira, Fernandez-Gonzalez, Rizos, Ramirez, Perez-Crespo and Gutierrez-Adan (2005) for one report describing inadvertent SMGT of bacterial DNA following murine ICSI]. A role for RT in spermatozoa is more difficult to posit. According to the activities highlighted above, one wonders what would stop endogenous spermatozoal mRNAs from being successfully stabilised as autonomously replicating DNAs. Perhaps this does indeed occur under certain circumstances? If it does, what effect might it have on any egg that is fertilised by that spermatozoon?

Unfertilised murine oocytes also appear to have a higher level of RT activity than somatic cells; however, two-cell murine embryos contain levels that are at least ten times higher (Evsikov, De Vries, Peaston, Radford, Fancher, Chen, Blake, Bult, Latham, Solter and Knowles 2004; and see below) suggesting that it is not uniquely introduced with the spermatozoon. Interestingly, proviral particles, a rich source of RT enzyme, have been observed in mature human oocytes by transmission electron microscopy (Larsson, Nilsson, Sundstrom and Widehn, 1981; Nilsson, Kattstrom, Sundstrom, Jaquemin and Larsson, 1992). They have also been observed in murine epididymides (Kiessling, Crowell and Fox, 1989). Hence endogenous provirus and RT may be general features of reproductive cells and tissue that are required to perform some particular as yet unknown post-fertilisation function. If LINE RT is indeed the source of active enzyme in spermatozoa and oocytes, it must be a special property of germ cells, because in all somatic cells (with the notable exception of some malignant cell lines), LINE expression is almost completely repressed by DNA methylation (Thayer, Singer and Fanning, 1993).

Paternal RNA mediated epigenetic effects may be co-opted from 'selfish' elements

Role of RT in embryogenesis

The latest evidence demonstrating the transmission of an epigenetic trait to the zygote via spermatozoal RNA begs the question of whether this is only ever seen under extreme experimental conditions or whether it is an occasional or general natural phenomenon. From the murine data, it appears as though a role for RT either in mediating RNA to DNA conversion within the spermatozoon itself prior to fertilisation or following fertilisation is certainly possible, particularly if this new DNA construct is somehow involved in fertilisation events. Evidence for such a mechanism has come from experiments where RT activity was inhibited in mouse zygotes. In such embryos, irreversible developmental arrest was observed at the two to four cell stage, coincident with the time of activation of the embryonic genome (ZGA) (Beraldi, Pittoggi, Sciamanna, Mattei and Spadafora, 2006). Beraldi *et al.* prevented activation using antisense morpholinos specifically targeting the LINE1 RT. An illustration of this is provided in Figure 4: Fig. 4A is similar to Fig. 3A, except that the sperm factor (RT?) may interact directly either with the female pronucleus before syngamy or the zygotic nucleus after syngamy (Fig. 4B) priming them for ZGA. In conjunction with a cleavage stage RT derived from the zygotic nucleus (possibly after priming by the spermatozoal factor), ZGA goes ahead. In Fig. 4C, the spermatozoal factor primes the maternal pronucleus but ZGA is prevented by inhibition of RT.

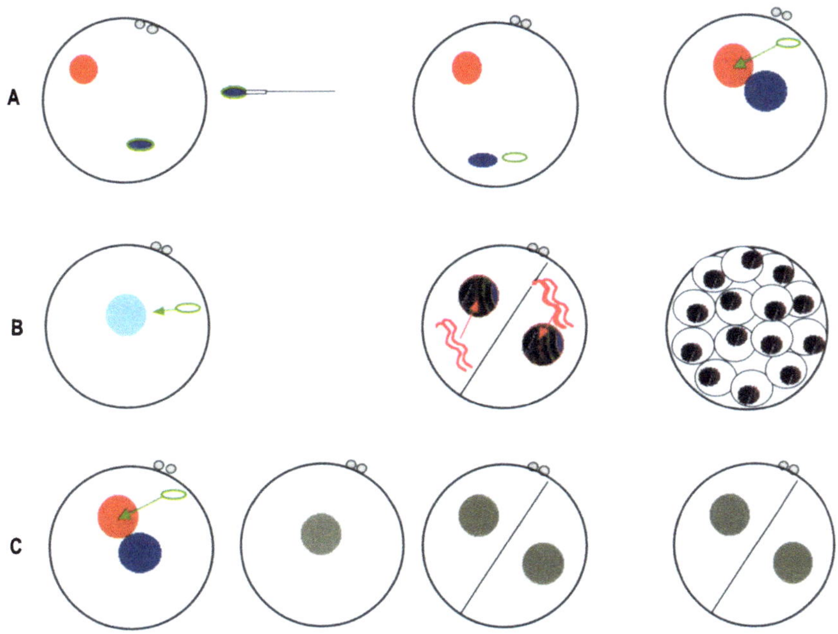

Figure 4. Proposed requirement for an RNA mediated interaction that primes zygote genome activation (ZGA). A spermatozoon carries with it an RNA signal (green nuclear periphery) that is introduced into the egg at fertilisation (**A**). This signal may affect the female pronucleus (red) and possibly primes ZGA. Alternatively, the introduced RNA may act at syngamy, priming the one-cell embryo prior to cleavage (**B**). Following cleavage in mouse an oocyte cytoplasmic factor (reverse transcriptase) facilitates the final switch. If priming is permissive through fertilisation but the RT is inhibited, ZGA does not occur and embryo development arrests (**C**). It follows that inhibiting the proposed priming stage in the one cell embryo should also abolish ZGA.

Understanding what RT may be doing under these circumstances and why its abrogation is capable of abolishing ZGA is partially helped by our knowledge of what the enzyme does during the normal replication of LINE 1 elements. We know, for example that these elements can replicate because occasional pathological insertional inactivation of essential genes is observed from time to time (Kazazian, Wong, Youssoufian, Scott, Phillips and Antonarakis, 1988). The mechanism of replication is also reasonably well understood and we can infer from this mechanism and observations of insertional events generated *in vitro* that L1-mediated exon shuffling is a potentially powerful force for reshaping the genome (Moran, DeBerardinis and Kazazian, 1999). Moreover, the LINE element is also known to be expressed in the testis through the expression of the p40 ribonucleoprotein encoded by the element's ORF1 (Branciforte and Martin, 1992). As indicated above, LINE 1 transcripts are also present in spermatozoa and in the ooplasm although there is apparently a much greater load of retroelement RNA in early cleavage stage embryos. Indeed one study estimated that over 10% of the RNA in these embryos is retroelement in origin (Evsikov *et al.*, 2004). All mammals carry a proviral load, the expression of which appears to be under strictly controlled regulation. Expression is occasionally observed under permissive conditions in reproductive cells and tissues and there is good evidence that these tissues have taken advantage of their presence in the distant past. For example, the syncytin protein that mediates the fusion of placental cytotrophoblast into syncytiotrophoblast has been co-opted to this task from an endogenous retroviral ancestor and it is tempting to speculate that the *quid pro quo* for this co-option is the permissive replication of proviral particles (Mi, Lee, Li, Veldman, Finnerty, Racie, LaVallie, Tang, Edouard, Howes, Keith, and McCoy, 2000).

An hypothesis for the origin of spermatozoal RNA mediated epigenetic effects

Now that spermatozoal RNA has been shown to be transferred to the oocyte and that, under special conditions, it can modify the genome via an epigenetic pathway, it is also conceivable that this same mechanism could serve the transmission of both male benefit and selfish elements. The male benefit elements could take the form of small inhibitory RNAs, antisense RNAs or even episomal elements that help ensure the successful transmission of the paternal genome. The flip side of this coin is the occasional or possibly routine transmission of selfish elements including proviral and LINE intermediates. Indeed, spermatogenesis is probably the best time for such opportunistic elements to be transmitted because of the much greater proliferation of spermatogonia compared with oogonia and the wider window of opportunity afforded by a constantly replicating genome during a male's reproductive lifetime. Once in the ooplasm, these elements may be free to replicate, hence the occasional observation of proviral particles in oocytes; alternatively, they could either integrate silently (or maliciously) into the host genome or remain as autonomously replicating episomes. Anyone who concludes that this is wildly imaginative should consider the parasitism of insect eggs. These are frequently subject to infection by a rickettsia-like bacterium, *Wolbachia,* that ensures its vertical transmission to offspring by only permitting development of female embryos (Hurst and McVean, 1996). A similar sex ratio distortion phenomenon exists in mice carrying the *t*-haplotype where the spermatids of heterozygous animals that are not carrying the *t*-allele are somehow 'poisoned' and fail to develop in favour of spermatids that do carry the allele. Females are more likely to be born and transmit the trait because male mice that are homozygous for the *t*-allele usually die *in utero* or are born infertile if they survive to term (Olds-Clarke, 1997).

The repression of ZGA by inhibitors of reverse transcriptase could arise either through interference with RT derived from the spermatozoon or ZGA-associated retroelements. One possible requirement for RT may be related to the exon shuffling that is observed as a consequence of LINE activity. Perhaps a juxtaposition of distant DNA elements (possibly a master promoter or enhancer placed close to a master transcriptional regulator) is required to 'switch on' the embryonic genome and that this arises through a re-arrangement that is RNA mediated, using a similar mechanism that drives episome formation in SMGT (Smith and Spadafora, 2005) (i.e. DNA to RNA to DNA). While highly speculative, this hypothesis is attractive for a number of reasons, first and foremost because it suggests an alternative reason for the poor efficiency of somatic cell cloning beyond epigenetic reprogramming. Since they are derived from a normally fertilised egg, it follows that all somatic cell nuclei must already have this master switch in place. As indicated above, resetting such a switch would be energetically unfavourable; hence quite apart from the need to reprogramme the nuclear epigenome, the recipient ooplasm would have to work very efficiently to reverse the ZGA switch to a point where it was also reset. Moreover, any switch normally requiring an epigenetic element derived from an incoming spermatozoon would be very difficult to circumvent with similar outcomes for cloning efficiency.

The latter possibility is particularly appealing because it fits well with the notion of the paternal (selfish?) contribution making itself indispensable to the zygote and hence ensuring its own transmission by whatever means. The male genome is certainly dispensable as demonstrated recently by the successful (albeit inefficient) production of gynogenetic mice (Kono *et al.*, 2004). Such a mechanism also concords with the occasional opportunistic transmission of selfish elements from one generation to the next and the favourable co-opting of selfish components for use by the host. Unless selfish elements cause obvious phenotypes such as disease states, they will normally be transmitted unnoticed. Only human intervention through the widespread adoption of assisted conception technologies like ICSI is potentially able to change the balance of this relationship.

The concept of a ZGA switch is illustrated in Figure 5 and is loosely based on the exon shuffling potential for LINE elements described by Moran *et al.* (1999) where two elements are brought together via a DNA to RNA to DNA process that occurs in cis (Fig. 5A) or trans (Fig. 5B). In a cis acting mechanism, transcription produces an RNA intermediate with the two elements separated by an intron-like spacer. This self-splices itself out and the two elements are ligated. The RT then converts the new element back to DNA where it is integrated into the genome forming the final switch. The mechanism could be similar to the one used by LINE 1 and could lead to the juxtaposition of the newly spliced element next to a promoter. In the trans-acting mechanism, the two distant elements are physically brought together by DNA bending proteins (possibly HMG related), permitting transcription of a switch component that is then reverse transcribed back to DNA and integrated as before. This mechanism is appealing because it provides a way for elements in the two pronuclei to interact prior to syngamy.

Conclusions

The past four years has seen some major advances in spermatozoal RNA research. There is now a comprehensive array based data set cataloguing the complexity of human spermatozoal RNA that is now supported by a SAGE map. Other array-based surveys are expected in due course that should include the murine. Fresh developments in the field have challenged the view that spermatozoa are transcriptionally and translationally silent cells with reports demonstrating both. The most recent discovery of a spermatozoal RNA mediated transmission of an epigenetic trait is perhaps the most compelling evidence yet for a role of spermatozoal RNA in the zygote. Taken together with the work arising from the fields of sperm mediated gene transfer and the expression of proviral and LINE elements in reproductive cells and tissues, we are beginning to see a pattern that paints a much broader picture for the spermatozoon's role beyond mere delivery of the paternal genome. Of particular interest is the possibility that spermatozoa use an ancient mechanism to effect epigenetic alterations that has evolved in tandem (and relatively peacefully) with selfish genetic elements. Both host and selfish elements stand to gain from this interaction through mutually beneficial cooperation; the host (male) ensures delivery of his genome to the egg while the selfish element catches a free ride along the way. Occasionally, the two may interact in new and interesting ways and the increasing use of ICSI procedures may inadvertently be increasing the chances of such interaction.

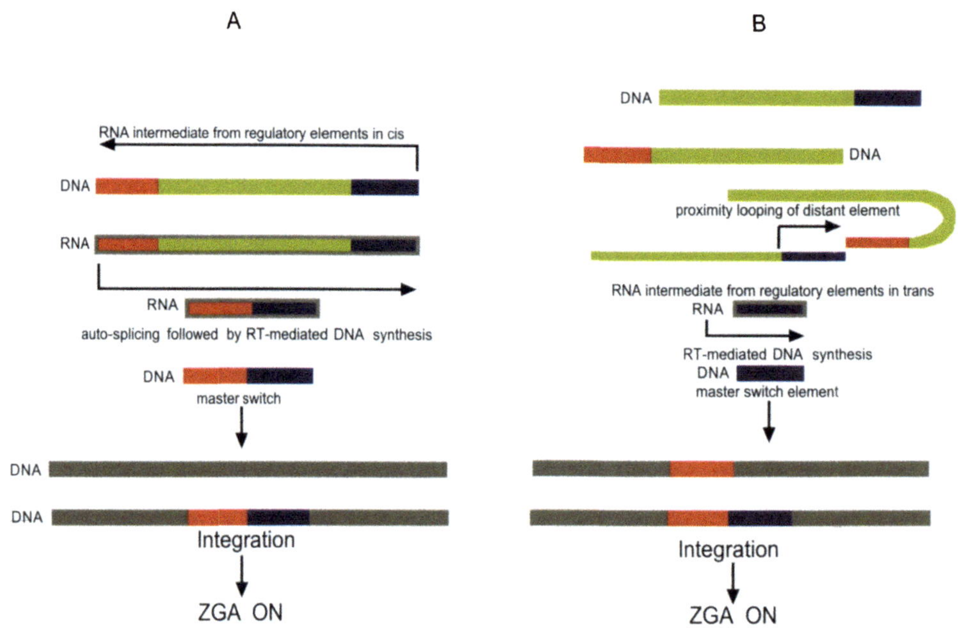

Figure 5. Proposed mechanisms for zygote genome activation (ZGA) switch. Assuming that the switch uses regulatory elements that are spaced apart, there are two possible mechanisms that could bring those elements together. If the elements are in cis but separated by intron-like sequences (**A**), then transcription along the full length of the sequence would produce an intermediate RNA containing the introns. These introns could be automatically spliced out bringing the elements together and then reverse transcribed and 'slotted into place'. Alternatively, a special DNA bending factor could bring two elements (an enhancer and a coding region in trans) together by a looping mechanism. This would enable the transcriptional machinery to create an RNA copy of the master switch which is then reverse transcribed into DNA and 'slotted into place' as before. The principal difference between these two mechanisms is that critically interacting elements are brought together either as RNA (**A**) or as DNA (**B**).

Late breaking news

Since this manuscript went to press, two papers (Aravin, Gaidatzis, Pfeffer, Lagos-Quintana, Landgraf, Iovino, Morris, Brownstein, Kuramochi-Miyagawa, Nakano, Chien, Russo, Ju, Sheridan, Sander, Zavolan and Tuschl, 2006; Girard, Sachidanandam, Hannon and Carmell, 2006) have reported on the presence of populations of small RNAs including a novel population that differs from both miRNA and siRNA in the murine, bovine and human testes. This new 30nt class of small RNAs binds the PIWI group of Argonaute proteins, is testis-specific and is essential for spermatogenesis. While evidence was presented for their absence in murine epididymal spermatozoa, it is my view that the detection method was too insensitive to detect either these RNAs or small RNAs of the other testis-expressed classes.

Acknowledgements

The author thanks Charles Ostermeier (The Jackson Laboratory, Technology Evaluation and Development, Reproductive Technology Resources, Bar Harbour, ME, USA) for assistance with illustrations.

References

Abraham, K.A. and Bhargava, P.M. (1963). Nucleic acid metabolism of mammalian spermatozoa. *Biochemical Journal* **86:** 298-307

Aoki, V.W., Liu, L. and Carrell, D.T. (2006). A novel mechanism of protamine expression deregulation highlighted by abnormal protamine transcript retention in infertile human males with sperm protamine deficiency. *Molecular Human Reproduction* **12:** 41-50

Aravin, A., Gaidatzis, D., Pfeffer, S., Lagos-Quintana, M., Landgraf, P., Iovino, N., Morris, P., Brownstein, M.J., Kuramochi-Miyagawa, S., Nakano, T., Chien, M., Russo, J.J., Ju, J., Sheridan, R., Sander, C., Zavolan, M. and Tuschl, T. (2006). A novel class of small RNAs bind to MILI protein in mouse testes. *Nature* **442:** 203-207

Bache, I., Assche, E.V., Cingoz, S., Bugge, M., Tumer, Z., Hjorth, M., Lundsteen, C., Lespinasse, J., Winther, K., Niebuhr, A. *et al.* (2004). An excess of chromosome 1 breakpoints in male infertility. *European Journal of Human Genetics* **12:** 993-1000

Balhorn, R., Brewer, L., Corzett, M. and Cosman, J. (1999). Protamine-mediated condensation of DNA and the subunit structure of sperm chromatin. *Biology of Reproduction* **60:** M34

Beraldi, R., Pittoggi, C., Sciamanna, I., Mattei, E. and Spadafora, C. (2006). Expression of LINE-1 retroposons is essential for murine preimplantation development. *Molecular Reproduction and Development* **73:** 279-287

Betlach, C.J. and Erickson, R.P. (1976). 28 s and 18 s ribonucleic acid from mammalian spermatozoa. *Journal of Experimental Zoology* **198:** 49-55

Bhargava, P.M. (1957). Incorporation of radioactive amino-acids in the proteins of bull spermatozoa. *Nature* **179:** 1120-1121

Branciforte, D. and Martin, S.L. (1992). Expression of Line-1 RNA and protein in mouse testis. *Molecular Biology of the Cell* **3:** A101

Brito, M., Concha, I.I., Pessot, C.A., Mezzano, S. and Burzio, L.O. (1992). Small nuclear ribonucleoprotein polypeptides in rat sperm. *Comparative Biochemistry and Physiology B-Biochemistry and Molecular Biology* **103:** 197-200

Brouha, B., Schustak, J., Badge, R.M., Lutz-Prigge, S., Farley, A.H., Moran, J.V. and Kazazian, H.H., Jr. (2003). Hot L1s account for the bulk of retrotransposition in the human population. *Proceedings of the National Academy of Sciences USA* **100:** 5280-5285

Caldwell, K.A. and Handel, M A (1991). Protamine transcript sharing among postmeiotic spermatids. *Proceedings of the National Academy of Sciences USA* **88:** 2407-2411

Chiang, M.H., Steuerwald, N., Lambert, H., Steinleitner, A. and Main, E.K. (1993). Detection of HLA-G Messenger-RNA in Human Sperm Using RT-PCR. *Journal of Immunology* **150:** A283

Engel, M.L., Chaboud, A., Dumas, C. and McCormick, S. (2003). Sperm cells of *Zea mays* have a complex complement of mRNAs. *Plant Journal* **34:** 697-707

Evsikov, A.V., De Vries, W.N., Peaston, A.E., Radford, E.E., Fancher, K.S., Chen, F.H., Blake, J.A., Bult, C.J., Latham, K.E., Solter, D. and Knowles, B.B. (2004). Systems biology of the 2-cell mouse embryo. *Cytogenetics and Genome Research* **105:** 240-250

Girard, A., Sachidanandam, R., Hannon, G.J. and Carmell, M.A. (2006). A germline-specific class of small RNAs binds mammalian Piwi proteins. *Nature* **442:** 199-202

Grunewald, S., Paasch, U., Glander, H.J. and Anderegg, U. (2005). Mature human spermatozoa do not transcribe novel RNA. *Andrologia* **37:** 69-71

Gur ,Y. and Breitbart, H. (2006). Mammalian sperm translate nuclear-encoded proteins by mitochondrial-type ribosomes. *Genes and Development* **20:** 411-416

Hecht, N.B. and Williams, J.L. (1978). Synthesis of RNA by separated heads and tails from bovine spermatozoa. *Biology of Reproduction* **19:** 573-579

Hurst, L.D. and Mcvean, G.T. (1996). Clade selection, reversible evolution and the persistence of selfish elements: The evolutionary dynamics of cytoplasmic incompatibility. *Proceedings of the Royal Society of London Series B-Biological Sciences* **263:** 97-104

Kazazian, H.H. Jr, Wong, C., Youssoufian, H., Scott, A.F., Phillips, D.G. and Antonarakis S. (1988). Haemophilia A resulting from *de novo* insertion of L1 sequences represents a novel mechanism for mutation in man. *Nature* **332:** 164-166

Kerjean, A., Dupont, J.M., Vasseur, C., Le Tessier, D., Cuisset, L., Paldi, A., Jouannet, P. and Jeanpierre M. (2000). Establishment of the paternal methylation imprint of the human H19 and MEST/PEG1 genes during spermatogenesis. *Human Molecular Genetics* **9:** 2183-2187

Kiessling, A.A., Crowell, R. and Fox C. (1989). Epididymis is a principal site of retrovirus expression in the mouse. *Proceedings of the National Academy of Sciences USA* **86:** 5109-5113

Kleene, K.C. (1996). Patterns of translational regulation in the mammalian testis. *Molecular Reproduction and Development* **43:** 268-281

Kono, T., Obata, Y., Wu, Q., Niwa, K., Ono, Y.,

Yamamoto, Y., Park, E.S., Seo, J.S. and Ogawa, H. (2004). Birth of parthenogenetic mice that can develop to adulthood. *Nature* **428:** 860-864

Krawetz, S.A. (2005). Paternal contribution: new insights and future challenges. *Nature. Reviews Genetics* **6:** 633-642

Kumar, G., Patel, D. and Naz R.K. (1993). c-MYC mRNA is present in human sperm cells. *Cell Molecular Biology Research* **39:** 111-117

Lambard, S., Galeraud-Denis, I., Martin, G., Levy, R., Chocat, A. and Carreau S. (2004) Analysis and significance of mRNA in human ejaculated sperm from normozoospermic donors: relationship to sperm motility and capacitation. *Molecular Human Reproduction* **10:** 535-541

Larsson, E., Nilsson, B.O., Sundstrom, P. and Widehn S. (1981). Morphological and microbiological signs of endogenous C-virus in human oocytes. *International Journal of Cancer* **28:** 551-557

Maclaughlin, J. and Terner, C. (1973). Ribonucleic acid synthesis by spermatozoa from the rat and hamster. *Biochemical Journal* **133:** 635-639

Markewitz, M., Graff, S. and Veenema R.J. (1967). Absence of RNA synthesis in shed human spermatozoa. *Nature* 2**14:** 402-403

Martin, F. and Brachet J. (1959). Autoradiographic studies on the incorporation of amino acids into spermatozoa. *Experimental Cell Research* **17:** 399-404

Mi, S., Lee, X., Li, X., Veldman, G.M., Finnerty, H., Racie, L., LaVallie, E., Tang, X.Y., Edouard, P., Howes, S., Keith, J.C. Jr and McCoy, J.M. (2000). Syncytin is a captive retroviral envelope protein involved in human placental morphogenesis. *Nature* **403:** 785-789

Miller, D., Briggs, D., Snowden, H., Hamlington, J., Rollinson, S., Lilford, R. and Krawetz, S.A. (1999). A complex population of RNAs exists in human ejaculate spermatozoa: implications for understanding molecular aspects of spermiogenesis. *Gene* **237:** 385-392

Miller, D., Ostermeier, G.C. and Krawetz, S.A. (2005). The controversy, potential and roles of spermatozoal RNA. *Trends in Molecular Medicine* **11:** 156-163

Miller, D., Tang, P.Z., Skinner, C. and Lilford, R. (1994). Differential RNA fingerprinting as a tool in the analysis of spermatozoal gene expression. *Human Reproduction* **9:** 864-869

Moran, J.V., DeBerardinis R.J. and Kazazian, H.H. Jr. (1999). Exon shuffling by L1 retrotransposition. *Science* **283:** 1530-1534.

Moreira, P.N., Fernandez-Gonzalez, R., Rizos, D., Ramirez, M., Perez-Crespo, M. and Gutierrez-Adan, A. (2005). Inadvertent transgenesis by conventional ICSI in mice. *Human Reproduction* **20:** 3313-3317

Nilsson, B.O., Kattstrom, P.O., Sundstrom, P., Jaquemin, P. and Larsson, E. (1992). Human oocytes express murine retroviral equivalents. *Virus Genes* **6:** 221-227

Olds-Clarke, P. (1997). Models for male infertility: the t haplotypes. *Reviews of Reproduction* **2:** 157-164

Ostermeier, G.C., Dix, D.J., Miller, D., Khatri, P. and Krawetz S.A. (2002). Spermatozoal RNA profiles of normal fertile men. *The Lancet* **360:** 772-777

Ostermeier, G.C., Goodrich, R.J., Moldenhauer, J.S., Diamond, M.P. and Krawetz, S.A. (2005). A suite of novel human spermatozoal RNAs. *Journal of Andrology* **26:** 70-74

Ostermeier, G.C., Miller, D., Huntriss, J.D., Diamond, M.P. and Krawetz, S.A. (2004). Reproductive biology: delivering spermatozoan RNA to the oocyte. *Nature* **429:** 154

Palermo, G.D., Colombero, L.T. and Rosenwaks, Z. (1997). The human sperm centrosome is responsible for normal syngamy and early embryonic development. *Reviews in Reproduction* **2:** 19-27

Pessot, C.A., Brito, M., Figueroa, J., Concha, I.I, Yanez, A. and Burzio, L.O. (1989). Presence of RNA in the sperm nucleus. *Biochemical and Biophysical Research Communications* **158:** 272-278

Pittoggi, C., Beraldi, R., Sciamanna, I., Barberi, L., Giordano, R., Magnano, A.R., Torosantucci, L., Pesarmona, E. and Spadafora, C. (2006). Generation of biologically active retro-genes upon interaction of mouse spermatozoa with exogenous DNA. *Molecular Reproduction and Development* **73:** 1239-1246

Pittoggi, C., Renzi, L., Zaccagnini, G., Cimini, D., Degrassi, F., Giordano, R., Magnano, A.R., Lorenzini, R., Lavia, P. and Spadafora C. (1999). A fraction of mouse sperm chromatin is organized in nucleosomal hypersensitive domains enriched in retroposon DNA. *Journal of Cell Science* **112:** 3537-3548

Pittoggi, C., Zaccagnini, G., Giordano, R., Magnano, A.R., Baccetti, B., Lorenzini, R. and Spadafora, C. (2000). Nucleosomal domains of mouse spermatozoa chromatin as potential sites for retroposition and foreign DNA integration. *Molecular Reproduction and Development* **56:** 248-251

Premkumar, E. and Bhargava, P.M. (1972). Transcription and translation in bovine spermatozoa. *Nature New Biology* **240:** 139-143

Rassoulzadegan, M., Grandjean, V., Gounon, P., Vincent, S., Gillot, I. and Cuzin, F. (2006). RNA-mediated non-mendelian inheritance of an epigenetic change in the mouse. *Nature* **441:** 469-474

Rejon, E., Bajon, C., Blaize, A. and Robert, D. (1988). RNA in the nucleus of a motile plant spermatozoid: characterization by enzyme-gold cytochemistry and in situ hybridization. *Molecular Reproduction and Development* **1:** 49-56

Saunders, C.M., Larman, M.G., Parrington, J., Cox, L.J., Royse, J., Blayney, L.M., Swann, K. and Lai F.A. (2002). PLC zeta: a sperm-specific trigger of Ca^{2+} oscillations in eggs and embryo development. *Development* **129:** 3533-3544

Sciamanna, I., Barberi, L., Martire, A., Pittoggi, C., Beraldi, R., Giordano, R., Magnano, A.R., Hogdson, C. and Spadafora, C. (2003). Sperm endogenous reverse transcriptase as mediator of new genetic information. *Biochemical and Biophysical Research Com-*

munications **312:** 1039-1046

Smith, K. and Spadafora, C. (2005). Sperm-mediated gene transfer: applications and implications. *Bioessays* **27:** 551-562

Sotolongo, B., Lino, E. and Ward, W.S. (2003). Ability of hamster spermatozoa to digest their own DNA. *Biology of Reproduction* **69:** 2029-2035

Steger, K., Fink, L., Failing, K., Bohle, R.M., Kliesch, S., Weidner, W. and Bergmann, M. (2003). Decreased protamine-1 transcript levels in testes from infertile men. *Molecular Human Reproduction* **9:** 331-336

Sutovsky, P. (2003). Ubiquitin-dependent proteolysis in mammalian spermatogenesis, fertilization, and sperm quality control: killing three birds with one stone. *Microscopy Research Techniques* **61:** 88-102

Thayer, R.E., Singer, M.F. and Fanning, T.G. (1993). Undermethylation of specific Line-1 sequences in human-cells producing a Line-1-encoded protein. *Gene* **133:** 273-277

Wang, H., Zhou, Z., Xu, M., Li, J., Xiao, J., Xu, Z.Y. and Sha, J. (2004). A spermatogenesis-related gene expression profile in human spermatozoa and its potential clinical applications. *Journal of Molecular Medicine* **82:** 317-324

Ward, M.A. and Ward, W.S. (2004). A model for the function of sperm DNA degradation. *Reproduction Fertility and Development* **16:** 547-554

Witkin, S.S. and Bendich, A. (1977). DNA synthesizing activity in normal human sperm: location and characterization of the endogenous reaction. *Experimental Cell Research* **106:** 47-54

Witkin, S.S., Traganos, F. and Bendich, A. (1977). Isolation of a nuclear DNA synthesizing complex from human sperm. *Biochemical and Biophysical Research Communications* **77:** 1404-1410

Wykes, S.M., Visscher, D.W. and Krawetz S.A. (1997). Haploid transcripts persist in mature human spermatozoa. *Molecular Human Reproduction* **3:** 15-19

Zalenskaya, I.A., Bradbury, E.M. and Zalensky, A.O. (2000). Chromatin structure of telomere domain in human sperm. *Biochemical and Biophysical Research Communications* **279:** 213-218

Zhao, Y., Li, Q., Yao, C., Wang, Z., Zhou, Y., Wang, Y., Liu, L., Wang, Y., Wang, L. and Qiao Z. (2006). Characterization and quantification of mRNA transcripts in ejaculated spermatozoa of fertile men by serial analysis of gene expression. *Human Reproduction* **21:** 1583-1590

Spermatology. SRF Vol. 65. ERS Roldan and M Gomendio (eds) Nottingham University Press, Nottingham

Protein translation in mammalian sperm

Yael Gur and Haim Breitbart

The Mina & Everard Goodman Faculty of Life Sciences, Bar-Ilan University, Ramat-Gan 52900, Israel

Our recent work has shown for the first time the translation of nuclear-encoded proteins in mammalian spermatozoa. We demonstrate that several sperm proteins are degraded after a relatively long period of incubation, and some of these proteins are resynthesised when sperm were incubated under capacitation conditions. In the present study using specific antisenses we demonstrated a significant reduction in the level of several sperm proteins and inhibition of sperm motility, actin polymerization and the acrosome reaction. Thus, this study further supports our previous notion regarding the synthesis of proteins during sperm capacitation.

Introduction

It is generally known that the nuclear-encoded protein synthesis does not occur in spermatozoa. However, it was shown that mammalian sperm contain mRNAs originated from the nucleus (Wykes, Vischer and Krawetz, 1997; Miller, Briggs, Snowden, Hamlington, Rollinson, Lilford and Krawetz, 1999; Ostermeier, Dix, Miller, Khatri and Krawetz, 2002). Spermatozoa of mammals reside in the female reproductive tract for several hours, first as sperm reservoir (Suarez, 2002) and later on for capacitation. Sperm capacitation includes several biochemical changes that must occur before sperm can effectively bind to the oocyte (reviewed in Breitbart, 2003).

We hypothesised, that proteins are degraded during the residence of the sperm in the female, and there is resynthesis of these proteins which are important for the capacitation processes. We recently showed that several nuclear-encoded proteins which are important for the processes occur in sperm capacitation, are degraded and resynthesised when sperm are incubated under *in vitro* capacitation conditions (Gur and Breitbart, 2006). In the present study we provide new data based on the use of several antisenses which further support this notion.

Results

In situ hybridization of sperm mRNA transcripts with antisense transcripts is highly specific (see Gur and Breitbart, 2006). This hybridization may inhibit protein synthesis, because binding to the active mRNA sequence interferes with the translation process. The inhibition of translation of a specific mRNA by antisense is called antisense orientation (Crooke, 1999).

We first examined whether hybridization of a specific antisense prevents the translation of the protein for which the complementary sense strand codes. As a control, we also examined the effect of treatment with an antisense to an mRNA whose protein is not present in mammals.

The anti-bovine-AT1-R, anti-mouse-PKA-Cs, and the non-specific anti-EGFP antisense transcripts were added to the capacitation medium with the transfection agent fugene 6. Fugene 6 coats the DNA fragments and allows the antisense transcripts to penetrate the plasma membrane. We also incubated the cells with fugene 6 alone to make sure that it does not have a non-specific inhibitory effect on the translation of the proteins of interest. We calibrated the concentration of fugene 6 (1:200) so that there was no influence of the fugene 6 itself on the cells, but the DNA was successfully inserted.

Corresponding author E-mail: breith@mail.biu.ac.il

Protein synthesis inhibition was examined by Western blot analysis in order to find changes in the total amount of protein in each treatment condition.

Sperm were incubated with antisense transcripts against the AT1-R or PKA-Cs sense sequences, and the amount of bovine AT1-R and mouse PKA-Cs, respectively, was examined (Fig. 1). In each experiment, the levels of AT1-R or PKA-Cs proteins in sperm that were incubated with transcripts against the AT1-R or PKA-Cs sense sequences were compared to sperm incubated with the non-specific antisense transcript to the EGFP mRNA. EGFP served as a control, because it does not normally exist in mammalian cells. Translation of each protein was inhibited by its specific antisense transcript and was unaffected by the EGFP antisense.

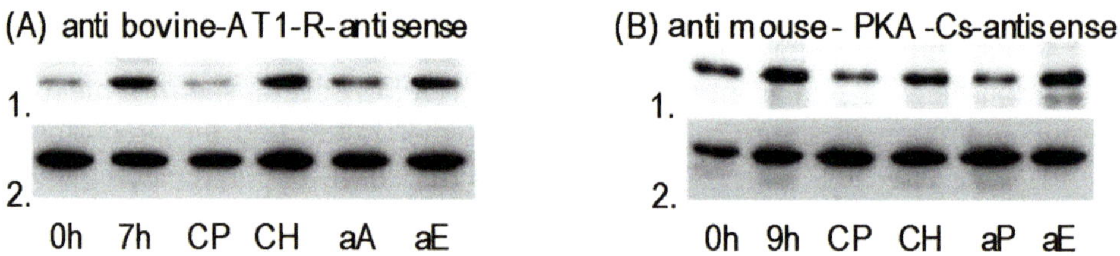

Figure 1. Reduction of protein levels by antisense application. Bovine and mouse sperm were incubated in capacitation medium for 7 and 9 h, respectively, in the presence of fugene 6, the antisense transcripts, and with or without D-chloramphenicol (CP; 100 μg/ml) or cyclohexidine (CH; 1 mg/ml). At the end of the capacitation incubation period, the proteins were extracted, run on a gel and exposed to either anti-AT1-R or anti-PKA-Cs antibodies. (A1) The gel was probed with anti-AT1-R or (B1) anti- PKA-Cs antibody; aA - anti-AT1-R antisense; aP - anti-PKA-Cs antisense; and aE - anti-EGFP antisense. (A2) (B2) The gels were probed with anti-tubulin for the total protein that was loaded in each lane.

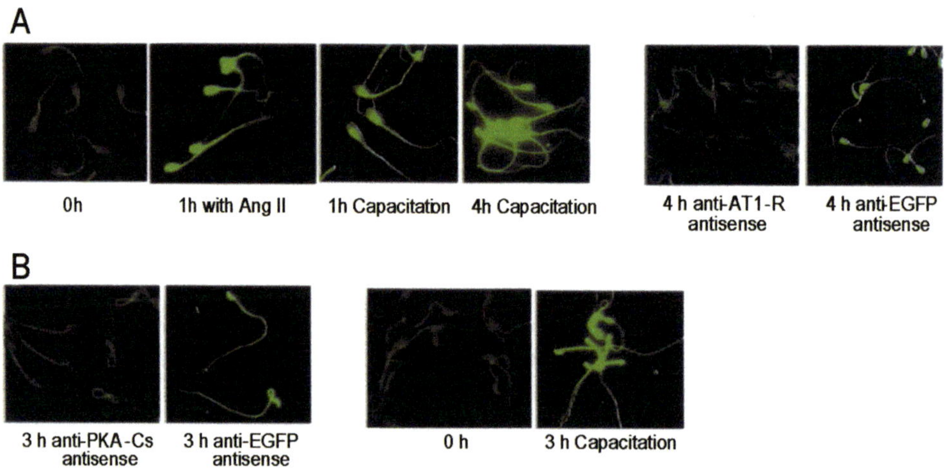

Figure 2. Actin polymerization in bovine and mouse sperm. (A) Bovine sperm cells were incubated in capacitation medium in the absence or presence of Angiotensin II (1 nM) to induce actin polymerization. Cells were also incubated with anti-AT1-R antisense (10 μg/100 μl) in the presence of angiotensin II (1 nM) or anti-EGFP antisense (10 μg/100 μl). (B) Mouse sperm cells were incubated in capacitation medium and in the presence of anti-PKA-Cs and anti-EGFP antisense transcripts (10 μg/100 μl). Cells were stained for F- actin by FITC-phalloidin (3 μM) and observed under fluorescent microscope.

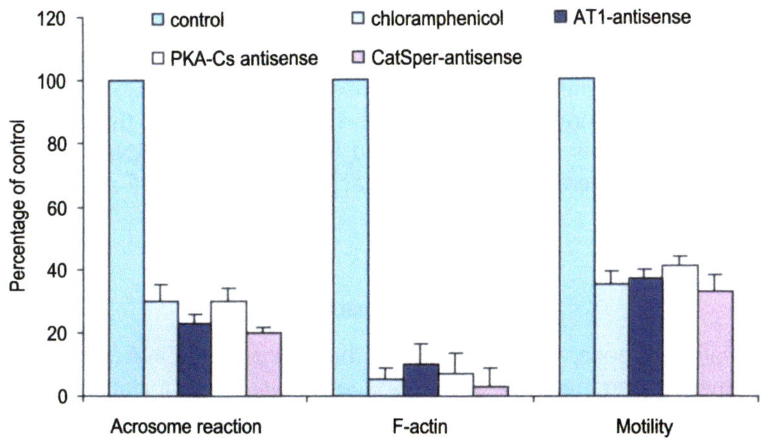

Figure 3. Inhibition of sperm function by specific antisense transcripts. Bovine and mouse sperm were incubated in capacitation medium with or without D-chloramphenicol (CP; 100 μg/ml), bovine-AT1-R, mouse PKA-Cs or mouse CatSper (10 μg/100 μl) antisense transcripts and with fugene 6 (1:200). The cells were preincubated with Angiotensin II (1 nM) for 1 h for the examination of the function of the AT1-R. The rate of motility, actin polymerization and the acrosome reaction were determined. Samples labelled control were incubated under capacitation conditions without antisense. Incubation with an irrelevant antisense (anti-EGFP (10 μg/100 μl)) did not show significant effect (not shown). The control represents the maximal activity (100%) for each time period: acrosome reaction for 6, 4, 6, and 6 h for CP, AT1-R, PKA-Cs and CatSper antisenses respectively, F-actin for 4, 4, 3 and 3 h for CP, AT1-R, PKA-Cs and CatSper antisenses respectively, motility for 6 h for CP and all the antisenses. For each measurement, treatment with AT1-R, PKA-Cs, and CatSper antisense were compared with treatment with the control antisense (anti-EGFP or anti-SL). Data are presented as the mean percentage of control $\pm$ SD of 3-4 experiments for each treatment.

Our earlier *in situ* hybridization experiments examined the levels of AT1-R, PKA-Cs, and CatSper mRNA (Gur and Breitbart, 2006). The proteins that are translated from these mRNAs are known to be involved in sperm capacitation and acrosome reaction. Inhibition of the synthesis of those proteins by incubating the sperm with the specific antisense transcripts revealed a significant inhibition of sperm capacitation as determined by measuring motility, actin polymerization and the acrosome reaction (Fig. 3).

For each measurement, treatment with AT1-R, PKA-Cs, and CatSper antisense were compared with treatment with a non-specific antisense (anti-EGFP or anti-SL).

To examine the inhibitory effect of anti-AT1-R antisense on the acrosome reaction and actin polymerization, we first treated the cells with Angiotensin II. We found that Angiotensin II can induce the acrosome reaction (Gur, Breitbart, Lax, Rubinstein and Zamir, 1998), but it was necessary to find out whether Angiotensin II can induce actin polymerization known to occur during sperm capacitation (Brener, Rubinstein, Cohen, Shternall, Rivlin and Breitbart, 2003; Cohen, Rubinstein, Gur and Breitbart, 2004) and can be used as a marker for the determination of this process. We showed that indeed Angiotensin II induces actin polymerization within 1 h of incubation under capacitation conditions (Fig. 2). This actin polymerization is inhibited by specific antisense to AT1-R but not EGFP (Fig. 2). The quantitation of the inhibition of actin polymerization by antisense treatment is presented in Figure 3. The incubation of sperm cells with AT1-R, PKA-Cs and CatSper antisense transcripts inhibits several measures of sperm function, as can be observed in Figure 3.

In addition, the effectiveness of these antisense transcripts in inhibiting sperm function is comparable to that of D-chloramphenicol (Fig. 3). The results indicate that the AT1, PKA-Cs and CatSper antisense transcripts successfully bound to the specific mRNA sequence, interfered with its translation, and thereby inhibited the acrosome reaction, F-actin levels and sperm motility.

Discussion

Antisense orientation (Crooke, 1999) is a method where exogenous DNA fragments are hybridised to the complementary endogenous mRNA sequences, creating a stable DNA-RNA double-strand. The hybridization, by removing the mRNA that is available for translation, interferes with the production of the protein that is related to the mRNA. The same antisense sequences that were used for *in situ* hybridizations (see Gur and Breitbart, 2006) were used for antisense orientation, because of their specificity for the gene of interest. The ability of applied antisenses to inhibit the synthesis of the specific protein related to the mRNA was examined. The effect of AT1-R and PKA-Cs specific antisenses on the amount of the AT1-R and PKA-Cs protein was evaluated by Western blot (Fig. 1).

Unlike in the *in situ* hybridization experimental procedure (Gur and Breitbart, 2006), the antisenses were added and transfected to capacitating live cells with fugene 6 to allow the exogenous DNA to penetrate the cell. As a control for the specific antisenses, two antisenses to an mRNA whose proteins are not present in mammals (the irrelevant antisenses anti-EGFP or anti-SL) were applied under the same conditions. The total amount of the proteins AT1-R and PKA-Cs were reduced in the presence of their specific mRNA antisense (Fig. 1).

The second step was to evaluate the physiological impact of this inhibition. The effect of a single antisense on the levels of cellular proteins is similar to a hypothetical inhibition of a single protein by D-chloramphenicol - the existing proteins in the cell are still functional until they are degraded, while the resynthesis of new proteins is blocked. Further, the physiological effect of each of the detected proteins was determined and compared to their effects in other published studies.

Each of the three proteins that were examined belongs to a different protein family and is involved in different protein cascades in sperm. The sperm functions were evaluated every hour for up to 7 hours. Due to the large amount of data, we summarised the maximal effect of each antisense for each function in Figure 3.

Each of the examined proteins is a key protein in a different stage of fertilisation, starting from motility, then capacitation and finally with the acrosome reaction: CatSper is involved in motility, PKA-Cs and AT1-R are involved in motility, capacitation and acrosome reaction.

CatSper is a sperm-specific calcium channel that regulates calcium-dependent sperm motility (Ren, Navarro, Perez, Jackson , Hsu, Shi, Tilly and Clapham, 2001). However, as a calcium channel its activity might be important for actin polymerization which occur in the capacitation process and depend on intracellular calcium (Cohen *et al.*, 2004). This can explain why antisenses to CatSper blocks actin polymerization (Figs. 2 and 3). CatSper$^{-/-}$ mice are an example of infertility that is caused by one among the numerous proteins that are involved in sperm fertilization. The sperm of CatSper$^{-/-}$ mice have poor motility. For example, the path velocity, progressive velocity and track speed of CatSper$^{-/-}$ mouse sperm are significantly impaired. As a result of the impaired motility, CatSper$^{-/-}$ sperm cannot penetrate the zona pellucida and fertilise the egg (Ren *et al.*, 2001). Our new finding here suggests that CatSper is also involved in sperm capacitation, which further supports the previous results on its importance for fertilisation.

In our work, CatSper antisense was incubated in capacitation medium with mouse sperm for up to 6 hours, even though the typical time period of mouse capacitation *in vitro* is 3 hours. This protocol allowed the remaining CatSper protein to be maximally degraded. Motility was affected by application of CatSper antisense within 3 hours. The delayed effect of CatSper antisense on motility may be attributed to the relatively high number of CatSper molecules at zero time and the relative stability of the protein. This could also explain the relatively low distribution of CatSper mRNA and proteins inside the mitochondria in the *in situ* localizations (Gur and Breitbart, 2006). Although our results regarding motility and infertility using antisense were similar to those from CatSper knockout mice, we could not directly compare their effects. In our study, the sperm cells have an initial amount of CatSper molecules that remained from spermatogenesis and can still function until degradation. In the CatSper knockout mice, on the other hand, the sperm cells do not have any CatSper protein.

PKA-Cs is a sperm-specific protein and was the second protein that was examined. PKA plays a central role in sperm capacitation, motility and acrosome reaction (reviewed by Breitbart, 2003). All of the members of the PKA signalling system are found in sperm, particularly in the vicinity of the mitochondria (San Agustin, Leszyk, Nuwaysir and Witman, 1998). Thus, PKA regulates capacitation and actin polymerization, motility and acrosome reaction. Our results indicate that application of PKA-Cs antisense inhibits motility, acrosome reaction and actin polymerization within 2 hours from the beginning of capacitation. The relatively fast response can be attributed to the large amount of PKA-Cs molecules that are found in spermatozoa at zero time (San Agustin, Leszyk, Nuwaysir and Witman, 1998). Furthermore, PKA-Cs mRNA and protein were relatively abundant in *in situ* localizations (Gur and Breitbart, 2006). This indicates that PKA-Cs mRNA is being translated intensively during capacitation, and the interruption of this process produces a relatively fast response. Furthermore, the anchor protein for PKA, the AKAP, is present in sperm and the fact that mRNA for AKAP 110 is found in the polysomal fraction in sperm (Gur and Breitbart, 2006) support our notion regarding AKAP synthesis which is necessary for the activation of PKA during capacitation.

AT1-R is a hormone receptor, part of the RAS (Rennin-Angiotensin-System) family in somatic and gamete cells, and has a major role in the induction of the acrosome reaction (Gur *et al.*, 1998). The male reproductive system has several sites of intrinsic RAS activity. The roles of the RAS in local processes at the testis, vas deferens, prostate and semen are still not fully understood, although there is evidence that they are involved in tubular contractility, spermatogenesis, sperm maturation, capacitation, acrosomal exocytosis and fertilisation (reviewed in Tan, Ng, Lim and Leung, 2001).

Angiotensin II is needed to activate the AT1-R. We found that Angiotensin II induces actin polymerization (Fig. 2) indicating that AT1-R is involved in sperm capacitation. The induction of sperm motility by AT1-R was shown in other studies (Vinson, Ho and Puddefoot, 1995; Vinson, Mehta, Evans, Matthews, Puddefoot, Saridogan, Holt and Djahanbakhch, 1996), and was verified in our hands in the present study.

AT1-R antisense inhibits motility, acrosome reaction and actin polymerization within 2 hours of capacitation (Fig. 3). We attribute the fast effect to the relatively large number of AT1-R molecules in the cell. AT1-R amount was low at zero time and increased dramatically after 4 hours of capacitation (Gur and Breitbart, 2006). The immunogold labelling also showed a high concentration of AT1-R mRNA and protein in bovine sperm sections (Gur and Breitbart, 2006).

Similar to our results using AT1-R antisense, the specific AT1-R inhibitor losartan completely inhibits the acrosome reaction (Gur *et al.*, 1998) and ion transport (Leung and Sernia, 2003).

In summary, we conclude that AT1-R, CatSper and PKA-Cs are central proteins in the spermatozoal processes that lead to fertilisation. We demonstrated that they are synthesised during sperm capacitation using several complementary techniques. We also provided evidence that AT1-R, CatSper and PKA-Cs proteins are associated with the mitochondrial-translation machinery. Finally, the physiological role and importance of these proteins was demonstrated using specific inhibition by antisenses.

Based on our results, we can conclude that antisenses are a powerful tool to research general phenomenon at the single protein level. The antisense technique allows us to evaluate specific pathways in the cell with an inhibitor that has no side effects and whose specificity is undisputed. Furthermore, this technique can be used to study the role of a specific protein in a signalling cascade without interfering with the other proteins in the cascade.

Based on the work presented here, we present a model for a previously unknown phenomenon in mammalian sperm. Mammalian sperm begin capacitation with a limited amount of proteins. During the final maturation process called capacitation, sperm cells resynthesized proteins that are necessary for motility, capacitation and fertilisation. Some of these proteins are encoded by nuclear genes, but their synthesis is mediated by mitochondrial-type ribosomes (Gaur and Breitbart, 2006). Although the localization of mitochondrial ribosomes in sperm is not clear, several nuclear-encoded mRNA transcripts and their related proteins were localised to the mitochondria. Thus, one of two mechanisms is possible: Proteins may be translated inside the mitochondria and then translocated to their target site (Fig. 4A), or proteins may be translated by mitochondrial-type ribosomes in the cytoplasm and then translocated to their active sites (Fig. 4B). Further research of colocalization of the translating ribosomes with the translated mRNAs could elucidate which of the mechanisms is active.

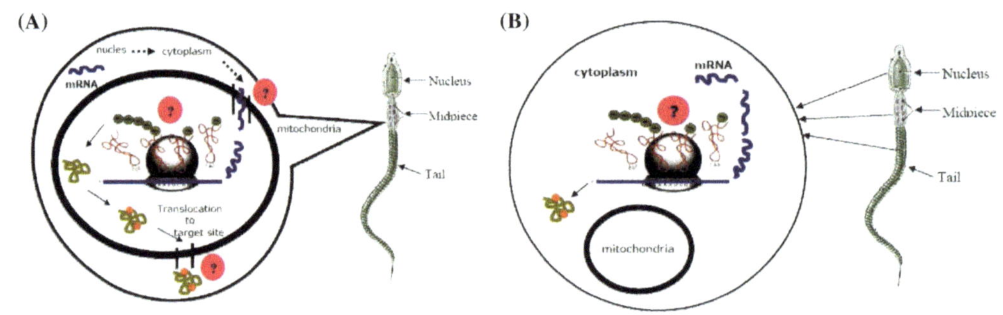

Figure 4. A suggested model for protein translation in sperm. Model (**A**): mRNAs translocate from the nucleus, through the cytoplasm into the mitochondria, then translated by mitochondrial ribosomes. The translated proteins are exported from the mitochondria to their target sites in the cell. Model (**B**): nuclear encoded mRNAs are translated by mitochondrial-type ribosomes in the cytoplasm and then the translated proteins are translocated to their active sites.

These studies provide evidence that protein synthesis is physiologically important to the sperm for its final maturation and successful fertilisation.

This new concept that protein translation is significant to the process of capacitation in mature sperm and the possibility that nuclear-encoded proteins are translated on mitochondrial ribosomes are breakthroughs in our understanding of the molecular mechanisms that are involved in protein translation during the maturation of sperm cells, as well as somatic cells.

These breakthroughs may provide new approaches to understanding sperm-egg interactions, treatment of infertility problems, and the development of new contraceptives.

References

Breitbart, H. (2003). Signaling pathways in sperm capacitation and acrosome reaction. *Cell and Molecular Biology* **49**: 321-327

Brener, E., Rubinstein, S., Cohen, G., Shternall, K., Rivlin, J. and Breitbart, H. (2003). Remodeling of the actin cytoskeleton during mammalian sperm capacitation and acrosome reaction. *Biology of Reproduction* **68**: 837-845

Cohen, G., Rubinstein, S., Gur, Y. and Breitbart, H. (2004). Crosstalk between protein kinase A and C regulates phospholipase D and F-actin formation during sperm capacitation. *Developmental Biology* **267**: 230-241

Crooke, S.T. (1999). Molecular mechanisms of action of antisense drugs. *Biochimica et Biophysica Acta* **1489**: 31-44

Gur, Y. and Breitbart, H. (2006). Mammalian sperm translate nuclear-encoded proteins by mitochondrial-type ribosomes. *Genes and Development* **20**: 411-416

Gur, Y., Breitbart, H., Lax, Y., Rubinstein, S. and Zamir, N. (1998). Angiotensin II induces acrosomal exocytosis in bovine spermatozoa. *American Journal of Physiology* **275**: E87-E93

Leung, P.S. and Sernia, C. (2003). The renin-angiotensin system and male reproduction: new functions for old hormones. *Journal of Molecular Endocrinology* **30**: 263-270

Miller, D., Briggs, D., Snowden, H., Hamlington, J., Rollinson, S., Lilford, R. and Krawetz, S.A. (1999). A complex population of RNAs exists in human ejaculate spermatozoa: implications for understanding molecular aspects of spermiogenesis. *Gene* **237**: 385-392

Ostermeier, G.C., Dix, D.J., Miller, D., Khatri, P. and Krawetz, S.A. (2002). Spermatozoal RNA profiles of normal fertile men. *The Lancet* **360**: 772-777

Ren, D., Navarro, B., Perez, G., Jackson, A.C., Hsu, S., Shi, Q., Tilly, J.L. and Clapham, D.E. (2001). A sperm ion channel required for sperm motility and male fertility. *Nature* **413**: 603-609

San Agustin, J.T., Leszyk, J.D., Nuwaysir, L.M. and Witman, G.B. (1998). The catalytic subunit of the cAMP-dependent protein kinase of ovine sperm flagella has a unique amino-terminal sequence. *Journal of Biological Chemistry* **273**: 24874-24883

Suarez, S.S. (2002). Formation of a reservoir of sperm in the oviduct. *Reproduction in Domestic Animals* **37**: 140-143

Tan, I., Ng, C.H., Lim, L. and Leung, T. (2001). Phosphorylation of a novel myosin binding subunit of protein phosphatase 1 reveals a conserved mechanism in the regulation of actin cytoskeleton. *Journal of Biological Chemistry* **276**: 21209-21216

Vinson, G.P., Ho, M.M. and Puddefoot, J.R. (1995). The distribution of angiotensin II type 1 receptors, and the tissue renin-angiotensin systems. *Molecular Medicine Today* **1**: 35-39

Vinson, G.P., Mehta, J., Evans, S., Matthews, S., Puddefoot, J.R., Saridogan, E., Holt, W.V. and Djahanbakhch, O. (1996). Angiotensin II stimulates sperm motility. *Regulatory Peptides* **67**: 131-135

Wykes, S.M., Visscher, D.W. and Krawetz, S.A. (1997) Haploid transcripts persist in mature human spermatozoa. *Molecular Human Reproduction* **3**: 15-19

Spermatology. SRF Vol. 65. ERS Roldan and M Gomendio (eds) Nottingham University Press, Nottingham

Sperm mitochondria and fertilisation

Justin C. St. John[1], Emma J. Bowles[1] and Alexandra Amaral[1,2]

[1]*The Mitochondrial and Reproductive Genetics Group, The Medical School, University of Birmingham, Birmingham B15 2TT, UK;* [2]*Center for Neuroscience and Cell Biology, Department of Zoology, University of Coimbra, 3004-517 Coimbra, Portugal*

Human mitochondrial DNA (mtDNA) is an extranuclear genome that encodes 13 of the polypeptides associated with the process of oxidative phosphorylation (OXPHOS). The role and importance of OXPHOS in sperm motility and function has been debated over the last few years. Here, we argue that sperm OXPHOS is important to sperm function in the light of clinical based evidence in the human where pathogenic mutations have also been described in sperm and are associated with varying degrees of male subfertility. We also discuss the importance of maintaining maternal inheritance of mtDNA and how sperm mtDNA might be eliminated during early embryogenesis in a manner similar to the process which decreases oocyte mtDNA to extremely low levels once it reaches the blastocyst stage of preimplantation development. Finally, we discuss the role of sperm mtDNA replication and why it may be prudent to considerably reduce sperm mtDNA numbers during the transition from spermatogenesis to spermiogenesis.

Introduction

Mitochondria are organelles of bacterial origin that invaded eukaryotic cells to form a symbiotic relationship. Within each of these organelles are one or more copies of a bacterial type genome, namely mitochondrial DNA (mtDNA), which contributes extensively to cellular metabolic function. However, the role that the sperm mitochondria play in sperm metabolism and function prior to fertilisation, and their subsequent fate post-fertilisation, are highly contentious issues that have perplexed geneticists, biochemists and cell biologists. This article discusses many of these issues and perpetuates the discussion further.

The generation of cellular ATP

A cell can generate ATP through various intracellular processes. These include anaerobic mechanisms such as the Krebs cycle and ß-oxidation, which precede the aerobic production of oxidative phosphorylation (OXPHOS). Indeed, OXPHOS is the cell's major generator of ATP contributing 32 molecules of ATP to every 2 generated through glycolysis (Pfeiffer *et al.*, 2001). OXPHOS takes place within the mitochondrion using an intracellular apparatus known as the electron transfer chain (ETC). The ETC consists of five protein complexes which pass electrons from one complex to the next, while protons are pumped out of the complexes to generate the mitochondrial membrane potential (see Fig. 1). Returning these protons into the mitochondria through the fifth complex results in ATP generation.

Corresponding author E-mail: j.stjohn.1@bham.ac.uk

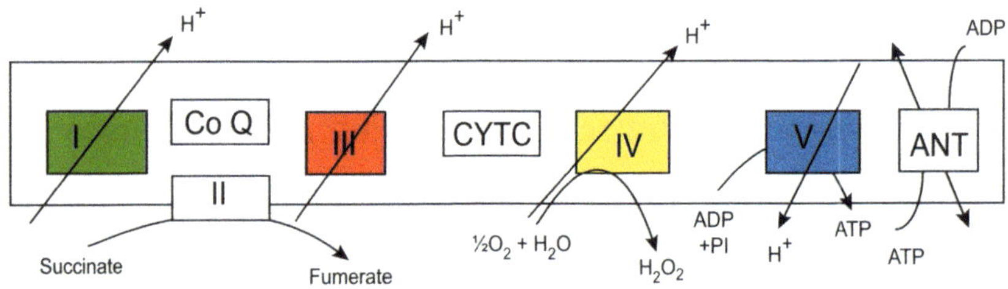

Complex	I	II	III	IV	V
nDNA subunits	>18	4	8	10	10
mtDNA subunits	7	0	1	3	2

Figure 1. The electron transfer chain. Protons (H^+) are released into the surrounding environment which establishes an electrochemical gradient. This results in sufficient energy being generated to support ATP synthesis. As the electrons are transmitted along the ETC, they interact with the protons and react with molecular oxygen which produces H_2O. The vast majority of the subunits for the complexes are encoded by nDNA. However, each of the complexes, except for complex II, is dependent on mtDNA genes for the generation of one or more of their subunits. Abbreviations: CoQ = Coenzyme Q; CYTC + cytochrome c; ANT = adenine nucleotide transporter; nDNA = nuclear DNA; mtDNA = mitochondrial DNA.

The mitochondrion

In the vast majority of cases mitochondria are located in the cytoplasm. However, in the mature sperm, they are isolated in the midpiece where 22 to 28 of these organelles are packed in a helical formation (Otani *et al.*, 1998). This relocation takes place during the process of spermiogenesis when other organelle structures undergo conformational changes, such as the Golgi apparatus which gives rise to the acrosome (Sinowatz and Wrobel, 1981). At this point, the mitochondria are translocated so that they are located between the centrioles and the flagellum. It appears that these mitochondria are packaged in such a manner that their isolation to perform classical and specific ATP studies is a rate-limiting step. However, this decrease in the number of mitochondria present would suggest that OXPHOS-derived ATP is not critical for sperm function based on the assumption that the number of mitochondria per cell indicates that cell's requirement for ATP (Moyes *et al.*, 1998). For example, the female gamete possesses from 10,000 to over 700,000 depending on its ability to fertilise and support preimplantation development (Almeida Santos *et al.*, 2006).

The sperm OXPHOS debate

The inability to isolate mitochondria from mature sperm has led to a series of contradictory conclusions being drawn when classical OXPHOS inhibitors were used. For example, analysis of a non-fractionated sperm population determined a poor association between sperm motility and mitochondrial function (Ford and Harrison, 1981). However, the significance of OXPHOS-derived ATP for sperm motility was demonstrated in a clinical scenario where a patient harbouring a maternally inherited mitochondrial gene defect in Complex I of the ETC exhibited increased sperm motility when his sperm were cultured in a substrate for Complex II thus by-passing the mutagenic effect (Folgero *et al.*, 1993). Just as convincing are the data demonstrating an association between an increase in the concentration of individual OXPHOS inhibitors for respiratory Complexes I, III and IV and the decrease in sperm motility and flagellum movement (Ruiz-Pesini et *al.*, 2000). Other studies by these investigators have also demonstrated a link between these complexes and asthenozoospermia, i.e. sperm with poor motility characteristics when compared to normal controls (Ruiz-Pesini et *al.*, 1998).

In an attempt to concentrate on sperm mitochondrial ATP generation, other investigators have ruptured cytoplasmic membranes (Storey, 1975, 1978, 1980; Storey and Kane, 1975, 1977, 1980). In this instance, they have analysed respiration rates and oxygen consumption to determine whether OXPHOS is critical for ATP production and thus motility in spermatozoa. The findings tended to be species specific, where the level of ATP generated through OXPHOS was indicative of the glucose concentration of the female reproductive tract (Quinn *et al.*, 1995). More recently, the role of glycolysis has been debated in the context of a paracrine mechanism for the delivery of ATP to the flagellum and experiments where sperm have been cultured in sugar-free media and shown to be motile (Ford, 2006) strongly suggesting a role for OXPHOS. In this respect, we have previously argued that glucose concentration would support glycolysis and thus allow these cells to use anaerobic metabolism with the inevitable outcome that loss of OXPHOS function was therefore masked (St. John *et al.*, 2005a). Here, we were able to demonstrate that, in the environment of a lower glucose concentration similar to that of the female reproductive tract, OXPHOS inhibitors were able to reduce the motility of good quality fractionated sperm.

Mitochondrial DNA

Located within the mitochondrion is the mitochondrial genome (Fig. 2). In mammals, this circular genome is approximately 16.5kb in size and double-stranded, consisting of a Heavy and Light strand and encodes 13 of the polypeptides of the ETC (Fig. 1). Each of the complexes of the ETC has one or more of its genes encoded by mtDNA except for Complex II which like the other non-mtDNA encoded genes are derived from chromosomal DNA (Anderson *et al.*, 1981). Specifically, mtDNA codes for seven subunits of NADH dehydrogenase (Complex I), one component of Complex III, three subunits of cytochrome C oxidase (Complex IV) and two subunits of ATP synthase (Complex V). All other components of the ETC are encoded by the nuclear genome (see Fig. 1). MtDNA also encodes 22 tRNAs and 2 rRNAs again indicating the semi-autonomous nature to the function of this genome. MtDNA is further unlike chromosomal DNA as this highly compact genome contains no introns between the coding regions (Anderson *et al.*, 1981) and some of the coding regions even overlap (ATPase 6 and ATPase 8, ND4 and ND4L; see Anderson *et al.*, 1981). Some of the other genes do not possess termination codons, as these sequences are generated through post-transcriptional polyadenylation (Ojala *et al.*, 1981).

Transcription and replication of mtDNA

Transcription and replication of mtDNA are under the control of nuclear-encoded transcription and replication factors that are translocated to the mitochondria (Clayton, 1998). However, mtDNA transcription and replication take place at significantly greater rates than for nuclear genes (Bogenhagen and Clayton, 1977) demonstrating, despite being under nuclear control, a strong link to original bacterial regulation rather than synchrony with the cell cycle of the nucleus. The process of replication, which ensures sufficient mtDNA is present, is mediated by

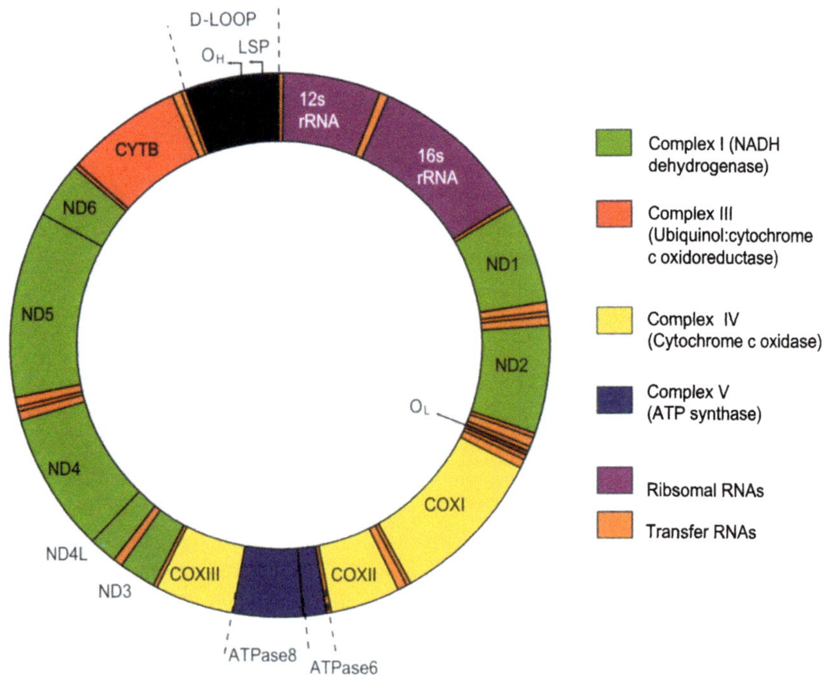

Figure 2. The human mitochondrial genome. This genome encodes 13 of the polypeptides for 4 out of the 5 complexes of the ETC. This double stranded genome consists of the heavy (H) strand, which encodes 12 of these proteins along with 14 tRNAs and the 2 rRNAs, whilst the light (L) strand encodes one enzymatic protein (NADH 6) and eight tRNAs. Within the D-loop, the H-strand origin of replication (O_H), the H- and L-strand promoters (LSP) and conserved sequence boxes are located. The D-loop is the only region of the genome that does not undergo transcription. However, it has two hypervariable regions (HV1 and HV2) that are used for molecular fingerprinting of individuals which define maternal inheritance. O_L = strand origin of replication.

several key factors (see Fig. 3). However, the initiation of replication is dependent on transcription having proceeded beforehand.

The key factors included in mtDNA transcription are mitochondrial transcription factor A (TFAM; Fisher and Clayton 1985, 1988), either mitochondrial transcription factor B1 (TFB1M) or B2 (TFB2M; Falkenberg *et al.*, 2002; McCulloch *et al.*, 2002) and the mitochondrial specific RNA polymerase (mtRNApol; Tiranti *et al.*, 1997). Transcription can commence from two initiation sites on the heavy (H) strand (H_1 and H_2) and one site located on the light strand (L) (Montoya *et al.*, 1982, 1983; Bogenhagen *et al.*, 1984). The H_1 and L transcription initiation sites are located within promoter regions, named the Heavy Strand Promoter (HSP_1) and the Light Strand Promoter (LSP), respectively. Initiation from H_1 and L necessitates TFAM to interact with an enhancer element (Parisi and Clayton, 1991; Dairaghi *et al.*, 1995) that permits mtRNApol to bind and trigger transcription in the presence of either TFB1M or TFB2M. Transcription from H_1 produces the two rRNAs, 12S and 16S, and the Phe and Val tRNAs. Initiation of transcription at H_2 generates a polycistronic molecule containing the majority of the H strand and is independent of TFAM activity (Montoya *et al.*, 1983). Furthermore, transcription from this site occurs 20 times less frequently (Montoya *et al.*, 1983).

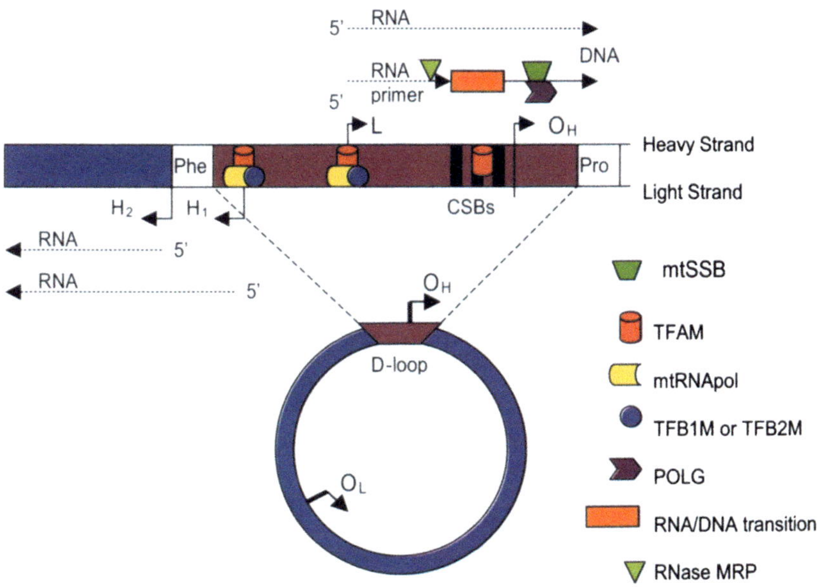

Figure 3. Transcription and replication of mtDNA. Transcription and replication of mtDNA is dependent on a number of nuclear-encoded factors being exported to the mitochondrion and their subsequent interaction. These include: mitochondrial transcription factor A (TFAM); mitochondrial transcription factor B1 (TFB1M); mitochondrial transcription factor B2 (TFB2M); the mitochondrial specific RNA polymerase (mtRNApol); RNA-processing endoribonuclease (RNase MRP); mitochondrial single strand binding protein (mtSSB). Other abbreviations: conserved sequence blocks (CSBs); L (Light strand); H (Heavy strand); Initiation sites (H_1, H_2, L_1). Pro (Proline); Phenylalinine (Phe). See text for details.

Initiation of transcription from the initiation site L results in the generation of a polycistronic molecule for the L-strand and the primer necessary for mtDNA replication. This primer is then processed by RNA-processing endoribonuclease (RNase MRP) which permits the initiation of mtDNA replication from the origin of replication for the heavy strand (O_H) through DNA polymerisation (Lee and Clayton, 1996; Xu and Clayton, 1996). DNA polymerisation is mediated by the mitochondrial specific polymerase gamma (POLG). Transition from the RNA primer to the newly synthesised DNA is mediated through a series of conserved sequence blocks (CSBs). Once H-strand replication reaches the origin of light strand replication (O_L), this factor is exposed and L-strand synthesis commences in an anti-clockwise manner (Wong and Clayton, 1985; Hixson *et al.*, 1986). Once replication of both strands has been completed, the RNA primers are discarded and the newly formed molecule takes on its customary tertiary structure after ligation of the contributing strands of DNA (Lecrenier and Foury, 2000). More recently, an additional mechanism for mtDNA replication has been described. This proposes *coupled leading-* and *lagging-strand* synthesis (Holt *et al.*, 2000). Here, both H and L strands are replicated from multiple initiation sites in close proximity to the O_H and in a simultaneous manner (Yang *et al.*, 2002; Bowmaker *et al.*, 2003).

The role of mitochondrial DNA transcription and replication factors in mtDNA type disease

In the human, PolG is located on chromosome 15q24-15q26 (Ropp and Copeland, 1996; Walker et al., 1997) whilst in the mouse it has been mapped to chromosome 7 (Zullo et al., 1997). POLG comprises a 140KDa catalytic subunit (POLGA; see Figure 4) and a 54KDa accessory subunit (POLGB; Gray and Wong, 1992). The catalytic unit functions as both the polymerase and a 3′-5′ exonuclease, for both proofreading and DNA repair purposes (Graves et al., 1998; see Fig. 3). The accessory unit promotes enzymatic DNA binding, processivity and fidelity (Lim et al., 1999; Longley et al., 2001) and potentially the recognition of RNA primers to mediate mtDNA replication through recruitment of POLG to the primer-template junction (Fan et al., 1999). Dysfunction to POLG can arise through missense mutations which are associated with Progressive External Opthalmoplegia (PEO) and mitochondrial neurogastrointestinal encephalomyopathy (Cormier et al., 1991; Van Goethem et al., 2001, 2003) and testicular cancer (Nowak et al., 2005). Additionally, mutations in PolG have been associated with Alper's disease (Naviaux et al., 2004, 2005 ; Davidzon et al., 2005 ; Nguyen et al., 2005, 2006), Parkinsonism and premature menopause (Luoma et al., 2004; Pagnamenta et al., 2006).

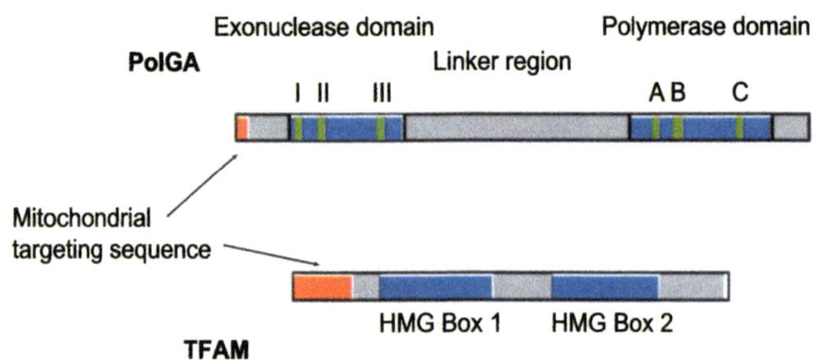

Figure 4. Properties of PolGA and Tfam. Both genes possess mitochondrial targeting sequences allowing their translocation to the mitochondria. HMG = High Mobility Group.

POLG is also particular sensitive to nucleoside analogue reverse transcriptase inhibitors (NRTIs), commonly used in the treatment of HIV. Consequently, HIV positive patients administered such drugs can present with mitochondrial myopathies and neuropathies (Dalakas et al., 2001). The power of such agents is perhaps best observed through cell culture studies where, for example 2',3'-dideoxycytidine (ddC), can result in near mtDNA depletion within a few days of culture (Ashley et al., 2005). Indeed, sperm samples from HIV postive men treated with NRTIs, present with mtDNA depletion after 18 months administration (White et al, 2001).

Unlike other species, it appears that human PolG encodes a series of trinucleotide CAG repeats that gives rise to a polyglutamine tract (Ropp and Copeland, 1996). Interestingly, the length variation of the CAG repeat region of the androgen receptor gene has been suggested to play a role in male infertility (reviewed in Gottlieb et al., 2005) and thus makes it an interesting region to study in PolG. These repeats are located at the 5′ end. This particular region has been subject to much debate within the last few years in terms of sperm quality and function. A series of contradictory reports have questioned its role in idiopathic subfertility. An association between the absence of the common allele and male infertility was first suggested by Rovio et al. (2001) who found the homozygous mutant genotype (not 10/not 10) at an elevated frequency in infertile patients compared with fertile men (particularly in patients with moderate oligozoospermia; and excluding extreme oligozoospermia and azoospermia; Rovio et al., 2001). A link between the lack of the common allele and idiopathic infertility in a Danish population has also been identified (Jensen et al., 2004). However, in a subsequent Italian study, the frequency of the homozygous mutant genotype was found to be similar both in infertile and normozoospermic fertile men (Krausz *et al.*, 2004). Similar outcomes were obtained in a large study in a French cohort. Importantly, more than half of these homozygous mutant men fathered either naturally or after assisted reproductive techniques (ART; Aknin-Seifer *et al.*, 2005). Moreover, these authors suggested recalculations of the results obtained both by Rovio *et al.* and Jensen *et al.* as the final conclusions of these studies excluded a number of samples. In agreement with these results, Brusco *et al.* (2006) found no significant differences in the frequency of each genotype when comparing oligozoospermic Italian men with normozoospermic men. In conclusion, at least in the Italian and French populations, there is no relationship between the polymorphic CAG repeat region in the PolG gene and male infertility.

If the presence of the mutant PolG genotype (either in homozygosity or heterozygosity) has, in fact, any effect in PolG enzymatic properties, leading to a decrease in mtDNA replication efficiency, one should expect an accumulation of mtDNA mutations and/or a decrease in mtDNA copy number in men with the mutant genotype. Results of a recent study showed no differences in the frequencies of mtDNA nucleotide substitutions for the different PolG genotypes in both normozoospermic and non-normozoospermic men (Harris *et al.*, 2006). These authors hypothesised that the results obtained by Rovio *et al.* and Jensen *et al.* could be caused by the presence of mutations somewhere else in the PolG gene. To test that possibility, they analysed the three exonuclease motifs of PolG, but no mutations were found in all the samples tested. Moreover, Harris and colleagues found the genotypic frequencies of pooled heterozygous and homozygous mutant PolG to be significantly higher in normozoospermic men when compared with non-normozoospermics. This is contrary to that anticipated should PolG CAG repeat polymorphisms have an effect on sperm quality. However, this unexpected statistically significant difference could result from the number of normozoospermic samples used being less than half those of the non-normozoospermic cohort.

TFAM belongs to the High Mobility Group (HMG) of proteins and in the human is located to chromosome 10q21. Its 204 amino acids largely consist of 2 HMG boxes and a mitochondrial targeting sequence (see Fig. 4; Parisi and Clayton, 1991). The exact role of TFAM is still being elucidated. It is thought to be a regulator of mtDNA copy number and this has been supported by mouse knockout studies where heterozygous knockouts suffered from decreased mtDNA copy number and myocardial OXPHOS deficiency (Larsson *et al.*, 1998). Homozygous mice had severe mtDNA depletion and abolished OXPHOS with their embryos surviving only as far as gastrulation. Other investigators have analysed cell culture systems and depleted their cultures of mtDNA to varying degrees, and observed reduced levels of TFAM and mtRNApol, though the relationship was not linear (Seidel-Rogol and Shadel, 2002; Lloyd *et al.*, 2006). Others regard TFAM as a regulator of mitochondrial gene expression (Gensler *et al.*, 2001). However, when expressed in excess, TFAM binds and saturates the DNA template, resulting in complete inhibition of transcription from the LSP *in vitro* (Dairaghi *et al.*, 1995) which obviously also impacts on mtDNA replication.

It has become increasingly evident that TFAM requires moderation from TFB1M and TFB2M (Falkenberg *et al.*, 2002) and Nuclear Respiratory Factors 1 (NRF-1) and 2 (NRF-2). TFAM's promoter encodes recognition sites for these latter two factors, (Evans and Scarpulla, 1989; Virbasius *et al.*, 1993). Interestingly, these sites possess CpG islands which, following investigation, have been proposed as sites for DNA methylation control of mtDNA transcription and replication (Choi *et al.*, 2004).

As with POLG, it is evident that decreased expression of TFAM in humans is associated with mtDNA-type depletion syndromes which manifest as severely debilitating mtDNA diseases such as mitochondrial myopathy and encephalomyopathy disorders (Larsson *et al.*, 1994; Poulton *et al.*, 1994, Spelbrink *et al.*, 1998; Siciliano *et al.*, 2000). Although TFAM has not so far been implicated in sperm dysfunction, it is evident that during murine spermatogenesis, a nuclear isoform is expressed (Larsson *et al.*, 1996) that is characterised by a 10-fold reduction in mtDNA copy to approximately 70 to 80 mtDNA copies per cell as the spermatocyte differentiates into a round spermatid (Hecht and Liem, 1984). In the human, TFAM expression in the testis is merely down-regulated without the specific expression of a testis-specific isoform (Larsson *et al.*, 1997). Consequently, one could speculate that, in subfertile men, TFAM is poorly regulated during spermatogenesis and spermiogenesis so that one could account for the greater number of mtDNA molecules observed in men requiring infertility treatment (Díez-Sanchez *et al.*, 2003; May-Panloup *et al.*, 2003). However, these findings require further verification as there was considerable variability in mtDNA copy between the observations of these two groups. The average mtDNA/Beta-globin ratio of sperm was determined by May-Panloup *et al.* (2003) using real-time quantitative PCR. In sperm collected from 100% density layers, the mean ratios were 1.4 for normozoospermic samples, 6.1 for samples with one abnormal sperm parameter and 9.1 for samples presenting with two or three abnormal sperm criteria. Sperm collected from 40% gradient layers showed a mean ratio of 17.1. However, through slot-blot hybridization, Díez-Sanchez *et al.* (2003) estimated the number of copies of mtDNA per cell to be at 700 and 1200 in progressive and non-progressive spermatozoa, respectively. In addition, one other group has produced observations indicating the contrary (Kao *et al.*, 2004). These authors compared the relative mtDNA content between samples with low and normal motility and found means of 7.2 for the asthenozoospermic individuals and 74 for the normozoospermics. Consequently, some consensus is required to verify whether mtDNA copy number is a defining characteristic of good and poor quality sperm. If those observations reporting abnormal regulation of TFAM during human spermiogenesis are indicative of a cause of male subfertility, one could speculate that this would also be the case for POLG.

MtDNA inheritance

Transmission of mtDNA tends to be solely through the oocyte in mammals (Birky, 1995, 2001), and especially so in humans (Giles *et al.*, 1980). This arises as a restriction event takes place during early oogenesis that ensures only a few molecules are available in primordial germ cells (Jansen and de Boer, 1998) for transmission to the subsequent generation (Hauswirth and Laipis, 1982). These molecules are clonally expanded throughout the process of oocyte maturation so that the oocytes giving rise to the next generation of offspring will possess over a 100,000 copies of a single population of mtDNA (Marchington *et al.*, 1997), otherwise referred to as homoplasmy.

However, for other species this is not always the case. For example, mussels transmit male- and female-type mtDNA genomes to male offspring and female only type genomes to female offspring (Fisher and Skibinski, 1990; Hoeh *et al.*, 1991). Both intra- and inter-specific crosses of fruit fly can transmit sperm and oocyte mtDNA to subsequent offspring (Kondo *et al.*, 1992). Intra- and inter-specific crossing also influences mammalian mtDNA patterns of inheritance. It is evident from early studies in mouse that interspecific crosses are capable of transmitting sperm mtDNA to offspring but only at very low levels (Gyllensten *et al.*, 1991; Shitara *et al.*, 1998) with transmission being restricted to the first generation (Shitara *et al.*, 1998). Sperm mtDNA transmission has also been observed in interspecific crosses in sheep (Zhao *et al.*, 2004) and non-human primates (St. John and Schatten, 2004). Similarly, sperm mitochondria are observed at later stages in mouse (Kaneda *et al.*, 1995), bovine and rhesus macaque (Sutovsky *et al.*, 1999) preimplantation embryos, a phenomenon not observed in intraspecific crosses. In the case of intraspecific crosses, a ubiquitin-mediated elimination process has been proposed to account for such outcomes (Sutovsky *et al.*, 1996, 1999). Here, spermatogonial cells are labelled with ubiquitin which is then recognised once fertilisation has taken place and the mitochondria are thus eliminated. More recently, active digestion of sperm mtDNA has been demonstrated in Japanese medaka embryos just after fertilisation and this takes place before the destruction of the organelle itself (Nishimura *et al.*, 2006).

The mechanism for eliminating sperm mtDNA, although rigorous in many circumstances, has proven to permit leakage. This has certainly been the case for a male patient diagnosed with a mitochondrial myopathy where the muscle mtDNA was inherited from his father's mtDNA. However, the 2bp deletion giving rise to his myopathy was *de novo* (Schwartz and Vissing, 2002). Additionally, it appears that poor quality oocytes allow sperm mtDNA to persist through to blastocyst in a few cases (St. John *et al.*, 2000a).

In the pursuit of attempting to identify mechanisms for the elimination of sperm mtDNA, it is also valid to consider the fate of oocyte mtDNA during preimplantation development. Based on earlier data, it has always been assumed that mtDNA levels remained constant (Piko and Taylor, 1987). However, the use of more sensitive PCR assays have indicated that bovine oocyte mtDNA is reduced to approximately 40% of its original content at fertilisation by the 8-cell stage (May-Panloup *et al.*, 2005). Although these data require verification in other species, this presents an interesting scenario, namely that the oocyte mediates a mass programme of elimination perhaps to preserve a few copies of the genome that would be transmitted to the immediate offspring as well as preserving the maternal mtDNA lineage through the oocytes of subsequent generations. This large-scale degradation of embryonic mtDNA would account for the very few mitochondria observed in undifferentiated embryonic stem cells (Sathananthan et *al.*, 2002; St. John et *al.*, 2005b) which are derived from the inner cell mass of the blastocyst. Above all, this low level of mtDNA transmitted through to the next generation could act as a mechanism to ensure against the transmission of mtDNA type disease, especially as reports suggest that mtDNA deletions accumulate in the later stages of *in vitro* maturation (Gibson et *al.*, 2005).

MtDNA disease

MtDNA disease arises from rearrangements in the mtDNA genome. These diseases have been categorised by their association with maternally-inherited point mutations into either rRNAs or tRNAs such as for mitochondrial myopathy, encephalopathy, lactic acidosis, and stroke-like episodes (MELAS – Kobayashi *et al.*, 1990); or by their presence in coding genes as observed for neurogenic weakness, ataxia, retinitis pigmentosa (NARP – Fryer *et al.*, 1994), mitochondrial encephalomyopathy (Clark *et al.*, 1999) and Leber's hereditary optic neuropathy (LHON – Brown *et al.*, 1992). Also important, are the single large-scale deletions that are indicative of Kearns-Sayre syndrome (Schon *et al.*, 1989), and the multiple large-scale deletions observed with ageing (Melov *et al.*, 1995). Various mechanisms have been proposed to account for these rearrangements and these are discussed at length in St. John *et al.* (2000b).

Most mtDNA diseases are observed in a heteroplasmic manner, namely the ratio of wild type to mutated/deleted molecule. Consequently, for certain disease states, a critical threshold appears to exist before the phenotypic onset of the disease is observed. In patients suffering, from LHON, it is anticipated that the onset of disease requires >60% mutant mtDNA load before being observed (Chinnery *et al.*, 2001). For other cases, such as myoclonic epilepsy and ragged red fibres (MERRF), 85-90% mutant molecule is projected for the phenotypic onset to be observed (Boulet *et al.*, 1992).

Many studies related to mtDNA mutations and deletions in male infertility have been undertaken to determine whether any of these mutations or deletions are indicative of one or more forms of subfertility. Consequently, many studies have relied on the separation of sperm using, for example, Percoll gradients, to characterise the degree of motility to mutant/deletion load. Others have based their analysis on a World Health Organisation (WHO) classification of infertility, for example the World Health Organisation Guidelines (1992, 1999), which are, at best, subjective. That having been said, some interesting associations have been formulated.

As with mtDNA disease in somatic tissues, a maternally inherited point mutation, A3243G, has been described where the greater proportion of mutant was identified to be in the lower density Percoll fractions (range = 54 to 42% from 0% to 80% fractions - Spiropoulos *et al.*, 2002), suggesting that OXPHOS and therefore mitochondrial ATP are mediators of sperm motility. Less definitive but equally interesting are the two reports of mutations at nt 9055 and nt 11719. These were indicative of poor semen quality through their greater incidence in men with poor semen parameters (11 to 12%) compared to men with normal semen characteristics (0% to 1.3%; Holyoake *et al.*, 1999, 2001).

The case for large-scale mtDNA deletions and their association with male subfertility has also been made. Although there are subtle differences between outcomes, an association exists between such deletions and poor semen quality. In terms of assessing motility though Percoll gradients and drawing a direct link to asthenozoospermia, the 4977bp 'common' deletion has been identified with increasing regularity in lower density Percoll fractions (Kao *et al.*, 1995). It should, however, be pointed out that the levels of deleted molecule compared to wild type mtDNA were very low (0.0032% for the 80% Percoll fractions to 0.0708% for the <50% Percoll fractions). Additionally, this work was not verified by two subsequent studies, one analysing sperm only samples (St. John *et al.*, 2001) and the other semen samples (Cummins *et al.*, 1998a). A further report indicated 2 other deletions, 7345 and 7599 bp, that were also characteristic of poor motile sperm (Kao *et al.*, 1998). However, the persistence of a similar sized deletion was not deemed to be indicative of such a trend (St. John *et al.*, 2001).

Large-scale deletions can also manifest themselves in multiple forms. Interestingly, multiple deletions identified in muscle fibres from a patient presenting with ptosis were also identified in his sperm (Lestienne *et al.*, 1997), again demonstrating that an OXPHOS disorder affecting one tissue was also characteristic of poor sperm motility. Interestingly though, two further studies have identified multiple mtDNA rearrangements in a range of patients. The first observed such deletions in both normozoospermic and oligozoospermic men (Reynier *et al.*, 1998) and the other confirmed these findings (St. John *et al.*, 2001). However, the latter study also demonstrated a greater accumulation of multiple deletions in poorer quality sperm samples, with a greater number being observed in oligoasthenoteratozoospermic men (St. John *et al.*, 2001). Although some normozoospermic samples possess a number of large-scale deletions, this does not imply that all sperm within a sample are likely to carry mutated molecules. Indeed, in those samples with large concentrations of sperm, one could predict that even if 50% of them harboured mtDNA deletions, the remainder would still be sufficient to comply with their categorisation as normozoospermic, as defined by the WHO. Interestingly though, sperm separated into high-density (90%) fractions are significantly more likely to possess more wild-type mitochondrial molecule than 45% fractioned sperm (O'Connell *et al.*, 2003). Coupled to this, is the greater frequency of single deletions than two or more deletions for the 90% fraction.

Analysis of testicular and epididymal sperm has provided some interesting outcomes. There appears to be significantly more wild-type mtDNA present in testicular sperm than in epididymal sperm and that epididymal sperm possess more large-scale deletions than testicular sperm (O'Connell *et al.*, 2002a). Equally interesting, the mean size of the deletions in testicular sperm from fertile men is significantly less than those from men with obstructive azoospermia (O'Connell *et al.*, 2002b). These findings allow interesting hypotheses to be posed. For example, although transcription and replication are completed at the late spermatocyte-early spermatid stage (Hecht and Liem, 1984; Larsson *et al.*, 1996, 1997), is it possible that sperm mtDNA transcription and replication are not entirely complete? We know from oogenesis that an increase in mtDNA deletion accumulates post germinal vesicle breakdown (Gibson *et al.*, 2005) and more interestingly that TFAM transcripts have been identified in mature ejaculated sperm (Zhao *et al.*, 2006). Consequently, one might hypothesise that sperm mtDNA replication is perhaps an on-going process in those sperm which may not have completed maturation in time or a process of self-renewal is taking place. This might arise as mtDNA replication is not synchronous with the cell cycle. Indeed, one can again observe similarities in oocytes where incompetent oocytes delay completion of mtDNA replication and are unlikely to have completed this before they are *in vitro* fertilised, which likely accounts for their arrest during preimplantation development (Spikings and St. John, 2006).

MtDNA haplotype

MtDNA haplotypes are the collection of individuals characterised by single or short variations in the D-Loop region of the mtDNA genome. These haplotypes are indicative of ethnic origin and a predisposition to a particular phenotypic onset of disease (see Wallace, 1995). MtDNA haplotype has been associated with several traits of enhanced performance. Such examples include increased growth and physical performance in mice (Nagao *et al.*, 1998); milk quality in Holstein cows (Schutz *et al.*, 1994), fertility in beef cattle (Sutarno *et al.*, 2002) and in pigs (El Shourbagy *et al.*, 2006), and increased preimplantation development following *in vitro* fertilisation in cattle (Tamassia *et al.*, 2004). Specifically related to sperm, certain mtDNA haplotypes have been associated with poor sperm motility in Caucasians. Here, there is a tendency for haplotype H male patients not to be associated with asthenozoospermia, whilst haplotype T are predisposed to asthenozoospermia (Ruiz-Pesini *et al.*, 2000). These outcomes were supported through sperm motility and biochemical analysis of Complex I and IV enzymatic activity in the ETC. In a recent study, differences in both sperm motility and vitality were found in a number of sublineages of haplogroup U, which might be caused by highly conserved missense mutations in the COX III and cytochrome B genes (Montiel-Sosa *et al.*, 2006).

ART and mtDNA transmission

We have argued that the more diverse the ART procedure the greater the potential for two or more populations of mtDNA to be transmitted (Spikings *et al.*, 2006). As studies related to germinal vesicle transfer (GVT), pronuclear transfer (PNT), and nuclear transfer undergo more rigorous investigation, it is evident that mtDNA accompanying the donor nuclei can also be transmitted. However, this can be at extremely low levels. In somatic cell nuclear transfer (SCNT) the donor cell is still primed to drive mtDNA replication which accounts for the low levels of donor cell mtDNA which persist up to blastocyst (Lloyd *et al.*, 2006). In GVT, PNT and embryonic cell (EC)NT, the donor nucleus is embryonic and therefore not primed to drive mtDNA replication. However, evidence from PNT (Meirelles and Smith, 1997, 1998; Sato *et al.*, 2005) derived embryos suggest preferential replication of the donor nuclei's own mtDNA population prior to blastocyst. In ECNT, sperm mtDNA, which contributed to the blastomeres that were subsequently used to generate the only two cloned non-human primates, was detected in both offspring (St. John and Schatten, 2004). This most likely arises due to the interspecific crossings that firstly generated the donor blastomeres and then the subsequent NT-embryos. Consequently, as with cytoplasmic transfer (Brenner *et al.*, 2000), the introduction of a 'foreign' population of mtDNA can result in its transmission.

However, one of the initial concerns associated with intracytoplasmic sperm injection (ICSI) was that sperm mtDNA could be potentially transmitted. Indeed, studies analysing the fate of murine round spermatid injections into oocytes demonstrated a strong tendency to eliminate sperm mitochondria, although there was low level persistence in the 8-cell or later staged embryos (1%; Cummins *et al.*, 1998b). This however has so far not been the case following sperm injection in humans with two definitive studies, with appropriate levels of sensitivity, reporting that these sperm molecules are not transmitted to live human offspring (Danan *et al.*, 1999; Marchington *et al.*, 2002). Furthermore, sperm mtDNA appears not to be present in 3-pronuclei human blastocysts following ICSI (St. John *et al.*, 2000a), which one might expect otherwise, considering their abnormal nature. These outcomes are refreshing considering that, should the contrary outcome be the case, the risk of transmission would be even greater due to the increased number of molecules being present in poorer sperm samples (Díez-Sanchez *et al.*, 2003; May-Panloup *et al.*, 2003) with increased numbers of mutations (Spiropoulos *et al.*, 2002) and deletions present (St. John *et al.*, 2001).

Conclusions

The presence of mtDNA mutations and deletions associated with loss of sperm function suggests a role for OXPHOS function in sperm. However, the precise nature of OXPHOS function in sperm function still requires clarification. As we begin to understand the role of sperm mtDNA replication in more detail and also clarify whether this has indeed terminated prior to spermiogenesis as previously thought, we will have a clearer idea of how post-testicular modification of sperm can alter the mtDNA genome and affect its ability to transmit chromosomal DNA and the male dominant centrioles to the oocyte for fertilisation. Although we have not managed to mediate the transmission of the highly mutagenic sperm mtDNA molecule to the offspring through ART, we must bear this outcome in mind as we seek to introduce more sophisticated ARTs into the clinical setting. However, the use of such techniques to develop animal models to study the physiological effects of such transmission would be a useful tool and would allow heteroplasmic disease models to be established.

References

Aknin-Seifer, I.E., Touraine, R.L., Lejeune, H., Jimenez, C., Chouteau, J., Siffroi, J.P., McElreavey, K., Bienvenu, T., Patrat, C. and Levy, R. (2005). Is the CAG repeat of mitochondrial DNA polymerase gamma (*POLG*) associated with male infertility? A multi-centre French study. *Human Reproduction* **20:** 736-740

Almeida Santos, T., El Shourbagy. S. and St John, J. (2006). Mitochondrial content reflects oocyte variability and fertilization outcome. *Fertility and Sterility* **85:** 584-591

Anderson, S., Bankier, A.T., Barrell, B.G., de Bruijn, M.H., Coulson, A.R., Drouin, J., Eperon, I.C., Nierlich, D.P., Roe, B.A. and Sanger, F. (1981). Sequence and organisation of the human mitochondrial genome. *Nature* **290:** 457-465

Ashley, N., Harris, D. and Poulton, J. (2005). Detection of mitochondrial DNA depletion in living human cells using PicoGreen staining. *Experimental Cell Research* **303:** 432-446

Birky, C.W. Jr. (2001). The inheritance of genes in mitochondria and chloroplasts: laws, mechanisms, and models. *Annual Review of Genetics* **35:** 125-148

Birky, C.W. Jr. (1995). Uniparental inheritance of mitochondrial and chloroplast genes: mechanisms and evolution. *Proceedings of the National Academy of Sciences USA* **92:** 11331-11338

Bogenhagen, D. F., Applegate, E. F. and Yoza, B. K. (1984). Identification of a promoter for transcription of the heavy strand of human mtDNA: in vitro transcription and deletion mutagenesis. *Cell* **36:** 1105-1113

Bogenhagen, D. and Clayton, D.A. (1997). Mouse L cell mitochondrial DNA molecules are selected randomly for replication throughout the cell cycle. *Cell* **11:** 719-727

Boulet, L., Karpati, G. and Shoubridge, E.A. (1992). Distribution and threshold expression of the tRNA (Lys) mutation in skeletal muscle of patients with myoclonic epilepsy and ragged red fibres (MERRF). *American Journal of Human Genetics* **51:** 1187-1200

Bowmaker, M., Yang, M. Y., Yasukawa, T., Reyes, A., Jacobs, H. T., Huberman, J. A. and Holt, I. J. (2003). Mammalian mitochondrial DNA replicates bidirectionally from an initiation zone. *Journal of Biological Chemistry* **278:** 50961-50969

Brenner, C.A., Barritt, J.A., Willadsen, S. and Cohen J. (2000). Mitochondrial DNA heteroplasmy after human ooplasmic transplantation. *Fertility and Sterility* **74:** 573-578

Brown, M.D., Voljavec, A.S., Lott, M.T., Torroni, A., Yang, C.C. and Wallace, D.C. (1992). Mitochondrial DNA complex I and III mutations associated with Leber's hereditary optic neuropathy. *Genetics* **130:** 163-173

Brusco, A., Michielotto, C., Gatta, V., Foresta, C., Matullo, G., Zeviani, M.,Ferrari G., Dragone, E., Calabrese, G., Rossato M., Stuppia, L. and Migone, N. (2006). The polymorphic polyglutamine repeat in the mitochondrial DNA polymerase gamma gene is not associated with oligozoospermia. *Journal of Endocrinological Investigation* **29:** 1-4

Chinnery, P.F., Andrews, R.M., Turnbull, D.M. and Howell, N.N. (2001). Leber hereditary optic neuropathy: Does heteroplasmy influence the inheritance and expression of the G11778A mitochondrial DNA mutation? *American Journal of Medical Genetics* **98:** 235-243

Choi, Y.S., Kim, S., Kyu Lee, H., Lee, K.U. and Pak, Y.K. (2004). *In vitro* methylation of nuclear respiratory factor-1 binding site suppresses the promoter activity of mitochondrial transcription factor A. *Biochemical and Biophysical Research Communications* **314:** 118-122

Clark, K.M., Taylor, R.W., Johnson, M.A., Chinnery, P.F., Chrzanowska-Lightowlers, Z.M., Andrews, R.M., Nelson, I.P., Wood, N.W., Lamont, P.J., Hanna, M.G., Lightowlers, R.N. and Turnbull, D.M. (1999). An mtDNA mutation in the initiation codon of the cytochrome C oxidase subunit II gene results in lower levels of the protein and a mitochondrial encephalomyopathy. *American Journal of Human Genetics* **64:** 1330-1339

Clayton, D.A. (1998). Nuclear-mitochondrial intergenomic communication. *Biofactors* **7:** 203-205

Cormier, V., Rotig, A., Tardieu, M., Colonna, M., Saudubray, J.M. and Munnich, A. (1991). Autosomal dominant deletions of the mitochondrial genome in a case of progressive encephalomyopathy. *American Journal of Human Genetics* **48:** 643-648

Cummins, J.M., Jequier, A.M., Martin, R., Mehmet, D. and Goldblatt, J. (1998a). Semen levels of mitochondrial DNA deletions in men attending an infertility clinic do not correlate with phenotype. *International Journal of Andrology* **21:** 47-52

Cummins, J.M., Wakayama, T. and Yanagimachi, R. (1998b). Fate of microinjected spermatid mitochondria in the mouse oocyte and embryo. *Zygote* **6:** 213-222

Dairaghi, D.J., Shadel, G.S. and Clayton, D.A. (1995). Addition of a 29 residue carboxyl-terminal tail converts a simple HMG box-containing protein into a transcriptional activator. J *Molecular Biology* **249:** 11-28

Dalakas, M. C., Semino-Mora, C. and Leon-Monzon, M. (2001). Mitochondrial alterations with mitochondrial DNA depletion in the nerves of AIDS patients with peripheral neuropathy induced by 2'3'-dideoxycytidine (ddC). *Laboratory Investigation* **81:** 1537-1544

Danan, C., Sternberg, D., Van Steirteghem, A. Cazeneuve, C., Duquesnoy, P., Besmond, C., Goossens, M., Lissens, W. and Amselem S. (1999). Evaluation of parental mitochondrial inheritance in neonates born after intracytoplasmic sperm injection. *American Journal of Human Genetics* **65:** 463-473

Davidzon, G., Mancuso, M., Ferraris, S., Quinzii, C., Hirano, M., Peters, H.L., Kirby, D., Thorburn, D.R. and DiMauro, S. (2005). POLG mutations and Alpers syndrome. *Annuals of Neurology* **57:** 921-923

Díez-Sanchez, C., Ruiz-Pesini, E., Lapeña, A., Montoya, J., Pérez-Martos, A., Enríquez, J.A. and López-Pérez, M.J. (2003). Mitochondrial DNA content of human spermatozoa. *Biology of Reproduction* **68:** 180-185

El Shourbagy, S.H., Spikings, E.C., Freitas, M. and St John, J.C. (2006). Mitochondria directly influence fertilisation outcome in the pig. *Reproduction* **131:** 233-245

Evans, M.J. and Scarpulla, R.C. (1989). Interaction of nuclear factors with multiple sites in the somatic cytochrome c promoter. Characterization of upstream NRF-1, ATF, and intron Sp1 recognition sequences. *Journal of Biological Chemistry* **264:** 14361-14368

Falkenberg, M., Gaspari, M., Rantanen, A., Trifunovic, A., Larsson, N.G. and Gustafsson, C.M. (2002). Mitochondrial transcription factors B1 and B2 activate transcription of human mtDNA. *Nature Genetics* **31:** 289-294

Fan, L., Sanschagrin, P.C., Kaguni, L.S. and Kuhn, L.A. (1999). The accessory subunit of mtDNA polymerase shares structural homology with aminoacyl-tRNA synthetases: implications for a dual role as a primer recognition factor and processivity clamp. *Proceedings of the National Academy of Sciences USA* **96:** 9527-9532

Fisher, R.P. and Clayton, D.A. (1985). A transcription factor required for promoter recognition by human mitochondrial RNA polymerase. Accurate initiation at the heavy- and light-strand promoters dissected and reconstituted *in vitro. Journal of Biological Chemistry* **260:** 11330-11388

Fisher, R.P. and Clayton, D.A. (1988). Purification and characterization of human mitochondrial transcription factor 1. *Molecular and Cellular Biology* **8:** 3496-3509

Fisher, C. and Skibinski, D.O.F. (1990). Sex-biased mitochondrial DNA heteroplasmy in the marine mussel Mytilus. *Proceedings of the Royal Society of London series B* **242:** 149-156

Folgero, T., Bertheussen, K., Lindal, S., Torbergsen, T. and Oian P. (1993). Mitochondrial disease and reduced sperm motility. *Human Reproduction* **8:** 1863-1868

Ford, W.C. (2006). Glycolysis and sperm motility: does a spoonful of sugar help the flagellum go round? *Human Reproduction Update* **12:** 269-274

Ford, W.C. and Harrison, A. (1981). The role of oxidative phosphorylation in the generation of ATP in human spermatozoa. *Journal of Reproduction and Fertility* **63:** 271-278

Fryer, A., Appleton, R., Sweeney, M.G., Rosenbloom, L. and Harding, A.E. (1994). Mitochondrial DNA 8993 (NARP) mutation presenting with a heterogeneous phenotype including 'cerebral palsy'. *Archives of the Disabled Child* **71:** 419-422

Gensler, S., Weber, K., Schmitt, W. E., Perez-Martos, A., Enriquez, J. A., Montoya, J. and Wiesner, R. J. (2001). Mechanism of mammalian mitochondrial DNA replication: import of mitochondrial transcription factor A into isolated mitochondria stimulates 7S DNA synthesis. *Nucleic Acids Research* **29:** 3657-3663

Gibson, T.C., Kubisch, H.M. and Brenner, C.A. (2005). Mitochondrial DNA deletions in rhesus macaque oocytes and embryos. *Molecular Human Reproduction* **11:** 785-789

Giles, R.E., Blanc, H., Cann, H.M. and Wallace, D.C. (1980). Maternal inheritance of human mitochondrial DNA. *Proceedings of the National Academy of Sciences USA* **77:** 6715-6719

Graves, S.W., Johnson, A.A. and Johnson, K.A. (1998). Expression, purification, and initial kinetic characterization of the large subunit of the human mitochondrial DNA polymerase. *Biochemistry* **37:** 6050-6058

Gray, H. and Wong, T.W. (1992). Purification and identification of subunit structure of the human mitochondrial DNA polymerase. *Journal of Biological Chemistry* **267:** 5835-5841

Gottlieb B., Lombroso R., Beitel L.K. and Trifiro, M.A. (2005). Molecular pathology of the androgen receptor in male (in)fertility. *Reproductive Biomedicine Online* **10:** 42-48

Gyllensten, U., Wharton, D., Josefsson, A. and Wilson, A.C. (1991). Paternal inheritance of mitochondrial DNA in mice. *Nature* **352:** 255-267

Hauswirth, W.W. and Laipis, P.J. (1982). Mitochondrial DNA polymorphism in a maternal lineage of Holstein cows. *Proceedings of the National Academy of Sciences USA* 79: 4686-4690

Harris, T.P., Gomas, K.P., Weir, F., Holyoake, A.J., McHugh, P., Wu, M., Sin, Y., Sin, I.L. and Sin, F.Y.T. (2006). Molecular analysis of polymerase gamma gene and mitochondrial polymorphism in fertile and subfertile men. *International Journal of Andrology* **29:** 421-433

Hecht, N.B. and Liem, H. (1984). Mitochondrial DNA is synthesized during meiosis and spermiogenesis in the mouse. *Experimental Cell Research* **154**, 293-298

Hixson, J.E., Wong, T.W. and Clayton, D.A. (1986). Both the conserved stem-loop and divergent 5'-flanking sequences are required for initiation at the human mitochondrial origin of light-strand DNA replication. *Journal of Biological Chemistry* **261:** 2384-2390

Hoeh, W.M., Blakley, K.H. and Brown, W.M. (1991). Heteroplasmy suggests limited biparental inheritance of Mytlius mitochondrial DNA. *Science* **251:** 1488-1490

Holt, I. J., Lorimer, H. E., and Jacobs, H. T. (2000). Coupled leading- and lagging-strand synthesis of mammalian mitochondrial DNA. *Cell* **100:** 515-524

Holyoake, A.J., McHugh, P., Wu, M., O'Carroll, S., Benny, P., Sin, I.L. and Sin, F.Y. (2001). High incidence of single nucleotide substitutions in the mitochondrial genome is associated with poor semen parameters in men. *International Journal of Andrology* **24:** 175-182

Holyoake, A.J., Sin, I.L., Benny, P.S. and Sin, F.Y. (1999). Association of a novel human mtDNA ATPase6 mutation with immature sperm cells. *Andrologia* **31:** 339-345

Jansen, R.P. and de Boer, K. (1998). The bottleneck: mitochondrial imperatives in oogenesis and ovarian follicular fate. *Molecular and Cellular Endocrinology* **145:** 81-88

Jensen, M., Leffers, H., Petersen, J.H., Andersen, A.N., Jørgensen, N., Carlsen, E., Jensen, T.K., Skakkebæk, N.E. and Meyts, E.R.-D. (2004). Frequent polymorphism of the mitochondrial DNA polymerase gamma gene (POLG) in patients with normal spermiograms and unexplained subfertility. *Human Reproduction* **19:** 65-70

Kaneda, H., Hayashi, J.I., Takahama, S., Taya, C., Lindahl, K.F. and Yonekawa, H. (1995). Elimination of paternal mitochondrial DNA in intraspecific crosses during early mouse embryogenesis. *Proceedings of the National Academy of Sciences USA* **92:** 4542-4546

Kao S.H., Chao H.T., Liu H.W., Liao T.L. and Wei, Y.H. (2004). Sperm mitochondrial DNA depletion in men with asthenospermia. *Fertility and Sterility* **82:** 66-73

Kao, S.H., Chao, H.T. and Wei, Y.H. (1995). Mitochondrial deoxyribnucleic acid 4977 bp deletion is associated with diminished fertility and motility of human sperm. *Biology of Reproduction* **52:** 729-736

Kao, S.H., Chao, H.T and Wei, Y.H. (1998). Multiple deletions of mitochondrial DNA are associated with the decline of motility and fertility of human spermatozoa. *Molecular Human Reproduction* **4:** 657-666

Kobayashi, Y., Momoi, M.Y., Tominaga, K., Momoi, T., Nihei, K., Yanagisawa, M., Kagawa, Y. and Ohta, S. (1990). A point mutation in the mitochondrial tRNA(Leu)(UUR) gene in MELAS (mitochondrial myopathy, encephalopathy, lactic acidosis and stroke-like episodes). *Biochemical and Biophysical Research Communications* **173:** 816-822

Kondo, R., Matsuura, E.T. and Chigusa, S.I. (1992). Further observation of paternal transmission of Drosophila mitochondrial DNA by PCR selective amplification method. *Genetic Research* **59:** 81-84

Krausz, C., Guarducci, E., Becherini, L., degl'Innocenti, S., Gerace, L., Balercia, G. and Forti, G. (2004). The clinical significance of the POLG gene polymorphism in male infertility. *Journal of Clinical Endocrinology & Metabolism* **89:** 4292-4297

Larsson, N.G., Garman, J.D., Oldfors. A., Barsh, G.S. and Clayton, D.A. (1996). A single mouse gene encodes the mitochondrial transcription factor A and a testis-specific nuclear HMG-box protein. *Nature Genetics* **13:** 296-302

Larsson, N.G., Oldfors, A., Garman, J.D., Barsh, G.S. and Clayton, D.A. (1997). Down-regulation of mitochondrial transcription factor A during spermatogenesis in humans. *Human Molecular Genetics* **6:** 185-191

Larsson, N.G., Oldfors, A., Holme, E. and Clayton, D.A. (1994). Low levels of mitochondrial transcription factor A in mitochondrial DNA depletion. *Biochemical and Biophysical Research Communications,* **200:** 1374-1381

Larsson, N.G., Wang, J., Wilhelmsson, H., Oldfors, A., Rustin, P., Lewandoski, M., Barsh, G.S. and Clayton, D.A. (1998). Mitochondrial transcription factor A is necessary for mtDNA maintenance and embryogenesis in mice. *Nature Genetics* **18:** 231-236

Lecrenier, N. and Foury, F. (2000). New features of mitochondrial DNA replication system in yeast and man. *Gene* **246:** 37-48

Lee, D.Y. and Clayton, D.A. (1996). Properties of a primer RNA-DNA hybrid at the mouse mitochondrial DNA leading-strand origin of replication. *Journal of Biological Chemistry* **271:** 24262-24269

Lestienne, P., Reynier, P., Chretien, M.-F., Penisson-Besnier, I., Malthiery, Y. and Rohmer, V. (1997). Oligoasthenospermia associated with multiple mitochondrial DNA rearrangements. *Molecular Human Reproduction* **3:** 811-814

Lim, S.E., Longley, M.J. and Copeland, W.C. (1999). The mitochondrial p55 accessory subunit of human DNA polymerase gamma enhances DNA binding, promotes processive DNA synthesis, and confers N-ethylmaleimide resistance. *Journal of Biological Chemistry* **274:** 38197-38203

Lloyd, R.E., Lee, J.H., Alberio, R., Bowles, E.J., Ramalho-Santos, J., Campbell, K.H. and St. John, J.C. (2006). Aberrant nucleo-cytoplasmic cross-talk results in donor cell mtDNA persistence in cloned embryos. *Genetics* **172:** 2515-2527

Longley, M.J., Nguyen, D., Kunkel, T.A., and Copeland, W.C. (2001). The fidelity of human DNA polymerase gamma with and without exonucleolytic proofreading and the p55 accessory subunit. *Journal of Biological Chemistry* **276:** 38555-38562

Luoma, P., Melberg, A., Rinne, J.O., Kaukonen, J.A., Nupponen, N.N., Chalmers, R.M., Oldfors, A., Rautakorpi, I., Peltonen, L., Majamaa, K., Somer, H. and Suomalainen, A. (2004). Parkinsonism, premature menopause, and mitochondrial DNA polymerase gamma mutations: clinical and molecular genetic study. *Lancet* **364**: 875-882

Marchington, D.R., Hartshorne, G.M., Barlow, D. and Poulton, J. (1997). Homopolymeric tract heteroplasmy in mtDNA from tissues and single oocytes: support for a genetic bottleneck. *American Journal of Human Genetics* **60:** 408-416

Marchington, D.R., Scott Brown, M.S., Lamb, V.K., van Golde, R.J., Kremer, J.A., Tuerlings, J.H., Mariman, E.C., Balen, A.H. and Poulton, J. (2002). No evidence for paternal mtDNA transmission to offspring or extra-embryonic tissues after ICSI. *Molecular Human Reproduction* **8:** 1046-1049

May-Panloup, P., Chretien, M.F., Savagner, F., Vasseur, C., Jean, M., Malthiery, Y. and Reynier, P. (2003). Increased sperm mitochondrial DNA content in male infertility. *Human Reproduction* **18:** 550-56

May-Panloup, P., Vignon, X., Chretien, M. F., Heyman, Y., Tamassia, M., Malthiery, Y. and Reynier, P. (2005). Increase of mitochondrial DNA content and transcripts in early bovine embryogenesis associated with upregulation of mtTFA and NRF1 transcription factors. *Reproductive Biology and Endocrinology* **3:** 65-72

McCulloch, V., Seidel-Rogol, B. L. and Shadel, G. S. (2002). A human mitochondrial transcription factor is related to RNA adenine methyltransferases and binds S-adenosylmethionine. *Molecular and Cellular Biology* **22:** 1116-1125

Meirelles, F.V. and Smith, L.C. (1997). Mitochondrial genotype segregation in a mouse heteroplasmic lineage produced by embryonic karyoplast transplantation. *Genetics* **145:** 445-451

Meirelles, F. V. and Smith, L. C. (1998). Mitochondrial genotype segregation during preimplantation development in mouse heteroplasmic embryos. *Genetics* **148:** 877-883

Melov, S., Shoffner, J.M., Kaufman, A. and Wallace, D.C. (1995). Marked increase in the number and variety of mitochondrial DNA rearrangements in aging human skeletal muscle. *Nucleic Acids Research* **23:** 4122-4126

Montiel-Sosa F., Ruiz-Pesini, E., Enriquez, J.A., Marcuello, A., Diez-Sanchez, C., Montoya, J., Wallace, D.C. and Lopez-Perez, M.J. (2006). Differences of sperm motility in mitochondrial DNA haplogroup U sublineages. *Gene* **368:** 21-27

Montoya, J., Christianson, T., Levens, D., Rabinowitz, M. and Attardi, G. (1982). Identification of initiation sites for heavy-strand and light-strand transcription in human mitochondrial DNA. *Proceedings of the National Academy of Sciences USA* **79:** 7195-7199

Montoya, J., Gaines, G. L. and Attardi, G. (1983). The pattern of transcription of the human mitochondrial rRNA genes reveals two overlapping transcription units. *Cell* **34:** 151-159

Moyes, C.D., Battersby, B.J and Leary, S.C. (1998).

Regulation of muscle mitochondrial design. *Journal of Experimental Biology* **201:** 299-307

Nagao, Y., Totsuka, Y., Atomi, Y., Kaneda, H., Lindahl, K. F., Imai, H. and Yonekawa, H. (1998). Decreased physical performance of congenic mice with mismatch between the nuclear and the mitochondrial genome. *Genes and Genetic Systems* **73:** 21-27

Naviaux, R.K. and Nguyen, K. V. (2004). POLG mutations associated with Alpers' syndrome and mitochondrial DNA depletion. *Annuals of Neurology* **55:** 706-712

Naviaux, R.K. and Nguyen, K.V. (2005). POLG mutations associated with Alpers syndrome and mitochondrial DNA depletion. *Annuals of Neurology* **58**: 491

Nguyen, K., Ostergaard, V.E., Ravn, S.H., Balslev, T., Danielsen, E.R., Vardag, A., McKiernan, P.J., Gray, G. and Naviaux, R.K. (2005). POLG mutations in Alpers syndrome. *Neurology* **65:** 1493-1495

Nguyen, K.V., Sharief, F.S., Chan, S.S., Copeland, W.C. and Naviaux, R.K. (2006). Molecular diagnosis of Alpers syndrome. *Journal of Hepatology* **45:** 108-116

Nishimura, Y., Yoshinari, T., Naruse, K., Yamada, T., Sumi, K., Mitani, H., Higashiyama, T. and Kuroiwa, T. (2006). Active digestion of sperm mitochondrial DNA in single living sperm revealed by optical tweezers. *Proceedings of the National Academy of Sciences USA* **103:** 1382-1387

Nowak, R., Zub, R., Skoneczna, I., Sikora, K. and Ligaj, M. (2005). CAG repeat polymorphism in the DNA polymerase gamma gene in a Polish population: an association with testicular cancer risk. *Annals of Oncology* **16:** 1211-1212

O'Connell, M., McClure, N. and Lewis, S.E. (2002a) Mitochondrial DNA deletions and nuclear DNA fragmentation in testicular and epididymal human sperm. *Human Reproduction* **17:** 1565-1570

O'Connell, M., McClure, N. and Lewis, S.E. (2002b) A comparison of mitochondrial and nuclear DNA status in testicular sperm from fertile men and those with obstructive azoospermia. *Human Reproduction* **17:** 1571-1517

O'Connell, M., McClure, N., Powell, L.A., Steele, E.K. and Lewis, S.E (2003). Differences in mitochondrial and nuclear DNA status of high-density and low-density sperm fractions after density centrifugation preparation. *Fertility and Sterility* **79:** 754-762

Ojala, D., Montoya, J. and Attardi, G. (1981). tRNA punctuation model of RNA processing in human mitochondria. *Nature* **290:** 470-474

Otani, H., Tanaka, O., Kasai, K. and Yoshioka, T. (1988). Development of mitochondrial helical sheath in the middle piece of the mouse spermatid tail: regular dispositions and synchronized changes. *Anatomical Record* **222:** 26-33

Pagnamenta, A.T., Taanman , J.W., Wilson, C.J., Anderson, N.E., Marotta, R., Duncan, A.J., Bitner-Glindzicz, M., Taylor, R.W., Laskowski, A., Thorburn, D.R. and Rahman, S. (2006). Dominant inheritance of premature ovarian failure associated with mutant mitochondrial DNA polymerase gamma *Human Reproduction* **21:** 2467-2473

Parisi, M.A. and Clayton, D.A. (1991). Similarity of human mitochondrial transcription factor 1 to high mobility group proteins. *Science* **252:** 965-969

Pfeiffer, T., Schuster, S. and Bonhoeffer, S. (2001). Cooperation and competition in the evolution of ATP-producing pathways. *Science* **292:** 504-507

Piko, L. and Taylor, K.D. (1987). Amounts of mitochondrial DNA and abundance of some mitochondrial gene transcripts in early mouse embryos. *Developmental Biology* **123:** 364-374

Poulton, J., Morten, K., Freeman-Emmerson, C., Potter, C., Sewry, C., Dubowitz, V., Kidd, H., Stephenson, J., Whitehouse, W. and Hansen, F.J. (1994). Deficiency of the human mitochondrial transcription factor h-mtTFA in infantile mitochondrial myopathy is associated with mtDNA depletion *Human Molecular Genetics* **3:** 1763-1769

Quinn, P., Kerin, J.F. and Warnes, G.M. (1985). Improved pregnancy rate in human in vitro fertilization with the use of a medium based on the composition of human tubal fluid. *Fertility and Sterility* **44:** 493-498

Reynier, P., Chretien, M.F., Savagner, F, Larcher, G., Rohmer, V., Barriere, P. and Malthiery, Y. (1998). Long PCR analysis of human gamete mtDNA suggests defective mitochondrial maintenance in spermatozoa and supports the bottleneck theory for oocytes. *Biochemical and Biophysical Research Communications* **252:** 373-377

Ropp, P.A. and Copeland, W.C. (1996). Cloning and characterization of the human mitochondrial DNA polymerase, DNA polymerase gamma. *Genomics* **36:** 449-458

Rovio, A.T., Marchington, D.R., Donat, S., Schuppe, H.C., Abel, J., Fritsche, E., Elliott, D.J., Laippala, P., Ahola, A.L., McNay, D., Harrison, R.F., Hughes, B., Barrett, T., Bailey, D.M., Mehmet, D., Jequier, A.M., Hargreave, T.B., Kao, S.H., Cummins, J.M., Barton, D.E., Cooke, H.J., Wei Y.H., Wichmann, L., Poulton, J. and Jacobs HT. (2001). Mutations at the mitochondrial DNA polymerase (POLG) locus associated with male infertility. *Nature Genetics* **29:** 261-262

Ruiz-Pesini, E., Diez, C., Lapena, A.C., Perez-Martos, A., Montoya, J., Alvarez, E., Arenas, J. and Lopez-Perez, M.J. (1998). Correlation of sperm motility with mitochondrial enzymatic activities. *Clinical Chemistry* **44:** 1616-1620

Ruiz-Pesini, E., Lapeña, A.C., Diez-Sanchez, C., Perez-Martos, A., Montoya, J., Alvarez, E., Diaz, M., Urries, A., Montoro, L., Lopez-Perez, M.J. and Enriquez, J.A. (2000). Human mtDNA haplogroups associated with high or reduced spermatozoa motility. *American Journal of Human Genetics* **67:** 682-696

Sathananthan, H., Pera, M. and Trounson, A. (2002). The fine structure of human embryonic stem cells. *Reproductive Biomedicine Online* **4:** 56-61

Sato, A., Kono, T., Nakada, K., Ishikawa, K., Inoue, S., Yonekawa, H. and Hayashi, J. (2005). Gene therapy for progeny of mito-mice carrying pathogenic mtDNA by nuclear transplantation. *Proceedings of the Na-*

tional Academy of Sciences USA **102:** 16765-16770

Schon, E.A., Rizzuto, R., Moraes., C.T., Nakase, H., Zeviani, M. and DiMauro, S. (1989). A direct repeat is a hotspot for large-scale deletion of human mitochondrial DNA. *Science* **244:** 346-349

Schutz, M. M., Vanraden, P. M. and Wiggans, G. R. (1994). Genetic variation in lactation means of somatic cell scores for six breeds of dairy cattle. *Journal of Dairy Science* **77:** 284-293

Seidel-Rogol, B. L. and Shadel, G. S. (2002). Modulation of mitochondrial transcription in response to mtDNA depletion and repletion in HeLa cells. *Nucleic Acids Research* **30:** 1929-1934

Schwartz, M. and Vissing, J. (2002). Paternal inheritance of mitochondrial DNA. *New England Journal of Medicine* **347:** 576-580

Shitara, H., Hayashi, J.I., Takahama, S., Kaneda, H. and Yonekawa, H. (1998). Maternal inheritance of mouse mtDNA in interspecific hybrids: segregation of the leaked paternal mtDNA followed by the prevention of subsequent paternal leakage. *Genetics* **148:** 851-857

Siciliano, G., Mancuso, M., Pasquali, L., Manca, M. L., Tessa, A. and Iudice, A. (2000). Abnormal levels of human mitochondrial transcription factor A in skeletal muscle in mitochondrial encephalomyopathies. *Neurological Sciences* **21:** S985-987

Sinowatz, F. and Wrobel, K.H. (1981). Development of the bovine acrosome. An ultrastructural and cytochemical study. *Cell Tissue Research* **219:** 511-524

Spelbrink, J.N., Van Galen, M.J., Zwart, R., Bakker, H.D., Rovio, A., Jacobs, H.T. and Van den Bogert, C. (1998). Familial mitochondrial DNA depletion in liver: haplotype analysis of candidate genes. *Human Genetics* **102:** 327-331

Spiking, E., Alderson, J. and St. John, J.C. (2006) Transmission of mitochondrial DNA following assisted reproduction and nuclear transfer. *Human Reproduction Update* **12:** 401-415

Spikings, E and St. John J. (2006). Delayed TFAM and PolG activity reduces oocyte competence. *Society for Reproduction and Fertility Annual Conference.* Leeds, UK. 4 – 5 July, 2006.

Spiropoulos, J., Turnbull, D.M. and Chinnery, P.F. (2002). Can mitochondrial DNA mutations cause sperm dysfunction? *Molecular Human Reproduction* **8:** 719-721

St. John, J.C., Jokhi, R.P. and Barratt, C.L. (2001). Men with oligoasthenoteratozoospermia harbour higher numbers of multiple mitochondrial DNA deletions in their spermatozoa, but individual deletions are not indicative of overall aetiology. *Molecular Human Reproduction* **7:** 103-111

St. John, J.C., Ramalho-Santos, J., Gray, H.L., Petrosko, P., Rawe, V.Y., Navara, C.S., Simerly, C.R. and Schatten G.P. (2005b). The expression of mitochondrial DNA transcription factors during early cardiomyocyte in vitro differentiation from human embryonic stem cells. *Cloning and Stem Cells* **7:** 141-153

St. John, J., Sakkas, D., Dimitriadi, K., Barnes, A., Maclin, V., Ramey, J., Barratt, C. and De Jonge C. (2000a). Abnormal human embryos show a failure to eliminate paternal mitochondrial DNA. *Lancet* **355:** 200

St. John, J.C., Sakkas, D. and Barratt, C.L.R. (2000b). A role for mitochondrial DNA in sperm survival. *Journal of Andrology* **21:** 189-199

St. John, J.C., Jokhi, R.P. and Barratt C.L.R. (2005a) The impact of mitochondrial genetics on male infertility. *International Journal of Andrology* **28:** 65-73

St John, J. C. and Schatten, G. (2004). Paternal mitochondrial DNA transmission during nonhuman primate nuclear transfer. *Genetics* **167:** 897-905

Storey, B.T. (1975). Energy metabolism of spermatozoa. IV. Effect of Calcium on Respiration of Mature Epididymal Sperm of the Rabbit. *Biology of Reproduction* **13:** 1-9

Storey, B.T. (1978). Effect of ionophores and inhibitors and uncouplers of oxidative phosphorylation on sperm respiration. *Archives in Andrology* **1:** 169-177

Storey, B.T. (1980). Strategy of oxidative metabolism in bull spermatozoa. *Journal of Experimental Zoology* **212:** 61-67

Storey, B.T. and Kayne, F.J. (1975). Energy metabolism of spermatozoa. V. The Embden-Myerhof pathways of glycolysis: Activities of pathway enzymes in Hypotonically Treated Rabbit Epididymal Spermatozoa. *Fertility and Sterility* **26:** 1257-1265

Storey, B.T. and Kayne, F.J. (1977). Energy metabolism of spermatozoa. VI. Direct intramitochondrial lactate oxidation by rabbit sperm mitochondria. *Biology of Reproduction* **16:** 549-556

Storey, B.T. and Kayne, F.T. (1980). Properties of pyruvate kinase and flagellar ATPase in rabbit spermatozoa: relation to metabolic strategy of the sperm cell. *Journal of Experimental Zoology* **211:** 361-367

Sutarno, Cummins, J. M., Greeff, J. and Lymbery, A. J. (2002). Mitochondrial DNA polymorphisms and fertility in beef cattle. *Theriogenology* **57:** 1603-1610

Sutovsky, P., Moreno, R.D., Ramalho-Santos, J., Dominko, T., Simerly, C. and Schatten, G. (1999). Ubiquitin tag for sperm mitochondria. *Nature* **402:** 371-372

Sutovsky, P., Navara, C.S. and Schatten, G. (1996). Fate of the sperm mitochondria, and the incorporation, conversion, and disassembly of the sperm tail structures during bovine fertilization. *Biology of Reproduction* **55:** 1195-1205

Tamassia, M., Nuttinck, F., May-Panloup, P., Reynier, P., Heyman, Y., Charpigny, G., Stojkovic, M., Hiendleder, S., Renard, J. P. and Chastant-Maillard, S. (2004). In vitro embryo production efficiency in cattle and its association with oocyte adenosine triphosphate content, quantity of mitochondrial DNA, and mitochondrial DNA haplogroup. *Biology of Reproduction* **71:** 697-704

Tiranti, V., Savoia, A., Forti, F., D'Apolito, M.F., Centra, M., Rocchi, M. and Zeviani, M. (1997). Identification of the gene encoding the human mitochondrial RNA polymerase (h-mtRPOL) by cyberscreening of the Expressed Sequence Tags database. *Human Molecular Genetics* **6:** 615-625

Van Goethem, G., Dermaut, B., Lofgren, A., Martin, J.J. and Van Broeckhoven, C. (2001). Mutation of POLG is associated with progressive external ophthalmoplegia characterized by mtDNA deletions. *Nature Genetics* **28:** 211-212

Van Goethem, G., Schwartz, M., Lofgren, A., Dermaut, B., Van Broeckhoven, C. and Vissing, J. (2003). Novel POLG mutations in progressive external ophthalmoplegia mimicking mitochondrial neurogastrointestinal encephalomyopathy. *European Journal of Human Genetics* **11:** 547-549

Virbasius, C.A., Virbasius, J.V. and Scarpulla, R.C. (1993). NRF-1, an activator involved in nuclear-mitochondrial interactions, utilizes a new DNA-binding domain conserved in a family of developmental regulators. *Genes and Development* **7:** 2431-2445

Walker, R.L., Anziano, P. and Meltzer, P.S. (1997). A PAC containing the human mitochondrial DNA polymerase gamma gene (POLG) maps to chromosome 15q25. *Genomics* **40:** 376-378

Wallace, D.C. (1995). Mitochondrial DNA variation in human evolution, degenerative disease, and aging. *American Journal of Human Genetics* **57:** 201-223

White, D.J., Mital, D., Taylor, S. and St. John, J.C. (2001). Sperm mitochondrial DNA deletions as a consequence of long term highly active antiretroviral therapy. *AIDS* **15**: 1061-1062

World Health Organization (1992) *WHO Laboratory Manual for the Examination of Human Semen and Semen-Cervical Mucus Interaction.* 3rd edn. Cambridge University Press, Cambridge

World Health Organization (1999) *WHO Laboratory Manual for the Examination of Human Semen and Semen-Cervical Mucus Interaction.* 4th edn. Cambridge University Press, Cambridge

Wong, T.W. and Clayton, D.A. (1985). In vitro replication of human mitochondrial DNA: accurate initiation at the origin of light-strand synthesis. *Cell* **42:** 951-958

Xu, B. and Clayton, D.A. (1996). RNA-DNA hybrid formation at the human mitochondrial heavy-strand origin ceases at replication start sites: an implication for RNA-DNA hybrids serving as primers. *EMBO Journal* **15:** 3135-3143

Yang, M.Y., Bowmaker, M., Reyes, A., Vergani, L., Angeli, P., Gringeri, E., Jacobs, H.T., and Holt, I.J. (2002). Biased incorporation of ribonucleotides on the mitochondrial L-strand accounts for apparent strand-asymmetric DNA replication. *Cell* **111:** 495-505

Zhao, X., Li, N., Guo, W., Hu, X., Liu, Z., Gong, G., Wang, A., Feng, J. and Wu, C. (2004). Further evidence for paternal inheritance of mitochondrial DNA in the sheep (*Ovis aries*). *Heredity* **93:** 399-403

Zhao, Y., Li, Q., Yao, C., Wang, Z., Zhou, Y., Wang, Y., Liu, L., Wang, Y., Wang, L. and Qiao, Z. (2006). Characterization and quantification of mRNA transcripts in ejaculated spermatozoa of fertile men by serial analysis of gene expression. *Human Reproduction* **21:** 1583-1590

Zullo, S.J., Butler, L., Zahorchak, R.J., Macville, M., Wilkes, C. and Merril, C.R. (1997). Localization by fluorescence in situ hybridization (FISH) of human mitochondrial polymerase gamma (POLG) to human chromosome band 15q24—>q26, and of mouse mitochondrial polymerase gamma (Polg) to mouse chromosome band 7E, with confirmation by direct sequence analysis of bacterial artificial chromosomes (BACs). *Cytogenetics and Cellular Genetics* **78:** 281-284

Spermatology. SRF Vol. 65. ERS Roldan and M Gomendio (eds) Nottingham University Press, Nottingham

Sperm cryopreservation in endangered felids: developing linkage of *in situ-ex situ* populations

W.F. Swanson, M.A. Stoops, G.M. Magarey and J.R. Herrick

Center for Conservation and Research of Endangered Wildlife, Cincinnati Zoo & Botanical Garden, 3400 Vine Street, Cincinnati, OH 45220, USA

Many of the world's cat species face growing threats to their continued survival in nature. For some species, managed captive populations may provide a reservoir for future reintroduction or genetic augmentation. Because most zoo populations are derived from small founder sizes and are subject to loss of genetic variation over time, periodic infusion of founder alleles is necessary to avoid the dire consequences of inbreeding. Collection and freezing of semen from free-living nondomestic felids offers a viable option for introducing founder genes into captive populations without removal of animals from the wild. The effective application of this strategy requires established protocols for safely capturing and anaesthetising wild cats coupled with suitable methods of semen recovery, processing and cryopreservation under field conditions. In small-sized non-domestic felids, the general feasibility of this approach is being explored in two studies of black-footed cats and Pallas' cats. Two factors – relatively low sperm numbers per ejaculate and compromised status of frozen-thawed cat spermatozoa – suggest that *in vitro* fertilisation (IVF) and embryo transfer present the most efficient use of this limiting resource in small-sized cats. Our studies with captive felids have explored alternative methods of sperm cryopreservation that are adaptable to field situations and shown that frozen-thawed spermatozoa from Pallas' cats, ocelots, and fishing cats exhibit adequate function to fertilise heterologous and/or homologous oocytes *in vitro*. Most recently, we investigated the fertilising capacity of frozen-thawed spermatozoa obtained from wild Pallas' cats in Mongolia. Combined with improved methods for embryo culture and transfer in small cat species, sperm banking *in situ* will facilitate introduction of new founders into captive populations without causing further depletion of their wild counterparts. As one component of holistic conservation programs, including ongoing support of field ecology studies in range countries, this reproductive strategy serves to further strengthen linkages among imperiled *ex situ* and *in situ* cat populations.

Corresponding author E-mail: bill.swanson@cincinnatizoo.org

Conservation of the forgotten felids – the small cats

Among carnivores, the 36 extant species of non-domestic felid have been fairly resilient over their recent evolutionary history. Although several cat populations, most notably three subspecies of tigers, have been extirpated by man in the past century, there have been no species level extinctions of wild cats in at least several thousand years. However, persistent habitat loss, combined with poaching and the escalating threats associated with global climate change, have created an altered environment in which several cat species are now in imminent danger of disappearance (Nowell and Jackson, 1996). The International Union for Conservation of Nature (IUCN, 2003) includes 24 species of the world's wild cats on its Red List under some degree of risk for extinction (i.e., critically endangered, endangered, vulnerable, near threatened; Table 1).

Table 1. *In situ* and *ex situ* conservation status of the 7 large-sized and 17 small-sized nondomestic felid species classified under some degree of risk on the IUCN Red List.

Cat species				
Common name	Latin name	*In situ* status[a]	*Ex situ* population[b]	Captive management[c]
Cheetah	*Acinonyx jubatus*	Vu	Yes	SSP/EEP
Clouded leopard	*Neofelis nebulosa*	Vu	Yes	SSP/EEP
Lion	*Panthera leo*	Vu	Yes	SSP/EEP
Jaguar	*Panthera onca*	NT	Yes	SSP
Puma	*Puma concolor*	NT	Yes	-
Snow leopard	*Uncia uncia*	En	Yes	SSP/EEP
Tiger	*Panthera tigris*	En	Yes	SSP/EEP
African golden cat	*Profelis aurata*	Vu	No	-
Andean mountain cat	*Oreailurus jacobita*	En	No	-
Asian golden cat	*Catopuma temmickii*	Vu	Yes	EEP
Black-footed cat	*Felis nigripes*	Vu	Yes	SSP/EEP
Bornean bay cat	*Catopuma badia*	En	No	-
Chinese desert cat	*Felis bieti*	Vu	No	-
Fishing cat	*Prionailurus viverrinus*	Vu	Yes	SSP/EEP
Flat-headed cat	*Prionailurus planiceps*	Vu	Yes	-
Geoffroy's cat	*Oncifelis geoffroyi*	NT	Yes	EEP
Iberian lynx	*Lynx pardinus*	Cr	Yes	-
Kodkod	*Oncifelis guigna*	Vu	No	-
Marbled cat	*Pardofelis marmorata*	Vu	Yes	-
Oncilla	*Leopardus tigrinus*	NT	Yes	-
Pallas' cat	*Otocolobus manul*	NT	Yes	SSP/EEP
Pampas cat	*Oncifelis colocolo*	NT	Yes	-
Rusty-spotted cat	*Prionailurus rubiginosus*	Vu	Yes	-
Sand cat	*Felis margarita*	NT	Yes	SSP/EEP

[a]IUCN classification categories (Cr, critically endangered; En, endangered; Vu, vulnerable; NT, near threatened) (IUCN, 2003)
[b]Presence or absence of the species in captivity, based on the most recent ISIS data (ISIS, 2006)
[c]SSP, Species Survival Plan of the Association of Zoos and Aquariums (AZA); EEP, European Endangered Species Program of the European Association of Zoos and Aquaria (EAZA)

Most of the world's cat species are small in size, with individuals of 28 species averaging less than 20 kg in body mass (Sunquist and Sunquist, 2002). Compared to the eight large (> 20 kg body mass) cat species, small-sized cats have been relatively neglected in both conservation and science circles. In particular, comprehensive *in situ* studies have been virtually non-existent in the majority of small cat species, creating a dearth of reliable information on the population status and ecology of these forgotten felids (Nowell and Jackson, 1996; Sunquist and Sunquist, 2002). Based on limited *in situ* data, the IUCN (2003) has listed 17 of the 28 small cat species as facing varying degrees of risk for extinction in the wild (Table 1). Most small cats have been only sporadically maintained in zoos and few species are currently managed effectively in captive breeding programmes. Of the 24 total cat species included on the IUCN Red List, the International Species Inventory System (ISIS, 2006) indicates that the 7 large cat species and 12 of the 17 small cat species currently are housed in the world's zoos. Although all large cats species, with the exception of the puma (*Puma concolor*), are intensively managed within cooperative breeding programmes in North American and/or European zoos, only a handful of small cat species are provided with similar oversight by these regional zoo associations.

In North American zoos accredited by the Association of Zoos and Aquariums (AZA), just five small cat species – the Brazilian ocelot (*Leopardus pardalis mitis*), fishing cat (*Prionailurus viverrinus*), Pallas' cat (*Otocolobus manul*), Arabian sand cat (*Felis margarita harrisoni*) and black-footed cat (*Felis nigripes*) - are managed by Species Survival Plans (SSPs) (Table 1). European zoos that are members of the European Association of Zoos and Aquaria (EAZA) have comparable management programmes (European Endangered Species Programs, EEPs) in place for six small cat species, including four of the SSP species (Table 1). SSPs and EEPs represent the most intensive management programmes available within the world's zoos, with programme coordinators using demographic and genetic analyses of studbook data to make breeding recommendations, limit inbreeding and maximise genetic diversity within participating institutions. Achieving these management goals are a difficult proposition in small cats, given the limited founder numbers, small population sizes, short generation intervals and reproductive life spans, and breeding challenges that are typical of most of these species (Swanson, 2006).

The need for *in situ* linkage

Captive felid populations managed by SSPs or EEPs are potentially important as genetic reservoirs for possible reintroduction of living animals into nature, provided that suitable habitat remains *in situ* and the tremendous difficulties of integrating captive-born individuals into a hostile wild environment can be overcome (Beck, Rapaport, Stanley Price and Wilson, 1994). The number of species, however, that can be effectively managed in zoos within conventional breeding programmes is limited. In a seminal paper on the carrying capacity of zoos, Conway (1986) calculated that all of the worlds' zoos, working in concert, could only maintain a total of 330 mammalian species in captivity, given space restrictions in zoos and the need for minimal population sizes for genetic viability. Included in his calculations was the yearly cost of caring for individual animals of various species, illustrating the financial obligations incurred with large zoo populations. These costs have only increased in the past twenty years, with zoos investing more money in building naturalistic exhibits while continually striving to improve the husbandry of their charges by providing imaginative forms of enrichment, higher quality diets, and more sophisticated veterinary care. It might be illustrative to revisit the genetic constraints and maintenance costs for zoo animals in the twenty-first century, specifically using the Pallas' cat as an example for the small cats that are the primary focus of this paper.

In North American zoos, SSP population managers use computer software programmes to collate studbook data (ISIS, 2003) and conduct pedigree analysis and population modeling (Ballou and Foose, 1996; Pollack, Lacy and Ballou, 2001). From these analyses, we find that the Pallas' cat SSP population currently consists of 65 animals derived from 24 founders (i.e., primarily wild-born cats, presumably unrelated) imported into North America in the 1990's

(Caron, 2006). Taking into consideration the total cage spaces available for all small cats in North American zoos, this species has been allocated 100 exhibit spaces. Over the past ten years, this population has grown at an average rate of 5.7% per year with an effective breeding population size (Ne) equal to ~24% of the total population (N) and a generation interval of 4.7 years. At present, the SSP population retains 93.6% of the species' existing genetic diversity (GD) with the goal being to maintain 90% GD over the next 50 years. However, population modeling indicates that 90% GD is sustainable for only 6 more years and, after 50 years, this value will have declined to 74%. With a moderate improvement in breeding success (i.e., Ne/N = 50%; maximal growth rate = 20%), 90% GD for 50 years can be achieved but the maximal population size must be allowed to grow to 392 cats (Fig. 1A). If the maximal population size is held at 100 cats, our genetic goal is sustainable for just 16 years and GD declines to 84% by 50 years (Fig. 1B).

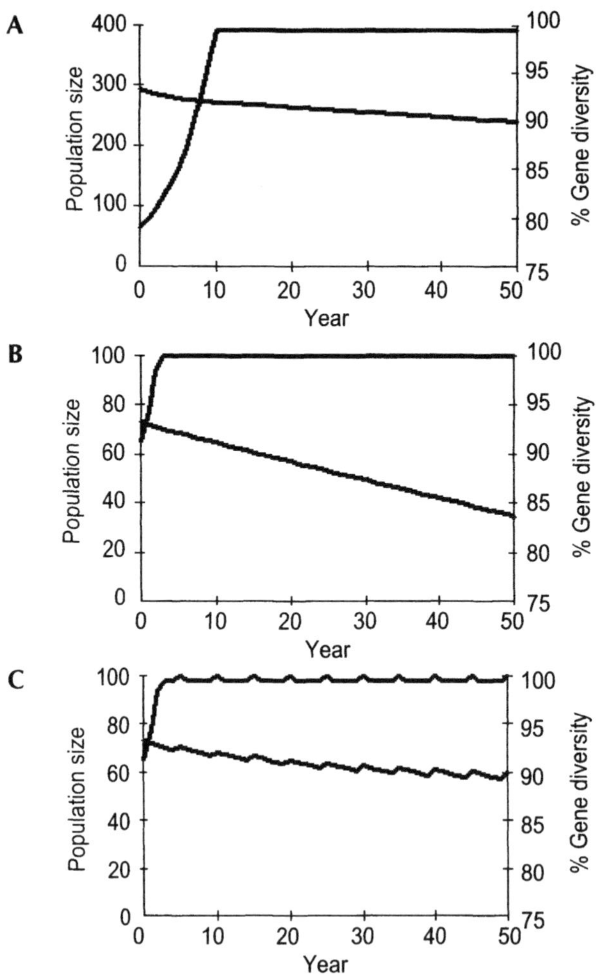

Figure 1. Number of captive Pallas' cats required to maintain 90% genetic diversity over a 50 year period under three scenarios: (**A**) unlimited total population size with improved management to increase the effective population size (Ne) and maximal population growth rate; (**B**) improved management with a total population size limited to 100 cats; and (**C**) improved management of 100 total cats combined with introduction of 2 new founders from the wild every 5 years.

Even if ~400 cage spaces somehow could be reserved for Pallas' cats in zoos, the financial implications of managing 392 cats *versus* 100 cats over 50 years are profound. At the Cincinnati Zoo, we estimate that housing, feeding and caring for each of our Pallas' cats costs $2,300 per year (Table 2). Accordingly, to maintain 392 cats over 50 years, North American zoos would need to provide a total of $45,080,000 in financial support. In contrast, housing 100 cats over 50 years would require $11,500,000 in total funding for a projected cost savings of almost $33,500,000 or ~$700,000 per year. One means to realise this cost savings while still meeting the SSP's genetic goals is through the periodic introduction of new founders into the captive population. By adding two new founders every 5 years for the next 50 years, the SSP population can be sustained at 100 total cats with 90% GD (Fig. 1C). Traditionally, zoos have acquired founders through the capture and importation of free-living wild-born animals, as with the establishment of the Pallas' SSP population. In most cases, the removal of a handful of animals from a vibrant wild population is unlikely to have any long-lasting detrimental impact on species' survival. For example, over the past 10 years, 105 wild-born Pallas' cats have been brought into captivity internationally (Caron, 2006). Although ecological studies have just begun to assess wild Pallas' cats (Munkhtsog and Ross, in press), it is unlikely that this limited off-take from multiple geographic sites has had any significant adverse effects.

Table 2. Financial costs (USD, $) for maintaining a captive population of Pallas' cats over the next 50 years based on a projected total population size of 392 cats versus 100 cats.

Parameter	Cost/cat/yr	Cost/cat/50 yr	Cost/392 cats/50 yr	Cost/100 cats/50 yr
Labour	$1,750	$87,500	$34,300,000	$8,750,000
Food	$310	$15,500	$6,076,000	$1,550,000
Veterinary care	$100	$5,000	$1,960,000	$500,000
Facility overhead	$140	$7,000	$2,744,000	$700,000
Total cost	$2,300	$115,000	$45,080,000	$11,500,000

For many species, the periodic capture of wild founders remains as one option for zoos interested in genetic augmentation of captive populations. In essence, this practice represents one variation on the concept of ecosystem services in which intact functional environments provide essential biological services to humankind - in this case, providing genetic resources to zoos (Costanza, d'Arge, de Groot, Farber, Grasso, Hannon, Limburg, Naem, O'Neill, Paruelo, Raskin, Sutton and van den Belt, 1997; Pimental, Wilson, McCullum, Huang, Dwen, Flack, Tran, Saltman and Cliff, 1997). As zoos have evolved to focus more on conservation as their primary mission, the need for zoos to support habitat preservation has received greater emphasis (Hutchins, 2003; Miller, Conway, Reading, Wemmer, Wildt, Kleiman, Monfort, Rabinowitz, Armstrong and Hutchins, 2004). One direct incentive for supporting *in situ* conservation is the notion of zoo reserves, in which zoos provide financial support for wildlife reserve management in exchange for occasional removal of animals for zoo collections (Conway, 1999). In the Pallas' cat example, the substantial financial savings gained by managing a smaller captive population could be invested in support of Pallas' cat reserves, providing major funding for habitat preservation in the range countries of Central Asia.

However, if zoos are sincere in their intent to become true conservation organisations and be perceived as such by other conservationists and the public, then they should continue to distance themselves from the traditional "bring 'em back alive" method of founder acquisition whenever possible. Even if sustainable from an ecological perspective, capturing any animal in the wild for placement in zoos harkens back to an earlier era when zoos were major extractive and destructive forces on wild animal populations. Although nearly 90% of zoo animals today are born in captivity (ISIS, 2006), this perception of zoos as consumers of wildlife persists and, with each transfer of a formerly free-living animal into captivity, is reinforced. Many zoo professionals, including the authors, struggle with this apparent contradiction of exhorting society to conserve endangered species *in situ* while simultaneously advocating the removal of these same species from the wild. Times change, science advances, and alternatives do exist.

In situ genetic resource banks

Beginning in the 1980's, Conway (1986) and others (Seager, 1983; Benirschke, 1984; Wildt, Schiewe, Schmidt, Goodrowe, Howard, Phillips, O'Brien and Musch, 1986; Holt and Moore, 1988) noted that the carrying capacity of zoos could be substantially enhanced by the application of assisted reproductive technologies, including the use of sperm cryopreservation, in combination with artificial insemination (AI) or *in vitro* fertilisation (IVF) and embryo transfer (ET). Twenty years later, the true potential of assisted reproduction for population management has yet to be realised but we are progressing incrementally closer to being able to apply these technologies as genetic management tools (Swanson,2006). In theory, assisted reproductive technologies would allow gene flow to be established between disjunct locations, including linking of *in situ* and ex *situ* populations (Lasley, Loskutoff and Anderson, 1994; Johnston and Lacy, 1995; Wildt, Rall, Critser, Monfort and Seal, 1997; Wildt and Roth, 1997). In the near future, unidirectional movement of genetic material from the wild into captivity presents the most viable option for creating this initial linkage. In fact, the general feasibility of this approach already has been demonstrated in cheetahs (*Acinonyx jubatus*), using frozen-thawed spermatozoa collected from wild males in Namibia for AI, culminating in the production of three litters of offspring in North American zoos (Wildt *et al.*, 1997; Howard, Marker, Pukazhenthi, Roth, Swanson, Grisham and Wildt, 2002).

In our laboratory at the Cincinnati Zoo's Center for Conservation and Research of Endangered Wildlife (CREW), one of the goals is to develop and apply similar reproductive strategies of genetic linkage for small cat SSP species. In effect, small cat populations in the wild would serve as living *in situ* genetic resource banks with the periodic withdrawal of founder genes in the form of frozen spermatozoa to augment the genetic diversity of captive populations. In return, a portion of the substantial cost savings accrued by zoos from reducing captive population sizes, as seen with the Pallas' cat example, would be redirected to support *in situ* research and conservation efforts with wild populations. Substantial progress already has been made in developing the basic and applied scientific knowledge and establishing the *in situ* collaborative partnerships that will be necessary for implementing this *in situ*-ex *situ* linkage strategy with small cats.

Semen collection and cryopreservation *in situ*

For *in situ*-ex *situ* genetic linkage to be a realistic option, it is imperative that the wild cats can be physically located in their natural habitat within a defined time period and then safely captured, anaesthetised and subjected to electroejaculation before being released back into the wild. Recovered semen must be evaluated, processed and cryopreserved under field conditions in a manner that allows recovery of adequate post-thaw sperm function for IVF or AI procedures. Lastly, frozen semen must be transported internationally with proper government permits and awareness of airport security screening hazards. Each of these steps present logistical challenges that require some flexibility in experimental design and methodology. Comparative studies with domestic cats and captive nondomestic felids are essential for establishing effective protocols for semen collection, analysis and cryopreservation before venturing into the field to study cats in the wild.

Capture and restraint

Earlier studies of semen collection in wild felids focused on two African plains species, the lion (*Panthera leo*) and the cheetah (Wildt, O'Brien, Howard, Caro, Roelke, Brown and Bush, 1987a; Wildt, Bush, Goodrowe, Packer, Pusey, Brown, Joslin and O'Brien, 1987b). Because these populations already were acclimated to the presence of field researchers, capture prior to anaesthesia was unnecessary and the cats could be easily approached and darted with anaesthetic agents delivered remotely from a nearby vehicle. Subsequent studies with wild Florida panthers and Brazilian jaguars (*Panthera onca*) proved more challenging, requiring the use of trained dogs to track the cats and force the animals to climb trees for subsequent darting (Barone, Roelke, Howard, Brown, Anderson and Wildt, 1994; Morato, Conforti, Azevedo, Jacomo, Silveira, Sana, Nunes, Guimaraes and Barnabe, 2001). With small-sized cats, animals typically have been captured either in baited live traps (Shindle and Tewes, 2000; Grassman, Austin, Tewes and Silvy, 2004; Sliwa, 2004) or found by visual sighting and then physically restrained after a short chase and netting or cornering in an underground burrow (Sliwa, 2004; Brown, Lappin, Brown, Munkhtsog and Swanson, 2005; Munkhtsog and Ross, in press). In an optimal situation, the wild cats will have been captured initially at an earlier date and fitted with radiocollars as part of ongoing field ecology studies. This approach in Mongolia allowed us to relocate and recapture four radiocollared Pallas' cats over a two day period, greatly shortening the time commitment needed for biological sampling and reducing field expedition expenses.

Anaesthesia

Among wildlife species, felids are among the safest to anaesthetise using injectible anaesthetic agents alone or injectible anaesthetics combined with inhalant gas anaesthesia (Swanson, 2003). Traditionally, many biologists working with wild cats in the field have conducted anaesthesia without veterinary support, using ketamine, sometimes combined with xylazine, or tiletamine-zolezapam to achieve adequate anaesthetic depth for blood sampling, physical measurements and/or placement of radiocollars (Franklin, Johnson, Sarno and Iriate, 1999; Shindle and Tewes, 2000; Grassman *et al.*, 2004). Ideally, an experienced veterinarian should be present to administer the anaesthetic agents and monitor the animal's vital signs throughout the sampling period, especially when a greater depth of anaesthesia is required for stimulatory procedures such as electroejaculation. To the authors' knowledge, veterinarians have been responsible for conducting anaesthesia in all previous and ongoing studies involving semen collection of cats in the wild (Wildt *et al.*, 1987ab; Barone *et al.*, 1994; Morato *et al.*, 2001; Herrick, Lamberski, Wilson, Anderson, Swanson and Sliwa, 2005; unpublished). Ketamine alone has been used for semen collection procedures in some species, including the cheetah and Pallas' cat (Wildt *et al.*, 1987a; Swanson et al., 1996), but combination regimens are now preferred for providing a greater level of sedation and analgesia during electroejaculation. Common injectible agents used for electroejaculation in field situations include tiletamine-zolazepam (Wildt *et al.*, 1987a; Morato *et al.*, 2001) and ketamine-medetomidine (Herrick *et al.*, 2005), but each combination has drawbacks, such as prolonged recovery times with the former anaesthetic drug and blood pressure fluctuations with the latter. In some species, inhalant anaesthesia has proven feasible in a field situation, such as in our ongoing study of wild Pallas' cats in Mongolia. One advantage is that the dosage of injectible anaesthetic may be reduced, allowing for quicker recovery times after cessation of gas anaesthesia. The main disadvantage is the need to acquire oxygen tanks and compatible flow regulators in the study region and transport this bulky equipment into the field. A recent study, however, has shown that a non-rebreathing system attached to a flow-over anaesthestic vapouriser may be used with ambient air alone, eliminating the need for compressed oxygen gas cylinders (Lewis, 2004). With continual monitoring of oxygen saturation using pulsoximetry, this approach provides an adequate margin of safety for effective field anaesthesia in a variety of species, including felids.

Electroejaculation and semen analysis

Semen collection from wild felids requires the electroejaculation of anaesthetised individuals. Most felids are easily collected in this manner (Howard, Bush and Wildt, 1986; Howard, 1993), although some species, such as ocelots, fishing cats and jaguarundis, are prone to urination during electrostimulation. In field situations, a mobile laboratory approach is required (Swanson, 2003), including transport of electroejaculation equipment to the study site and provision of a suitable power source. In some of the earlier field studies, a standard alternating current (AC) - powered electrostimulator (PTE Electronics, Boring OR USA) was used with electricity provided from a portable gas generator capable of generating a minimum of 300-500 watts (Wildt *et al.*, 1987ab; Barone *et al.*, 1994). In more remote study locations where the transport of heavy equipment must be minimised, battery-powered direct current (DC) electrostimulators (Lane Manufacturing, Denver CO, USA; Eletrovet Electronic Equipment, Sao Paulo, Brazil) have proven to be adequate for semen collection (Barone *et al.*, 1994; Morato *et al.*, 2001; unpublished).

Semen analysis in the field entails evaluation of most of the same parameters that are assessed in a normal laboratory setting, specifically semen volume, sperm concentration, sperm progressive motility and rate of progressive movement, sperm morphology and acrosomal integrity (Howard *et al.*, 1986). If a gas-powered electrical generator is available and equipment transport is not an issue, then a standard-sized phase contrast laboratory microscope may be used to assess sperm concentration, motility and morphology (Wildt *et al.*, 1987ab; Barone et *al.*; 1994). Alternatively, sperm motility parameters may be evaluated using a small, lightweight phase-contrast field microscope (Swift Instruments, San Jose CA, USA) with a small flashlight as the source of illumination (Morato *et al.*, 2001; Swanson, 2003). Sperm concentration may be determined by fitting the field microscope eyepiece with a calibrated optical grid and counting the total number of spermatozoa within the grid area on a thin counting chamber slide (Microcell, Conception Technologies, San Diego CA, USA). Since sperm morphology assessments require higher magnification (>400x), aliquots of raw semen usually are fixed in glutaraldehyde for later evaluation with a standard laboratory microscope. Lastly, sperm acrosome evaluation may be conducted in the field using stable colorimetric stains, such as a rose bengal-fast green stain (Pope, Zhang and Dresser, 1991) or a Coomassie blue stain (Thiangtum, Swanson, Howard, Tunwattana, Tongthainan, Wichasilpa, Patumrattanathan and Pinyopoommintr, 2006; Crosier, Pikazhenthi, Henghali, Howard, Dickman, Marker and Wildt, 2006). With both stains, spermatozoa are fixed in either alcohol or paraformaldehyde so acrosomal degradation prior to microscopic evaluation is of minimal concern.

Most knowledge of basal seminal parameters in felids is based on studies of captive populations (Howard, 1993; Wildt, Brown, Bush, Barone, Cooper, Grisham and Howard, 1993; Morais, Mucciolo, Gomes, Lacerda, Moraes, Moreira, Graham, Swanson and Brown, 2002; Swanson,

Johnson, Cambre, Citino, Quigley, Brousset, Morais, Moreira, O'Brien and Wildt, 2003; van Dorrser and Strick, 2005; Thiangtum *et al.*, 2006). Although these data are valuable for assessing the reproductive potential of our breeding populations, their validity as true species norms, especially in relation to wild individuals, is unknown for most cats. Multiple factors associated with the captive environment, including levels of genetic heterozygosity and inbreeding, nutrition, seasonal and latitudinal temperature and photoperiod fluctuations, and physiological or psychological stressors, may affect measured reproductive characteristics of felids in zoos. Comparative seminal data from wild felid populations presently exist for only a few large cat species (cheetah, Wildt et al., 1987a; lion, Wildt et al., 1987b; puma, Barone et al., 1994; jaguar, Morato et al., 2001). Our ongoing studies of black-footed cats in South Africa and Pallas' cats in Mongolia are providing the first comparative information about the reproductive status of small cat species in the wild. In black-footed cats and Pallas' cats, data collection has been limited thus far to analysis of semen from two adult males of each species but reproductive values for the wild cats appear similar to those of captive individuals (Swanson *et al.*, 1996; Herrick *et al.*, 2005; Herrick *et al.*, in press; Table 3). Notably, wild Pallas' cats evaluated in October produced minimal sperm numbers, similar to that seen in this seasonally breeding species in captivity during the nonbreeding season (Swanson *et al.*, 1996), but sperm quality was still adequate for cryopreservation for future use with IVF procedures.

Table 3. Mean (± SEM) values for seminal and testicular traits of wild South African black-footed cats and Mongolian Pallas' cats versus captive black-footed cats and Pallas' cats in North American zoos.

	Black-footed cats		Pallas' cats	
Parameter	Captive[a]	Wild	Captive[b]	Wild
No. cats	5	2	1	2
No. ejaculates	12	2	5	2
Semen volume (ml)	0.25 ± 0.02	0.13 ± 0.03	0.19 ± 0.01	0.18 ± 0.06
Total sperm/ejaculate (x 10^6)	28.4 ± 6.3	26.6 ± 4.6	0.7 ± 0.3	1.7 ± 0.2
Progressively motile sperm (%)	80.4 ± 2.5	82.5 ± 7.5	78.0 ± 2.0	67.5 ± 2.5
Normal sperm morphology (%)	48.8 ± 3.3	46.3 ± 4.3	34.6 ± 4.9	46.5 ± 4.5
Testicular volume (cm^3)	1.64 ± 0.08	2.19 ± 0.46	1.6 ± 0.1	2.6 ± 0.3

[a]Herrick *et al.*, in press; [b]Swanson *et al.*, 1996 (for the non-breeding season only)

Sperm cryopreservation

Effective techniques for cryopreservation of domestic cat spermatozoa are still under development (Pukazhenthi, Pelican, Wildt and Howard, 1999; Pukazhenthi, Spinder, Wildt, Bush and Howard, 2002) and methods used at present are far from ideal for retaining maximal post-thaw sperm motility and intact acrosomal membranes. Variables such as cooling rate, glycerol concentration and timing of addition, packaging method, and freezing and thawing rate all can affect post-thaw semen parameters (Watson, 2000). Extrapolating these cryopreservation methods to nondomestic cats introduces the further complication of interspecies variability in sperm cryobiological properties (Holt, 2000). For example, in fishing cats, cooling the semen sample prior to glycerol addition benefits post-thaw motility and function whereas the temperature of glycerol addition has little impact on cheetah spermatozoa (Crosier *et al.*, 2006; Thiangtum *et al.*, 2006). For *in situ* banking of spermatozoa from nondomestic cats, cryopreservation techniques that are adaptable to a field environment also are a necessity. In some situations, anaesthetised males, such as black-footed cats, have been transported to an indoor mobile lab facility for semen collection (Herrick *et al.*, 2005) or semen recovered at a remote location has been transferred to a more conventional laboratory for further processing and freezing, as in cheetahs (Crosier *et al.*, 2006). At other study sites, semen collection, processing and freezing all have been conducted under more primitive conditions in the field, such as with Pallas' cats in Mongolia. Based on current knowledge, the most 'field-friendly' cryopreservation method involves adding glycerol at room temperature, packaging the semen in sealed straws, cooling slowly in an ice chest or refrigerator and then freezing over liquid nitrogen vapour or directly in a dry shipper. In a recent study, we have shown that ocelot spermatozoa can be processed using similar methods and frozen either on dry ice, over liquid nitrogen vapour or in a dry shipper, resulting in minimal differences among treatments in post-thaw sperm longevity and *in vitro* function (unpublished data).

All of these factors in combination create major challenges for being able to freeze cat spermatozoa in the field. In small cats, an additional consideration is the relatively low number of spermatozoa typically recovered in each ejaculate (Table 4). The amount of frozen spermatozoa banked from small cats in the wild represents the major variable dictating the use of this limiting resource. For assisted reproduction, AI is relatively uncomplicated and easy to perform (Howard, 1999; Wildt, 2003), but IVF represents a more economical choice for assisted reproduction when sperm numbers are at a premium and post-thaw sperm parameters are compromised. Insemination of each IVF drop requires only ~50,000 motile sperm as compared to a minimum of 10,000,000 motile sperm for each AI procedure. With small cats, cryopreservation of each ejaculate would permit 1 - 7 AI attempts; in contrast, 300 - 1500 IVF drops could be inseminated with the same number of motile spermatozoa (Table 4). Although IVF adds the additional challenge of transferring either freshly-produced or frozen-thawed embryos to synchronised recipients, recent research progress has led to improved ET success with some cat species (Pope, 2000; Swanson and Brown, 2004; Pope, Gomez and Dresser, 2006; Swanson, 2006).

Table 4. Relative merit of using frozen-thawed spermatozoa for artificial insemination *versus in vitro* fertilisation for assisted reproduction in small cat species.

Cat species	No. cats (no. ejaculates)	Total sperm/ ejaculate (x 10^6)*	Pre-freeze motility (%)*	AI doses/ ejaculate**	IVF doses/ ejaculate***
Fishing cat	8 (8)[a]	56.4 ± 17.7	73.0 ± 4.0	3	600
Ocelot	10 (10)[b]	129.4 ± 39.4	76.9 ± 5.6	7	1470
Black-footed cat	5 (12)[c]	28.4 ± 6.3	81.3 ± 2.3	1	345
Sand cat	7 (10)[d]	43.5 ± 11.0	77.0 ± 2.3	2	495
Pallas' cat	1 (5)[e]	25.2 ± 3.2	78.0 ± 2.0	1	290

*Mean (± SEM).
**Based on 10 x 10^6 motile sperm per AI with an estimated 20% decline in post-thaw motility.
***Based on 5 x 10^4 motile sperm per IVF drop with an estimated 20% decline in post-thaw motility.
[a]Thiangtum *et al.* (2006); [b]Stoops, unpublished; [c]Herrick *et al.*, in press; [d]Herrick *et al.* (2006); [e]Swanson *et al.* (1996) (breeding season only).

Given the potential pitfalls associated with effective banking of semen from wild felids in the field, assessing the functionality of frozen-thawed cat spermatozoa before applied usage is a necessity. Since routine collection of oocytes from non-domestic cats for fertility evaluations is not feasible, heterologous IVF of viable domestic cat oocytes has been developed as an alternative. Freshly-collected spermatozoa from at least nine nondomestic cat species, including three of five small cat SSP species, have proven capable of fertilising viable domestic cat oocytes, allowing evaluation of all sequential aspects of sperm function from capacitation to syngamy (Swanson, McRae, Davidson, Levens and Campbell, 1998; unpublished). With frozen-thawed spermatozoa from Pallas' cats (Swanson, Maggs, Clarke, Newell, Bond, Bateman and Kennedy-Stoskopf, 2006), ocelots (unpublished) and fishing cats (Thiangtum *et al.*, 2006), heterologous IVF percentages have ranged from 33% to 46%, including 36% fertilisation using frozen-thawed spermatozoa from two wild Pallas' cats in Mongolia (Table 5; Fig. 2). Because *in vitro* matured (IVM) oocytes frequently have reduced nuclear maturation compared to *in vivo* matured oocytes, fertilisation percentages may be adjusted based on mature oocytes only. For frozen-thawed fishing cat and ocelot spermatozoa, assessing fertilisation of mature oocytes only increases heterologous IVF percentages to 59 - 62% (Table 5). Direct comparison to values from homologous IVF with frozen-thawed spermatozoa is difficult, given limited data using mobile laboratory techniques for sperm cryopreservation and/or both homologous and heterologous IVF (Magarey, Herrick, Thiangtum, Tunwattana and Swanson, 2006; Swanson *et al.*, 2006). In a conventional laboratory setting, homologous IVF with frozen-thawed fishing cat and caracal spermatozoa has resulted in high (65-71%) fertilisation success (Pope *et al.*, 2006).

Table 5. Homologous *in vitro* fertilisation (IVF) of conspecific oocytes and heterologous IVF of domestic cat oocytes with frozen-thawed spermatozoa from small-sized nondomestic felids.

	Homologous IVF			Heterologous IVF		
Cat Species	Gonadotropin treatment**	Total oocytes inseminated	Fertilisation percentage	Oocyte source**	Total oocytes inseminated***	Fertilisation percentage
Pallas' cat	eCG/hCG	30	66.7%*[a]			
				eCG/hCG	45	35.6%*[b]
				eCG/hCG	201	46.1%*[c]
Ocelot				IVM	298	36.0%*[d]
					(157)	(59.1%)
Fishing cat	eCG/hCG	29	37.9%*[e]			
	FSH/LH	313	70.6%[f]			
				IVM	110	32.7%*[g]
					(58)	(62.1%)
Caracal	FSH/LH	392	64.8%[f]			

*Sperm cryopreservation and/or IVF procedures conducted using a mobile laboratory approach.
**Oocytes for homologous IVF were recovered from eCG/hCG or FSH/LH treated conspecific females whereas domestic cat oocytes for heterologous IVF were either recovered from eCG/hCG treated females or after *in vitro* maturation (IVM) of oocytes recovered after spaying.
***For IVM oocytes, fertilisation percentage also was calculated based on the total number of oocytes (in parentheses) that had completed nuclear maturation.
[a]Swanson and Kennedy-Stoskopf, unpublished; [b]Swanson, unpublished (semen from wild Pallas' cats); [c]Swanson *et al.* (2006); [d]Stoops, unpublished; [e]Magarey *et al.* (2006); [f]Pope *et al.* (2006); [g]Thiangtum *et al.* (2006).

Disease and transportation concerns

Two other considerations with *in situ* sperm banking of wild cats are the potential for disease transmission in frozen cat semen and the impact of airport security screening on sperm viability. The scientific literature contains little to no information on these two concerns in cats but either could have a significant impact on the applied usage of this genetic resource. In other mammalian species, infectious viruses have received the most attention as pathogens in semen (van Oirschot, 1995; Guerin and Pozzi, 2005). Of the common viruses found in cats (feline immunodeficiency virus, FIV; herpes virus, FHV; leukemia virus, FLV; corona virus, FCV; panleukopenia virus, FPV), only FIV has been thoroughly investigated as a potential contaminant of cat semen. In domestic cats, FIV is found in semen of infected males and may be

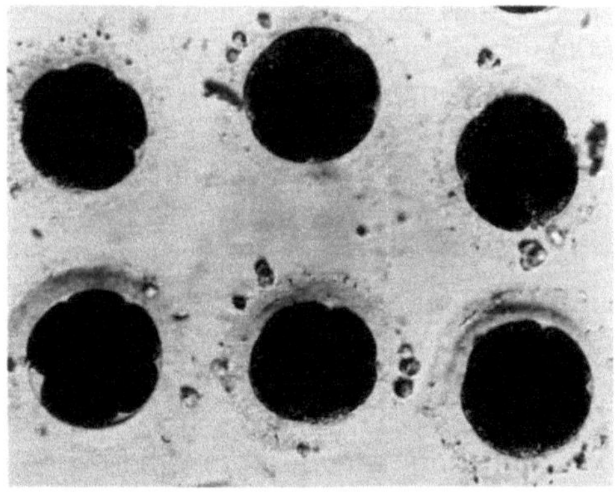

Figure 2. Pallas' cat - domestic cat hybrid embryos (48 h post-insemination; 5-8 cell stage) produced from heterologous IVF of domestic cat oocytes using frozen-thawed spermatozoa from wild Mongolian Pallas' cats.

transmitted via laparoscopic AI (Jordan, Howard, Barr, Kennedy-Stoskopf, Levy and Tompkins, 1998), but no comparative studies have been conducted in non-domestic felids. To our knowledge, FHV is the only other virus that has been assessed for presence in cat semen. In FHV-infected Pallas' cats, viral DNA was not detectable by PCR in either seminal fluid or frozen sperm cells, although conjunctival biopsies showed recurrent viral shedding (Swanson *et al.*, 2006). Given the limited data available, the safest strategy for banking semen from wild felids would be to screen blood samples from all sperm donors for viral exposure and freeze samples in sealed cryostraws for added biosecurity.

For *ex situ* usage, frozen cat semen typically must be transported internationally in liquid nitrogen dewars (or dry shippers) using commercial airlines as the means of conveyance. With recent terrorist activity, airline security has become more restrictive and now requires that all checked luggage undergo x-ray screening. A recent study has shown that a single screening procedure causes a significant decline in post-thaw cat sperm motility, possibly due to mitochondrial injury, and that multiple screenings induce double-stranded DNA breakage (Gloor, Winget and Swanson, 2006). For the near term, transporting dry shippers via overnight carrier services (such as Fedex or DHL) on cargo aircraft is one safe option; however, in some countries, these carriers either are not available or ship their cargo on passenger aircraft, necessitating x-ray exposure. These companies currently are developing high-intensity x-ray screening methods that can be applied to large cargo containers which, when implemented, will effectively eliminate overnight carriers as a transport option.

Holistic conservation programs

Linkage of *in situ* and *ex situ* populations is one essential step in the development of holistic conservation programs that incorporate captive breeding, habitat preservation and environmental education. This process involves identifying native collaborators in range countries, forming cooperative partnerships with local conservation organizations and engaging the local human communities in each region in these conservation efforts. An integral component of these partnerships has been scientific capacity building, in disciplines such as field ecology and reproductive biology, so that resident scientists acquire the skills to independently study indigenous wildlife and apply their new knowledge to species management and preservation (Rudran and Wemmer, 2001; Swanson and Brown, 2004). Zoos have played an integral role in these research training programs and increasingly are providing funding to support field ecology studies and habitat preservation (Hutchins, 2003; Miller *et al.*, 2004).

Among small cats, field ecology projects are ongoing in all five SSP species, including camera trap surveys of fishing cats in Thailand, radiotelemetry studies of Pallas' cats in Mongolia and sand cats in Saudi Arabia, distribution and biomedical surveys of black-footed cats in South Africa and habitat restoration for ocelots in Brazil. The Cincinnati Zoo provides financial support for each of these projects with other North American zoos contributing funding as well. *In situ* sperm banking already has been incorporated into field ecology studies in Pallas' cats and black-footed cats and likely will become a component of these other field projects within the next few years. We anticipate producing offspring in several of these small cat species using frozen-thawed spermatozoa from wild males in the near future, laying the foundation for more effective genetic management of captive cat populations. Provided that this linkage strategy shows adequate efficiency, the intriguing but even more challenging prospect of reversing gene flow (i.e., from ex *situ* to *in situ*) then may be explored for future management of genetically impoverished, geographically isolated cat populations in the wild.

Acknowledgements

The authors gratefully acknowledge the following individuals for assistance with *in situ* studies: Dr. Meredith Brown, Bariushaa Munkhtsog, Steve Ross and Dr. Amanda Fine (Mongolia) and Dr. Alex Sliwa, Beryl Wilson, Morgan Hauptfleisch, Dr. Corne Anderson and Dr. Nadine Lamberski (South Africa). Our appreciation also is extended to the multiple zoos and universities that collaborated on research in ocelots (African Wildlife Safari, Carnivore Preservation Trust, Dallas Zoo, Fort Worth Zoo, Nashville Zoo, Oglebay's Good Zoo), fishing cats (Feline Conservation Center, Kasetsart University, Khao Kheow Open Zoo, Oklahoma City Zoological Park, San Francisco Zoological Gardens), black-footed cats (Central Florida Zoological Park, Oklahoma City Zoological Park, Omaha's Henry Doorly Zoo, Riverbanks Zoo & Garden, San Antonio Zoological Gardens), sand cats (Erie Zoo, Feline Conservation Center, Living Desert Zoo, Saint Louis Zoo) and Pallas' cats (North Carolina State University). At the Cincinnati Zoo, the authors also thank Dr. Mark Campbell, Dr. Greg Levens, Jennifer Bond, Helen Bateman and Jackie Newsom for help with animal procedures and John Dinon and Barbara Lintzenich for estimates of Pallas' cat maintenance costs. Funding for these studies was provided, in part, by grants from the National Institutes of Health (RR001588) and the Morris Animal Foundation (D04ZO-72).

References

Ballou, J.D. and Foose, T.J. (1996). Demographic and genetic management of captive populations. In: *Wild Mammals in Captivity: Principles and Techniques*. Edited by Kleiman, D.G., Allen, M.A., Thompson, K.V. and Lumpkin, S., University of Chicago Press, Chicago, IL, pp. 263-283

Barone, M.A., Roelke, M.E., Howard, J.G., Brown, J.L., Anderson, A.E. and Wildt, D.E. (1994). Reproductive characteristics of male Florida panthers: comparative studies from Florida, Texas, Colorado, Latin America, and North American zoos. *Journal of Mammalogy* **75:** 150-162

Beck, B.B., Rapaport, L.G., Stanley Price, M.R., and Wilson, A.C. (1994). Reintroduction of captive-born animals. In: *Creative Conservation: Interactive Management of Wild and Captive Animals*. Edited by Olney, P.J.S., Mace, G.M. and Feistner, A.T.C., Chapman and Hall, London, pp. 287-303

Benirschke, K. (1984). The frozen zoo concept. *Zoo Biology* **3:** 325-328

Brown, M., Lappin, M.R., Brown, J.L., Munkhtsog, B. and Swanson, W.F. (2005). Exploring the ecological basis for extreme susceptibility of Pallas' cats (*Otocolobus manul*) to fatal toxoplasmosis. *Journal of Wildlife Diseases* **41:** 691-700

Caron, M. (2006). International Pallas' Cat (*Otocolobus manul*) Studbook. Minnesota Zoological Garden, Minneapolis, MN

Conway, W.G. (1998). Zoo reserves: a proposal. *Proceedings of the Annual Conference of the American Zoo and Aquarium Association*, pp. 54-58

Conway, W.G. (1986). The practical difficulties and financial implications of endangered species breeding programs. *International Zoo Yearbook* **24/25:** 210-219

Costanza, R., d'Arge, R., de Groot, R.S., Farber, S., Grasso, M., Hannon, B., Limburg, K., Naeem, S., O'Neill, R.V., Paruelo, J., Raskin, R.G., Sutton, P. and van den Belt, M. (1997). The value of the world's ecosystem services and natural capital. *Nature* **387:** 253-260

Crosier, A.E., Pukazhenthi, B.S., Henghali, J.N., Howard, J.G., Dickman, A.J., Marker, L. and Wildt, D.E. (2006) Cryopreservation of spermatozoa from wild-born Namibian cheetahs (*Acinonyx jubatus*) and influence of glycerol on cryosurvival. *Cryobiology* **52:** 169-181

Franklin, W.L., Johnson, W.E., Sarno, R.J. and Iriate, J.A. (1999). Ecology of the Patagonia puma (*Felis concolor patagonica*) in southern Chile. *Biological Conservation* **90:** 33-40

Gloor K.T., Winget, G.D. and Swanson, W.F. (2006). Conservation science in a terrorist age: the impact of airport security screening on the viability and DNA integrity of frozen felid spermatozoa. *Journal of Zoo and Wildlife Medicine* **37:** 327-335

Grassman, L.I. Jr., Austin, S.C., Tewes, M.E. and Silvy, N.J. (2004). Comparative immobilization of wild felids in Thailand. *Journal of Wildlife Diseases* **40:** 575-578

Guerin, B. and Pozzi, N. (2005). Viruses in boar semen: detection and clinical as well as epidemiological consequences regarding disease transmission by artificial insemination. *Theriogenology* **63:** 556-572

Herrick, J., Lamberski, N., Wilson, B., Anderson, C., Swanson, B. and Sliwa, A. (2005). Linking ex situ and in situ in black-footed cats. In: *North American Felid Taxon Advisory Group (TAG) 2005 Annual Report*. Edited by Swanson, W.F. and Flecthall, N., Felid TAG official website (*www.felidtag.org*), pp. 15-17

Herrick, J.R., Leiske, K., Magarey, G.M. and Swanson, W.F. (2006). Basal seminal traits and *in vitro* fertilization in the sand cat (*Felis margarita*). *Reproduction, Fertility and Development* **18:** 218-219 (abstract 221)

Herrick, J.R., West, G., Benson, K., Coke, R., Campbell, M. and Swanson, W.F. Semen evaluation and *in vitro* sperm function in the black-footed cat (*Felis nigripes*). *Proceedings of the Annual Meeting of the Society for the Study of Reproduction* (in press)

Holt, W.V. (2000). Fundamental aspects of sperm cryobiology: the importance of species and individual differences. *Theriogenology* **53:** 47-58

Holt, W.V. and Moore, H.D.M. (1988). Semen banking – is it now feasible for captive endangered species? *Oryx* **22:** 172-178

Howard, J.G. (1993). Semen collection and analysis in carnivores. In: *Zoo and Wild Animal Medicine, Current Therapy 3*. Edited by Fowler, M.E., W.B. Saunders Company, Philadelphia, PA, pp. 390-399

Howard, J.G. (1999). Assisted reproductive techniques in nondomestic carnivores. In: *Zoo & Wild Animal Medicine, Current Therapy 4*. Edited by Fowler, M.E. and Miller, R.E., W.B. Saunders Company, Philadelphia, PA, pp. 449-457

Howard, J.G., Bush, M. and Wildt, D.E. (1986). Semen collection, analysis and cryopreservation in nondomestic mammals. In: *Current Therapy in Theriogenology*. Edited by Morrow, D., W.B. Saunders Company, Philadelphia, PA, pp. 1047-1053

Howard, J.G., Marker, L., Pukazhenthi, B.S., Roth, T.L., Swanson, W.F., Grisham, J. and Wildt, D.E. (2002). Genome resource banking and successful artificial insemination with cryopreserved sperm in the cheetah. *Proceedings of the 9th International Symposium on Spermatology*, p. 70 (abstract PL15)

Hutchins, M. (2003). Zoo and aquarium animal management and conservation: current trends and future challenges. *International Zoo Yearbook* **38:** 14-28

ISIS [International Species Inventory System]. (2003). *Single Population and Animal Record Keeping System (SPARKS), Version 1.53.* ISIS, Apple Valley, MN

ISIS [International Species Inventory System]. (2006). Global inventory of mammalian species in captivity. ISIS, Apple Valley, MN. ISIS official website (*www.isis.org*)

IUCN [International Union for Conservation of Nature and Natural Resources]. (2003). IUCN Red List of Threatened Species. IUCN, Gland, Switzerland. IUCN official website (*www.redlist.org*)

Johnston, L.A. and Lacy, R.C. (1995). Genome resource banks for species conservation: selection of sperm donors. *Cryobiology* **32:** 68-77

Jordan, H.L., Howard, J., Barr, M.C., Kennedy-Stoskopf, S., Levy, J.K. and Tompkins, W.A. (1998). Feline immunodeficiency virus is shed in semen from experimentally and naturally infected cats. *AIDS Research and Human Retroviruses* **14:** 1087-1092

Lasley, B.L., Loskutoff, N.M. and Anderson, G.B. (1994). The limitation of conventional breeding programs and the need and promise of assisted reproduction in nondomestic species. *Theriogenology* **41:** 119-132

Lewis, J.C.M. (2004). Field use of isoflurane and air anesthetic equipment in wildlife. *Journal of Zoo and Wildlife Medicine* **35:** 303-311

Magarey ,G.M., Herrick, J., Thiangtum, K., Tunwattana, W. and Swanson, W.F. (2006). Comparative fertility of freshly-collected versus frozen-thawed spermatozoa for *in vitro* fertilization in the fishing cat (*Prionailurus viverrinus*). *Reproduction, Fertility and Development* **18:** 249 (abstract 284)

Miller, B., Conway, W., Reading, R.P., Wemmer, C.,

Wildt, D.E., Kleiman, D., Monfort, S., Rabinowitz, A., Armstrong, B. and Hutchins, M. (2004). Evaluating the conservation mission of zoos, aquariums, botanical gardens, and natural history museums. *Conservation Biology* **18:** 86-93

Morais, R.N., Mucciolo, R.G., Gomes, M.L.F., Lacerda, O., Moraes, W., Moreira, N., Graham, L.H., Swanson, W.F. and Brown, J.L. (2002). Seasonal analysis of seminal characteristics, serum testosterone and fecal androgens in the ocelot (*Leopardus pardalis*), margay (*L. wiedii*) and tigrina (*L. tigrinus*). *Theriogenology* **57:** 2027-2041

Morato, R.G., Conforti, V.A., Azevedo, F.C., Jacomo, A.T.A., Silveira, L., Sana, D., Nunes, A.L.V., Guimaraes, M.A.B.V. and Barnabe, R.C. (2001). Comparative analyses of semen and endocrine characteristics of free-living versus captive jaguars (*Panthera onca*). *Reproduction* **122:** 745-751

Munkhtsog, B. and Ross, S. (2006). Pallas' cat (*Otocolobus manul*) foraging ecology and conservation in central Mongolia. In: *North American Felid Taxon Advisory Group (TAG) 2006 Annual Report.* Edited by Swanson, W.F. and Flecthall, N., Felid TAG official website (www.felidtag.org) (in press)

Nowell, K. and Jackson, P. (1996). *Wild Cats: Status Survey and Conservation Action Plan.* IUCN/SSC Cat Specialist Group, Gland, Switzerland

Pimentel, D., Wilson, C., McCullum, C., Huang, R., Dwen, P., Flack, J., Tran, Q., Saltman, T. and Cliff, B. (1997). Economic and environmental benefits of biodiversity. *Bioscience* **47:** 747-757

Pollack, J.P., Lacy, R.C. and Ballou, J.D. (2001). *Population Management 2000 (PM2000).* Chicago Zoological Society, Brookfield, IL

Pope, C.E. (2000). Embryo technology in conservation efforts for felids. *Theriogenology* **53:** 163-174

Pope, C.E., Gomez, M.C., Dresser, B.L. (2006). *In vitro* embryo production and embryo transfer in domestic and nondomestic cats. *Theriogenology* **66:** 59-71

Pope, C.E., Zhang, Y.Z. and Dresser, B.L. (1991). A simple staining method for evaluating acrosomal status of cat spermatozoa. *Journal of Zoo and Wildlife Medicine* **22:** 87-95

Pukazhenthi, B., Pelican, K., Wildt, D.E., and Howard, J.G. (1999). Sensitivity of domestic cat (*Felis catus*) sperm from normospermic versus teratospermic donors to cold-induced acrosomal damage. *Biology of Reproduction* **61:** 135-141

Pukazhenthi, B., Spindler, R., Wildt, D., Bush, L.M., and Howard, J.G. (2002). Osmotic properties of spermatozoa from felids producing different proportions of pleiomorphisms: influence of adding and removing cryoprotectant. *Cryobiology* **44:** 288-300

Rudran, R. and Wemmer, C. (2001). Teaching conservation biology to developing country nationals: the National Zoo's experience. In: *The AZA Field Conservation Resource Guide.* Edited by Conway, W.G., Hutchins, M., Souza, M., Kapetanakos, Y. and Paul, E., Zoo Atlanta, Atlanta, GA, pp. 240-244

Seager, S.W.J. (1983). The breeding of captive wild species by artificial means. *Zoo Biology* **2:** 235-239

Shindle, D.B. and Tewes, M.E. (2000). Immobilization of wild ocelots with tiletamine and zolazepam in southern Texas. *Journal of Wildlife Diseases* **36:** 546-550

Sliwa, A. (2004). Home range size and social organization of black-footed cats (*Felis nigripes*). *Mammalian Biology* **69:** 96-107

Swanson, W.F. (2003). Research in nondomestic species: experiences in reproductive physiology research for conservation of endangered felids. *ILAR Journal* **44:** 307-316

Swanson, W.F. (2006). Application of assisted reproduction for population management in felids: the potential and reality for conservation of small cats. *Theriogenology* **66:** 49-58.

Swanson, W.F. and Brown, J.L. (2004). International training programs in reproductive sciences for conservation of Latin American felids. *Animal Reproduction Science* **82:** 21-34

Swanson, W.F., Johnson, W.E., Cambre, R.C., Citino, S.B., Quigley, K.B., Brousset, D.M., Morais, R.N., Moreira, N., O'Brien, S.J. and Wildt, D.E. (2003). Reproductive status of endemic felid species in Latin American zoos and implications for *ex situ* conservation. *Zoo Biology* **22:** 421-441

Swanson, W.F., Maggs, D.J., Clarke, H.E., Newell, A.E., Bond, J.B., Bateman, H.L. and Kennedy-Stoskopf, S. (2006). Assessment of viral presence in semen and reproductive function of frozen-thawed spermatozoa from Pallas' cats (*Otocolobus manul*) infected with feline herpesvirus. *Journal of Zoo and Wildlife Medicine* **37:** 336-346

Swanson, W.F., McRae, M.A., Davidson, J.L., Levens, G.P. and Campbell, M.K. (1998). Comparative sperm function and gamete interaction among small felid species from divergent evolutionary lineages. *Biology of Reproduction* **58 (Suppl. 1):** 176 (abstract)

Swanson, W.F., Brown, J.L., and Wildt, D.E. (1996). Influence of seasonality on reproductive traits in the male Pallas' cat (*Felis manul*) and implications for captive management. *Journal of Zoo and Wildlife Medicine* **27:** 234-240.

Thiangtum, K., Swanson, W.F., Howard, J.G., Tunwattana, W., Tongthainan, D., Wichasilpa, W., Patumrattanathan, P. and Pinyopoommintr, T. (2006). Assessment of basic seminal characteristics, sperm cryopreservation and heterologous *in vitro* fertilization in the fishing cat (*Prionailurus viverrinus*). *Reproduction, Fertility and Development* **18:** 373-382

van Dorrser, F.J.H. and Strick, J.A. (2005). Semen characteristics and sperm morphology in the Arabian leopard (*Panthera pardus nimr*) and how these vary with age and season. *Reproduction, Fertility and Development* **17:** 675-682

van Oirschot, J.T. (1995). Bovine herpes virus 1 in semen of bulls and the risk of transmission: a brief review. *Veterinary Quarterly* **17:** 29-33

Watson, P.F. (2000). The causes of reduced fertility with cryopreserved spermatozoa. *Animal Reproduction*

Science **60-61:** 481-492

Wildt, D.E. (2003). The role of reproductive technologies in zoos: past, present and future. *International Zoo Yearbook* **38:** 111-118

Wildt, D.E., Brown, J.L., Bush, M., Barone, M.A., Cooper, K.A., Grisham, J. and Howard, J.G. (1993). Reproductive status of cheetahs (*Acinonyx jubatus)* in North American zoos: the benefits of physiological surveys for strategic planning. *Zoo Biology* **12:** 45-80

Wildt, D.E., Bush, M., Goodrowe, K.L., Packer, C., Pusey, A.E., Brown, J.L., Joslin, P. and O'Brien, S.J. (1987b). Reproductive and genetic consequences of founding isolated lion populations. *Nature* **329:** 328-331

Wildt, D.E., O'Brien, S.J., Howard, J.G., Caro, T.M., Roelke, M.E., Brown, J.L. and Bush, M. (1987a). Similarity in ejaculate-endocrine characteristics in captive versus free-living cheetahs of two subspecies. *Biology of Reproduction* **36:** 351-360

Wildt, D.E., Rall, W.F., Critser, J.K., Monfort, S.L. and Seal, U.S. (1997). Genome resource banks - living collections for biodiversity conservation. *Bioscience* **47:** 689-698

Wildt, D.E. and Roth, T.L. (1997). Assisted reproduction for managing and conserving threatened felids. *International Zoo Yearbook* **35:** 164-172

Wildt, D.E., Schiewe, M.C., Schmidt, P.M., Goodrowe, K.L., Howard, J.G., Phillips, L.G., O'Brien, S.J. and Bush, M. (1986). Developing animal model systems for embryo technologies in rare and endangered wildlife. *Theriogenology* **25:** 33-51

Spermatology. SRF Vol. 65. ERS Roldan and M Gomendio (eds) Nottingham University Press, Nottingham

Challenges in cryopreserving endangered mammal spermatozoa: Morphology and the value of acrosomal integrity as markers of cryo-survival

Budhan Pukazhenthi[1], Rachel Santymire*, Adrienne Crosier[2], JoGayle Howard[1] and David E. Wildt[2]

[1]*Department of Reproductive Sciences, Smithsonian's National Zoological Park, 3001 Connecticut Avenue NW, Washington DC 20008-2598, USA;* [2]*Smithsonian's National Zoological Park, Conservation & Research Center, Front Royal, VA 22630, USA;* **Davee Center for Epidemiology and Endocrinology, Lincoln Park Zoo, Conservation & Science, 2001 N. Clark Street, Chicago, IL 60614, USA*

The science of cryobiology is essential to the effective, practical use of semen for assisted breeding to help manage small populations of rare wildlife species. In this review, we describe challenges associated with cryopreserving gametes from wild fauna. Based on more than 25 years of experience across a diversity of mammals, it appears that the primary driving force dictating cryo-survival of a spermatozoon is its initial pre-freeze quality and morphology, especially having a morphologically normal, intact acrosome. This assertion is supported through extensive studies of three animal groups that routinely ejaculate semen containing (1) normal sperm/acrosomal quality (examples, Eld's deer, *Cervus eldi* and giant panda, *Ailuropoda melanoleuca*), (2) normal acrosomal quality, but from teratospermic donors (>70% pleiomorphic sperm; cheetah, *Acinonyx jubatus* and black-footed ferret, *Mustela nigripes*) and (3) abnormal acrosomal quality and general teratospermia (clouded leopard, *Neofelis nebulosa*). Data revealed that species producing high quality sperm with >70% normal, intact acrosomes were best able to survive cryopreservation (~80% intact acrosomes post-thaw). Species that were teratospermic, but with high proportions of intact acrosomes (72 to 88%) in ejaculates varied significantly (4 to 55% intact acrosomes post-thaw) in sperm survival to freeze-thawing. Spermatozoa from the clouded leopard (that was both teratospermic while producing only 11% normal acrosomes in fresh semen) failed to survive cryopreservation despite using an array of conventional and unconventional freezing approaches. These observations (combined with zona penetration assays and artificial insemination results) suggest that proportions of malformed sperm and especially initial structural integrity of the acrosome are more important predictors of sperm survivability post-thaw than initial sperm motility scores.

Corresponding author E-mail: pukazhenthib@si.edu

Introduction

The utility of artificial insemination (AI) and sperm cryopreservation in the propagation and genetic management of several endangered species has been demonstrated (Wildt, Pukazhenthi, Brown, Monfort, Howard and Roth, 1995; Wildt and Roth, 1997; Howard, 1999; Howard, Marinari and Wildt, 2003; Pukazhenthi and Wildt, 2004). Modern zoo breeding programmes rely on matching appropriate males with females for the purposes of retaining maximal genetic diversity and producing healthy, viable offspring. These intensive breeding programmes require moving animals between zoos, or, in the case of assisted breeding, transporting sperm for artificial insemination (AI). For captive (as well as wild) populations, a repository of frozen spermatozoa (or a genome resource bank) also serves as important insurance for the future (Wildt, Rall, Critser, Monfort and Seal, 1997; Holt, 2001). For example, valuable genotypes in zoos have been lost forever from unexpected catastrophe (such as fire or epizootics). Most animal populations in nature are under significant pressures (the primary threat being habitat fragmentation and loss) that can lead to reduced levels of genetic diversity that, in turn, can have dramatic effects on fitness and viability (Roelke, Martenson and O'Brien, 1993). Periodically banking sperm from genetically valuable males can provide a means of protecting and saving the genes of individuals, populations or entire species. For zoo programmes, the ability to freeze and store sperm reduces the number of males needed in captivity, which is important because too little space exists in zoos to preserve much biodiversity. The use of viable cryopreserved sperm also can increase the generation interval of a species indefinitely, allowing genes to be reintroduced into living populations decades from now, long after the death of the current donor (Wildt, 1992; Holt, 2001).

In recognition of these benefits, some zoos and academic institutions have begun systematically banking sperm samples from various wildlife species, either from captive collections or even those free-living in nature. These efforts can be challenging, because such research often is not conducted under modern laboratory environments. In some cases, such work is done under rigorous, remote field conditions that require the development of novel cryopreservation technologies (Crosier, Pukazhenthi, Henghali, Howard, Dickman, Marker and Wildt, 2006). Nonetheless, advances in conventional cryobiology have substantially increased our understanding of the fundamental mechanisms and events to be expected in the cooling, freezing and thawing of spermatozoa. This scholarly knowledge has had some direct application and commonality across a broad spectrum of species (Gao, Mazur and Critser, 1997; Holt, 2000a; Mazur, Katkow, Katkova and Critser, 2000; Ball and Vo, 2001; Pukazhenthi, Spindler, Wildt, Bush and Howard, 2002a; Songsasen, Yu, Murton, Paccamonti, Eilts, Godke and Leibo, 2002). For example, there are numerous studies on the cytotoxicity of cryoprotectants, membrane permeability and lipid/fatty acid composition and the influences of cooling and warming rates on sperm viability and functionality (Rana, 1995; Gao *et al.*, 1997; Songsasen and Leibo, 1997; Roth, Bush, Wildt and Weiss, 1999; Holt, 2000b; Songsasen *et al.*, 2002; Garde, Soler, Cassinello, Crespo, Malo, Espeso, Gomendio and Roldan, 2003; Herrick, Bartels and Krisher, 2004; Woods, Benson, Agca and Critser, 2004; Harris, Baumgard, Arns and Webel, 2005; Rota, Rota, Martini, Milani and Romagnoli, 2005; Waterhouse, Hofmo, Tverdal and Miller, 2006). Throughout such comparative *in vitro* studies, the end-point question for deciding success (or failure) after thawing generally is: Are the sperm motile and structurally intact? In some instances, additional efforts have been made to examine functionality of the thawed spermatozoon by testing its ability to undergo capacitation and fertilise conspecific or heterologous oocytes (Roth, Weiss, Buff, Wildt and Bush, 1998; Pena, Barrio, Becerra, Quintela and Herradon, 2004; Moore, Squires and Graham, 2005). Of course, the optimal method for evaluating achievement is the consistent production of living offspring.

Of the many factors that can influence a spermatozoon's ability to withstand the rigours of cooling, freezing and thawing, its biophysical properties (especially membrane stability and permeability) certainly play a major role in dictating cellular survival or death (Gao *et al.*, 1997). What seems even more undervalued is that these traits are highly taxon-specific, that is, that sperm cryopreservation protocols generally must be developed for each species of interest. In retrospect, such biological variation should not seem surprising because freeze-thawing success can vary among donors within a species and even within individuals from ejaculate to ejaculate (Songsasen and Leibo, 1997; Songsasen *et al.*, 2002; Waterhouse *et al.*, 2006).

For more than 25 years, our laboratory has had the unique opportunity of exploring the cryobiology of sperm from 'unconventional' mammals - wildlife species that have received little, if any, previous research attention. Collectively, these studies have demonstrated the vast array of inherently different ejaculate and sperm traits, even among species within the same family (Wildt, 1994). Our basic investigations also have provided valuable insight into (1) existing and marked variations in sperm cryosensitivity, (2) the need for developing alternative cryo-methodologies to meet quality standards that should ensure offspring production after AI and (3) identifying appropriate donor candidates for genome resource banking for use in the near- or long-term. Because resources are limited, our goal is not to cryopreserve sperm from 'marginal' ejaculates, that is from donors in which post-thaw viability would be too inadequate to result in offspring. Thus, a high priority is to explore markers in freshly ejaculated spermatozoa that are most indicative of potential freeze-thawing success.

A particularly interesting feature of ejaculates from many wildlife species is an unusually high incidence of sperm pleiomorphisms that, in many cases, appears due to a historic loss in genetic diversity (Wildt, Bush, Howard, O'Brien, Meltzer, van Dyk, Ebedes and Brand, 1983; O'Brien, Wildt, Goldman, Merril and Bush, 1983; Roelke *et al.*, 1993). This phenomenon of teratospermia (production of <40% morphologically normal sperm/ejaculate) has been especially prevalent in certain felids (Howard, 1993; Pukazhenthi, Wildt and Howard, 2001; Pukazhenthi, Neubauer, Jewgenow, Howard and Wildt, 2006c) as well as species derived from small numbers of founders (Wolf, Wildt, Vargas, Marinari, Kreeger, Ottinger and Howard, 2000; Wildt, Ellis and Howard, 2001). In some cases, these sperm deformities include the acrosome, whereas in others, this critical structure remains intact, even in the presence of other structural malformations associated with the head, midpiece and flagellar regions. The functionality of spermatozoa from teratospermic ejaculates has been thoroughly studied, especially in felids where such cells are known to be compromised in ability to undergo capacitation, penetrate zona-intact oocytes and regulate protein tyrosine phosphorylation, all while containing increased amounts of defective DNA (Howard, Donoghue, Johnston and Wildt, 1993; Pukazhenthi et *al.*, 2006c). Spermatozoa from teratospermic donors also appear hyper-sensitive to rapid cooling to 5°C (Pukazhenthi, Pelican, Wildt and Howard, 1999) as well as hypo-osmotic stress (Pukazhenthi *et al.*, 2002a). Although some species and populations experience high levels of teratospermia, others produce mostly structurally normal spermatozoa, much like that observed in heterozygous domesticated livestock.

For this paper, we analysed our database on five wild mammal species to conclude that sperm morphology (especially acrosomal shape and structural integrity) is the primary factor dictating sperm cryosurvival. Species were divided into three sperm quality types: (1) those that produce mostly structurally normal cells with intact acrosomes (e.g., giant panda, *Ailuropoda melanoleuca* and Eld's deer, *Cervus eldi thamin*); (2) those that are teratospermic, but produce high proportions of normal, intact acrosomes (e.g., cheetah, *Acinonyx jubatus* and black-footed ferret, *Mustela nigripes*); and (3) a species that produces unusually high proportions of pleiomorphic sperm that also include a marked incidence of deranged acrosomes (e.g., clouded leopard, *Neofelis nebulosa*).

Ejaculate traits and sperm morphology

Common electroejaculate traits for the five studied species are provided in Table 1. With obvious exceptions in seminal volume and sperm density, these species express somewhat similar sperm motility traits (both total percentages of motile cells and forward progression scores). The giant panda (endemic to China) and the Eld's deer (whose natural range is Thailand, Myannmar, Laos, Cambodia and India) are species typical of producing semen with a high incidence of structurally normal (>85%) spermatozoa, including intact acrosomes (Spindler *et al.*, 2004). In contrast, the African cheetah and the North American black-footed ferret are examples of species that consistently ejaculate low proportions of normally shaped spermatozoa (~18% and ~29%, respectively) (Wildt, Brown and Bush, 1993; Wolf *et al.* 2000; Santymire, Marinari, Kreeger, Wildt and Howard, 2006). Interestingly, in both of these cases, despite a majority of the cells being malformed, the acrosomes of these spermatozoa remain largely intact and normal in appearance (Table 1). In contrast, ejaculates of the clouded leopard (a species that ranges widely throughout Southeast Asia), commonly contain few morphologically normal sperm, with about 85% of these cells having an abnormal acrosome, largely expressed as knobbed or folded acrosomes (Fig. 1) (Pukazhenthi, Laroe, Crosier, Bush, Spindler, Pelican, Bush, Howard and Wildt, 2006b).

Table 1. Seminal traits of giant pandas, Eld's deer, cheetahs, black-footed ferrets and clouded leopards.

Traits	Giant panda (n = 15 males)	Eld's deer (n = 23 males)	Cheetah (n = 97 males)	Black-footed ferret (n = 16 males)	Clouded leopard (n = 5 males)
Ejaculate volume (ml)	1.6 ± 0.3	3.5 ± 0.4	2.1 ± 0.4	0.03 ± 0.0	1.02 ± 0.1
Sperm concentration (x 10^6/ml)	1950 ± 3.0	459 ± 2.0	22.2 ± 1.6	558.6 ± 121.5	43 ± 2.1
Sperm motility (%)	80 ± 2.5	85 ± 2.0	69 ± 1.1	62.2 ± 2.3	73.2 ± 4.0
Progressive sperm motility[a]	3.7 ± 0.2	4.0 ± 0.0	3.3 ± 0.1	2.9 ± 0.1	3.5 ± 0.1
Structurally normal spermatozoa (%)	88.3 ± 0.8	85.7 ± 5.4	18.4 ± 0.8	15.3 ± 3.0	18.5 ± 2.5
Normal, intact acrosomes (%)	85.2 ± 1.5	84 ± 2.3	73.6 ± 1.5	72.9 ± 3.7	15.8 ± 2.1

Values are means ± SEM.
[a]Scale ranges from 0 to 5; 0 = no forward movement and 5 = rapid forward linear movement.
Adapted from Spindler *et al.* (2004), Monfort *et al.* (1993), Crosier *et al.* (2006), Santymire *et al.* (2006) and Pukazhenthi *et al.* (2006).

Assessment of acrosomal integrity

Our laboratory routinely relies on an array of assays for assessing seminal quality. These include sperm motility (phase-contrast optics), morphology (phase-contrast optics) and acrosomal integrity (phase-contrast optics, dye-based staining techniques, flourescein conjugated lectin staining techniques and/or flow cytometry). Generally and especially for the species described here, phase-contrast microscopy alone fails to allow adequate assessment of acrosomal integrity. An exception is the black-footed ferret where phase-contrast optics can be used effectively to classify acrosomal integrity into four categories: normal apical ridge, damaged, missing or loose acrosomal cap (Howard, 1993; Santymire *et al.*, 2006) (Fig. 2). In contrast, evaluating acrosomal integrity in the giant panda (Spindler, Huang, Howard, Wang, Zhang, Zhang and Wildt, 2004), Eld's deer (Harnal, Spindler, Monfort, Pukazhenthi, Bird and Wildt, 2001), cheetah (Crosier *et al.* 2006) and clouded leopard (Pukazhenthi *et al.*, 2006b) requires a dye-based

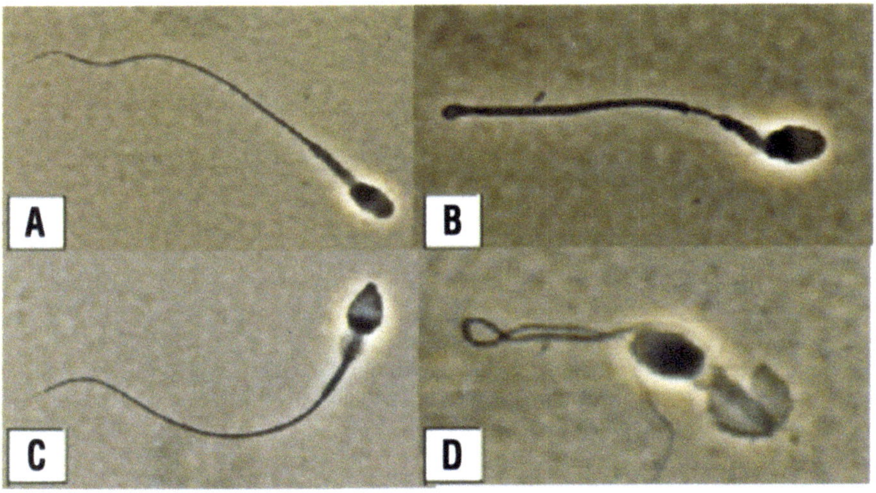

Figure 1. Common sperm morphological abnormalities consistently observed in the clouded leopard electroejaculate: (**A**) normal; (**B**) partial midpiece aplasia; (**C**) abnormal acrosome and retained proximal cytoplasmic droplet; and (**D**) bent midpiece and retained cytoplasmic droplet. Adapted from Pukazhenthi *et al.* (2006).

or fluorescein-based staining technique. Acrosomal integrity then can be determined microscopically and classified as intact, damaged or missing [for the Eld's deer (Fig. 3A) or cheetah] or these same categories plus the loose acrosomal cap defect (for the giant panda). In the case of the clouded leopard, the consistently high proportion of abnormal acrosomes mandates that we classify stained cells as intact normal, intact abnormal, damaged abnormal and missing (Fig. 3B).

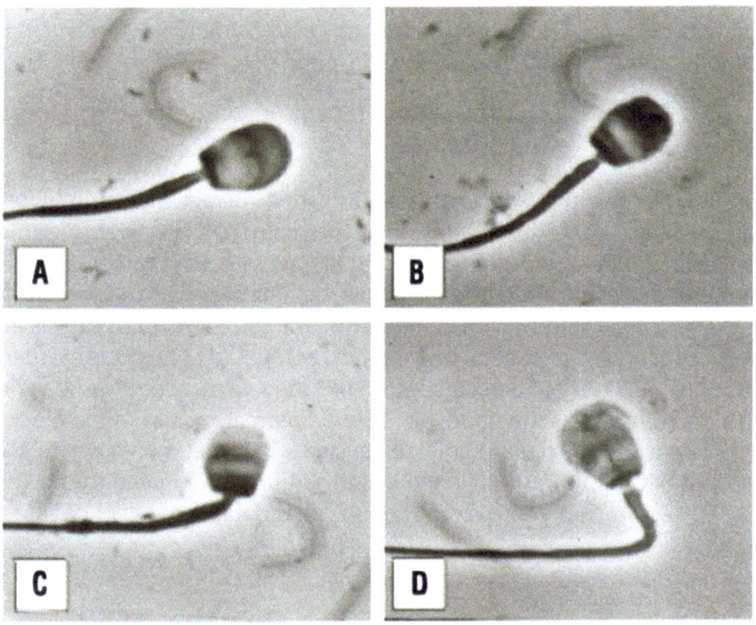

Figure 2. Classification of sperm acrosomal morphology in the black-footed ferret: (**A**) normal apical ridge; (**B**) damaged apical ridge; (**C**) missing apical ridge; and (**D**) loose acrosomal cap. Adapted from Howard *et al.* (1991).

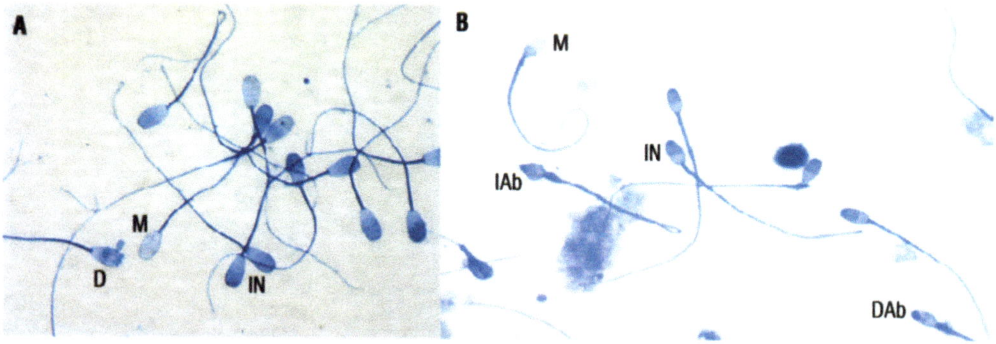

Figure 3. Coomassie blue staining of Eld's deer (**A**) and clouded leopard (**B**) spermatozoa for assessment of acrosomal morphology including intact (IN), damaged (D), missing (M), intact abnormal (IAb) and damaged abnormal (DAb).

Impact of cryofactors and teratospermia on sperm post-thaw survival and acrosomal integrity

During sperm cryopreservation, ejaculates are extended in a cryodiluent (in the absence or presence of cryoprotectant), cooled to 5°C, mixed with a cryoprotectant and then loaded in plastic straw containers that are cryopreserved over dry-ice or liquid nitrogen vapour (Howard, Bush and Wildt, 1986; Howard, Zhang, Li, Huang, Zhang, Hou, Ye, Li, Zhang, Huang, Zhang, Zhang, Spindler and Wildt, 2006b). The specifics of each protocol generally are species dependent with explicit steps developed systematically. In some cases, protocols must be developed and/or modified to accommodate the need to cool/freeze semen under field conditions in the absence of a standard laboratory environment and resources such as dry-ice. The design and implementation of specific protocols generally involve addressing four primary challenges that include determining: (1) the appropriate cooling rate to temperatures above 0°C; (2) ideal cryoprotectant and dilution; (3) freezing approach (e.g., using dry-ice *versus* liquid nitrogen vapour; and (4) the influence and mitigation of osmotic stress.

Cooling rate

Our suspicion that morphology and acrosomal integrity interact significantly with the above factors was derived from early observations that spermatozoa from teratospermic felid ejaculates were especially vulnerable to rapid cooling (above freezing temperatures) (Pukazhenthi *et al.*, 1999; Pukazhenthi *et al.*, 2006c). For example, cooling spermatozoa from the teratospermic domestic cat to 0°C (ultra-rapid; ~14°C/min) or 5°C (rapid; ~4°C/min) causes marked reductions in proportions of intact acrosomal membranes, compared to their normospermic counterparts (Pukazhenthi *et al.*, 1999). Such damage is abolished when sperm suspensions are cooled more slowly (0.5°C/min) (Pukazhenthi *et al.*, 1999). Spermatozoa from the black-footed ferret are similarly cryo-sensitive, also undergoing extensive acrosomal derangement after cooling to 0°C (ultra-rapid) or 5°C (rapid). And like the cat, this injury is mitigated by slower cooling, for this species at 0.2°C/min (Santymire *et al.*, unpublished data).

Cryoprotectant

Studies of domesticated species have shown that the adverse effects of cryoprotectant type and timing of supplementation are revealed at thawing (rather than during cooling) (Gao *et al.*, 1997). Our studies of wildlife species have demonstrated taxonomic differences as well as (for some species) an impact of temperature at the time of glycerol addition to sperm suspensions. Some species are amazingly resistant to cryoprotectant exposure. For example, post-thaw motility and acrosomal integrity of giant panda spermatozoa are unaffected ($P > 0.05$) after pre-freeze glycerol exposure at 37°C versus 4°C. Likewise, for the cheetah, neither sperm motility nor acrosomal integrity is influenced immediately after thawing in aliquots where glycerol is added pre-freeze at 22°C versus 5°C (Crosier *et al.*, 2006). However, sperm motility *in vitro* eventually is poorer ($P < 0.05$) in samples exposed initially to glycerol at 22°C (~48%) compared to 5°C (~55%) and then diluted after thawing. This indicates that toxic effects of glycerol (no doubt due to induced osmotic stress) can occur eventually as a result of inappropriate cryoprotectant dilution. However, it is worth emphasising that post-thaw sperm motility in this study was sufficiently adequate (with almost half the sperm being motile) even when glycerol is added at 22°C. This is important because this investigation was conducted under rather primitive field conditions in Africa, where resources were so limited that it was impossible to add cryoprotectant at 5°C. Furthermore, the eventual post-thaw motility was consistent with sperm quality used to produce cheetah offspring by AI with fresh or thawed semen (Howard, Donoghue, Barone, Goodrowe, Blumer, Snodgrass, Starnes, Tucker, Bush and Wildt, 1992; Howard, Marker, Pukazhenthi, Roth, Swanson, Grisham and Wildt, 2002). Thus, this inherent ability of cheetah sperm to avoid lethal damage from glycerol addition at ambient temperature means that one can indeed initiate field programmes to bank spermatozoa from wild cheetahs captured in nature (Crosier *et al.*, 2006). The variability across species for this capacity is well illustrated by contrasting data for the black-footed ferret. For this animal, glycerol exposure at ambient temperature or 5°C causes a sharp, irreparable decline in sperm motility and acrosomal integrity (Santymire *et al.*, unpublished data). Thus, it clear that wildlife species vary markedly in susceptibility to glycerol toxicity.

Freezing approach

Although sperm cryopreservation technologies have evolved significantly for common animals and humans, these approaches generally are impractical for conditions under which most wildlife species are found (in captivity or nature). Among the limitations are sheer cost of equipment (zoos often having only small budgets for research), its difficulty in transport and a dependence on electricity. For these reasons, we have focused efforts on developing 'field-friendly' sperm cryopreservation technologies that require economical and easily available/transportable resources (Fig. 4). These techniques typically have involved freezing diluted sperm suspensions in standard, plastic, sealed straws on dry-ice, over liquid nitrogen vapour or by direct placement in a liquid nitrogen dry-shipper.

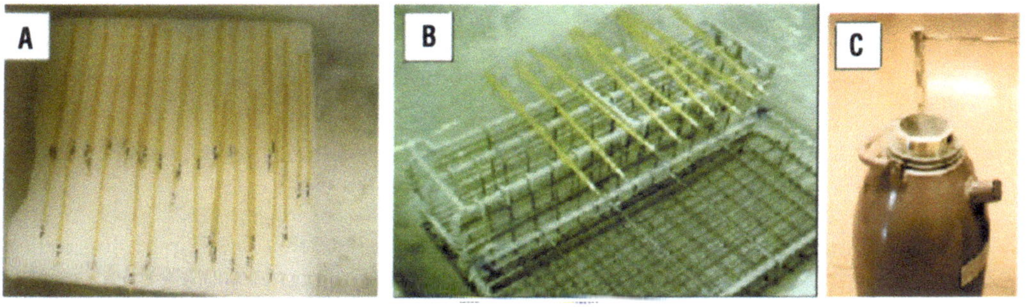

Figure 4. Sperm cryopreservation approaches used for wildlife species: (**A**) dry-ice; (**B**) two-step method; and (**C**) dry-shipper method for freezing spermatozoa.

For the dry-ice option, a block of this material is placed in an insulated container (with a lid), and the straws with the extended semen placed directly on the ice for 2 to 10 min before plunging into liquid nitrogen (Fig. 4A). Alternatively, the same type packaged samples are placed over liquid nitrogen vapour using a 'two-step' method: initially 7.62 cm on a metal rack above the liquid for 1 min, then 2.54 cm above the liquid for an additional 1 min before plunging (Fig.4B) (Spindler *et al.*, 2004; Crosier *et al.*, 2006). The dry-shipper method has been developed for situations where there is minimal access to liquid nitrogen (Roth *et al.*, 1999). Straws containing diluted semen are placed in a pre-cooled goblet attached to a cane and rapidly dropped into a charged (fully saturated with liquid nitrogen per manufacturer instructions) dry shipping container (Fig. 4C).

Influence and mitigation of osmotic or centrifugation stress

Spermatozoa from teratospermic cats are more susceptible to hypotonic stress during the initial phases of cooling than normospermic counterparts (Pukazhenthi, Noiles, Pelican, Donoghue, Wildt and Howard, 2000). However, spermatozoa from both populations are equally susceptible to extensive membrane damage when initially exposed to hypertonic conditions and rapidly returned to an isotonic environment (Pukazhenthi *et al.*, 2000, 2002a). Membrane damage caused by osmotic stress can be minimised by slowing the return of spermatozoa from a hyperosmotic to isosmotic state (Gao *et al.*, 1997). We have reported that step-wise removal of permeating cryoprotectants (in the absence of cryopreservation) causes minimal ($P > 0.05$) damage to membrane integrity in the domestic cat (normospermic and teratospermic), the normospermic serval (*Felis serval*) and the teratospermic clouded leopard (Pukazhenthi *et al.*, 2002a; Pukazhenthi, Crosier, Howard, Marker and Wildt, 2002b). However, at least in the clouded leopard, step-wise removal of cryoprotectants fails to improve acrosomal integrity in cryopreserved samples (Pukazhenthi *et al.*, 2006b). Recently, we assessed the beneficial effects of non-permeating cryoprotectants on cryo-survival of cheetah and clouded leopard spermatozoa. Interestingly, although raffinose contributes minimally to the osmolarity of the cryodiluent, it is unable to abolish dilution (osmotic stress) induced membrane disruption in these species (Pukazhenthi *et al.*, 2006b). However, we have determined that centrifuging thawed cheetah spermatozoa through an Accudenz discontinuous density gradient (Crosier, Pukazhenthi, Henghali, Howard, Marker and Wildt, 2004) enhances ($P < 0.05$) the proportions of intact acrosomes (control, ~37% versus Accudenz, ~42%) with minimal impact on sperm motility (Crosier *et al.*, 2004). Although the improvement in acrosomal integrity has only been moderate, these early findings are incentive for exploring other options for minimising the damage that arises from normal osmotic stresses associated with thawing and cryoprotectant removal.

Successes and challenges in sperm cryopreservation

The giant panda is a normospermic species that appears highly amenable to these low-tech cryopreservation approaches. A recent study reports that spermatozoa from this species readily survive freezing over liquid nitrogen vapour using the two-step method (Spindler *et al.*, 2004). Immediately after thawing, more than 55% of sperm are motile and >81% (Fig. 5) retain intact acrosomes. Although post-thaw sperm motility at 0 h is less than that found in fresh ejaculate (~71%), by 3 h of *in vitro* incubation motility scores are the same ($P>0.05$) for both groups. Most importantly, giant panda cubs have been produced after AI with sperm that have the same quality scores as produced in this freeze-thawing study (Howard, Huang, Wang, Li, Zhang, Hou, Zhang, Durrant, Spindler, Zhang, Zhang and Wildt, 2006a). Ungulate spermatozoa, including those of the Eld's deer, also appear to nicely withstand the vapour freezing approach (Fig. 5). For this normospermic species, 80% of spermatozoa have retained an intact acrosome after thawing and diluting out of the cryoprotectant. In contrast, sperm from the teratospermic cheetah exposed to the two-step method produces interesting results (Crosier *et al.*, 2006). Typically, raw cheetah electroejaculates contain spermatozoa of good motility (~80%) with mostly intact acrosomes (~88%). However, immediately upon thawing, sperm motility (~40%) and acrosomal integrity (~65%) decline ($P<0.05$) sharply. Although sperm motility improves slightly (~44%) upon dilution of cryoprotectant, acrosomal integrity declines ($P<0.05$) further to ~55% (Fig. 5). At the end of a 3 h incubation, only ~32% of spermatozoa are motile with 35% retaining an intact acrosome.

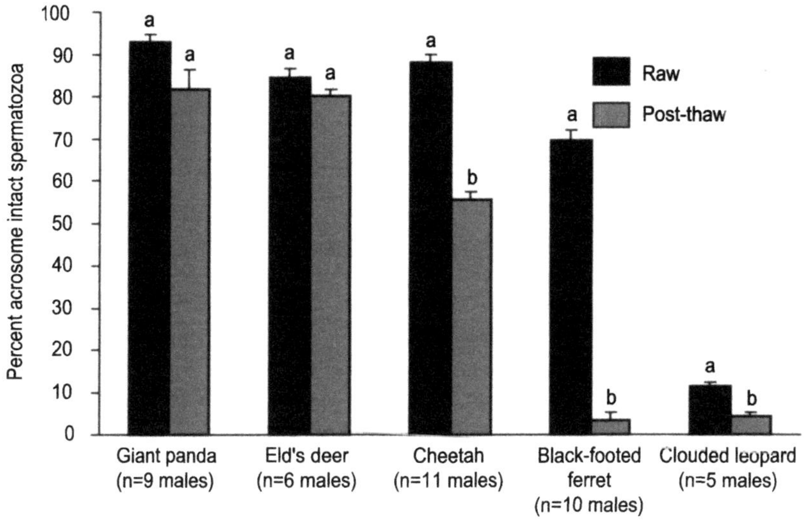

Figure 5. Effect of cryopreservation on sperm acrosomal integrity. Post-thaw values represent acrosomal integrity of post-thawed samples that were washed by centrifugation and resuspended in fresh medium. Values are mean ± S.E.M. Within a species, bars with different superscripts are different ($P < 0.05$). Data adapted from Spindler *et al.* (2004), Crosier *et al.* (2006), Pukazhenthi *et al.* (2006) and Santymire *et al.* (2006).

The teratospermic black-footed ferret is a particularly challenging species for sperm cryopreservation. Our studies have benefited by comparative investigations involving the domesticated ferret (*Mustela putorious furo*) that produces mostly structurally normal spermatozoa (Howard, Bush, Morton, Morton and Wildt, 1991). Preliminary sperm cryo-studies clearly reveal distinctive differences between these two taxonomically-related species (Santymire *et al.*, unpublished data). Sperm motility in fresh, raw ejaculate is ~75% in the domestic ferret and ~60% in the black-footed counterpart ($P<0.05$). Likewise, the proportion of spermatozoa exhibiting a normal apical ridge is higher ($P<0.05$) in the domestic (~95%) compared to the black-footed ferret (~70%). Interestingly, both species are markedly influenced by cryopreservation using the pellet technique or 2-step method, resulting in a significant ($P<0.05$) decline in sperm motility (domestic, ~37% versus black-footed ferret, ~15%) immediately upon thawing. The proportion of spermatozoa with normal apical ridge also is sharply reduced ($P<0.05$; domestic, ~23%; black-footed ferret, <3%). Thus, although sperm from both species suffer major damage from cryopreservation, the process of freeze-thawing virtually destroys all acrosomal integrity in the teratospermic black-footed ferret.

Of the species compared here, sperm from the clouded leopard are the most resistant to successful cryopreservation. This animal is an excellent example of the need to first understand 'normal' acrosomal traits prior to conducting experiments. This is because light microscopy fails to detect what is an unusually high incidence of acrosomal anomalies (>68%) that is only detectable after staining with dyes (such as Coomassie blue). We have assessed the influence of glycerol (4% final concentration) or raffinose (final concentration 4 or 8% with or without 4% glycerol) on the ability of clouded leopard sperm to withstand cryopreservation (Pukazhenthi *et al.*, 2006b). Additionally, we have examined post-thaw survival using a conventional approach (washed spermatozoa resuspended at room temperature in cryodiluent containing 4% glycerol and cooled at 5 °C for 30 min) *versus* a slow cooling method (Spindler *et al.*, 2004) followed by freezing using the 2-step method. Compared to the raw controls, sperm motility declined ($P < 0.05$) dramatically in all cryodiluents immediately upon thawing. Diluting out cryoprotectants immediately at thawing failed to improve sperm motility long-term, and, regardless of treatment, <26% of sperm retained motility by the end of a 4 h *in vitro* culture interval. In general, freezing and thawing clouded leopard sperm had a further detrimental impact on the percentages of intact acrosomes, with (on average) only 3% of cells retaining acrosomal viability after 4 h of *in vitro* culture. Thus, we encounter a species that is a poor candidate for sperm cryopreservation, largely because fresh ejaculates contains so many acrosomal abnormalities. It then may become necessary to attempt to recover viable sperm after cryopreservation at a much different scale. Rather than targeting the recovery of tens of percentages of motile sperm with intact acrosomes, it may require tinkering with methods to allow salvaging a few percentages of intact cells that have the potential of fertilising ova. Although such samples have negligible value in conventional AI protocols, there is an array of emerging technologies that may have application in propagating critically small populations of endangered species or genetically valuable males unable to breed naturally (e.g., intracytoplasmic sperm injection, among others) (Pukazhenthi, Comizzoli, Travis and Wildt, 2006a). Using sophisticated, *in vitro* injection or fertilisation methods may be all that is needed to artificially generate embryos.

Conclusions and perspectives

Studying sperm cryosensitivity in wild animals provides a renewed appreciation for the diversity of reproductive mechanisms in nature. Sperm cells, even among taxonomically-related species, differ amazingly in response to cryoprotectant, cooling, freezing and the many osmotic stresses associated with the entire cryo-process. Although this variation is due to as yet rather ambiguous factors, it is clear (and rather logical) that initial pre-freeze seminal quality plays a major role in ensuring whether a given spermatozoon survives cryopreservation. Hopefully, this review has been useful because it emphasises the significance of sperm morphology – the better the structural integrity, the more likely that cells show post-thaw evidence of survival. However, we believe one also can responsibly conclude that the incidence of acrosomal integrity in the original ejaculate is an excellent predictor of subsequent sperm vigour after thawing. Certainly, the usual variables (e.g., cryoprotectant type, cooling rate, post-thaw handling) can be manipulated in innumerable ways. Yet, if the original semen sample has adequate proportions of sperm with intact acrosomes, then there is a high likelihood of recovering viable sperm after cryopreservation.

This conclusion is important from a practical perspective as zoos with limited resources attempt to identify species that can benefit from using assisted breeding (including genome resource banks and AI) for small population management. Our experiences and scientific data would suggest that, in most cases, efforts for *applied studies* should be directed at species and populations that have extensive genetic diversity and/or those that consistently produce good-to-excellent quality sperm from male donors. At the same time and from a scholarly perspective, it indeed is fascinating that certain species, populations or even individuals produce a high incidence of teratospermia. These males provided unlimited opportunities for *basic research studies*, especially understanding the aetiology and significance of diverse sperm morphotypes as well as developing technical strategies to circumvent and 'rescue' what might be invaluable DNA from such cells. Therefore, we are advocates for both types of long-term investigations, although the number of candidate species that require study is overwhelming. For example, almost 98% of the world's mammalian species have gone unstudied in the reproductive sciences, with the percentages even higher for birds, reptiles, amphibians and fish (Wildt *et al.,* 2001). These species deserve our attention.

More specifically, we would argue that there is a need to strenuously characterise the seminal quality of all available species. In addition to identifying new traits and reproductive mechanisms, this information is critical-first-step data before proceeding with further basic or applied studies. For example, as aptly demonstrated by the clouded leopard example presented here, conventional assessments of sperm traits likely would have missed the usually low incidence of acrosomal integrity. Without the latter finding, we would have remained perplexed and frustrated about the observed failures of post-thaw sperm viability. Secondly, fundamental and comparative spermatology studies across diverse species will continue to answer important questions as well as provide direction for how reproductive science is relevant to the genetic management of endangered species. For example, from the review here, we now have three criteria for helping to prioritise species for cryo-studies depending on research and/or management needs (i.e. percentage of sperm motility, the incidence of teratospermia and the presence or absence of a normal complement of acrosomes). Such information is essential to designing studies or even determining if limited resources should be redirected at other species. Ideally, however, no species 'should be left behind'. Therefore, we will continue to encourage our fellow spermatologists to become directly involved, taking advantage of the vast opportunities afforded in the study of wildlife species.

Acknowledgements

This work was supported, in part, by grants from the Friends of the National Zoo, National Center for Research Resources of the National Institutes of Health, and the Morris Animal Foundation.

References

Ball, B.A. and Vo, A. (2001). Osmotic tolerance of equine spermatozoa and the effects of soluble cryoprotectants on equine sperm motility, viability, and mitochondrial membrane potential. *Journal of Andrology* **22:** 1061-1069

Crosier, A., Pukazhenthi, B.S., Henghali, J.N., Howard, J.G., Dickman, A.J., Marker, L. and Wildt, D.E. (2006). Cryopreservation of spermatozoa from wild-born Namibian cheetahs (*Acinonyx jubatus*) and influence of glycerol on cryosurvival. *Cryobiology* **52:** 169-181

Crosier, A.E., Pukazhenthi, B.S., Henghali, J.N., Howard, J.G., Marker, L.L. and Wildt, D.E. (2004). Improved acrosomal integrity of cryopreserved cheetah (*Acinonyx jubatus*) sperm after centrifugation through Accudenz. *Biology of Reproduction* **185:** Suppl abstract 404

Gao, D.Y., Mazur, P. and Critser, J.K. (1997). Fundamental cryobiology of mammalian spermatozoa. In: *Reproductive Tissue Banking*. Edited by Karow A.M. and Critser J.K., Academic Press, San Diego, pp. 263-328

Garde, J.J., Soler, A.J., Cassinello, J., Crespo, C., Malo, A.F., Espeso, G., Gomendio, M. and Roldan, E.R.S. (2003). Sperm cryopreservation in three species of endangered gazelles (*Gazella cuvieri, G. dama mhorr,* and *G. dorcas neglecta*). *Biology of Reproduction* **69:** 602-611

Harnal, V.K., Spindler, R., Monfort, S.L., Pukazhenthi, B., Bird, D.M. and Wildt, D.E. (2001). Sperm capacitation *in vitro* in the eld's deer. *Theriogenology* **56:** 399-413

Harris, M.A., Baumgard, L.H., Arns, M.J. and Webel, S.K. (2005). Stallion spermatozoa membrane phospholipid dynamics following dietary n-3 supplementation. *Animal Reproduction Science* **89:** 234-237

Herrick, J.R., Bartels, P. and Krisher, R.L. (2004). Postthaw evaluation of *in vitro* function of epididymal spermatozoa from four species of free-ranging African bovids. *Biology of Reproduction* **71:** 948-958

Holt, W.V. (2000a). Basis aspects of frozen storage of semen. *Animal Reproduction Science* **62:** 3-22

Holt, W.V. (2000b). Fundamental aspects of sperm cryobiology: the importance of species and individual differences. *Theriogenology* **53:** 47-58

Holt, W.V. (2001). Genetic resource banking and maintaining biodiversity. In: *Cryobanking the Genetic Resource*. Edited by Watson, P.F. and Holt, W.V., Taylor & Francis, London, pp. 11-19

Howard, J.G. (1993). Semen collection and analysis in carnivores. In: *Zoo and Wildlife Medicine: Current Therapy III*. Edited by Fowler ME, W.B. Saunders Co., Philadelphia, pp. 390-399

Howard, J.G. (1999). Assisted reproductive techniques in nondomestic carnivores. In: *Zoo and Wild Animal Medicine IV*. Edited by Fowler ME and Miller RE, W.B. Saunders Co. Philadelphia, pp. 449-457

Howard, J.G., Bush, M., Morton, C., Morton, F. and Wildt, D.E. (1991). Comparative semen cryopreservation in ferrets (*Mustela putorious furo*) and pregnancies after laparoscopic intrauterine insemination with frozen-thawed spermatozoa. *Journal of Reproduction and Fertility* 109-118

Howard, J.G., Bush, M. and Wildt, D.E. (1986). Semen collection, analysis and cryopreservation in nondomestic mammals. In: *Current Therapy in Theriogenology*. Edited by Morrow D, W.B. Saunders, Philadelphia, pp. 1047-1053

Howard, J.G., Donoghue, A.M., Barone, M.A., Goodrowe, K.L., Blumer, E., Snodgrass, K., Starnes, D., Tucker, M., Bush, M. and Wildt, D.E. (1992). Successful induction of ovarian activity and laparoscopic intrauterine artificial insemination in the cheetah (*Acinonyx jubatus*). *Journal of Zoo and Wildlife Medicine* **23:** 288-300

Howard, J.G., Donoghue, A.M., Johnston, L.A. and Wildt, D.E. (1993). Zona pellucida filtration of structurally abnormal spermatozoa and reduced fertilization in teratospermic cats. *Biology of Reproduction* **49:** 131-139

Howard, J.G., Huang, Y., Wang, P., Li, D., Zhang, G., Hou, R., Zhang, Z., Durrant, B.S., Spindler, R.E., Zhang, H., Zhang, A. and Wildt, D.E. (2006a). Role and efficiency of artificial insemination and genome resource banking. In: *Giant Panda: Biology, Veterinary Medicine and Management*. Edited by Wildt D.E., Zhang, A., Zhang, H., Janssen, D.L. and Ellis, S. Cambridge University Press, Cambridge, pp. 469-494

Howard, J.G., Marinari, P.E. and Wildt, D.E. (2003). Black-footed ferret: model for assisted reproductive technologies contributing to *in situ* conservation. In: *Reproductive Science and Integrated Conservation*. Edited by Holt WV, Pickard AR, Roger JC and Wildt DE, Cambridge University Press, Cambridge, pp. 249-266

Howard, J.G., Marker, L., Pukazhenthi, B.S., Roth, T.L., Swanson, W.F., Grisham, J. and Wildt, D.E. (2002). Genome resource banking and successful artificial insemination with cryopreserved sperm in the cheetah. *Proceedings of the 9th International Symposium on Spermatology*, Cape Town, South Africa, p. 70

Howard, J.G., Zhang, Z., Li, D., Huang, Y., Zhang, M., Hou, R., Ye, Z., Li, G., Zhang, J., Huang, S., Zhang, H., Zhang, J., Spindler, R.E. and Wildt, D.E. (2006b). Male reproductive biology in giant pandas. In: *Giant Panda: Biology, Veterinary Medicine and Management*. Edited by Wildt, D.E., Zhang, A., Zhang, H., Janssen, D.L. and Ellis, S. Cambridge University Press, Cambridge, pp.159-197

Mazur, P., Katkov, I.I., Katkova, N. and Critser, J.K. (2000). The enhancement of the ability of mouse sperm to survive freezing and thawing by the use of high concentrations of glycerol and the presence of an *Escherichia coli* membrane preparation (Oxyrase) to lower the oxygen concentration. *Cryobiology* **40:** 187-209

Monfort, S.L., Brown, J.L., Bush, M., Wood, T.C., Wemmer, C., Vargus, A., Williamson, L.R., Montali, R. and Wildt, D.E. (1993). Circannual interrelationships among reproductive hormones, gross morphometry, bahaviour, ejaculate characteristics and testicular histology in Eld's deer stags (*Cervus eldi thamin*). *Journal of Reroduction and Fertility* **98:** 471-480

Moore, A.I., Squires, E.L. and Graham, J.K. (2005). Adding cholesterol to the stallion sperm plasma membrane improves cryosurvival. *Cryobiology* **51:** 241-249

O'Brien, S.J., Wildt, D.E., Goldman, D., Merril, C.R. and Bush, M. (1983). The cheetah is depauperate in genetic variation. *Science* **221:** 459-462

Pena, A.I., Barrio, M., Becerra, J.J., Quintela, L.A. and Herradon, P.G. (2004). Zona pellucida binding ability and responsiveness to ionophore challenge of cryopreserved dog spermatozoa after different periods of capacitation *in vitro*. *Animal Reproduction Science* **84:** 193-210

Pukazhenthi, B., Comizzoli, P., Travis, A. and Wildt, D.E. (2006a). Applications of emerging technologies to the study and conservation of threatened and endangered species. *Reproduction Fertility Development* **18**: 77-90

Pukazhenthi, B.S., Crosier, A.E., Howard, J.G., Marker, L. and Wildt, D.E. (2002b). Cheetah (*Acinonyx jubatus*) sperm survive cryopreservation in the absence of permeating cryoprotectant. *Proceedings of the 9th International Symposium on Spermatology*, Cape Town, South Africa, p. 73

Pukazhenthi, B.S., Laroe, D., Crosier, A., Bush, L.M., Spindler, R., Pelican, K., Bush, M., Howard, J.G. and Wildt, D.E. (2006b). Challenges in cryopreservation of clouded leopard (*Neofelis nebulosa*) spermatozoa. *Theriogenology* (in press)

Pukazhenthi, B.S., Neubauer, K., Jewgenow, K., Howard, J. and Wildt, D.E. (2006c). The impact and potential etiology of teratospermia in the domestic cat and its wild relatives. *Theriogenology* **66:** 112-121

Pukazhenthi, B., Noiles, E., Pelican, K., Donoghue, A., Wildt, D. and Howard, J.G. (2000). Osmotic effects on feline spermatozoa from normospermic versus teratospermic donors. *Cryobiology* **40:** 139-150

Pukazhenthi, B., Pelican, K., Wildt, D. and Howard, J.G. (1999). Sensitivity of domestic cat (*Felis catus*) sperm from normospermic versus teratospermic donors to cold-induced acrosomal damage. *Biology of Reproduction* **61:** 135-141

Pukazhenthi, B., Spindler, R., Wildt, D., Bush, L.M. and Howard, J.G. (2002a). Osmotic properties of spermatozoa from felids producing different proportions of pleiomorphisms: influence of adding and removing cryoprotectant. *Cryobiology* **44:** 288-300

Pukazhenthi, B.S. and Wildt, D.E. (2004). Which reproductive technologies are most relevant to studying, managing and conserving wildlife? *Reproduction Fertility Development* **16:** 33-46

Pukazhenthi, B.S., Wildt, D.E. and Howard, J.G. (2001). The phenomenon and significance of teratospermia in felids. *Journal of Reproduction and Fertility Supplement* **57:** 423-433

Rana, K.J. (1995). Cryopreservation of fish spermatozoa. *Methods in Molecular Biology* **38:** 151-165

Roelke, M.E., Martenson, J.S. and O'Brien, S.J. (1993). The consequences of demographic reduction and genetic depletion in the endangered Florida panther. *Current Biology* **3:** 340-350

Rota, A., Rota, A., Martini, M., Milani, C. and Romagnoli, S. (2005). Evaluation of dog semen quality after slow (biological freezer) or rapid (nitrogen vapours) freezing. *Reproduction Nutrition Development* **45:** 29-37

Roth, T.L., Bush, L.M., Wildt, D.E. and Weiss, R.B. (1999) Scimitar-horned oryx (*Oryx dammah*) spermatozoa are functionally competent in a heterologous bovine *in vitro* fertilization system after cryopreservation on dry ice, in a dry shipper, or over liquid nitrogen vapor. *Biology of Reproduction* **60:** 493-498

Roth, T.L., Weiss, R.B., Buff, J.L., Buff, L.M., Wildt, D.E. and Bush, M. (1998) Heterologous in vitro fertilization and sperm capacitation in an endangered African antelope, the Scimitar-horned oryx (*Oryx dammah*). *Biology of Reproduction* **58:** 475-482

Santymire, R.M., Marinari, P.E., Kreeger, J.S., Wildt, D.E. and Howard, J.G. (2006). Sperm viability in the black-footed ferret (*Mustela nigripes*) is influenced by seminal and medium osmolality. *Cryobiology* **53:** 37-50

Songsasen, N. and Leibo, S.P. (1997). Cryopreservation of mouse spermatozoa II: Relationship between survival after cryopreservation and osmotic tolerance of spermatozoa from three strains of mice. *Cryobiology* **35:** 255-269

Songsasen, N., Yu, I., Murton, S., Paccamonti, D.L., Eilts, B.E., Godke, R.A. and Leibo, S.P. (2002). Osmotic sensitivity of canine spermatozoa. *Cryobiology* **44:** 79-90

Spindler, R.E., Huang, Y., Howard, J.G., Wang, P., Zhang, H., Zhang, G. and Wildt, D.E. (2004). Acrosomal integrity and capacitation are not influenced by sperm cryopreservation in the giant panda. *Reproduction* **127:** 547-556

Waterhouse, K.E., Hofmo, P.O., Tverdal, A. and Miller, R.R., Jr. (2006). Within and between breed differences in freezing tolerance and plasma membrane fatty acid composition of boar sperm. *Reproduction* **131:** 887-894

Wildt, D.E. (1992) Genetic resource banks for conserving wildlife species: justification, examples, and becoming organized on a global basis. *Animal Reproduction Science* **28:** 247-257

Wildt, D.E. (1994) Endangered species spermatozoa: diversity, research, and conservation. In: *Functions of Somatic Cells in the Testis*. Edited by Bartke A, Springer-Verlag, New York, pp. 1-24

Wildt, D.E., Brown, J.L. and Bush, M. (1993) Reproductive status of cheetahs (*Acinonyx jubatus*) in North American Zoos: the benefits of physiological surveys for strategic planning. *Zoo Biology* **12:** 45-80

Wildt, D.E., Bush, M., Goodrowe, K.L., Packer, C., Pusey, A.E., Brown, J.L., Joslin, P. and O'Brien, S.J. (1987). Reproductive and genetic consequences of founding isolated lion populations. *Nature* **329:** 328-331

Wildt, D.E., Bush, M., Howard, J.G., O'Brien, S.J., Meltzer, Dyk, A.V., Ebedes, H. and Brand, D.J. (1983). Unique seminal quality in the South African cheetah and a comparative evaluation in the domestic cat. *Biology of Reproduction* **29:** 1019-1025

Wildt, D., Ellis, E. and Howard, J. (2001). Linkage of reproductive sciences: from 'quick fix' to 'integrated' conservation. *Journal of Reproduction and Fertility Supplement* **57:** 295-307

Wildt, D., Pukazhenthi, B., Brown, J., Monfort, S., Howard, J. and Roth, T. (1995). Spermatology for understanding, managing and conserving rare species. *Reproduction Fertility and Development* **7:** 811-824

Wildt, D.E., Rall, W.F., Critser, J.K., Monfort, S.L. and Seal, U.S. (1997). Genome Resource Banks: Living collections for biodiversity conservation. *BioScience* **47:** 689-698

Wildt, D.E., Roth, T.L. (1997). Assisted reproduction for managing and conserving threatened felids. *International Zoo Yearbook* **35:** 164-172

Wolf, K.N., Wildt, D.E., Vargas, A., Marinari, P.E., Kreeger, J.S., Ottinger, M.A. and Howard, J.G.

(2000). Age-dependent changes in sperm production, semen quality, and testicular volume in the black-footed ferret (*Mustela nigripes*). *Biology of Reproduction* **63:** 179-187

Woods, E.J., Benson, J.D., Agca, Y. and Critser, J.K. (2004). Fundamental cryobiology of reproductive cells and tissues. *Cryobiology* **48:** 146-156

Spermatology. SRF Vol. 65. ERS Roldan and M Gomendio (eds) Nottingham University Press, Nottingham

Transplantation of germ cells and testis tissue for the study and preservation of fertility

Ina Dobrinski

Center for Animal Transgenesis and Germ Cell Research, 145 Myrin Bldg., New Bolton Center, School of Veterinary Medicine. University of Pennsylvania, 382 West Street Rd., Kennett Square, PA 19348, USA

The germ line stem cells in the mammalian testis form the basis of male fertility. When these stem cells are transplanted from the testis of a fertile donor animal to the testis of an infertile recipient they can establish donor-derived spermatogenesis in the recipient testis, and the resulting sperm can transmit the genotype of the donor to the offspring of the recipient. Germ cell transplantation provides a system to study the biology of germ line stem cells, to explore stem cell isolation and culture, to examine defects in spermatogenesis and to overcome male infertility. Although most widely studied in rodents, germ cell transplantation is now applied to larger mammals, including primates. Germ cell transplantation can preserve fertility from valuable animals and potentially restore fertility in patients that underwent cytotoxic treatments for cancer. In addition, genetic manipulation of germ cells prior to transplantation provides a new approach to germ line modification and transgenesis. As an alternative to transplantation of isolated germ cells to a recipient testis, ectopic grafting of testis tissue from diverse mammalian donor species, including primates, into a mouse host represents a model to study spermatogenesis, to investigate the effects of substances with the potential to enhance or suppress male fertility, and to produce fertile sperm from immature donors. Therefore, transplantation of germ cells or testis tissue are uniquely valuable approaches for the study, preservation and manipulation of male fertility in mammalian species.

Introduction

Mammalian spermatogenesis is an organised sequence of cell proliferation and differentiation resulting in the continuous production of virtually unlimited numbers of spermatozoa throughout the life of the male (Russell, Ettlin, Sinha Hikim and Clegg, 1990). At the foundation of this system is the male germ line stem cell which has the potential for both self-renewal and production of differentiated daughter cells which will ultimately form spermatozoa (Huckins, 1971; Clermont, 1972; Meistrich and van Beek, 1993). In 1994, Brinster and colleagues reported that transplantation of germ cells from fertile donor mice to the testes of infertile recipient mice resulted in donor-derived sperm production by the recipient animal (Brinster and Zimmermann, 1994) and sperm arising from transplanted donor germ cells are capable of

Corresponding author E-mail: dobrinsk@vet.upenn.edu

fertilisation *in vivo* and *in vitro* (Brinster and Avarbock, 1994; Goossens, Frederickx, De Block, Van Steirteghem and Tournaye, 2003; Honaramooz, Behboodi, Megee, Overton, Galantino-Homer, Echelard and Dobrinski, 2003b). Subsequently, germ cell transplantation was also successfully performed between rats (Jiang and Short, 1995; Ogawa, Dobrinski and Brinster, 1999a; Zhang, Renfree and Short, 2003). The germ cell transplantation system in rodents became instrumental to define the stem cell niche in the testis and to characterise putative spermatogonial stem cells (Parreira, Ogawa, Avarbock, Franca, Brinster and Russell, 1998; Nagano, Avarbock and Brinster, 1999; Ventela, Ohta, Parvinen and Nishimune, 2002). For example, transplantation experiments recently demonstrated that mouse primordial germ cells, the precursor cells of spermatogonial stem cells, already possess the potential to initiate spermatogenesis when transplanted into a post-natal testis (Chuma, Kanatsu-Shinohara, Inoue, Ogonuki, Miki, Toyokuni, Hosokawa, Nakatsuji, Ogura, and Shinohara, 2005). Serial transplantation experiments have been employed to study stem cell proliferation and homing efficiency after transplantation into recipient mouse testes (Ogawa, Ohmura, Yumura, Sawada and Kubota, 2003; Nagano, 2003; Kanatsu-Shinohara, Toyokuni, Morimoto, Matsui, Honjo and Shinohara, 2003c). A morphological characterization of stem cells and the development of the stem cell niche have also been presented for the rat (Orwig, Shinohara, Avarbock and Brinster, 2002a; Ryu, Orwig, Avarbock and Brinster, 2003; Ryu, Orwig, Kubota, Avarbock and Brinster, 2004).

Characterization of germ line stem cells

It has been estimated that there are only about 2 x 10^4 stem cells in 10^8 cells of a mouse testis (Meistrich and van Beek, 1993; Tegelenbosch and de Rooij, 1993). To study the biology of these stem cells, enriched populations of stem cells need to be obtained from testis cell preparations. Using colonization of recipient testes after transplantation as a functional endpoint it was shown that selection of cells that express α_6-and β_1-integrin but not the c-kit receptor, as well as collection of cells from experimentally induced cryptorchid testes resulted in a significant enrichment for spermatogonial stem cells (Shinohara, Avarbock and Brinster, 1999; Shinohara, Avarbock and Brinster, 2000a; Shinohara, Orwig, Avarbock and Brinster, 2000b). Subsequent studies further characterised the antigenic phenotype of putative spermatogonial stem cells showing that expression of Thy-1, CD9 or Egr3, surface proteins expressed also on embryonic or bone marrow stem cells, can serve as markers for enrichment of mouse germ line stem cell populations (Kubota, Avarbock and Brinster, 2003, 2004; Kanatsu-Shinohara, Toyokuni and Shinohara, 2004a; Hamra, Schultz, Chapman, Grellhesl, Cronkhite, Hammer and Garbers, 2004). Relatively little information is available for species other than rodents. Expression of protein gene product 9.5 (PGP 9.5) has been reported to be a robust marker for undifferentiated spermatogonia in cattle and pigs (Wrobel, 2000; Luo, Megee, Rathi, Dobrinski, 2006).

Germ cell culture

Nagano, Avarbock, Leonida, Brinster and Brinster (1998) showed first that stem cells could be maintained in culture for a long period of time. Since then, significant progress has been made in improving culture conditions for mouse germ cells (Kanatsu-Shinohara, Ogonuki, Inoue, Miki, Ogura, Toyokuni and Shinohara, 2003b; Jeong, McLean and Griswold, 2003). Co-culture with embryonic fibroblast or bone marrow stromal cells, but not Sertoli cell lines, and addition of several growth factors known to be beneficial for culture of other stem cell types or primordial germ cells, such as glial cell-line derived neurotrophic factor (GDNF), leukemia inhibitory factor (LIF), epidermal growth factor (EGF), and basic fibroblast growth factor (bFGF), successfully maintained mouse germline stem cells in culture for varying periods of time (Nagano *et al.*, 1998; Kanatsu-Shinohara *et al.*, 2003b; Nagano, Ryu, Brinster, Avarbock and Brinster, 2003b). Efficient long-term culture systems for mouse spermatogonial stem cells have been described (Kubota *et al.*, 2004; Kanatsu-Shinohara, Miki, Inoue, Ogonuki, Toyokuni, Ogura and Shinohara, 2005). Recently, the developmental plasticity of male germ cells was highlighted by reports that pluripotent stem cells could be isolated from cultures of neonatal and adult mouse testis cells (Kanatsu-Shinohara, Inoue, Lee, Yoshimoto, Ogonuki, Miki, Baba, Kato, Kazuki, Toyokuni, Toyoshima, Niwa, Oshimura, Hike, Nakahata, Ishino, Ogura and Shinohara, 2004b; Guan, Nayernia, Maier, Wagner, Dresel, Lee, Nolte, Wolf, Li, Engel and Hasenfuss, 2006) and by the demonstration that adult male germ line stem cells can even give rise to fertilisation competent eggs when transplanted into an undifferentiated gonad in fish (Okutsu, Suzuki, Takeuchi, Takeuchi and Yoshizaki, 2006).

The majority of the work to date was performed with primary cultures of putative male germ line stem cells. While progress in this area has been significant, availability of immortalised cell lines would provide tremendous potential for the study and manipulation male germ cells *in vitro*. To date, there are reports of immortalised germ cell lines from rat and mouse (van Pelt, Roepers-Gajadien, Gademan, Creemers, de Rooij and van Dissel-Emiliani, 2002; Feng, Chen, Dettin, Reijo Pera, Herr, Goldberg and Dym, 2002; Hofmann, Braydich-Stolle, Dettin, Johnson and Dym, 2005).

Application of germ cell transplantation to characterise phenotypes of infertility

Mammalian spermatogenesis requires close interaction between germ cells and the supporting testicular somatic cells as well as the endocrine and paracrine environment of the testis. When presented with a phenotype of male infertility with a defect in spermatogenesis, it is often difficult to determine whether the defect originates in the germ cells, or in the somatic components of the testis. Germ cell transplantation in rodents has become a standard approach to address this question. Initially, we demonstrated that transplantation of germ cells from infertile Steel mutant mice (Sertoli cell defect) to infertile W/W^v mutant mice (germ cell defect) could restore fertility (Ogawa, Dobrinski, Avarbock and Brinster, 2000). In this model donor cells collected from an infertile donor could restore fertility when transplanted into a permissive testis environment. Because donor cells were collected from adult infertile mice, it also demonstrated that stem cells can retain their ability to support spermatogenesis for a long period of time even in the absence of adequate environmental stimuli. This finding is particularly important for potential applications of this approach to adult patients presenting with infertility. In the above experiment, the molecular mechanisms underlying the defects in spermatogenesis were known. However, to characterise an unknown defect, standard experimental design now includes reciprocal transplantation of germ cells from affected donors to wild-type testes and vice versa. If donor cells from affected animals successfully colonise wild-type testes, the defect is localised to the somatic compartment of the testis. In that case donor cells from fertile wild-type animals will not be able to support spermatogenesis in affected recipient animals. If, on the other hand, affected donor cells fail to develop in wild-type testes and wild-type germ cells can colonise affected testes, the defect is localised to the germ cells. Accordingly, germ cell transplantation was successfully applied to characterise the role of the c-kit receptor and its ligand stem cell factor in regulation of germ cell proliferation (Ohta, Yomogida, Dohmae and Nishimune, 2000), to show that the defect associated with the juvenile spermatogonial depletion *(jsd)* mutation was inherent to the germ cells (Boettger-Tong, Higbstibm

Russell, Griswold and Bishop, 2000; Ohta, Yomogida, Tadokoro, Tohda, Dohmae and Nishimune, 2001), that germ cells do not require oestrogen receptors (Mahato, Goulding, Korach and Eddy, 2000) or androgen receptors (Johnston, Russell, Friel and Griswold, 2001) for development, and that germ cell differentiation is regulated by GDNF (Creemers, Meng, den Ouden, van Pelt, Izadyar, Santgoro, de Rooij and Sariola, 2002; Yomogida, Yagura, Tadokoro and Nishimune, 2003) and CREM function (Wistuba, Schlatt, Cantauw, von Schonfeldt, Nieschlag and Behr, 2002). Transplantation of wild-type germ cells into the testes of *Dazl* null mice established that the somatic compartment of the Dazl null testes remains functional (Rilianawati, Speed, Taggart and Cooke, 2003). More recently, *plzf* has been shown by reciprocal transplantation experiments using mutant and wild-type strains to be essential for stem cell self-renewal in the mouse (Buaas, Kirsch, Sharma, McLean, Morris, Griswold, de Rooij and Braun, 2004; Costoya, Hobbs, Barna, Cattoretti, Manova, Sukhwani, Orwig, Wolgemuth and Pandolfi, 2004). In the rat, germ cell transplantation experiments elucidated the defect underlying the as-mutation (Noguchi, Toyama, Yuasa, Kukuchi and Kaneko, 2002).

Transplantation of germ cells from non-rodent mammalian species

Cross-species transplantation

In 1996, production of rat sperm in mouse testes was achieved following cross-species spermatogonial transplantation from rats to mice (Clouthier, Avarbock, Maika, Hammer and Brinster, 1996) and transplantation was subsequently also successful from mice to rats (Ogawa *et al.*, 1999a; Zhang *et al.*, 2003). Transplantation of germ cells between rats and mice established that the cell cycle during spermatogenesis is controlled by the germ cell and not the Sertoli cell (Franca, Ogawa, Avarbock, Brinster and Russell, 1998). Hamster spermatogenesis could also occur in the mouse (Ogawa, Dobrinski, Avarbock and Brinster, 1999b); however, with increasing phylogenetic distance between donor and recipient species, complete spermatogenesis could no longer be achieved in the mouse testis. Transplantation of germ cells from non-rodent donors ranging from rabbits, dogs, pigs, bulls, horses and ultimately non-human primates and humans, resulted in colonization of the mouse testis but spermatogenesis became arrested at the stage of spermatogonial expansion (Dobrinski, Avarbock and Brinster, 1999, 2000; Nagano, McCarrey and Brinster, 2001a; Nagano, Patrizio and Brinster, 2002a). Therefore, the initial steps of germ cell recognition by the Sertoli cells, localization to the basement membrane and initiation of cell proliferation are conserved between evolutionary divergent species. However, when donor and recipient species are phylogeneticaly more distant than rodents, the recipient testicular environment appears to become unable to support spermatogenic differentiation and meiosis. This incompatibility of donor germ cells and recipient testicular environment could be overcome by co-transplantation of germ cells and Sertoli cells (Shinohara, Orwig, Avarbock and Brinster, 2003) or by testis tissue transplantation as described below (Honaramooz, Snedaker, Boiani, Schöler, Dobrinski and Schlatt, 2002a). Although cross species spermatogonial transplantation did not have the envisioned immediate practical application, it nonetheless provides a bioassay for stem cell potential of germ cells isolated from other species (Dobrinski *et al.*, 1999; 2000; Izadyar, Matthijs-Rijsenbilt, den Ouden, Creemers, Woelders, and de Rooij, 2002).

Germ cell transplantation in non-rodent species

Application of germ cell transplantation technology to species other than rodents has so far been reported in pigs, goats, cattle, monkeys and recently fish (Honaramooz Megee and Dobrinski, 2002b; Honaramooz, Behboodi, Hausler, Blash, Ayres, Azuma, Echelard and Dobrinski, 2003a; Honaramooz *et al.*, 2003b; Izadyar, den Ouden, Stout, Stout, Coret, Lankveld, Spoormakers, Colenbrander, Oldenbroek, Van der Ploeg, Woelders, Kal and de Rooij, 2003; Schlatt, Foppiani, Rolf, Weinbauer and Nieschlag, 2002a; Takeuchi, Yoshizaki and Tekeuchi, 2003; Yoshizaki, Tago, Takeuchi, Sawatari, Kobayashi and Takeuchi, 2005; Okutsu *et al.*, 2006). While direct injection of donor cells into rodent seminiferous tubules is possible via the efferent ducts, this is not feasible in larger species. Instead, a combination of ultrasound-guided cannulation of the centrally located rete testis with delivery of germ cells by gravity flow (Honaramooz *et al.*, 2002b, 2003b) was shown to be a successful approach in large animals. When germ cells from transgenic donor goats were transplanted into the testes of immunocompetent, prepubertal recipient animals, the recipients produced sperm carrying the donor haplotype and transmitted the donor genetic makeup to the offspring. This provided proof-of-principle that germ cell transplantation results in donor-derived sperm production and fertility in a non-rodent species (Honaramooz *et al.*, 2003a).

Choice and preparation of suitable recipients for germ cell transplantation

The success of germ cell transplantation requires the availability of a stem cell niche in the recipient testis. Using young mice and treating recipients with GnRH-agonists to suppress high intratesticular testosterone levels improved donor cell colonization (Shinohara, Orwig, Avarbock and Brinster, 2001, Ogawa, Dobrinski, Avarbock and Brinstger, 1998; Ogawa *et al.*, 1999a; Dobrinski, Ogawa, Avarbock and Brinster, 2001). The efficiency of colonization of seminiferous tubules by the transplanted germ cells can be improved if the recipient testes have little or no endogenous spermatogonia. Busulfan, a DNA alkylating agent that destroys proliferating cells, is frequently used in rodents to deplete recipient germ cells prior to germ cell transplantation. However, the sterilizing dose of busulfan is species- and strain-specific and treatment can be lethal due to severe bone marrow depression (Ogawa *et al.*, 1999a; Brinster, Ryu, Avarbock, Karagenc, Brinster and Orwig, 2003). As an alternate approach, *in utero* treatment of pigs during a period of high proliferation of fetal gonocytes resulted in depletion of germ cells with no observed adverse effects on the piglets or on sow health or fertility (Honaramooz, Behboodi, Hausler, Blash, Ayres, Azuma, Echelard and Dobrinski, 2005). As an alternative to cytotoxic treatment irradiation of the testes will also result in depletion of endogenous germ cells (Creemers *et al.*, 2002; Schlatt *et al.*, 2002a). Provided a suitable radiation source is available, local testicular irradiation appears to be the method of choice in species where the anatomical position of the testes facilitates shielding of the body from irradiation, thereby minimizing systemic effects. Fractionated testicular irradiation resulted in depletion of germ cells in goats and cats (Honaramooz *et al.*, 2005; Kim, Selvaraj, Dobrinski, Lee, McEntee and Travis, 2006) and reduction of germ cell numbers was also reported after testicular irradiation in rams (Oatley, Tibary, de Avila, Wheaton, McLean and Reeves, 2005). Interestingly, germ cell transplantation in rodents and perhaps also in older cattle requires that donor and recipients are closely related or that recipient animals are immuno-suppressed (Kanatsu-Shinohara, Ogonuki, Inoue, Ogura, Toyokuni, Honjo and Shinohara, 2003a; Zhang *et al.*, 2003; Izadyar *et al.*, 2003), whereas germ cell transplantation in pigs, goats and young bulls was successful also between unrelated individuals (Honaramooz *et al.*, 2002a, 2003a,b; Hill and Dobrinski, 2006). The testis is considered to be an immune privileged site, but it is unclear why transplantation between unrelated, immunocompetent animals is possible in domestic animal species but not in rodents. Nonetheless, this makes the technique infinitely more applicable in non-rodent species.

Transplantation of testis tissue

As described above, cross-species germ cell transplantation did not result in complete sperm production in species other than rodents, likely due to an incompatibility between donor germ cells and recipient testicular environment. Co-transplantation of the donor germ cells with their surrounding testicular tissue into a mouse host would preserve the testicular environment and still allow experimentation in a small rodent. We therefore developed ectopic grafting of testicular tissue under the back skin of immunodeficient mice as an alternate approach for the maintenance and propagation of male germ cells that can be more readily applied to different mammalian species (Honaramooz *et al.*, 2002a). Xenografting of testis tissue from newborn pigs and goats resulted in production of normal, functional sperm in a mouse host. This was the first report of complete, functional cross-species spermatogenesis from species other than rodents and it also represented the first time that sperm could be obtained from neonatal donors. Sperm recovered from allografts (mouse to mouse) and xenografts (monkey to mouse) supported embryo development when injected into oocytes (Schlatt, Honaramooz, Boiani, Schöler and Dobrinski, 2003; Honaramooz, Li, Penedo, Meyers and Dobrinski, 2004) and following embryo transfer, mouse sperm from allografts sired normal, fertile progeny (Schlatt *et al.*, 2003). The onset of spermatogenesis in xenografted pig testis tissue occurred slightly earlier than in the donor species (Honaramooz *et al.*, 2002a) and testicular maturation and sperm production in rhesus macaque testis tissue was significantly accelerated by exposure to the endocrine environment of the castrated adult mouse host (Honaramooz *et al.*, 2004). This shortened time to sperm production is due to accelerated maturation of the testicular somatic cells whereas the length of the spermatogenic cycle remains unchanged (Zeng, Avelar, Rathi, Franca and Dobrinski, 2006). Recently, we also established that human testis tissue can survive when grafted into host mice. While germ cell loss was noted in tissue with active spermatogenesis at the time of grafting, germ cells survived in tissue from patients with quiescent germinal epithelium (Schlatt, Honaramooz, Ehmcke, Goebell, Rubben, Dhir, Dobrinski and Patrizio, 2006; Geens, De Block, Goossens, Frederickx, Van Steirteghem and Tournaye, 2006).

Application of transplantation of germ cells and testis tissue to the preservation of fertility

Mouse spermatogonial stem cells can be cryopreserved for prolonged periods of time before transplantation, can still establish spermatogenesis in the recipient testis (Avarbock, Brinster and Brinster, 1996) and live offspring resulted from spermatozoa produced after transplantation of frozen-thawed mouse germ cells (Kanatsu-Shinohara, Ogonuki, Inoue, Ogura, Toyokuni and Shinohara, 2003d). Therefore, germ cell transplantation could serve to restore male fertility after an insult to the testis. In human cancer patients, germ cells could be frozen prior to irradiation or chemotherapy treatment for cancer as these treatments often lead to temporary or permanent destruction of spermatogenesis. Re-introduction of autologous germ cells could then restore fertility in the patient once the previous illness has been overcome. This approach was demonstrated in principle in the monkey (Schlatt *et al.*, 2002a) and its application in humans has been discussed (Radford, Shalet and Lieberman, 1999; Radford, 2003; Brook, Radford, Shalet, Joyce and Gosden, 2001), but so far no definitive data have been reported. Re-introduction of donor cells collected before treatment carries the risk that cancerous cells could also be re-introduced into the patient as demonstrated in leukemic rats (Jahnukainen, Hou, Petersen, Setchell and Soder, 2001). Cell sorting technology could potentially be employed to overcome this risk (Fujita, Ohta, Tsujimura, Takao, Miyagawa, Takada, Matsumiya, Wakayama and Okuyama, 2005). Germ cell transplantation has an advantage over the cryopreservation of sperm prior to treatment, in that it could be applied to pre-pubertal males where sperm cannot be obtained or to adult males rendered azoospermic or teratozoospermic by the disease. The technique is also of great interest in domestic or endangered animals for its potential to preserve genetic material from immature males that are lost before they reach puberty. Even when cryopreservation of sperm is possible from adult individuals, the preserved sperm will provide a finite resource. In contrast, cryopreserved germ cells will undergo genetic recombination after transplantation, thereby virtually preserving the entire genetic potential of the donor male. This will provide an invaluable advantage for application of germ cell preservation to the conservation of genetic diversity.

Another important application of germ cell transplantation especially in animals other than mice is transgenesis through the male germ line using transplantation of transfected germ cells. This approach has tremendous potential in species where embryonic stem cell technology is not available and current options to generate transgenic animals are inefficient. Transgenic mice and rats have been generated by viral transduction of germ cells prior to transplantation (Nagano, Shinohara, Avarbock and Brinster, 2000; Nagano, Brinster, Orwig, Ryu, Avarbock and Brinster, 2001b; Nagano, Watson, Ryu, Wolfe and Brinster, 2002b; Orwig, Avarbock and Brinster, 2002b; Hamra, Gatlin, Chapman, Grellhesl, Garcia, Hammer and Garbers, 2002). Recently, the advances in mouse germ cell culture even allowed for gene targeting and production of knock out mice using germ cell transplantation (Kanatsu-Shinohara, Ikawa, Takehashi, Ogonuki, Miki, Inoue, Kazuki, Lee, Toyokuni, Oshimura, Ogura and Shinohara, 2006).

While transplantation of germ cells has tremendous potential for the preservation of fertility, it is technically difficult and not easily adapted to diverse species. In contrast, ectopic grafting of testis tissue to a mouse host has been shown to be widely applicable to species ranging from mice to men. The accessibility of the tissue in the mouse host makes it possible to manipulate spermatogenesis and steroidogenesis in a controlled manner that is not feasible in the donor animal and certainly not in humans. This in turn will allow detailed analysis of the effects of toxins and compounds to enhance or suppress male fertility in an *in vivo* system without extensive experimentation in the target species. Ectopic testis tissue grafting also represents a new option for male germ line preservation. Similar to isolated germ cells, testicular tissue can be stored frozen prior to grafting while retaining its developmental potential (Honaramooz *et al.*, 2002a; Schlatt, Kim and Gosden, 2002b). As it provides a source of male gametes even from immature gonads, grafting of fresh or preserved testis tissue offers an invaluable tool for the conservation of fertility. Transplantation of gonadal tissue can provide genetic rescue for neonatal lethal phenotypes in rodents (Ohta and Wakayama, 2005) and potentially can preserve the genetic material of rare or endangered animals that are lost before puberty. Unlike autologous transplantation of isolated germ cells to restore fertility in a patient following cancer therapy, xenologous grafting and use of the resultant sperm for assisted fertilisation would eliminate the potential risk of tumor cell transmission. The potential applications of germ cell transplantation and tissue grafting to preserve male fertility are illustrated in Figure 1.

However, safety and ethical issues associated with the use of a xenogeneic system would have to be addressed before application to human patients is considered.

Conclusions

Transplantation of germ cells and testis tissue are complementary approaches to explore basic biological aspects of male germ line stem cells and testis function as well as potential causes of male infertility. Practical applications include the introduction of genetic modifications into the germline of domestic animals, and the preservation of fertility in rare and endangered animals and in patients undergoing potentially sterilizing treatments for cancer therapy. Transplantation

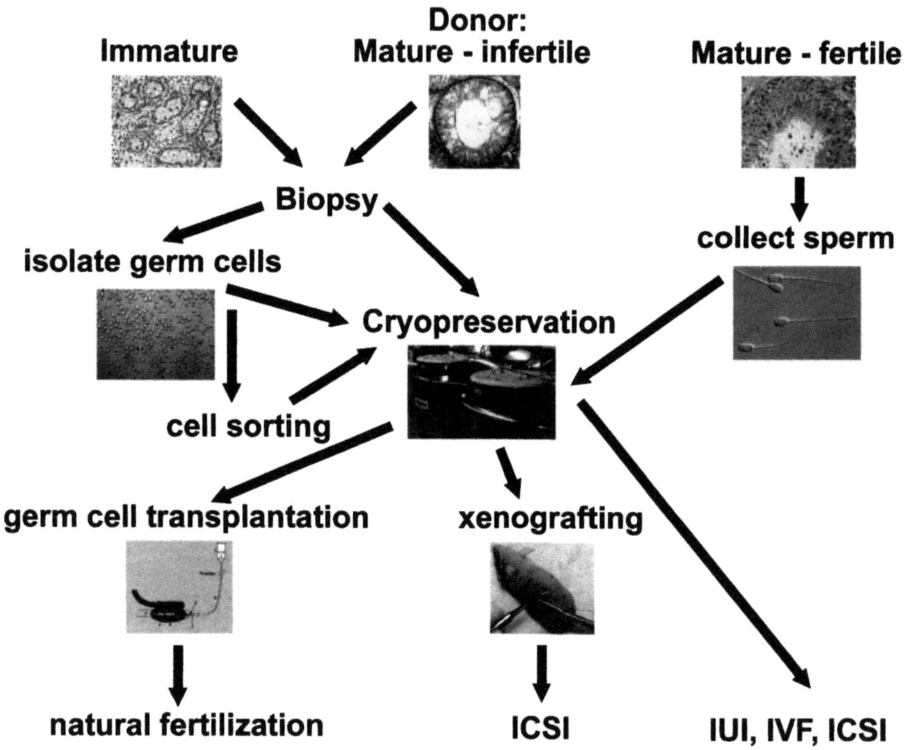

Figure 1. Schematic representation of strategies for preservation of fertility.

of germ cells or testis tissue will continue to significantly enhance our understanding of testis function and our ability to control and preserve male fertility.

Acknowledgements

Work from the author's laboratory presented here was supported by 5R01RR017359-04, 1R21RR017419-01 (NCRR/NIH), and 1R42 HD044780-01 (NICHD/NIH) and USDA/CSREES/NRICGP (2003-35203-13486).

References

Avarbock, M.R., Brinster, C.J. and Brinster, R.L. (1996). Reconstitution of spermatogenesis from frozen spermatogonial stem cells. *Nature Medicine* **2:** 693-696

Boettger-Tong H.L., Johnston, D.S., Russell, L.D., Griswold, M.D. and Bishop, C.E. (2000). Juvenile spermatogonial depletion (jsd) mutant seminiferous tubules are capable of supporting transplanted spermatogenesis. *Biology of Reproduction* **63:** 1185-1191

Brinster, R.L. and Avarbock, M.R. (1994). Germline transmission of donor haplotype following spermatogonial transplantation. *Proceedings of the National Academy of Sciences USA* **91:** 11303-11307

Brinster, R.L. and Zimmermann, J.W. (1994). Spermatogenesis following male germ-cell transplantation. *Proceedings of the National Academy of Sciences USA* **91:** 11298-11302

Brinster, C.J., Ryu, B.-Y., Avarbock, M.R., Karagenc, L., Brinster, R.L. and Orwig, K.E. (2003). Restoration of fertility by germ cell transplantation requires effective recipient preparation. *Biology of Reproduction*

69: 412-420

Brook, P.F., Radford, J.A., Shalet, S.M., Joyce, A.D. and Gosden, R.G. (2001). Isolation of germ cells from human testicular tissue for low temperature storage and autotransplantation. *Fertility and Sterility* **75:** 269-274

Buaas, F.W., Kirsh, A.L., Sharma, M., McLean, D.J., Morris, J.L., Griswold, M.D., de Rooij, D.G. and Braun, R.E. (2004). Plzf is required in adult male germ cells for stem cell self-renewal. *Nature Genetics* **36:** 647-652.

Chuma, S., Kanatsu-Shinohara, M., Inoue, K., Ogonuki, N., Miki, H., Toyokuni, S., Hosokawa, M., Nakatsuji, N., Ogura, A. and Shinohara, T. (2005). Spermatogenesis from epiblast and primordial germ cells following transplantation into postnatal mouse testis. *Development* **132:** 117-122

Clermont, Y. (1972). Kinetics of spermatogenesis in mammals: seminiferous epithelium cycle and spermatogonial renewal. *Physiological Reviews* **52:** 198-236

Clouthier, D.E., Avarbock, M.R., Maika, S.D., Hammer, R.E. and Brinster, R.L. (1996). Rat spermatogenesis in mouse testis. *Nature* **381:** 418-421

Costoya, J.A., Hobbs, R.M., Barna, M., Cattoretti, G., Manova, K., Sukhwani, M., Orwig, K.E., Wolgemuth, D.J. and Pandolfi, P.P. (2004). Essential role of Plzf in maintenance of spermatogonial stem cells. *Nature Genetics* **36:** 653-659

Creemers, L.B., Meng, X., den Ouden, K. van Pelt, M.M., Izadyar, F., Santoro, M., de Rooij, D.G. and Sariola, H. (2002). Transplantation of germ cells from glial cell line-derived neurotrophic factor-overexpressing mice to host testes depleted of endogenous spermatogenesis by fractionated irradiation. *Biology of Reproduction* **66:** 1579-1584

Dobrinski, I., Avarbock, M.R. and Brinster, R.L. (1999). Transplantation of germ cells from rabbits and dogs into mouse testes. *Biology of Reproduction* **61:** 1331-1339

Dobrinski I, Avarbock, M.R. and Brinster, R.L. (2000). Germ cell transplantation from large domestic animals into mouse testes. *Molecular Reproduction and Development* **57:** 270-279

Dobrinski, I., Ogawa, T., Avarbock, M.R. and Brinster, R.L. (2001). Effect of the GnRH-agonist leuprolide on colonization of recipient testes by donor spermatogonial stem cells after transplantation in mice. *Tissue and Cell* **33:** 200-207

Feng, L.-X., Chen, Y., Dettin, L., Reijo Pera, R.A., Herr, J.C., Goldberg, E. and Dym, M. (2002). Generation and in vitro differentiation of a spermatogonial cell line. *Science* **297:** 392-395

Franca, L.R., Ogawa, T., Avarbock, M.R., Brinster, R.L. and Russell, L.D. (1998). Germ cell genotype controls cell cycle during spermatogenesis in the rat. *Biology of Reproduction* **59:** 1371-1377

Fujita, K., Ohta, H., Tsujimura, A., Takao, T., Miyagawa, Y., Takada, S., Matsumiya, K., Wakayama, T. and Okuyama, A. (2005). Transplantation of spermatogonial stem cells isolated from leukemic mice restores fertility without inducing leukemia. *Journal of Clinical Investigation* **115:** 1855-1861

Geens, M., De Block, G., Goossens, E., Frederickx, V., Van Steirteghem, A. and Tournaye, H. (2006). Spermatogonial survival after grafting human testicular tissue to immunodeficient mice. *Human Reproduction* **21:** 390-396

Goossens, E., Frederickx, V., De Block, G., Van Steirteghem, A. and Tournaye, H. (2003). Reproductive capacity of sperm obtained after germ cell transplantation in a mouse model. *Human Reproduction* **18:** 1874-1880

Guan, K., Nayernia, K., Maier, L.S., Wagner, S., Dressel, R., Lee, J.H., Nolte, J., Wolf, F., Li, M., Engel, W. and Hasenfuss, G. (2006) Pluripotency of spermatogonial stem cells from adult mouse testis. *Nature* **440:** 1190-1203

Hamra, F.K., Gatlin, J., Chapman, K.M., Grellhesl, D.M., Garcia, J.V., Hammer, R.E. and Garbers, D.L. (2002). Production of transgenic rats by lentiviral transduction of male germ-line stem cells. *Proceedings of the National Academy of Sciences USA* **99:** 14931-14936

Hamra, F.K., Schultz, N., Chapman, K.M., Grellhesl, D.M., Cronkhite, J.T., Hammer, R.E. and Garbers, D.L. (2004). Defining the spermatogonial stem cell. *Developmental Biology* **269:** 393-410

Hill, J.R. and Dobrinski, I. (2006). Male germ cell transplantation in livestock. *Reproduction Fertility and Development* **18:** 13-18

Hofmann, M.-C., Braydich-Stolle, L., Dettin, L., Johnson, E. and Dym, M. (2005). Immortalization of mouse germ line stem cells. *Stem Cells* **23:** 200-210

Honaramooz, A., Behboodi, E., Blash, S., Megee, S.O. and Dobrinski, I. (2003a). Germ cell transplantation in goats. *Molecular Reproduction and Development* **64:** 422-428

Honaramooz, A., Behboodi,E., Hausler, C.L., Blash, S., Ayres, S.L., Azuma, C., Echelard, Y. and Dobrinski, I. (2005). Depletion of endogenous germ cells in male pigs and goats in preparation for germ cell transplantation. *Journal of Andrology* **26:** 698-705

Honaramooz, A., Behboodi, E., Megee, S.O., Overton, S.A., Galantino-Homer, H.L., Echelard, Y. and Dobrinski, I. (2003b). Fertility and germline transmission of donor haplotype following germ cell transplantation in immunocompetent goats. *Biology of Reproduction* **69:** 1260-1264

Honaramooz, A., Li, M.-W., Penedo, C.T., Meyers, S.A. and Dobrinski, I. (2004). Accelerated maturation of primate testis by xenografting into mice. *Biology of Reproduction* **70:** 1500-1503

Honaramooz, A., Megee, S.O. and Dobrinski, I. (2002b). Germ cell transplantation in pigs. *Biology of Reproduction* **60:** 21-28

Honaramooz, A., Snedaker, A., Boiani, M., Schöler, H.R., Dobrinski, I. and Schlatt, S. (2002a). Sperm from neonatal mammalian testes grafted in mice. *Nature* **418:** 778-781

Huckins, C. (1971). The spermatogonial stem cell population in adult rats. I. Their morphology, proliferation, and maturation. *Anatomical Record* **160:** 533-558

Izadyar, F., den Ouden, K., Stout, T.A.E., Stout, J., Coret, J., Lankveld, D.P.K., Spoormakers, T.J.P., Colenbrander, B., Oldenbroek, J.K., Van der Ploeg, K.D., Woelders, H., Kal, H.B. and de Rooij, D.G. (2003). Autologous and homologous transplantation of bovine spermatogonial stem cells. *Reproduction* **126:** 765-774

Izadyar, F., Matthijs-Rijsenbilt, J.J., den Ouden, K., Creemers, L.B., Woelders, H. and de Rooij, D.G. (2002). Development of a cryopreservation protocol for type A spermatogonia. *Journal of Andrology* **23:** 537-545

Jahnukainen, K., Hou, M., Petersen, C., Setchell, B. and Soder, O. (2001). Intratesticular transplantation of testicular cells from leukemic rats causes transmission of leukemia. *Cancer Research* **61:** 706-710

Jeong, D., McLean, D.J. and Griswold, M.D. (2003). Long-term culture and transplantation of murine testicular germ cells. *Journal of Andrology* **24:** 661-669

Jiang,F.X. and Short, R.V. (1995). Male germ cell transplantation in rats: apparent synchronization of spermatogenesis between host and donor seminiferous epithelia. *International Journal of Andrology* **18:** 326-330

Johnston, D.S., Russell, L.D., Friel, P.J. and Griswold, M.D. (2001). Murine germ cells do not require functional androgen receptors to complete spermatogenesis following spermatogonial stem cell transplantation. *Endocrinology* **142:** 2405-2408

Kanatsu-Shinohara, M., Ikawa, M., Takehashi, M., Ogonuki, N., Miki, H., Inoue, K., Kazuki, Y., Lee, J., Toyokuni, S., Oshimura, M., Ogura, A. and Shinohara, T. (2006). Production of knockout mice by random or targeted mutagenesis in spermatogonial stem cells. *Proceedings of the National Academy of Sciences USA* **103:** 8018-8023

Kanatsu-Shinohara, M., Inoue, K., Lee, J., Yoshimoto, M., Ogonuki, N., Miki, H., Baba, S., Kato, T., Kazuki, Y., Toyokuni, S., Toyoshima, M., Niwa, O., Oshimura, M., Heike, T., Nakahata, T., Ishino, F., Ogura, A. and Shinohara, T. (2004b). Generation of pluripotent stem cells from neonatal mouse testis. *Cell* **119:** 1001-1012

Kanatsu-Shinohara, M., Miki, H., Inoue, K., Ogonuki, N., Toyokuni, S., Ogura, A. and Shinohara, T. (2005). Long-term culture of mouse male germline stem cells under serum- or feeder-free conditions. *Biology of Reproduction* **72:** 985-991

Kanatsu-Shinohara, M., Ogonuki, N., Inoue, K., Miki, H., Ogura, A., Toyokuni, S. and Shinohara, T. (2003b). Long-term proliferation in culture and germline transmission of mouse male germline stem cells. *Biology of Reproduction* **69:** 612-616

Kanatsu-Shinohara, M., Ogonuki, N., Inoue, K., Ogura, A., Toyokuni, S., Honjo, T. and Shinohara, T. (2003a). Allogeneic offspring produced by male germ line stem cell transplantation into infertile mouse testis. *Biology of Reproduction* **68:** 167-173

Kanatsu-Shinohara, M., Ogonuki, N., Inoue, K., Ogura, A., Toyokuni, S. and Shinohara, T. (2003d). Restoration of fertility in infertile mice by transplantation of cryopreserved male germline stem cells. *Human Reproduction* **18:** 2660-2667

Kanatsu-Shinohara, M., Toyokuni, S., Morimoto, T., Matsui, S., Honjo, T. and Shinohara, T. (2003c) Functional assessment of self-renewal activity of male germline stem cells following cytotoxic damage and serial transplantation. *Biology of Reproduction* **68:** 1801-1807

Kanatsu-Shinohara, M., Toyokuni, S. and Shinohara, T. (2004a). CD9 is a surface marker on mouse and rat male germline stem cells. *Biology of Reproduction* **70:** 70-75

Kim, Y., Selvaraj, V., Dobrinski, I,, Lee, H., McEntee, M.C. and Travis, A.J. (2006). Recipient preparation and mixed germ cell isolation for spermatogonial stem cell transplantation in domestic cats. *Journal of Andrology* **27:** 248-256

Kubota, H., Avarbock, M.R. and Brinster, R.L. (2003). Spermatogonial stem cells share some, but not all, phenotypic and functional characteristics with other stem cells. *Proceedings of the National Academy of Sciences USA* **100:** 6487-6492

Kubota, H., Avarbock, M.R. and Brinster, R.L. (2004). Growth factors essential for self-renewal and expansion of mouse seprmatogonial stem cells. *Proceedings of the National Academy of Sciences USA* **101:** 16489-16494

Luo, J., Megee, S., Rathi, R. and Dobrinski, I. (2006). Protein gene product 9.5 is a spermatogonia-specific marker in the pig testis: Application to enrichment and culture of porcine spermatogonia. *Molecular Reproduction and Development* **73:** 1531-1540

Mahato, D., Goulding, E.H., Korach, K.S. and Eddy, E.M. (2000). Spermatogenic cells do not require estrogen receptor-α for development or function. *Endocrinology* **141:** 1273-1276

Meistrich, M.L. and van Beek, M.E.A.B. (1993). Spermatogonial stem cells. In: *Cell and Molecular Biology of Testis*. Edited by Desjardins C. and Ewing L.L., Oxford University Press, New York, pp. 266-295

Nagano, M. (2003). Homing efficiency and proliferation kinetics of male germ line stem cells following transplantation in mice. *Biology of Reproduction* **69:** 701-707

Nagano, M., Avarbock, M.R. and Brinster, R.L. (1999). Pattern and kinetics of mouse donor spermatogonial stem cell colonization in recipient testes. *Biology of Reproduction* **60:** 1429-1436

Nagano, M., Avarbock, M.R., Leonida, E.B., Brinster, C.J. and Brinster, R.L. (1998). Culture of mouse spermatogonial stem cells. *Tissue and Cell* **30:** 389-397

Nagano, M., Brinster, C.J., Orwig, K.E., Ryu, B.-Y., Avarbock, M.R. and Brinster, R.L. (2001b). Transgenic mice produced by retroviral transduction of male germ-line stem cells. *Proceedings of the National Academy of Sciences USA* **98:** 13090-13095

Nagano, M., McCarrey, J.R. and Brinster, R.L. (2001a).

Primate spermatogonial stem cells colonize mouse testes. *Biology of Reproduction* **64:** 1409-1416

Nagano, M., Patrizio, P. and Brinster, R.L. (2002a). Long-term survival of human spermatogonial stem cells in mouse testes. *Fertility and Sterility* **78:** 1225-1233

Nagano, M., Ryu, B.-Y., Brinster, C.J., Avarbock, M.R. and Brinster, R.L. (2003b). Maintenance of mouse male germ line stem cells in vitro. *Biology of Reproduction* **68:** 2207-2214

Nagano, M., Shinohara, T., Avarbock, M.R. and Brinster, R.L. (2000). Retrovirus-mediated gene delivery into male germ line stem cells. *FEBS Letters* **475:** 7-10

Nagano, M.,Watson, D.J., Ryu, B.-Y., Wolfe, J.H. and Brinster, R.L. (2002b). Lentiviral vector transduction of male germ line stem cells in mice. *FEBS Letters* **524:** 111-115

Noguchi, J., Toyama, Y., Yuasa, S., Kikuchi, K. and Kaneko, M. (2002). Hereditary defects in both germ cells and the blood-testis barrier system in as-mutant rats: Evidence from spermatogonial transplantation and tracer-permeability analysis. *Biology of Reproduction* **67:** 880-888

Oatley, J.M., Tibary, A., de Avila, D.M., Wheaton, J.E., McLean, D.J. and Reeves, J.J. (2005). Changes in spermatogenesis and endocrine function in the ram testis due to irradiation and active immunization against luteinizing hormone-releasing hormone. *Journal of Animal Science* **83:** 604-612

Ogawa, T., Dobrinski, I., Avarbock, M.R. and Brinster, R.L. (1998). Leuprolide, a gonadotropin-releasing hormone agonist, enhances colonization after spermatogonial transplantation into mouse testes. *Tissue and Cell* **30:** 583-588

Ogawa, T., Dobrinski, I., Avarbock, M.R. and Brinster, R. (1999b). Xenogeneic spermatogenesis following transplantation of hamster germ cells to mouse testes. *Biology of Reproduction* **60:** 515-521

Ogawa T, Dobrinski, I., Avarbock, M.R. and Brinster, R.L. (2000). Transplantation of male germ line stem cells restores fertility in infertile mice. *Nature Medicine* **6:** 29-34

Ogawa, T., Dobrinski, I. and Brinster, R.L. (1999a). Recipient preparation is critical for spermatogonial transplantation in the rat. *Tissue and Cell* **31:** 461-472

Ogawa, T., Ohmura, M., Yumura, Y., Sawada, H. and Kubota, Y. (2003). Expansion of murine spermatogonial stem cells through serial transplantation. *Biology of Reproduction* **68:** 316-322

Ohta, H., Yomogida, K., Dohmae, K. and Nishimune, Y. (2000). Regulation of proliferation and differentiation in spermatogonial stem cells: the role of c-kit and its ligand SCF. *Development* **127:** 2125-2131

Ohta, H., Yomogida, K., Tadokoro,Y., Tohda, A., Dohmae, K. and Nishimune, Y. (2001). Defect in germ cells, not supporting cells, is the cause of male infertility in the jsd mutant mouse: proliferation of spermatogonial stem cells without differentiation. *International Journal of Andrology* **24:** 15-23

Ohta, H. and Wakayama, T. (2005). Generation of normal progeny by intracytoplasmic sperm injection following grafting of testicular tissue from cloned mice that dies postnatally. *Biology of Reproduction* **73:** 390-395

Okutsu, T., Suzuki, K., Takeuchi, Y., Takeuchi, T. and Yoshizaki, G. (2006). Testicular germ cells can colonize sexually undifferentiated embryonic gonad and produce functional eggs in fish. *Proceedings of the National Academy of Sciences USA* **103:** 2725-2729

Orwig, K.E., Avarbock, M.R. and Brinster, R.L. (2002b). Retrovirus-mediated modification of male germline stem cells in rats. *Biology of Reproduction* **67:** 874-879

Orwig, K.E., Shinohara, T., Avarbock, M.R. and Brinster, R.L. (2002a). Functional analysis of stem cells in the adult rat testis. *Biology of Reproduction* **66:** 944-949

Parreira, G.C., Ogawa, T., Avarbock, M.R., Franca, L.R., Brinster, R.L. and Russell, L.D. (1998). Development of germ cell transplants in mice. *Biology of Reproduction* **59:** 1360-1370

Radford, J.A., Shalet, S.M. and Lieberman, B.A. (1999). Fertility after treatment for cancer. *British Medical Journal* **319:** 935-936

Radford, J. (2003). Restoration of fertility after treatment for cancer. *Hormone Research* **59:** (Supp1) 21-23

Rilianawati, Speed, R., Taggart, M. and Cooke, H.J. (2003). Spermatogenesis in testes of Dazl null mice after transplantation of wild-type germ cells. *Reproduction* **126:** 599-604

Russell, L.D., Ettlin, R.A., Sinha Hikim, A.P. and Clegg, E.D. (1990). Mammalian spermatogenesis. In: *Histological and Histopathological Evaluation of the Testis.* Edited by Russell, L.D., Ettlin, R.A., SinhaHikim, A.P. and Clegg, E.D., Cache River Press, Clearwater, pp 1-40

Ryu, B.-Y., Orwig, K.E., Avarbock, M.R. and Brinster, R.L. (2003). Stem cell and niche development in the postnatal rat testis. *Developmental Biology* **263:** 253-263

Ryu, B.-Y., Orwig, K.E., Kubota, H., Avarbock, M.R. and Brinster, R.L. (2004). Phenotypic and functional characteristics of spermatogonial stem cells in rats. *Developmental Biology* **274:** 158-170

Schlatt, S., Foppiani, L., Rolf, C., Weinbauer, G.F. and Nieschlag, E. (2002a). Germ cell transplantation into X-irradiated monkey testes. *Human Reproduction* **17:** 55-62

Schlatt, S., Honaramooz, A., Boiani, M., Schöler, H.R. and Dobrinski, I. (2003). Progeny from sperm obtained after ectopic grafting of neonatal mouse testes. *Biology of Reproduction* **68:** 331-335

Schlatt, S., Honaramooz, A., Ehmcke, J., Goebell, P.J., Rubben, H., Dhir, R., Dobrinski, I. and Patrizio, P. (2006). Limited survival of adult human testicular tissue as ectopic xenograft. *Human Reproduction* **21:** 384-389

Schlatt, S., Kim, S.S. and Gosden, R. (2002b). Spermatogenesis and steroidogenesis in mouse, hamster and monkey testicular tissue after cryopreservation and heterotopic grafting to castrated hosts. *Reproduction* **124:** 339-346

Shinohara, T., Avarbock, M.R. and Brinster, R.L. (1999)

β_1- and α_6-integrin are surface markers on mouse spermatogonial stem cells. *Proceedings of the National Academy of Sciences USA* **96:** 5504-5509

Shinohara, T., Avarbock, M.R. and Brinster, R.L. (2000a). Functional analysis of spermatogonial stem cells in Steel and cryptorchid infertile mouse models. *Developmental Biology* **220:** 401-411

Shinohara, T., Orwig, K.E., Avarbock, M.R. and Brinster, R.L. (2000b). Spermatogonial stem cell enrichment by multiparameter selection of mouse testis cells. *Proceedings of the National Academy of Sciences USA* **97:** 8346-8351

Shinohara, T., Orwig, K.E., Avarbock, M.R and Brinster, R.L. (2001). Remodeling of the postnatal mouse testis is accompanied by dramatic changes in stem cell number and niche accessibility. *Proceedings of the National Academy of Sciences USA* **98:** 6186-6191

Shinohara, T., Orwig, K.E., Avarbock, M.R. and Brinster, R.L. (2003). Restoration of spermatogenesis in infertile mice by Sertoli cell transplantation. *Biology of Reproduction* **68:** 1064-1071

Takeuchi, Y., Yoshizaki, G. and Takeuchi, T. (2003). Generation of live fry from intraperitoneally transplanted primordial germ cells in rainbow trout. *Biology of Reproduction* **69:** 1142-1149

Tegelenbosch, R.A.J. and de Rooij, D.G. (1993). A quantitative study of spermatogonial multiplication and stem cell renewal in the C3H/101 F_1 hybrid mouse. *Mutation Research* **290:** 193-200

van Pelt, M.M., Roepers-Gajadien, H.L., Gademan, I.S., Creemers, L.B., de Rooij, D.G. and van Dissel-Emiliani, F.M.F. (2002). Establishment of cell lines with rat spermatogonial stem cell characteristics. *Endocrinology* **143:** 1845-1850

Ventela, S., Ohta, H., Parvinen, M. and Nishimune, Y. (2002). Development of stages of the cycle in mouse seminiferous epithelium after transplantation of green fluorescent protein-labeled spermatogonial stem cells. *Biology of Reproduction* **66:** 1422-1429

Wistuba, J., Schlatt, S., Cantauw, C., von Schonfeldt, V., Nieschlag, E. and Behr, R. (2002).Transplantation of wild-type spermatogonia leads to complete spermatogenesis in testes of cyclic 3',5'-adenosine monophosphate response element modulator-deficient mice. *Biology of Reproduction* **67:** 1052-1057

Wrobel, K.-H. (2000). Preseprmatogenesis and spermatogoniogenesis in the bovine testis. *Anatomical Embryology* **202:** 209-222

Yomogida, K., Yagura, Y., Tadokoro, Y. and Nishimune, Y. (2003). Dramatic expansion of germinal stem cells by ectopically expressed human glial cell line-derived neurotrophic factor in mouse Sertoli cells. *Biology of Reproduction* **69:** 1303-1307

Yoshizaki, G., Tago, Y., Takeuchi, Y., Sawatari, E., Kobayashi, T. and Takeuchi, T. (2005). Green fluorescent protein labeling of primordial germ cells using a nontransgenic method and its application for germ cell transplantation in salmonidae. *Biology of Reproduction* **73:** 88-93

Zeng, W., Avelar, G.F. Rathi, R., Franca, L.R. and Dobrinski, I. (2006). The length of the spermatogenic cycle is conserved in porcine and ovine testis xenografts. *Journal of Andrology* **27:** 527-533

Zhang, Z., Renfree, M.B. and Short, R.V. (2003). Successful intra- and interspecific male germ cell transplantation in the rat. *Biology of Reproduction* **68:** 961-967

Spermatology. SRF Vol. 65. ERS Roldan and M Gomendio (eds) Nottingham University Press, Nottingham

Sperm-mediated gene transfer: mechanisms and implications

Corrado Spadafora

Istituto Superiore di Sanità, Viale Regina Elena 299, 00161 Rome, Italy

The finding that mature spermatozoa act as vectors of genetic material, not only for their own genome, but also for exogenous DNA molecules, has suggested a strategy for animal transgenesis alternative to DNA microinjection. Exploiting this ability, protocols for sperm-mediated gene transfer have been developed in a variety of animal species with extremely variable results. In the last years, we have devoted much effort to clarify the mechanism of sperm/DNA interaction, with the dual aim to improve the protocols for transgenesis and to expand our understanding of the molecular mechanism(s) underlying sperm-mediated gene transfer. The results from these studies challenge the traditional view that spermatozoa are metabolically inert cells, showing that the binding of exogenous DNA to sperm cells, and its nuclear internalization, are not casual events but steps of a regulated process mediated by specific factors. The binding of exogenous sequences triggers enzymatic functions that are otherwise repressed in mature spermatozoa under normal conditions. Among those, we have identified an endogenous retrotransposon-encoded reverse transcriptase activity able to reverse transcribe exogenous RNA molecules in cDNA copies that can be delivered to embryos at fertilisation. The reverse-transcribed molecules have peculiar features: they are propagated in tissues as low copy extrachromosomal structures, mosaic distributed, transcriptionally competent, able to inducing phenotypic variations in positive tissues, and sexually transferred from one generation to the next in a non-mendelian fashion. Taken as a whole, these results suggest that sperm-mediated gene transfer is a retrotransposon-mediated phenomenon: thus, the sperm endogenous retrotransposon machinery can be regarded as a novel source of genetic information.

Background and state of the art

It is now a well established notion that mature spermatozoa of virtually all animal species have the spontaneous ability to bind exogenous DNA molecules and deliver them to oocytes at fertilisation. The evidence that spermatozoa can behave as vectors of genetic material not only contained in their own genome, but also derived from foreign nucleic acids sequences, suggested long ago the idea that sperm cells may play a role in animal transgenesis. The first protocol for Sperm-Mediated Gene Transfer (SMGT) in mice was developed in our laboratory

Corresponding author E-mail: cspadaf@tin.it

(Lavitrano, Camaioni, Fazio, Dolci, Farace, and Spadafora, 1989), and was subsequently extended and adapted, by our and other laboratories to numerous animal species, as schematically represented in Fig.1 (for a recent review see Smith and Spadafora, 2005). Since its very first appearance, however, SMGT has been characterised by a great deal of controversy, because various laboratories reported their failure to reproduce our results (Brinster, Sandgren, Behringer, and Palmiter, 1989; Smith, 1999): the reason for the discrepancy between our and others results remained obscure at that time. In spite of that initial controversy, however, more than seventy reports have appeared in peer-reviewed journals since then, which describe the application of SMGT protocols to a variety of animal species. These reports show an extremely broad spectrum of results. First, it emerges that the frequency of positive individuals is extremely variable among experiments, ranging in some cases from 0% to 100% (Maione, Lavitrano, Spadafora and Kiessling, 1998). Second, the specific organization of the transgene sequences is also extremely variable: they can be found integrated in the host genome or maintained as extrachromosomal structures, transcriptionally competent or silent, transmittable to the next generation or altogether absent from the germ line; moreover, in some cases, transgenes are only detected during embryogenesis but not in adult animals while in others they are stably present in born animals (for a review see Smith and Spadafora, 2005). A turning point occurred in the SMGT story when a novel protocol, termed TransgenICSI, was successfully applied in mice (Perry, Wakayama, Kishikawa, Kasai, Okabe, Toyoda and Yanagimachi, 1999). In this procedure, demembranated spermatozoa, after a short exposure to exogenous DNA, are microinjected in oocytes following the well-known ICSI (Intra Cytoplasmic Sperm Injection) procedure and embryos are subsequently implanted into foster mothers. TransgenICSI improved not only the efficiency but, most importantly, the reproducibility of the sperm-mediated transgenesis procedure (Kato, Ishikawa, Kaneko, Yagi, Hochi and Hirabashi, 2004; Naruse, Ishikawa, Kawano, Ueda, Kurome, Miyazaki, Endo, Sawasaki, Nagashima and Makuuchi, 2005; Hirabayashi, Kato, Ishikawa, Kaneko, Yagi and Hochi, 2005; Kurome, Ueda, Tomii, Naruse and Nagashima, 2006), with the transgene being clearly integrated in the host genome, even when exogenous DNA fragments as large as 200 kb were used (Osada, Toyoda, Moisyadi, Akutsu, Hattori, Sakaki and Yanagimachi, 2005). These data thus suggest that when foreign DNA bypasses the sperm plasma membrane following the ICSI procedure, genuinely transgenic animals are generated with a high reproducibility and efficiency; in contrast, the binding of the DNA with the sperm membrane on intact cells triggers alternative pathways that lead mainly to the generation of DNA extra-chromosomal structures. To my knowledge, only one single group has reported the generation of transgenic pigs with high efficiency after direct sperm/DNA interaction (Lavitrano, Bacci, Forni, Lazzereschi, Di Stefano, Fioretti, Giancotti, Marfe, Pucci, Renzi *et al.*, 2002; Webster, Forni, Bacci, Giovannoni, Razzini, Fantinati, Zannoni, Fusetti, Dalpra, Bianco *et al.*, 2005), although in that work the claimed integration events are not unambiguously demonstrated or characterised by cloning and sequencing. By and large, SMGT studies have yielded relevant outcomes to both applied and basic research for at least three reasons: (i) a reliable and reproducible variant protocol for SMGT has been established, based on the exposure of demembranated spermatozoa to DNA rather than on direct interaction between DNA and intact spermatozoa; (ii) the central role of sperm cells in the process of transgenesis has been confirmed, in that it has been clarified that transgenic animals can be generated when the complex demembranated sperm cells/ exogenous DNA is microinjected in oocytes, but not when sperm and DNA are sequentially microinjected in oocytes as individual components; and (iii) the biological relevance of specific DNA-binding substrates located on the sperm plasma membrane has been elicited. In the latter respect, indeed, it is important to point out that the final fate of the exogenous sequences depends entirely on their interaction with the sperm membrane: this conveys the exogenous sequences towards alternative fates, either as transgenes integrated in the host genome - if the membrane is bypassed - or as autonomously replicating extrachromosomal sequences - if the sperm membrane is encountered. These results therefore highlight the notion that the very first step of the SMGT process, i.e. the sperm/DNA interaction, is most crucial to determine the fate of the foreign DNA.

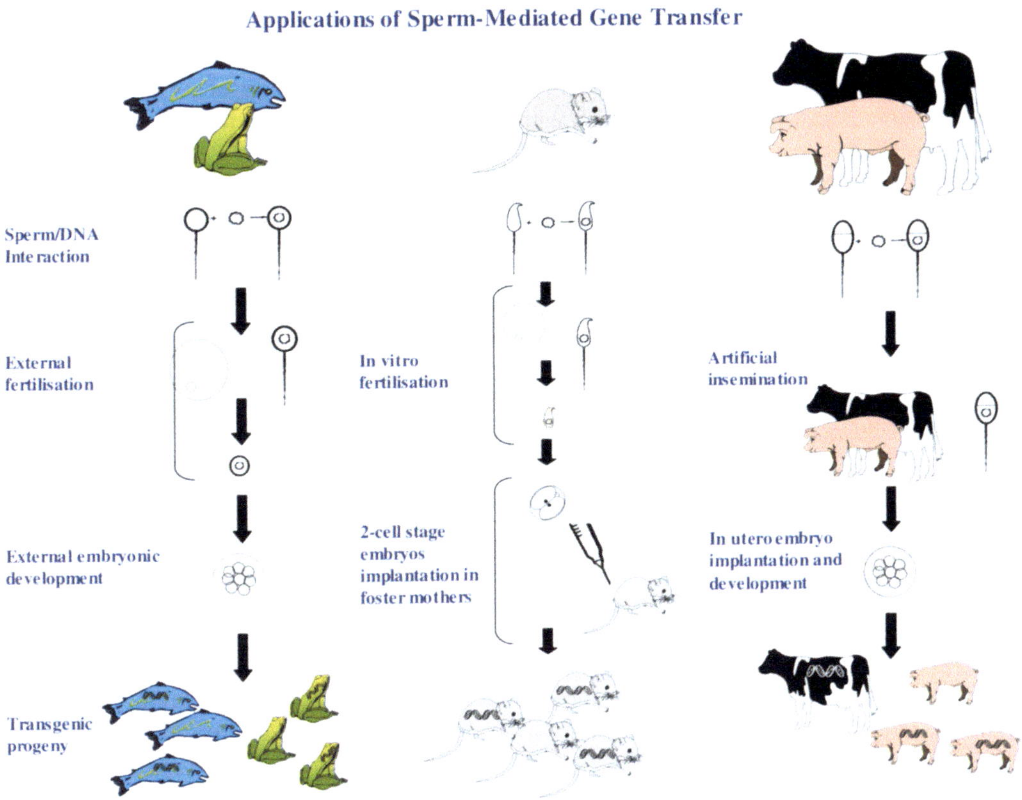

Figure 1: Routes to the production of genetically modified animals via SMGT.

The binding between sperm cells and exogenous DNA

Over the past seventeen years, we have concentrated our efforts to dissect and clarify the molecular steps of the earliest and most fundamental event of the whole SMGT process, i.e. the interaction between spermatozoa and exogenous DNA. We have found that this binding is not a random event but is a well-regulated process mediated by specific factors present in the sperm cells. Briefly, we have found that the binding of exogenous DNA molecules to sperm cells is mediated by a specific class of DNA-binding proteins (Zani, Lavitrano, French, Lulli, Maione, Sperandio and Spadafora, 1995), and is modulated by a conserved glycoprotein abundant in the seminal fluid of mammals, called IF1 (inhibitory factor 1) (Lavitrano, French, Zani, Frati and Spadafora, 1992; Zani *et al.*, 1995). A portion of sperm-bound DNA is further internalised within sperm nuclei (Francolini, Lavitrano, Lora-Lamia, French, Frati, Cotelli and Spadafora, 1993) in a process mediated by CD4 molecules (Lavitrano, Maione, Forte, Francolini, Sperandio, Testi and Spadafora, 1997). The internalised exogenous DNA reaches the nuclear scaffold, in close contact with the chromosomal DNA and is heavily rearranged under the action of sperm endogenous nucleases, whose activity is triggered by the exogenous DNA itself (Maione, Pittoggi, Achene, Lorenzini and Spadafora, 1997). The exogenous sequences eventually undergo recombination events that cause their integration in the sperm genome (Zoraqi and Spadafora, 1997); evidence in support of such integration events was obtained from the molecular analysis of a genomic library constructed with the DNA extracted from sperm cells after incubation with pSV2CAT plasmid. Two clones were selected and fully characterised, which showed that integration occurs in one, or few, preferential sites of the sperm chromatin (Zoraqi and Spadafora, 1997). From the cloning data, integration emerges as a very unfrequent event, likely to occur only in a small subpopulation of spermatozoa. Consistent with this observation is the finding that the binding of exogenous DNA to sperm cells triggers the activation of endogenous nucleases as a DNA dose-dependent response (Maione *et al.*, 1997); the activated nucleases can locally degrade a small nucleohistone component that is present within the sperm nucleus (Pittoggi, Renzi, Zaccagnini, Cimini, Degrassi, Giordano, Magnano, Lorenzini, Lavia and Spadafora, 1999). This observation thus suggests that chromatin of sperm cells is not a uniform compact structure, but harbours discrete sites of hypersensitivity within the overall tight packaging of the chromatin bulk: these nuclease-sensitive domains may be the preferential integration sites for foreign DNA sequences (Pittoggi, Zaccagnini, Giordano, Magnano, Lorenzini and Spadafora, 2000). These results have been assembled in an integrated model (Spadafora, 1998), according to which the basic mechanism of sperm/DNA interaction and nuclear internalization can be rationalised.

Reverse transcriptase activity in mature spermatozoa

The sperm nucleohistone chromatin component was subsequently characterised thoroughly and found to be highly enriched in retrotransposon-derived sequences, most abundantly in the reverse transcriptase-encoding ORF2 of LINE-1 retrotransposable elements, and organised in a structural conformation that closely resembles that of transcriptionally competent chromatin of somatic nuclei (Pittoggi *et al.*, 1999). That finding prompted us to investigate whether an endogenous reverse transcriptase (RT) activity is present in mature spermatozoa. To this end, we reasoned that the most direct evidence for a sperm RT activity might come from testing the ability of sperm cells to reverse transcribe an exogenous RNA into cDNA copies. In a first set of experiments, mouse epididymal spermatozoa were incubated with human poliovirus RNA molecules. This virus was chosen in order to avoid any potential artifact due to DNA contamination, since the poliovirus chromosomal RNA replicates through a RNA (–) strand with no DNA replicative intermediate, differently from retroviruses. We found that the exogenous RNA was indeed taken up by sperm cells, internalised in nuclei and reverse transcribed in cDNA fragments that were further transferred to two-cell stage embryos in *in vitro* fertilisation (IVF) essays (Giordano, Magnano, Zaccagnini, Pittoggi, Moscufo, Lorenzini and Spadafora, 2000). RT molecules were localised by immunoelectronmicroscopy on the sperm nuclear scaffold. To determine whether reverse transcription generates inert products or transcriptionally competent sequences, we incubated spermatozoa with a RNA population transcribed from a ß-gal-containing DNA construct. RNA-loaded spermatozoa were used in IVF essays and embryos were implanted into foster mothers to generate F0 founders, which in turn generated an F1 progeny by natural breeding. Randomly selected individuals from both F0 and F1 populations were screened for the presence of ß-gal-containing DNA sequences by both Southern blot and direct PCR analysis of genomic DNA samples from various organs. The results showed that ß-gal cDNAs newly synthesised in sperm cells had the following features: (i) low abundance in tissues (< 1copy/genome); (ii) mosaic distributed in adult animals; (iii) sexually transmitted from F0 to the F1 progeny and (iv) again underrepresented

(<1 copy/genome) and mosaic distributed in the tissues of F1 individuals. Due to their low abundance, the reverse-transcribed sequences are below the level of resolution of Southern blot analysis and are only detectable by PCR amplification (Sciamanna, Barberi, Martire, Pittoggi, Beraldi, Giordano, Magnano, Hogdson and Spadafora, 2003). A considerable amount of experimental effort was invested to clarify if these sequences were integrated in the host genome or remained as extrachromosomal structures. All our pursued approaches, including the construction and screening of two genomic libraries and ligation-mediated PCR genomic analysis, failed to demonstrate a covalent link between the mouse genome and the exogenous cDNA sequences. Together, these features suggest that the reverse transcribed sequences are not integrated in the host genome, but are rather maintained and propagated in adult animals as extrachromosomal structures. In addition, ß-gal protein expression was detected in various tissues, thus confirming that the reverse transcribed cDNAs are indeed transcriptionally competent sequences, able to induce phenotypic variations in both F0 and F1 individuals.

As a whole, these results provided the first evidence that "transgenic" mice can be generated using exogenous RNA rather than DNA molecules and, most importantly, suggested that new genetic information, besides that present in the genome, can be generated through a RT-mediated mechanism active in mature spermatozoa and propagated in born animals and further to their progeny. We have called this phenomenon Sperm-Mediated "Reverse" Gene Transfer.

SMGT as a retrotransposon-dependent phenomenon: proposed mechanism

Figure 2 schematically illustrates the RT-dependent mechanism responsible for the genesis of reverse transcribed new genetic information. Sperm-bound exogenous RNA molecules are internalised in the nucleus, in close contact with the nuclear matrix, and therein are taken up by the RT-dependent retrotransposon machinery and are reverse-transcribed in cDNA copies by the endogenous RT activity. At this point, the newly synthesised cDNAs have two possible fates: (i) a proportion of the cDNAs, probably the majority of them, remain as extrachromosomal structures; (ii) a minor proportion of cDNA is integrated in one, or few specific sites present in the nucleohistone chromatin fraction of the sperm which, in our model, connects two adjacent nucleoprotamine domains. The integrated cDNA copies behave as actively transcribing proviruses whose RNA is in turn reverse transcribed in cDNA copies, thus further amplifying the generation of extrachromosomal structures. The extrachromosomal structures are transferred from spermatozoa to embryos at fertilisation and are then destined to be further propagated throughout embryogenesis and in born animals. This model is based on the results briefly reported above, but some crucial aspects are still unclear and several questions remain unanswered at this time. Specifically, it is not clear why integration events only occur in a minor population of spermatozoa: this is in sharp contrast with the widespread ability to bind DNA, which is instead a general feature of virtually all spermatozoa. A second aspect requiring more investigation is the structure of the extrachromosomal sequences and the mechanism through which they propagate in animals. How these structures replicate, and whether they propagate as naked DNA or assemble in some kind of nucleoprotein complexes, has remained elusive thus far.

Risks for human assisted reproduction

Spermatozoa depleted of seminal fluid become extremely reactive towards foreign nucleic acids and behave as vectors of both exogenous and newly reverse-transcribed DNA molecules,

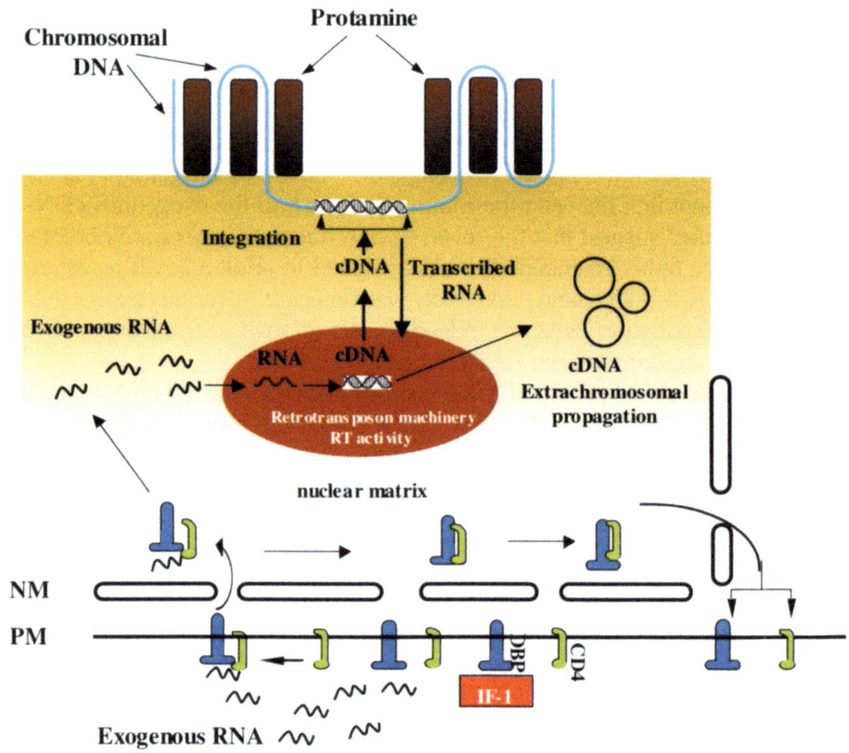

Figure 2. Hypothetical model for the genesis of new genetic information through reverse transcription. Exogenous RNA molecules are taken up by sperm cells, internalised in nuclei and reverse-transcribed in cDNA copies by the endogenous matrix-associated RT activity (represented by the brown oval structure). A small proportion of the resulting cDNA may eventually integrate into "accessible" sites of the sperm chromatin which retain a nucleosomal organization and are probably located between adjacent nucleoprotamine domains. The majority of cDNA copies instead appear to propagate in tissues of the offspring as extrachromosomal structures (blue circles). Extrachromosomal structures can be generated either by direct reverse transcription of the exogenous RNA, or from transcription of the integrated cDNA copy into RNA. DBP: DNA-binding protein; PM: plasma membrane; NM: nuclear membrane; IF1: inhibitory factor 1.

both of which can then be delivered to embryos in IVF essays. Our work on RT activity has been carried out using essentially mouse epididymal spermatozoa, but unpublished evidence suggest that human spermatozoa are also endowed with the same enzymatic activity. Therefore, when used in assisted reproduction, this constitutive ability of mature spermatozoa may represent a high risk factor for human health, because exogenous DNA or RNA molecules can be easily taken up, internalised and eventually become integrated in the sperm genome. The risk is evidently higher when spermatozoa from HIV-positive individuals, or from men carrying other sexually-transmitted viral diseases, are used in IVF. It is known that viral particles responsible for sexually-transmitted diseases, such as HIV (Baccetti, Benedetto, Burrini, Collodel, Costantino Ceccarini, Cris, Di Caro, Garbuglia, Massacesi, Piomboni and Sollazzo, 1994: Baccetti, Benedetto, Collodel, Di Caro, Garbuglio and Piomboni, 1998) and herpes (Bobrosky, Bagasra, Patel, Saikumari, Memoli, Abbey, Wood, Sosa and Bagasra, 1998), can be taken up by sperm cells and transferred to oocytes at fertilisation. Under these circumstances, spermatozoa may represent a potential threat as vectors of viral transmission from the father directly to the offspring. Moreover, given that sperm endogenous RT can reverse-transcribe cDNA copies from viral RNA, the possibility may be envisaged that HIV cDNA molecules are propagated to the offspring. If so, an additional IVF-associated risk, besides the danger of spreading serum-positivity in the offspring, may be associated to the disruption of the genetic integrity caused by the integration of the reverse-transcribed viral sequences. In the latter case, the risk is not serum positivity per se, but rather that spermatozoa may potentially lead to the generation of a "HIV-transgenic" individual. These considerations call for a high degree of awareness of the conditions and experimental control under which IVF procedures are operated.

Implications for evolution

The work summarised in this review shows that sperm cells incubated with exogenous RNA molecules trigger the synthesis of reverse-transcribed, transcriptionally competent, new genetic information that can be transmitted to the progeny at fertilisation and thereafter to the next generation. Paradoxically, the new phenotypic traits are not related to any identifiable transgene integration in the host genome, but rather to autonomously replicating extrachromosomal sequences, whose structural and functional features remain as yet poorly understood.

As a whole, these results reveal that mature spermatozoa are endowed with a previously unknwon RT-mediated mechanism that can promote the genesis of new genetic traits transmittable to the progeny in a non-mendelian fashion. This mechanism is not yet fully characterised, nor is its functional role clear during development.

It would seem plausible, however, that the spontaneous ability of spermatozoa to generate (retro)genes physically unlinked to those present in the nuclear genome, but biologically active and inheritable, has had, and possibly still has, an impact on evolutionary processes. Although still hypothetical and requiring experimental confirmation, the results reported thus far suggest that the expression of some phenotypic traits may not depend uniquely on the chromosomal genome but may also derive from a more subtle flow of genetic information generated by the retrotransposon/ endogenous retroviral system.

Conclusions

An increasing body of published results now challenges the traditional view of spermatozoa as metabolically inert cells, and suggests instead that they are actively involved in previously unsuspected functional roles. In this review, I have described the novel functional aspects of these cells and have attempted to place them in the framework of a unifying model. First, the spontaneous ability of spermatozoa of all species to take up, internalise and deliver exogenous DNA molecules to oocytes at fertilisation is now well established. This ability confers an inherently mutagenic potential to sperm cells, which may be viewed, in this perspective, both as a threat to human health when used in assisted reproduction and as a source of genetic variation with relevant implications in evolutionary processes. A second key aspect of more recent emergence is the preferential expression of genes of retrotransposon/retroviral origin in the germ line and the "storage" of functional RT enzymatic activity encoded by retrotransposons in mature sperm cells. That finding revealed an unsuspected link between sperm cells and the retrotransposon/retroviral gene machinery. The combination of these features has far-reaching implications, not only because it increases the overall potential of mature spermatozoa as inducers of genetic variability, given that both exogenous DNA or RNA sequences can act as suitable substrates, but, most importantly, because it offers an alternative possibility to generate new genetic traits and induce phenotypic variation in adult individuals.

Parallel lines of research in my and other laboratories indicate that endogenous RT also plays unsuspected roles in other fundamental biological processes, including embryo development, cell transformation and tumorigenesis (for recent reviews see Spadafora, 2004; Sinibaldi-Vallebona, Lavia, Garaci and Spadafora, 2006). Indeed, we found that LINE-1-encoded RT is essential for mouse preimplantation development (Pittoggi, Sciamanna, Mattei, Beraldi, Lobascio, Mai, Quaglia, Lorenzini and Spadafora, 2003; Beraldi, Pittoggi, Sciamanna, Mattei and Spadafora, 2006); conversely, RT inhibition results in reduced proliferation and induction of differentiation in somatic cells (Mangiacasale, Pittoggi, Sciamanna, Careddu, Mattei, Lorenzini, Travaglini, Landriscina, Barone, Nervi, Lavia and Spadafora, 2003; Landriscina, Fabiano, Altamura, Bagala, Piscazzi, Cassano, Spadafora, Giorgino, Barone and Cignarelli, 2005), and, furthermore, antagonises tumor progression in animal models (Sciamanna, Landriscina, Pittoggi, Quirino, Mearelli, Beraldi, Mattei, Serafino, Cassano, Sinibaldi-Vallebona, Garaci, Barone and Spadafora, 2005).

References

Baccetti, B., Benedetto, A., Burrini, A.G., Collodel, G., Costantino Ceccarini, E., Cris, N., Di Caro A., Garbuglia, A.R., Massacesi, A., Piomboni, P. and Sollazzo, D. (1994). HIV-1 particles in spermatozoo of patients with AIDS and their transfer into the oocytes. *Journal of Cell Biology* **127**: 903-914

Baccetti, B., Benedetto, A., Collodel, G., Di Caro, A., Garbuglio, A.R. and Piomboni, P. (1998). The debite on the presence of HIV-1 in human gametes. *Journal of Reproductive Immunology* **41**: 41-67

Beraldi, R., Pittoggi, C., Sciamanna, I., Mattei, E. and Spadafora, C. (2006). Expression of LINE-1 retroposons is essential for murine preimplantation development. *Molecular Reproduction and Development* **73**: 279-287

Bobroski, L., Bagasra, A.U., Patel, D., Saikumari, P., Memoli, M., Abbey, M.V., Wood, C., Sosa, C. and Bagasra, O. (1998). Localization of human herpesvirus type 8 (HHV-8) in the Kaposi's sarcoma tissues and the semen specimen of HIV-1 infected and uninfected individuale by utilizing in situ polymerase chain reaction. *Journal of Reproductive Immunology* **41**: 149-160

Brinster, R.L., Sandgren, E.P., Behringer, R.R. and Palmiter, R.D. (1989). No simple solution for making transgenic mice. *Cell* **59**: 239-241

Francolini, M., Lavitrano, M., Lora-Lamia, C., French, D., Frati, L., Cotelli, F. and Spadafora, C. (1993). Evidence for nuclear internalization of exogenous DNA into mammalian sperm cells. *Molecular Reproduction and Development* **34**: 133-139

Giordano, R., Magnano, A.R., Zaccagnini, G., Pittoggi, C., Moscufo, N., Lorenzini, R. and Spadafora, C. (2000). Reverse transcriptase activity in mature spermatozoa of mouse. *Journal of Cell Biology* **148**: 1107-1113

Hirabayashi, M., Kato, M., Ishikawa, A., Kaneko, R., Yagi, T. and Hochi, S. (2005). Factors affecting production of transgenic rats by ICSI-mediated DNa transfer: effects of sonication and freeze-thawing of spermatozoa, rat strains for sperm and oocyte donors, and different constructs of exogenous DNA. *Molecular Reproduction and Development* **70**: 422-428

Kato, M., Ishikawa, A., Kaneko, R., Yagi, T., Hochi, S. and Hirabayashi, M. (2004). Production of transgenic rats by ooplasmic injection of spermatogonic cells exposed to exogenous DNA: a preliminary study. *Molecular Reproduction and Development* **69**: 153-158

Kurome, M., Ueda, H., Tomii, R., Naruse, K. and Nagashima, H. (2006). Production of transgenic-clone pigs by the combination of ICSI-mediated gene transfer with somatic cell nuclear transfer. *Transgenic Research* **15**: 229-240

Landriscina, M., Fabiano, A., Altamura, S., Bagala, C., Piscazzi, A., Cassano, A., Spadafora, C., Giorgino, F., Barone, C. and Cignarelli, M. (2005). Reverse transcriptase inhibitors downregulate cell proliferation *in vitro* and *in vivo* and restore TSH signaling and iodine uptake in human thyroid anaplastic carcinoma. *Journal of Clinical Endocrinology and Metabolism* **90**: 5663-5671

Lavitrano, M., Camaioni, A., Fazio, V., Dolci, S., Farace, M.G. and Spadafora, C. (1989). Sperm cells as vectors for introducing foreign DNA into eggs: genetic transformation of mice. *Cell* **57**: 717-723

Lavitrano, M., French, D., Zani, M., Frati, L. and Spadafora, C. (1992). The interaction between exogenous DNA and sperm cells. *Molecular Reproduction and Development* 31: 161-169

Lavitrano, M., Maione, B., Forte, E., Francolini, M.,

Sperandio, S., Testi, R. and Spadafora, C. (1997). The interaction of sperm cells with exogenous DNA: A role of CD4 and major histocompatibility complex class II molecules. *Experimental Cell Research* **233**: 56-62

Lavitrano, M., Bacci, M.L., Forni, M., Lazzereschi, D., Di Stefano, C., Fioretti, D., Giancotti, P., Marfe, G., Pucci, L., Renzi, L., Wang, H., Stoppacciaro, A., Stassi, G., Sargiacomo, M., Sinibaldi, P., Turchi, V., Giovannoni, R., Della Casa, G., Seren, E. and Rossi, G. (2002). Efficient production by sperm-mediated gene transfer of human decay accelerating factor (hDAF) transgenic pigs for xenotransplantation. *Proceedings of the National Academy of Sciences USA* **99**: 14230-14235

Maione, B., Pittoggi, C., Achene, L., Lorenzini, R. and Spadafora, C. (1997). Activation of endogenous nucleases in mature sperm cells upon interaction with exogenous DNA. *DNA and Cell Biology* **16**: 1087-1097

Maione, B., Lavitrano, M., Spadafora, C. and Kiessling, A.A. (1998). Sperm-mediated gene transfer in mice. *Molecular Reproduction and Development* **50**: 406–409

Mangiacasale, R., Pittoggi, C., Sciamanna, I., Careddu, A., Mattei, E., Lorenzini, R., Travaglini, L., Landriscina, M., Barone, C., Nervi, C., Lavia, P. and Spadafora, C. (2003). Exposure of normal and transformed cells to nevirapine, a reverse transcriptase inhibitor, reduces cell growth and promotes differentiation. *Oncogene.* **22**: 2750-2761

Naruse, K., Ishikawa, H., Kawano, H., Ueda, H., Kurome, M., Miyazaki, K., Endo, M., Sawasaki, T., Nagashima, H. and Makuuchi, M. (2005). Production of transgenic pigs expressing human albumin and enhanced green fluorescent protein. *Journal of Reproduction and Development* **51**: 539-546

Osada, T., Toyoda, A., Moisyadi, S., Akutsu, H., Hattori, M., Sakaki, Y. and Yanagimachi, R. (2005). Production of inbred and hybrid transgenic mice carrying large (> 200 kb) foreign DNA fragments by intracytoplasmic sperm injection. *Molecular Reproduction and Development* **72**: 329-335

Perry, A.C.F., Wakayama, T., Kishikawa, H., Kasai, T., Okabe, M., Toyoda, Y. and Yanagimachi R. (1999). Mammalian transgenesis by intracytoplasmic sperm injection. *Science* **284**:1180-1183

Pittoggi, C., Renzi, L., Zaccagnini, G., Cimini, D., Degrassi, F., Giordano, R., Magnano, A.R., Lorenzini, R., Lavia, P. and Spadafora, C. (1999). A fraction of mouse sperm chromatin is organized in nucleosomal hypersensitive domains enriched in retroposon DNA. *Journal of Cell Science* **112**: 3537-3548

Pittoggi, C., Zaccagnini, G., Giordano, R., Magnano, A. R., Lorenzini, R. and Spadafora, C. (2000). Nucleosomal domains of mouse spermatozoa chromatin as potential sites for retroposition and foreign DNA integration. *Molecular Reproduction and Development* **56**: 248-251

Pittoggi, C., Sciamanna, I., Mattei, E., Beraldi, R., Lobascio, A.M., Mai, A., Quaglia, M.G., Lorenzini, R. and Spadafora, C. (2003). A role of endogenous reverse transcriptase in murine early embryo development. *Molecular Reproduction and Development* **66**: 225-236

Sciamanna, I., Barberi, L., Martire, A., Pittoggi, C., Beraldi, R., Giordano, R., Magnano, A.R., Hogdson, C. and Spadafora, C. (2003). Sperm endogenous Reverse Transcriptase as mediator of new genetic information. *Biochemical and Biophysical Research Communications* **312**: 1039-1046

Sciamanna, I., Landriscina, M., Pittoggi, C., Quirino, M., Mearelli, C., Beraldi, R., Mattei, E., Serafino, A., Cassano, A., Sinibaldi-Vallebona, P., Garaci, E., Barone, C. and Spadafora, C. (2005). Inhibition of endogenous Reverse Transcriptase antagonizes human tumor growth. *Oncogene* **24**: 3923-3931

Sinibaldi-Vallebona, P., Lavia, P., Garaci, E. and Spadafora, C. (2006). A role for endogenous reverse transcriptase in tumorigenesis and as a target in differentiating cancer therapy. *Genes Chromosomes and Cancer.* **45**: 1-10

Smith, K.R. (1999). Sperm cell mediated transgenesis: A review. *Animal Biotechnology* **10**: 1–13

Smith, K. and Spadafora, C. (2005). Sperm-mediated gene transfer: Applications and implications. *BioEssays* **27**: 551-562

Spadafora, C. (2004). Endogenous Reverse Transcriptase: a mediator of cell proliferation and differentiation. *Cytogenenetics and Genome Research* **105**: 346 – 350

Spadafora, C. (1998). Sperm cells and foreign DNA: a controversial relation. *BioEssays* **20**: 955-964.

Webster, N.L., Forni, M., Bacci, M.L., Giovannoni, R., Razzini, R., Fantinati, P., Zannoni, A., Fusetti, L., Dalpra, L., Bianco, M. R., Papa, M., Seren, E., Sandrin, M.S., Mc Kenzie, I.F. and Lavitrano, M. (2005). Multi-transgenic pigs expressing three fluorescent proteins produced with high efficiency by sperm mediated gene transfer. *Molecular Reproduction and Development* **72**: 68-76

Zani, M., Lavitrano, M., French, D., Lulli, V., Maione, B., Sperandio, S. and Spadafora, C. (1995). The mechanism of binding of exogenous DNA to sperm cells: factors controlling the DNA uptake. *Experimental Cell Research* **217**: 57-64

Zoraqi, G. and Spadafora, C. (1997). Integration of foreign DNA sequences into mouse sperm genome. *DNA and Cell Biology* **16**: 291-300

Spermatology. SRF Vol. 65. ERS Roldan and M Gomendio (eds) Nottingham University Press, Nottingham

In vivo gene transfer into the testis by electroporation and viral infection – A novel way to study testis and sperm function

J. Parrington[1], K. Coward[1], O. Hibbitt[1], H. Kubota[1], C. Young[1], J. McIlhinney[1] and O. Jones[2]

[1]*Department of Pharmacology, University of Oxford, Oxford, UK;* [2]*School of Biological Sciences, The University of Manchester, Manchester, UK*

The study of gene function in testis and sperm has been greatly assisted by creation of transgenic mice by injection of a transgene into the fertilised egg. However this approach is costly and laborious and is not applicable to other species of importance for the study of sperm function, such as the hamster. We have investigated alternative ways of expressing transgenes in mouse and hamster testis and sperm by *in vivo* gene transfer. DNA expression constructs were introduced into the testis by injection of DNA followed by electroporation, or by injection of a lentiviral vector. Expression of fluorescent proteins was assessed by fluorescence microscopy. *In vivo* gene transfer by electroporation led to expression of a fluorescent reporter protein and a fluorescently tagged version of sperm protein phospholipase C zeta in hamster and mouse testis and epididymal sperm. *In vivo* gene transfer by lentiviral infection led to high level expression of a fluorescent reporter protein in male germ cells. In conclusion, *in vivo* gene transfer offers a novel way to study gene function in testis and sperm and may also have potential as a way of creating transgenic versions of important model organisms such as the hamster.

Introduction

Transgenic mice are usually created by injection of DNA constructs into the pronucleus of a fertilised egg and have greatly facilitated studies of gene function in the testis and sperm. However, the production of transgenic mice is a costly and laborious process. In addition, there has been no success in using this approach to create transgenic versions of mammalian species like the hamster, that have been used extensively for studying reproduction.

Here we explored the potential of *in vivo* gene transfer into the testis as a way of introducing transgenes into male germ cells. Previous studies in mice have shown that it is possible to use electroporation to express reporter genes in spermatogenic cells (Muramatsu *et al.*, 1997; Yamazaki *et al.*, 1998; Huang *et al.*, 2000; Yamazaki *et al.*, 2000; Ike *et al.*, 2004; Kubota *et al.*, 2005), suggesting that *in vivo* gene transfer into the testis using electroporation may be an alternative way to study gene function in testis and sperm and generate transgenic animals. However, doubts have been raised about the usefulness of the technique compared to standard transgenic approaches (Widlak *et al.*, 2003).

Corresponding author E-mail: john.parrington@pharmacology.oxford.ac.uk

We investigated whether *in vivo* gene transfer into the testis by electroporation could be applied to mammals other than mice. We chose to study the hamster because of its important role as a model organism for reproductive biology (Yanagimachi, 1994). Another key aim was to see whether this approach could be used to generate mouse sperm expressing fluorescently labelled versions of endogenous sperm proteins as a way to study their localization properties. Finally, we have begun exploring alternative *in vivo* gene transfer approaches, such as the use of lentiviral vectors.

Experimental Approaches

Animals

Adult male Golden Syrian hamsters and mice (MF1 strain), both aged 4-6 weeks, were obtained from Harlan (UK). All animal procedures were in accordance with Home Office guidelines as appropriate to the UK Animals (Scientific Procedures) Act (1986).

DNA constructs and lentiviral vectors

A mammalian expression construct (EYFP-Mito), designed to generate enhanced yellow fluorescent protein (EYFP), was obtained from Professor Norio Nakatsuji, (Kyoto University, Japan) (Huang *et al.*, 2000). A full length clone of mouse phospholipase C zeta (PLCζ) was subcloned into the EYFP-C1 vector (Clontech, UK). This created a fusion of full length mouse PLCζ cDNA with EYFP at the N-terminus. For *in vivo* gene transfer, DNA constructs were resuspended in an appropriate volume of HEPES buffered saline (HBS) to obtain a 3 μg/μl working solution, with trypan blue (Sigma, UK) added to a final concentration of 0.04% to monitor micropipette position and injection accuracy/volume during gene transfer. A self-inactivating lentiviral vector was created that was designed to generate green fluorescent protein (GFP). In all these cases gene expression was controlled by the constitutive CMV promoter.

In vivo *gene transfer by electroporation and lentiviral infection*

Animals were anaesthetised and provided with analgesia peri-operatively by sub-cutaneous injection. For each animal, the right testis was treated experimentally whilst the left testis remained completely untouched throughout the procedure to act as a control. DNA constructs or lentivirus were injected into the rete testis using a CellTram Oil manual injector and glass microcapillary pipette, until the majority (80%) of the seminiferous tubules visible on the surface of the testis displayed trypan blue. For electroporation an electrical current was applied to the right testis using an Electrosquare Porator ECM830 (BTX, USA) (voltage 50 V; pulse length 10 msec). The testis was replaced, muscle and skin layers sutured and the animal allowed to recover. Animals were maintained for varying periods of time prior to killing dependent upon experimental requirement.

Analysis of EYFP and GFP expression

Testes from killed animals were treated such that both thin histological sections and mature epididymal sperm could be analysed for expression of fluorescent proteins. The epididymis was separated from each testis and the testis snap frozen in liquid nitrogen to await histological analysis. Epididymal sperm were obtained by partially macerating the epididymis in PBS and allowing sperm to swim free. Epididymal sperm were counter-stained with propidium iodide (1 μg/μl) and analysed using a fluorescence microscope (Leica, UK) equipped with an excitation light source and

Texas Red (for propidium iodide staining), YFP and GFP filter sets. Images were captured using Leica IM50 Image Manager software. Serial frozen sections (10 µm in thickness) were prepared from frozen testis, transferred onto glass slides coated with 0.1% w/v poly-L-lysine, fixed for 20 minutes in Bouin's solution, counterstained with neutral red (0.01% w/v neutral red, 0.01% w/v sodium azide, 0.04% v/v glacial acetic acid) and analysed using a fluorescence microscope as described above.

Expression of a fluorescent reporter gene in hamster testis and sperm following *in vivo* gene transfer by electroporation

We investigated whether a transgene could be expressed in hamster testis and epididymal sperm following *in vivo* gene transfer by electroporation. We injected the right testis of hamsters with a DNA construct that expresses EYFP-Mito protein under the control of a constitutive promoter, and used electroporation to drive the DNA into the cells of the testis. We found that 20 days after the procedure, seminiferous tubules of the treated testis showed clusters of fluorescent cells that were not evident in tubules from the control testis (Hibbitt *et al.*, 2006). Histological sections of testis demonstrated that these fluorescent cells were mostly situated adjacent to the basement membrane of the seminiferous tubules (Fig. 1A) (Hibbitt *et al.*, 2006). This pattern of expression suggested that at this time-point the transgene was primarily being expressed in the youngest male germ cells. We also investigated whether *in vivo* gene transfer into the testis by electroporation results in expression of EYFP-Mito in sperm. When we looked at sperm from the right epididymis at 40 days following the procedure, we saw sperm expressing EYFP fluorescence in their midpiece (Fig. 1B) (Hibbitt *et al.*, 2006), the expected site of localization of EYFP-Mito as mitochondria are situated here. This pattern of fluorescence was not evident in sperm from control epididymis (Hibbitt *et al.*, 2006).

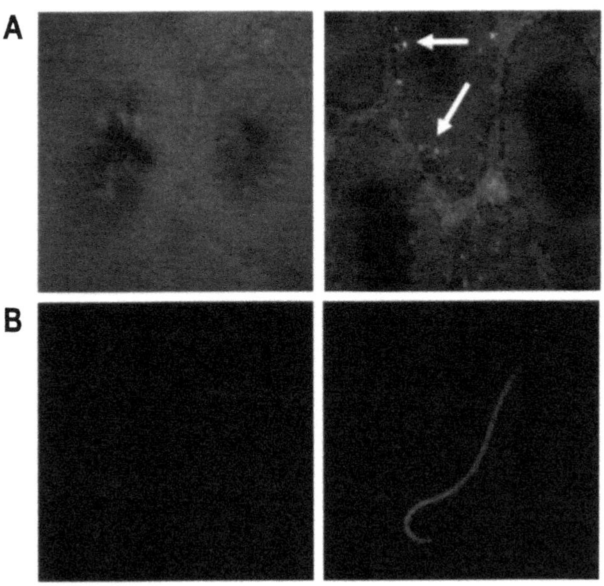

Figure 1. (**A**) Expression of EYFP-Mito in hamster testis following *in vivo* gene transfer using electroporation of the testis. Left panel: control testis. Right panel: treated testis. Arrows indicate cells showing EYFP fluorescence. (**B**) Expression of EYFP-Mito in hamster epididymal sperm following *in vivo* gene transfer using electroporation of the testis. Left panel: sperm from control epididymis. Right panel: sperm from treated epididymis.

Expression of PLCζ-EYFP protein in mouse testis and epididymal sperm

We next looked to see whether *in vivo* gene transfer into the testis by electroporation resulted in the expression of a fluorescently labelled recombinant version of a sperm protein, phospholipase C zeta (PLCζ) (Saunders *et al.*, 2002). We created an expression construct in which PLCζ was fused to EYFP with the fluorescent moiety at the N-terminus, and driven by a constitutive promoter. Expression of PLCζ-EYFP protein was first assessed in cultured mammalian cells. We showed that a fluorescent fusion protein of the expected size was generated from the construct (Coward *et al.*, 2006). After introducing this construct into mouse testis by *in vivo* gene transfer using electroporation, epididymal sperm were collected from animals 40 days later and analysed for PLCζ-EYFP expression by fluorescence microscopy. A proportion of epididymal sperm obtained from testis injected with PLCζ-EYFP-C1 followed by electroporation 40 days previously, clearly expressed EYFP fluorescence (Fig. 2) (Coward *et al.*, 2006). This was not evident in sperm collected from testis that had not been subjected to the gene transfer procedure (Fig. 2) (Coward *et al.*, 2006). Expression of PLCζ-EYFP protein was evident predominantly in the sperm head and to a lesser extent, also in the mid-piece.

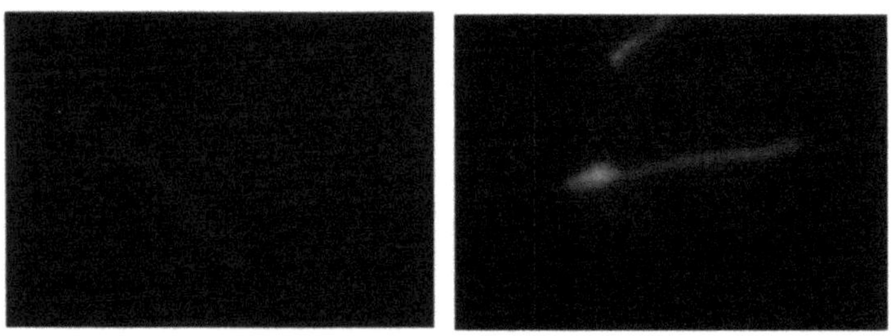

Figure 2. Expression of PLCζ-EYFP in mouse sperm following *in vivo* gene transfer using electroporation of the testis. Left panel: sperm from control epididymis. Right panel: sperm from treated epididymis.

Expression of GFP protein in hamster testis following *in vivo* gene transfer using lentiviral infection

To see if we could improve the efficiency of *in vivo* gene transfer into the testis and integration of transgenes into the sperm genome of hamsters, we used a lentiviral vector that expressed GFP from a constitutive promoter. The infectivity and ability to express GFP of this lentiviral vector were assessed in cultured mammalian cells and the titre of the virus determined (data not shown). Subsequently, varying amounts of the GFP lentivirus were injected into the right testis of hamsters, leaving the left testis as a control. After 3 days GFP expression in testis sections was assessed. We found that there was high level expression of GFP in male germ cells in treated, but not in control, testis (Fig. 3), after injection of ~20,000,000 colony forming units.

Implications for spermatology

In vivo gene transfer into the testis by electroporation has been put forward as an alternative way of studying testicular and sperm function and as a novel way of creating transgenic animals. We have shown that *in vivo* gene transfer into the rete testis by electroporation can be

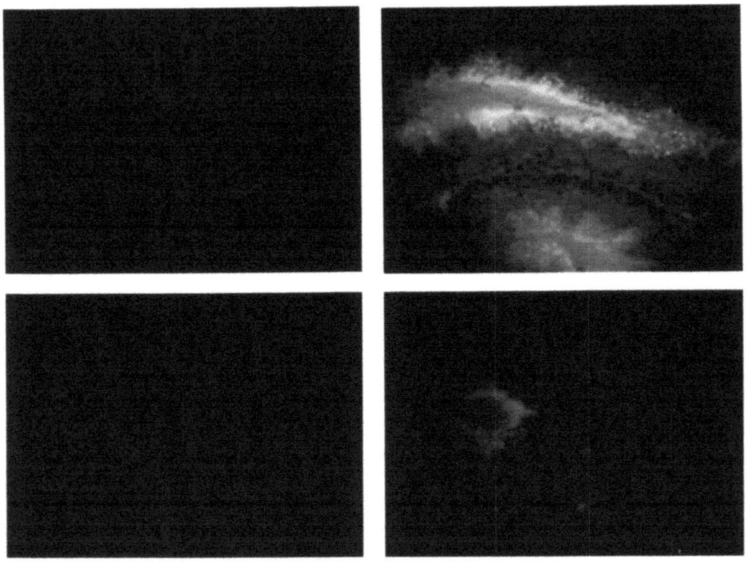

Figure 3. Expression of GFP in hamster testis following *in vivo* gene transfer using lentiviral vectors. Left panels: control testis. Right panels: treated testis.

used to achieve long-lived, high level expression of a fluorescent reporter protein in male germ cells of the hamster, including in mature, epididymal sperm (Hibbitt *et al.*, 2006).

In mice, we have shown that this approach can be used to express a fluorescently labelled version of sperm protein PLCζ in epididymal sperm (Coward *et al.*, 2006). The recombinant PLCζ-EYFP protein appears to localise primarily to the sperm head. This pattern of localization appears to be specific to PLCζ-EYFP, as a control construct expressing EYFP protein alone, failed to show any specific localization to the sperm head (Coward *et al.*, 2006). Moreover, this pattern of localization is in line with an immunofluorescence study that showed that PLCζ in mice appears to localise to a specific part of the sperm head (Fujimoto *et al.*, 2004). These findings suggest that *in vivo* gene transfer into the testis may offer a novel way to study the localization of sperm proteins.

Our demonstration that a lentiviral vector can be used to express a fluorescent reporter protein in male germ cells of the hamster is the first demonstration that such an approach can be used to introduce transgenes into the male germ cells of a mammal other than mice. Previous studies have shown that viral vectors can be used to introduce genes into mouse and rat male germ cells *in vitro* (Nagano *et al.*, 2001; Hamra *et al.*, 2002). Most recently, *in vivo* gene transfer into the testis using a retroviral vector resulted in successful expression of a transgene in mouse testis and germline transmission and creation of transgenic mice (Kanatsu-Shinoahara *et al.*, 2004). These findings combined suggest that *in vivo* gene transfer into the testis may offer an important new way of creating transgenic versions of mammals where standard approaches have failed.

Acknowledgements

This work was supported by an MRC Non-Clinical Senior Fellowship and a Royal Society Research Grant awarded to J.P.

References

Coward, K., Kubota, H., Hibbitt, O., McIlhinney, J., Kohri, K. and Parrington, J. (2006). Expression of a fluorescent recombinant form of sperm protein phospholipase C zeta in mouse epididymal sperm by in vivo gene transfer into the testis. *Fertilility and Sterility* **85**: Suppl 1: 1281-1289

Fujimoto, S., Yoshida, N., Fukui, T., Amanai, M., Isobe, T., Itagaki, C., Izumi, T. and Perry, A.C. (2004). Mammalian phospholipase Czeta induces oocyte activation from the sperm perinuclear matrix. *Developmental Biology* **274:** 370-383.

Hamra, F.K., Gatlin, J., Chapman, K.M., Grellhesl, D.M., Garcia, J.V., Hammer, R.E. and Garbers, D.L. (2002). Production of transgenic rats by lentiviral transduction of male germ-line stem cells. *Proceedings of the National Academy of Sciences USA* **99:** 14931-14936

Hibbitt, O., Coward, K., Kubota, H., Prathalingham, N., Holt, W., Kohri, K. and Parrington, J. (2006). In vivo gene transfer by electroporation allows expression of a fluorescent transgene in hamster testis and epididymal sperm and has no adverse effects upon testicular integrity or sperm quality. *Biology of Reproduction* **74:** 95-101

Huang, Z., Tamura, M., Sakurai, T., Chuma, S., Saito, T. and Nakatsuji N. (2000). In vivo transfection of testicular germ cells and transgenesis by using the mitochondrially localized jellyfish fluorescent protein gene. *FEBS Letters* **487:** 248-251

Ike, A., Ohta, H., Onishi, M., Iguchi, N., Nishimune, Y. and Nozaki, M. (2004). Transient expression analysis of the mouse ornithine decarboxylase antizyme haploid-specific promoter using in vivo electroporation. *FEBS Letters* **559:** 159-164

Kanatsu-Shinohara, M., Toyokuni, S. and Shinohara, T. (2004). Transgenic mice produced by retroviral transduction of male germ line stem cells *in vivo*. *Biology of Reproduction* **71:** 1202-1207

Kubota, H., Hayashi, Y., Kubota, Y., Coward K. and Parrington J. (2005). Comparison of two methods of *in vivo* gene transfer by electroporation. *Fertility and Sterility* **83:** 1310-1318

Muramatsu, T., Shibata, O., Ryoki, S., Ohmori, Y. and Okumura, J. (1997). Foreign gene expression in the mouse testis by localized *in vivo* gene transfer. *Biochemical and Biophysical Research Communications* **233:** 45-49

Nagano, M., Brinster, C.J., Orwig, K.E., Ryu, B.Y., Avarbock, M.R. and Brinster, R.L. (2001). Transgenic mice produced by retroviral transduction of male germ-line stem cells. *Proceedings of the National Academy of Sciences USA* **98:** 13090-13095

Saunders, C.M., Larman, M.G., Parrington, J., Cox, L.J., Royse, J., Blayney, L.M., Swann, K. and Lai FA. (2002). PLC zeta: a sperm-specific trigger of Ca^{2+} oscillations in eggs and embryo development. *Development* **129:** 3533-3544

Widlak, W., Scieglinska, D., Vydra, N., Malusecka, E. and Krawczyk, Z. (2003). *In vivo* electroporation of the testis versus transgenic mice model in functional studies of spermatocyte-specific hst70 gene promoter: A comparative study. *Molecular Reproduction and Developement* **65:** 382-388

Yamazaki, Y., Fujimoto, H., Ando, H., Ohyama, T., Hirota, Y. and Noce, T. (1998). *In vivo* gene transfer to mouse spermatogenic cells by deoxyribonucleic acid injection into seminiferous tubules and subsequent electroporation. *Biology of Reproduction* **59:** 1439-1444

Yamazaki, Y., Yagi, T., Ozaki, T. and Imoto, K. (2000). *In vivo* gene transfer to mouse spermatogenic cells using green fluorescent protein as a marker. *Journal of Experimental Zoology* **286:** 212-218

Yanagimachi, R. (1994). Mammalian fertilization. In: *The Physiology of Reproduction*, 2nd edition, Edited by Knobil, E. and Neill, J.D. Raven Press, New York, pp. 189-317

Spermatology. SRF Vol. 65. ERS Roldan and M Gomendio (eds) Nottingham University Press, Nottingham

XY sperm separation and use in artificial insemination and other ARTs

David G. Cran

XY (UK) Ltd., Scottish Agriculture College, Aberdeen AB21 7TN, Scotland, UK

Many tens of thousands of calves resulting from artificial insemination (AI) have been born worldwide after XY sperm separation and commercial production is underway in several countries. Accuracy of sex selection is some 90% and can be achieved both in research facilities and at AI studs in rural locations. Most facilities sort X- sperm which have also be utilised for superovulation and embryo transfer projects and for *in vitro* fertilisation (IVF) as well as AI. Sort rates of some 15 x 10^6 sperm/h are currently achievable and are used for low dose insemination, generally at 2 x 10^6 frozen sperm per dose, at a minimum of 35% post thaw motility. Pregnancy rates are some 70% to 80% of normal "high" dose unsexed inseminates. Good herd management is essential for high pregnancy rates with sexed sperm. In addition to cattle, offspring of predetermined sex have been born in the human, sheep, pig, horse, rabbit, elk, buffalo, cat and dolphin. Each species has its own challenge with regard to sperm handling and insemination procedure. In pigs, horses and sheep, the available dose of sexed sperm is very considerably lower than that which is used for conventional AI and special approaches have been devised for each species. In the pig a flexible catheter has been used to deliver a small dose of semen as close to the site of ovulation as possible. In the horse, hysteroscopic insemination at the utero tubal junction has resulted in fertilisation and, in the sheep, laparoscopic insemination into the uterus is the standard procedure for both sexed and unsexed sperm. Further advances in the efficiency of sorting together with improvements in sperm handling should result in acceptable pregnancy rates in these species.

Introduction

In spite of strenuous efforts over many years, the only effective method for achieving sex selection before birth is by means of sorting X- and Y- chromosome bearing sperm by flow cytometry. Such sorted sperm are then subsequently used for AI or IVF with subsequent transfer of the embryos (ET). In addition, a combination of AI and ET and other ARTs have also been used.

Developmental work on the technique of differentiation between X- and Y-populations by flow cytometry occurred during the 1970's and 1980's (Fulwyler, 1977; Dean, Pinkel and Mendelson, 1978; Van Dilla, Gledhill, Lake, Dean, Gray, Jachel, Barlogie and Gohde,1977;

Corresponding author E-mail: dcran@xyinc.com

Pinkel, Gledhill, Lake, Stephenson, and Van Dilla, 1982a; Pinkel, Lake, Gledhill, Van Dilla, Stephensen and Watchmaker, 1982b; Garner, Gledhill, Pinkel, Lake, Stephenson, Van Dilla, and Johnson,1983; Gledhill, Pinkel, Lake, Garner, and Johnson, 1984) during which time the problems associated with the asymmetric shape of the sperm cell were examined. The historical perspective of the development of sperm sorting has been described in a review by Garner (2001). Until the late 1980's, preparative methods for staining and analysing mammalian sperm were destructive and resulted in non viable cells. The use of the membrane permeable DNA binding dye Hoechst 33342 allowed for the use of live sperm for analysis and sorting (Johnson, Flook and Look, 1987). The first live births using this stain as a marker for DNA (Johnson, Flook and Hawk, 1989) were rabbits with litters containing a total of more than 50 individuals being born.

At present the main driving force behind research and development of sperm sorting is the commercial benefit which may be accrued from it, primarily in cattle. Wheeler, Rutledge, Fischer-Brown, VanEtten, Malusky and Beebe (2006) recently reported that in Pennsylvania in the USA there was a premium of between $340 and $380 for Holstein heifer calves as opposed to bull calves and there is a similar driving force in many countries. In addition, there are also advantages in terms of faster genetic progress and benefits in livestock management and efficiencies in production. This predominance of interest in sorting sperm from cattle is reflected in the number of sorters world-wide which are devoted solely to this purpose. At the time of writing there are 66 flow cytometers specifically adapted to sort sperm. All apart from four are primarily devoted to sorting bull sperm with the remaining four also sorting sperm from a variety of species.

The process of sperm sorting is a patented technology held by the United States Department of Agriculture (USDA) (Johnson, 1991a) and two licenses have been granted by them for non-human mammals. One was granted to a British company, Mastercalf Ltd, in 1991 and a second to XY Inc. based in America and founded in 1996. The two companies joined forces in 1997 and XY Inc. continues to develop the technology in a number of species with the aim reaching the point where it is sufficiently robust to be applied in the field. Since the birth of the first sex determined calves in 1992 (Cran, Johnson, Miller, Cochrane and Polge, 1993) many tens of thousands of sexed calves have been produced in some 13 countries and the technology has been the subject of a number of reviews (Cran and Johnson, 1996; Amann, 1999; Hohenboken, 1999; Johnson and Welch, 1999; Lu, Cran and Seidel, 1999; Schenk, Suh, Cran and Seidel, 1999; Seidel, Schenk, Herickhoff, Doyle, Brink, Green and Cran 1999; Seidel, 1999, Johnson, 2000; Garner, 2001; Seidel and Garner, 2002; Seidel, 2003; Cran, 2003; Maxwell, Evans, Hollinshead, Bathgate, De Graaf, Eriksson, Gillan, Morton and O'Brien 2004; Garner, 2006). The nature of sperm sorting places constraints on how it may be used. This paper will describe the sorting technique and its effects on sperm and progress being made towards it becoming part of standard farm procedures using AI techniques, particularly in cattle.

Flow cytometric sperm sorting

The challenge of sperm sorting is to obtain pure populations of X- and Y-sperm while at the same time maintaining physiological function. Sperm sorting is based on the principle that the DNA difference between X- and Y sperm is constant. In general, the greater the DNA difference, the easier it is to obtain good separation of the populations. The asymmetry of sperm cells is a major factor determining the rate at which sperm may be sorted since discrimination between X and Y populations can only be made if the laser hits the sperm surface at right angles to the flat side (Fig.1). The population so aligned is termed the orientated population. Sperm from species economically of interest (mainly cattle, sheep and pigs) have paddle-shaped sperm which will orientate well with respect to the laser. Other species such as the human have bullet-shaped sperm which makes for poorer orientation. The DNA difference for domestic livestock (cattle, pigs, sheep, and horses) varies from 3.6 to 4.2% (Johnson, 1992; Johnson and Welch, 1999). With such species it is now relatively straightforward to obtain purities of 90% or over for both X- and Y-sperm. For a species such as the human which orientates poorly and has a low DNA difference (2.8%), it is difficult to obtain good resolution and while good purity can routinely be achieved for X-sperm, a much lower purity is obtained for Y-sperm (Stern, Karabinus, Wiley, Deresh, Matken and Blauer, 2003). It would appear that the sperm of most species are capable of being sorted and species ranging from elephants, gorilla and giraffe to dogs and cats have been successfully stained and analysed for DNA difference (Garner, 2006). Provided that resolution of the two populations can be detected, pure X- and Y-sperm samples should be capable of being sorted. With the current technology, the rate at which sorting will take place to give a set purity will depend largely on resolution, sperm concentration and sperm viability.

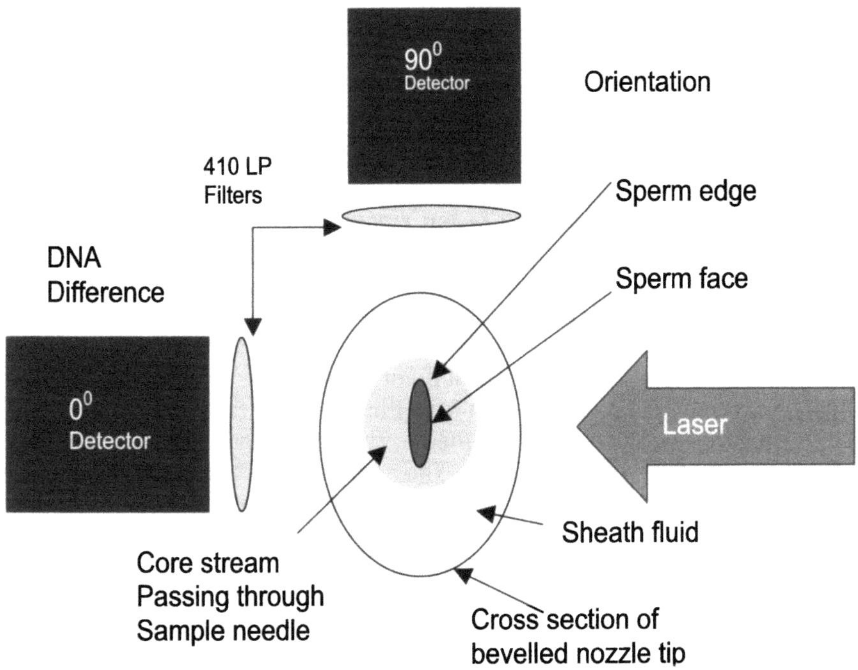

Figure 1. Transection of the nozzle tip demonstrating the sperm orientation necessary to obtain resolution of X- and Y- sperm populations.

The ability to differentiate X- and Y-sperm is determined by the positioning of photomultiplier tubes set at right angles to one another – one to determine that percentage of the population which is orientated at right angles to the incident laser beam and the other to distinguish DNA difference within this orientated population (Fig. 1). This configuration is the basis of the USDA patent (Johnson, 1991a). There are also several other consequent patents either granted or applied for by XY Inc. Using a standard fluidics delivery, only some 10% of the sperm population is appropriately orientated with respect to the forward detector. An early adaptation was to modify the injection needle such that the end was bevelled. This had the effect of producing a ribbon-like sheer area immediately in front of the needle and increased the orientated population to some 30%. Until the mid 1990's sort rates were in the region of 100 sperm/sec and therefore at that time although the purities obtained were often in excess of 90% for X- and Y- sperm, the numbers of sperm sorted were too low for use with AI, and therefore pregnancies with sexed sperm were obtained either with surgical AI into the uterus or oviduct (Johnson *et al.*, 1989; Johnson, 1991), ICSI (Catt, Catt, Gomez and Evans, 1996; Hamano, Li, Qian, Funanchi, Furadate, and Minato, 1999), conventional laparoscopic intrauterine AI (Cran, McKelvey, King, Dolman, McEvoy and Robinson, 1997) or IVF and ET (Cran *et al.*, 1993; Cran, Cochrane, Wei, Lu and Polge, 1994; Rath, Johnson, Dobrinsky, Welch and Neimann, 1997; Spinaci, De Ambrogi, Volpe, Galeati, Tamanini and Seren, 2004).

The major advance which allowed for the possibility of the application of sperm sorting to AI was the development of the high speed cell sorter. A high speed cell sorter was, in fact, developed in 1985 (Peters, Branscomb, Dean, Merill, Pinkel, Van Dilla and Gray, 1985) but due to the very high pressures involved (~200 p.s.i.) to obtain a high throughput of cells, not surprisingly, sperm cells did not survive. A subsequent version (van den Engh and Stokdijk, 1989) operated under much lower pressures and the sperm survived and retained their fertilising capacity. The design was licensed by Cytomation (now Dako) and marketed from 1996 under the trade name of MoFlo®. There has been subsequent development of this instrument for sperm sorting which is sold under the name MoFlo-SX®. With the original configuration of the bevelled injection needle, and with a ten-fold increase in the number of sperm passing through the instrument (an increase from ~2,500 to ~25,000), there was an associated increase in the rate at which sperm could be sorted into X- and Y-populations from ~100/sec to ~1000/sec. The orientating effect of the bevelled needle has subsequently been enhanced by an orientating bevelled nozzle tip (Fig.1). This was initially developed for a Coulter sorter (Rens, Welch and Johnson, 1998; Rens, Welch and Johnson, 1999; Johnson and Welch, 1999) by sculpting the interior of the nozzle tip to produce a tapering ellipse which had the effect of maintaining orientation of the sperm close to the intercept point of the laser. With the bevelled needle the orientating forces were much further distant from this point and thus many of the cells had lost orientation by the time sort decisions were made. This refinement had the effect of more than doubling orientation and therefore sort rates. The nozzle tip has been further developed by XY Inc. such that it (termed Cytonozzle™) is now a standard part of the MoFlo-SX® sperm sorter. Further developments relating to the resolution of the populations and refinements as to the determination of the percentage of the population which can be selected and software to abolish drift (Cytrack™) have further increased sort rates. It is now routinely possible to obtain sort rates in excess of 4000 sperm/sec (14.4×10^6/sperm/h) and on occasion sort rates of X-sperm can be as high as 6000 sperm/sec while maintaining about 90% purity.

There are limits as to the number of sperm which may be sorted from a single nozzle assembly. At present, some 30% are lost as a result of malorientation and a further ~15% are lost due to coincidence. Coincidence occurs when sperm are passing through the flow cytometer too closely for the detection system to recognise them as a signal designated either X or Y. Lower coincidence can be achieved by lowering the concentration of sperm being sorted but this will in itself result in a lower sort rate. Currently the concentration of sperm in the sort tube is 100×10^6/ml. A further ~2% are discarded due to aborts when sperm are positioned at the edge of droplets such that they can not be distinguished and a further 10% or more are lost due to the overlap in fluorescence signal between X- and Y-sperm. Little can be done with regard to coincidence or aborts. By increasing the number of droplets it is theoretically possible to increase the sort rate, but by doing so the size of the droplets decrease and there is a concomitant likelihood that there will be sperm damage. Increasing the spatial separation of the two populations could, in theory, increase the sort rate as would improving the orientation further. However, as has been pointed out by Seidel and Garner (2002) the theoretical limit to sorting sperm may be in the region of 10,000 sperm/sec of each sex and even this rate may be optimistic and not compatible with maintenance of optimum physiology. Monsanto claim to have developed a multinozzle sorter which has a multiplier effect marketed under the name Decisive™, but as yet there is no published extensive scientific trial data indicating the merit of this development.

The effect of sorting on sperm

Before a straw of frozen sorted semen can be prepared, the sperm undergo various steps which may potentially result in cellular damage which, in turn, may affect fertilising ability and embryonic development. To date, most information is available for cattle and the following largely pertains to this species. In most sites ejaculates are collected close to the location of the sperm sorters and in most commercial situations, sorters are located on bull studs. However, bull sperm can be transported chilled (5°C to 18°C) and although the sperm can be sorted, some damage does occur and further work is required on this. In the laboratory, the ejaculate is diluted to 200 x 10^6 sperm/ml in a modified Tyrode's albumin lactate pyruvate medium (TALP) (Schenk *et al.*, 1999) and stained by the addition of Hoechst 33342 for 45 min at 34°C after which it is diluted to 100 x 10^6 sperm/ml by the addition of an equal volume of TALP containing 4% clarified egg yolk and food dye (FD&C#40, Sensient Colours, USA). The sperm are passed through the cell sorter where each sperm is interrogated by the 351 and 364 nm lines of an argon ion laser and collected in a collection fluid comprising Tris buffer containing egg yolk. Sorting continues until about 15 ml of fluid has accumulated at which point the egg yolk concentration is some 3% and the concentration of sperm is some 800,000/ml. The sperm are then cooled and frozen by conventional methods. From this brief description of the process it is clear that there are various potentially damaging steps – hydrodynamic forces, UV irradiation, high dilution, etc., which must be taken into account with regard to sperm function. One such example is the effect of pressure on sperm function. To obtain high speed cell sorting it is necessary that there be high pressure within the cell sorter. Until recently this had typically been 50 p.s.i. and this had resulted in viable sperm and numerous pregnancies. However, studies by Suh, Schenk and Seidel (2005) have clearly shown that by reducing the pressure to 40 p.s.i there is a significant improvement in sperm quality without any detriment to sperm sorting speeds. Campos-Chillon and de la Torre (2003) have also shown a beneficial effect of a reduction in pressure. In a study comparing sperm sorted at 50 p.s.i. with those sorted at 40 p.s.i. they demonstrated that there was a higher cleavage and blastocyst development rate at the lower pressure.

The Hoechst stain permeates cell membranes and binds selectively to A-T bases by hydrogen bonding, van der Waals forces and electrostatic interactions, and unlike many other DNA binding dyes, is intercalative. Teng, Usman, Frederick and Wang (1988) have suggested that for some somatic cells Hoechst 33342 may give some irradiation protection to DNA damage. A study by Garner *et al.* (2001) indicated that dye and laser caused a small non-significant rise in DNA damage but concluded that the centrifugation step necessary to concentrate the sperm prior to freezing was more damaging. In addition Guthrie, Johnson, Garett, Welch and Dobrinsky (2002) examined the effect of different laser levels and found no effect on sperm function, and Catt, Sakkas, Bizzaro, Bianchi, Maxwell, and Evans (1997) found no increase in endogenous

DNA nicks after exposure of sperm to high concentrations of Hoechst 33342. An investigation by Boe-Hansen, Morris, Ersboll, Greve, and Chrisensen (2005) compared the effect of sorting on DNA integrity and found that there was greater DNA damage in control sperm but that there was a higher standard deviation in sorted sperm. In a study in the pig, Parrilla, Vazquez, Cuello, Gil, Roca, Di Bernardino and Martinez (2004) found that controls were the same as sperm subjected to sorting and staining with regard to sister chromatid exchanges or chromosome or chromatid breaks. It is possible that the sperm from some bulls may be more susceptible to damage by sorting than others. It is well known that the sperm from some bulls will freeze better than those from others and this effect may well be magnified by the sorting process. In a limited study it has been demonstrated (Zhao *et al.*, unpublished) that after sorting the sperm from some bulls will freeze as well as those which have not been sorted, but that sperm from other bulls will not freeze satisfactorily. This would indicate the desirability for testing bulls before their inclusion in a sorting programme at the onset.

The studies described indicate that there is little evidence to suggest that there is DNA damage which might potentially manifest itself in embryonic loss or foetal abnormalities. A study by Tubman, Brink, Suh and Seidel (2004), which examined calves born following AI with sexed sperm, indicated no abnormalities and, in addition, of the many tens of thousands of calves that have now been born, there has been no indication of a deleterious effect of the sorting process on development. Offspring have also been born in several other species (pig, sheep, horse, elk, cat, dolphin, rabbit, buffalo and human) with no contraindications. This is not to say, however, that the use of Hoechst is totally benign. It has been consistently demonstrated that there is an effect of the stain concentration on post thaw sperm motility *in vitro* and although there are indications that high stain levels may depress pregnancy rates, these do not reach significance (Schenk *et al.*, unpublished). The standard stain level is 112.5 μM, a level which ensures good resolution. However, with well aligned instruments this may often be lowered to 63.0 μM. Further, the current generation of continuous wave UV lasers is being replaced by pulsed lasers which, for a given output, give superior resolution resulting in the potential for a further lowering of stain concentration.

It is clearly essential to have a reliable method for estimating the purity of the samples which have been sorted. This is most readily carried out by removing the tails of an aliquot of sorted sperm and incubating the heads with a low concentration of Hoechst to ensure that any quenching due to irradiation with the laser is compensated for. The sperm are slowly passed through the flow cytometer and the purity of sperm is determined based on the DNA difference and the histograms analysed (Johnson *et al.*, 1987) by computer fitting to double Gaussian peaks (Welch and Johnson, 1999). This re-sort is the form of analysis which is generally carried out but other means of analysing the data by flow cytometry are available (Cran and Johnson, 1996). Because the analysis of purity is carried out by flow cytometric analysis, the same process by which the X- and Y-populations are analysed, several studies have been carried out to independently verify the accuracy of the procedure. No difference has been found between sort reanalysis and fluorescence in situ hybridization (FISH) using sex specific probes (Kawarasaki, Welch, Long, Yoshida, and Johnson, 1998; Parrilla, Vazquez, Oliver-Bonet, Navarro, Roca, and Martinez, 2003) and chromosome painting (Rens, Welch and Johnson, 2001). However, the most pressing argument for the validity of the procedure is the sex ratio of offspring following insemination with sexed sperm. The original cattle trials (Cran, Johnson and Polge, 1995) to the publications in recent years (Seidel, Schenk, Herickhoff, Doyle, Brink, Green and Cran, 1999; Crichton, Huffman, McSweeney and Schenk, 2006) have consistently demonstrated a sex bias of about 90% for both male and female calves.

Artificial insemination of cattle with sexed sperm

The most immediate practical application of the integration of sperm sexing and AI is in the cattle industry and there are clear commercial imperatives for its use in this species. There have been several studies examining the economic value of the use of sexed sperm in the cattle industry (Van Vleck, 1981; Hohenboken, 1999; Doyle, 2000; Seidel, 2003) and a detailed discussion is beyond the scope of this paper. The difference in value of a heifer over a male calf in a dairy herd is generally higher than for the reverse in a beef herd, and most attention has been placed on the sorting of X- sperm for the former sector. The first commercial licensee started operations in the UK in 2000 and there are now a further 8 commercial licensees (2 in the USA, 2 in Brazil, 2 in Japan and 1 each in Argentina and Mexico) and several other precommercial licensees are expected to take up commercial licenses within the next few years.

Even with the advances that have been made to improve the speed of sperm sorting, it is nevertheless a relatively slow procedure and requires that every sperm is individually analysed. As a consequence, considerable attention has been placed on attempting to optimise the conditions under which sperm are sorted, processed and frozen. This is particularly relevant when it is considered that the "standard" AI dose of sexed sperm is some ten fold lower than that which is used for conventional AI.

The initial studies in which fresh sorted sperm were used for insemination (Seidel, Allen, Johnson, Holland, Brink, Welch, Graham and Catell, 1997; Seidel, Herickhoff, Schenk, Doyle and Green, 1998) indicated that pregnancies could be generated following AI with sorted sperm. These initial experiments were speedily supplanted with the use of frozen sexed sperm when it was apparent that the quality of sexed sperm was very high with excellent post sort motility and a high percentage of acrosome intact cells. This is, in part, due to the fact that only viable sperm are sorted, the dead cells having been gated out by fluorescence quenching with food dye. The capacity to use sorted frozen bovine sperm to generate pregnancies was first demonstrated by Seidel, Cran, Herickhoff, Schenk, Doyle and Green (1999) and this formed the basis of all subsequent handling procedures (Schenk *et al.*, 1999, Seidel *et al.*, 1999).

The vast majority of inseminations in cattle have taken place in heifers due to their higher intrinsic fertility than cows. The management of cattle is perhaps that factor which determines above all else whether there will be a successful outcome following AI. This is, of course, true both for unsexed as well as for sexed sperm. However, due to the lower dose used and an indication that sorted sperm may suffer some damage caused by processing, the fertilising lifespan of sorted sperm is shorter than for unsorted. In addition, in the ram and boar capacitation-like changes have been observed (Maxwell and Johnson; 1997, Maxwell, Welch and Johnson, 1997; Maxwell, Long, Johnson, Dobrinsky, and Welch, 1998; Maxwell and Johnson, 1999), although no clear-cut changes of this nature have been reported in cattle. However, Moce, Graham and Schenk (2006) have shown that the membranes of cryopreserved sorted bovine sperm have been altered such that they undergo the acrosome reaction more readily than do those from similarly treated non-sorted sperm. In view of this, detailed attention to how sorted sperm are thawed and subsequently handled is essential. And, of course it should be reiterated that unless there is careful on-farm management, poor fertility data will follow.

The site of deposition of the sperm when using low doses of sexed semen might intuitively be considered to be particularly important. However, several studies have shown that there is no advantage to uterine horn insemination compared with that in the uterine body (Table 1). Several trials have compared the site of deposition and in very few has there been any indication of a benefit of using horn insemination.

Table 1. Percentage of heifers pregnant following insemination into the body or horn of the uterus. After Schenk (2000).

Treatment	Number inseminated	Pregnancy rate (%)
Unsexed body insemination (Control at 20 x 10^6 sperm)	112	67^a
Sexed body insemination	105	54^b
Sexed horn insemination	105	53^b

^{a,b}Means without common superscripts differ ($P < 0.05$).

The explanation is likely to be that deep insemination as used in the embryo transfer industry is a very skilled operation and there is a possibility of damage occurring in the uterine lining caused by inexperienced operators. All inseminations are now carried out into the uterine body in a manner similar to that used in a standard AI operation.

There have also been numerous studies in relation to the optimal dose to be used with sexed sperm. Although sort rates are now relatively high, nevertheless it is necessary to utilise sorted sperm in as efficient a manner as possible and to this end low dose insemination is a necessity. But what constitutes an effective low dose? In practice this will vary with bull fertility. Den Daas, de Jong, Lansbergen and van Wagendonk-De Leeuw (1998) conducted a trial into the effect of reducing the dose (non-sexed) from a range of bulls and found that there was no decline in fertility for highly fertile bulls but that there was a reduction for others. Such a study would indicate the necessity for conducting an investigation into whether a low dose will give an acceptable pregnancy rate with bulls chosen for use with sexed sperm (Table 2). Doses from as low as 1 million sperm to as high as 6 million sorted sperm have been used, and within this range there has been no observable benefit of increasing the dose over the standard of 2 million. The bulls used in the trials were of unknown fertility which may explain, in part, the difference in pregnancy rate with the "high dose" control. It has been observed at several locations that it is possible, with 2 million sexed sperm, to obtain pregnancy rates which do not differ significantly from 20 x 10^6 control levels. In these instances, it was invariably the case that there were high quality management practices in place.

Table 2. The effect of differing doses of sexed semen on pregnancy rate. After Seidel and Schenk (2002).

Treatment	Number of inseminations	Pregnancy rate (%)
20 x 10^6	263	62^a
6 x 10^6	246	41^b
1.5 x 10^6	288	43^b

^{a,b}Means without common superscripts differ ($P < 0.05$).

The use of competitive heterospermic insemination may be a useful approach to determining the fertility of bulls without the need for large scale trials. Thus, using sex as a marker, viable sperm sorted for X- sperm from bull A would be mixed pre- or post freeze with an equal number of Y- sperm from bull B and used for half of the inseminations and the converse for the other half (Seidel, Brink and Schenk, 2003). Fertility is assessed by sexing embryos or using ultrasound to check the sex of the foetuses. This approach although expensive, is less costly than that incurred when using the many hundreds of inseminations that would be necessary to obtain statistical significance in a field trial. This approach also has the advantage of testing procedures which may be potentially damaging to sorted sperm. One such recent example relates to the effect of the laser used to discriminate between X- and Y- sperm. Until recently, a continuous wave (CW) argon ion ultraviolet laser has been used. This has required water coolant, an expensive consideration. Recently a quasi CW or pulsed laser has become available (Vanguard 350-HDM, Spectra Physics, Mountain View, CA, USA). This laser is air cooled and has a considerably greater functional lifetime than the CW water cooled laser. Concerns were raised regarding the potential harmful effects of the heat generated within sperm from the incident laser and a competitive heterospermic assay was performed to determine whether a difference in fertility existed between the two. Sperm were sorted into X- and Y-sperm with the two lasers and 1 x 10^6 each of the X- and Y-sperm mixed from both laser treatments. Pregnancies marked by foetal sex with competitive fertilisation gave a ratio of 52 (pulsed):48 (CW) ratio (n = 179) (Schenk, Brink and Suh, 2005) indicating no deleterious effect of the pulsed laser. In addition, no difference was seen from the control unsexed sperm which were inseminated at a dose of 10 x 10^6 cells.

To date, the use of sexed sperm has been limited for use in heifers and the vast majority have been used to produce females. It would clearly increase the scope of the technology if it were possible to inseminate lactating dairy cows. To date there have been very limited data available and this has been of a variable and unreliable nature. A trial with commercially produced sexed sperm on commercial herds (Andersson, Taponen, Kommeri and Dahlbom, 2006) has compared 2 x 10^6 sorted sperm with unsexed sperm at 15 x 10^6 inseminated into unsynchronised lactating cows. They obtained a 21% pregnancy rate with the sorted sperm compared with 46% with the unsexed. A limited trial has recently been carried out by XY Inc. in which two doses (10 and 2 x 10^6) of X-sorted sperm from two bulls were used and compared with 10 x 10^6 unsexed sperm (Table 3). There was no difference between the two dose levels of sexed sperm or between sexed sperm and the control. This was a very limited trial with small numbers of animals and the cows were highly selected in terms of potential reproductive performance. XY Inc. is currently conducting a large field trial under commercial conditions to establish the practical relevance of this study and to establish the parameters for cattle management. If these data hold for trials held at multiple sites with a variety of bulls, the application of sperm sexing in the cattle industry would be markedly widened.

Table 3. Pregnancy rate in dairy cows. From Crichton *et al.* (2006)

Insemination Dose	Treatment	Percentage pregnant	Percentage female
10 x 10^6	Unsexed	55.6	51.5
10 x 10^6	Sexed	43.9	95.2
2 x 10^6	Sexed	40.4	90.9

Superovulation and IVF

The use of sexed semen together with superovulation is a niche market that remains to be exploited. Often, embryo sexing has been used in conjunction with embryo flushing and transfer but this has the potential for causing damage to the embryo. There are potential logistical problems in applying sexing to superovulation, in that as a result of ovarian stimulation and increased numbers of ovulations, the eggs may be ovulated over an extended period and thus there is the potential for asynchrony between sperm and eggs. To overcome this, AI, often with increased numbers of sperm, may be carried out over an extended period. Panarace, Medina, Cattaneo, Caballero, Cerrate, Dalla Lasta and Kaiser (2003) investigated this by inseminating cows and heifers at 0h, 12h and 24h after the onset of heat with 10 x 10^6 sperm at each time point. Control animals were inseminated with conventional unsexed semen. They obtained a mean of 4.2 ± 3.0, 5.5 ± 4.6 and 5.4 ± 4.4 viable embryos for sexed sperm placed in cows, heifers and unsexed control animals respectively. While there was no difference from controls, and while the same bulls were not always used in the control group, nevertheless, this trial demonstrated that it is possible to obtain high numbers of sexed transferable embryos from this approach. However, the use of 30 x 10^6 sorted sperm per animal is an expensive use of resources. Schenk, Suh and Seidel (2006) have taken a different approach and have inseminated at one time point with 20 or 10 x 10^6 sorted sperm. These treatments were compared with controls of 40 x 10^6 unsorted sperm. From 98 inseminations a mean of 2.0, 1.3 and 3.2 respectively, transferable embryos were obtained. These data indicated that this procedure may be a valid use of resources provided that the genetic value of the embryos is justified.

A third approach that has been used fairly extensively with cattle is that of IVF and this has been the subject of a recent review (Wheeler *et al.*, 2006). The advantage of using IVF is that considerably fewer sperm are required and, if required, a higher purity obtained. Much of the original work with sorted sperm used IVF (Cran *et al.*, 1993; Lu *et al.*, 1999). There has been somewhat of a resurgence in interest in its use over recent years, particularly since the large calf syndrome (McEvoy, Sinclair, Wilmut and Robinson, 2000) which was an impediment to large scale use of *in vitro* produced embryos appears to have been largely controlled. Several groups have observed that while sorting has no significant effect on cleavage, there was reduction in the number of transferable blastocysts (Merton, Haring, Stap, Hoebe, and Aten, 1997; Lu *et al.*, 1999; Wilson, Weigel, Fricke, Rutledge, Lebfried-Rutledge, Matthews and Schutzkus, 2005; Wheeler *et al.*, 2006). As has been observed for non sexed sperm, the sire used can have an impact on development rates and this may be modulated to some extent by changes in the concentration of heparin within the culture media (Lu and Seidel, 2004). In a study by Fischer-Brown, Barquero, Clark, Ferguson, Ireland, Jensen, Lane, Lindsey, Lopes, Monson, Northey, Reeder, Rutledge, Wheeler and Kesler (2005) it was found that twin transfer gave a higher percentage of foetuses (55%) than single transfers (30%). There was, however, a higher rate of foetal loss for the twin transfer group and a better understanding of both culture conditions required to produce a higher proportion of high grade embryos together with an understanding of factors leading to foetal loss would help make this approach more widespread.

Sheep

The application of sperm sexing in the sheep raises challenges not found in cattle. The ovine cervix does not dilate during oestrus and as a consequence transcervical insemination is not possible without a high degree of manipulation. While Halbert, Dobson, Walton, Sharpe and Bickrell (1980) reported acceptable conception rates by this approach others did not. The most common approach to AI in the sheep is by intrauterine insemination using a fine pipette, and conception rates of 50-80% have been reported (Maxwell and Hewitt, 1986). Practical application of sexed sperm in this species will use this route.

The first lamb born using sperm sorted by flow cytometry was produced by intracytoplasmic sperm injection (ICSI) with a fresh spermatozoon (Catt *et al.*, 1996). Not surprisingly, as with cattle (Hamano *et al.*, 1999), this approach has not been repeated. The first lambs conceived from intrauterine insemination were born in 1996 (Cran *et al.*, 1997). A dose of 10 x 10^5 fresh sperm was inseminated resulting in the birth of six sex selected lambs. The pregnancy rate was 50% that following insemination of unsexed sperm. Hollinshead, O'Brien, Maxwell and Evans (2002) subsequently inseminated with 2-4 x 10^6 frozen thawed sorted sperm, a similar dose to that used in cattle, and obtained a conception rate of 20% compared with 54% for that of a high dose control. Maxwell, Hollinshead, Rath, O'Brien and Evans (2003) showed that, with the then current techniques, a dose of some 20 x 10^6 sperm was required to obtain a pregnancy rate similar to control levels. Attempts to shorten the time of ovulation and also inseminate at about the time of ovulation did not improve conception rates. This group have also demonstrated the use of IVF procedures to produce selected embryos and offspring both using standard abattoir ovaries (Hollinshead, Evans, Evans, Catt, Maxwell and O'Brien, 2004) and also from oocytes obtained from hormonally stimulated juvenile lambs (Morton, Catt, Hollinshead, Maxwell and Evans, 2004).

The location of the sperm sorter in relation to that of the sperm source can prove to be logistically difficult. The majority of sorters for bull semen are located at bull studs. However, on occasion, semen needs to shipped over considerable distances and techniques for transporting chilled bovine semen without significant loss of viability have been developed. For sheep, work has been carried out on the potential for sorting previously frozen sperm. Frozen sperm from cattle were the first to be used in this manner (Stap, Hoebe, Merton, Haring, Bakker and Aten, 1998; Lu *et al.*, 1999). It was found that such sperm were viable and capable of fertilisation *in vitro* (Lu *et al.*, 1999). However, the presence of egg yolk in the diluent posed a problem to the sorting of sperm since it necessitated the use of high stain levels and resolution of the X- and Y- populations was compromised. Hollinshead *et al.* (2004) re-examined the potential to sort frozen sheep sperm and successfully produced IVF embryos. They also went one step further and used frozen/thawed sorted sperm to obtain IVF embryos and offspring following ET. The rates of pregnancy were similar to controls indicating that the freezing steps had not impaired fertilising capacity. de Graaf, Evans, Maxwell, Cran and O'Brien (2007) have taken this procedure one step further and have used frozen sorted and refrozen sperm in conjunction with AI by laparoscopy. Ewes have been inseminated with either fresh sorted sperm which were subsequently frozen and thawed or with sperm which were frozen, thawed and then sorted and frozen and thawed again. These two groups were compared with unsorted controls. Lambs were born from all three groups with no statistical difference between them. These are the first recorded births in any species after the insemination of frozen-thawed, sex sorted and refrozen sperm and demonstrate that the sperm maintain viability and fertilising capacity. The data did, however, tend to confirm that of Maxwell *et al.* (2003) that a relatively high number of motile sperm are required to give a pregnancy rate similar to that of controls. While this number of sperm requires a long sorting time, improvements in post thaw motility have decreased the total number of sperm required from 2 X the number of motile sperm to 1.4 X (de Graaf, Gillan, Evans, Maxwell and O' Brien, 2004) and current preliminary work in Sydney (De Graff *et al.*, unpublished) indicates that doses of sexed sperm which may be commercially acceptable may potentially be possible.

Pigs

The first farm animal to be born as a result of insemination with sorted sperm was the pig (Johnson, 1991b). These pigs were surgically inseminated into the fallopian tubes. The standard method of AI in pigs uses large numbers of sperm – typically some 3 x 10^9 sperm are used two or three times during each oestrus which is clearly impractical for use with sorted sperm. Two approaches to this problem have been investigated – the use of IVF and deep intrauterine insemination. Groups in the USA and Germany have been successful in producing offspring of predetermined sex by IVF (Rath *et al.*, 1997; Abeydeera, Johnson, Welch, Wang, Boquest, Cantley, Rieke, and Day 1998). In addition, Probst and Rath (2003) have successfully carried out ICSI on pig eggs with flow cytometrically sorted sperm and obtained offspring following transfer.

Neither IVF nor ICSI present a practical approach to the commercial use of sexed sperm in the pig. To address this problem a group in Spain are making progress in developing a method for the deep intrauterine delivery of relatively small numbers of sperm (Vazquez, Martinez, Parrilla, Cuello, Gil, Lucas, Roca, Vazquez, Didion and Day 2001). The initial work was encouraging with good farrowing rates with an inseminate of 50×10^6 sperm. Subsequent work led to the development of a robust 1.3m intrauterine insemination catheter with which acceptable pregnancy rates were achieved. A further improved version of the catheter has been developed and recent work by this group (Vazquez, Martinez, Parrilla, Cuello, Gil, Garcia, Caballero, Alminana, Roca and Vazquez, 2006) would indicate that follicle status is important for the successful use of sexed sperm in this species. It was observed that high levels of monospermy were obtained following intrafallopian insemination of sexed sperm in animals with preovulatory follicles as opposed to those with ovulating or post ovulatory ovaries. However, as with sheep, practical commercial application will require further development.

Horses

The conventional AI dose in horses is in the region of 500×10^6 (Pickett and Voss, 1975). Thus as with pigs and sheep, there is a considerable challenge to very markedly reduce the number of sperm in an inseminate if sexed sperm are to be used successfully. Work by Morris, Hunter and Allen (2000) using low dose inseminations of unsexed sperm indicated that by using transcervical videoendoscopic hysteroscopy it was possible to obtain an acceptable pregnancy rate with as low as 1×10^6 sperm with highly fertile mares. Buchanan, Seidel, McCue, Schenk, Herickhoff and Squires (2000) achieved a pregnancy rate of 40% with 25×10^6 sorted sperm inseminated into the tip of the uterine horn ipsilateral to the preovulatory ovary and Lindsey, Schenk, Graham, Bruemner and Squires (2002) and Lindsey, Morris, Allen, Schenk, Squires and Bruemner (2002) obtained pregnancies with both fresh sorted and frozen sorted sperm with doses as low as 5×10^6 sperm. Pregnancy rates varied from 13% to over 50% but the numbers involved were small and many more inseminations will be required before the approach can be considered to be robust enough for common practice. One challenge to be overcome will be the development of protocols which will allow for the reliable transport of stallion semen from centres distant from the sorting facility. Initial results from experiments in which semen has been cooled and stored overnight and then used for sorting have given promising results.

Other species

There are obvious advantages for the application of sperm sexing as an adjunct to the management of captive wildlife particularly in species such as the elephant or gorilla where groups are dominated by a single sex structure (O'Brien, Crichton, Evans, Schenk, Stojanov, Evans, Maxwell and Loskutoff, 2002). The challenges are great in that sperm sorting would be imposed on what are, in many cases, rather poorly understood male and female reproductive systems. An example of a successful development is the characterisation of the oestrus cycle of the bottlenose dolphin (Robeck, Steinman, Yoshioka, Jensen, O'Brien, Katsumata, Gili, McBain, Sweeney and Monfort, 2005) and the use of AI in this species. This has been followed by the sorting of dolphin sperm and AI resulting in the first birth of a marine mammal of predetermined sex (O'Brien and Robeck, 2006).

Perhaps those species whose reproductive characteristics are somewhat similar to their domestic counterparts may be more amenable to the application of sperm sexing. Thus, for example, in the elk (*Cervus elaphus*) the sperm can be sorted and treated in a manner similar to that which has been used for cattle. In a small trial in this species 11/17 (64.7%) were pregnant following insemination with sexed sperm, a number similar to that of controls (Schenk and DeGrofft, 2003). In a somewhat similar fashion sperm from the Italian buffalo (*Bubalus bubalis*) have been sorted and a low dose inseminated near the utero tubal junction (Presicce, Verberckmoes, Senatore, Klinc and Rath 2005). A 43% pregnancy rate was obtained which is the same as that for control sperm. A group from China is also investigating the potential of sperm sexing in the buffalo but using an IVF approach (Zhang, Liang, Lu and Lu, 2006). They found no difference in blastocyst development rate between sexed or unsexed sperm. Recently, a litter of cats has been born following insemination with sexed sperm (Crichton *et al.*, unpublished) thus opening up the possibility of the use in large cats. Attempts at sperm sexing have also been made for a number of non-human primates (O'Brien, Hollinshead, Evans, Evans and Maxwell; 2004 O'Brien, Stojanov, Crichton, Evans, Leigh, Maxwell, Evans and Loskutoff, 2004) and motile intact sperm have been recovered. In species such as this other approaches such as ICSI are likely to be those of choice due to the paucity of sorted sperm available and its rather poor quality.

Concluding remarks

Advances in approaches to bovine sperm sorting and sperm treatment have resulted in a technology which is sufficiently robust to be placed in a commercial situation. With a ready supply of good quality semen, provided protocols are closely adhered to and acceptable farm practices are maintained, it is possible to obtain commercially acceptable pregnancy rates at any location. Sperm sorters are currently deployed in rural locations in the Americas, Europe, and the Far East and in each location, sex predetermined calves are routinely produced. Further developments, particularly in relation to the sperm sorting hardware will expand the applicability to this species in the field.

There has been a successful outcome in sheep, although it is likely that the dose will have to be further lowered to make it commercially viable. Although there has been much progress, further advances remain to be made such that sperm sexing could be a routine application in the horse and pig. Application in other species will require considerable investment to understand the effects of sorting on their sperm and to develop AI approaches which may be quite dissimilar from those currently practiced in the domestic species.

Acknowledgements

The author is grateful to Dr Elizabeth Crichton and John Schenk for reading the manuscrript and providing valuable comments and constructive suggestions.

References

Abeydeera, L.R., Johnson, L.A., Welch, G.R., Wang, W.H., Boquest, A.C., Cantley, T.C., Rieke, A. and Day, B.N. (1998). Birth of piglets preselected for gender following in vitro fertilization of in vitro matured pig oocytes by X and Y chromosome bearing sperm by high speed flow cytometry. *Theriogenology* **50**: 981-988

Amann, R.P. (1999). Issues affecting commercialization of sexed semen. *Theriogenology* **52:** 1441-1457

Andersson, M., Taponen, J, Kommeri, M. and Dahlbom, M. (2006). Pregnancy rates in lactating Holstein-Friesian cows after artificial insemination with sexed sperm. *Reproduction in Domestic Animals* **41:** 95-97

Boe-Hansen, G., Morris, I.D., Ersboll A.K., Greve, T. and Christensen, P. (2005). DNA integrity in sexed bull sperm assessed by neutral Comet assay and sperm chromatin assay. *Theriogenology* **63:** 1789-1802

Buchanan, B.R., Seidel, G.E., McCue, P.M., Schenk, J.L., Herickhoff L.A. and Squires E.L. (2000). Insemination of mares with low numbers of either unsexed or sexed spermatozoa. *Theriogenology* **53:** 1309-1321

Campos-Chillon, L.F. and de la Torre, J.F. (2003). Effect of concentration of sexed bovine sperm sorted at 40 and 50 PSI on developmental capacity of in vitro produced embryos. *Theriogenology* **59:** 506

Catt, S.L., Catt, J.W., Gomez, M.C. and Evans, G. (1996). Birth of a male lamb derived from in vitro matured oocyte fertilized by intracytoplasmic injection of a single "male" sperm. *Veterinary Record* **139:** 494-495

Catt, S.L. Sakkas, D., Bizzaro, D., Bianchi, P.G., Maxwell, W.M.C. and Evans, G. (1997) Hoechst staining and exposure to UV laser during flow cytometric sorting does not affect the frequency of detected endogenous DNA nicks in abnormal and normal human sperm. *Molecular Human Reproduction* **3:** 821-825

Cran, D.G. (2003). Current developments in sperm sorting in farm animals. *Proceedings of the 19th meeting A.E.T.E. Rostock*, 113-123

Cran, D.G. and Johnson, L.A. (1996). The predetermination of embryonic sex using flow cytometrically sorted X- and Y- spermatozoa. *Human Reproduction Update* **2:** 355-363

Cran, D.G. Cochrane, D.J., Wei, H., Lu, K.H. and Polge, C.E. (1994). Separation of X- and Y- chromosome bearing sperm by flow cytometry for use in IVF. *Theriogenology* **41:** 183

Cran, D.G., Johnson, L.A., Miller, N.G.A., Cochrane, D. and Polge, C.E. (1993). Production of bovine calves following separation of X- and Y- bearing sperm and in vitro fertilization. *Veterinary Record* **132:** 40-41

Cran, D.G., Johnson, L.A. and Polge, C.E. (1995). Sex preselection in cattle: a field trial. *Veterinary Record* **136:** 495-496

Cran, D.G., McKelvey, W.A.C.., King, M.E., Dolman, D.P., McEvoy, T.G. and Robinson, J.J. (1997). Production of lambs by low dose intrauterine insemination of flow cytometrically sorted and unsorted semen. *Theriogenology* **47:** 267

Crichton, E., Huffman, S., McSweeney, K. and Schenk, J. (2006). Artificial insemination of lactating Holstein cows with sexed sperm. *Reproduction, Fertility and Development* **18:** 281

de Graaf, S.P., Evans, G., Maxwell, W.M.C., Cran, D.G. and O'Brien, J.K. (2007). Birth of offspring of predetermined sex after artificial insemination of frozen-thawed, sex-sorted and re-frozen ram spermatozoa. *Theriogenology* **67:** 391-398.

de Graaf, S.P., Gillan, L., Evans, G. , Maxwell, W.M.C. and O'Brien J.K. (2004). Effect of flow cytometric sorting and freezing diluent on in vitro quality of ram spermatozoa. *Proceedings of the 15th International Congress on Animal Reproduction, Porto Seguro, Brazil*, p. 477

Dean, P.N.D., Pinkel, D., and Mendelson, M.L. (1978). Hydrodynamic orientation of sperm heads for flow cytometry. *Cytogenetics and Cell Genetics* **29:** 189-197

Den Daas, J.H.G., de Jong, G., Lansbergen, L.M.T.E. and van Wagendonk-De-Leeuw, A.M. (1998). The relationship between the number of spermatozoa inseminated and reproductive efficiency of individual dairy bulls. *Journal of Dairy Science* **81:** 1714-1723

Doyle, S.P. (2000). *Impacts of sexed semen on commercial beef production.* Ph.D. dissertation, Colorado State University

Fischer-Brown, A., Barquero, G., Clark, S., Ferguson, C., Ireland, F., Jensen, N., Lane, S., Lindsey, B., Lopes, P., Monson, R., Northey, D., Reeder, A., Rutledge, J., Wheeler, M. and Kesler, D. (2005). Twin vs single transfer Holstein embryos to beef recipients. *Reproduction, Fertility and Development* **17:** 230

Fulwyler, M.J. (1977). Hydrodynamic orientation of cells. *Journal of Histochemistry and Cytochemistry* **25:** 781-783

Garner, D.L. (2001). Sex-sorting mammalian sperm – concept to application in animals. *Journal of Andrology* **22:** 519-526

Garner,D.L. (2006). Flow cytometric sexing of mammalian sperm. *Theriogenology* **65:** 943-957

Garner, D.L., Gledhill, B.L., Pinkel, D., Lake, S., Stephenson, D., Van Dilla, M.A. and Johnson, L.A. (1983). Quantification of the X- and Y- chromosome – bearing spermatozoa of domestic animals by flow cytometry *Biology of Reproduction* **28:** 312-321

Garner, D., Schenk. J.L. and Seidel, G. (2001). Chromatin stability in sex sorted sperm. In *Proceedings of the VIIth International Congress of Andrology Montreal. Andrology in the twenty first century.* Canada Short Communcations. Edited by Robaire, B., Chemes, H. and Morales, C.R., Medimond, Englewood, NJ, pp. 3-7

Gledhill, B.L., Pinkel, D., Lake, S., Garner, D.L. and Johnson, L.A. (1984). Identifying and separating X- and Y- chromosome-bearing mammalian sperm by flow cytometry. *Proceedings of the 10th International Congress of Animal Reproduction and Artificial Insemination* Urbana-Champaign, IL, USA, pp55-57

Guthrie, H.D., Johnson, L.A. Garett, W.M., Welch, G.R. and Dobrinsky, J.R. (2002). Flow cytometric sperm sorting: effects of varying laser power on embryo development in swine. *Proceedings of the Society of Reproductive Biology* **32:** 20

Halbert, G.W , Dobson, H., Walton, J.S., Sharpe, P. and Bickrell, B.C. (1980). Field evaluation of a technique for transcervical intra-uterine insemination of ewes. *Theriogenology* **33:** 1231-1234

Hamano, K., Li, X., Qian, X.Q, Funanchi, K., Furadate, M. and Minato, Y. (1999). Gender preselection in cattle with intracytoplasmically injected flow cytometrically sorted sperm heads. *Biology of Reproduction* **60:** 1194-1197

Hohenboken, W.D. (1999). Applications of sexed semen in cattle production. *Theriogenology* **52:** 1421-1433

Hollinshead, F.K., O'Brien, J.K., Maxwell, W.M.C. and Evans, G. (2002). Production of lambs of predetermined sex after the insemination of ewes with low numbers of frozen-thawed sorted X- or Y- chromosome bearing spermatozoa. *Reproduction, Fertility and Development* **14:** 503-508

Hollinshead, F.K., Evans, G., Evans, K.M., Catt, S.L., Maxwell, W.M.C. and O'Brien, J.K. (2004). Birth of lambs of a pre-determined sex after in vitro production of embryos using frozen-thawed sex-sorted ram spermatozoa. *Reproduction* **127:** 557-568

Johnson, L.A. (1991a) Method to preselect the sex of offspring. *United States Patent #5* **135:** 759.

Johnson, L.A. (1991b). Sex selection in swine: Altering sex ratios in offspring following surgical insemination with flow sorted X- and Y- bearing sperm. *Reproduction in Domestic Animals* **26:** 195-204

Johnson, L.A. (1992). Gender preselection in domestic animals using flow cytometrically sorted sperm. *Journal of Animal Science* **70** (Suppl. **2**): 8-18

Johnson, L.A. (2000). Sexing mammalian sperm for production of offspring: the state of the art. *Animal Reproduction Science* **60-61**: 93-107

Johnson, L.A., Flook, J.P. and Hawk, H.W. (1989). Sex preselection in rabbits: Live births from X and Y sperm separated by DNA and cell sorting. *Biology of Reproduction* **41**:199-203

Johnson, L.A., Flook, J.P. and Look, M.V. (1987). Flow cytometry of X- and Y- chromosome – bearing sperm for DNA using an improved preparation method and staining with Hoechst 33342. *Gamete Research* **17:** 203-212

Johnson, L.A. and Welch, G.R. (1999). Sex preselection: high speed flow cytometric sorting of X and Y sperm for maximum efficiency. *Theriogenology* **52:** 1323-1341

Kawarasaki, T., Welch, G.R., Long, C.R., Yoshida, M. and Johnson, L.A. (1998). A verification of flow cytometrically-sorted X- and Y- bearing porcine spermatozoa for DNA content using fluorescence in situ hybridization (FISH) technique. *Theriogenology* **50:** 625-635

Lindsey, A.C., Morris, L.H.A., Allen, W.R., Schenk, J.L., Squires, E.L. and Bruemner, J.E. (2002). Hysteroscopic insemination of mares with low numbers of non-sorted or flow sorted spermatozoa. *Equine Veterinary Journal* **34:** 128-132

Lindsey, A.C., Schenk, J.L., Graham, J.K., Bruemner, J.E. and Squires, E.L. (2002). Hysteroscopic insemination of low numbers of flow sorted fresh and frozen/thawed stallion spermatozoa. *Equine Veterinary Journal* **34:** 121-127

Lu, K.H., Cran, D.G. and Seidel, G.E. (1999). In vitro fertilization with flow cytometrically sorted bovine spermatozoa. *Theriogenology* **52:** 1393-1405

Lu, K.H. and Seidel, G.E. (2004). Effects of heparin and sperm concentration on cleavage rates of bovine oocytes inseminated with flow-cytometrically sorted bovine sperm. *Theriogenology* **62:** 819-830

Maxwell, W.M.C. and Hewitt, L.J. (1986). A comparison of vaginal, cervical and intrauterine insemination in sheep. *Journal of Agricultural Science, Cambridge* **106:** 191-193

Maxwell, W.M.C. and Johnson, L.A. (1997). Chlortetracycline analysis of boar spermatozoa after incubation, flow cytometric sorting, cooling or cryopreservation. *Molecular Reproduction and Development* **46:** 408-418

Maxwell, W.M.C. and Johnson, L.A. (1999). Physiology of spermatozoa at high dilution rates: The influence of seminal plasma. *Theriogenology* **52:** 1353-1362

Maxwell, W.M.C., Evans, G., Hollinshead, F.K., Bathgate, R., De Graaf, S.P., Eriksson, B., Morton, K. and O'Brien, J.K. (2004) Integration of sperm sexing technology into the ART toolbox. *Animal Reproduction Science* **82-83:** 79-95

Maxwell, W.M.C., Hollinshead, F.K., Rath, D., O'Brien, J.K. and Evans, G. (2003). Effect of dose of sperm processed for sex-sorting and cryopreservation on fertility in ewes. *Theriogenology* **59:** 511

Maxwell, W.M.C., Long, C.R., Johnson, L.A. Dobrinsky, J.R. and Welch, G. (1998). The relationship between membrane status and fertility of boar spermatozoa after flow cytometric sorting in the presence or absence of seminal plasma. *Reproduction, Fertility and Development* **10:** 433-40

Maxwell, W.M.C., Welch, G.R. and Johnson, L.A. (1997). Viability and membrane integrity of spermatozoa after dilution and flow cytometric sorting in the presence or absence of seminal fluid. *Reproduction, Fertility and Development* **8:** 1165-1178

McEvoy, T.G., Sinclair, K.D., Wilmut, I. and Robinson, J.J. (2000). Large offspring syndrome and other consequences of ruminant embryo culture in vitro: relevance to blastocyst culture in human ART. *Human Fertility* **3:** 238-246

Merton, J.S., Haring, R.M., Stap, J., Hoebe R.A. and Aten, J.A. (1997). Effect of flow cytometrically sorted frozen thawed semen on success rate of in vitro bovine embryo production. *Theriogenology,* **47:** 295

Moce, E., Graham, J.K. and Schenk, J.L. (2006). Effect of sex-sorting on the ability of fresh and cryopreserved sperm to undergo an acrosome reaction. *Theriogenology* **66:** 929-936

Morris, L. H.A., Hunter, R.H.F. and Allen, W.A. (2000). Hysteroscopic insemination of small numbers of spermatozoa at the uterotubal junction of preovulatory mares. *Journal of Reproduction and Fertilty* **118:** 95-100

Morton, K.M., Catt, S.L., Hollinshead, F.K., Maxwell, W.M.C. and Evans, G. (2004). Lambs born after in vitro embryo production from prepubertal lamb oocytes and frozen-thawed unsorted and sex - sorted spermatozoa. *Reproduction, Fertility and Development* **16:** 260

O'Brien J.K., Crichton, E.G., Evans, K.M., Schenk, J.L., Stojanov, T., Evans, G., Maxwell, W.M.C. and Loskutoff, N.M. (2002). Sex ratio modification using sperm sorting and assisted reproductive technology – a population strategy. *The 2nd International Symposium on Assisted Reproductive Technology for the Conservation and Genetic Management of Wildlife, Omaha, Nebraska, USA* 28-29 Sept. 2002 pp. 224-231. Henry Doorly Zoo, Omaha, USA.

O'Brien, J.K., Hollinshead, F.K., Evans, K.M., Evans, G., and Maxwell, W.M. (2004). Flow cytometric sorting of frozen-thawed spermatozoa in sheep and non-human primates. *Reproduction, Fertility and Development* **15:** 367-375

O'Brien J.K. and Robeck, T.R. (2006). Development of sperm sexing and associated assisted reproductive technology for sex preselection of captive bottlenose dolphins (*Tursiops truncatus*). *Reproduction, Fertility and Development* **18:** 319-329

O'Brien , J.K., Stojanov, T.,Crichton, E.G., Evans, K.M., Leigh, D., Maxwell, W.M.C. Evans, G. and Loskutoff, N.M. (2004). Flow cytometric sorting of fresh and frozen-thawed spermatozoa in the western lowland gorilla (*Gorilla gorilla gorilla*). *American Journal of Primatology* **66:** 297-315

Paranace, M., Medina, M., Cattaneo, L., Caballero, J., Cerrate, H., Dalla Lasta, M. and Kaiser G. (2003). Embryo production using sexed semen in superovulated cows and heifers. *Theriogenology* **59:** 513

Parrilla, I., Vazquez, J.M., Cuello, C., Gil, M.A., Roca, J., Di Bernadino, D. and Martinez, E.A. (2004). Hoechst 33342 stain and u.v. laser exposure do not induce genotoxic effects in flow-sorted boar spermatozoa. *Reproduction* **128:** 615-621

Parrilla, I., Vazquez, J.M., Oliver-Bonet, M., Navarro, J., Roca, J. and Martinez, E.A. (2003). Fluorescence in situ hybridization in diluted and flow cytometrically sorted boar spermatozoa using specific probes labelled by nick translation. *Reproduction* **126:** 317-325

Peters, D., Branscomb, E., Dean, P., Merrill, T., Pinkel, D. Van Dilla, M. and Gray, J.W. (1985). The LLNL high speed sorter: Design features, operational characteristics, and biological utility. *Cytometry* **6:** 290-301

Pickett, B.W. and Voss, J.L. (1975). The effect of semen extenders and sperm number on mare fertility *Journal of Reproduction and Fertility Supplement* **23:** 95-98

Pinkel, D., Gledhill, B.L., Lake, S., Stephenson, D. and Van Dilla, (1982a). Sex preselection in mammals? Separation of Y and "O" chromosomes in the vole *Microtus oregoni*. *Science* **218:** 904-905

Pinkel, D., Lake, S., Gledhill, B.L., Van Dilla, M.A., Stephenson, D., and Watchmaker, G. (1982b). High resolution DNA content measurements of mammalian sperm. *Cytometry* **2:** 1-9

Presicce, G.A., Verberckmoes, S., Senatore, E.M., Klinc, P. and Rath, D. (2005). First established pregnancies in Mediterranean Italian Buffalos (*Bubalus bubalis*) following deposition of sexed spermatozoa near the utero tubal junction. *Reproduction in Domestic Animals* **40:** 73-75

Probst, S. and Rath, D. (2003). Production of piglets using intracytoplasmic sperm injection (ICSI) with flow cytometrically sorted boar semen and artificially activated oocytes. *Theriogenology* **59:** 961-973

Rath, D., Johnson, L.A., Dobrinsky, J.R., Welch, G.R. and Neimann, H. (1997). Production of piglets preselected for sex following in vitro fertilisation with X- and Y- bearing spermatozoa sorted by flow cytometry. *Theriogenology* **47:** 795-800

Rens, W., Welch, G.R. and Johnson, L.A. (1998). A novel nozzle for more efficient sperm orientation to improve sperm sorting efficiency of X and Y chromosome bearing sperm. *Cytometry* **33:** 125-128

Rens, W., Welch, G.R., and Johnson, L.A. (1999). Improved flow cytometric sorting of chromosome bearing sperm: substantial increase in yield of sexed semen. *Molecular Reproduction and Development* **52:** 50-56

Rens , W., Welch, G.R. and Johnson, L.A. (2001). An X-Y paint set and sperm FISH protocol that can be used for cattle sperm separation procedures. *Reproduction* **121:** 541-546

Robeck, T.R., Steinman, K., Yoshioka, M., Jensen, E., O'Brien, J.K., Katsumata, E., Gili, C., McBain, J.F., Sweeney, and J.and Monfort, S.L. (2005). Estrous cycle characterisation and artificial insemination using frozen–thawed spermatozoa in the bottlenose dolphin (*Tursiops truncatus*). *Reproduction* **129:** 659-674

Schenk, J.L., Brink, Z. and Suh, T.K. (2005). Use of competitive insemination to evaluate a simpler laser for flow cytometric sexing of bovine sperm. *Reproduction, Fertility and Development* **17:** 306

Schenk, J.L. and DeGrofft, D.L. (2003). Insemination of cow elk with sexed frozen semen. *Theriogenology* **59:** 514

Schenk, J.L (2000). Applying sexing technology to the AI Industry. *Proceeding of the 18th Technical Conference on Artificial Insemination and Reproduction* 73-78

Schenk, J.L., Suh, T.K., Cran, D.G. and Seidel, G.E. (1999). Cryopreservation of flow-sorted bovine spermatozoa. *Theriogenology* **52:** 1375-1391

Schenk, J.L., Suh, T.K. and Seidel, G.E. (2006). Embryo production from superovulated cattle following insemination of sexed sperm. *Theriogenology* **65:** 299-307

Seidel, G.E. (1999). Sexing mammalian sperm and embryos – state of the art. *Journal of Reproduction and Fertility Supplement* **54:** 475-485

Seidel, G.E. (2003). Economics of selecting for sex: the most important genetic trait. *Theriogenology* **59:** 585-598

Seidel, G.E. Jr., Allen, C.H., Johnson, L.A., Holland, M.D., Brink, Z., Welch, G.R., Graham, J.K. and Cattell, M.B. (1997). Uterine horn insemination of heifers with very low numbers of nonfrozen and sexed spermatozoa. *Theriogenology* **48:** 1255-1264

Seidel, J.R., Brink, Z. and Schenk, J.L. (2003). Use of heterospermic insemination with fetal sex as the genetic marker to study fertility of sexed sperm.

Theriogenology **59:** 515

Seidel, G.E., Cran, D.G., Herickhoff, L.A., Schenk, J.L., Doyle, S.P. and Green, R.D. (1999). Insemination of heifers with sexed frozen or sexed liquid semen. *Theriogenology* **59:** 400

Seidel G.E. and Garner, D.L. (2002). Current status of sexing mammalian spermatozoa. *Reproduction* **124:** 733-743

Seidel, G.E., Herickhoff, L.A., Schenk, J.L., Doyle, S.P. and Green, R.D. (1998). Artificial insemination of heifers with unfrozen sexed semen. *Theriogenology* **49:** 365

Seidel, G.E. and Schenk, J.L. (2002). Field trials with sexed frozen bovine semen. *Proceedings of the 19th Technical Conference on Artificial Insemination and Reproduction* 64-69

Seidel, J.L., Schenk, J.L., Herickhoff, L.A., Doyle, S.P., Brink, Z., Green, R.D. and Cran, D.G. (1999). Insemination of heifers with sexed sperm. *Theriogenology* **52:** 1407-1420

Spinaci, M., De Ambrogi, S., Volpe, S. Galeati, G., Tamanini, C. and Seren, E. (2005). Effect of staining and sorting on boar membrane integrity, mitochondrial activity and in vitro blastocyst development. *Theriogenology* **64:** 191-201

Stap, J., Hoebe, R.A., Merton, J.S., Haring, R.M., Bakker, P.J.M. and Aten, J.A. (1998). Improving the resolution of crypreserved X- and Y- sperm during DNA flow cytometric analysis with the addition of Percoll to quench the fluorescence of dead sperm. *Journal of Animal Science* **76:** 1896-1902

Stern, H., Karabinus, D., Wiley, S., Deresh, D., Matken, R. and Blauer, K. (2003). Microsort® babies 1994-2003: Preliminary postnatal follow-up results. *American Society of Reproductive Medicine,* San Antonio, Texas.

Suh, T.K., Schenk, J.L. and Seidel, G.E. (2005). High pressure flow cytometric sorting damages sperm. *Theriogenology* **54:** 1035-1048

Teng, M., Usman, N., Frederick, C.A. and Wang, A.H.J. (1988). Molecular structure of the complex Hoechst 33258 and the DNA dodecamer d(CGCGAATTCCGCG). *Nucleic Acids Research* **16:** 2671-2690

Tubman, L.M., Brink, Z., Suh, T.K. and Seidel , G.E. (2004). Characteristics of calves produced with sperm sexed by flow cytometry/cell sorting. *Journal of Animal Science* **82:** 1029-1036

Van den Engh, G. and Stokdijk, W. (1989). Parallel processing data acquisition system for multilaser flow cytometry and cell sorting. *Cytometry* **10:** 282-293

Van Dilla, M.A., Gledhill, B.L., Lake, S., Dean, P.N., Gray, J.W. Jachel, V., Barlogie, B. and Gohde, W. (1977). Measurement of mammalian sperm deoxyribonucleic acid by flow cytometry – problems and approaches. *Journal of Histochemistry and Cytochemistry* **25:** 763-773

Van Vleck, L.D. (1981). Potential genetic impact of artificial insemination, sex selection, embryo transfer, cloning and selfing in dairy cattle. In *New Technologies in Animal Breeding*. Edited by Brackett, B.G., Seidel, G.E. and Seidel, S.M. San Diego, Academic Press, pp 222-242

Vazquez, J.M., Martinez, E.A., Parrilla, L., Cuello, C., Gil, M.A. Lucas, X., Roca, J., Vazquez, J.L. Didion, B.A. and Day, B.N. (2001). Deep intrauterine insemination in natural post-weaning estrus sows. *Proceedings of the Sixth International Conference on Pig Reproduction*, University of Missouri, Columbia, MO, p134

Vazquez, J.M., Martinez, I., Parrilla, I., Cuello, C., Gil, M.A., Garcia, E., Caballero, I., Alminana, C., Roca, J. and Vazquez, J.L. (2006). Improving the efficiency of laparoscopic intraoviductal insemination with sex-sorted spermatozoa. *Reproduction, Fertility and Development* **18:** 283

Welch, G.R. and Johnson, L.A. (1999). Sex preselection: laboratory validation of the sperm sex ratio of flow sorted X- and Y- sperm by sort reanalysis for DNA. *Theriogenology* **52:** 1343-1352

Wheeler, M.B., Rutledge, JJ, Fischer-Brown, A., VanEtten, T. Malusky, S. and Beebe, D.J. (2006). Application of sexed semen technology to *in vitro* embryo production in cattle. *Theriogenology* **65:** 219-227

Wilson, R.D., Weigel, K.A., Fricke, P.M., Rutledge, J.J., Lebfried-Rutledge, M.L., Matthews, D.L. and Schutzkus, V.R. (2005). In vitro production of Holstein embryos using sex-sorted sperm and oocytes from selected cull cows. *Journal of Dairy Science* **88:** 776-782

Zhang, M., Liang, X. W., Lu, Y.Q. and Lu K.H. (2006). In vitro fertilisation with flow-sorted buffalo sperm. *Reproduction, Fertility and Development* **18:** 283

Spermatology. SRF Vol. 65. ERS Roldan and M Gomendio (eds) Nottingham University Press, Nottingham

Sperm cryopreservation in fish and shellfish

Terrence R. Tiersch[1], Huiping Yang[1,2], Jill A. Jenkins[3] and Qiaoxiang Dong[4]

[1]*Aquaculture Research Station, Louisiana State University Agricultural Center, Louisiana Agricultural Experiment Station, Baton Rouge, Louisiana, USA;* [2]*Ocean College, Hainan University, Haikou, People's Republic of China;* [3]*U.S. Geological Survey, National Wetlands Research Center, Lafayette, Louisiana, USA;* [4]*School of Public Health, Wenzhou Medical College, Wenzhou, People's Republic of China*

Initial success in sperm cryopreservation came at about the same time for aquatic species and livestock. However, in the 50-plus years since then cryopreserved sperm of livestock has grown into a billion-dollar global industry, while despite work in some 200 species with well over 200 published reports, cryopreservation of aquatic species sperm remains essentially a research activity with little commercial application. Most research has focused on large-bodied culture and sport fishes, such as salmonids, carps, and catfishes, and mollusks such as commercially important oyster and abalone species. However, only a handful of studies have addressed sperm cryopreservation in small fishes, such as zebrafish, and in endangered species. Overall, this work has yielded techniques that are being applied with varying levels of success around the world. Barriers to expanded application include a diverse and widely distributed literature base, technical problems, small sperm volumes, variable results, a general lack of access to the technology, and most importantly, the lack of standardization in practices and reporting. The benefits of cryopreservation include at least five levels of improvements for existing industries and for creation of new industries. First, cryopreservation can be used to improve existing hatchery operations by providing sperm on demand and simplifying the timing of induced spawning. Second, frozen sperm can enhance efficient use of facilities and create new opportunities in the hatchery by eliminating the need to maintain live males, potentially freeing resources for use with females and larvae. Third, valuable genetic lineages such as endangered species, research models, or improved farmed strains can be protected by storage of frozen sperm. Fourth, cryopreservation opens the door for rapid genetic improvement. Frozen sperm can be used in breeding programs to create improved lines and shape the genetic resources available for aquaculture. Finally, cryopreserved sperm of aquatic species will at some point become an entirely new industry itself. A successful industry will require integrated practices for sample collection, refrigerated storage, freezing, thawing, rules for use and disposal, transfer agreements, and database development. Indeed the development of this new industry is currently constrained by factors including the technical requirements for scaling-

Corresponding author E-mail: ttiersch@agctr.lsu.edu

up to commercial operations during the transition from research, and the absence of uniform quality control practices, industry standards, marketing and price structures, and appropriate biosecurity safeguards.

Introduction

The scientific underpinnings of cryobiology can be traced back to the 1950s after the discovery of the cryoprotective qualities of glycerol for fowl sperm (Polge, Smith and Parkes, 1949). The first studies of fish sperm cryopreservation were published soon thereafter (Blaxter, 1953), and since then more than 200 species of fish have been studied (Rana, 1995; Tiersch, 2000). Cryopreservation is a process where biological materials such as cells and tissues are preserved by cooling to ultra-low temperatures, typically -196°C (for liquid nitrogen), yet remain viable after subsequent warming to temperatures above 0°C. For sperm cryopreservation, this process typically includes gamete collection, suspension of sperm in an extender, quality assessment, addition of cryoprotectants, equilibration, freezing, thawing and fertilisation, and the development of early life stages for assessment of cryopreservation success (Tiersch, 2000).

Conventional cryopreservation of sperm cells often involves the use of cryoprotectants and slow freezing to produce cellular dehydration and shrinkage. Cryoprotectants are chemicals used to protect cells from damage during freezing and thawing. Although their mechanisms of action are as yet not completely understood, permeating cryoprotectants such as dimethyl sulfoxide are believed to lower the freezing point of the solution, minimise osmotic shock by replacing the water inside the cell, and reduce formation of destructive intracellular ice (Doebbler, 1966; Rowe, 1966; Leung, 1991). Non-permeating cryoprotectants such as sugars and polymers are believed to help stabilise the membrane during cryopreservation (Meryman, 1971). Too little cryoprotectant entering the cell before cooling reduces effectiveness (Taylor, Adams, Boardman and Wallis, 1974), whereas too much causes osmotic swelling and rupture during thawing and dilution. In addition, cryoprotectants are often toxic to cells, and thus choice of cryoprotectant and their optimal concentration (a balance between protection and toxicity) has been the focus for numerous studies. Optimal equilibration before freezing is necessary to allow permeating cryoprotectants the time necessary to penetrate the sperm while minimising toxicity and can be influenced by temperature.

The choice of optimal cooling rate for freezing has been another major focus of numerous studies of sperm cryopreservation. To be considered as optimal, a rate should be slow enough to minimise the amount of intracellular ice (below a damaging level) and yet be rapid enough to minimise the length of time cells are exposed to what is referred to as the "solution effect" (the concentration and precipitation of solutes that occurs when solubility limits are exceeded during the dehydration caused by ice formation). In general, rapid thawing is preferred to minimise the damage associated with recrystalization (the coalescence of small ice crystals into large crystals during thawing). Numerous studies in sperm cryopreservation have been devoted to optimising specific components of cryopreservation procedures. However, aside from those factors mentioned above, other factors such as sample density, freezing container, starting temperatures, final temperatures (before plunging into liquid nitrogen), dilution, and cryoprotectant removal after thawing can also affect results (Leibo, 2000). Therefore, cryopreservation procedures must be tailored for each species and cell type, and be based on a thorough understanding of cryobiological properties (e.g., Mauger, Le Bail and Labbe, 2006).

The term "fish" is an artificial collective of more than 25,000 species characterised more by their differences than by their similarities. To broaden this grouping to aquatic species entails further inclusion of invertebrates such as oysters, abalone, and corals (e.g., Hagedorn, Carter,

Steyn, Krupp, Leong, Lang and Tiersch, 2006). A review of cryopreservation within aquatic species is thus a balancing act of attempting to generalise observations into basic principles while acknowledging the considerable diversity that exists across these organisms. For example, consider Figure 1 wherein the relationship of sperm motility to osmotic pressure is illustrated in three groupings of aquatic organisms. Sperm of freshwater species generally become active and increasingly motile in response to a lowering of osmotic pressure (hypotonic to blood plasma). Sperm of marine species behave in an opposite manner with motility increasing with osmotic pressure (hypertonic). The point being that these organisms are adapted to spawn externally at the osmotic pressure of the external environment. A third group displays sperm motility activation in an intermediate range (isotonic). This group, live-bearing fishes of the genus *Xiphophorus*, have evolved internal fertilisation, and as such, the internal anatomy of the female has become the working environment for the gametes with a corresponding change in the osmotic pressure of sperm activation (Huang, Dong, Walter and Tiersch, 2004b; Yang, Hazelwood, Walter and Tiersch, 2006).

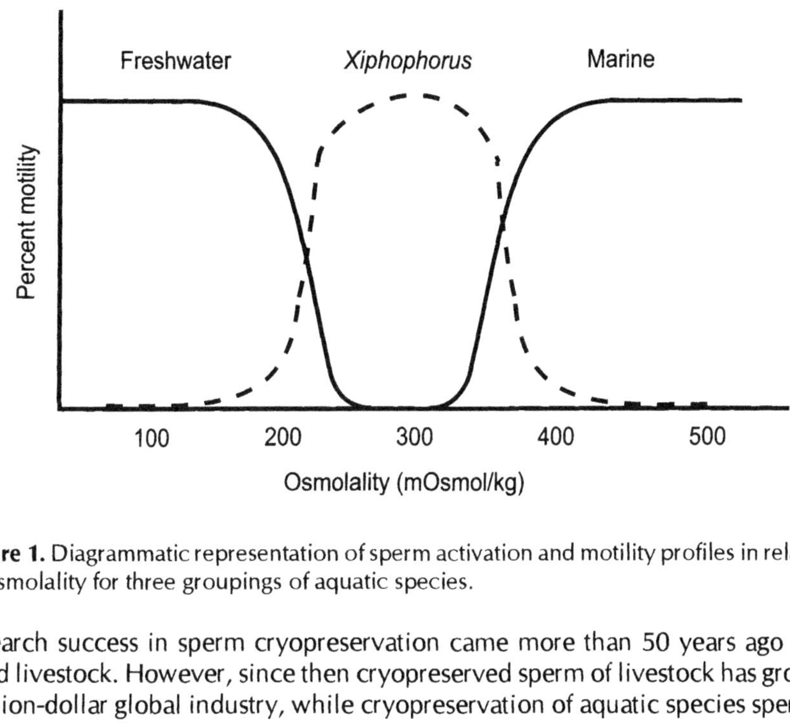

Figure 1. Diagrammatic representation of sperm activation and motility profiles in relation to osmolality for three groupings of aquatic species.

Initial research success in sperm cryopreservation came more than 50 years ago for aquatic species and livestock. However, since then cryopreserved sperm of livestock has grown into an annual billion-dollar global industry, while cryopreservation of aquatic species sperm remains essentially a research activity with little commercial application. Most research work has focused on large-bodied culture and sport fishes, such as salmonids, carps, and catfishes. Other groups such as mollusks, represented by commercially important oyster and abalone species, have received a fair amount of attention as well. However, only a handful of studies have addressed sperm cryopreservation in aquarium fishes, such as zebrafish (*Danio rerio*), which are becoming increasingly important in biomedical research (Tiersch, 2001), and in endangered species. Overall, this work has yielded techniques that are being applied with varying levels of success around the world. However, the barriers to expanded application include a diverse and widely distributed literature base, technical problems, small sperm volumes, variable results, a general lack of access to the technology, and most importantly, the lack of standardization in practices and reporting.

This review will focus on the needs and opportunities for future research and application of cryopreservation in aquatic species. To do this we will follow the general thematic approach afforded by the groupings shown in Figure 1. First, because several excellent review articles are already available that focus on sperm cryopreservation in freshwater fishes (primarily large-bodied commercially important species) we will direct the reader elsewhere for this information (Table 1). We will use the two other groupings to serve as models for aquatic species in general. Accordingly, second, we will present an overview of aquatic species cryopreservation research through examination of live-bearing fishes of the genus *Xiphophorus*, as representatives of the neglected but important aquarium fishes. Third, we will present the need for standardization by an overview of the existing oyster literature (a marine invertebrate; marine fishes have been reviewed by Suquet, Dreanno, Fauvel, Cosson and Billard, 2000). We will finish with a look to the future of application of cryopreservation in aquatic species.

Table 1. Examples of review articles, books, and special journal issues addressing cryopreservation of large-bodied freshwater fishes.

Short Title	Citation
Biology, handling and storage of salmonid spermatozoa	Scott and Baynes, 1980
Fish gamete preservation and spermatozoan physiology	Stoss, 1983
Live preservation of fish gametes	Leung and Jamieson, 1991
Cryopreservation of fish spermatozoa	Rana, 1995
Cryopreservation of finfish and shellfish gametes	Chao, Lin, Chen, Hsu and Liao, 1997
Cryopreservation in aquatic species (book)	Tiersch and Mazik (editors), 2000
Cryopreservation of gametes in aquatic species (special issue)	Lahnsteiner (editor), 2000
Techniques of genetic resource banking in fish	Billard and Zhang, 2001
Cryopreservation of gametes and embryos of aquatic species	Zhang, 2004
Cryopreservation and short-term storage of sturgeon sperm	Billard, Cosson, Noveiri and Pourkazemi, 2004
Semen cryopreservation in catfish species	Viveiros, 2005

Sperm cryopreservation in live-bearing fishes of the genus *Xiphophorus*

Viviparous fishes of the genus *Xiphophorus* (family Poecilliidae) are valuable for biomedical research because hybrid crosses between certain species within this genus have provided well-studied spontaneous and induced tumor models (Anders, 1967; Schartl, 1995; Walter and Kazianis, 2001). In 1928, a melanoma model for *Xiphophorus* was established by Gordon and Kosswig, demonstrating that certain cancers were inherited diseases (Anders, 1991). Since then, numerous studies have used *Xiphophorus* in medical research. Specific strains have been developed (Kazianis and Walter, 2002) and inbred lines have been created for study of gene structure and expression (e.g., David, Mitchell and Walter, 2004; Heater, Oehlers, Rains and Walter, 2004; Kazianis, Khanolkar, Naim, Rains, Trono, Garcia, Williams and Walter, 2004; Meierjohann, Schartl and Volff, 2004; Oehlers, Heater, Rains, Wells, David and Walter, 2004). Additionally, ornamental strains with specific body coloration have been developed by commercial farmers (e.g, www.ekkwill.com).

Although of great importance for conservation of valuable species and lines, sperm cryopreservation was essentially unexplored in *Xiphophorus* or in live-bearing fishes as a group before 2004 (Tiersch, 2001). The genus *Xiphophorus* can serve as a representative of both viviparous fish (with reproduction by internal fertilisation) and aquarium fish. As for the aquarium fishes, sperm cryopreservation research has been limited to zebrafish (Harvey, Norman and Ashwood-Smith, 1982; Morris, Berghmans, Zahrieh, Neuberg, Kanki and Look, 2003; Draper,

McCallum, Scout, Slade and Moens, 2004) and medaka (*Oryzias latipes*) (Aoki, Okamoto, Tatsumi and Ishikawa, 1997) which are also medical research models. Internal fertilisation is an atypical mode of reproduction in fishes and has involved specialised adaptations in morphology, physiology, and biochemistry in males and females. These features of *Xiphophorus* fishes, in combination with their small body size (2-5 cm) and associated small testis (5-12 mg), make sperm cryopreservation considerably more challenging than in fish species with external fertilisation.

Sperm characteristics of fishes from the genus Xiphophorus

Due to viviparity, sperm from *Xiphophorus* fishes possess specialised characteristics distinct from those found in sperm of fishes with external fertilisation (Fig. 2): (1) mature sperm have a cylindrical nucleus, a short, thickened midpiece, and an elongated axoneme, and within the testis sperm form discrete head-head agglutination bundles (Jonas-Davies, Winfrey and Olson, 1983); (2) after insemination, the spermatozoa remain in the head-head agglutination bundles within the female reproductive tract (Hoar, 1969; Grier, 1975); (3) inseminated sperm can be stored live for months in the female reproductive tract before fertilisation of mature oocytes, and females may produce broods at approximately 30-day intervals over 4-5 months after a single insemination (Tavolga, 1949); (4) motility can be activated by suspension of sperm obtained from crushed bundles in buffer isotonic to *Xiphophorus* plasma (Yang *et al.*, 2006); (5) suspended sperm can remain continuously motile for 4-7 days of storage at 4°C (Huang *et al.*, 2004b), and (6) spermatozoa possess glycolytic activity comparable to mammalian species (Kallman, 1975; Gardiner, 1978). In the following section, recent studies of sperm cryopreservation in *Xiphophorus* are reviewed. These studies could assist research in other viviparous fishes. Also, due to internal fertilisation, this group could serve as a model for studies of sperm cryopreservation and reproduction in other taxa including mammals.

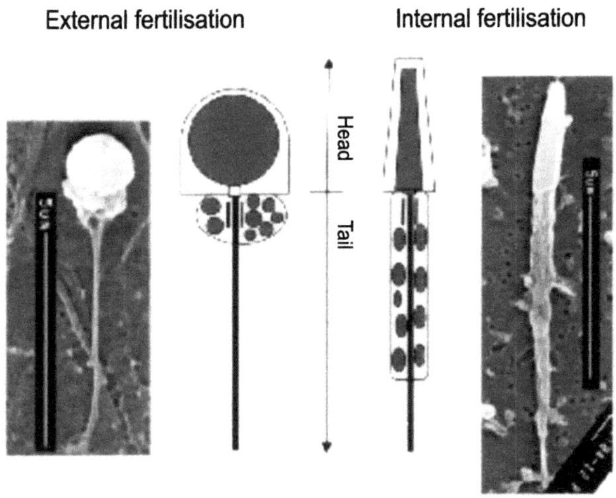

Figure 2. Representative scanning electron micrographs and diagrammatic views of sperm cells typical of fishes that employ external fertilisation (left, Nile tilapia, *Oreochromis niloticus*) or internal fertilisation (right, green swordtail *Xiphophorus helleri*). Figure reprinted with permission from Huang *et al.* (2004c).

Development of protocols for sperm cryopreservation in Xiphophorus

Generally, sperm cryopreservation comprises a series of steps including: (1) sperm collection; (2) extension of sperm in extender; (3) addition of cryoprotectant; (4) packing of sperm samples; (5) freezing; (6) thawing, and (7) estimation of fertility (Tiersch, 2000). Protocol establishment involves evaluation and optimization of multiple factors at each step (e.g., the type and concentration for each cryoprotectant), and recognition of the interactions among the steps (e.g., between cryoprotectant and cooling rate).

Sperm collection and dilution

Usually fish sperm can be obtained by either stripping or crushing of dissected testis. Because of the small body size of *Xiphophorus*, the availability of sperm by stripping is limited to ~1 microliter. To maximise the volume of available sperm, crushing of dissected testis was chosen for sperm collection in the following studies (Huang, Dong and Tiersch, 2004a; Huang et *al.*, 2004b; Huang, Dong, Walter and Tiersch, 2004c; Yang et *al.*, 2006). Dilution after sperm collection was necessary to maximise the volume available for subsequent experiments of various factor optimizations. However, extreme dilution of sperm samples has been found to reduce sperm motility in mammals, fishes, and oysters (e.g., Paniagua-Chavez, Buchanan and Tiersch, 1998). Thus, dilution ratios of 1:50, 1:100, 1:200, 1:400, 1:800, and 1:1600 (v:v) were evaluated with sperm of *X. helleri* (Huang *et al.*, 2004b). Motility of samples diluted at 1:50, 1:100 and 1:200 was not significantly different before freezing and after thawing, but the motility declined significantly at ratios of higher than 1:200.

Evaluation of extender solutions

The term "extender" refers to a solution of salts, sometimes including organic compounds such as sugars that help to maintain sperm viability prior to and during the freezing process (e.g., Hanks' balanced salt solution). The use of extenders is based on the control of pH and osmolality as well as the supply of energy, and can extend the functional life and fertilising capability of sperm. As stated above, in most fishes with external fertilisation, unless controlled by the concentration of specific ions, sperm motility can be activated by either hypotonic media (for freshwater fish) or hypertonic media (marine species), and have a short life span (1-2 min) upon being activated (Morisawa and Suzuki, 1980; Morisawa, Suzuki, Shimizu, Morisawa and Yasuda, 1983). Thus, sperm typically need to be maintained in an extender with proper osmolality (usually nearly isotonic to the plasma osmolality) to inhibit undesired sperm activation during refrigerated storage or cryopreservation (e.g., Bates, Wayman and Tiersch, 1996).

In contrast to fishes with external fertilisation, the sperm of *Xiphophorus* fishes exhibit different behaviours (Yang *et al.*, 2006). When sperm of *X. helleri* were suspended at 11 different osmotic pressures (24–500 mOsmol/kg) of Hanks' balanced salt solution (HBSS), motility was observed between 116 and 425 mOsmol/kg, with highest motility observed for samples suspended in HBSS at 310 mOsmol/kg, which was similar to that of the blood plasma (~320 mOsmol/kg). Sperm were not motile when the osmolality was lower than 116 or higher than 425 mOsmol/kg. Motility of the immobilised (non-motile) sperm could be activated by changing the osmotic pressure to 291–316 mOsmol/kg, and motility of immobilised sperm from hypertonic HBSS (> 425 mOsmol/kg) was significantly higher than that from hypotonic HBSS (< 145 mOsmol/kg) after 48 h of storage.

Based in part on these observations, HBSS at 310 mOsmol/kg was chosen as an initial extender for sperm cryopreservation and motility as high as 70% was obtained in thawed sperm after cryopreservation (Huang *et al.*, 2004a,b,c). Further study revealed that fresh sperm immobilised in HBSS 500 mOsmol/kg retained motility during refrigerated (non-frozen) storage significantly longer than did sperm stored in a motile (non-immobilised) state in isotonic HBSS (Yang *et al.*, 2006). The immobilised sperm suspended at 500 mOsmol/kg in HBSS were also cryopreserved, and high motility (~55%) was obtained after thawing. Subsequent parallel testing of cryopreservation of sperm from individual fish showed that motile sperm (in isotonic extender) and immobilised sperm (in hypertonic extender) showed the same motility after thawing with activation in isotonic HBSS (Yang *et al.*, 2006). By immobilising sperm prior to freezing, the use of hypertonic HBSS was hypothesised to conserve energy capacity, and consequently to provide an advantage for artificial insemination compared to sperm cryopreserved using isotonic HBSS.

Cryoprotectant selection and equilibration

Cryoprotectants used in screening for cryopreservation in *Xiphophorus* include dimethyl sulfoxide (DMSO), N-dimethyl acetamide, glycerol, propylene glycol, methanol, and sucrose. These are commonly used cryoprotectants for aquatic species. Each was evaluated with final concentrations of 6% and 10% (v/v). Results with *X. helleri* indicated that DMSO and glycerol were suitable cryoprotectants, and further evaluation with different concentrations showed that glycerol was better than DMSO in retaining motility and prolonging storage time of sperm after thawing. The most effective concentration for glycerol was 14% for non-immobilised sperm of *X. helleri* (Huang *et al.*, 2004b,c). Similarly, 14% glycerol showed the best effect in cryopreservation for another species, *X. couchianus* (Huang *et al.*, 2004a), and in immobilised sperm of *X. helleri* with concentrations of 10% and 15% (Yang *et al.*, 2006).

Prior to freezing, an equilibration period is necessary to allow permeating cryoprotectants to penetrate the sperm. Equilibration times of 10, 20, 30, 60, and 120 min at 4 °C before freezing were evaluated for sperm of *X. helleri* and *X. couchianus*. An equilibration time less than 30 min yielded the highest post-thaw motility in both species, but there were no consistent differences across the range of 10 to 120 min. Thus, equilibration time was considered to be not critical when glycerol was used as the cryoprotectant for *Xiphophorus* fishes.

Packing of samples for freezing

In cryopreservation, packing of samples for freezing and storing is important to standardise cooling and thawing rates, and to assure sample identification. Because of the small volume of sperm available, glass capillary tubes or cryovials have been employed in cryopreservation in aquarium fish (Harvey *et al.*, 1982; Aoki *et al.*, 1997; Morris *et al.*, 2003; Draper *et al.*, 2004). For *Xiphophorus* fishes, standard French straws (0.25 ml), used for cryopreservation of livestock sperm, were chosen for sperm packaging. The use of French straws offers the advantages of efficient and reliable sample identification by permanent printing on colored straws, sample safety by complete sealing of the straws, and standardization of the cooling and thawing processes.

Determination of cooling rate for freezing

Cooling rate is an important factor in sperm cryopreservation. Theoretically, it should be fast enough to minimise the exposure time of sperm to concentrated extracellular solutions and yet it should be slow enough to minimise, by cellular dehydration, the formation of intracellular ice to below a damaging level. The optimum cooling rates vary with different cryoprotectants and species, and can be determined empirically by experimentation, or by theoretical calculation using techniques such as differential scanning calorimetry, a technique used to measure two heat releases from the same cell suspension during freezing of live cells and dead cells (Devireddy, Raha and Bischof, 1998). For *Xiphophorus*, rates of 5, 15, 20, 25, 30, 35 and 45°C per min were tested to cool sperm from 5°C to –80°C before plunging into liquid nitrogen. The results showed that 20-30°C per min were best when sperm were cryopreserved with 14% glycerol (Huang *et al.*, 2004a,b). The prediction by differential scanning calorimetry agreed closely with this result in *X. helleri* (Thirumala, Huang, Dong, Tiersch and Devireddy, 2005) but not in *X. maculatus* (Pinisetty, Huang, Dong, Tiersch and Devireddy, 2005).

Overall, using post-thaw motility as an indicator of overall gamete quality, the optimised protocol for sperm cryopreservation in *X. helleri* and *X. couchianus* is as follows: collection of sperm by crushing of dissected testis in a volume of HBSS equal to 20-100 times the testis weight (yielding a sperm concentration of $\sim 10^8$ cells per ml) at an osmolality of 310 or 500 mOsmol/kg, holding on ice, the addition of chilled HBSS-glycerol as cryoprotectant to the sperm suspension to yield a final concentration of 14%, equilibration for 10-120 min, loading into 250-ml French straws, freezing with a cooling rate of 20-25°C per minute from 5°C to -80°C, plunging of the samples into liquid nitrogen for storage, and thawing at 40°C for 7 seconds before use for insemination and motility estimation.

Artificial insemination

After insemination, sperm of *Xiphophorus* fishes can reside within the female reproductive tract for months prior to fertilisation (Tavolga, 1949). This creates significant problems for monitoring the fate of sperm after insemination and for evaluating fertility. In initial studies, despite high motility in thawed samples, artificial insemination with cryopreserved sperm failed to produce live young. This resulted in the development of several hypotheses, including: (1) sperm have limited energy stores, and in the process of cryopreservation, motility in the extender could shorten the storage time of sperm in the female reproductive tract prior to fertilisation; (2) in the process of cryopreservation, the suspension of sperm in extender reduces the sperm density below that necessary for fertilisation, or (3) the toxicity of cryoprotectants used in cryopreservation may shorten the sperm lifetime, or affect inseminated females, and consequently reduce fertilisation (Yang *et al.*, 2006). To address these hypotheses, immobilisation of sperm by hypertonic HBSS was used to maintain potential energy stores; centrifugation was used to increase the concentration of sperm density (Yang *et al.*, 2006), and washing of cryoprotectant from thawed sperm was also accomplished by centrifugation, which improved retention of motility to 72 hours. The ionic composition of the extender, and addition of glucose and fetal bovine serum were also studied for the purposes of increasing sperm motility after thawing and improving effectiveness with the female reproductive tract for internal fertilisation (Dong, Huang and Tiersch, 2006b).

Based on these efforts, successful fertilisation and offspring production by cryopreserved sperm has been obtained in *X. helleri* (Yang, Hazlewood, Heater, Guerrero, Walter and Tiersch, 2007). This claim was supported by use of artificial insemination between two species that yield distinct hybrid offspring to verify paternity via cryopreserved sperm by both phenotype and genotype. Virgin females of *X. maculatus* were used for insemination with cryopreserved sperm from *X. helleri*. Motile and osmotically immobilised sperm each produced confirmed offspring. In addition, washed and non-washed sperm each produced confirmed offspring, suggesting that the toxicity of the glycerol concentrations tested did not affect production of offspring.

Overall, the optimised protocol to prepare sperm samples after thawing for artificial insemination is as follows: transfer of thawed samples into 1.5-ml microcentrifuge tubes, centrifugation at 1000 x g for 5 minutes, decanting of most of the supernatant, and re-suspension of the sperm pellet into 15 μl of fresh HBSS before use for insemination. Care was taken in these trials to ensure that a concentration of at least 10^9 sperm cells per ml was attained before insemination. These protocols represent the first report of successful fertilisation and offspring production by cryopreserved sperm in a live-bearing fish. Future studies will focus on the refinement of current protocols, and their application to other species of live-bearing fishes.

The need for standardization

For sperm cryopreservation to become a reliable, cost-effective tool for genetic banking in aquatic species, the overall process needs to be improved, and the approach needs to be integrated into an efficient large-scale platform that links with genetic and biological databases, long-term storage capabilities, inventory management, quality control, sample distribution pathways, biosecurity assurance, utilization and disposal practices, and a sound cryobiological foundation (Tiersch and Mazik, 2000). At present, a student of the aquatic species cryopreservation literature would conclude that the available procedures are plagued with extraordinary variability in results and reporting (Tiersch, 2000). The sources of this underlying variability derive from the empirical approach used for protocol development, the diversity of training background and experience of the researchers involved, a lack of standardization and definition of procedures (e.g., Leibo, 2000), and unexplainable results in terms of classical cryobiological principles.

The importance of estimating sperm concentration

It is likely that a lack of control of sperm concentration is one of the main reasons for the inconsistency observed among various studies, especially in relation to the concentration of cryoprotectant that offers best protection. This is because most cryopreservation studies in aquatic species do not standardise sperm concentrations prior to freezing, and instead have used methods such as dilution ratios based on volume. For example, in a review of the most studied group of aquatic organisms, the salmonid fishes, examination of 27 published reports (between 1978 and 2003) revealed the use of dilution ratios (mainly 1:3 of sperm-to-extender) for all studies (Dong and Tiersch, unpublished). The variations in sperm concentration that are possible by use of a set dilution ratio for a population of males are sufficient to span across and beyond the effective range of the available concentration of cryoprotectant. This effectively acts as a significant uncontrolled variable within otherwise well designed studies and can produce considerable variability in results.

Oysters, the most studied group of invertebrates, offer a relatively self-contained literature base that is representative of aquatic species in general (reviewed by Dong 2005). Currently, there are approximately 26 reports directly related to sperm cryopreservation in oysters since the first study some 35 years ago (Lannan, 1971). These reports include 16 peer-reviewed journal articles, 1 abstract, 2 book chapters, 2 conference proceedings, 1 thesis, 1 dissertation, 1 technical report, and 2 review articles. Nineteen of the 24 research reports (~80%) were produced for sperm from a single species, the Pacific oyster, *Crassostrea gigas*. A review of this literature identified the lack of sperm concentration standardization in 75% of the 16 peer-reviewed journal articles. These studies in particular were characterised by inconsistent or conflicting results even with the same cryoprotectants used at the same nominal concentrations. In other words it is difficult to find agreement in results with particular cryoprotectants or other variables in any two studies within a single species.

Based on these observations, a recent study was performed to illustrate the need for standardisation of sperm concentration by research of sperm agglutination in Pacific oysters (Dong *et al.*, 2006c). Sperm agglutination after thawing is a relatively frequent phenomenon observed for aquatic species, especially when sub-optimal cryopreservation protocols are used. The study evaluated various factors affecting sperm agglutination of thawed samples from diploid and tetraploid oysters. It was found that agglutination in thawed samples was mainly due to the lack of sufficient cryoprotectant for a specific sperm concentration. The exact mechanism of sperm agglutination remains unclear. However, morphological examination by microscopy of cross sections indicated at least two forms of agglutination (formed with and without cryoprotectant) that could be used as an important tool to understand the cryopreservation process within the micro-environment of the straw. Furthermore, the fact that the level of sperm agglutination was directly determined by sperm concentration, in addition to the type of cryoprotectant, cryoprotectant concentration, and cooling and thawing methods emphasised the importance of procedural standardization and systematic optimization and integration of protocols involving multiple factors. The results presented in this study call attention to the requirement to standardise sperm concentrations prior to cryopreservation; otherwise the reporting of cryoprotectant concentration or molarity offers little value and can be misleading. Sperm concentrations can be estimated readily by use of spectrophotometric methods (e.g., Dong, Huang and Tiersch, 2006c).

Other areas for standardization

The need for standardization extends elsewhere, as inconsistency exists in other components of cryopreservation technology among and within studies, such as in initial sperm quality, gamete collection methods, extender formulation, cryoprotectant choice, cooling rate and method, thawing rate and method, insemination protocols, and evaluation of post-thaw sperm quality (Rana, 1995; Gwo, 2000; Tiersch, 2000). Again we can look to the oyster literature review described above as an illustrative example for other aquatic species. Cryopreservation of oyster sperm involves variables ranging from broodstock condition to larval development, and for each step, various procedures have been used among different studies. For gamete collection, the most commonly used methods were dry stripping (which kills the oyster), aspiration using a pipette or syringe of gonad material through a hole drilled in the shell (without killing the oyster), or induced spawning. Few studies indicated what part of the gonad was sampled, although a recommendation has been made that no more than half of the gonad volume be extracted to avoid including immature or nutritive cells (McFadzen, 1995). Milt from individual males or pooled milt from several males has been used for various studies. Fewer than half of the reports indicated a sperm quality assessment prior to freezing, and when quality was assessed, motility was the sole criterion used. A similar state of within and among study variability exists for most factors that are studied. This lack of standardization places a substantial burden on the aquatic cryopreservation research community.

Furthermore, with respect to cryobiology, it is important to note that even simple cells such as sperm have a high degree of internal complexity. The various structures within a sperm cell represent different functional compartments that can each require different optimal conditions, and thus can each exhibit differential responses to cryopreservation. This can cause a variety of damages and outcomes. For example, damage to the tail could interfere with motility whereas damage to the head could interfere with embryonic development. Sperm quality is a generic term that encompasses proper functionality of a combination of cellular structures (such as the nucleus, plasma membrane, or axoneme) that can be evaluated individually by specific assays such as viability staining combined with flow cytometry (e.g., DeBaulny, LeVern, Kerboeuf and Maisse, 1997; Segovia, Jenkins, Paniagua-Chavez and Tiersch, 2000; Paniagua-Chavez, Jenkins, Segovia and Tiersch, 2006), motility estimates, Comet assay (e.g., Zilli, Schiavone, Zonno, Storelli and Vilella, 2003), computer-assisted sperm analysis (e.g., Kime, Van Look, McAllister, Huyskens, Rurangwa and Ollevier, 2001), or in aggregate by examining factors such as the capacity of sperm to fertilise eggs that hatch and develop normally. Sperm quality can be affected by a number of factors, and damage can occur to multiple structures. These would not all be detectable with a single assay (such as motility estimates). Accordingly, it is important to realise that it will be necessary to integrate several parameters of sperm quality into screening panels. As indicated above, several assays already exist for fish, but no comprehensive screening exists for aquatic species. Research is necessary not only to identify and further develop assays, but also to understand the principles and mechanisms that underlie results.

Misinterpretation of osmotic effects

A final example of systematic error that a lack of standardization can introduce into cryopreservation research in aquatic species, especially marine organisms, comes again from oysters. In another recent study (Dong, Huang, Henk and Tiersch, 2006a) spermatozoa from Pacific oysters were examined after non-isotonic fixation using glutaraldehyde. Morphological anomalies, such as membrane rupture, detached tails, and the formation of tail vesicles (typically associated with damage attributable to procedures such as cryopreservation) were observed. Osmolalities within the range of 800 to 1,086 mOsmol/kg were found to be functionally isotonic to sperm of Pacific oysters, but osmolalities below or above this caused severe cell damage. Although the damage associated with non-isotonic fixation was evident in all parts of the cells, the most vulnerable locations were the plasma membrane and flagellar motor apparatus. Hypotonic fixation was associated with the formation of tail vesicles (Fig. 3). Oyster sperm became swollen in hypotonic fixative, and bending or coiling of the axoneme within the tail vesicles led to the appearance of multiple axonemal structures in cross sections when observed by transmission electron microscopy. This phenomenon might be generally misinterpreted as the presence of double tails. This and other fixation artifacts can lead to the misinterpretation of damage caused by cryopreservation in ultrastructure studies of sperm of aquatic species. These findings indicate that caution is required in the interpretation of previously published results addressing ultrastructural changes in relation to cryopreservation. Future studies should ensure that samples are prepared with isotonic fixation, especially in marine species, and that the formation of tail vesicles be scrutinised objectively before claims are made linking them to treatment effects (e.g., cryopreservation) or biological variation (e.g., double tails).

To summarise, a lack of standardization is typically observed within the scientific literature for aquatic species in each step involved in the cryopreservation process. Comparisons among different studies are difficult to perform and could be invalid in most cases due to the procedural and reporting variations across studies. The points made in the section above call attention to the requirement for researchers to standardise sperm concentration in particular and protocols in general during cryopreservation research and reporting. Optimization of protocols without standardization offers little value for the improvement of existing methods and results, especially for the future development of commercial application. Controversy and inconsistency would be reduced if more congruent approaches were utilised and results among various studies could be compared directly. Suggestions for improvement include the creation and

Figure 3. Scanning electron micrographs of spermatozoa from Pacific oysters. Left panel: Samples fixed in 4% glutaraldehyde at 356 mOsmol/kg (hypotonic). Right: Samples fixed in 4% glutaraldehyde supplemented with sufficient sucrose to bring the final osmolality to 967 mOsmol/kg (isotonic). Arrows indicate tail vesicles formed after hypotonic fixation. Figure adapted with permission from Dong *et al.* (2006a).

widespread acceptance of standard references to assist in harmonising terminology (e.g., Tiersch and Mazik, 2000; Fuller, Lane and Benson, 2004) and the development and utilization of standardised educational programmes. Standardization of research practices and reporting could be accomplished through establishment of guidelines for publication of results. These guidelines could be developed and approved by societies and organizations that are involved in cryopreservation, and they could be applied gradually in phases to allow researchers to adjust methodologies. Once in place the guidelines could be made available to journal editors and reviewers to assist in evaluation of research reports.

The future prospects for application of cryopreservation in aquatic species

With respect to commercialization, the benefits of cryopreservation include at least five levels of improvements that address existing industries and the creation of new industries. First, cryopreservation, at a minimum, can be used to improve existing hatchery operations by providing sperm on demand, and greatly simplifying the timing of induced spawning. This prevents the problem, for example, of collecting ripe eggs, but not having sperm available to fertilise them. Second, frozen sperm can greatly enhance efficient use of facilities and create new opportunities in the hatchery by eliminating the need to maintain live males. Potentially all of the resources in a hatchery, which are typically limited, could be diverted to use for females and larvae. Third, valuable genetic lineages that currently exist, such as endangered species, research models, or improved farmed strains can be protected by storage of frozen sperm. This could be very important for marine species such as shellfish in which valuable broodstocks are often stored in natural waters. Fourth, cryopreservation opens the door for rapid genetic improvement. Frozen sperm can be used in breeding programs to create new improved lines and shape the genetic resources available for aquaculture operations. A dramatic example of this potential opportunity is provided by the dairy industry, which relies almost entirely upon cryopreserved sperm to produce improvements in milk yields. Finally, cryopreserved sperm of aquatic species will at some point, likely within the coming decade, become an entirely new industry itself. The global market for livestock sperm is around a billion dollars each year. Large, highly valuable global markets for cryopreserved sperm of aquatic species are now on the horizon.

Most of the technological innovations that have advanced the field of germplasm cryopreservation arose from a sound understanding of the mechanisms of cryodamage and cryoprotection (Mazur, 1970, 1984). Successful cryopreservation of germplasm must address intrinsic biophysical properties (e.g., osmotic tolerance limits), and cryopreservation procedures based on these biophysical properties are necessary to minimise cryodamage and maximise survival (Rall, 1993). The sections above call attention to the importance of the cumulative effects arising from all activities in cryopreservation and the importance of standardization for the future potential commercialization of cryopreserved sperm in aquatic species.

Cryopreservation research and application each require consideration of an interconnected series of activities and this involves more than simple freezing of samples. A successful programme involves integrated practices for sample collection, refrigerated storage, freezing, thawing, rules for use and disposal, transfer agreements, and database development. Sperm cryopreservation in aquatic species is only beginning to find application on a commercial scale. The development of this new industry is constrained by a number of factors including the technical requirements for scaling-up to commercial operations during the transition from research. This problem has been addressed by previous research that documents the feasibility of utilising commercial dairy cryopreservation facilities to provide a jumpstart for industry-scale cryopreservation in aquatic species such as catfish (Lang, Riley, Chandler and Tiersch, 2003) and oysters (e.g., Dong, Endeline, Huang, Allen and Tiersch, 2005). Although market price structures for cryopreserved sperm are yet to emerge for aquatic species, they will be resolved as the industry develops in response to the existing and growing strong demand for genetic resources (Caffey and Tiersch, 2000). Thus, overall, the largest practical constrains to realization of a cryopreservation industry for aquaculture are at present the absence of uniform quality control practices, industry standards, and appropriate biosecurity safeguards. The lack of quality control and biosecurity (Tiersch and Jenkins, 2003) cast a dark shadow over industry development and could discourage use of cryopreservation technology in general.

Acknowledgements

This work was supported in part by funding from USDA-SBIR, USDA-CSREES, the National Institutes of Health, and the National and Louisiana Sea Grant College Program. This manuscript was approved for publication by the Director of the Louisiana Agricultural Experiment Station as number 06-11-0364.

References

Anders, F. (1967). Tumour formation in platyfish-swordtail hybrids as a problem of gene regulation. *Experientia* **23**: 1-10

Anders, F. (1991). Contributions of the Gordon-Kosswig melanoma system to the present concept of neoplasia. *Pigment Cellular Research* **4**: 7-29

Aoki, K., Okamoto, M., Tatsumi, K. and Ishikawa, Y. (1997). Cryopreservation of medaka spermatozoa. *Zoological Science* **14**: 641-644

Bates, M.C., Wayman, W.R. and Tiersch, T.R. (1996). Effect of osmotic pressure on the activation and storage of channel catfish sperm. *Transactions of the American Fisheries Society* **125**: 798-802

Billard, R., Cosson, J., Noveiri, S.B. and Pourkazemi, M. (2004). Cryopreservation and short-term storage of sturgeon sperm, a review. *Aquaculture* **236**: 1-9

Billard, R. and Zhang, T. (2001). Techniques of genetic resource banking in fish. In: *Cryobanking the Genetic Resource. Wildlife Conservation for the Future?* Edited by Watson, P.F. and Holt, W.V., Taylor and Francis, London, pp. 156-170

Blaxter, J.H.S. (1953). Sperm storage and cross-fertilization of spring and autumn spawning herring. *Nature* **172**: 1189-1190

Caffey, R.H. and Tiersch, R.T. (2000). Economics and marketing of cryopreserved fish sperm. In: *Cryopreservation in Aquatic Species*. Edited by Tiersch T.R. and Mazik, P.M., World Aquaculture

Society, Baton Rouge, LA, pp. 388-408

Chao, N.H., Lin, T.T., Chen, Y.J., Hsu, H.W. and Liao, H.W. (1997). Cryopreservation of late embryos and early larvae in the oyster and hard clam. *Aquaculture* **155**: 31-44

David, W.M., Mitchell, D.L. and Walter, R.B. (2004). DNA repair in hybrid fish of the genus *Xiphophorus*. *Comparative Biochemistry and Physiology C* **138**: 301-309

DeBaulny, B.O., LeVern, Y., Kerboeuf, D. and Maisse, G. (1997). Flow cytometric evaluation of mitochondrial activity and membrane integrity in fresh and cryopreserved rainbow trout (*Oncorhynchus mykiss*) spermatozoa. *Cryobiology* **34**: 141-149

Devireddy, R.V., Raha, D. and Bischof, J.C. (1998). Measurement of water transport during freezing in cell suspensions using a differential scanning calorimeter. *Cryobiology* **36**: 124-155

Doebbler, G.F. (1966). Cryoprotective compounds - review and discussion of structure and function. *Cryobiology* **3**: 2-11

Dong, Q. (2005). *Comparative studies of sperm cryopreservation of diploid and tetraploid Pacific oysters*. PhD thesis, Louisiana State University, Baton Rouge, Louisiana.

Dong, Q., Eudeline, B., Huang, C., Allen, J. and Tiersch, T.R. (2005). Commercial-scale sperm cryopreservation of diploid and tetraploid Pacific oysters, *Crassostrea gigas*. *Cryobiology* **50**: 1-16

Dong, Q., Huang, C., Henk, M.C. and Tiersch, T.R. (2006a). Fixation methods can produce misleading artifacts in sperm cell ultrastructure of diploid and tetraploid Pacific oysters, *Crassostrea gigas*. *Cell and Tissue Research* **324**: 335-345

Dong, Q., Huang, C. and Tiersch, T.R. (2006b). Post-thaw amendment of cryopreserved sperm for use in artificial insemination of a viviparous fish, the green swordtail *Xiphophorus helleri*. *Aquaculture* (in press)

Dong, Q., Huang, C. and Tiersch, R.T. (2006c). Control of sperm concentration is necessary for standardization of sperm cryopreservation in aquatic species: evidence from sperm agglutination in oysters. *Cryobiology* (in press)

Draper, B.W., McCallum, C.M., Scout, J.L., Slade, A.J. and Moens, C.B. (2004). A high-throughput method for identifying *N*-ethyl-*N*-nitrosourea (ENU)-induced point mutations in zebrafish. In: *The Zebrafish Genomics, and Informatics. Methods in Cell Biology*, volume 77. Edited by Detrich III, H.W., Westerfield, M. and Zon, L.I., Elsevier Press, San Diego, pp. 91-112

Fuller, B.J., Lane, N. and Benson, E.E. (2004). *Life in the Frozen State*. CRC Press, New York, pp. 1-672

Gardiner, D.M. (1978). Utilization of extracellular glucose by spermatozoa of 2 viviparous fishes. *Comparative Biochemistry and Physiology A* **59**: 165-168

Grier, H.J. (1975). Aspects of germinal cyst and sperm development in *Poecilia-latipinna* (Teleostei-Poeciliidae). *Journal of Morphology* **146**: 229-249

Gwo, J.C. (2000). Cryopreservation of aquatic invertebrate semen: a review. *Aquaculture Research* **31**: 259-271

Hagedorn, M., Carter, V.L., Steyn, R.A., Krupp, D., Leong, J.C., Lang, R.P. and Tiersch, T.R. (2006). Preliminary studies of sperm cryopreservation in the mushroom coral, *Fungia scutaria*. *Cryobiology* **52**: 454-458

Harvey, B., Norman, K.R. and Ashwood-Smith, M.J. (1982). Cryopreservation of zebrafish spermatozoa using methanol. *Canadian Journal of Zoology* **60**: 1867-1870

Heater, S.J., Oehlers, L.P., Rains, J.D. and Walter, R.B. (2004). DNA polymerase mRNA and protein expression in *Xiphophorus* fish. *Comparative Biochemistry and Physiology C* **138**: 325-334

Hoar, W.S. (1969). Reproduction. In: *Fish Physiology*. Edited by Hoar, W.S. and Randall, D.J., Academic Press, New York, pp. 1-72

Huang, C., Dong, Q. and Tiersch, T.R. (2004a). Sperm cryopreservation of a live-bearing fish, the platyfish *Xiphophorus couchianus*. *Theriogenology* **62**: 971-989

Huang, C., Dong, Q., Walter, R.B. and Tiersch, T.R. (2004b). Initial studies on sperm cryopreservation of a live-bearing fish, the green swordtail *Xiphophorus helleri*. *Theriogenology* **62**: 179-194

Huang, C., Dong, Q., Walter, R.B. and Tiersch, T.R. (2004c). Sperm cryopreservation of green swordtail *Xiphophorus helleri*, a fish with internal fertilization. *Cryobiology* **48**: 295-308

Jonas-Davies, J.A.C., Winfrey, V. and Olson, G.E. (1983). Plasma-membrane structure in spermatogenic cells of the swordtail (Teleostei, *Xiphophorus helleri*). *Gamete Research* **7**: 309-324

Kallman, K.D. (1975). The platyfish, *Xiphophorus maculatus*. In: *Handbook of Genetics*. Edited by King, R.C., Plenum Publishing Corporation, New York, pp. 81-132

Kazianis, S., Khanolkar, V.A., Naim, R.S., Rains, J.D., Trono, D., Garcia, R., Williams, E.L. and Walter, R.B. (2004). Structural organization, mapping, characterization and evolutionary relationships of CDKN2 gene family members in *Xiphophorus* fishes. *Comparative Biochemistry and Physiology C* **138**: 291-299

Kazianis, S. and Walter, R.B. (2002). Use of platyfishes and swordtails in biological research. *Laboratory Animal* **31**: 46-52

Kime, D.E., Van Look, K.J.W., McAllister, B.G., Huyskens, G., Rurangwa, E. and Ollevier, F. (2001). Computer-assisted sperm analysis (CASA) as a tool for monitoring sperm quality in fish. *Comparative Biochemistry and Physiology C* **130**: 425-433

Lahnsteiner, F. (2000). Introduction to the special issue on 'Cryopreservation of gametes in aquatic species'. *Aquaculture Research* **31**: 229-229

Lang, R.P., Riley, K.L., Chandler, J.E. and Tiersch, T.R. (2003). The use of dairy protocols for sperm cryopreservation of blue catfish *Ictalurus furcatus*. *Journal of the World Aquaculture Society* **34**: 66-75

Lannan, J.E. (1971). Experimental self-fertilization of

Pacific oyster, *Crassostrea-gigas*, utilizing cryopreserved sperm. *Genetics* **68**: 599-601

Leibo, S.P. (2000). Sources of variation is cryopreservation. In: *Cryopreservation in Aquatic Species*. Edited by Tiersch, T.R. and Mazik, P.M., World Aquaculture Society, Baton Rouge, Louisiana, pp. 75-83

Leung, L.K.P. (1991). Principle of biological cryopreservation. In: *Fish Evolution and Systmatics: Evidence form Spermatozoa*. Authored by Jamieson, B.G.M., Cambridge University Press, Cambridge, pp. 231-244

Leung, L.K.P. and Jamieson, B.G.M. (1991). Live preservation of fish gametes. In: *Fish Evolution and Systmatics: Evidence form Spermatozoa*. Authored by Jamieson, B.G.M., Cambridge University Press, Cambridge, pp. 245-269

Mauger, R.E., Le Bail, P.Y. and Labbe, C. (2006). Cryobanking of fish somatic cells: optimizations of fin explant culture and fin cell cryopreservation. *Comparative Biochemistry and Physiology B* **144**: 29-37

Mazur, P. (1970). Cryobiology - freezing of biological systems. *Science* **168**: 939-949

Mazur, P. (1984). Freezing of living cells - mechanisms and implications. *American Journal of Physiology* **247**: C125-C142

McFadzen, I.R.B. (1995). Cryopreservation of the sperm of Pacific oyster *Crassostrea gigas*. In: *Methods in Molecular Biology: Cryopreservation and Freeze-Drying Protocols*. Edited by Day, J.G. and McCellan M.R., Humana Press, Totowa, pp. 145-149

Meierjohann, S., Schartl, M. and Volff, J.N. (2004). Genetic, biochemical and evolutionary facets of Xmrk-induced melanoma formation in the fish *Xiphophorus*. *Comparative Biochemistry and Physiology C* **138**: 281-289

Meryman, H.T. (1971). Cryoprotective agents. *Cryobiology* **8**: 173-183

Morisawa, M. and Suzuki, K. (1980). Osmolality and potassium ion - their roles in initiation of sperm motility in teleosts. *Science* **210**: 1145-1147

Morisawa, M., Suzuki, K., Shimizu, H., Morisawa, S. and Yasuda, K. (1983). Effects of osmolality and potassium on motility of spermatozoa from fresh water cyprinid fishes. *Journal of Experimental Biology* **107**: 95-103

Morris, J.P., Berghmans, S., Zahrieh, D., Neuberg, D.S., Kanki, J.P. and Look, A.T. (2003). Zebrafish sperm cryopreservation with N,N-dimethylacetamide. *Biotechniques* **35**: 956-968

Oehlers, L.P., Heater, S.J., Rains, J.D., Wells, M.C., David, W.M. and Walter, R.B. (2004). Gene structure, purification and characterization of DNA polymerase beta from *Xiphophorus maculatus*. *Comparative Biochemistry and Physiology C* **138**: 311-324

Paniagua Chavez, C.G., Buchanan, J.T. and Tiersch, T.R. (1998). Effect of extender solutions and dilution on motility and fertilizing ability of Eastern oyster sperm. *Journal of Shellfish Research* **17**: 231-237

Paniagua-Chavez, C.G., Jenkins, J., Segovia, M. and Tiersch, T.R. (2006). Assessment of gamete quality for the eastern oyster (*Crassostrea virginica*) by use of fluorescent dyes. *Cryobiology* **53**: 128-138

Pinisetty, D., Huang, C., Dong, Q., Tiersch, T.R. and Devireddy, R.V. (2005). Subzero water permeability parameters and optimal freezing rates for sperm cells of the southern platyfish, *Xiphophorus maculatus*. *Cryobiology* **50**: 250-263

Polge, C., Smith, A.U. and Parkes, A.S. (1949). Revival of spermatozoa after vitrification and dehydration at low temperatures. *Nature* **164**: 666

Rall, W.F. (1993). Advances in the cryopreservation of embryos and prospects for the application to the conservation of salmonid fishes. In: *Genetic Conservation of Salmonid Fishes*. Edited by Thorgaard, G.H. and Cloud, J.G., Plenum Press, New York, pp. 137-158

Rana, K.J. (1995). Cryopreservation of fish spermatozoa. In: *Methods in Molecular Biology: Cryopreservation and Freeze-Drying Protocols*. Edited by Day, J.G. and McCellan M.R., Humana Press, Totowa, pp. 151-165

Rowe, A.W. (1966). Biochemical aspects of cryoprotective agents in freezing and thawing. *Cryobiology* **3**: 12-18

Schartl, M. (1995). Platyfish and swordtails - a genetic system for the analysis of molecular mechanisms in tumor-formation. *Trends in Genetics* **11**: 185-189

Scott, A.P. and Baynes, S.M. (1980). A review of the biology, handling and storage of Salmonid spermatozoa. *Journal of Fish Biology* **17**: 707-739

Segovia, M., Jenkins, J.A., Paniagua-Chavez, C. and Tiersch, T.R. (2000). Flow cytometric evaluation of antibiotic effects on viability and mitochondrial, function of refrigerated spermatozoa of Nile tilapia. *Theriogenology* **53**: 1489-1499

Stoss, J. (1983). Fish gamete preservation and spermatozoan physiology. In: *Fish Physiology, Volume IX, Part B, Behavior and Fertility Control*. Edited by Hoar, W.S., Randall, D.J. and Donaldson, E.M., Academic Press, Orlando, FL, pp. 305-350

Suquet, M., Dreanno, C., Fauvel, C., Cosson, J. and Billard, R. (2000). Cryopreservation of sperm in marine fish. *Aquaculture Research* **31**: 231-243

Tavolga, W.N. (1949). Embryonic development of the platyfish (*Platypoecilus*), the swordtail (*Xiphophorus*) and their hybrids. *Bulletin of the American Museum of Natural History* **94**: 161-230

Taylor, R., Adams, G.D.J., Boardman, C.F. and Wallis, R.G. (1974). Cryoprotection permeant vs nonpermeant additives. *Cryobiology* **11**: 430-438

Thirumala, S., Huang, C.J., Dong, Q.X., Tiersch, T.R. and Devireddy, R.V. (2005). A theoretically estimated optimal cooling rate for the cryopreservation of sperm cells from a live-bearing fish, the green swordtail *Xiphophorus helleri*. *Theriogenology* **63**: 2395-2415

Tiersch, T.R. (2000). Introduction. In: *Cryopreservation in Aquatic Species*. Edited by Tiersch T.R. and Mazik, P.M., World Aquaculture Society, Baton Rouge, LA, pp. xix-xxvi

Tiersch, T.R. (2001). Cryopreservation in aquarium fishes. *Marine Biotechnology* **3**: S212-S223

Tiersch, T.R. and Jenkins, J.A. (2003). Biosecurity considerations for cryopreserved gametes and early life stages of aquatic species. In: *Biosecurity in Aquaculture Production Systems: Exclusion of Pathogens and Other Undesirables*. Edited by Lee, C.-S. and O'Bryen, P.J., World Aquaculture Society, Baton Rouge, Louisiana

Tiersch, T.R. and Mazik, P.M. (2000). *Cryopreservation in Aquatic Species*. World Aquaculture Society, Baton Rouge, LA

Viveiros, A.T.M. (2005). Semen cryopreservation in catfish species, with particular emphasis on the African catfish. *Animal Breeding Abstracts* **73**: 1N-9N

Walter, R.B. and Kazianis, S. (2001). *Xiphophorus* interspecies hybrids as genetic models of induced neoplasia. *Journal of Institute for Laboratory Animal Research* **42**: 299-322

Yang, H., Hazlewood, L., Heater, S.J., Guerrero, P.A., Walter, R.B. and Tiersch, T.R. (2007). Production of F1 interspecies hybrid offspring with cryopreserved sperm from a live-bearing fish, the swordtail *Xiphophorus helleri*. *Biology of Reproduction* **76**: 401-406

Yang, H., Hazelwood, L., Walter, R.B. and Tiersch, T.R. (2006). Effect of osmotic immobilization on refrigerated storage and cryopreservation of sperm from a viviparous fish, the green swordtail *Xiphophorus helleri*. *Cryobiology* **2**: 209-218

Zhang, T. (2004). Cryopreservation of gametes and embryos of aquatic species. In: *Life in the Frozen State*. Edited by Fuller, B., Lane, N. and Benson, E., CRC Press, New York, pp. 415-435

Zilli, L., Schiavone, R., Zonno, V., Storelli, C. and Vilella, S. (2003). Evaluation of DNA damage in *Dicentrarchus labrax* sperm following cryopreservation. *Cryobiology* **47**: 227-235

Spermatology. SRF Vol. 65. ERS Roldan and M Gomendio (eds) Nottingham University Press, Nottingham

M540 bodies and their impact on flow cytometric analyses of human spermatozoa

Sara Marchiani, Lara Tamburrino, Gianni Forti, Elisabetta Baldi and Monica Muratori

Department of Clinical Physiopathology, Andrology Unit and Center for Research, Transfer and High Education MCIDNENT, University of Florence, Italy

M540 bodies are membrane-surrounded round bodies occurring in semen of sub-fertile men. They appear variable in size and density, virtually devoided of chromatinic material and especially frequent in oligoasthenoteratozoospermic patients. Up to now, data collected by our group suggest that they may be apoptotic bodies that somehow escaped from testicular/epididymal phagocytosis. Indeed, they promptly stain with Merocyanine 540, a probe detecting the changes occurring in the membrane of apoptotic cells. In addition, they exhibit many of the apoptotic markers occurring in testicular apoptosis, including caspases activity, Fas receptor, p53 and DNA fragmentation (the latter detected by TUNEL assay). Due to the similarity in size and density between head sperm and M540 bodies, traditional protocols of sample preparation fail to yield sperm population completely free of M540 bodies, except for swim-up selection. In flow cytometry, it is possible to distinguish sperm from bodies by labelling samples with nuclear probes, because the latter fail to stain M540 bodies. Occurrence of M540 bodies in semen has been revealed only recently, at least in quantitative studies, and ,hence, many flow cytometric studies have not accounted for them. We summarise two studies (one on sperm ubiquitination and another on sperm DNA fragmentation) in which flow cytometric analyses were conducted both including and excluding M540 bodies from the sperm population. We found that in both cases M540 bodies largely affected the results. The study on sperm ubiquitination reveals that the direct correlation between sperm ubiquitination and good semen parameters is unmasked only after exclusion of M540 bodies from the analysis. The study on sperm DNA fragmentation shows that the amount of DNA damage in sub-fertile patients is more dramatic than expected from past investigations that included M540 bodies in the analysis.

The occurrence of M540 bodies in human semen

Human semen is a heterogeneous biological fluid in which both germ cells with different degree of maturity (other than mature spermatozoa) and somatic cells may be present. Re-

Corresponding author E-mail: e.baldi@dfc.unifi.it

cently, we have reported the occurrence of membrane-surrounded round bodies with variable dimensions in semen of sub-fertile subjects (Muratori Porazzi, Luconi, Marchiani, Forti and Baldi, 2004). We termed such structures "M540 bodies" due to their bright staining with Merocyanine 540 (M540). M540 is a lipophilic molecule, sensitive to changes in the membrane architecture, including those occurring during somatic apoptosis (Aussel, Bernard, Breittmayer, Pelassy, Zoccola and Bernard, 1993; Mower, Peckham, Illera, Fishbaugh, Stunz and Ashman, 1994) and capacitation of several mammalian sperm (Gadella and Harrison 2000; Rathi, Colenbrander, Bevers and Gadella, 2001), with the exception of human spermatozoa (Muratori *et al.*, 2004). Staining of M540 bodies with M540 indicates that their membrane is differently packaged from normal cell membranes. In addition, the failure of the nuclear stain Yo-Pro-1 to label M540 bodies, suggests their virtual lack of chromatinic material (Muratori *et al.*, 2004). Finally, M540 bodies migrate in different fractions of discontinuous density gradients, suggesting that they are heterogeneous elements in size and density (Muratori *et al.*, 2004). We have also showed that M540 bodies are ubiquitinated and that the amount of semen ubiquitination due to M540 bodies (excluding sperm ubiquitination) was sharply associated to poor quality semen (Muratori, Marchiani, Forti and Baldi, 2005).

Up to now, the nature of M540 bodies is unknown. Their characteristics prompted us to hypothesise that they might be apoptotic bodies escaping from phagocytosis. Indeed, abortive apoptosis has been hypothesised as a process that starts in the testis, fails to complete and to delete apoptotic germ cells and their residues, which appear in the human ejaculates of subfertile patients (Sakkas, Mariethoz, Manicardi, Bizzaro, Bianchi and Bianchi, 1999). M540 bodies do not expose phosphatydilserine (PS) on their surface (Muratori *et al.*, 2004). Such finding might appear to contradict our hypothesis on the nature of M540 bodies, since translocation of PS from the inner to the outer plane of the membrane is an early step of apoptotic programme and one of the main signals for recognition of apoptotic cells by phagocytes (Williamson and Schlegel, 2002). On the other hand, the same finding might explain why they escape testis/epididymal phagocytosis, thus supporting the hypothesis that M540 bodies are apoptotic bodies of testicular origin. In addition, preliminary experiments conducted by our group and aimed to challenge such hypothesis, show that apoptotic markers, including caspase activity, p53, Fas receptor and small quantities of fragmented DNA, are actually present in M540 bodies (unpublished results).

The first report on the incidence of M540 bodies in semen of sub-fertile patients, as assessed by flow cytometry within the region characteristic of sperm (Muratori *et al.*, 2004), showed that M540 bodies are especially frequent in oligoasthenoteratozoospermic subjects (up to 26% of the total population of sperm and bodies, $n = 7$). However, studies in progress conducted in a larger group of patients show that M540 bodies are much more frequent than the originally reported values (up to 53.3% of the total population of sperm and bodies, $n = 14$) and may be occasionally found in high percentages also in the other categories of patients (unpublished results).

The variable incidence of M540 bodies in semen and the possibility that they are (apoptotic) residues of germ cells, suggest that studies on ejaculated sperm have to take them into account, especially those studies conducted by flow cytometry that can not directly distinguish between sperm and bodies (see below). Indeed, their presence may be somehow confusing when analysing sperm populations, as evidenced by our own experience (see below). On the other hand, it is not easy to separate sperm from M540 bodies before the analysis due to their close similarity in size and density. Among the protocols of sample preparation, only selection by swim-up can yield a sperm population completely devoid of M540 bodies (Muratori *et al.*, 2004).

Flow cytometry is a suitable technique to studying sperm due to the ability to perform rapid, objective and reliable measures in large number of cells. The accuracy of cytometric results relies on several parameters, including the possibility to perform the analysis of fluorescence in the population of interest by gating it in the FSC/SSC dot plots (i.e., based on characteristics of size and density). When studying sperm by flow cytometry, we established a "flame-shaped region" in the FSC/SSC dot plot, characteristic of sperm and excluding debris, large cells and aggregates (Fig. 1A) (Muratori, Piomboni, Baldi, Filimberti, Pecchioli, Moretti, Gambera, Baccetti, Biagiotti, Forti and Maggi, 2000; Muratori, Maggi, Spinelli, Filimberti, Forti and Baldi, 2003). However most of M540 bodies are included in such FSC/SSC region of sperm, due to the above-mentioned similarity between sperm and M540 bodies. Since M540 bodies do not label with nuclear probes, we discriminate M540 bodies from sperm by labelling samples with a nuclear stain, i.e, propidium iodide (PI) or Yo-Pro1 (Fig.1B). Then we analyse the fluorescence signal after excluding the unstained elements (i.e., M540 bodies) from the region containing both sperm and M540 bodies (Fig. 1C).

In the next sections, we report two examples showing that the occurrence of M540 bodies can yield different, and even opposite, results depending on their inclusion or exclusion from the flow cytometric analysis of sperm.

Ubiquitination in sperm

Protein ubiquitination in sperm seems to be associated both to normal and to defective sperm. It has been proposed that ubiquitination in defective cells serves to mark and commit sperm to their elimination by phagocytosis by epididymal epithelial cells (Sutovsky, Moreno, Ramalho-Santos, Dominko, Thompson and Schatten, 2001). Such hypothesis partially relies on flow cytometric studies showing that sperm ubiquitination is associated to poor quality semen, including abnormal morphology (Sutovsky, Hauser and Sutovsky, 2004) and DNA fragmentation of sperm (Sutovsky, Neuber and Schatten, 2002). Since our group had found that M540 bodies were ubiquitinated, we studied sperm ubiquitination after excluding bodies from the flow cytometric analysis (Muratori *et al.*, 2005). We found that the results of correlation analyses between ubiquitination and standard semen parameters were quite different depending on what population was considered for the analyses. When the population formed solely by sperm was investigated, we found that the amount of ubiquitination correlates to good quality of semen, including normal sperm morphology. This result is in sharp contrast with previous reports (Sutovsky *et al.*, 2004; Rawe, Olmedo, Benmusa, Shiigi, Chemes and Sutovsky, 2002). In addition, when we examined ubiquitination due solely to M540 bodies, and correlated it to standard semen parameters, we found that ubiquitin bound to bodies is associated to poor semen quality (Muratori *et al.*, 2005). This last finding is consistent with our previous data demonstrating that the occurrence of M540 bodies in the ejaculate is a sign of poor semen quality (Muratori *et al.*, 2004). In addition, the finding suggests that the negative correlation previously found by other authors (Sutovsky *et al.*, 2004) between global semen ubiquitination and good quality semen (especially normal morphology) is mainly driven by components different from sperm and including M540 bodies.

DNA fragmentation in sperm

Sperm DNA fragmentation is a genomic anomaly frequently detected in sub-fertile men. This phenomenon has attracted the attention of researchers in andrology for two main reason. First, because concerns have been raised about the possibility that a sperm with fragmented DNA may fertilise the oocyte when assisted reproduction techniques and, in particular,

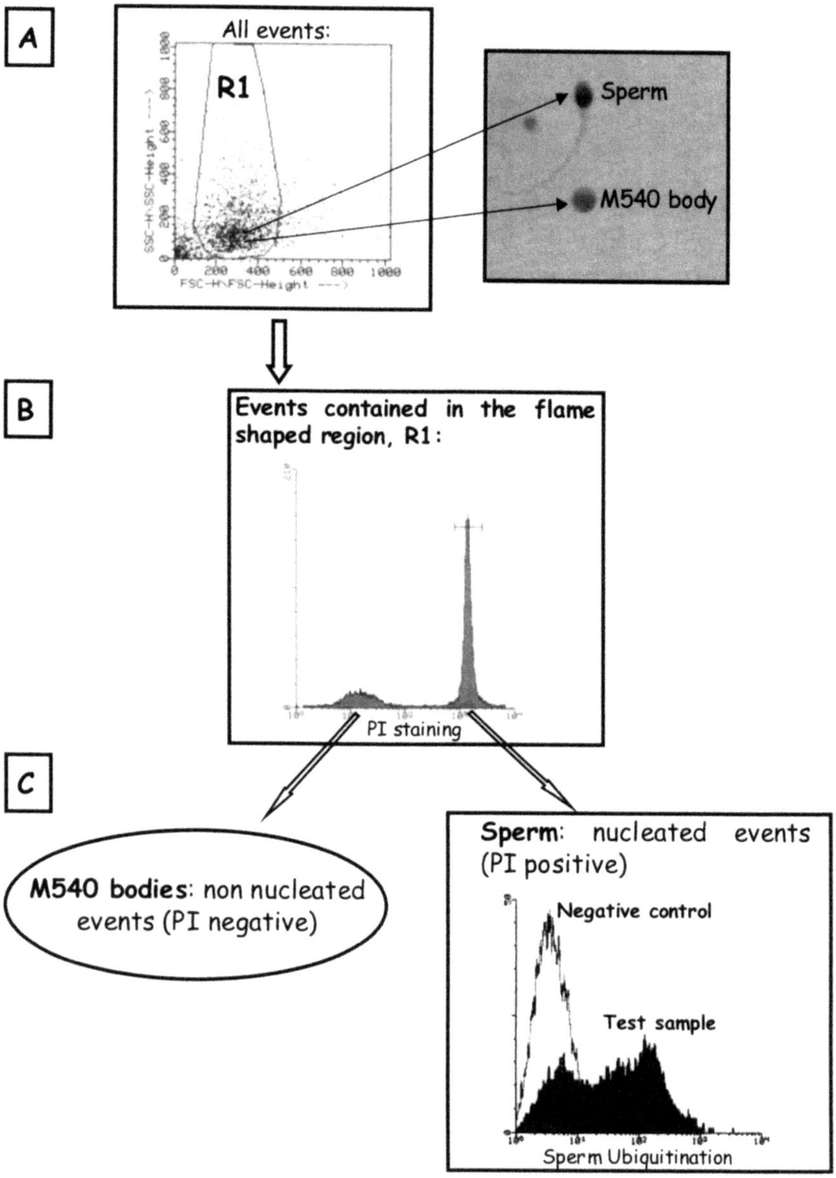

Figure 1. Strategy for sperm analysis by flow cytometry with exclusion of M540 bodies. (**A**) Left panel: FSC/SSC dot plot obtained from a semen sample. A "flame-shaped region" (R1) is established to exclude debris, large cells and aggregates. Right panel: Micrograph of a May Grunwald-Giemsa-stained sample, examined by light microscopy, showing the two kinds of elements (M540 body and sperm) contained in the R1 region. (**B**) Propidium iodide (PI) fluorescence histogram obtained within the R1 region and showing the nucleated (PI positive) and non-nucleated (PI negative) elements. (**C**) Analysis of fluorescence due to sperm ubiquitination, conducted within the PI positive population (sperm). The open histogram represents the negative control (absence of the primary antibody against ubiquitin); the solid histogram represents the test sample.

intracytoplasmatic sperm injection are applied. Secondly, because andrologists are presently searching for new semen parameters that are predictive of the fertilisation potential in men. In this light, sperm DNA fragmentation might represent a good option. Indeed, percentages of DNA-fragmented sperm correlate, although not exactly overlap, the extent of poor quality semen (reduced count and motility, abnormal morphology) (Lopes, Sun, Jurisicova, Meriano and Casper, 1998; Irvine, Twigg, Gordon, Fulton, Milne and Aitken, 2000; Muratori *et al.*, 2000). In addition, the occurrence of DNA fragmentation has been reported to negatively correlate to the outcome of *in vitro* fertilisation (Sun, Jurisicova and Casper, 1997; Lopes *et al.*, 1998; Virro, Larson-Cook and Evenson, 2004) and a threshold value of DNA fragmentation seems to exist over which the probability of pregnancy is dramatically reduced (Evenson, Larson and Jost, 2002).

Our group has extensively studied DNA fragmentation in human sperm (Muratori *et al.*, 2000, 2003; Muratori, Marchiani, Maggi, Forti and Baldi, 2006). The occurrence of M540 bodies in human semen together with the finding that they contain a small quantity of (intensively fragmented) DNA (unpublished results), prompted us to reconsider the amount of DNA fragmentation in sperm of sub-fertile patients in the light of this result. By flow cytometry, we determined DNA fragmentation by simultaneously labelling sperm using the TUNEL assay and PI in a group of 25 sub-fertile patients. Then we calculated the percentages of DNA-fragmented sperm both including and excluding M540 bodies from the analyses and compared results obtained (unpublished results). When M540 bodies were excluded from the analysis, we found that: (i) the percentage of sperm DNA fragmentation was higher than that found when excluding M540 bodies (difference: 8.6 $\pm$ 13.9%, mean $\pm$ SE; range: from -5.0 to +64.8%); (ii) sperm DNA fragmentation (calculated after exclusion of bodies) in this population of subjects correlates to poor semen parameters (at variance with DNA fragmentation in sperm plus bodies, which does not correlate with any semen parameter); (iii) two semen samples (out of 25) in which the sperm population was almost entirely DNA fragmented were unmasked. Overall, these data indicate that sperm DNA fragmentation and its impact on the outcome of *in vitro* fertilisation should be reconsidered in the light of the existence of M540 bodies.

References

Aussel, C., Bernard, G., Breittmayer, J.P., Pelassy, C., Zoccola, D. and Bernard, A. (1993). Monoclonal antibodies directed against the E2 protein (MIC2 gene product) induce exposure of phosphatidylserine at the thymocyte cell surface. *Biochemistry* **32:** 10096–10101

Evenson, D.P., Larson, K.L. and Jost, L.K. (2002). Sperm chromatin structure assay: its clinical use for detecting sperm DNA fragmentation in male infertility and comparisons with other techniques. *Journal of Andrology* **23:** 25–43

Gadella, B.M. and Harrison, R.A. (2000). The capacitating agent bicarbonate induces protein kinase A-dependent changes in phospholipid transbilayer behaviour in the plasma membrane. *Development* **127:** 2407–2420

Irvine, D.S., Twigg, J.P., Gordon, E.L., Fulton, N., Milne, P.A. and Aitken, R.J. (2000). DNA integrity in human spermatozoa: relationships with semen quality. *Journal of Andrology* **21:** 33-44

Lopes, S., Sun, J.G., Jurisicova, A., Meriano, J. and Casper, R.F. (1998). Sperm deoxyribonucleic acid fragmentation is increased in poor-quality semen samples and correlates with failed fertilization in intracytoplasmic sperm injection. *Fertility and Sterility* **69:** 528-532

Mower, D.A. Jr., Peckham, D.W., Illera, V.A., Fishbaugh, J.K., Stunz, L.L. and Ashman, R.F. (1994). Decreased membrane phospholipid packing and decreased cell size precede DNA cleavage in mature mouse B cell apoptosis. *Journal of Immunology* **152:** 4832–4842

Muratori, M., Marchiani, S., Maggi, M., Forti, G. and Baldi, E. (2006). Origin and biological significance of DNA fragmentation in human spermatozoa. *Frontiers in Biosciences* **11:** 1491-1499.

Muratori, M., Maggi, M., Spinelli, S., Filimberti, E., Forti, G. and Baldi, E. (2003). Spontaneous DNA fragmentation in swim-up selected human spermatozoa during long term incubation. *Journal of Andrology* **24:** 253–262

Muratori, M., Marchiani, S., Forti, G. and Baldi, E. (2005).

Sperm ubiquitination positively correlates to normal morphology in human semen. *Human Reproduction* **20:** 1035-1043

Muratori, M., Piomboni, P., Baldi, E., Filimberti, E., Pecchioli, P., Moretti, E., Gambera, L., Baccetti, B., Biagiotti, R., Forti, G. and Maggi, M. (2000). Functional and ultrastructural features of DNA-fragmented human sperm. *Journal of Andrology* **21:** 903-912

Muratori, M., Porazzi, I., Luconi, M., Marchiani, S., Forti, G. and Baldi, E. (2004). Annexin V binding and merocyanine staining fail to detect human sperm capacitation. *Journal of Andrology* **25:** 797–810

Rathi, R., Colenbrander, B., Bevers, M.M. and Gadella, B.M. (2001). Evaluation of in vitro capacitation of stallion spermatozoa. *Biology of Reproduction* **65:** 462–470

Rawe, V.Y., Olmedo, S.B., Benmusa, A., Shiigi, S.M., Chemes, H.E. and Sutovsky, P. (2002). Sperm ubiquitination in patients with dysplasia of the fibrous sheath. *Human Reproduction* **17:** 2119–2127

Sakkas, D., Mariethoz, E., Manicardi, G., Bizzaro, D., Bianchi, P.G. and Bianchi, U. (1999). Origin of DNA damage in ejaculated human spermatozoa. *Reviews of Reproduction* **4:** 31–37

Sun, J.G., Jurisicova, A. and Casper, R.F. (1997). Detection of deoxyribonucleic acid fragmentation in human sperm: correlation with fertilization in vitro. *Biology of Reproduction* **56:** 602-607

Sutovsky, P., Hauser, R. and Sutovsky, M. (2004). Increased levels of sperm ubiquitin correlate with semen quality in men from an andrology laboratory clinic population. *Human Reproduction* **19:** 628–638

Sutovsky, P., Moreno, R., Ramalho-Santos, J., Dominko, T., Thompson, W.E. and Schatten, G. (2001). A putative, ubiquitin-dependent mechanism for the recognition and elimination of defective spermatozoa in the mammalian epididymis. *Journal of Cell Science* **114:** 1665–1675

Sutovsky, P., Neuber, E. and Schatten, G. (2002). Ubiquitin-dependent sperm quality control mechanism recognizes spermatozoa with DNA defects as revealed by dual ubiquitin-TUNEL assay. *Molecular Reproduction and Development* **61:** 406–413

Virro, M.R., Larson-Cook, K.L. and Evenson, D.P. (2004). Sperm chromatin structure assay (SCSA) parameters are related to fertilization, blastocyst development, and ongoing pregnancy in in vitro fertilization and intracytoplasmic sperm injection cycles. *Fertility and Sterility* **81:** 1289-1295

Williamson, P. and Schlegel, R.A. (2002). Transbilayer phospholipid movement and the clearance of apoptotic cells. *Biochimica et Biophysica Acta* **1585:** 53–63

Sperm selection in assisted reproductive techniques

Uwe Paasch, Sonja Grunewald and Hans-Juergen Glander

European Academy of Andrology Center, University of Leipzig, Philipp-Rosenthal-Str. 23/25, 04103 Leipzig, Germany

The application of assisted reproduction techniques (ART) has provided help to many men seeking to father a child, although the current success rates of these procedures remain suboptimal. Selection of human spermatozoa prior to assisted reproduction techniques is currently based on criteria as viability, motility and morphology. The first sperm separation methods developed embraced washing procedures with subsequent resuspension of the male germ cells. Double density gradient centrifugation (DGC) and the swim-up procedure are currently used as standard preparation techniques while glass wool filtration (GWF) is also known to provide sperm samples with comparable recovery rates, motility, morphology, and fertilising capacity. Advanced protocols allow to select sperm according to their ultrastructural morphology or surface charges by electrophoresis. New insights into the molecular biology of spermatozoa prompted the development of molecular selection strategies. Those developments comprise hyaluronic acid mediated sperm selection, the annexin V magnetic activated cell separation (MACS), and annexin V molecular glass wool filtration. Sperm preparation that combines MACS with double density centrifugation provides spermatozoa of higher quality in terms of motility, viability and apoptosis indices compared with other conventional sperm preparation methods. Furthermore, sperm prepared according to this protocol showed improved ability to fertilise eggs using the hamster oocyte penetration assay.

Introduction

The application of assisted reproduction techniques (ART) has provided help to many men seeking to father a child, although the current success rates of these procedures remain suboptimal (Society for Assisted Reproductive Technology and American Society for Reproductive Medicine, 2004). Since raw semen cannot be used in most of the current techniques of ART a workup of the ejaculate is needed to extract those sperm that are capable to fertilise the egg.

The first sperm separation methods developed embraced one- or two-step washing procedures with subsequent resuspension of the male germ cells (Edwards, Bavister and Steptoe, 1969). Double density gradient centrifugation (DGC) and the swim-up procedure are currently used as standard preparation techniques while glass wool filtration (GWF) is also known to provide sperm samples with comparable recovery rates, motility, morphology, and fertilising capacity (Holmgren and Jeyendran, 1993; Johnson, Confino and Jeyendran, 1996; Van Den, Revelard, Bertrand, Biramane, Vanin and Englert, 1997; Henkel and Schill, 2003).

Corresponding author E-mail: uwe.paasch@medizin.uni-leipzig.de

To overcome the limitations of these classical separation procedures advanced protocols allow to select sperm according to their ultrastructural morphology (Bartoov, Berkovitz, Eltes, Kogosowski, Menezo and Barak, 2002) or surface charges by electrophoresis (Ainsworth, Nixon and Aitken, 2005). New insights into the molecular biology of spermatozoa prompted the development of molecular selection strategies. Those developments comprise hyaluronic acid mediated sperm selection (Huszar, Ozkavukcu, Jakab, Celik-Ozenci, Sati and Cayli, 2006), the annexin V magnetic activated cell separation (MACS), and annexin V molecular glass wool filtration (Grunewald, Paasch and Glander, 2001; Paasch, Grunewald, Fitzl and Glander, 2003).

Conventional selection strategies

Optimization of assisted conception outcomes involves the development of rapid, safe, effective techniques for the isolation of functional human spermatozoa free from significant DNA damage. Over the last decades a variety of standard procedures have been developed with certain modifications (conventional selection strategies). These sperm selection techniques can be classified as based on centrifugation, filtration, and migration of sperm. Among the centrifugation techniques the density gradient centrifugation (DGC) has been proposed to be a kind of gold standard in sperm preparation. The latest developments in the area of sperm selection are targeting the sperm surface with or without combination of a standard preparation protocols (molecular selection strategies).

Centrifugation techniques

Simple sperm wash. Today the one-step washing technique is considered a good alternative for processing certain compromised samples (Srisombut, Morshedi, Lin, Nassar and Oehninger, 1998). Since simple centrifugation causes definite harm to the cells it should not be applied for routine sperm processing.

Density gradient centrifugation (DGC). The semen sample is placed on top of a density gradient and centrifuged for 15–30 minutes. During this procedure, highly motile spermatozoa move actively in the direction of the sedimentation gradient and can therefore reach lower areas quicker than poorly motile or immotile cells. Finally highly purified motile sperm cells are enriched in the soft pellet at the bottom. The methods allows to enrich motile sperm (multiplicative factor of 1.2 to 2.1) and morphologically normal sperm (multiplicative factor of 1.2 to 1.8) (Glander, 1993). Continuous or discontinuous gradients are in use for DGC (Bolton and Braude, 1984; Pousette, Akerlof, Rosenborg and Fredricsson, 1986). The former involves a gradual increase in density towards the bottom of the vial, whereas the latter is characterised by clear boundaries between the different densities of the gradient.

For DGC, different separation media like Ficoll® (Harrison, 1976; Bongso, Ng, Mok, Lim, Teo, Wong and Ratnam, 1989), Dextran-Visotrast (Glander, Schaller and Ladusch, 1990), Nycodenz (Gellert-Mortimer, Clarke, Baker, Hyne and Johnston, 1988) and polyvinylpyrrolidone (PVP)-coated silica particles Percoll® (Gorus and Pipeleers, 1981) have been introduced. The Percoll®-DGC has been demonstrated to be beneficial for all methods of assisted reproduction (Mortimer and Mortimer, 1992). Comparison of the discontinuous Percoll® gradient with other techniques for separation of motile sperm indicated that the discontinuous Percoll® gradient has advantages in terms of recovery, enhancement of motility, and increased ability to penetrate zona-free hamster ova (Berger, Marrs and Moyer, 1985). Because of the risk of contaminations with endotoxins, possible membrane alterations and inflammatory responses that could be induced by the insemination of sperm populations contaminated with Percoll® it has been withdrawn from the market for clinical use in assisted reproduction in October 1996 (Pickering, Fleming, Braude, Bolton and Gresham, 1989; Pharmacia Biotech, 1996) and replacement products (e.g. IxaPrep®, PureSperm®, SilSelect®, Isolate®) have recently been introduced (Makkar, Ng, Yeung and Ho, 1999, Sills, Wittkowski, Tucker, Perloe, Kaplan and Palermo, 2002). All these less toxic media contain silane-coated silica particles. The quality of sperm prepared with these media seems to be equivalent to those prepared with Percoll® after assessing recovery rate, motility, viability, normal sperm morphology and velocity (Claassens, Menkveld and Harrison, 1998; Chen and Bongso, 1999; Soderlund and Lundin, 2000) although there are conflicting reports for motility (McCann and Chantler, 2000). Since nitric oxide production has been found to be lower using IxaPrep® as replacement, and the fact that NO adversely affects sperm motility, sperm-zona binding and embryonic development, there is an overall advantage of those new replacing media (Wu, Huang, Tai, Lui and Liu, 2004). Recently, samples processed by DGC have been shown to be low in signs known to be present in apoptotic somatic cells such as caspase activation, disruption of mitochondrial membrane potential (MMP), DNA-fragmentation, and externalised phosphatidylserine (PS) (Larson, Brannian, Timm, Jost and Evenson, 1999; Ollero, Gil-Guzman, Lopez, Sharma, Agarwal, Larson, Evenson, Thomas and Alvarez, 2001; Paasch, Agarwal, Gupta, Sharma, Grunewald, Thomas and Glander, 2003; Barroso, Taylor, Morshedi, Manzur, Gavino and Oehninger, 2006).

To further improve the quality of sperm a combination of washing procedures or DGC and swim-up are widely used (Scott, Oehninger, Menkveld, Veeck and Acosta, 1989). The combined use of DGC and a revised swim-up procedure was recently demonstrated to remove completely HIV-1 RNA and pro-viral DNA (Kato, Hanabusa, Kaneko, Takakuwa, Suzuki, Kuji, Jinno, Tanaka, Kojima, Iwashita, Yoshimura and Tanaka, 2006).

Filtration techniques

Glass wool filtration (GWF). The principle of glass wool filtration has been very early introduced to prepare sperm (Paulson and Polakoski, 1977; Paulson, Polakoski and Leto, 1979). It is of great advantage that the whole ejaculate can be processed followed by an additional centrifugation step to remove the seminal plasma. Highly motile spermatozoa are separated from immotile sperm cells by means of densely packed glass wool fibers using gravitational forces and the self-propelled motion of the cells. The efficacy of each filtration run depends on the properties of the glass wool used (Sanchez, Concha, Ichikawa, Henkel and Schill, 1996). The chemical nature of the glass (i.e. borate glass, silicate glass or quartz glass), the surface structure, charge, and thickness of the glass wool fibers influence the efficacy of filtration (Ford, McLaughlin, Prior, Rees, Wardle and Hull, 1992). Potential risks of the technique are damages to the membrane and acrosome or the transmission of glass particles into the filtrate (Sherman, Paulson and Liu, 1981). The glass wool, code number 112, from Manville Fiber Glass Corp. (Denver, CO, USA) available as SpermFertil® columns (TransMIT GmbH, Giessen, Germany) has been tested extensively. Glass-wool filtration separates human spermatozoa according to motility and size of the sperm head. The size of the sperm head closely correlated with the chromatin condensation quality and DNA fragmentation as measured by the Sperm Chromatin Structure Assay (Henkel, Franken, Lombard and Schill, 1994; Larson *et al.*, 1999). Reactive oxygen species are known to induce DNA fragmentation in sperm which in turn are directly and negatively correlated to the outcome of ART (Greco, Romano, Iacobelli, Ferrero, Baroni, Minasi, Ubaldi, Rienzi and Tesarik, 2005). GWF eliminates up to 90% of leukocytes (Sanchez *et al.*, 1996), leading to a significant reduction of reactive oxygen species in the sample and, therefore, contributes to prevent sperm DNA damage (Henkel and Schill, 1998). In comparison to Percoll-DGC, glass wool filtration yielded more functionally intact spermatozoa of superior quality (Johnson *et al.*, 1996) without affecting the fertilisation rate and embryo quality in an intracytoplasmic sperm injection (ICSI) programme (Van Den *et al.*, 1997). However these positive effects were related, in part, to the initial sperm concentration (Rhemrev, Jeyendran, Vermeiden and Zaneveld, 1989). Compared to the swim-up techniques GWF yields a higher recovery of viable and potentially fertile spermatozoa (Van der Ven, Jeyendran, Al Hasani, Tunnerhoff, Hoebbel, Diedrich, Krebs and Perez-Pelaez, 1988; Holmgren and Jeyendran, 1993; Coetzee, Erasmus, Kruger, Menkveld and Lombard, 1994). At the molecular level, samples processed by GWF were shown to be low in caspase activation, disruption of MMP, DNA-fragmentation, and externalised PS (Larson *et al.*, 1999; Rasch, Miska, Grunewald, Glander and Paasch, 2005).

Glass bead filtration (GBF). Since selected glass surfaces are able to sort out leukocytes and sperm with damaged membranes the idea has been developed to use these features in a system which provides a rather smart surface instead of the unpredictable glass wool surface. Spheres made of glass arranged in a column meet these criteria (Daya, Gwatkin and Bissessar, 1987). The glass bead filtration system filters spermatozoa that have significantly better progressive motility and viability and also have enhanced fertilising capacity (Daya and Gwatkin, 1987). Samples processed with GBF provided higher counts of motile sperm than with swim-up preparation (Lechtzin, Garside, Heyner and Hillman, 1991), although with lower velocity and linearity (Calamera, Quiros, Brugo and Nicholson, 1991). GBF before cryopreservation significantly enhanced post-thaw sperm quality (Klinc, Kosec and Majdic, 2005). However, concerns regarding the possibility of transfer of glass spheres into the filtrate prevented a wider acceptance.

Sephadex columns. Sephadex bead sperm preparation has been introduced to ART in the early nineties (Drobnis, Zhong and Overstreet, 1991). The gel filtration is based on a chromatographic media composed of macroscopic beads synthetically derived from the polysaccharide dextran. SpermPrep®, a commercial kit, has been intensively investigated. Compared to migration-sedimentation and swim-up it provides significantly higher yields (Gabriel and Vawda, 1993) and more motile spermatozoa than obtained by DGC (Ohashi, Saji, Wakimoto, Kato, Tsutsui and Tanizawa, 1992). In comparison to swim-up, morphologically normal sperm are enriched in the filtrate and significantly higher pregnancy rates for intrauterine insemination are reported (Zavos and Centola, 1992). However, after DGC spermatozoa showed a significantly higher percentage of normally chromatin-condensed and morphologically normal spermatozoa although the fertilisation rates were similar (Hammadeh, Zavos, Rosenbaum and Schmidt, 2001).

Migration techniques

For all migration methods, the self-propelled movement of spermatozoa is an essential requisite and a guarantee for a very clean workup.

Swim-up procedure. The swim-up procedure uses the active motion of spermatozoa. Intact moving cells are swimming out of a pellet derived by a simple washing step into an overlaid media for 30-60 minutes. Highly motile, morphologically intact spermatozoa are enriched in the absence of other cells, proteins and debris within the supernatant (Dominguez, Burgos and Fornes, 1999). The antigravitational centrifugation has been proposed to shorten the duration of preparation (Babbo, Hecht and Jeyendran, 1999).

The method may have limitations due to the fact that a close sperm-sperm contact, debris and other substances may lead to extensive reactive oxygen species production and subsequent DNA damage (Twigg, Irvine, Houston, Fulton, Michael and Aitken, 1998), although this could not be verified (Younglai, Holt, Brown, Jurisicova and Casper, 2001). The swim-up performed without pre-wash directly from the ejaculate, together with other technical modifications, are attempts to reduce reactive oxygen species production and to increase the total number of recovered spermatozoa, which can then also be used for ICSI (Al Hasani, Kupker, Baschat, Sturm, Bauer, Diedrich and Diedrich, 1995). The use of highly purified preparation of hyaluronic acid at 1 mg/ml in culture medium (SpermSelect™, Pharmacia, Uppsala, Sweden) resulted in a significantly higher recovery of motile spermatozoa and higher pregnancy rates in a clinical *in vitro* fertilisation programme (Wikland, Wik, Steen, Qvist, Soderlund and Janson, 1987). When compared with gradient centrifugation, the clearing efficacy after swim-up is lower for disomies and higher for diploidies (Henkel *et al.*, 1994; Jakab, Kovacs, Zavaczki,Borsos, Bray-Ward, Ward and Huszar, 2003). Sperm vitality in cryopreserved oligozoospermic samples was improved by the swim-up prior to freezing in a comparable level to DGC (Counsel, Bellinge and Burton, 2004).

Migration-sedimentation. This method combines a direct swim-up procedure with an additional sedimentation step requiring a special glass or plastic tube with an inner cone (Tea, Jondet and Scholler, 1983). Although modifications have been proposed and the spermatozoa obtained are of good quality in terms of their motility pattern, the method has not been established widely in ART due to its low yield (Yener, Matghur and Parent, 1990; Yavetz, Hauser, Homonnai, Paz, Lessing, Amit and Yogev, 1996). However, as sperm characteristics correlated significantly with their fertilising capacity, the method could be used in the evaluation of sperm function (Hauser, Homonnai, Paz, Yavetz, Amit, Lessing, Peyser and Yogev, 1992).

Transmembrane migration. This sperm preparation method uses filtration through a membrane filter with cylindrical pores at right angles to the plane of the membrane (Chijioke, Crocker, Gilliam, Owens and Pearson, 1988). Due to a low ratio of the total cross-sectional area of the pores to the overall membrane area the yield is extremely low. L4-membranes selective for spermatozoa with normal membrane integrity have been introduced; its use also leads to a simultaneous increase in motility and significant depletion of leukocytes (Agarwal, Manglona and Loughlin, 1992). Despite these improvements the transmembrane migration technique has never come into practical clinical use for human assisted reproduction.

Advanced sperm selection methods

Ultrastructural sperm selection. The ultrastructural morphology of the sperm head components have been correlated with sperm fertilising capacity *in vitro* (Mashiach, Fisch, Eltes, Tadir, Ovadia and Bartoov, 1992). The examination is performed in real time using an inverted light microscope equipped with high-power Nomarski optics enhanced by digital imaging to achieve a magnification of up to 6300x. The motile sperm organelle morphology examination (MSOME) method was significantly and positively associated with both fertilisation rate and pregnancy outcome (Bartoov, Berkovitz, Eltes, Kogosovsky, Yagoda, Lederman, Artzi, Gross and Barak, 2002). Intracytoplasmic sperm injection performed with selected spermatozoa with strictly defined morphologically normal nuclei improved significantly the incidence of pregnancy in couples with previous ICSI failures (Bartoov *et al.*, 2003) especially if the use of sperm with vacuoles is avoided.

Electrophoretic sperm isolation. A novel approach to sperm isolation, the electrophoretic sperm separation based on sperm size and charge, has been described recently (Ainsworth *et al.*, 2005). The electrophoresis-based Microflow® technology for the separation of spermatozoa by size and charge consists of two outer chambers separated from two inner chambers by polyacrylamide restriction membranes with a pore size of 15 kDa. The pore size allows directional movement of competent spermatozoa in the applied electric field and the size-exclusion of contaminating cell populations. Large numbers of spermatozoa that are viable and morphologically normal can be selected. The cell population provided shows low DNA damage and contaminating cells. It compared favorably with DGC in terms of the purity of the sperm population and the lack of reactive oxygen species generation, as well as in the viability and morphological integrity of the isolated cells. However, electrophoresis of spermatozoa is detrimental to their motility and therefore this procedure does not enrich for sperm motility (Ainsworth *et al.*, 2005).

Molecular selection strategies

Hyaluronic acid-mediated sperm selection. Hyaluronic acid or hyaluronan, which is a linear repeating polymer of disaccharides, increases the motility and viability of freshly ejaculated, as well as cryopreserved-thawed human spermatozoa when added to sperm preparation media (Huszar, Willetts and Corrales, 1990). Hyaluronic acid-mediated sperm selection is a novel technique that is comparable to sperm-zona pellucida binding. The presence of hyaluronic acid receptor on the plasma membrane of mature acrosome-intact sperm, coupled with hyaluronic acid coated glass or plastic surfaces, facilitates selection of single mature sperm (Huszar *et al.*, 2006). The frequencies of sperm with chromosomal disomy are reduced approximately fourfold to fivefold in hyaluronic acid-selected sperm compared with semen sperm compensating the increase in such abnormalities in ICSI offspring. Hyaluronic acid binding also excludes immature sperm with cytoplasmic extrusion, persistent histones, and DNA chain breaks (Huszar, Ozenci, Cayli, Zavaczki, Hansch and Vigue, 2003).

Annexin V MACS separation. Although the presence of apoptosis in ejaculated sperm, as compared with somatic cells, still remains to be demonstrated (Sakkas, Mariethoz and StJohn, 1999, Sakkas, Moffatt, Manicardi, Mariethoz, Tarozzi and Bizzaro 2002) ejaculates from infertility patients are of low quality as measured by standard semen parameters and apoptosis-like characteristics, e.g. activated caspases, externalization of phosphatidylserine (PS) to the outer plasma membrane, DNA-fragmentation, and disruption on the mitochondrial membrane potential (Comhaire, Vermeulen, Hinting and Schoonjans, 1988; Glander and Schaller, 1999; Oehninger, Morshedi, Weng, Taylor, Duran, and Beebe 2003, Paasch *et al.*, 2003; Grunewald, Paasch, Wuendrich and Glander, 2005).

The main apoptotic event detectable at the sperm surface is the externalization of PS, which is usually present on the inner leaflet of the sperm plasma membrane only (Vermes, Haanen, Steffens-Nakken and Reutelingsperger, 1995; Oosterhuis and Vermes, 2004). Annexin V is a phospholipid binding protein that has high affinity for PS and lacks the ability to pass through an intact sperm membrane (van Heerde, de Groot and Reutelingsperger, 1995). Therefore, annexin V binding to spermatozoa may be used to label sperm that have compromised membrane integrity and that are less capable to fertilise eggs (Glander and Schaller, 1999).

Annexin V-conjugated super-paramagnetic microbeads can effectively separate non-apoptotic spermatozoa from those with deteriorated plasma membranes based on the externalization of PS using magnetic-activated cell sorting (MACS). MACS separation of sperm yields 2 fractions:

annexin V-negative (intact membranes, non-apoptotic) and annexin V-positive (externalised PS, apoptotic) which is retained in the magnetic field (Grunewald *et al.*, 2001; Paasch *et al.*, 2003). Sperm preparation that combines MACS with double density centrifugation provides spermatozoa of higher quality in terms of motility, viability and apoptosis indices (caspase activation, MMP disruption and DNA fragmentation) compared with other conventional sperm preparation methods (Said, Grunewald, Paasch, Glander, Liang and Agarwal, 2005). Furthermore, sperm prepared according to this protocol showed improved ability to fertilise eggs using the hamster oocyte penetration assay (Said, Agarwal, Grunewald, Rasch, Baumann, Kriegel, Li, Glander, Thomas and Paasch, 2005).

Annexin V molecular glass wool separation. Although there is a major potential advantage of integrating annexin V-MACS in standard semen preparation protocols, to date the possibility of an accidental transmission of super-paramagnetic microbeads into eggs cannot be fully excluded. Therefore, a sperm preparation system using the binding properties of annexin V by avoiding free flotation of super-paramagnetic microbeads in a liquid phase buffer might help to avoid side effects. Towards the development of a molecular-based preparation system with enhanced impact on selection of vital spermatozoa with superior fertilising capacity the annexin V glass filtration technique is promising.

Recently we have demonstrated the feasibility to combine the classical glass wool filtration method with PS-binding properties of annexin V to a solid phase molecular filtration system. We investigated the apoptosis markers of sperm following a simple wash, glass wool filtration, molecular glass wool filtration, and MACS. It could be demonstrated that the application of annexin V-glass wool may improve the outcome of assisted reproductive techniques. Our data support the highly efficient filtration capacity of a solid phase annexin V coated glass wool filter.

Conclusion

Assisted conception techniques are now applied for 2-4% of new births in developed countries. While this technology has revolutionised the treatment of infertile couples, sperm preparation techniques still require improvements in evaluation of the classical and development of advanced and molecular selection strategies.

There is insufficient evidence to recommend any of the classical preparation techniques. Large, high quality randomised controlled trials, comparing the effectiveness of gradient-, swim-up-, wash-, and centrifugation-based techniques on clinical outcomes are lacking. Results from studies comparing semen parameters may suggest a preference for gradient technique, but firm conclusions cannot be drawn and the limitations should be taken into consideration (Boomsma, Heineman, Cohlen and Farquhar, 2004).

Another area to be improved is the safety of the sperm selection method. It is thought that the presence of DNA damage in the fertilising spermatozoon is responsible for early pregnancy loss and birth defects (Ainsworth *et al.*, 2005). Therefore there is an urgent need for optimised procedures to isolate spermatozoa for ART that maintain full functional competence by avoiding extrinsic harm due to reactive oxygen species production (Aitken, Gordon, Harkiss, Twigg, Milne, Jennings and Irvine, 1998; Zini, Nam, Mak, Phang and Jarvi, 2000).

The latest developments in advanced and molecular sperm selection strategies open a new field to develop customised sperm preparation techniques according to the primary quality of the ejaculate and the method of ART to be used.

References

Agarwal, A., Manglona, A. and Loughlin, K.R. (1992). Improvement in semen quality and sperm fertilizing ability after filtration through the L4 membrane: comparison of results with swim up technique. *Journal of Urology* **147:** 1539-1541

Ainsworth, C., Nixon, B. and Aitken, R.J. (2005). Development of a novel electrophoretic system for the isolation of human spermatozoa. *Human Reproduction* **20:** 2261-2270

Aitken, R.J., Gordon, E., Harkiss, D., Twigg, J.P., Milne, P., Jennings, Z. and Irvine, D.S. (1998). Relative impact of oxidative stress on the functional competence and genomic integrity of human spermatozoa. *Biology of Reproduction* **59:** 1037-1046

Al Hasani, S., Kupker, W., Baschat, A.A., Sturm, R., Bauer, O., Diedrich, C. and Diedrich, K. (1995). Mini-swim-up: a new technique of sperm preparation for intracytoplasmic sperm injection. *Journal of Assisted Reproduction and Genetics* **12:** 428-433

Babbo, C.J., Hecht, B.R. and Jeyendran, R.S. (1999). Increased recovery of swim-up spermatozoa by application of "antigravitational" centrifugation. *Fertility and Sterility* **72:** 556-558

Barroso, G., Taylor, S., Morshedi, M., Manzur, F., Gavino, F. and Oehninger, S. (2006). Mitochondrial membrane potential integrity and plasma membrane translocation of phosphatidylserine as early apoptotic markers: a comparison of two different sperm subpopulations. *Fertility and Sterility* **85:** 149-154

Bartoov, B., Berkovitz, A., Eltes, F., Kogosowski, A., Menezo, Y. and Barak, Y. (2002). Real-time fine morphology of motile human sperm cells is associated with IVF-ICSI outcome. *Journal of Andrology* **23:** 1-8

Bartoov, B., Berkovitz, A., Eltes, F., Kogosovsky, A., Yagoda, A., Lederman, H., Artzi, S., Gross, M. and Barak, Y. (2003). Pregnancy rates are higher with intracytoplasmic morphologically selected sperm injection than with conventional intracytoplasmic injection. *Fertility and Sterility* **80:** 1413-1419

Berger, T., Marrs, R.P. and Moyer, D.L. (1985). Comparison of techniques for selection of motile spermatozoa. *Fertility and Sterility* **43:** 268-273

Bolton, V.N. and Braude, P.R. (1984). Preparation of human spermatozoa for in vitro fertilization by isopycnic centrifugation on self-generating density gradients. *Archives of Andrology* **13:** 167-176

Bongso, A., Ng, S.C., Mok, H., Lim, M.N., Teo, H.L., Wong, P.C. and Ratnam, S. (1989). Improved sperm concentration, motility, and fertilization rates following Ficoll treatment of sperm in a human in vitro fertilization program. *Fertility and Sterility* **51:** 850-854

Boomsma, C.M., Heineman, M.J., Cohlen, B.J. and Farquhar, C. (2004). Semen preparation techniques for intrauterine insemination. *Cochrane Database of Systematic Reviews* CD004507

Calamera, J.C., Quiros, M.C., Brugo, S. and Nicholson, R.F. (1991). Comparison between swim-up and glass bead column techniques for the separation of human spermatozoa. *Andrologia* **23:** 259-261

Chen, M.J. and Bongso, A. (1999). Comparative evaluation of two density gradient preparations for sperm separation for medically assisted conception. *Human Reproduction* **14:** 759-764

Chijioke, P.C., Crocker, P.R., Gilliam, M., Owens, M.D. and Pearson, R.M. (1988). Importance of filter structure for the trans-membrane migration studies of sperm motility. *Human Reproduction* **3:** 241-244

Claassens, O.E., Menkveld, R. and Harrison, K.L. (1998). Evaluation of three substitutes for Percoll in sperm isolation by density gradient centrifugation. *Human Reproduction* **13:** 3139-3143

Comhaire, F.H., Vermeulen, L., Hinting, A. and Schoonjans, F. (1988). Accuracy of sperm characteristics in predicting the in vitro fertilizing capacity of semen. *Journal of In Vitro Fertilization and Embryo Transfer* **5:** 326-331

Counsel, M., Bellinge, R. and Burton, P. (2004). Vitality of oligozoospermic semen samples is improved by both swim-up and density gradient centrifugation before cryopreservation. *Journal of Assisted Reproduction and Genetics* **21:** 137-142

Daya, S. and Gwatkin,R.B. (1987). Improvement in semen quality using glass bead column. *Archives of Andrology* **18:** 241-244

Daya, S., Gwatkin, R.B. and Bissessar, H. (1987). Separation of motile human spermatozoa by means of a glass bead column. *Gamete Research* **17:** 375-380

Dominguez, L.A., Burgos, M.H. and Fornes, M.W. (1999). Morphometrical comparison of human spermatozoa obtained from semen and swim-up methodology. *Andrologia* **31:** 23-26

Drobnis, E.Z., Zhong, C.Q. and Overstreet, J.W. (1991). Separation of cryopreserved human semen using Sephadex columns, washing, or Percoll gradients. *Journal of Andrology* **12:** 201-208

Edwards, R.G., Bavister, B.D. and Steptoe, P.C. (1969). Early stages of fertilization *in vitro* of human oocytes matured in vitro. *Nature* **221:** 632-635

Erasmus, E.L., Kruger, T.F., Menkveld, R. and Lombard, C.J. (1994). Glass wool filter preparation of cryopreserved spermatozoa. *Andrologia* **26:** 33-34

Ford, W.C.L., McLaughlin, E.A., Prior, S.M., Rees, J.M., Wardle, P.G. and Hull, M.G. (1992). The yield, motility and performance in the hamster egg test of human spermatozoa prepared from cryopreserved semen by four different methods. *Human Reproduction* **7:** 654-659

Gabriel, L.K. and Vawda, A.I. (1993). Preparation of human sperm for assisted conception: a comparative study. *Archives of Andrology* **30:** 1-6

Gellert-Mortimer, S.T., Clarke, G.N., Baker, H.W., Hyne, R.V. and Johnston, W.I. (1988). Evaluation of Nycodenz and Percoll density gradients for the selection of motile human spermatozoa. *Fertility and*

Sterility **49:** 335-341

Glander, H.J. (1993). Effektivität der Methoden für die Spermienkollektion, -separation und -konzentrierung. *Hautnah Derm* **9:** 52-60

Glander, H.J. and Schaller, J. (1999). Binding of annexin V to plasma membranes of human spermatozoa: a rapid assay for detection of membrane changes after cryostorage. *Molecular Human Reproduction* **5:** 109-115

Glander, H.J., Schaller, J. and Ladusch, M. (1990). [A simple method for separating fresh and cryopreserved human sperm using dextran-visotrast density gradient centrifugation]. *Zentralblatt Gynakologie* **112:** 91-97

Gorus, F.K. and Pipeleers, D.G. (1981). A rapid method for the fractionation of human spermatozoa according to their progressive motility. *Fertility and Sterility* **35:** 662-665

Greco, E., Romano, S., Iacobelli, M., Ferrero, S., Baroni, E., Minasi, M.G., Ubaldi, F., Rienzi, L. and Tesarik, J. (2005). ICSI in cases of sperm DNA damage: beneficial effect of oral antioxidant treatment. *Human Reproduction* **20:** 2590-2594

Grunewald, S., Paasch, U. and Glander, H.J. (2001). Enrichment of non-apoptotic human spermatozoa after cryopreservation by immunomagnetic cell sorting. *Cell and Tissue Banking* **2:** 127-133

Grunewald, S., Paasch, U., Wuendrich, K. and Glander, H.J. (2005). Sperm caspases become more activated in infertility patients than in healthy donors during cryopreservation. *Archives of Andrology* **51:** 449-460

Hammadeh, M.E., Zavos, P.M., Rosenbaum, P. and Schmidt, W. (2001). Comparison between the quality and function of sperm after semen processing with two different methods. *Asian Journal of Andrology* **3:** 125-130

Harrison, R.A.P. (1976). A highly efficient method for washing mammalian spermatozoa. *Journal of Reproduction and Fertility* **48:** 347-353

Hauser, R., Homonnai, Z.T., Paz, G.F., Yavetz, H., Amit, A., Lessing, J.B., Peyser, M.R. and Yogev, L. (1992). Migration sedimentation technique as a predictive test for the fertilizing capacity of spermatozoa in an in-vitro fertilization programme. *International Journal of Andrology* **15:** 498-503

Henkel, R., Franken, D.R., Lombard, C.J. and Schill, W.B. (1994). Selective capacity of glass-wool filtration for the separation of human spermatozoa with condensed chromatin: a possible therapeutic modality for male-factor cases? *Journal of Assisted Reproduction and Genetics* **11:** 395-400

Henkel, R. and Schill, W.B. (1998). Sperm separation in patients with urogenital infections. *Andrologia* **30** *Supplement* **1:** 91-97

Henkel, R.R. and Schill, W.B. (2003). Sperm preparation for ART. *Reproductive Biology and Endocrinology* **1:** 108

Holmgren, W.J. and Jeyendran, R.S. (1993). Synergistic effect of TEST-yolk buffer treatment and glass wool filtration of spermatozoa on the outcome of the hamster oocyte penetration assay. *Human Reproduction* **8:** 425-427

Huszar, G., Ozenci, C.C., Cayli, S., Zavaczki, Z., Hansch, E. and Vigue, L. (2003). Hyaluronic acid binding by human sperm indicates cellular maturity, viability, and unreacted acrosomal status. *Fertility and Sterility* **79** *Supplement* **3:** 1616-1624

Huszar, G., Ozkavukcu, S., Jakab, A., Celik-Ozenci, C., Sati, G.L. and Cayli, S. (2006). Hyaluronic acid binding ability of human sperm reflects cellular maturity and fertilizing potential: selection of sperm for intracytoplasmic sperm injection. *Current Opinion in Obstetrics and Gynecology* **18:** 260-267

Huszar, G., Willetts, M. and Corrales, M. (1990). Hyaluronic acid (Sperm Select) improves retention of sperm motility and velocity in normospermic and oligospermic specimens. *Fertility and Sterility* **54:** 1127-1134

Jakab, A., Kovacs, T., Zavaczki, Z., Borsos, A., Bray-Ward, P., Ward, D. and Huszar, G. (2003). Efficacy of the swim-up method in eliminating sperm with diminished maturity and aneuploidy. *Human Reproduction* **18:** 1481-1488

Johnson, D.E., Confino, E. and Jeyendran, R.S. (1996). Glass wool column filtration versus mini-Percoll gradient for processing poor quality semen samples. *Fertility and Sterility.* **66:** 459-462

Kato, S., Hanabusa, H., Kaneko, S., Takakuwa, K., Suzuki, M., Kuji, N., Jinno, M., Tanaka, R., Kojima, K., Iwashita, M., Yoshimura, Y. and Tanaka, K. (2006). Complete removal of HIV-1 RNA and proviral DNA from semen by the swim-up method: assisted reproduction technique using spermatozoa free from HIV-1. *AIDS* **20:** 967-973

Klinc, P., Kosec, M. and Majdic, G. (2005). Freezability of equine semen after glass beads column separation. *Equine Veterinary Journal* **37:** 43-47

Larson, K.L., Brannian, J.D., Timm, B.K., Jost, L.K. and Evenson, D.P. (1999). Density gradient centrifugation and glass wool filtration of semen remove spermatozoa with damaged chromatin structure. *Human Reproduction* **14:** 2015-2019

Lechtzin, N., Garside, W., Heyner, S. and Hillman, N. (1991). Glass-bead column separation of motile and nonmotile human spermatozoa. *Journal of In Vitro Fertililzation and Embryo Transfer* **8:** 96-100

McCann, C.T. and Chantler, E. (2000). Properties of sperm separated using Percoll and IxaPrep density gradients. A comparison made using CASA, longevity, morphology and the acrosome reaction. *International Journal of Andrology* **23:** 205-209

Makkar, G., Ng, H.Y., Yeung, S.B. and Ho, P.C. (1999). Comparison of two colloidal silica-based sperm separation media with a non-silica-based medium. *Fertility and Sterility* **72:** 796-802

Mashiach, R., Fisch, B., Eltes, F., Tadir, Y., Ovadia, J. and Bartoov, B. (1992). The relationship between sperm ultrastructural features and fertilizing capacity in vitro. *Fertility and Sterility* **57:** 1052-1057

Mortimer, D. and Mortimer, S.T. (1992). Methods of sperm preparation for assisted reproduction. *Annals* of the Academy of Medicine of Singapore **21:** 517-524

Oehninger, S., Morshedi, M., Weng, S.L., Taylor, S., Duran, H. and Beebe, S. (2003). Presence and significance of somatic cell apoptosis markers in human ejaculated spermatozoa. *Reproductive Biomedicine Online* **7:** 469-476

Ohashi, K., Saji, F., Wakimoto, A., Kato, M., Tsutsui, T. and Tanizawa, O. (1992). Preparation of oligozoospermic and/or asthenozoospermic semen for intrauterine insemination using the SpermPrep semen filtration column. *Fertility and Sterility* **57:** 866-870

Ollero, M., Gil-Guzman, E., Lopez, M.C., Sharma, R.K., Agarwal, A., Larson, K., Evenson, D., Thomas, A.J., Jr. and Alvarez, J.G. (2001). Characterization of subsets of human spermatozoa at different stages of maturation: implications in the diagnosis and treatment of male infertility. *Human Reproduction* **16:** 1912-1921

Oosterhuis, G.J. and Vermes, I. (2004). Apoptosis in human ejaculated spermatozoa. *Journal of Biological Regulators and Homeostatic Agents* **18:** 115-119

Paasch, U., Agarwal, A., Gupta, A.K., Sharma, R.K., Grunewald, S., Thomas, A.J., Jr. and Glander, H.J. (2003). Apoptosis signal transduction and the maturity status of human spermatozoa. *Annals of the New York Academy of Science* **1010:** 486-489

Paasch, U., Grunewald, S., Fitzl, G. and Glander, H.J. (2003). Deterioration of plasma membrane is associated with activation of caspases in human spermatozoa. *Journal of Andrology* **24:** 246-252

Paulson, J.D. and Polakoski, K.L. (1977). A glass wool column procedure for removing extraneous material from the human ejaculate. *Fertility and Sterility* **28:** 178-181

Paulson, J.D., Polakoski, K.L. and Leto, S. (1979). Further characterization of glass wool column filtration of human semen. *Fertility and Sterility* **32:** 125-126

Pharmacia Biotech (1996). *Important notice: Percoll® NOT to be used in Assisted Reproduction Technologies in Humans*. Pharmacia Biotech, Uppsala, Sweden

Pickering, S.J., Fleming, T.P., Braude, P.R., Bolton, V.N. and Gresham, G.A. (1989). Are human spermatozoa separated on a Percoll density gradient safe for therapeutic use? *Fertility and Sterility* **51:** 1024-1029

Pousette, A., Akerlof, E., Rosenborg, L. and Fredricsson, B. (1986). Increase in progressive motility and improved morphology of human spermatozoa following their migration through Percoll gradients. *International Journal of Andrology* **9:** 1-13

Rasch, M., Miska, W., Grunewald, S., Glander, H.J. and Paasch, U. (2005) In: *8th International Congress of Andrology*. Edited by Kim, S.-C., Grootegoed, J.A. and Chemes, H., Medimond, Seoul (Korea), pp. 1-4

Rhemrev, J., Jeyendran, R.S., Vermeiden, J.P. and Zaneveld, L.J. (1989). Human sperm selection by glass wool filtration and two-layer, discontinuous Percoll gradient centrifugation. *Fertility Sterility* **51:** 685-690

Said, T.M., Agarwal, A., Grunewald, S., Rasch, M., Baumann, T., Kriegel, C., Li, L., Glander, H.J., Thomas Jr, A.J. and Paasch, U. (2005). Selection of nonapoptotic spermatozoa as a new tool for enhancing assisted reproduction outcomes: An in vitro model. *Biology of Reproduction* **74:** 530-537

Said, T.M., Grunewald, S., Paasch, U., Glander, H.J., Liang, L. and Agarwal, A. (2005). Advantage of combining magnetic cell separation with sperm preparation technique. *Reproductive Biomedicine Online* **10:** 740-746

Sakkas, D., Mariethoz, E. and StJohn, J.C. (1999). Abnormal sperm parameters in humans are indicative of an abortive apoptotic mechanism linked to the Fas-mediated pathway. *Experimental Cell Research* **251:** 350-355

Sakkas, D., Moffatt, O., Manicardi, G.C., Mariethoz, E., Tarozzi, N. and Bizzaro, D. (2002). Nature of DNA damage in ejaculated human spermatozoa and the possible involvement of apoptosis. *Biology of Reproduction* **66:** 1061-1067

Sanchez, R., Concha, M., Ichikawa, T., Henkel, R. and Schill, W.B. (1996). Glass wool filtration reduces reactive oxygen species by elimination of leukocytes in oligozoospermic patients with leukocytospermia. *Journal of Assisted Reproduction and Genetics* **13:** 489-494

Scott, R.T., Jr., Oehninger, S.C., Menkveld, R., Veeck, L.L. and Acosta, A.A. (1989). Critical assessment of sperm morphology before and after double wash swim-up preparation for in vitro fertilization. *Archives of Andrology* **23:** 125-129

Sherman, J.K., Paulson, J.D. and Liu, K.C. (1981). Effect of glass wool filtration on ultrastructure of human spermatozoa. *Fertility and Sterility* **36:** 643-647

Sills, E.S., Wittkowski, K.M., Tucker, M.J., Perloe, M., Kaplan, C.R. and Palermo, G.D. (2002). Comparison of centrifugation- and noncentrifugation-based techniques for recovery of motile human sperm in assisted reproduction. *Archives of Andrology* **48:** 141-145

Society for Assisted Reproductive Technology and American Society for Reproductive Medicine (2004). Assisted reproductive technology in the United States: 2000 results generated from the American Society for Reproductive Medicine/Society for Assisted Reproductive Technology Registry. *Fertility and Sterility* **81:** 1207-1220

Soderlund, B. and Lundin,K. (2000). The use of silane-coated silica particles for density gradient centrifugation in in-vitro fertilization. *Human Reproduction* **15:** 857-860

Srisombut, C., Morshedi, M., Lin, M.H., Nassar, A. and Oehninger, S. (1998). Comparison of various methods of processing human cryopreserved- thawed semen samples. *Human Reproduction* **13:** 2151-2157

Tea, N.T., Jondet, M. and Scholler, R. (1983) In: *Proceedings of the World Congress on Fertility and Sterility*. Edited by Harrison, F. Dublin

Twigg, J., Irvine, D.S., Houston, P., Fulton, N., Michael,

L. and Aitken, R.J. (1998). Iatrogenic DNA damage induced in human spermatozoa during sperm preparation: protective significance of seminal plasma. *Molecular Human Reproduction* **4:** 439-445

Van Den, B.M., Revelard, P., Bertrand, E., Biramane, J., Vanin, A.S. and Englert, Y. (1997). Glass wool column filtration, an advantageous way of preparing semen samples for intracytoplasmic sperm injection: an auto-controlled randomized study. *Human Reproduction* **12:** 509-513

Van der Ven, H.H., Jeyendran, R.S., Al Hasani, S., Tunnerhoff, A., Hoebbel, K., Diedrich, K., Krebs, D. and Perez-Pelaez, M. (1988). Glass wool column filtration of human semen: relation to swim-up procedure and outcome of IVF. *Human Reproduction* **3:** 85-88

van Heerde, W.L., de Groot, P.G. and Reutelingsperger, C.P. (1995). The complexity of the phospholipid binding protein Annexin V. *Thrombosis and Haemostasis* **73:** 172-179

Vermes, I., Haanen, C., Steffens-Nakken, H. and Reutelingsperger, C.P. (1995). A novel assay for apoptosis: Flow cytometric detection of phosphatidylserine expression of early apoptotic cells using fluorescein labelled Annexin V. *Journal of Immunological Methods,* **184:** 39-51

Wikland, M., Wik, O., Steen, Y., Qvist, K., Soderlund, B. and Janson, P.O. (1987). A self-migration method for preparation of sperm for in-vitro fertilization. *Human Reproduction* **2:** 191-195

Wu, T.P., Huang, B.M., Tsai, H.C., Lui, M.C. and Liu, M.Y. (2004). Effects of nitric oxide on human spermatozoa activity, fertilization and mouse embryonic development. *Archives of Andrology* **50:** 173-179

Yavetz, H., Hauser, R., Homonnai, Z.T., Paz, G.F., Lessing, J.B., Amit, A. and Yogev, I. (1996). Separation of sperm cells by sedimentation technique is not suitable for in vitro fertilization purposes. *Andrologia* **28:** 3-6

Yener, C., Mathur, S. and Parent, B. (1990). Comparison of two sperm preparation techniques using automated sperm motion analysis: migration sedimentation versus swim-up. *Archives of Andrology* **25:** 17-20

Younglai, E.V., Holt, D., Brown, P., Jurisicova, A. and Casper, R.F. (2001). Sperm swim-up techniques and DNA fragmentation. *Human Reproduction* **16:** 1950-1953

Zavos, P. M. and Centola, G.M. (1992). Methods of semen preparation for intrauterine insemination and subsequent pregnancy rates. *Tohoku Journal of Experimental Medicine* **168:** 583-590

Zini, A., Nam, R.K., Mak, V., Phang, D. and Jarvi, K. (2000). Influence of initial semen quality on the integrity of human sperm DNA following semen processing. *Fertility and Sterility* **74:** 824-827

Spermatology. SRF Vol. 65. ERS Roldan and M Gomendio (eds) Nottingham University Press, Nottingham

Proteomics of human spermatozoa, protamine content and assisted reproduction outcome

Rafael Oliva[1], Juan Martínez-Heredia[1], Sara de Mateo[1], Cristina Gázquez[1], Josep Oriola[1], Josep Maria Estanyol[2], Marta Guimerà[3], Juan Balasch[3] and José Luis Ballescà[3]

[1]Human Genetics Research Group, IDIBAPS, Faculty of Medicine, University of Barcelona, 08036-Barcelona, Spain and Genetics Service, Hospital Clínic i Provincial, 08036-Barcelona, Spain; [2]Unitat de Proteòmica, Faculty of Medicine, University of Barcelona, Spain; [3]Institut Clínic de Ginecologia, Obstetricia i Neonatología, Hospital Clínic i Provincial, 08036-Barcelona, Spain

It is well known that alterations in the expression of the major sperm nuclear proteins (protamines) are related to infertility in man. In addition, other minor proteins extracted from human spermatozoa are being analysed by 2-dimensional polyacrylamide gel electrophoresis (2-D PAGE) and identified by MALDI-TOF MS analysis. The function of the identified proteins turns out to be energy production, transcription, protein synthesis, transport, folding and turnover, cell cycle, apoptosis and oxidative stress, signal transduction, cytoskeleton, flagella and cell movement, cell recognition, metabolism and unknown function. Many of the proteins identified using MALDI-TOF had not been yet been described as being expressed in human spermatozoa. Substantial differences have been detected in the levels of some of the newly identified human sperm proteins in the different groups of infertile patients. Present research efforts are targeting the potential correlations among changes in the proteomic composition, protamine content, DNA integrity and assisted reproduction outcome.

Introduction

The human spermatozoon is a highly specialised cell with extremely marked compositional, morphological and functional differences, as compared to other somatic or germinal cells (Oliva and Dixon, 1991; Kierszenbaum and Tres, 2004; Pixton *et al.*, 2004). So far conventional protein separation strategies have led to the identification of nuclear proteins (Balhorn *et al.*, 1988; Oliva and Dixon, 1991; Yoshii *et al.*, 2005), to proteins required for sperm motility, and membrane proteins required for capacitation, egg interaction and fertilisation (Naaby-Hansen *et al.*, 2002). Also anomalies of the expression of the most abundant nuclear sperm proteins have been found associated to infertility (Balhorn *et al.*, 1988; de Yebra *et al.*, 1993; Mengual *et al.*, 2003; Aoki *et al.*, 2005). But the majority of causes of male infertility still remains unknown. This is at least in part due to the relative lack of knowledge concerning the protein composition of spermatozoa. Present proteomic approaches offer the possibility to boost research in this field. The present article briefly and partially summarises some recent progress in the understanding of the proteomic composition of human spermatozoa and its implications for fertility with a particular focus on sperm nuclear proteins.

Corresponding author E-mail: roliva@ub.edu

The nucleohistone to nucleoprotamine transition

In the final stage of spermatogenesis the nucleosomal structure is progressively disassembled and replaced by transition proteins and finally by protamines (Oliva and Dixon, 1991; Kierszenbaum and Tres, 2004). Protamines represent the major human sperm proteins extracted with 0.5M HCl from the sperm nucleus (Fig. 1). Two types of protamines are present in the human sperm: the P1 protamine and the family of P2 proteins (Fig. 1). Due to its extremely high arginine content (~ 50% in human protamines) they form a highly compact complex with the negatively charged DNA. In addition, the formation of inter-disulfide bonds between adjacent protamine molecules further stabilises the nucleoprotamine complex. However, despite substantial amount of information available on the abundant protamines, the human sperm nucleus also contains thousands of additional minor proteins with an intriguing function (Fig. 1). These minor proteins are now being targeted with the present proteomic tools (Martínez-Heredia *et al.*, 2006).

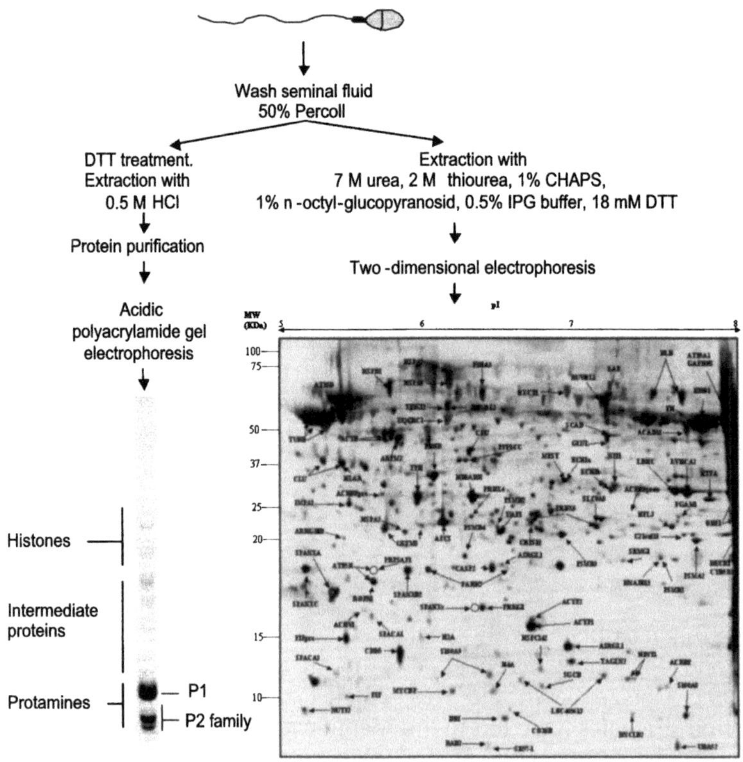

Figure 1. Extraction and electrophoresis of human sperm proteins. Before protein extraction, human spermatozoa are usually washed and purified through a Percoll gradient in order to remove seminal fluid and potential contaminating cells present in semen. After this step two direct extraction procedures are carried out. As shown on the left side of the figure, the basic sperm proteins are first treated with DTT, in order to reduce the disulfide bridges between adjacent protamines, and are extracted with 0.5M HCl, purified, separated by electrophoresis in an acetic-urea polyacrylamide gel and are stained with Coomassie blue (Mengual *et al.*, 2003). On the right side of the figure, a direct lysis and extraction of all sperm proteins is shown, followed by separation using two dimensional gel electrophoresis and visualization of the proteins using silver staining. The indicated proteins have been subsequently excised from the gel and identified by mass spectrometry (Martinez-Heredia *et al.*, 2006).

Proteomic composition of spermatozoa

The mature male germ cell proteome has been studied in model species such as plants (Holmes-Davis *et al.*, 2005) or worms (Hozumi *et al.*, 2004). In man specific approaches to target sub-cellular organelles or compartments have been successfully used (Naaby-Hansen *et al.* 2002). The human sperm proteome has also been analysed after separation of the proteins by liquid chromatography and MS/MS although the individual proteins identified have not been reported (Johnston *et al.*, 2005). A previous case report has already provided some evidence of the potential of the present proteomic tools to identify proteins involved in infertility (Pixton *et al.*, 2004). More recently the postranslational modifications during sperm maturation in the rat have also been studied using fluorescent 2-D PAGE. Recently the most abundant proteins extracted from human sperm cells fractionated in Percoll gradients, have been identified through 2-D PAGE and subsequent mass spectrometry (Martinez-Heredia *et al.*, 2006). In this study over 1000 protein spots were resolved by 2-D PAGE, corresponding to sperm proteins extracted from normozoospermic human samples (Fig. 1). A total of 145 spots ranging from a MW mass of 10 to 70 KDa and a pI of 5-8 were excised from the gels. After MALDI-TOF MS analysis 98 different proteins were identified (Fig. 1). The most abundant group corresponds to the proteins involved in energy production (23%). The next most abundant group of proteins identified corresponds to those involved in transcription, protein synthesis and turnover, protein transport and folding (23%). The identification of these proteins represented the first step towards the detection of expression differences in different types of infertile patients.

Relevance of alterations of the protamines and proteomic content for male fertility and assisted reproduction outcome

There have been several studies that analysed the protamines and detected an increased P1/P2 ratio in infertile patients (Balhorn *et al.*, 1988; de Yebra *et al.*, 1993; Mengual *et al.*, 2003). Also a reduction of the P1/P2 ratio results in a marked reduction of the *in vitro* fertilisation index in comparison with the patients with a normal or an increased P1/P2 ratio (Aoki *et al.*, 2005). Protamine deficient animal models indicate that the integrity of the DNA decreases upon spermatozoan passage from the epididymis affecting subsequent embryo development (Suganuma *et al.*, 2005).

Since differences in the expression of the major sperm nuclear proteins (protamines) are related to fertility it is also reasonable to expect differences in the expression of other minor sperm proteins in different groups of infertile patients. A previous case report already provided clear evidence of the potential of the present proteomic tools to identify proteins involved in infertility (Pixton *et al.*, 2004). One of our ongoing projects is targeting the identification of proteins differentially expressed in different groups of infertile patients as compared to controls. Specifically, we have studied the expression pattern of proteins separated by 2-D PAGE in groups of patients classified by the motility of their sperm cells, the protamine content, or the integrity of their DNA. The results indicate that up to 9% of the proteins are over-expressed or under-expressed in asthenozoospermic patients as compared to normozoospermic patients or controls (Martínez-Heredia, J., de Mateo, S., Vidal-Taboada, J., Ballescà, J.L. and Oliva, R., manuscript in preparation). Also different additional proteins are detected as over- or under-expressed when the samples are stratified according to the protamine content or to the integrity of DNA in the sperm nucleus (de Mateo, S., Martínez-Heredia, J., Estanyol, J.M., Ballescà, J.L. and Oliva, R., manuscript in preparation). It can be expected that the results obtained will have important diagnostic and prognostic clinical applications and provide important basic information concerning the function of the newly identified proteins.

Acknowledgements

Supported by grants from Ministerio de Ciencia y Tecnología BMC2003-03937, European Regional Development Fund (ERDF), Ministerio de Sanidad y Consumo V-2003-REDC07A-O and by Generalitat de Catalunya 2001SGR00382.

References

Aoki, V.W., Liu, L. and Carrell, D.T. (2005). Identification and evaluation of a novel sperm protamine abnormality in a population of infertile males. *Human Reproduction* **20:** 1298-1306

Balhorn, R., Reed, S. and Tanphaichitr, N. (1988). Aberrant protamine 1/protamine 2 ratios in sperm of infertile human males. *Experientia* **44:** 52-55

de Yebra, L., Ballesca, J.L., Vanrell, J.A., Bassas, L. and Oliva, R. (1993). Complete selective absence of protamine P2 in humans. *Journal of Biological Chemistry* **268:** 10553-10557

Holmes-Davis, R., Tanaka, C. K., Vensel, W. H., Hurkman, W. J. and McCormick, S. (2005). Proteome mapping of mature pollen of Arabidopsis thaliana. *Proteomics* **5:** 4864-4884

Hozumi, A., Satouh, Y., Ishibe, D., Kaizu, M., Konno, A., Ushimaru, Y., Toda, T. and Inaba, K. (2004). Local database and the search program for proteomic analysis of sperm proteins in the ascidian *Ciona intestinalis. Biochemical and Biophysical Research Communications* **319:** 1241-1246

Johnston, D.S., Wooters, J., Kopf, G.S., Qiu, Y. and Roberts, K.P. (2005). Analysis of the human sperm proteome. *Annals of the New York Academy of Science* **1061:** 190-202.

Kierszenbaum, A.L. and Tres, L.L. (2004). The acrosome-acroplaxome-manchette complex and the shaping of the spermatid head. *Archives of Histology and Cytology* **67:** 271-284

Martínez-Heredia, J., Estanyol, J.M., Ballescà, J.L. and Oliva, R. (2006). Proteomic identification of human sperm proteins. *Proteomics* **6:** 4356-3469

Mengual, L., Ballesca, J.L., Ascaso, C. and Oliva, R. (2003). Marked differences in protamine content and P1/P2 ratios in sperm cells from Percoll fractions between patients and controls. *Journal of Andrology* **24:** 438-447

Naaby-Hansen, S., Mandal, A., Wolkowicz, M.J., Sen, B., Westbrook, V.A., Shetty, J., Coonrod, S.A., Klotz, K.L., Kim, Y.H., Bush, L.A., Flickinger, C.J. and Herr, J.C. (2002). CABYR, a novel calcium-binding tyrosine phosphorylation-regulated fibrous sheath protein involved in capacitation. *Developmental Biology* **242:** 236-254

Oliva, R. and Dixon, G.H. (1991). Vertebrate protamine genes and the histone-to-protamine replacement reaction. *Progress in Nucleic Acids Research and Molecular Biology* **40:** 25-94

Pixton, K.L., Deeks, E.D., Flesch, F.M., Moseley, F.L., Bjorndahl, L., Ashton, P.R., Barratt, C.L. and Brewis, I.A. (2004). Sperm proteome mapping of a patient who experienced failed fertilization at IVF reveals altered expression of at least 20 proteins compared with fertile donors: case report. *Human Reproduction* **19:** 1438-1447

Suganuma, R., Yanagimachi, R. and Meistrich, M.L. (2005). Decline in fertility of mouse sperm with abnormal chromatin during epididymal passage as revealed by ICSI. *Human Reproduction* **20:** 3101-3108

Yoshii, T., Kuji, N., Komatsu, S., Iwahashi, K., Tanaka, Y., Yoshida, H., Wada, A. and Yoshimura, Y. (2005). Fine resolution of human sperm nucleoproteins by two-dimensional electrophoresis. *Molecular Human Reproduction* **11:** 677-681

Single nucleotide polymorphisms: discovery of the genetic causes of male infertility

H. Tanaka[1], M. Hirose[1], K. Tokuhiro[1], Y. Matsuoka[1], Y. Miyagawa[2], A. Tsujimura[2], A. Okuyama[2] and Y. Nishimune[1]

[1]*TANAKA Project, Center for Advanced Science and Innovation, and* [2]*Department of Urology, Osaka University, 3-1 Yamadaoka, Suita, Osaka 565-0871, Japan*

We carried out single nucleotide polymorphism (SNP) and mutation analyses of haploid germ cell-specific genes. An analysis of 13 genes associated with male infertility in approximately 300 infertile male patients and approximately 300 male volunteers with proven fertility revealed two mutations that might produce male infertility, and three SNP/mutations associated with male infertility in 13 germ cell-specific genes. These findings strongly support the hypothesis that dysfunction of germ cell-specific genes causes idiopathic human male infertility.

Introduction

In Japan, as in many European countries, more than one tenth of married couples are troubled with infertility (Skakkebaek *et al.*, 2006). In Japan, the birth rate is declining, while the proportion of elderly people is increasing rapidly. Infertility is one of the important factors causing the decrease in the birth rate. We have cloned and characterised germ cell-specific genes in the mouse, together with human orthologues (Nishimune and Tanaka, 2006). It was shown that deficiency of genes specifically expressed during spermatogenesis lead to male infertility in animal experiments. Advances in medical technology have allowed infertile couples to have children. Nevertheless, more than half of all infertility is idiopathic. Within this group, some causes of infertility must have a high incidence. To understand infertility, it is important to identify gene mutations involved in human spermiogenesis. We assessed the prevalence of single nucleotide polymorphisms (SNPs) in germ cell-specific genes by direct sequencing of PCR-amplified DNA from male patients who were undergoing fertility evaluation.

Isolation and characterisation of germ cell-specific molecules

We cloned and characterised germ cell-specific genes of mice using germ cell-specific antibodies and cDNA subtraction methods (Nishimune and Tanaka, 2006). A library was generated by subtracting the mRNA from 17-day-old mouse testes, which lack haploid germ cells, from the cDNA of 35-day-old mouse testes (Fujii *et al.*, 2002). The identification of germ cell-specific genes showed that many genes are expressed in differentiated spermatids, although the nuclei are compacted and most of the cytoplasm is lost during spermiogenesis. Since almost all mouse male germ cell-specific genes have human orthologues, we isolated and characterised the counterparts of human genes using mouse cDNA probes with a human testis cDNA library or computer analysis.

Corresponding author E-mail: tanaka@biken.osaka-u.ac.jp

Male infertility

Approximately 15% of couples attempting to conceive over a two-year period are unsuccessful. The causal frequency of male and female infertility is equal. More than half of male infertility has uncertain causes and is deemed idiopathic. *In vitro* fertilisation (IVF) and intracytoplasmic sperm injection (ICSI) are used to treat infertility, even when the cause is uncertain. Recently, testicular sperm extraction (TESE) with ICSI (TESE-ICSI) has become the first-line treatment for azoospermic male infertility in both non-obstructive and obstructive azoospermia (Tsujimura *et al.*, 2002). However, knowing the cause of infertility is important in order to develop novel and better remedies. Molecular techniques are much better able to identify these microdeletions than cytogenetic approaches. In nature, an infertility mutation on the Y chromosome is not transmitted to the next generation, because pregnancy does not occur. However, these mutations can be transmitted to male offspring through the use of assisted reproductive technologies (ART), either IVF or ICSI, and would cause male infertility in the offspring. If the effects of an autosomal mutation are recessive, the mutation can be inherited to the next generation, and under certain conditions, induce infertility. Even a dominant negative autosomal mutation that affects spermiogenesis can, in theory, be transferred to the next generation via females. Mutations in housekeeping genes would result in a loss of function, thereby affecting embryogenesis or somatic cell functions, which could lead to defects or illness. By contrast, even if the germ cell-specific genes were totally defective, the phenotype would be limited to germ cells and the only repercussion would be infertility. Therefore, we examined SNPs or detected mutations using direct DNA sequencing of the PCR products of chromosomal DNA from blood or semen samples.

SNP analysis of germ cell-specific genes

Azoospermic and oligospermic patients were examined for SNPs in an analysis of genes expressed specifically in spermiogenesis (Table 1). The analysis of human *PRM1* and *PRM2* gene sequences of infertile male patients and proven-fertile male volunteers revealed four SNPs in the *PRM1* coding region, which did not cause any amino acid substitutions, and one mutation in the *PRM2* haplotype, which terminated translation (Tanaka *et al.*, 2003). Translation termination may result in male infertility, due to haploinsufficiency of *PRM2* (Cho *et al.*, 2001). Analysis of the human *TNP1* and *TNP2* gene sequences revealed five amino acid substitutions caused by SNPs in the open reading frame of *TNP2*. In addition, a 15-nucleotide deletion encompassing the CRE motif was found in the 5'-promoter region of the *TNP1* haplotype in infertile men (Miyagawa *et al.*, 2005). This deletion mutation should greatly depress *TNP1* gene expression. Although TNP1 protein has been shown to play important roles in mouse spermiogenesis (Shirley *et al.*, 2004), its functions in human sperm formation are not yet fully understood. The 15-nucleotide deletion in the promoter region would reduce *TNP1* expression, and might be associated with human male infertility. HANP1 protein is essential for nucleus formation in functional spermatozoa, and is specifically involved in the replacement of histones with protamines during spermiogenesis (Tanaka *et al.*, 2005). The analysis of human *Hanp1* sequences revealed five SNPs in the coding region that introduce amino acid changes (Tanaka *et al.*, 2006). However, the prevalence of these five SNPs did not differ significantly between our infertile and proven-fertile clinical cases. Some human cases with deficient HANP1 will likely suffer teratozoospermia in the same way as knockout mice. We also identified mutations in the genomic sequence coding genes of Haspin and TSSK2 kinases in infertile men. These mutations terminated translation. Although the functions of these kinases in human sperm formation are not yet fully understood, it is possible to induce infertility by combining additional poly-

Table1. Frequency of SNPs/mutations in germ cell-specific genes

Gene	Number of SNPs in CDS	Number of nonsynonymous SNPs	Coding sequence length (bp)	Persons examined
PRM1	4	0	156	496
PRM2	0	0	309	496
(one nonsense mutation in infertile group)				
TNP1	2	2	168	552
(one mutation of 15 nucleotides deletion in promoter region in infertile group)				
TNP2	5	5	417	548
HANP1	5	5	705	530
t-actin 1 (ACTL7B)	5	1	1248	370
t-actin 2 (ACTL7A)	6	2	438	381
tPAP	7	4	1914	378
Gsg3	1	1	900	350
TSSK1	10	2	1098	378
TSSK2	13	7	1077	378
(one mutation of one nucleotide deletion in ORF in infertile group)				
Haspin	10	6	2265	544
(one mutation of one nucleotide addition in ORF in infertile group)				
Scot-t	3	3	1554	516

morphisms in other genes. Succinyl CoA:3-oxo acid CoA transferase (SCOT; EC 2.8.3.5) is a key enzyme in the metabolism of ketone bodies and its testicular isotype (SCOT-t) is localised exclusively to the cytoplasm of spermatids and the midpiece of ejaculated sperm where mitochondria exist. It has been suggested that SCOT-t plays important roles in spermiogenesis and sperm function (Koga *et al.*, 2000). The analysis of *Scot-t* revealed three SNPs causing amino acid substitutions and one in the non-coding 3′-terminal end. The prevalence of homozygous SNPs causing amino acid substitution was significantly higher in infertile patients than in fertile control men.

We found that a mutation in germ cell-specific genes might cause male infertility and some genes in germ cell-specific genes have SNPs/mutations that appear at considerably higher frequencies in infertile patients than in fertile control men. SNP analysis of germ cell-specific genes is useful for understanding the cause and medical treatment of male infertility. It is strongly hoped that the development of gene diagnosis and auxiliary reproductive medicine will follow our understanding of the molecular mechanism of spermatogenesis and the cause of infertility.

References

Cho, C., Willis, W.D., Goulding, E.H., Jung-Ha, H., Choi, Y.C., Hecht, N.B. and Eddy, E.M. (2001) Haploinsufficiency of protamine-1 or -2 causes infertility in mice. *Nature Genetics* **28**: 82-86

Fujii, T., Tamura, K., Masai, K., Tanaka, H., Nishimune, Y. and Nojima, H. (2002) Use of stepwise subtraction to comprehensively isolate mouse genes whose transcription is up-regulated during spermiogenesis. *EMBO Reports* **3**: 367-372

Koga, M., Tanaka, H., Yomogida, K., Nozaki, M., Tsuchida, J., Ohta, H., Nakamura, Y., Masai, K., Yoshimura, Y., Yamanaka, M., Iguchi, N., Nojima, H., Matsumiya, K., Okuyama, A. and Nishimune, Y. (2000) Isolation and characterization of a haploid germ-cell-specific novel complementary deoxyribonucleic acid; testis-specific homologue of succinyl CoA:3-oxo acid CoA transferase. *Biology of Reproduction* **63**: 1601-1609

Miyagawa, Y., Nishimura, H., Tsujimura, A., Matsuoka, Y., Matsumiya, K., Okuyama, A., Nishimune, Y. and Tanaka, H. (2005) Single-nucleotide polymorphisms and mutation analyses of the TNP1 and TNP2 genes of fertile and infertile human male populations. *Journal of Andrology* **26**: 779-786

Nishimune, Y. and Tanaka, H. (2006) Infertility caused by polymorphisms or mutations in spermatogenesis-specific genes. *Journal of Andrology* **27**: 326-334

Shirley, C.R., Hayashi, S., Mounsey, S., Yanagimachi, R. and Meistrich, M.L. (2004) Abnormalities and reduced reproductive potential of sperm from Tnp1- and Tnp2-null double mutant mice. *Biology of Reproduction* **71**: 1220-1229

Skakkebaek, N.E., Jorgensen, N., Main, K.M., Rajpert-De Meyts, E., Leffers, H., Andersson, A.M., Juul, A., Carlsen, E., Mortensen, G.K., Jensen, T.K. and Toppari, J. (2006) Is human fecundity declining? *International Journal of Andrology* **29**: 2-11

Tanaka, H., Iguchi, N., Isotani, A., Kitamura, K., Toyama, Y., Matsuoka, Y., Onishi, M., Masai, K., Maekawa, M., Toshimori, K., Okabe, M. and Nishimune, Y. (2005) HANP1/H1T2, a novel histone H1-like protein involved in nuclear formation and sperm fertility. *Molecular and Cellular Biology* **25**: 7107-7119

Tanaka, H., Matsuoka, Y., Onishi, M., Kitamura, K., Miyagawa, Y., Nishimura, H., Tsujimura, A., Okuyama, A. and Nishimune, Y. (2006) Expression profiles and single-nucleotide polymorphism analysis of human HANP1/H1T2 encoding a histone H1-like protein. *International Journal of Andrology* **29**: 353-359

Tanaka, H., Miyagawa, Y., Tsujimura, A., Matsumiya, K., Okuyama, A. and Nishimune, Y. (2003) Single nucleotide polymorphisms in the protamine-1 and -2 genes of fertile and infertile human male populations. *Molecular Human Reproduction* **9**: 69-73

Tsujimura, A., Matsumiya, K., Miyagawa, Y., Tohda, A., Miura, H., Nishimura, K., Koga, M., Takeyama, M., Fujioka, H. and Okuyama, A. (2002) Conventional multiple or microdissection testicular sperm extraction: a comparative study. *Human Reproduction* **17**: 2924-2929

Spermatology. SRF Vol. 65. ERS Roldan and M Gomendio (eds) Nottingham University Press, Nottingham

Eppin: a molecular strategy for male contraception

Zengjun Wang[1,2], E. E. Widgren[1], R. T. Richardson[1] and M. G. O'Rand[1]

[1]*Department of Cell & Developmental Biology, University of North Carolina at Chapel Hill, Chapel Hill, North Carolina 27599, USA;* [2]*Department of Urology, The First Affiliated Hospital of Nanjing Medical University, People's Republic of China*

New male contraceptives, both hormonal and non-hormonal, have many obstacles to overcome before they reach the market as a product. For hormonal contraceptives the long-term efficacy of oligospermia in a large population of unselected men remains to be determined. For non-hormonal contraception target selection remains a primary goal. Immunocontraception, which showed great promise for many years, has recently lost its appeal. Nevertheless, immunocontraception can be utilised as a strategy, particularly in primates, to discern the function of target molecules in the male. As an example, we discuss Eppin, an epididymal protease inhibitor that coats the surface of human spermatozoa. Because Eppin is predicted to be a serine protease inhibitor with chymotrypsin-like specificity and binds semenogelin, the natural substrate of PSA (prostate specific antigen, a serine protease), we investigated whether Eppin would modulate PSA activity and the hydrolysis of semenogelin. Additionally, because antibodies to Eppin provide contraception in male monkeys, we investigated whether antibodies to Eppin would inhibit the PSA hydrolysis of semenogelin. Eppin is a specific inhibitor of PSA activity that requires leucine 87, Eppin's P1 reactive site. Although Eppin modulates the hydrolysis of semenogelin by PSA, antibodies to Eppin do not inhibit PSA activity.

Introduction

Over the last 50 years contraception has had a major impact on human society and influenced the world wide distribution of family sizes and the variability of fertility rates (Bongaarts and Watkins, 1996; Bongaarts, 1997). This impact can be largely attributed to female contraceptive methods, their availability and economic and social costs. Male contraception on the other hand has had much less of a global impact, being largely limited to condoms and vasectomy (Nass and Strauss, 2004). Female hormonal contraceptives work through the mechanism of anovulation and the goal of male hormonal contraceptive research is analogous, namely the suppression of spermatogenesis to produce azoospermia. However, achievement of this goal in a reliable way for a diverse population of men is still many years away (Potts, 1996; Grimes, Gallo, Grigorieva, Nanda and Schulz, 2005). Even further away is the dream of a non-hormonal male contraceptive in which it may be envisioned that spermatozoa do not develop, or do not swim, or do not fertilise or some combination of these spermatozoan catastrophes. Numerous contraceptive targets abound and several of these targets are worthy of further exploratory work, including blocking transmembrane ion currents (Brenton, Smith, Lui and Brown, 1996;

Corresponding author E-mail: morand@unc.edu

Kirichok, Navarro and Clapham., 2006), disrupting Sertoli-germ cell adhesions (Cheng, Mo, Grima, Saso, Tita, Mruk and Silvestrini, 2002; Cheng, Mruk, Silvestrini, Bonanomi, Wong, Siu, Lee, Lui and Mo 2005) and disruption of spermiogenesis by imino sugars (Walden, Butters, Dwek, Platt and van der Spoel, 2006). And finally there is that enigmatic idea of immunocontraception, a non-hormonal approach that has developed into an entire field of its own.

Immunocontraception

Immunocontraception, the inhibition of fertility through manipulation of the immune system, has been a goal of researchers for many years (Metz, 1979; O'Rand, 1980). The concept is simple. Block an essential reproductive function with an antibody or cellular immune response and fertility will be lost. And the origin of the concept is old; it dates from Landsteiner (1899) and studies on the immunological properties of spermatozoa (Henle, Henle and Chambers, 1938; Tyler, 1948). Early workers were interested in the immunogenicity of spermatozoa because they could identify sperm antigens using the specificity of the resulting antisera (Tyler, 1961), utilizing monovalent and polyvalent antibodies as probes for molecular function (O'Rand and Metz, 1974). Eventually clinical observations of infertility patients with antisperm antibodies (Rümke and Hellinga, 1959) were cross-pollinated with basic research observations and descriptions of sperm antigens. The bench to bedside cross-pollination of research led to early attempts, with funding from international agencies (e.g. WHO, Population Council), to find sperm antigens that would serve as contraceptives in women (Munoz and Metz, 1978; Morton and McAnulty, 1979). During the next ten years (1980-1990) support for the development of a contraceptive vaccine grew and the key issue became selecting the appropriate target. Would antibodies to a sperm surface antigen really work or would an anti-hormone antibody be more effective? A successful phase I study of an anti-hCG vaccine supported by WHO was conducted without side-effects (Jones, Bradley, Judd, Denholm, Ing, Mueller, Powell, Griffin and Stevens, 1988). Successful reversible immunocontraception by the immunization of guinea pigs with a sperm antigen was reported by the end of the decade (Primakoff, Lathrop, Woolman, Cowan and Myles, 1988). Clearly in the last decade of the 20th century there was every expectation that a contraceptive vaccine was just around the corner (Griffin, 1990; Aitken, Paterson and Koothan, 1992). What went wrong?

Research in immunocontraception for men lagged behind interest in contraceptives for women primarily due to fear of an autoimmune response to the testis. Studies on guinea pigs and humans with Freund's adjuvant in association with an injury or infection of the testis generated early interest in the development of autoimmune orchitis (Mancini, 1976). However without a clear autosensitization procedure (such as injury or infection), the development of antisperm antibodies has not proven to be a clinical complication in men as evidenced by the widespread use of vasectomy for male contraception. Most vasectomised men have antisperm antibodies and suffer no long-term systemic dysfunctions (Beagley, Wu, Pomering and Jones, 1998). Consequently fears concerning the formation of immune complexes leading to autoimmune orchitis and epididymitis in male immunocontraception seem unduly alarmist and somewhat disingenuous (Nieschlag and Henke, 2005). Indeed both hormonal and immunocontraceptive approaches to male contraception have many obstacles to overcome before a product reaches the market, not the least of which is the problem of low responders, men who do not reach azoospermia or whose antibody titer remains low after immunization. The long-term efficacy of oligospermia in a large population of unselected men remains to be determined as does the long-term effect of androgen and progestin treatment on HDL levels and cardiovascular function.

The strategy

Unlike male hormonal contraceptive research, which is supported by the pharmaceutical industry (Nieschlag and Henke, 2005), immunocontraceptive research is not currently supported. It is very unlikely that significant progress beyond the "proof of principle" stage for immunocontraception will occur without major new commitments. However, major new commitments would demand new levels of confidence: not only in the selected target, but in the efficacy and safety of vaccine development as well. That confidence has been squandered over the last decade. While the concept is simple and it may seem intuitive that contraception would be possible if only the 'right' target could be found (O'Rand and Lea, 2002), it is not obvious that the same immunogen in different immunological environments (e.g. MHC context or T-cell tolerance) has profoundly different effects on fertility (Lea, van Lierop, Widgren, Grootenhuis, Wen, van Duin and O'Rand, 1998). Understanding the immunological arm of the immunocontraceptive axis will require additional research into the reasons for fertility differences between genetically identical individuals. The usefulness of immunocontraception is however not lost because of vaccine development problems. Consider the original reasons for interest in anti-sperm antibodies sixty years ago: they provide tools for discovering the function of sperm specific molecules and whether or not they have essential roles in fertility. This is particularly true in primates where creation of a null mutation is not possible.

Therefore a strategy for the development of new contraceptives is to utilise immunization of primates with specific sperm surface antigens and determine the effects of the immune response on the ejaculated spermatozoa of immunised males. Our recent work on Eppin, (SPINLW1; serine protease inhibitor-like, with Kunitz and WAP domains-1) provides an example of the utility of the immunocontraceptive approach (O'Rand, Widgren, Sivashanmugam, Richardson, Hall, French, VandeVoort, Ramachandra, Ramesh and Rao, 2004; Wang, Widgren, Sivashanmugam, O'Rand and Richardson, 2005; O'Rand, Widgren, Wang and Richardson, 2006, 2007). Our fertility study (O'Rand *et al.*, 2004) demonstrated that effective and reversible male immunocontraception in primates is an obtainable goal. We found that a high serum titer ($>$1:1000) sustained over several months achieves an effective level of contraception. Consequently, in addition to whatever conclusions one may wish to draw about the feasibility of using immunocontraception, we can definitely conclude from these studies that Eppin has an essential role in fertility. Our strategy has been successful; further studies focused on Eppin function led to the demonstration that in seminal plasma and on human spermatozoa following ejaculation Eppin is bound to semenogelin I (Sg) (Wang *et al.*, 2005). The mechanism of action of the antibody appears to prevent normal Eppin-Sg interaction (O'Rand *et al.*, 2006, 2007).

Eppin inhibits prostate specific antigen (PSA)

In addition to Eppin's interaction with Sg, Eppin interacts with PSA (O'Rand *et al.*, 2006, 2007) via the Kunitz-type trypsin inhibitor consensus sequence in its C-terminal half (amino acids 73-128). The C-terminal contains leucine reside 87 (leu87), the putative inhibitor P1 reactive site that is indicative of a specificity for chymotrypsin-like enzymes (Laskowski and Kato, 1980). Modelling studies (O'Rand *et al.*, 2006, 2007) indicate that leu87 is likely to be exposed on the outer binding surface of Eppin, similar to other P1 reactive sites in Kunitz-type trypsin inhibitors. To determine if the Kunitz-type inhibitor P1 site (leu87) is necessary for Eppin's inhibition of PSA, leu87 was changed to gly87 in Eppin (Δ87Eppin). As shown in Figure 1, Δ87Eppin had no effect on recombinant (r) PSA activity, demonstrating that leu87 is necessary for Eppin inhibition of rPSA activity. For comparison, Figure 1 also demonstrates that 5 mM Zn^{++}, a strong and specific non-competitive inhibitor of PSA, inhibited rPSA.

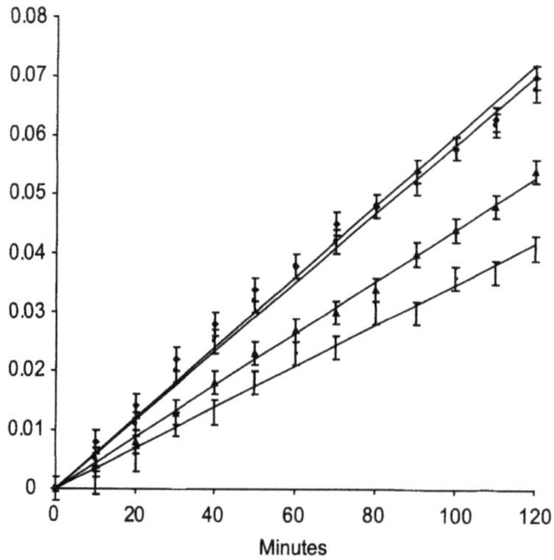

Figure 1. Eppin inhibition of rPSA activity is prevented by mutation of leucine 87 (Δ87Eppin). Zn^{++} (5 mM) is a specific non-competitive inhibitor of PSA. Key: — = Zn^{++}; ▲ = Eppin; ◆ = rPSA only; ● = Δ87Eppin

Eppin modulates hydrolysis of semenogelin

During ejaculation seminal fluid mixes with prostatic secretions aggregating Sg and fibronectin, and activating PSA (reviewed in Robert and Gagnon, 1999). PSA, one of the most abundant proteins in the secretions of the normal human prostate, is a serine protease in the human tissue kallikrein family that is produced as a zymogen. In seminal plasma PSA activity is inhibited by Zn^{++}, spermine, and spermidine (Carvalho, Sanz, Barettino, Romero, Calvete and Romao, 2002) and a serine residue at position S1 of the specificity pocket predicts a chymotrypsin-like specificity (Carvalho *et al.*, 2002). This chymotrypsin-like specificity is supported by experimental evidence and in particular PSA hydrolysis of its natural substrate, semenogelin (Sg), occurs C-terminally to leucine or tyrosine residues (Robert, Gibbs, Jacobson and Gagnon, 1997), although histidine and glutamine residues may occasionally substitute (Lovgren, Airas and Lilja, 1999). Activated PSA leads to liquefaction of semen (hydrolysis of the coagulum proteins), release of Sg proteolytic fragments and release of progressively motile spermatozoa (Robert *et al.*, 1997).

Because Eppin is predicted to be a serine protease inhibitor with chymotrypsin-like specificity that also binds PSA's natural substrate, Sg, we investigated whether Eppin would modulate PSA activity. Sg on the sperm surface is bound to Eppin (Wang *et al.*, 2005) and therefore the cleavage of sperm surface Sg by PSA must occur while Sg is bound to Eppin. We compared *in vitro* the digestion of Sg by rPSA in the presence or absence of Eppin. As shown in Figure 2A, digestion of Sg by rPSA initially produces several lower molecular weight fragments (< 12 kDa) and a 15-16 kDa fragment, which is further hydrolysed to lower molecular weight fragments by 12 hours. In the presence of Eppin the 15-16 kDa fragment persists for at least 24 hours as a protected fragment (Fig. 2B, lane 6), indicating that Eppin has modulated rPSA's hydrolysis of Sg. However when Sg is digested by rPSA for 24 hours in the presence of Δ87Eppin, there is no 15-16 kDa fragment detected (Fig. 2B, lane 3), indicating that Eppin's modulation of rPSA activity requires an active P1 inhibitor site. On far-Western blots Eppin binds to Sg (Wang *et al.*, 2005) and as shown in Figure 2C, Δ87Eppin binds to Sg. Therefore loss of the 15-16 kDa Sg fragment when Δ87Eppin was incubated with Sg and rPSA was not due to Δ87Eppin's failure to bind Sg. Finally we tested

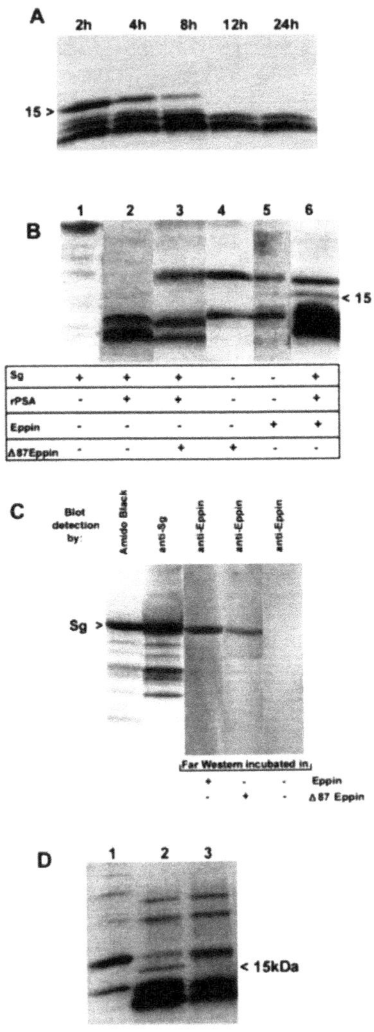

Figure 2. (**A**) Amido black (protein) stained blot following SDS-PAGE of the digestion of semenogelin (Sg) by rPSA in the absence of Eppin for 2-24 hours (h). Arrowhead indicates the 15-16 kDa Sg fragment. (**B**) Amido black (protein) stained blot following SDS-PAGE of the digestion of semenogelin (Sg) by rPSA in the presence or absence of Eppin for 24 hours. Arrowhead indicates the 15-16 kDa Sg fragment. Lane 1: Semenogelin (Sg). Lane 2: Sg digested with rPSA producing low molecular weight fragments <12 kDa. Lane 3: Sg digested with rPSA in the presence of Δ87Eppin. Lane 4: The leucine 87 mutant (Δ87Eppin). Lane 5: Eppin. Lane 6: Sg digested with rPSA in the presence of Eppin for 24 hours, a protected fragment is seen at 15-16 kDa. (**C**) Far-Western blot of Sg binding Eppin and Δ87Eppin, demonstrating that leucine 87 is not necessary for Eppin binding to Sg. Arrowhead indicates Sg band. Lane 1: Amido black (protein) stain of recombinant Sg (arrowhead). Lane 2: Anti-Sg binding to Sg on the blot. Lane 3: Eppin binding to Sg on the blot and detected with anti-Eppin antibody. Lane 4: Δ87Eppin binding to Sg on the blot and detected with anti-Eppin antibody. Lane 5: Control, no Eppin added, and no Δ87Eppin added detected with anti-Eppin antibody. (**D**) Amido black (protein) stained blot following SDS-PAGE of the digestion of semenogelin (Sg) or reduced and carboxymethylated Sg by rPSA in the presence of Eppin for 24 hours. Arrowhead indicates the 15-16 kDa Sg fragment, which is not present when Sg is reduced and carboxymethylated. Lane 1: Eppin. Lane 2: Digestion of Sg by rPSA in the presence of Eppin for 24 hours. Lane 3: Digestion of reduced and carboxymethylated Sg by rPSA in the presence of Eppin for 24 hours.

whether or not reduced and carboxymethylated Sg, which does not bind Eppin (Wang *et al.*, 2005), would produce a 15-16 kDa fragment when digested by PSA. As shown in Figure 2D, lane 3, a 15-16 kDa fragment was not produced. We conclude that Eppin's ability to modulate PSA's hydrolysis of its substrate, Sg, requires that Eppin bind to Sg and have an active P1 site residue at leu87.

Additionally, because antibodies to Eppin provide contraception in male monkeys (O'Rand *et al.*, 2004), we investigated whether antibodies to Eppin would affect the PSA hydrolysis of Sg. Previously anti-Eppin antibodies directed to the C-terminal were shown to prevent the formation of a 15-16 kDa fragment (O'Rand *et al.*, 2006, 2007), which could have been a result of blocking the active P1 site or the Sg binding site or both. This was tested by determining the effect of anti-Eppin antibodies on PSA's hydrolysis of the chromogenic substrate S-2586. As shown in Figure 3, anti-Eppin antibodies had no effect on PSA's ability to hydrolyse S-2586 in the presence of Eppin.

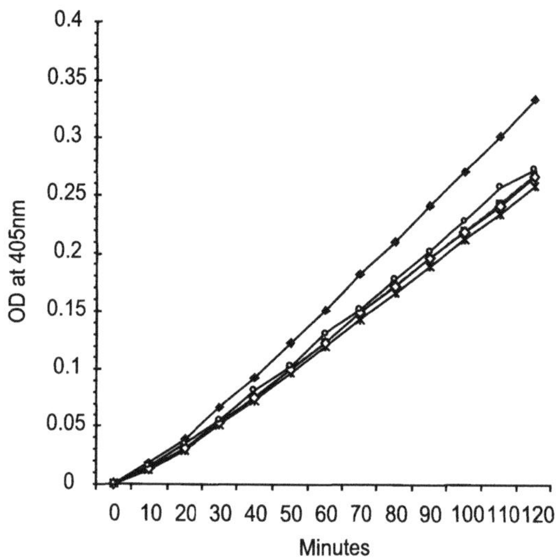

Figure 3. Inhibition of activated rPSA activity by Eppin is not affected by antibodies to Eppin. Normal rPSA activity = ◆ ; rPSA inhibited with 7 μg Eppin = *; rPSA + 7 μg Eppin + 14 mg rabbit IgG (control) =■; PSA + 7 μg Eppin + 14 mg anti-N terminal Eppin (Q20E) =◇; PSA + 7 mg Eppin + 14 mg anti-recombinant Eppin =○

While immunocontraception may have stalled in its path to a marketable product for male contraception, its power as a research tool, particularly for primates, remains intact. Immunization can be used as a molecular strategy to investigate new targets for male contraception and we have demonstrated the utility of this approach with Eppin. Eppin is an essential molecule in the pre-fertilisation preparation of human spermatozoa in the male reproductive tract and in the ejaculate coagulum. Continued study of its structure and function is likely to yield significant new information in the coming years.

Acknowledgements

Zengjun Wang was supported by a Fogarty Postdoctoral fellowship. Supported by grant CIG-96-06 from the CICCR Program of CONRAD (Contraceptive Research and Development Program), the Andrew W. Mellon Foundation, and by D43TW-HD00627, Program for International Training and Research in Population and Health from the Fogarty International Center and the National Institute of Child Health and Human Development (Grant HD 048843 to MO).

References

Aitken, R.J., Paterson, M. and Koothan, P.T. (1992). Risks and benefits of immunological contraception. In: *Spermatogenesis-Fertilization-Contraception*. Edited by Nieschlag, E., and Habenicht, U.-F., Springer-Verlag, Berlin, pp. 461-475

Beagley, K.W., Wu, Z.L., Pomering, M. and Jones, R.C. (1998). Immune responses in the epididymis: implications for immunocontraception. In: *The Epididymis: Cellular and Molecular Aspects*. Journal of Reproduction and Fertility Supplement 53. Edited by Jones, R.C., Holland, M.K. and Doberska, C., pp. 235-245

Bongaarts, J. and Watkins, S.C. (1996). Social interactions and contemporary fertility transitions. *Population and Development Review* **22**: 639- 682

Bongaarts, J. (1997). The role of family planning programmes in contemporary fertility transitions. In: *The Continuing Demographic Transition*. Edited by Jones, G.W., Douglas, R.M., Caldwell, J.C. and D'Souza, R.M. Clarendon Press, Oxford, pp. 422–443

Brenton, S., Smith, P.J.S., Lui, B. and Brown, D. (1996). Acidification of the male reproductive tract by a proton pumping (H^+)-ATPase. *Nature Medicine* **2**: 470-472

Carvalho,, A.L., Sanz, L., Barettino, D., Romero, A., Calvete, J.J. and Romao, M.J. (2002) Crystal structure of a prostate kallikrein isolated from stallion seminal plasma: a homologue of human PSA. *Journal of Molecular Biology* **322:** 325-337

Cheng, C.Y., Mo, M.Y., Grima, J., Saso, L., Tita, B., Mruk, D. and Silvestrini, B. (2002). Indazole carboxylic acids in male contraception. *Contraception* **65**: 265-268

Cheng, C.Y., Mruk, D., Silvestrini, B., Bonanomi, M., Wong, C.H., Siu, M.K.Y., Lee, N.P.Y., Lui, W.Y. and Mo, M.Y. (2005). AF-2364 [1-(2,4-dichlorobenzyl)-1H-indazole-3-carbohydrazide] is a potential male contraceptive: a review of recent data. *Contraception* **72**: 251-261

Griffin, D. (1990). Strategy of vaccine development. In: *Gamete Interaction. Prospects for Immunocontraception*. Edited by Alexander, N.J., Griffin, D., Spieler, J.M., and Waites, G.M.H., Wiley-Liss, New York, pp. 501-522

Grimes, D.A., Gallo, M.F., Grigorieva, V., Nanda, K. and Schulz, K.F. (2005). Steroid hormones for contraception in men: systematic review of randomized controlled trials. *Contraception* **71**: 89-94

Henle, W., Henle, G. and Chambers, L.S. (1938). Studies on the antigenic structure of some mammalian spermatozoa. *Journal of Experimental Medicine* **68**: 335-352

Jones, W.R., Bradley, J., Judd, S.J., Denholm, E.H., Ing, R.M., Mueller, U.W., Powell, J., Griffin, P.D. and Stevens, V.C. (1988). Phase I clinical trial of a World Health Organisation birth control vaccine. *Lancet* **1**: 1295-1298

Kirichok, Y., Navarro, B. and Clapham, D.E. (2006). Whole-cell patch-clamp measurements of spermatozoa reveal an alkaline-activated Ca^{2+} channel. *Nature* **439**: 737-740

Landsteiner, K. (1899). Zur kenntnis der spezifisch auf blütkorperchen wirkenden sera. *Centralblatt* **25**: 546-549

Laskowski, M. and Kato, I. (1980). Protein inhibitors of proteinases. *Annual Review of Biochemistry* **49**: 593-626

Lea, I.A., van Lierop, M.J.C., Widgren, E.E., Grootenhuis, Y., Wen, Y., van Duin, M. and O'Rand, M.G. (1998). A chimeric sperm peptide induces antibodies and strain-specific reversible infertility. *Biology of Reproduction* **59**: 527-536

Lovgren, J., Airas, K. and Lilja, H. (1999). Enzymatic action of human glandular kallikrein 2 (hK2). *European Journal of Biochemistry* **262**: 781-789

Mancini, R.E. (1976). Immunopathology of animal and human testes. In: *Human Semen and Fertility Regulation in Men*. Edited by Hafez, E.S.E., Mosby, St. Louis, USA, pp. 287-307

Metz, C.B. (1979). Immunological inhibition of sperm and egg function: Prospects for an antifertility vaccine from gametes. In: *Contraception: Science, Technology, and Application*. Proceedings of a Symposium, National Academy of Sciences, Washington DC, USA, pp. 180-220

Morton, D.B. and McAnulty, P.A. (1979). The effect on fertility of immunizing female sheep with ram sperm acrosin and hyaluronidase. *Journal of Reproductive Immunology* **1**: 61-73

Munoz, M.G. and Metz, C.B. (1978). Infertility in female rabbits isoimmunized with subcellular sperm fractions. *Biology of Reproduction* **18**: 669-678

Nieschlag, E. and Henke, A. (2005). Hopes for male contraception. *Lancet* **365**: 554-556

O'Rand, M.G. (1980). Antigens of spermatozoa and their environment. In: *Immunological Aspects of Infertility and Fertility Regulation*. Edited by Dhindsa, D.S. and Schumacher, G.F.B., Elsevier/North Holland, New York, USA, pp. 155-171

O'Rand, M.G. and Lea, I.A., (2002). Update on contraceptive vaccines. In: *Infertility and Reproductive Medicine Clinics of North America*. Volume 13(1). Edited by Arici A., W.B. Saunders, Harcourt, Philadelphia, USA, pp. 221-232

O'Rand, M.G. and Metz C.B. (1974). Tests for rabbit sperm surface iron-binding protein and hyaluronidase using the "exchange agglutination" reaction. *Biology of Reproduction* **11**: 326-334

O'Rand, M.G., Widgren, E.E., Sivashanmugam, P., Richardson, R.T., Hall, S.H., French, F.S., VandeVoort, C.A., Ramachandra, S.G., Ramesh, V. and Rao, A.J. (2004). Reversible immuno-contraception in male monkeys immunized with Eppin. *Science* **306**: 1189-1190

O'Rand, M.G., Widgren, E.E., Wang, Z. and Richardson, R.T. (2006). Eppin: an effective target for male contraception. *Molecular and Cellular Endocrinology* **250**: 157-162

O'Rand, M.G., Widgren, E.E., Wang, Z. and Richardson, R.T. (2007). Eppin: an epididymal protease inhibitor and a target for male contraception. In: *Gamete Biology: Emerging Frontiers in Fertility and Contraceptive Development*. Edited by Gupta, S., Nottingham University Press, Nottingham, pp. 445-454

Potts, M. (1996). The myth of a male pill. *Nature Medicine* **2**: 398-399

Primakoff, P., Lathrop, W., Woolman, L., Cowan, A. and Myles, D. (1988). Fully effective contraception in male and female guinea pigs immunized with the sperm protein PH20. *Nature* **335**: 543-546

Robert, M. and Gagnon, C. (1999). Semenogelin I: a coagulum forming, multifunctional seminal vesicle protein *Cellular and Molecular Life Science* **55** 944-960

Robert, M., Gibbs, B.F., Jacobson, E. and Gagnon, C. (1997). Characterization of prostate-specific antigen proteolytic activity on its major physiological substrate, the sperm motility inhibitor precursor/semenogelin I. *Biochemistry* **36**: 3811-3819

Rümke, P.H. and Hellinga, G. (1959). Autoantibodies against spermatozoa in sterile men. *American Journal of Clinical Pathology* **32**: 357-363

Tyler, A. (1948). Fertilization and immunity. *Physiology Reviews* **28**: 180-219

Tyler, A. (1961). Approaches to the control of fertility based on immunological phenomena. *Journal of Reproduction and Fertility* **2**: 473-506

Walden, C.M., Butters, T.D., Dwek, R.A., Platt, F.A. and van der Spoel, A.C. (2006). Long-term non-hormonal male contraception in mice using *N*-butyldeoxynojirimycin. *Human Reproduction* **21**: 1309-1315

Wang, Z., Widgren, E.E., Sivashanmugam, P., O'Rand, M.G. and Richardson, R.T. (2005). Association of Eppin with semenogelin on human spermatozoa. *Biology of Reproduction* **72**: 1064-1070

* 9 7 8 1 7 8 9 1 8 3 0 8 5 *

ND - #0108 - 050826 - C0 - 240/170/26 - PB - 9781789183085 - Matt Lamination